Industrial Applications of Advanced Oxidation Technologies: Past and Future

Industrial Applications of Advanced Oxidation Technologies: Past and Future

Editors

Gassan Hodaifa
Rafael Borja
Mha Albqmi

MDPI • Basel • Beijing • Wuhan • Barcelona • Belgrade • Manchester • Tokyo • Cluj • Tianjin

Editors

Gassan Hodaifa
Department of Molecular
Biology and Biochemical
Engineering
Pablo de Olavide University
Seville
Spain

Rafael Borja
Instituto de la Grasa
Spanish National
Research Council
Seville
Spain

Mha Albqmi
Chemistry Department
Jouf University
Sakaka
Saudi Arabia

Editorial Office
MDPI
St. Alban-Anlage 66
4052 Basel, Switzerland

This is a reprint of articles from the Special Issue published online in the open access journal *Catalysts* (ISSN 2073-4344) (available at: www.mdpi.com/journal/catalysts/special_issues/industrial_AOPs_catalysis).

For citation purposes, cite each article independently as indicated on the article page online and as indicated below:

LastName, A.A.; LastName, B.B.; LastName, C.C. Article Title. *Journal Name* **Year**, *Volume Number*, Page Range.

ISBN 978-3-0365-8193-4 (Hbk)
ISBN 978-3-0365-8192-7 (PDF)

Contents

About the Editors

Gassan Hodaifa

Dr. Gassan Hodaifa (H-Index of 26 in Web of Science) is a Food Engineer (University of Albaath, Syria) and Chemical Engineer (University of Granada). He received his doctoral degree from the University of Jaén. He joined the research group Bioprocesses TEP-138 (Junta de Andalucía), and works in the areas of microalgae, the treatment of wastewater, chemical oxidation, membrane technology, adsorption, pesticide removal, olive oil processing, enzyme biotechnology, and nematode biotechnology. He has participated in 27 research projects (8 as IP) and 4 European programs (1 PRIMA Project, and 3 in collaboration with M.A.P.A. and the INFAOLIVA and UNAPROLIVA companies), with which he has signed 16 research contracts (4 as IP). He has been contracted twice as Adviser for the mega Japanese company "Asahi Kasei Corporation", and has had six projects financed by J. Andalucía (two of them as IP). He has completed five projects with the National Research Plan, one project for the Madrid Community, and the rest he carried out with companies. As a result of his research work, he has had 64 articles published, with 56 articles published in journals included in the JCR (31 in Q1, 18 in Q2, and 7 in Q3). Of these articles, more than 10 articles have been published as a collaboration with authors from different countries on all continents. He has published six total articles in open access journals, four articles in journals with an impact factor different from that listed in the JCR, and five other articles in outreach magazines. He has two patents for the water purification of olive oil mill wastewater, and has written two books and 46 book chapters that have been published by prestigious publishers such as the Academic Press (Elsevier), CRC Press, Taylor and Francis Group, Springer International Publishing AG, etc. He was the Editor in Chief of three Special Issues in *J. Chemistry* and *Catalysts*. He has participated in 125 congresses. He is currently a member of 15 Editorial Boards of international journals such as *Catalysts* and *Heliyon*.

Rafael Borja

Rafael Borja is a Doctor (PhD) in Chemical Sciences. Since 1986, he has been working as a researcher at the Instituto de la Grasa (CSIC, Spanish National Research Council) in Seville, after obtaining an FPI grant funded by the Andalusian Government for his doctoral thesis. He received the San Alberto Magno Prize for Doctoral Theses, which was awarded by the National Association of Spanish Chemists (ANQUE), Delegation of Andalusia, in 1991. During the period of September 1992–September 1994, he carried out a postdoctoral stay at the "Environmental Technology Centre" belonging to the "Department of Chemical Engineering" of the University of Manchester's Institute of Science and Technology (UMIST) in the United Kingdom. In 1997 he obtained the position of Tenured Scientist of the CSIC, and since then he has developed several lines of research related to the use and treatment of wastewater and solid wastes from agro-food industries. In 2008 he obtained the position of Research Scientist at the CSIC. He has directed six doctoral theses and is a co-author of eight patents. He has published 302 articles in different scientific journals that are included in the "Science Citation Index (SCI)".

He has been the principal investigator of 12 projects funded by the R&D National Plan and by the European Union, as well as several R&D contracts developed at the Instituto de la Grasa (CSIC) in Seville. He currently is a member of the Editorial Board of the scientific journals *Process Biochemistry* (Elsevier), *Catalysts* (MDPI), and *Emerging Science Journal* (Ital Publication). During the period 2013–2015, he was the vice-president of the association, "International Association of the Mediterranean Agro-industrial Wastes (IAMAW)", which is located in Perugia, Italy, and from 2016 to 2019 he was a member of the "Audit Committee" of this association.

Mha Albqmi

Dr. Mha Albqmi currently holds the position of Assistant Professor at Jouf University and serves as the Director of the Olive Research Center at Jouf University, Saudi Arabia. Her educational background includes a Ph.D. in Chemistry from Oklahoma State University, Stillwater, Oklahoma, USA, a Master of Science Degree in Chemistry from Lamar University, Texas, USA, and a Bachelor of Science Degree in Chemistry from Taif University, Taif, Saudi Arabia.

Dr. Mha Albqmi's great passion lies in the field of catalysis, and studies its applications in various industrial chemical processes. Particularly, throughout her Ph.D. program, she made contributions to the development of new procedures that enabled the direct conversion of organic compounds into their resulting products, eliminating the requirement for solvents and effectively reducing waste generation.

Her current research interests lie in environmental chemistry, the recycling of waste materials, and the prevention of pollution. Additionally, she actively valorizes the olive waste residue on certain plants experiencing heavy-metal toxicity. She is involved in the study of nanotechnology applications, such as water purification and photocatalytic degradation processes. Furthermore, she is researching the direct conversion method of organic compounds using catalysts as well as the potential of natural products for the treatment of metal-overload diseases.

Preface to "Industrial Applications of Advanced Oxidation Technologies: Past and Future"

In recent years, climate change has become more evident due to the high values of the carbon footprints and the hydraulic footprints registered by industries. Additionally, the availability of drinking water is scarce in many countries around the world. Technological advances have permitted the establishment of large industries in the last century. Most of these industries are characterized by a high consumption of drinking water that is transformed into wastewater as it enters and leaves these industrial processes. Unfortunately, a large part of this wastewater is not recovered because it is difficult to treat due to the presence of a high organic load with persistent, toxic, and inhibitory compounds. The capabilities of conventional urban wastewater treatment plants (CUWTPs) make them unable to meet the standards set by local, national, or international legislation. On the other hand, society is unwilling to give up many of the comforts offered by technological developments, highlighting the need for effective solutions to minimize or reduce climate change, which has clearly made its presence felt. In response to this challenge, a green and sustainable pre-treatment, or the treatment of industrial wastewater prior to its discharge from CWWTPs, is mandatory for industries.

The use of Advanced Oxidation Technologies (AOTs) for wastewater treatment is an important area of research which has not yet been fully exploited at an industrial level and has significant potential in the disposal of many industrial effluents. In particular, this includes effluents that are difficult to treat by conventional biological treatment processes. Given the positive response from researchers to our first Special Issue entitled "Photocatalysis in the Wastewater Treatment", this new Special Issue represents a continuation and an extension of the work carried out in the previous Special Issue and aims to explain the importance of advanced oxidation technologies and how their incorporation into the industrial sector as green and clean technologies can improve the current situation of the Earth's ecosystems and environment. In this sense, it is worth mentioning treatments based on photolysis, TiO_2/solar light, ozone/ultraviolet irradiation, oxidant/ultraviolet irradiation, oxidant/catalyst/ultraviolet irradiation, high-energy electron beam irradiation (E-beam), sonication/photocatalysis, etc. This Special Issue includes works with new perspectives on catalytic ozonation for organic removal, photo-based advanced oxidation processes for pharmaceutical degradation, etc.

This reprint is addressed to all readers interested in the manufacture of catalysts and chemical oxidation-based treatment processes, whether they are academics or professionals. Finally, the Editors would like to thank the authors who have participated in the writing of this interesting book, which will be of great interest at the industrial, academic, and research level.

Gassan Hodaifa, Rafael Borja, and Mha Albqmi
Editors

 **catalysts**

Article

Comprehensive Study on Environmental Behaviour and Degradation by Photolytic/Photocatalytic Oxidation Processes of Pharmaceutical Memantine

Sandra Babić [1,*], Davor Ljubas [2], Dragana Mutavdžić Pavlović [1], Martina Biošić [1], Lidija Ćurković [2,*] and Dario Dabić [1,3]

[1] Faculty of Chemical Engineering and Technology, University of Zagreb, Trg Marka Marulića 19, 10000 Zagreb, Croatia
[2] Faculty of Mechanical Engineering and Naval Architecture, University of Zagreb, Ivana Lučića 1, 10000 Zagreb, Croatia
[3] Croatian Meteorological and Hydrological Service, Ravnice 48, 10000 Zagreb, Croatia
* Correspondence: sandra.babic@fkit.unizg.hr (S.B.); lidija.curkovic@fsb.hr (L.Ć.)

Abstract: Memantine is a pharmaceutical used to treat memory loss, one of the main symptoms of dementia and Alzheimer's disease. The use of memantine is expected to continue to grow due to the increasing proportion of the elderly population worldwide. The aim of this work was to conduct a comprehensive study on the behaviour of memantine in the environment and the possibilities of its removal from wastewater. Abiotic elimination processes (hydrolysis, photolysis and sorption) of memantine in the environment were investigated. Results showed that memantine is stable in the environment and easily leached from river sediment. Therefore, further investigation was focused on memantine removal by advanced oxidation processes that would prevent its release into the environment. For photolytic and photocatalytic degradation of memantine, ultraviolet (UV) lamps with the predominant radiation wavelengths of 365 nm (UV-A) and 254/185 nm (UV-C) were used as a source of light. TiO_2 in the form of a nanostructured film deposited on the borosilicate glass wall of the reactor was used for photocatalytic experiments. Photodegradation of memantine followed pseudo-first-order kinetics. The half-life of photocatalytic degradation by UV-A light was much higher (46.3 min) than the half-life obtained by UV-C light (3.9 min). Processes degradation efficiencies and evaluation of kinetic constants were based on the results of HPLC-MS/MS analyses, which also enable the identification of memantine oxidation products. The acute toxicity of the reaction mixture during the oxidation was evaluated by monitoring the inhibition of the luminescence of *Vibrio fischeri* bacteria. The results showed that memantine and its oxidation products were not harmful to *Vibrio fischeri*.

Keywords: memantine; hydrolysis; photolysis; sorption; photocatalysis; sol-gel TiO_2 film; degradation products; toxicity

Citation: Babić, S.; Ljubas, D.; Mutavdžić Pavlović, D.; Biošić, M.; Ćurković, L.; Dabić, D. Comprehensive Study on Environmental Behaviour and Degradation by Photolytic/ Photocatalytic Oxidation Processes of Pharmaceutical Memantine. *Catalysts* **2023**, *13*, 612. https://doi.org/10.3390/catal13030612

Academic Editors: Gassan Hodaifa, Rafael Borja and Mha Albqmi

Received: 20 February 2023
Revised: 15 March 2023
Accepted: 16 March 2023
Published: 17 March 2023

1. Introduction

With the increase in the aging population worldwide, Alzheimer's disease and other dementias have become a rapidly increasing public health concern, with an estimated 50 million people currently living with dementia [1]. The prevalence of Alzheimer's disease is approximately 0.6% at the age of 60 but it doubles every five years, so the prevalence is about 40% at the age of 90 [2]. Currently, no cure exists for Alzheimer's disease but there are treatments that temporarily slow down the development of symptoms and improve cognitive functions. In Europe and the USA two symptomatic treatments are approved, the use of acetylcholinesterase (AChE) inhibitors and *N*-methyl-*D*-aspartate receptor antagonist memantine (3,5-dimethyladamantan-1-amine) [1]. Once in the organism, memantine is poorly metabolized and most (57–82%) of the administered dose is excreted unchanged in urine [3]. Memantine is a compound highly soluble in water (29.4 mg/L, [4]).

In view of this, it is reasonable to expect that memantine will end up in wastewater and, without proper treatment, in environmental waters. Memantine was detected in rivers and sewage treatment plant (STP) influents and effluents in Japan at high frequency (>70%), with a maximal measured environmental concentration in seven rivers of 47.4 ng/L. The measured concentration of memantine in STP was lower than 1 µg/L with an average removal rate in three STPs lower than 20% [5]. Memantine was detected in effluent wastewater from a wastewater treatment plant (WWTP) located near Barcelona (Spain) at a concentration ranging from 0.028 to 0.134 µg/L for samples taken on 10 different days [6]. Memantine was also detected in sewage effluents in three Sweden STPs in concentrations ranging from 10 to 14 ng/L. It is also detected in the plasma of fishes exposed to sewage effluents (from <LOQ (0.5 ng/mL) to 2.3 ng/mL) [7]. Kårelid et al. [8] studied the adsorption of pharmaceuticals on granular activated carbon (GAC) and powdered activated carbon (PAC) at three Swedish wastewater treatment plants and observed that, among 22 investigated pharmaceuticals, only memantine showed removal lower than 95%. Despite the evidence that memantine is present in the environment, data on the fate and behaviour of memantine in the environmental are scarce.

The incomplete removal of pharmaceuticals in conventional wastewater treatment plants clearly indicates the need for the development of innovative technologies such as advanced oxidation processes (AOPs). AOPs have been proposed as a tertiary treatment for wastewater [9,10]. Among different AOPs, heterogeneous photocatalysis is a promising method for removing organic micropollutants (OMPs), including pharmaceuticals [11–13]. The most commonly used semiconductor photocatalyst is TiO_2 with the potential for the total mineralization of OMPs, resulting in the formation of non-toxic compounds (CO_2, H_2O and the corresponding mineral acids). TiO_2 can be used in the form of TiO_2 powder suspension (slurry) or it can be immobilized by different techniques on different substrates such as borosilicate glass [14,15], alumina foam [16], alumina ring and borosilicate ring. Immobilization of TiO_2 on the adequate reactor walls eliminates the need to separate the photocatalyst from the treated water. Among different deposition techniques (sol-gel, thermal treatment, pulsed laser deposition, reactive evaporation, physical vapour deposition (PVD), chemical vapour deposition (CVD), electrodeposition, sol-spray, hydrothermal deposition, etc.), the sol-gel technique offers many advantages: relatively low cost, low processing temperature, simple deposition, relatively simple control of composition, possibility of various forming processes, and ability to prepare nano-sized thin films and to produce fine structures [14,17]. Sol-gel films can be generally deposited by two methods—the dip coating and the spin coating technique [18].

The aim of this study was to investigate the environmental behaviour of memantine and possibility of its degradation by advanced oxidation processes. Environmental behaviour was investigated by studying abiotic elimination processes: hydrolysis, photolysis and sorption. For investigation of photolytic/photocatalytic oxidation of memantine, a photoreactor with UV-A and UV-C lamps, and TiO_2 in a form of a nanostructured film deposited on borosilicate glass wall of the reactor was used. Process degradation efficiencies and evaluation of kinetic constants were based on the results of HPLC-MS/MS analyses, which also enable identification and monitoring of memantine degradation products. In addition, the acute toxicity of the reaction mixture during the degradation experiment was evaluated by monitoring the inhibition of the luminescence of *Vibrio fischeri* bacteria.

2. Results and Discussion

2.1. Environmental Behaviour

2.1.1. Hydrolytic and Photolytic Degradation

Hydrolytic degradation of memantine was investigated according to the procedure described in OECD 111 [19]. The results of hydrolytic degradation experiments showed that memantine is persistent to hydrolytic degradation with the degree of hydrolytic degradation between 1.5% and 2.8% under the applied conditions (Figure 1A). Given that hydrolytic degradation of 10% at 50 °C corresponds to a half-life of approximately 30 days, which is equivalent to the half-life of 1 year at 25 °C [1], memantine was considered stable and no further investigation is required.

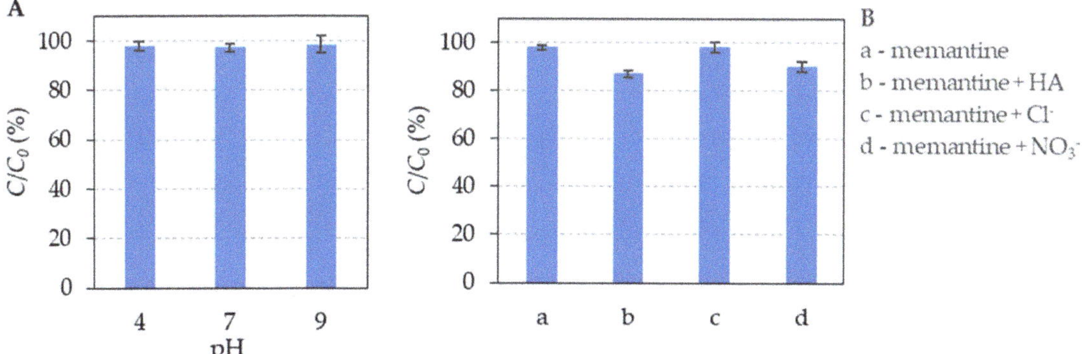

Figure 1. The degree of (**A**) hydrolytic degradation of memantine at 50 °C and (**B**) direct and indirect photolytic degradation (24 h exposure to artificial solar radiation, 10 mg/L memantine solution). Error bars refer to standard deviation.

Photolytic degradation was investigated with memantine solution in MilliQ water (direct photolysis) and in the presence of substances commonly present in environmental waters: humic acids (HA), chloride and nitrate (indirect photolysis). The concentrations of inorganic ions and humic acids were typical for the environment. The presence of HA and nitrates resulted in a lower concentration of memantine, while chlorides did not affect photolytic degradation of memantine. The photolysis due to the presence of HA and nitrate can be attributed to the formation of highly reactive hydroxyl radicals [20,21]. However, the observed decrease in the concentration of memantine was not significant (less than 15% after 24-h exposure to simulated solar radiation, Figure 1B), which points to the conclusion of its persistence during exposure to artificial solar radiation. Blum et al. [22] reported similar results of memantine photolytic persistence with a degree of degradation of less than 10%.

According to the available literature, similar environmental behaviour was not observed for other pharmaceuticals detected in environmental waters. They are usually susceptible to photolytic [20,21,23–25] or hydrolytic [26,27] degradation or to both elimination processes [28,29].

2.1.2. Sorption
Kinetics of Sorption and Desorption

Based on previously published papers [30,31], there is already some information about the tendency of memantine to sorption on soil and sediment particles, namely that sorption is definitely not a dominant process in its case. In this context, our goal was to compare from a kinetic point of view how much memantine is sorbed to the sediment particles and how much memantine is desorbed from the same sediment studied.

From Figure 2A, it can be seen that the "faster" sorption of memantine to the sediment sample occurs in the first 6 h, after which further sorption of memantine occurs slowly over the observed 24-h period. At the same time, desorption of the previously sorbed memantine takes place during the same time intervals (Figure 2B). It should be noted that, as the concentration of memantine increases, the amount of memantine sorbed and desorbed decreases, so that the largest difference between the amount sorbed and desorbed was obtained at the lowest initial concentration of memantine tested (0.1 mg/L). To investigate the control mechanisms of the sorption [32] and desorption processes, experiments were performed at different time periods, i.e., a kinetic study was performed at three concentration levels (0.1, 0.5 and 2.0 mg/L), in contrast to the other six concentrations at which the sorption experiments will be carried out.

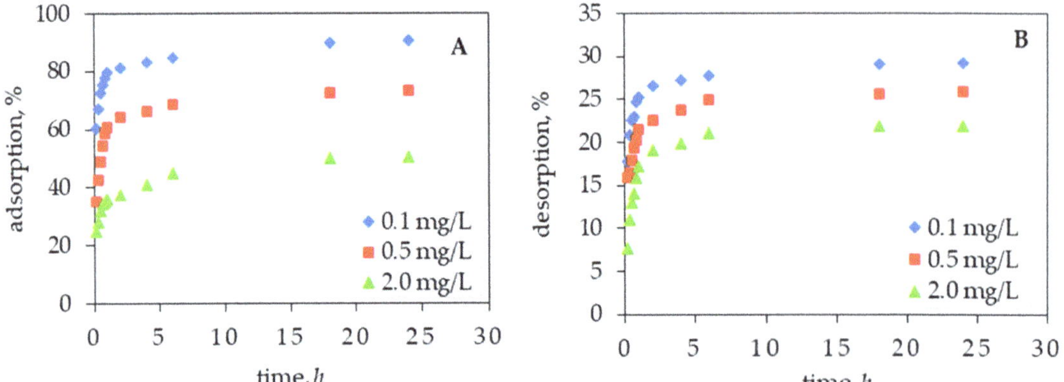

Figure 2. Kinetics of (**A**) sorption and (**B**) desorption for memantine on the sediment Studena, $T = 25\,°C$.

The sorption and desorption data were analysed using three different kinetic models; Lagergren pseudo-first-order, pseudo-second-order and the intraparticle diffusion (IPD) model. All kinetic models are presented in Table 1 where q_e and q_t are the amounts of memantine ($\mu g/g$) adsorbed/desorbed on investigated sediment samples at equilibrium and at time t; k_1 (1/min) is the rate constant of the pseudo-first-order adsorption; k_2 ($g/\mu g$ min) is the rate constant of the pseudo-second-order sorption and k_{pi} ($\mu g/g$ min$^{1/2}$) is the intraparticle diffusion rate parameter of stage i. C_i, the intercept of stage i, gives an indication of the thickness of the boundary layer, i.e., the larger the intercept, the larger the boundary layer effect.

Table 1. Kinetic models.

Kinetic Model	Linear Form
Lagergren pseudo-first-order	$\ln(q_e - q_t) = \ln q_e - k_1 t$
Ho's pseudo-second-order	$\frac{t}{q_t} = \frac{1}{k_2 q_e^2} + \frac{t}{q_e}$
IPD model	$q_t = k_{pi}\sqrt{t} + C_i$

The sorption and desorption rate constants k_1, k_2 and $q_{e,cal}$ as well as the correlation coefficients (R^2) for the pseudo-first and pseudo-second models are shown in Table 2.

Table 2. Sorption and desorption kinetic parameters of memantine on the sediment Studena.

Initial Concentration, mg/L		$q_{e,exp}$, $\mu g/g$	Pseudo-First-Order			Pseudo-Second-Order		
			$q_{e,calc}$, $\mu g/g$	k_1, 1/min	R^2	$q_{e,calc}$, $\mu g/g$	k_2, $g/\mu g$ min	R^2
Sorption process	2.0	10.11	13.33	$2.303·10^{-4}$	0.7666	10.24	0.0035	0.9990
	0.5	3.67	2.29	$4.606·10^{-4}$	0.5838	3.70	0.0162	0.9999
	0.1	0.91	0.25	$9.212·10^{-4}$	0.7508	0.91	0.0966	0.9998
Desorption process	2.0	4.40	3.74	$-2.303·10^{-4}$	0.8514	4.46	0.0107	1.000
	0.5	1.29	1.64	$-2.303·10^{-4}$	0.3369	1.30	0.0545	0.9999
	0.1	0.29	0.51	$-1.612·10^{-4}$	0.5115	0.29	0.3025	0.9999

From the results, it can be concluded that the pseudo-second-order kinetic model perfectly describes both the kinetics of memantine sorption and the kinetics of desorption of previously sorbed memantine, since very high correlation coefficients ($R^2 > 0.999$) were obtained in both cases. These results suggest that the sorption/desorption capacity is regulated by the number of available active sites on the sediment. According to this model, the maximum concentration absorbed at equilibrium (q_e) on the Studena sediment was approximately 10.11 µg/g, which corresponds to the maximum sorption capacity of this sediment for memantine in the experiments performed. However, if we consider the results of desorption according to the same model, it follows that the maximum concentration desorbed at equilibrium on the Studena sediment was about 4.40 µg/g, which practically corresponds to almost half of the amount of memantine previously sorbed. This ratio of sorbed/desorbed memantine from the sediment studied depends, of course, on the memantine concentration in contact with the sediment, so that Table 2 shows that, at concentrations of 0.1 mg/L and 0.5 mg/L, almost one-third is desorbed, which is less than the previously mentioned concentration of 2.0 mg/L, at which almost half was desorbed. The higher the concentration of memantine in contact with the sediment, the more memantine is washed out of the sediment, which is consistent with what was said before, i.e., that the sorption/desorption capacity is regulated by the number of available active sites on the sediment. The desorption rate constants according to the pseudo-second-order kinetic model are much higher than the sorption constants under the same conditions. Such a result is even more discouraging because, no matter how little memantine is sorbed on a sediment, it is still quite a lot and is rapidly desorbed, posing a risk of water contamination by memantine.

The kinetics of sorption and desorption can also be described from a mechanical point of view. The whole process of sorption and desorption can be controlled by one or more steps, such as surface diffusion, pore diffusion, external diffusion and sorption/desorption at the pore surface. When the sorbent is porous, as in the case of sediments, intraparticle diffusion often plays a major role. During rapid stirring, the diffusion mass transport can be related to the diffusion coefficient, which describes well the experimental sorption/desorption data. Results of the IPD model are shown in Table 3.

Table 3. Intraparticle diffusion model constants and correlation coefficients for memantine on the sediment Studena at different initial concentrations.

Initial Concentration, mg/L		Intraparticle Diffusion								
		First Phase			Second Phase			Third Phase		
		k_{p1}, µg/g min$^{1/2}$	C_1	R^2	k_{p2}, µg/g min$^{1/2}$	C_2	R^2	k_{p3}, µg/g min$^{1/2}$	C_3	R^2
Sorption process	2.0	0.5123	3.4451	0.9716	0.1853	5.4711	0.9819	0.0176	9.4375	1.000
	0.5	0.2895	0.8587	0.9934	0.0269	2.9168	0.9924	0.0062	3.4343	1.000
	0.1	0.0425	0.4792	0.9853	0.0046	0.7624	1.000	0.0014	0.8556	1.000
Desorption process	2.0	0.4058	0.3216	0.9915	0.0505	3.2505	0.9731	0.0009	4.3602	1.000
	0.5	0.0631	0.5692	0.9681	0.0142	0.9752	0.9975	0.0030	1.1806	1.000
	0.1	0.0156	0.1348	0.9673	0.0015	0.2490	0.9986	0.0003	0.2799	1.000

From these results, it is evident that the process of sorption and desorption of memantine on the studied sediment is multilinear, indicating that the sorption and desorption process occurs in three phases [33]: (i) initial boundary layer diffusion or adsorption/desorption at the outer surface, (ii) gradual intraparticle diffusion or diffusion in the pores, where the degree of intraparticle diffusion is rate-controlled, and (iii) equilibrium stage showing saturation of the sorbent surface.

This multilinearity of the sorption and desorption processes suggests that IPD was not the only rate-controlling step [34] but that multiple steps occur at this microlevel. All the kinetic results obtained are of great importance, especially the desorption information, which plays an important role in evaluating the behaviour of memantine in the environment.

Sorption Isotherms

The sorption of memantine was tested on three sediment samples (Pakra, Petrinjčica and Studena) and described by two sorption isotherms: the Linear and Freundlich sorption isotherms, and results are presented in Table 4. All presented values are expressed by the average value of three determinations. Achieved relative standard deviations are lower than 10%.

Table 4. Sorption coefficients (K_d), Freundlich and Dubinin-Radushkevich sorption isotherm parameters in 0.01 M $CaCl_2$ at initial pH values.

Sediment Samples	Linear			Freundlich		Dubinin-Radushkevich			
	K_d, mL/g	R^2	n	K_F, $(\mu g/g)(mL/\mu g)^{1/n}$	R^2	β, mol^2k/J^2	q_m, $\mu g/g$	E, kJ/mol	R^2
Pakra	1.4267	0.9917	1.8776	1.6372	0.9189	0.0435	1.6394	3.39	0.7082
Petrinjčica	2.9658	0.9906	2.0467	3.1362	0.8894	0.0286	2.7194	4.18	0.6642
Studena	0.9771	0.9933	1.5868	1.0290	0.8932	0.0477	1.0377	3.24	0.6451

From the obtained regression coefficients R^2, it can be seen that only the linear isotherm describes the sorption process with $R^2 > 0.99$ in all cases, while the range of regression coefficients for the Freundlich isotherm is 0.89–0.92. The values of the Freundlich exponent, n, range from 1 to 10, indicating favourable sorption [35]. The Dubinin–Radushkevich model shows the worst agreement with the experimental data since R^2 ranges from 0.6451 to 0.7082. Since the values of sorption energy, E (obtained from the D-R isotherm model), are from 3.24 to 4.18 kJ/mol for the investigated sediments, it could be said that the sorption of memantine on the investigated sediments is of a physical nature.

In addition to these three sorption models, many other sorption isotherm models were tried in the preliminary experiment to obtain more information about sorption of memantine, but without success. For example, when trying to applied the Langmuir isotherm we either obtained negative values for the Langmuir isotherm constants or the R^2 were extremely low ($R^2 < 0.25$). This indicates that the Langmuir sorption isotherm is not suitable to explain the sorption process of memantine on the sediment samples studied, since these Langmuir constants indicate the binding surface energy and monolayer coverage.

The values of the distribution coefficient K_d from the linear isotherm and the adsorption capacity K_F from the Freundlich isotherm indicate a slightly weaker binding of memantine to the sediments of Studena and Pakra compared to the sediment of Petrinjčica. However, it should be noted that high values of sorption coefficients were not expected at all, since, according to previous studies, a low tendency of memantine to sorption on soil and sediment particles was observed [30,31]. In any case, the results obtained in this study support the fact that memantine poses a major threat to natural waters because it is easily leached from sediment samples and thus has high mobility in these sediments.

Numerous previous studies clearly show that pH is one of the most important factors affecting the sorption mechanism and rate [36,37]. This is supported by the fact that whether the observed molecule behaves as a cation, anion or neutral molecule depends on the pH of the medium, but also that the activity/presence of metal ions present in the sediment changes depending on it. Since memantine as a molecule is characterized by a pK_a constant (10.27), it can be concluded that it also occurs in ionic form. However, based on the natural pH value of the sediments studied, it could be concluded that memantine occurs exclusively as a neutral molecule in all sediments studied [30].

In addition to the effect of pH on the distribution of the ionic/molecular species of the memantine under environmental conditions, the effect of pH on their sorption at three different pH values (pH 5, 7 and 9) was also investigated (Figure 3A).

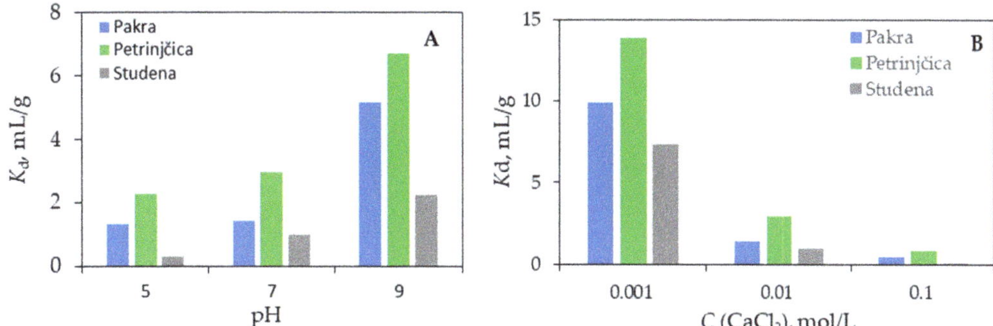

Figure 3. The influence of (**A**) pH and (**B**) ionic strength on the sorption capacity of memantine in studied sediments (T = 25 °C).

In these experiments, an inverse relationship between sorption and pH was observed for memantine [38,39], i.e., with higher acidity, the sorption coefficient decreases. In all three sediments studied, differences in K_d values are observed with the change in pH. While the changes in distribution coefficient from neutral to the alkaline pH range are easily visible, the change in distribution coefficient from the neutral to the acidic pH range is much less pronounced. From these results, it can be clearly concluded that the influence of pH on the sorption of memantine dominates.

In addition to the influence of pH, the influence of ionic strength on the sorption of memantine was also investigated. It was found that the sorption coefficients decreased with increasing ionic strength (Figure 3B). The obtained results indicate a possible surface complexity between the memantine and the studied sediments. For all sediments studied, the highest K_d coefficient values were obtained at the lowest concentration of $CaCl_2$ solution tested. The influence of ionic strength on sorption could be related to the fact that the thickness of the charged surface of the "electric double layer" is reduced, resulting in decrease in surface charge and fewer interactions between the ionic form of the drug and the sediment surface [40]. Of course, this theory is also supported by the fact that memantine is in the form of a neutral molecule [30] in all experiments performed, which makes a possible interaction even less likely. A similar trend was observed in a previous study of the sorption of memantine and in other studies of the sorption of pharmaceuticals [30,41–43].

2.2. Photolytic and Photocatalytic Oxidation of Memantine in Aqueous Solution

The photocatalytic activity of sol-gel TiO_2 film was evaluated through the degradation of memantine aqueous solution (10 mg/L) under ultraviolet (UV) lamps with the predominant radiation wavelengths of 365 nm (UV-A) and 254/185 nm (UV-C).

In order to investigate the kinetics of the photocatalytic degradation of memantine by photolytic and photocatalytic processes, the pseudo-first-order kinetic model was used. The linear form of the pseudo-first-order kinetic model is [44]:

$$ln\frac{C_0}{C_t} = -k_1 \cdot t \tag{1}$$

The half-life time $t_{1/2}$ was calculated using the following expression [16]:

$$t_{1/2} = \frac{ln\,(2)}{k} \tag{2}$$

where C_t (mg/L) is the concentration of memantine at time t (min), C_0 (mg/L) is the initial memantine concentration and k_1 (1/min) is the degradation rate constant.

The first-order degradation rate constant (k_1, 1/min) from equation 1 can be calculated by the slope of the straight line obtained from plotting linear regression of $-\ln\,(C_t/C_0)$ versus irradiation time (t) (Figure 4A). Table 5 shows the pseudo-first-order kinetic constant (k_1, 1/min), coefficient of determination (R^2), half-life time ($t_{1/2}$, min) and efficiency (η, %)

for memantine removal by photolysis and photocatalysis. It is noticed that the pseudo-first-order model has an $R^2 > 0.96$, which confirms that the memantine removal follows a pseudo-first-order model.

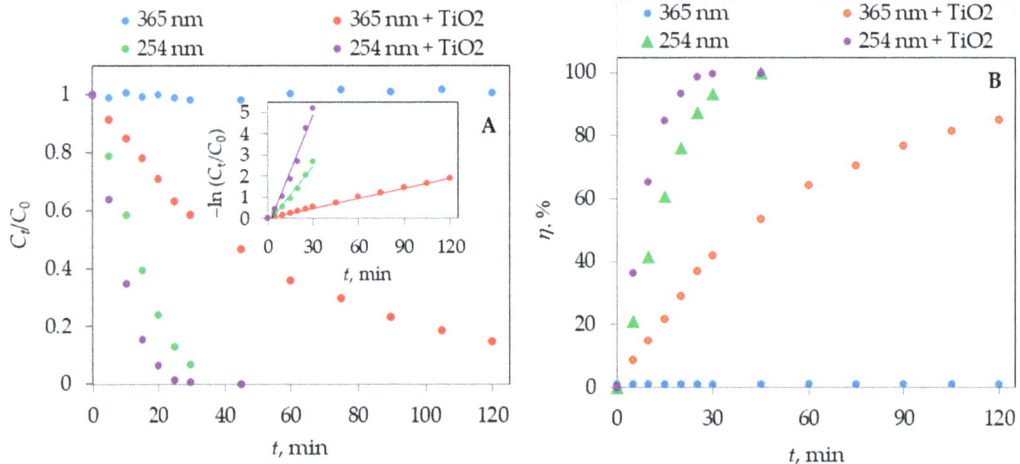

Figure 4. (**A**) Photolytic and photocatalytic degradation of memantine under UV-A (365 nm) and UV-C (254/185 nm) radiation by sol-gel nanostructured TiO$_2$ film as a function of irradiation time, C_0 (memantine) = 10 mg/L. Inset: linear transform of $-\ln (C_t/C_0)$ versus t. (**B**) Photolytic and photocatalytic degradation efficiency. All experiments were triplicated with the standard deviation from the average value ± 4%.

Table 5. Photolytic and photocatalytic degradation rate constants and half-lives of memantine.

Experiment	R^2	Regression Equation	k_1, 1/min	$t_{1/2}$, min	η, %
UV-A	–	–	–	–	0 (after 120 min)
UV-C (254/185 nm)	0.9686	y = 0.0908x − 0.2293	0.0908	7.6	100 (after 45 min)
TiO$_2$ film + UV-C (254/185 nm)	0.9693	y = 0.1779x − 0.4418	0.1779	3.9	100 (after 45 min)
TiO$_2$ film + UV-A (365 nm)	0.9984	y = 0.0159x + 0.0277	0.0159	46.3	85 (after 120 min)

From the photocatalytic experiments, it is observed that, under UV-A light (Figure 4A,B) after 45 min of irradiation, complete degradation of memantine is achieved (100% efficiency). Conversely, the photocatalytic experiments under UV-C light (Figure 4A,B) show that, after 120 min of irradiation, 85% of memantine removal is obtained. The photocatalytic degradation rate of memantine in the "UV-C + TiO$_2$ film" experiment is much faster (0.1779 min^{-1}) than that in the "UV-A + TiO$_2$ film" experiment (0.0159 min^{-1}). Similar behaviour was also observed in a recently published study of memantine oxidation [45]. In addition, photolytic oxidation of memantine by UV-A and UV-C light radiation was investigated. Memantine showed no degradation after 120 min of exposure to UV-A light alone, which is attributed to the low energy level of this type of UV light. Memantine degradation efficiency under exposure to UV-C light is the same as photocatalytic degradation by TiO$_2$ film irradiated with UV-C light (100% efficiency after 45 min, Table 5) but a rate constant of memantine degradation is two times lower for UV-C photolysis than for photocatalysis in the "TiO$_2$ film + UV-C" experiment (Table 5). The half-life of photolytic memantine degradation by UV-C light was almost two times higher (7.6 min) than the half-life obtained in the "TiO$_2$ film + UV-C" photocatalytic degradation experiment (3.9 min). It was found that the half-life of photocatalytic degradation by UV-A light was much higher (46.3 min) than the half-life obtained by UV-C light (3.9 min).

Švagelj et al. [16] published their findings on the photocatalytic degradation of memantine using a sol-gel TiO$_2$ film deposited on alumina foam substrate irradiated by UV-A light. The application of sol-gel TiO$_2$ film deposited on alumina foam substrate resulted in a larger specific surface area and therewith fast degradation of memantine can be obtained.

The diffuse reflectance spectroscopy (DRS) result and the Tauc plot are shown in Figure 5A,B. It was found that prepared TiO$_2$ only absorbs photons at wavelengths shorter than 400 nm. Based on the DRS, the Tauc plots can be obtained to determine the energy bandgap of TiO$_2$ (Figure 5B). It is observed that prepared TiO$_2$ presents a lower energy bandgap (2.99 eV) in comparison to commercial TiO$_2$ P25 (3.20 eV) [46].

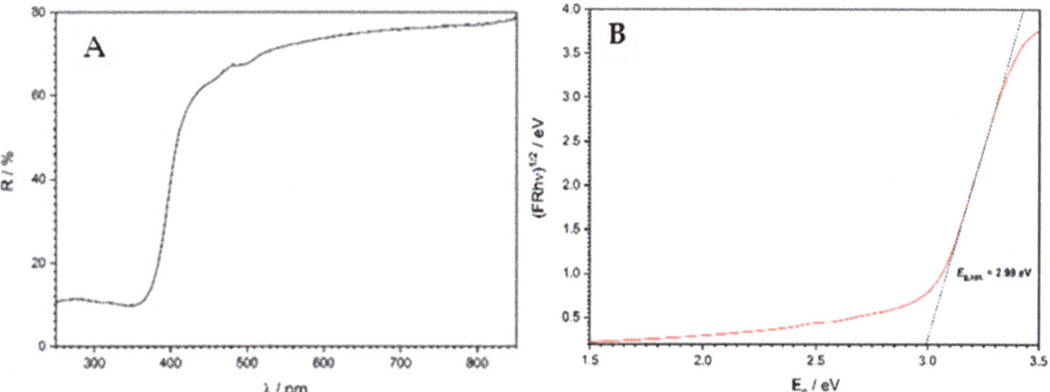

Figure 5. (**A**) DRS spectra and (**B**) Tauc plot for energy bandgap determination of TiO$_2$.

2.3. Oxidation Products of Memantine

Compared to the chromatogram of the memantine before oxidation, five new peaks were observed corresponding to the possible oxidation products of memantine. All five degradation products have lower retention times than memantine (Table 6), indicating that they are more polar. Tentative structures of memantine oxidation products (Table 6) were proposed based on the retention times, m/z-values and fragmentation patterns obtained through HPLC-MS/MS analysis. Mass spectra of memantine and its oxidation products are shown in Supplementary Materials, Figure S1.

Despite the different kinetics, all five degradation products were detected in all oxidation experiments, except photolysis UV-A light when degradation was not achieved. The same oxidation products after UV-C/H$_2$O$_2$ and UV-A/TiO$_2$ treatment were recently reported in [40].

2.4. Toxicity of the Mixture of Memantine and Its Degradation Products

Generally, some compounds do not show toxicity to a specific species, but this does not necessarily mean that they are not toxic or harmful to the environment or humans or another organism tested for toxicity [47,48]. Papac et. al. [45] determine that memantine was toxic to *Daphnia magna* (EC$_{50}$ = 7.19 mg/L). Blaschke et al. [49] demonstrate that chronic toxicity is not always more sensitive than acute toxicity. Keeping that in mind, assessment of the acute toxicity of memantine and its oxidation products during TiO$_2$ photocatalysis by UV-C light was carried out using *Vibrio fischeri* bacteria. Luminescence inhibition (%) was measured in triplicate for each tested sample and mean values and standard deviations (s) were calculated. The results presented in Table 7 indicate that memantine (10 mg/L) and its oxidation products were not harmful to *Vibrio fischeri* under the applied experimental conditions.

Table 6. Proposed chemical structures of memantine oxidation products.

Compound	t_R, min	Chemical Formula	Chemical Structure
memantine $[M+H]^+$ m/z 163 m/z 107	14.7	$C_{12}H_{22}N$ $C_{12}H_{19}$	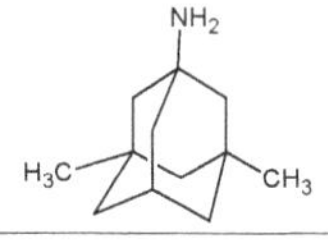
DP-1 $[M+H]^+$ m/z 179 m/z 135	2.87	$C_{12}H_{22}NO$ $C_{12}H_{19}O$ $C_{10}H_{15}$	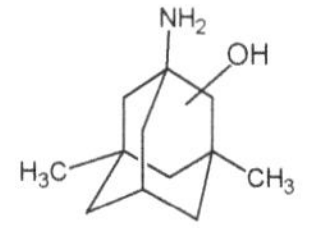
DP-2 $[M+H]^+$ m/z 177 m/z 149	4.77	$C_{12}H_{20}NO$ $C_{12}H_{17}O$ $C_{10}H_{13}O$	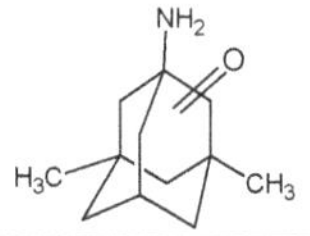
DP-3 $[M+H]^+$ m/z 193 m/z 164 m/z 135	2.29	$C_{12}H_{20}NO_2$ $C_{12}H_{17}O_2$ $C_{11}H_{18}N$ $C_{10}H_{15}$	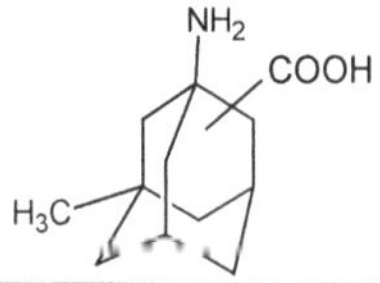
DP-4 $[M+H]^+$ m/z 195 m/z 179 m/z 135	4.11	$C_{12}H_{22}NO_2$ $C_{12}H_{19}O_2$ $C_{12}H_{19}O$ $C_{10}H_{15}$	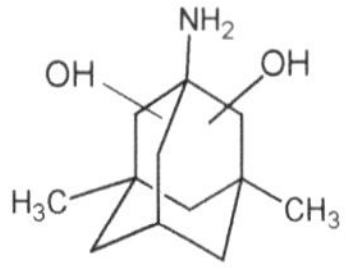
DP-5 $[M+H]^+$ m/z 209 m/z 193 m/z 135	1.53	$C_{12}H_{20}NO_3$ $C_{12}H_{17}O_3$ $C_{12}H_{17}O_2$ $C_{10}H_{15}$	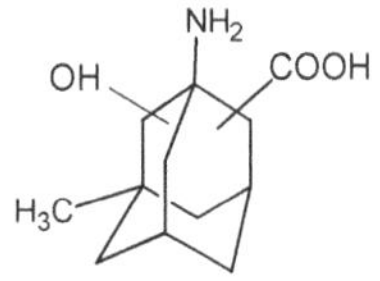

Table 7. Luminescence inhibition during the TiO_2 photocatalysis by UV-C light.

Exposure Time, min	0	10	20	30	45	60	180
Luminescence inhibition $\pm\ s$, %	0.86 ± 0.02	1.04 ± 0.03	1.40 ± 0.05	0.76 ± 0.03	0.67 ± 0.04	0.73 ± 0.04	1.98 ± 0.07

3. Environmental Relevance

The fate and behaviour of pharmaceuticals in the environment is controlled by their physicochemical properties and the characteristics of the environment. Once in the environment, the pharmaceutical can be distributed between different compartments of the environment (such as water, soil, air and biota) and be exposed to different biotic and abiotic elimination processes that can potentially lead to lowering of their environmental concentration. On the other hand, the results of elimination processes can lead to the formation of new compounds—degradation products—with different physicochemical and toxic properties.

Memantine is highly soluble in water (Table S1). From the results of in silico prediction of memantine biodegradability using different QSAR models provided by EPISuit [50], memantine can be considered a persistent compound because it does not biodegrade fast (biowin 2 < 0.5 or biowin 6 < 0.5) and its ultimate biodegradation timeframe prediction is longer than months (biowin 3 < 2.25) [51] (Table S1).

The results of our research showed that memantine is persistent to hydrolytic degradation. Although some photolytic degradation was observed in the presence of humic acids and nitrates, the degree of photodegradation was insignificant. Due to the absence of chromophores in the molecule of memantine, such results are expected. For persistent compounds, such as memantine, it is important to investigate their potential mobility in the environment determined by the compound's water solubility and sorption properties. Results of sorption/desorption experiments showed that memantine has a low tendency to sorption and is easily leached from river sediments. Considering this and its high solubility in water, it is possible to conclude that memantine will not be eliminated by natural processes and has the potential to be transported from the release site.

Such compounds, persistent and mobile in the environment, are of great concern for water quality since they are highly polar and are not removed from water by sorption. They can therefore end up in drinking water, posing a potential risk to human health [52]. Conventional wastewater treatment, based on microbial degradation and sorption, is expected to be ineffective for the removal of persistent and mobile compounds, since they are neither biodegradable nor sorbed substantially [52]. Therefore, it is of great importance to prevent their release into the environment by developing advanced and effective wastewater treatment processes.

Today, water treatment technology is trying to introduce processes that will be efficient and cheap, but also in accordance with ecological principles. One of the methods for degradation of such persistent and mobile compounds in water that is close to meeting these requirements is photocatalytic oxidation, with the use of titanium(IV) oxide (TiO_2) as a photocatalyst [53]. In addition to TiO_2, the presence of a suitable source of UV radiation that starts the process and oxygen dissolved in water are also necessary. It is a process that is included in the so-called advanced oxidation processes (AOPs). For the photocatalytic oxidation process, it is not necessary to add any additional chemicals except a solid photocatalyst (in the form of particles or nanostructured films on the reactor walls) and oxygen, while ensuring irradiance in UV spectra [54]. The use of solar radiation as a process activator (i.e., a source of UV radiation) and oxygen from the air around the reactor contribute to approaching the ecological principles of this technology. Photocatalytic oxidation of memantine resulted with the occurrence of five degradation products. According to the shorter chromatographic retention times compared to memantine, it is assumed that the oxidation products are more polar than memantine. This may indicate better solubility in water and a weaker tendency for sorption. Although oxidation products of memantine do not show inhibition of bioluminescence of *Vibrio Fischeri*, future studies on the current topic are suggested to assess the cytotoxicity and genotoxicity of memantine and its oxidation products in surface waters and wastewaters, as well as toxicological risks to ecosystems and human health. Furthermore, experimental data on the biodegradation of memantine and its oxidation products should be gathered.

4. Experimental Section

4.1. Materials and Chemicals

Analytical standard of memantine hydrochloride (CAS number: 41100-52-1) (Sigma Aldrich, St. Louis, MO, USA) of high purity (>98%) was used in this study. Memantine stock solution concentration of 1000 mg/L was prepared by weighing the accurate mass of memantine standard and dissolving it in methanol.

Acetonitrile, formic acid, citric acid, ascorbic acid and inorganic salts were of analytical grade and supplied by Kemika (Zagreb, Croatia). For buffer preparation, analytical grade reagents were used. Ultra-pure water was prepared using a Millipore Simplicity UV system (Millipore Corporation, Billerica, MA, USA).

For toxicity evaluation, freeze-dried and liquid luminescent bacteria *Vibrio fischeri* NRRL-B-11177 (LCK484, LUMINStox LUMISmini, Hach Lange, Varaždin, Croatia) were used. The bacterial reagents as well as reconstitution reagents were purchased from Kemika (Zagreb, Croatia).

For the preparation of TiO_2 sol (colloidal solution), the following components were used: titanium (IV) isopropoxide ($Ti(C_3H_5O_{12})_4$, TTIP, 97%, Sigma-Aldrich, St. Louis, MO, USA), i-propanol (C_3H_7OH, Grammol, Croatia), acetylacetone ($CH_3(CO)CH_2(CO)CH_3 \geq 99\%$, Honeywell, Charlotte, NC, USA), nitric acid (HNO_3, Carlo Erba Reagents, Barcelona, Spain) and polyethylene glycol ($H(OCH_2CH_2)nOH$, Mr = 5000–7000, Sigma-Aldrich, St. Louis, MO, USA). All these chemicals were analytical grade reagents.

4.2. Sediment Samples

Samples of river or fluvial sediments were collected on the territory of the Republic of Croatia and in the following areas: in Sisak-Moslavina County on the Petrinjčica River in the town of Petrinja, in Požega-Slavonia County on the Pakra River in the town of Pakrac and in Primorsko-Goranska County on the Studena River in the surroundings of the city Rijeka. In all locations, the samples were collected outside of human activities, which provides some assurance that the samples are not contaminated, especially in the case of pharmaceuticals, and in the summertime when it is easier to reach the area and collect samples due to dryness. All samples were air-dried, crushed, sieved through a 2-mm sieve and characterized according to the proposed procedure [55]. The physicochemical properties of used samples can be seen in the previously published work [31].

4.3. Hydrolytic Degradation Experiments

Hydrolytic degradation was conducted at $(50 \pm 0,1)$ °C (Incubator shakers KS 3000 i control, IKA, Staufen, Germany) for 5 days and at three pH values (1, 7 and 9) in capped glass vials under dark conditions. A buffer solution with a pH value of 4 was prepared by mixing 38.55 mL of 0.2 M K_2HPO_4 and 61.45 mL of 0.1 M citric acid. A pH value buffer solution of 7 was prepared by mixing 29.63 mL of 0.1 M NaOH, 50 mL of 0.1 M KH_2PO_4 and 20.37 mL of MilliQ water, and the pH 9 buffer solution was prepared by mixing 21.30 mL of 0.1 M NaOH, 50 mL of 0.1 M H_3BO_3 in 0.1 M KCl and 28.70 mL of MilliQ water. The pH of each buffer solution was checked with pH meter S20 SevenEasy (Mettler Toledo, Greifensee, Switzerland). Memantine solutions were prepared in appropriate buffer solutions at a concentration of 10 mg/L. Concentration of memantine solutions after hydrolytic degradation experiment were determined by HPLC-MS/MS. All experiments were performed in three replicates.

4.4. Photolytic Degradation Experiments at Environmentally Relevant Conditions

Direct photolysis experiment was performed with memantine solution in MilliQ water (10 mg/L). To test the possibility of indirect photolysis, solutions of memantine (10 mg/L) were prepared in solutions of Cl^- ions (10 mg/L), NO_3^- ions (3 mg/L) and humic acids (3 mg/L). Forty millilitres of test solution were irradiated in quartz vessels (diameter 4.6 cm) placed in Suntest CPS+ simulator (Atlas, Linsengericht, Germany). Suntest CPS+ is equipped with a temperature sensor and a xenon lamp as a source of artificial sunlight in the wavelength range of 300–800 nm. The distance between the liquid surface and the lamp was 14 cm. During the experiments, the radiation intensity was maintained at 500 W m^{-2} and the reaction temperature was kept at (25 ± 2) °C. In all photolysis experiments, dark control experiments were performed under the same conditions but protected from the radiation. Control samples with the same composition as test solutions were used to establish that memantine degradation was a consequence of the irradiation. Aliquots of irradiated memantine solution were analysed by HPLC-MS/MS. All experiments were performed in three replicates.

4.5. Sorption Experiments

Batch sorption experiments were performed according to the OECD 106 procedure [56]. The procedure is performed in triplicate by shaking on a laboratory shaker (Innova 4080 Incubator Shaker, New Brunswick Scientific, Edison, NJ, USA), which allows continuous contact of the sediment samples with the memantine solution. To avoid photolytic degradation, all experiments were performed in the dark and, to avoid microbiological activities, all sediment samples were sterilized beforehand.

It is very important to choose a good ratio between sediment (sorbent) and memantine solution. Since previous studies [30,31] show that memantine does not have excessive sorption potential to sediment or soil samples, all experiments were performed with a 1:10 (w/v) sediment/memantine solution ratio. In a previously published paper [30], it was determined that 24 h was sufficient for memantine to reach sorption equilibrium, so all experiments were conducted with 24 h of shaking. The procedure consisted of adding 10 mL known concentration (0.1–2.0 mg/L) of memantine solution in 50 mL of laboratory glass to 1 g of air-dried sediment. The prepared suspension was shaken in a shaker (at 200 rpm) for 24 h at room temperature (25 °C), filtered through a 0.45 μm syringe filter and transferred to HPLC vials. Blank samples containing the same amount of sediment and soil in contact with 10.0 mL of 0.01 M $CaCl_2$ solution were also included in the analysis. They served as controls to detect interfering compounds or contaminated sediment.

In order to investigate the influence of pH and ionic strength on the sorption of memantine on the sediments studied, a series of experiments were performed in which one of the factors was changed while the others remained constant. The effect of pH was monitored using three series of experiments with different pH values of the studied memantine solutions (pH values 5, 7 and 9). All these experiments were performed in a 0.01 M $CaCl_2$ solution. To determine the effect of ionic strength, the pH of the memantine solution must be adjusted to the initial value (pH 7.0). These experiments were performed with three different concentrations of $CaCl_2$ solution (0.001, 0.01 and 0.1 mol/L).

Since the experiments to determine the required contact time (24 h) were the basis for the determination of sorption kinetics, we were able to start immediately with the determination of the kinetics of memantine desorption from the sediments studied. Desorption kinetics were studied for three memantine solutions in 0.01 mol/L $CaCl_2$ (0.1, 0.5 and 2.0 mg/L) using a decanting and refilling technique. After 24 h of shaking, the memantine solutions in contact with the sediment samples were replaced with fresh 0.01 mol/L $CaCl_2$. Memantine solution was removed using a disposable glass pipette. The residual solution that could not be removed before the desorption experiment was determined gravimetrically, and the same amount of 0.01 mol/L $CaCl_2$ as the removed memantine solution was weighed and added to the residual solution. Samples were then shaken at 25 °C for various periods (10, 20, 30, 40 and 50 min, and 1, 2, 4, 6, 12, 18 and 24 h).

4.6. Photolytic and Photocatalytic Oxidation Experiments

All experiments were carried out with 10 mg/L memantine solution in two borosilicate glass tubes (200 mm in height, 30 mm in diameter, 0.11 L) with continuous purging with air (O_2), at (25 ± 0.2) °C: (i) with the TiO_2 film, for photocatalytic experiments and (ii) without the TiO_2 film, for photolytic experiments. UV lamps (UV-A and UV-C) were placed in the middle of each reactor. Detailed experimental set-up is published elsewhere [17]. UV lamps used in experiments were 15 W mercury UV lamps: (i) model Pen-Ray CPQ-7427, UV-A with λ_{max} = 365 nm and (ii) model Pen-Ray 90-0004-07, UV-C with λ_{max} = 254/185 nm, manufactured by UVP (Upland, CA, USA). Both lamps were used with the same electrical source PS-4, I = 0.54 A, from UVP, too. The experimental set-up was described in detail in [17]. The reaction temperature was controlled by the circulation of cooling water. Total irradiation time for each oxidation test was 120 min. Aliquots of 1 mL were collected in defined time intervals and stored in the dark at 4 °C until HPLC-MS/MS analysis.

Nanostructured TiO_2 film was deposited on a borosilicate glass substrate by the sol-gel process using the dip-coating method. TiO_2 colloidal solution (sol) was prepared by mixing titanium(IV) isopropoxide (Ti-iPrOH) as a precursor, i-propyl alcohol (iPrOH) as a solvent, acetylacetone (AcAc) as a chelating agent, nitric acid (NA, 0.5 M) as a catalyst and polyethylene glycol as an organic/polymer additive in the amount of 2 g. The molar ratio of these reactants was Ti-iPrOH:iPrOH:AcAc:NA=1:35:0.63:0.015. The film was dried at 100 °C for 1 h prior to the deposition of the next layer. After the deposition of the three layers, the deposited film was heat-treated at 550 °C for 4 h [14]. The procedure for the film preparation as well as its characterization was described in detail elsewhere [14,17].

Energy bandgap (Eg) of prepared TiO_2 was calculated from diffuse reflectance spectroscopy (DRS) measurements, which were performed on a QE Pro High-Performance Spectrometer (Ocean Insight, Orlando, FL, USA) equipped with an integrating sphere and a DH 2000 deuterium–halogen source in the analysis range 200–1000 nm with a resolu-tion of 1 nm and integration time of 10 s.

4.7. HPLC-MS/MS Analysis

Samples from photolytic and hydrolytic degradation experiments as well as samples from AOP experiments were analysed using an Agilent Series 1200 HPLC system (Santa Clara, CA, USA) coupled with an Agilent 6410 triple-quadrupole mass spectrometer equipped with an ESI interface (Santa Clara, CA, USA). Chromatographic separation was performed on an Kinetex C18 column (100 mm × 2.1 mm, 2.6 µm) (Phenomenex, Torrance, CA, USA) using mobile phase comprising MilliQ water with 0.1% formic acid as eluent A and acetonitrile with 0.1% formic acid as eluent B. The composition of 50% organic phase (B) was maintained at flow rate of 0.2 mL/min throughout the analysis. An injection volume of 5 µL was used in all analyses. The analyses were done in positive ion mode under the following conditions: drying gas temperature 350 °C; capillary voltage 4.0 kV; drying gas flow 11 L/min and nebulizer pressure 35 psi. Instrument control, data acquisition and evaluation were done with Agilent MassHunter 2003–2007 Data Acquisition for Triple Quad B.01.04 (B84) software (Santa Clara, CA, USA).

The residual concentration of memantine in the remaining liquid phase after sorption was analysed using UHPLC-MS (Agilent 6490 coupled with Agilent Infinity UHPLC System Triple Quadrupole Mass Spectrometer, Santa Clara, CA, USA) with electrospray ionization. The chromatographic column Shim pack XR ODS II (50 mm × 2 mm i.d., 1.6µm) (Shimadzu, Duisburg, Germany) was used at 30 °C with an injection volume of 1 µL. The mobile phase consisted of two eluents: eluent A (0.1% formic acid in MilliQ water) and B (0.1% formic acid in acetonitrile) and was performed in gradient elution mode. The gradient started with a 0.1-min linear gradient from 100% A to 10% B, followed by a 1.0-min linear gradient to 98% B, followed by a 0.5-min linear gradient back to 100% A held for 0.4 min. The flow rate was 0.2 mL/min. All analyses were performed in positive ion mode under the following parameters: drying gas temperature 200 °C, capillary voltage 3.0 kV, drying gas flow rate 15 L/min and nebulizer pressure 20 psi. Memantine was analysed by MRM, using the two highest characteristic precursor ion/product ion transitions (m/z 291.25→230.2; m/z 291.25→123.0).

4.8. Assessment of Acute Toxicity by Vibrio Fischeri

Acute toxicity assessment toward *Vibrio fischeri* culture was performed on standard solutions of memantine (10 mg/L), a mixture of memantine with its degradation products and finally the degradation products themselves without the detectable presence of memantine. Acute toxicity assessment was performed according to the method described in detail in [29]. In brief, sample solutions for toxicity measurements were prepared by serial dilutions in linear progression with addition of 2% NaCl. The experiments were conducted in a test tube by combining each volume of initial or diluted sample (1.5 mL) and 0.5 mL of bacterium *Vibrio Fischeri* suspension. The inhibition of luminescence was measured before and after 30 min of exposure of the sample to *Vibrio fischeri* on a luminometer (LUMIStox 300 Hach Lange, Düsseldorf, Germany) at 15 °C. In order to control bacteria performance, the reference substances $ZnSO_4$ x $(OH_2)_7$ (109.9 µg/mL) and $K_2Cr_2O_7$ (22.6 µg/mL) were used.

5. Conclusions

This research provides a comprehensive picture of memantine behaviour in the aquatic environment. The results showed that memantine is resistant to hydrolytic and photolytic (i.e., irradiated by solar light) degradation, has a low tendency to sorption and is easily desorbed from river sediments. For such compounds, which are persistent and mobile in the environment, it is of great importance to prevent their release into the environment by effective wastewater treatment.

Investigation of memantine oxidation by photolytic/photocatalytic oxidation by UV-A and UV-C light showed that memantine could be completely oxidized within 30 min during photocatalytic and within 50 min during photolytic oxidation processes by UV-C light. Photolytic degradation by UV-A light did not occur, while photocatalytic degradation by UV-A light did occur although the degradation rate was lower and memantine degradation was not completed even after 120 min. A kinetic study showed that the oxidation in all experiments, in which oxidation occurred, followed pseudo-first-order reaction kinetics.

As a result of the oxidation, five oxidation products were formed, which were identified using high performance liquid chromatography coupled with a triple quadrupole mass spectrometer. The same oxidation products were identified in all investigated processes in which oxidation occurred. The results of the acute toxicity assessment of memantine and its mixture with oxidation products indicate that the resulting products are not harmful.

Supplementary Materials: The following supporting information can be downloaded at: https://www. mdpi.com/article/10.3390/catal13030612/s1, Figure S1: Mass spectra of memantine and its oxidation products; Table S1: Results of EPISuite biodegradability prediction for memantine.

Author Contributions: Conceptualization, S.B. and D.L.; methodology, S.B., D.L., D.M.P., M.B., L.Ć. and D.D.; formal analysis, M.B., D.M.P. and D.D.; investigation, S.B., D.L., D.M.P., M.B., L.Ć. and D.D.; resources, S.B., D.L., D.M.P. and L.Ć.; data curation, S.B., D.L., D.M.P., M.B., L.Ć. and D.D.; writing—original draft preparation, S.B., L.Ć. and D.M.P.; writing—review and editing, S.B., D.L., D.M.P., M.B., L.Ć. and D.D.; visualization, L.Ć., D.M.P. and M.B.; supervision, S.B.; funding acquisition, S.B. All authors have read and agreed to the published version of the manuscript.

Funding: This research was funded by the Croatian Science Foundation under the project Fate of pharmaceuticals in the environment and during advanced wastewater treatment (PharmaFate) (IP-09-2014-2353).

Data Availability Statement: The data presented in this study are available upon reasonable request from the corresponding author.

Conflicts of Interest: The authors declare no conflict of interest.

References

1. Garcia, M.J.; Leadleya, R.; Lang, S.; Ross, J.; Vinand, E.; Ballard, C.; Gsteiger, S. Real-World Use of Symptomatic Treatments in Early Alzheimer's Disease. *J. Alzheimer's Dis.* **2023**, *91*, 151–167. [CrossRef] [PubMed]
2. Prevalence of Dementia in Europe. Available online: https://www.alzheimer-europe.org/dementia/prevalence-dementia-europe (accessed on 7 March 2023).
3. Food and Drug Administration, Approval Labeling Text NDA 21-487. Available online: https://www.accessdata.fda.gov/drugsatfda_docs/label/2003/021487lbl.pdf. (accessed on 7 March 2023).
4. PubChem, Memantine Hydrochloride. Available online: https://pubchem.ncbi.nlm.nih.gov/compound/181458#section=Molecular-Formula. (accessed on 7 March 2023).
5. Suzuki, T.; Kosugi, Y.; Watanabe, K.; Iida, H.; Nishimurab, T. Environmental Risk Assessment of Active Human Pharmaceutical Ingredients in Urban Rivers in Japan. *Chem. Pharm. Bull.* **2021**, *69*, 840–853. [CrossRef]
6. Gómez-Canela, C.; Edo, S.; Rodríguez, N.; Gotor, G.; Lacorte, S. Comprehensive Characterization of 76 Pharmaceuticals and Metabolites in Wastewater by LC-MS/MS. *Chemosensors* **2021**, *9*, 273. [CrossRef]
7. Fick, J.; Lindberg, R.H.; Parkkonen, J.; Arvidsson, B.; Tysklind, M.; Larsson, D.G.J. Therapeutic Levels of Levonorgestrel Detected in Blood Plasma of Fish: Results from Screening Rainbow Trout Exposed to Treated Sewage Effluents. *Environ. Sci. Technol.* **2010**, *44*, 2661–2666. [CrossRef]

8.	Kårelid, V.; Larsson, G.; Björlenius, B. Pilot-scale removal of pharmaceuticals in municipal wastewater: Comparison of granular and powdered activated carbon treatment at three wastewater treatment plants. *J. Environ. Manag.* **2017**, *193*, 491–502. [CrossRef] [PubMed]

9.	Mansouri, F.; Chouchene, K.; Roche, N.; Ksibi, M. Removal of Pharmaceuticals from Water by Adsorption and Advanced Oxidation Processes: State of the Art and Trends. *Appl. Sci.* **2021**, *11*, 6659. [CrossRef]

10.	McMichael, S.; Fernández-Ibáñez, P.; Byrne, J.A. A review of photoelectrocatalytic reactors for water and wastewater treatment. *Water* **2021**, *13*, 1198. [CrossRef]

11.	Ibhadon, A.O.; Fitzpatrick, P. Heterogeneous Photocatalysis: Recent Advances and Applications. *Catalysts* **2013**, *3*, 189–218. [CrossRef]

12.	Ma, D.; Yi, H.; Lai, C.; Liu, X.; Huo, X.; An, Z.; Li, L.; Fu, Y.; Li, B.; Zhang, M.; et al. Critical review of advanced oxidation processes in organic wastewater treatment. *Chemosphere* **2021**, *275*, 130104. [CrossRef]

13.	Gomes, J.; Lincho, J.; Domingues, E.; Quinta-Ferreira, R.; Martins, R. N–TiO$_2$ Photocatalysts: A Review of Their Characteristics and Capacity for Emerging Contaminants Removal. *Water* **2019**, *11*, 373. [CrossRef]

14.	Čizmić, M.; Ljubas, D.; Rožman, M.; Ašperger, D.; Ćurković, L.; Babíc, S. Photocatalytic degradation of azithromycin by nanostructured TiO$_2$ film: Kinetics, degradation products, and toxicity. *Materials* **2019**, *12*, 873. [CrossRef] [PubMed]

15.	Čizmić, M.; Ljubas, D.; Ćurković, L.; Škorić, I.; Babić, S. Kinetics and degradation pathways of photolytic and photocata-lytic oxidation of the anthelmintic drug praziquantel. *J. Hazard. Mater.* **2017**, *323*, 500–512. [CrossRef]

16.	Švagelj, Z.; Mandić, V.; Ćurković, L.; Biošić, M.; Žmak, I.; Gaborardi, M. Titania-Coated alumina foam photocatalyst for memantine degradation derived by replica method and sol-gel reaction. *Materials* **2020**, *13*, 227. [CrossRef] [PubMed]

17.	Ćurković, L.; Ljubas, D.; Šegota, S.; Bačić, I. Photocatalytic degradation of Lissamine Green B dye by using nanostructured sol-gel TiO$_2$ films. *J. Alloys Compd.* **2014**, *604*, 309–316. [CrossRef]

18.	Pant, B.; Park, M.; Park, S.-J. Recent Advances in TiO$_2$ Films Prepared by Sol-Gel Methods for Photocatalytic Degradation of Organic Pollutants and Antibacterial Activities. *Coatings* **2019**, *9*, 613. [CrossRef]

19.	OECD. *Test No. 111: Hydrolysis as a Function of pH, OECD Guidelines for the Testing of Chemicals, Section 1*; OECD Publishing: Paris, France, 2004. [CrossRef]

20.	Biošić, M.; Mitrevski, M.; Babić, S. Environmental behavior of sulfadiazine, sulfamethazine, and their metabolites. *Environ. Sci. Pollut. Res.* **2017**, *24*, 9802–9812. [CrossRef]

21.	Dabić, D.; Babić, S.; Škorić, I. The role of photodegradation in the environmental fate of hydroxychloroquine. *Chemosphere* **2019**, *230*, 268–277. [CrossRef]

22.	Blum, K.M.; Norström, S.H.; Golovko, O.; Grabic, R.; Järhult, J.D.; Koba, O.; Söderström Lindström, H. Removal of 30 active pharmaceutical ingredients in surface water under long-term artificial UV irradiation. *Chemosphere* **2017**, *176*, 175–182. [CrossRef]

23.	Bialk-Bielinska, A.; Stolte, S.; Matzke, M.; Fabianska, A.; Maszkowska, J.; Kolodziejska, M.; Liberek, B.; Stepnowski, P.; Kumirska, J. Hydrolysis of suplhonamides in aqueous solutions. *J. Hazard. Mater.* **2012**, *221–222*, 264–274. [CrossRef]

24.	Dabić, D.; Hanževački, M.; Škorić, I.; Žegura, B.; Ivanković, K.; Biošić, M.; Tolić, K.; Babić, S. Photodegradation, toxicity and density functional theory study of pharmaceutical metoclopramide and its photoproducts. *Sci. Total Environ.* **2022**, *807*, 150694. [CrossRef]

25.	Chiron, S.; Minero, C.; Evione, D. Photodegradation Processes of the Antiepileptic Drug Carbamazepine, Relevant to Estuarine Waters. *Environ. Sci. Technol.* **2006**, *40*, 5977–5983. [CrossRef] [PubMed]

26.	Babić, S.; Mutavdžić Pavlović, D.; Biošić, M.; Ašperger, D.; Škorić, I.; Runje, M. Fate of febantel in the aquatic environment—The role of abiotic elimination processes. *Environ. Sci. Pollut. Res.* **2018**, *25*, 28917–28927. [CrossRef]

27.	Tonski, M.; Dołżonek, J.; Stepnowski, P.; Białk-Bielinska, A. Hydrolytic stability of selected pharmaceuticals and their transformation products. *Chemosphere* **2019**, *236*, 124236. [CrossRef] [PubMed]

28.	Biošić, M.; Škorić, I.; Beganović, J.; Babić, S. Nitrofurantoin hydrolytic degradation in the environment. *Chemosphere* **2017**, *186*, 660–668. [CrossRef] [PubMed]

29.	Biošić, M.; Dabić, D.; Škorić, I.; Babić, S. Effects of environmental factors on nitrofurantoin photolysis in water and its acute toxicity assessment. *Environ. Sci. Process. Impacts* **2021**, *23*, 1385–1393. [CrossRef]

30.	Mutavdžić Pavlović, D.; Tolić Čop, K.; Barbir, V.; Gotovuša, M.; Lukač, I.; Lozančić, A.; Runje, M. Sorption of cefdinir, memantine, praziquantel and trimethoprim in sediment and soil samples. *Environ. Sci. Pollut. Res.* **2022**, *29*, 66841–66857. [CrossRef] [PubMed]

31.	Mutavdžić Pavlović, D.; Tolić Čop, K.; Prskalo, H.; Runje, M. Influence of organic matter on the sorption of cefdinir, memantine and praziquantel on different soil and sediment samples. *Molecules* **2022**, *27*, 8008. [CrossRef]

32.	Ozacar, M.; Sengýl, I.A. Two-stage batch sorber design using second-order kinetic model for the sorption of metal complex dyes onto pine sawdust. *Biochem. Eng. J.* **2004**, *21*, 39–45. [CrossRef]

33.	Mutavdžić Pavlović, D.; Ćurković, L.; Macan, J.; Žižek, K. Eggshell as a new biosorbent for the removal of the pharmaceuticals from aqueous solutions. *CLEAN–Soil Air Water* **2017**, *45*, 1700082-1–1700082-14. [CrossRef]

34.	Guo, X.; Yang, C.; Dang, Z.; Zhang, Q.; Li, Y.; Meng, Q. Sorption thermodynamics and kinetics properties of tylosin and sulfamethazine on goethite. *Chem. Eng. J.* **2013**, *223*, 59–67. [CrossRef]

35.	Samarghandi, M.R.; Hadi, M.; Moayedi, S.; Barjasteh Askari, F. Two-parameter isotherms of methyl orange sorption by pinecone derived activated carbon. *J. Environ. Health. Sci. Eng.* **2009**, *6*, 285–294.

36. Doretto, K.M.; Peruchi, L.M.; Rath, S. Sorption and desorption of sulfadimethoxine, sulfaquinoxaline and sulfamethazine antimicrobials in Brazilian soils. *Sci. Total Environ.* **2014**, *476–477*, 406–414. [CrossRef]
37. Mutavdžić Pavlović, D.; Glavač, A.; Gluhak, M.; Runje, M. Sorption of albendazole in sediments and soils: Isotherms and kinetics. *Chemosphere* **2018**, *193*, 635–644. [CrossRef]
38. Schaffer, M.; Boxberger, N.; Bornick, H.; Licha, T.; Worch, E. Sorption influenced transport of ionizable pharmaceuticals onto a natural sandy aquifer sediment at different pH. *Chemosphere* **2012**, *87*, 513–520. [CrossRef] [PubMed]
39. Kodešová, R.; Grabic, R.; Kočárek, M.; Klement, A.; Golovko, O.; Fér, M.; Nikodem, A.; Jakšík, O. Pharmaceuticals' sorptions relative to properties of thirteen different soils. *Sci. Total Environ.* **2015**, *511*, 435–443. [CrossRef] [PubMed]
40. Białk-Bielińska, A.; Maszkowska, J.; Mrozik, W.; Bielawska, A.; Kołodziejska, M.; Palavinskas, R.; Stepnowski, P.; Kumirska, J. Sulfadimethoxine and sulfaguanidine: Their sorption potential on natural soils. *Chemosphere* **2012**, *86*, 1059–1065. [CrossRef]
41. Cao, X.; Pang, H.; Yang, G. Sorption behaviour of norfloxacin on marine sediments. *J. Soils Sediments* **2015**, *15*, 1635–1643. [CrossRef]
42. Mutavdžić Pavlović, D.; Ćurković, L.; Grčić, I.; Šimić, I.; Župan, J. Isotherm, kinetic, and thermodynamic study of ciprofloxacin sorption on sediments. *Environ. Sci. Pollut. Res.* **2017**, *24*, 10091–10106. [CrossRef]
43. Tolić, K.; Mutavdžić Pavlović, D.; Židanić, D.; Runje, M. Nitrofurantoin in sediment and soils: Sorption, isotherms and kinetics. *Sci. Tot. Environ.* **2019**, *681*, 9–17. [CrossRef]
44. Gabelica, I.; Ćurković, L.; Mandić, V.; Panžić, I.; Ljubas, D.; Zadro, K. Rapid microwave-assisted synthesis of $Fe_3O_4/SiO_2/TiO_2$ core-2-layer-shell nanocomposite for photo-catalytic degradation of ciprofloxacin. *Catalysts* **2021**, *11*, 1136. [CrossRef]
45. Papac, J.; Garcia Ballesteros, S.; Tonković, S.; Kovačić, M.; Tomić, A.; Cvetnić, M.; Kušić, H.; Senta, I.; Terzić, S.; Ahel, M.; et al. Degradation of pharmaceutical memantine by photo-based advanced oxidation processes: Kinetics, pathways and environmental aspects. *J. Environ. Chem. Eng.* **2023**, *11*, 109334. [CrossRef]
46. Sanchez Tobon, C.; Ljubas, D.; Mandić, V.; Panžić, I.; Matijašić, G.; Ćurković, L. Microwave-Assisted Synthesis of N/TiO_2 Nanoparticles for Photocatalysis under Different Irradiation Spectra. *Nanomaterials* **2022**, *12*, 1473. [CrossRef] [PubMed]
47. Vulliet, E.; Emmelin, C.; Chovelon, J.M.; Chouteau, C.; Clement, B. Assessment of the toxicity of triasulfuron and its photoproducts using aquatic organisms. *Environ. Toxicol. Chem.* **2004**, *23*, 2837–2843. [CrossRef] [PubMed]
48. Froehner, K.; Backhaus, T.; Grimme, L.H. Bioassays with Vibrio fischeri for the assessment of delayed toxicity. *Chemosphere* **2000**, *40*, 821–828. [CrossRef] [PubMed]
49. Blaschke, U.; Paschke, A.; Rensch, I.; Schüürmann, G. Acute and chronic toxicity toward the bacteria Vibrio fischeri of organic narcotics and epoxides: Structural alerts for epoxide excess toxicity. *Chem Res Toxicol.* **2010**, *23*, 1936–1946. [CrossRef]
50. EPISuite US. *EPA Estimation Programs Interface Suite™ for Microsoft®Windows, v 4.11, United States Environmental Protection Agency*; EPISuite US: Washington, DC, USA, 2017.
51. ECHA European Chemicals Agency. Chapter R.11: PBT/vPvB Assessment. In Guidance on Information Requirements and Chemical Safety Assessment; Version 3.0. 2017. Available online: https://echa.europa.eu/documents/10162/13632/information_requirements_r11_en.pdf/a8cce23f-a65a-46d2-ac68-92fee1f9e54f (accessed on 7 March 2023).
52. Reemtsma, T.; Berger, U.; Arp, H.P.H.; Gallard, H.; Knepper, T.P.; Neumann, M.; Quintana, J.B.; de Voogt, P. Mind the Gap: Persistent and Mobile Organic Compounds—Water Contaminants That Slip Through. *Environ. Sci. Technol.* **2016**, *50*, 10308–10315. [CrossRef] [PubMed]
53. Hoffmann, M.R.; Martin, S.T.; Choi, W.; Bahnemann, D.W. Environmental applications of semiconductor photocatalysis. *Chem. Rev.* **1995**, *95*, 69–96.
54. Linsebigler, A.L.; Lu, G.; Yates, J.T., Jr. Photocatalysis on TiO_2 surfaces: Principles, Mechanisms and selected results. *Chem. Rev.* **1995**, *95*, 735–758. [CrossRef]
55. Mutavdžić Pavlović, D.; Ćurković, L.; Blažek, D.; Župan, J. The sorption of sulfamethazine on soil samples: Isotherms and error analysis. *Sci. Total Environ.* **2014**, *497–498*, 543–552. [CrossRef]
56. OECD. *Adsorption-Desorption Using a Batch Equilibrium Method. OECD Guideline for the Testing of Chemicals 106. Organization for Economic Cooperation and Development*; OECD: Paris, France, 2000.

Article

Development of Nanomedicine from Copper Mine Tailing Waste: A Pavement towards Circular Economy with Advanced Redox Nanotechnology

Amrita Banerjee [1,2], Ria Ghosh [3], Tapan Adhikari [4], Subhadipta Mukhopadhyay [1], Arpita Chattopadhyay [5,*] and Samir Kumar Pal [3,*]

1 Department of Physics, Jadavpur University, Kolkata 700032, India
2 Technical Research Centre, S. N. Bose National Centre for Basic Sciences, Block JD, Sector III, Salt Lake, Kolkata 700106, India
3 Department of Chemical and Biological Sciences, S. N. Bose National Centre for Basic Sciences, Block JD, Sector 3, Salt Lake, Kolkata 700106, India
4 Indian Institute of Soil Science Nabibagh, Bhopal 462038, India
5 Department of Basic Science and Humanities, Techno International New Town, Block—DG 1/1, Action Area 1 New Town, Rajarhat, Kolkata 700156, India
* Correspondence: arpita.chattopadhyay@tict.edu.in (A.C.); skpal@bose.res.in (S.K.P.)

Abstract: Copper, the essential element required for the human body is well-known for its profound antibacterial properties, yet salts and oxides of copper metals in the copper mine tailings are reported to be a big burden in the modern era. Among other copper oxides, CuO, in particular, is known to have beneficial effects on humans, while its slight nanoengineering viz., surface functionalization of the nanometer-sized oxide is shown to make some paradigm shift using its inherent redox property. Here, we have synthesized nanometer-sized CuO nanoparticles and functionalized it with a citrate ligand for an enhanced redox property and better solubility in water. For structural analysis of the nanohybrid, standard analytical tools, such as electron microscopy, dynamic light scattering, and X-ray diffraction studies were conducted. Moreover, FTIR and UV-VIS spectroscopy studies were performed to confirm its functionalization. The antibacterial study results, against a model bacteria (*S. hominis*), show that CuO nanohybrids provide favorable outcomes on antibiotic-resistant organisms. The suitability of the nanohybrid for use in photodynamic therapy was also confirmed, as under light its activity increased substantially. The use of CuO nanoparticles as antibiotics was further supported by the use of computational biology, which reconfirmed the outcome of our experimental studies. We have also extracted CuO nanogranules (top-down technique) from copper mine tailings of two places, each with different geographical locations, and functionalized them with citrate ligands in order to characterize similar structural and functional properties obtained from synthesized CuO nanoparticles, using the bottom-up technique. We have observed that the extracted functionalized CuO from copper tailings offers similar properties compared to those of the synthesized CuO, which provides an avenue for the circular economy for the utilization of copper waste into nanomedicine, which is known to be best for mankind.

Keywords: copper nanohybrid; citrate functionalized CuO; nano-medicine; *S. hominis* infection control; photodynamic therapy

Citation: Banerjee, A.; Ghosh, R.; Adhikari, T.; Mukhopadhyay, S.; Chattopadhyay, A.; Pal, S.K. Development of Nanomedicine from Copper Mine Tailing Waste: A Pavement towards Circular Economy with Advanced Redox Nanotechnology. *Catalysts* **2023**, *13*, 369. https://doi.org/10.3390/catal13020369

Academic Editors: Gassan Hodaifa, Rafael Borja and Mha Albqmi

Received: 10 January 2023
Revised: 1 February 2023
Accepted: 3 February 2023
Published: 7 February 2023

1. Introduction

Copper was the first metal discovered in the history of human civilization during the chalcolithic or copper age, around 6000 years ago. This marvel metal was well known in ancient times, not only for its ability to enhance the strength of the tools, but also for its impressive healing capacity. The ancient Indians, Greeks, and Egyptians all used copper containers for water purification, treatment of wounds, and lung ailments. Copper cooking utensils were employed throughout the Roman empire to stop the spread of disease [1]. In recent times, the US Environmental Protection Agency (EPA) has classified copper and its derivatives as antibacterial materials [2]. On the other hand, the presence of copper in the human body is essential for healthy development, cardiovascular and lung functionality, neovascularization, neuroendocrine function, and iron metabolism [3].

Metallic copper, cupric oxide (CuO), and cuprous oxide (Cu_2O) nanoparticles are attracting considerable research interest nowadays due to their widespread applications in catalysts and therapeutic domains [4–10] compared to their bulk counterparts, by virtue of their nano dimensional higher surface area to volume ratio. Specifically, copper oxide nanoparticles (CuO NPs) display broad-spectrum antibacterial and photocatalytic properties [11] and have the potential to be used as an alternative to antibiotics. The design and development of new compounds as antibiotics possess an emergent need, as the overuse and misuse of existing antibiotics are responsible for the growing episodes of antibiotic-resistant infections and deaths globally [12].

The antibacterial activity of CuO NPs against Gram-positive bacteria, such as *S. aureus* and *B. subtilis* as well as Gram-negative bacteria, such as *E. coli* and *P. aeruginosa* are reported in various studies [13–17]. Metallic Cu and CuO NPs are also found to exhibit multi-toxicity on multi-drug resistant bacterial species, such as the methicillin-resistant S. aureus (MRSA) [18]. In the case of CuO NPs, it is suggested that its antibacterial effect might be associated with cell membrane dissociation and reactive oxygen species (ROS) production [19]. Various simultaneous mechanisms of action of CuO nanoparticles against bacteria make it almost impossible for the microbes to develop resistance, as the bacterial cell would be required to generate multiple simultaneous gene mutations to develop this resistance [20]. Copper oxide nanoparticles are obtaining growing attention for their cheaper price and abundance in comparison with other noble and expensive metals, such as silver and gold, and their competent potential application as microbial agents [15,21,22].

The size, morphology, and solubility play a significant role in the antibacterial activities of Cu, CuO, and Cu_2O [23–25]. However, the major limitation of metallic CuO NPs in the nano-size range is the lack of significant stability in dispersions, due to their strong tendencies to aggregate and the formation of larger clusters that reduce the energy associated with their high surface area [26–28]. The formation of clusters results in sedimentation leading to loss of reactivity and antimicrobial performance, in which a nanometric size is essential [28]. Further surface modifications of CuO NPs using a post functionalization approach, not only enhances their colloidal stability, yet can also introduce unique physical and chemical properties, including the possible enhancement of their antimicrobial activities. Functionalization, or capping of an inorganic nanoparticle with an organic ligand-like citrate or folate, is evidenced to produce nanohybrids that have unique therapeutic potentials [29].

In the current study, we have explored the effect of citrate-capped CuO NPs on a Gram-positive *Staphylococcus hominis* (SH) bacterial strain. Indeed, *S. hominis* is a commensal bacteria that resides on human skin [30]. Although it is apparently harmless, it has been reported that one of its subspecies, novobiosepticus, is multidrug-resistant and causes nosocomial infections, such as sepsis, alongside bloodstream infections in neonates and immunocompromised patients [31], and various opportunistic infections of humans [32]. *S. hominis* is also well known for its ability to generate pungent body odor in humans [33]. In the present study, we have reported the synthesis or extraction, characterization, and antimicrobial activity of citrate functionalized CuO NPs on the Staphylococcus hominis bacterial strain. CuO NPs were synthesized and capped using a precipitation technique [34] and grafting method [35], respectively, in a bottom-up approach. Similarly in a top-down method, CuO NPs were extracted, and citrate functionalized from two types of copper-containing stones from copper mines. The structural properties of both synthesized and extracted CuO NPs were examined by X-ray diffraction (XRD), and field emission scanning electron microscopy (FESEM), which were equipped with an energy dispersive X-ray spectroscopy (EDS) and found to be similar. Dynamic light scattering (DLS) and zeta potential studies were also employed for estimating the hydrodynamic diameter and solubility assessment of the synthesized nanohybrid. The surface functionalization of CuO NPs by citrate ligands was confirmed by FTIR and UV-vis spectroscopy. The antimicrobial activity of citrate-CuO NPs was examined in the *S. hominis* bacteria strain. The citrate functionalized NPs were found to generate reactive oxygen species (ROS) upon photoexcitation, which is responsible for their antimicrobial action because ROS has the ability to destroy the active substances in the bacterial inner and outer membranes [36–38]. This phenomenon establishes the credentials of citrate CuO NPs for applications including antibacterial photodynamic therapy (PDT) with enhanced efficacy. We have also used computational biology strategy in order to rationalize the antibiotic-resistant bacterial remediation found in our experimental studies.

Immediately after the establishment of the CuO nanohybrid as a potential antibacterial agent, we also explored the use of copper mine tailings as a source of raw materials used for the nanohybrid synthesis, in order to simultaneously cater to low-cost antibiotics to a wider population across the globe and to remediate the burden from the copper mines. We have also developed a prototype FMCG product (talcum powder) for the remediation of bacteria *S. hominis* which are responsible for the generation of several human disorders.

2. Materials and Methods

2.1. Materials

Copper acetate, citric acid, sodium hydroxide, and sodium citrate were purchased from Sigma Aldrich (St. Louis, MO, USA), California. All solvents and all other used chemicals were procured from Merck (New Jersey, USA), unless otherwise stated. 2,7-dichlorodihydrofluorescein diacetate (DCFH-DA) was bought from Calbiochem to estimate the reactive oxygen species (ROS) production. Similarly, 2,2-diphenyl-1-picrylhydrazyl (DPPH) was obtained from Sigma (St. Louis, MO, USA) to monitor the antioxidant activity in the samples. All reagents were analytical grade and used without any further purification. Nanopore water, with a resistivity value ≥ 18 MΩ cm, from the Milli-Q system (Millipore GmbH, Germany) was used in all experiments. For bacterial studies, LB top agar and Luria broth (LB) medium were purchased from HIMEDIA. The Gram-positive bacteria, *Staphylococcus hominis* (*S. hominis*) strain, was procured from ATCC.

2.2. Synthesis of Functionalized CuO Nanoparticles

In this study, the CuO NPs were synthesized following the reported precipitation method by Zhu et al. [34]. The citrate ligands generated in the process provided satisfactory passivation against aggregation and sufficient stability to the NPs in colloidal suspension. Briefly, 150 mL of deionized water was used to dissolve 0.54 g of copper acetate. Next, 0.52 g of citric acid was added, and the mixture was vigorously stirred while being heated to boiling at 100 °C. Once the mixture's pH attained a value of 6–7, 0.7 gm or 0.015 mol of sodium hydroxide (NaOH) was added quickly, causing a significant amount of dark brown precipitate to form at the same time. The blue color solution was immediately converted to brown, indicating the production of CuO NPs. The liquid was cooled to room temperature while being stirred after 5 additional minutes of reflux. The CuO-NPs were subsequently separated by centrifugation (4000 rpm, 10 min), and washed twice with water and another two times with ethanol. The supernatant, which contains citrate-capped CuO NPs, was then separated.

Using the citric acid grafting procedure, the CuO NPs were further functionalized. In a water-to-ethanol ratio of 8:2, suspensions of 200 mM acetic acid and 65 mM CuO were performed. The produced citric acid solution was combined with the CuO suspension, and the pH of the resulting combination was raised to 12 by adding 6 M NaOH. Next, the mixture was refluxed for three hours. The product was then centrifuged and properly cleaned three times to remove the extra citric acid. To obtain citrate-capped CuO (C-CuO) from the two copper tailings in the mines of Peru and Bhopal (India), a similar procedure was followed. Only 160 mg of each stone dust was mixed initially with 150 mL of deionized water before the addition of citric acid and boiling occurred.

2.3. Characterization Tools and Techniques

Optical absorbance spectra of samples were measured in a double-beam UV-vis spectrophotometer (model UV-2600, Shimadzu, Japan) in the 200−800 nm wavelength range. The room temperature steady-state emission spectra were recorded using a Fluorolog Model LFI-3751 (Horiba-Jobin Yvon, Edison, NJ, USA) spectrofluorometer equipped with a microchannel plate−photomultiplier tube (MCP−PMT, Hamamatsu, Japan). All fluorescence spectra were corrected for variations with a wavelength in source intensity, photomultiplier response, and monochromator throughput. The X-ray diffraction (XRD) pattern of synthesized CuO nanoparticles was measured in a PANalytical X'PertPRO (Malvern Panalytical Ltd., Malvern, UK) diffractometer, with Cu Kα radiation (at 40 mA and 40 kV) generating at a rate of $0.02\ ^\circ\ s^{-1}$ in the 2θ range from 20 ° to 80 °. The liquid CuO nanoparticles were subjected to lyophilization for the XRD analysis. We have performed Fourier transform infrared spectroscopy (FTIR) on the liquid samples and the spectra were obtained using a JASCO FTIR-6300 spectrometer instrument (Oklahoma city, OK, USA). A NanoS Malvern (Zeta-seizer) instrument equipped with a 4 mW He:Ne laser (λ = 632.8 nm) and a thermostat coupled sample chamber was employed for dynamic light scattering (DLS) and ζ potential measurements. Quartz cuvettes of 10 mm path length were used to execute all spectroscopic experiments. The structural morphologies and chemical compositions of the synthesized citrate CuO nanohybrids from different sources were analyzed using scanning electron microscopy (SEM) and EDAX methods. Before scanning in a field emission scanning electron microscope (Quanta FEG 250: source of electrons, FEG source; operational accelerating voltage, 200 V to 30 kV; resolution, 30 kV under low vacuum conditions: 3.0 nm; detectors, large field secondary electron detector for the low vacuum operation), the coverslips containing samples were coated with gold.

2.4. Antioxidant Activity

The free radical scavenging capability of the samples of interest was determined using the DPPH assay method. A 0.15 mM DPPH solution was prepared in methanol and 0.5 mL citrate CuO nanoparticles of various concentrations were added to 2.5 mL of the freshly prepared DPPH solution. The characteristic absorption maxima of DPPH at 535 nm were selected to monitor the degradation process with DPPH in the presence and absence of light at room temperature. The absorption spectra of DPPH were recorded in the interval of 2 s for an hour using SPECTRA SUITE software provided by Ocean Optics.

2.5. Quantification and Characterization of ROS

For the purpose of quantifying the generated ROS, we used 2′,7′-dichlorofluorescein (DCFH), which is a well-known reagent. The DCFH was prepared via a de-esterification reaction from DCFH-DA at room temperature, following a standardized protocol described in previous studies [39,40]. The oxidation of DCFH, in the presence of light, leads to the production of DCF emitting fluorescence [41,42]. In this study, the ROS generated in the aqueous citrate functionalized CuO NPs convert DCFH into DCF, which has a characteristic emission maximum of 522 nm upon excitation at 488 nm. The DCF emissions were recorded in the Fluorolog Model LFI-3751 (Horiba-Jobin Yvon, Edison, NJ) spectrofluorometer. The ROS experiments were performed in the dark and in the light for 30 min.

2.6. Bacterial Strain and Culture Conditions

The antibacterial activity of the synthesized and extracted samples have been investigated against a strain of Staphylococcus hominis bacteria. The Gram-positive *S. hominis* strain was procured from ATCC. For the antibacterial assay, fresh *S. hominis* bacteria have been cultured using sterilized Luria−Bertani (LB) medium in a shaker incubator at a temperature of 37 °C for 24 h. All used glassware, suction nozzle, and culture medium were sterilized in an autoclave at a high pressure of 0.1 MPa and a temperature of 120 °C for 30 min prior to the experiments beginning. The treatment of bacteria was performed on LB agar plates using the colony forming unit (CFU) assay method under dark and light illumination conditions. The freshly grown original *S. hominis* bacterial suspension was firstly washed twice with phosphate-buffered saline (PBS; pH 7.4) solution and further diluted 10^6 times before the test samples were added. Then, they were incubated with the respective nanoparticles for 1 h. The resulting bacterial PBS suspensions (200 μL) were uniformly spread over gelatinous LB agar plates and cultured at 37 °C for 24 h to obtain the CFUs. To quantify the antibacterial activity, the number of survival colonies was manually counted and presented as a bar diagram.

Detection of microbial growth was also performed in the cuvette system for different citrate CuO NPs under various conditions. For this bacterial mortality study, a bacterial solution of 10^8 CFU/mL was considered and incubated with test solutions of 1 mM concentrations, initially for 3 h with photoactivation.

For the cuvette-based detection of microbial growth, the cells were cultured in an LB medium under an incubator shaker at 37 °C for 24 h. The optical density of the freshly grown overnight culture was fixed to 0.1 in LB medium initially. The culture was then put into a cuvette, mixed with the test samples, and incubated at 37 °C for 9 h with shaking under dark and illumination conditions. The absorbance was taken at 1 h intervals and plotted against time, with baseline correction for studying the growth curves.

For the microscopic studies, the bacteria cells after proper incubations with the nanoparticles were stained with DAPI and PI. DAPI stains all cells, while PI only stains the membrane-disrupted cells. The (1-red/blue) ratio was obtained to assess the viability of the *S. hominis*. The tests were repeated three times. The samples (15 μL) were observed under a fluorescence microscope (Leica digital inverted microscopes DMI8).

2.7. Statistical Analysis

All the data in this current work are represented as mean ± standard deviation (SD), unless otherwise stated. An unpaired 2-tailed *t*-test was used for comparison between the groups. A value of $p < 0.05$ was considered significant. GraphPad Prism (v8.0) software was used for all statistical tests.

2.8. Method of Computational Biology

The web resource STITCH (http://stitch.embl.de/; accessed on 24 December, 2022) was utilized to predict the chemical–protein (CP) interaction networks of CuO NP in *S. hominis*. The STITCH database can forecast around 960,000 proteins and 430,000 compounds from the 2031 eukaryotic and prokaryotic genomes [43,44]. The confidence score of a chemical–protein interaction can be used to predict the relationship, with a higher value indicating a stronger interaction. For the purposes of this investigation, a medium confidence score of 0.4 was taken into account.

3. Results and Discussion

X-ray diffraction (XRD) characterization of the synthesized and extracted CuO NPs was carried out to estimate the precise elemental composition, particle size, and superficial morphology, as depicted in Figure 1a. It was determined that all CuO NPs were in a monoclinic geometry with a space group of C2/C. No characteristic peaks of any other impurities were detected, suggesting the preparation of high-quality CuO NPs. Moreover, the obtained χ^2 value of 1.82 for the Le Bail fitting indicates excellent agreement with the previously reported literature [45,46]. The crystallite size is estimated from the XRD pattern using Debye Scherrer's Equation (1) [47]:

$$D = K\lambda/\beta \cos \theta \tag{1}$$

where $K = 0.94$ is the shape factor, λ is the X-ray wavelength of Cu Kα radiation (1.541 Å), θ is the Bragg diffraction angle, and β is the full width at half maxima (FWHM) of the respective diffraction peak. The crystallite size corresponding to the highest peak observed in XRD was found to be 34.4 nm. The presence of sharp structural peaks in XRD patterns and a crystallite size less than 100 nm corresponds to the nanocrystalline nature of synthesized CuO NPs. The peaks at 32.5, 35.4, 35.5, 38.7, 38.9, 46.2, 48.8, 51.3, 53.4, and 56.7 in 2θ correspond to the different CuO planes [48]. Similarly, as shown in Figure 1a, the CuO obtained from Peru and Bhopal mine samples exhibited peaks corresponding to the synthesized CuO NPs. The peaks at the same 2θ positions for the extracted CuO nanoparticles from both Peru and Bhopal mine samples are referred to here and confirm the extraction of CuO from the mine samples. The additional peaks appear due to the attribution of elements such as Au, Cr, etc., which are present in the mine samples, as revealed from the SEM EDAX analysis (Figure 2).

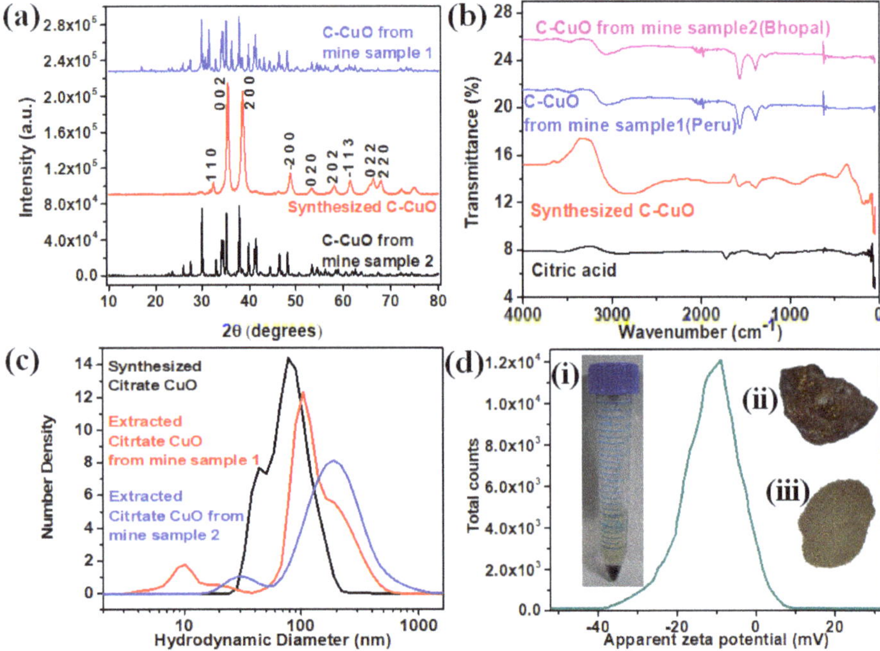

Figure 1. (**a**) XRD of the synthesized and extracted (from Peru and Bhopal samples) CuO nanoparticles. (**b**) FTIR spectra of different citrate-capped CuO samples and citric acid in the range between 0 and 4000 cm^{-1}, (**c**) DLS of the citrate-capped CuO from supernatants of C-CuO samples, (**d**) zeta potential distribution of the synthesized citrate functionalized CuO NP. The insets (i), (ii), and (iii) of Figure 1d are the synthesized C-CuO, the Peru mine tailing, and the Bhopal mine sludge, respectively.

To ensure the surface functionalization of the synthesized and extracted CuO NPs by citrate, FTIR analysis was performed on citric acid and the nanoparticles (Figure 1b). The absorption band, around 620.9 cm^{-1} in synthesized citrate CuO (C-CuO), C-CuO from mine sample 1 (Peru), and C-CuO from mine sample 2 (Bhopal) can be attributed to the vibrations of the Cu−O group [49,50]. Furthermore, the absence of other molecular vibrations, due to the calcination of the synthesized C-CuO, confirms the formation of a pristine surface. The citric acid was found to exhibit an absorption band occurring at around 1718 cm^{-1}, which is attributed to the C=O stretching, while the band occurring at 1394 cm^{-1} is attributed to the C−O stretching [51]. Moreover, the signature of the O−H stretching from the tertiary alcohol of citric acid can be witnessed at around 1099 cm^{-1}, and the bending of the O−H group from the carboxylic acid portion can be observed at 1383 cm^{-1}. Finally, the absorption band at 3299 cm^{-1} is due to the presence of the O−H bond from the citric acid. In the CuO FTIR spectrum, the C−O and C=O stretching can be seen at 1370 cm^{-1} and around 1600 cm^{-1}, while the tertiary alcohol band is at 1080 cm^{-1}. The shift in C=O and

C−O indicates that the CA is bonded to the CuO via the chemisorption of the carboxylate group. Therefore, the weakening of C=O causes a shift in frequency from 1700 to 1600 cm^{-1}. This characterization confirmed the success of the surface modification of CuO with citric acid. All the FTIR spectra shown in Figure 1b are baseline corrected; however, the apparent baseline shift may be due to the excess concentration of the synthesized citrate CuO nanoparticles in comparison with the extracted CuO NPs from the copper mine tailings.

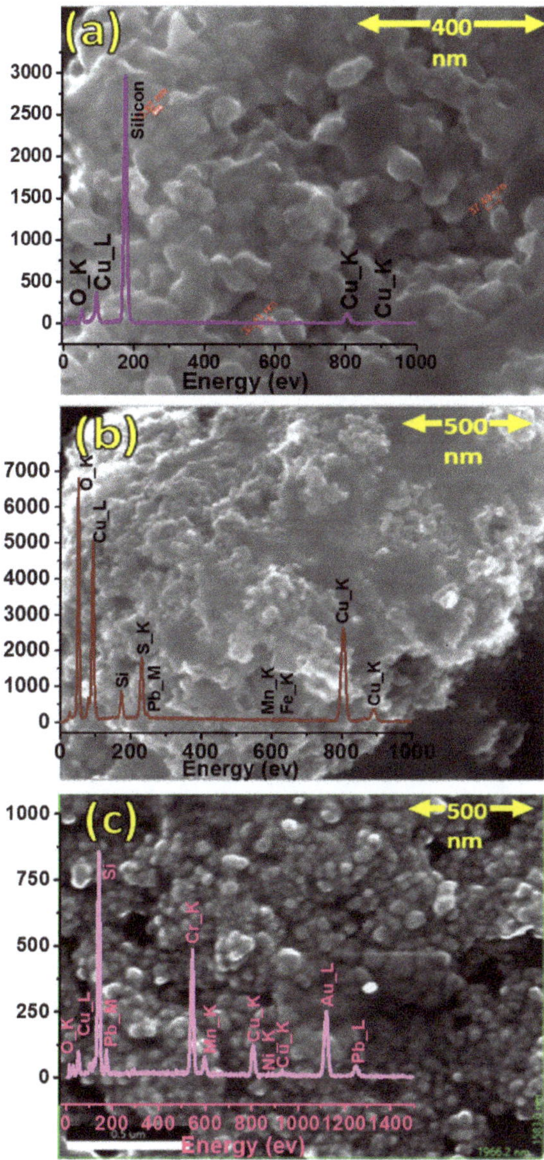

Figure 2. FESEM images of (**a**) synthesized citrate-capped CuO NPs. The inset shows the EDAX parameters of the nanoparticles. (**b**) Extracted and citrate-capped CuO from Peru mine stone sample. The inset shows the EDAX parameters of the nanoparticles. (**c**) Extracted and citrate-capped CuO from the Bhopal mine stone sample. The inset shows the EDAX parameters of the nanoparticles.

The hydrodynamic diameter of the synthesized CuO nanohybrid, C-CuO mine sample 1 (Peru) and 2 (Bhopal) were estimated to be 78.8 nm,106 nm and190 nm, respectively, from the dynamic light scattering (DLS) studies (Figure 1c), which are close enough to each other. The results of the DLS corroborate with the size obtained from the XRD (34.4 nm) and FESEM analyses (38.1 nm). Dynamic light scattering (DLS) accounts for the hydrodynamic diameter of the functionalized nanomaterial, which comprises core CuO and citrate ligands at the surface, along with associated water molecules and some possible aggregation of the functionalized nanoparticles in the aqueous solution. The electron micrograph of the nanoparticles shows the inorganic nanomaterial, as a whole, contains crystalline and amorphous inorganic core substrate. On the other hand, the size from the XRD accounts for only the crystalline materials in the core inorganic particles. Thus, it is obvious that DLS overestimates the size, while XRD reveals the crystalline nanoparticles in the core [52]. Additionally, the citrate CuO NPs exhibited a ζ-potential of magnitude -10.38 mV, assuring moderate solubility of the NPs (Figure 1d). The ζ-potential for the Peru and Bhopal samples were obtained to be -11.2 mV and -8 mV, respectively. These results lower the possibility of instability and particle agglomeration or precipitate tendencies of the CuO NPs out of the solutions. The full curve for a sample is only provided by the instrument software, while the titration experiment was performed because we obtained a synthesized pure citrate-capped CuO nanohybrid. However, for the extracted citrate CuO nanoparticles from Peru and Bhopal mine samples, only zeta potential measurements were carried out, for which only values in the mV unit are provided because, as mentioned previously, no full curves were available from the instrument. The insets in i, ii, and iii of Figure 1d are the synthesized C-CuO, the Peru mine tailing, and the Bhopal mine sludge, respectively.

Figure 2a–c shows the FESEM images of the C-CuO NPs. The inset of Figure 2a depicts the EDAX spectrum of the synthesized CuO NPs. The EDAX analysis of the extracted citrate CuO from the Peru and Bhopal mine tailings are shown in the insets of Figure 2b,c. The EDAX result shows that there are no other elemental impurities present in the prepared CuO NPs. However, trace amounts of Au and Cr, etc., were found in the extracted CuO NPs from the EDAX method. The peak of silicon appearing in the EDAX graphs of Figure 2 is attributed to the substrate. For EDAX analysis, the test sample was a drop cast on a Si wafer substrate and the thickness of the drop casted layer may not be uniform throughout the substrate. In the FESEM images of C-CuO, the synthesized nanohybrids are seen to consist of clustered spherical particulates with approximate diameters between 38 and 46 nm. The average diameter of the CuO NPs was calculated by measuring over 100 particles in a random field of FESEM view. The SEM-EDAX analysis demonstrated that the atomic compositions of the Cu and O elements were 38.41% and 61.59%, respectively. The mean ratio of Cu and O was, therefore, 38.41:61.5, and an accurate compound formula based on this atomic ratio of Cu and O can, thus, be given as Cu1.2O or CuO0.8. Therefore, it can be ensured that most of the synthesized nanoparticulate sample was indeed CuO.

The UV-vis spectra of the citrate-capped CuO samples (Figure 3a) exhibited a broad absorbance peak at 285 nm, which is characteristic of a surface plasmon resonance of the CuO nanoparticles [53]. A weaker peak at around 350 nm, signifies the d–d transition of CuO nanoparticles due to citrate functionalization and the quantum confinement effect of the CuO NPs on the citrate functionalization [54]. Similarly, the peaks observed at 285 nm and 350 nm, after the capping of Peru and the Bhopal samples, signified the presence of citrate-capped CuO NPs within the system.

The dark and photoinduced ROS generation capability of the citrate CuO NPs is illustrated using a well-known nonfluorescent probe: DCFH (Figure 3b). DCFH is oxidized to fluorescent dichlorofluorescein (DCF) by ROS, exhibiting an emission near 522 nm upon excitation at 488 nm. Thus, the enhancement of the ROS generation level is indicated by the increase in the emission intensity at 522 nm [37]. The oxidation of only DCFH control and CuO NPs are monitored for 35 min in the dark and then under irradiation of UV light (365 nm). In the dark, there is a considerable enhancement of emission intensity at 522 nm, indicating the presence of dark ROS generation. However, with light exposure, a greater increase in emission intensity is observed for the citrate CuO nanohybrid compared to the control (Figure 3b). This confirms the ROS generation capability of the synthesized nanohybrid under light exposure making it suitable to apply for photodynamic therapy.

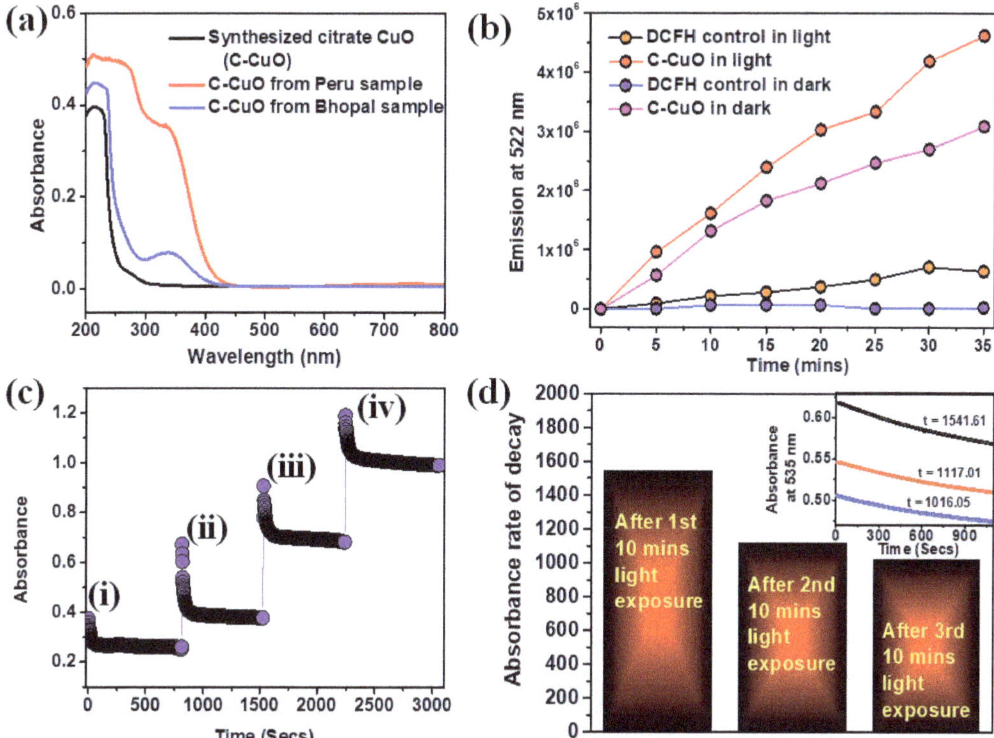

Figure 3. (**a**) Absorbance spectra of synthesized and extracted citrate-CuO nanohybrids. (**b**) DCFH oxidation (monitored at 522 nm) with time in the presence and absence of citrate CuO NPs under dark and illumination conditions. (**c**) The DPPH catalytic activity of the synthesized CuO NPs in dark conditions upon the addition of DPPH in periodic intervals of time (i) (ii) (iii) and (iv). (**d**) The DPPH assay of the synthesized CuO NPs under UV light illumination conditions. Inset shows the decay of absorbance of DPPH radical at 535 nm with multiple light exposures of 10 min intervals.

The antioxidant activity of the citrate functionalized CuO NPs was evaluated using the DPPH assay method (Figure 3c,d). The absorbance of DPPH at 535 nm was observed to decrease with time, establishing the presence of antioxidant properties of the CuO NPs under dark conditions and with a visual transition of the solution from purple to yellow. Now, upon further addition of 50 uM DPPH, after a periodic interval of time, it is observed that the free radical scavenging activity remains almost intact in the same 100 uM C-CuO sample. The study of antioxidant activity in the presence of UV light (365 nm) is a prerequisite to a full understanding of its potential antioxidant properties. As shown in Figure 3d, the rate of absorbance at 535 nm decreases (from 1541.51sec to 1117.01sec to 1016.05sec) with UV exposure of 10 mins for each case, keeping the concentrations of DPPH and C-CuO NPs to be constant. The results corroborate the enhancement of antioxidant activities with light exposure. The same sample of C-CuO NPs is capable of providing catalytic activities after getting recharged by light exposure.

The antibacterial activities of the C-CuO NPs, synthesized and extracted from nine samples were investigated against Gram-positive *S. hominis* bacteria by analyzing their growth curves. Figure 4a reveals the growth pattern of the bacteria in terms of absorbance at 600 nm with one-hour intervals. The synthesized C-CuO, under illumination as well as dark conditions, is observed to almost nullify the growth, with respect to the controls where only the bacteria are present in LB media. The C-CuO extracted from the Peru and Bhopal mine samples are observed to exhibit better antimicrobial activity in presence of light. The copper chloride salt used for the synthesis of CuO NPs, and sodium citrate used for the capping purpose were found to have almost no effect on bacterial growth, establishing the fact that the entire antimicrobial activities are solely generated by the C-CuO NPs.

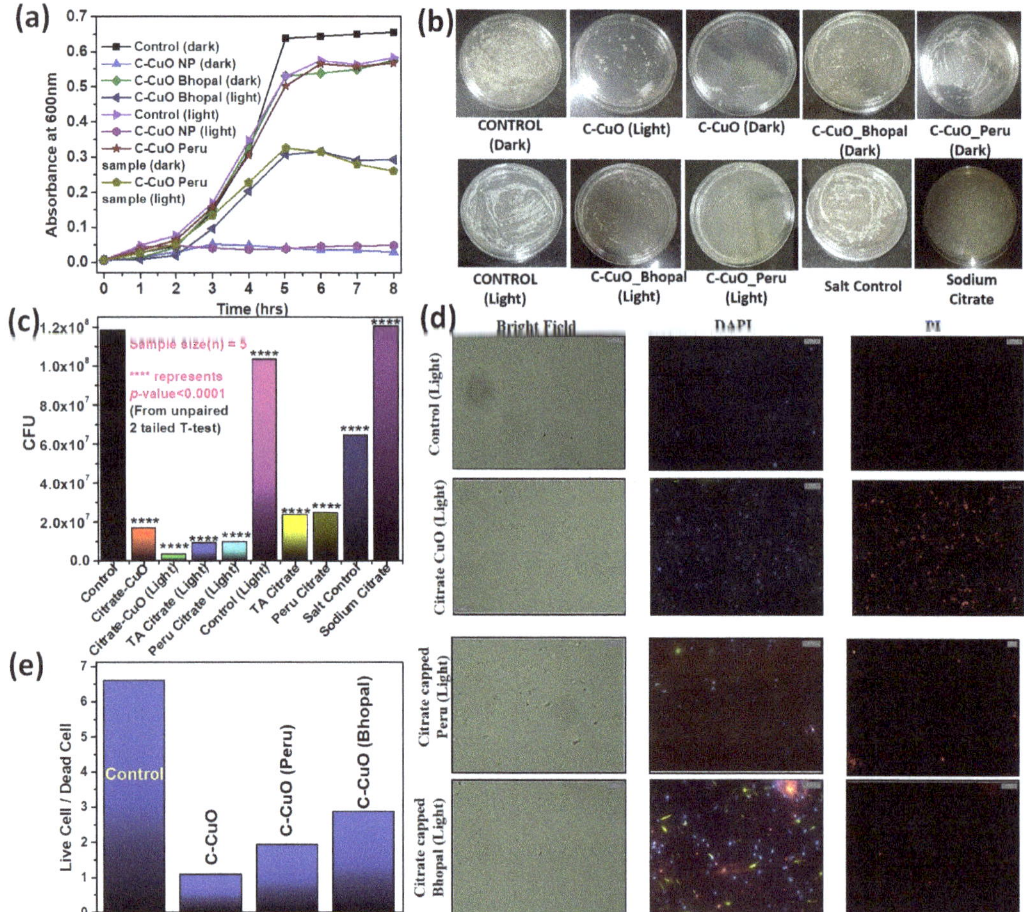

Figure 4. (**a**) Effect of C-CuO NPs on the growth curves of *S. hominis* bacteria under dark and illumination conditions. (**b**) Bacterial viability after treatment with different citrate CuO NPs in the presence and absence of light irradiation. (**c**) Agar plate colony count assay. '****' indicates *p*-value < 0.0001. (**d**) Bright-field microscopic images and fluorescence micrographs of bacteria stained with DAPI, and fluorescence micrographs of bacteria stained with PI. (**e**) The ratio of live and dead cells obtained from microscopy.

The antimicrobial activity of the synthesized and extracted CuO NPs after citrate functionalization was investigated against *S. hominis* growth to explore the antibiotic potential against bacterial infections in the agar plate assay method. To probe the antibacterial action of the nanoparticles, they were used for incubating the culture for 3 h. As shown in Figure 4b,c, minimal colonies were observed (the bacterial growth was found to have decreased by 96.9% in the CFU from the control plate) for the nanohybrid under UV light illumination conditions. The bacterial growth is found to be decreased by 85% only in the CFU under dark conditions. The effect of two different mine-extracted copper nanohybrids under dark and light irradiation conditions was then studied further upon the growth of *S. hominis* bacteria (Figure 4b). In the case of the Peru and Bhopal samples, the bacterial growth is found to be decreased by 79% CFU and 80% CFU, respectively, in dark conditions. On the other hand, a huge decrement in bacterial growth is observed for both the Peru and Bhopal samples after citrate-capping on UV illumination. The bacterial growth was reduced by 90% and 91% for the citrate-capped Peru and Bhopal samples, respectively, on the UV illumination. To nullify the effect of the salt and the capping agent (sodium citrate), their antibacterial effects were studied. No significant change in the number of colonies in the negative control group was found, indicating a low antibacterial effect of these salts on *S. hominis*. Thus, the antibacterial effect found is solely due to the citrate-capped CuO NP. From these results, it is evident that the nanohybrid is itself an antibacterial agent and its efficiency enhances many folds upon white light exposure, which triggers an overall huge antibacterial effect.

To study the effect of the citrate CuO nanohybrid, bacterial cultures were performed five times for each group (control in dark and light, citrate CuO in dark and light, different concentrations of the nanohybrid, etc.) and their differences were calculated to identify their significance levels. The sample size for each of the groups was five. The p-value was calculated using an unpaired 2-tailed t-test and a value of $p < 0.05$ was considered to be significant. The statistical difference between the control and treatment is designated by '*'. The '*' represents a p-value < 0.1 and '****' represents a p-value < 0.0001 [55].

The antimicrobial activity of the NPs was further investigated using optical microscopy. We used *S. hominis*, the Gram-positive bacteria, as the model biological system. *S. hominis* were incubated with both synthesized and extracted CuO samples for two hours in LB broth and then stained with DAPI and PI. While DAPI stains all the cells in the medium, PI is specific to dead cells as it cannot cross intact cell membranes. Therefore, the ratio of DAPI-stained cells to PI-stained cells reveals the viability of the cells. Figure 4d,e shows the results of the microscopic microbial studies. The first row of Figure 4d contains the microscopic images of the control cells after staining with DAPI and PI. As expected, the number of viable cells was more than the dead cells (ratio ~6.6). The microscopic images of the synthesized C-CuO NPs (Figure 4d, second row) incubated with bacteria showed a reduced ratio–i.e., the dead cells outnumbered the live cells (ratio ~2.1) compared to the control. Interestingly, the group of bacteria treated with C-CuO NPs extracted from the Bhopal and Peru mine samples (Figure 4d, rows 3 and 4) also showed higher viability compared to the control. The ratio for these C-CuO NP-treated groups was similar to that of the synthesized C-CuO-treated group, as shown in Figure 4e. Therefore, the microscopic studies revealed that the C-CuO NPs exerted toxic effects and effectively killed the *S. hominis*. The therapeutic action of the CuO NPs against *S. hominis* bacteria occurs primarily via the generation of reactive oxygen species, which mainly includes the formation of hydroxyl radicals ($HO.^{*}$) and superoxide radicals (O_2^{-*}). Initially, electron-hole (e^-/h^+) pairs are formed when the electromagnetic radiation of energy ($h\upsilon$) is either greater than or equal to the bandgap energy (E_g) of the CuO NPs. This phenomenon excites the electron from the valence band (VB) to the conduction band (CB), leaving holes behind in the VB. These photoexcited electrons reduce the surface adsorbed O_2 to $O_2^-.$, while the holes oxidize H_2O or HO- to OH^* [56–58], as described in the graphical abstract. This in situ production of the reactive radicals starts off by attacking the bacterial population and eliminates them through the production of toxic by-products, resulting from ROS-mediated damage to the cellular system.

$$CuO + h\upsilon = CuO\ (e^- + h^+)$$

$$O_2 + e^- = O_2^{-*}.$$

$$H_2O + h^+ = H^+ + HO^*.$$

In the absence of electromagnetic radiation, the antimicrobial activity may be attributed to the ligand-to-metal charge transfer (LMCT) of the C-CuO NPs. The LMCT bands originate due to the interaction of the $Cu^{1+}/^{2+}$ centers in the NP with the surface-bound citrate ligands. In the dark condition, the origin of the antibacterial activity might be due to the conversion of the Cu^{1+} to Cu^{2+} states at the center, accompanied by the direct injection of the electrons into the CB of the C-CuO NPs and followed by the reduction of the surface adsorbed O_2 to $O_2{}^-$.

$$Cu^{1+} - O = Cu^{+2} - O^* + e^-$$

$$O_2 + e^- = O_2{}^-$$

The enhanced efficacy of the citrate functionalized CuO NPs, in the presence of light for more bacterial remediation, establishes its credentials for the application of APDT (antibacterial photodynamic therapy).

The extraordinary effect of the synthesized copper nanohybrid may be hypothesized using predictive biological interactions. To compare our findings with the existing research and related predictive models, the previously mentioned STITCH database was used. It generated an interaction network (Figure 5) between the ligand CuO and its effect on various proteins of *S. hominis*. To understand how CuO affects the survival of the bacteria, a comprehensive table (Table 1) was populated with the protein names and their respective activities. The table suggests the negative impact of CuO on key proteins of *S. hominis*, which helps the organism deal with environmental stress, energy metabolism, GTP binding, and translation regulation, etc.

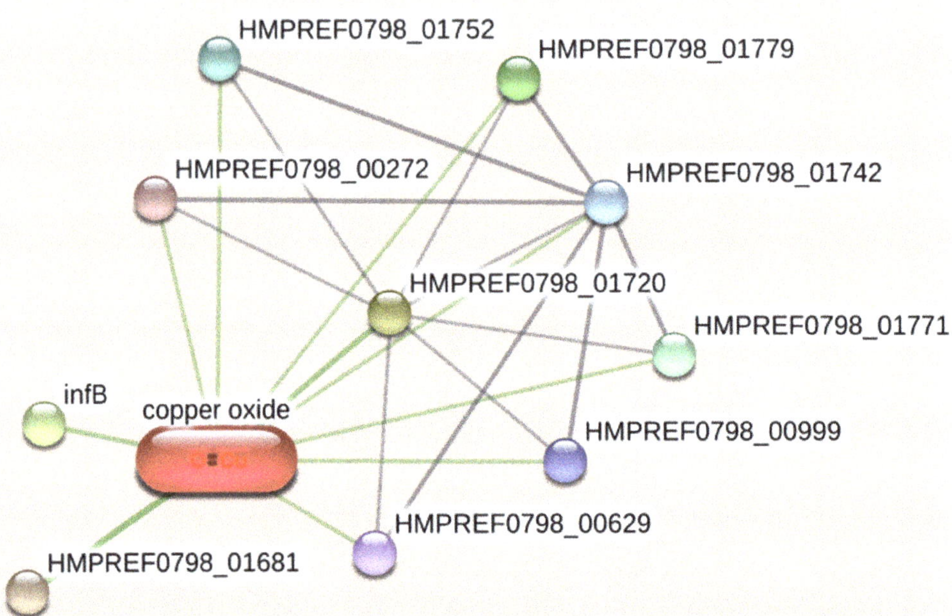

Figure 5. Compound protein interactions network of citrate CuO nanohybrid on *S. hominis* bacterial strain.

Table 1. Effect of citrate functionalized CuO NPs on *S. hominis* bacterial strain.

Identifier	Corresponding Protein Name	Activity
HMPREF0798_01720	Catalase	Catalase, an antioxidant enzyme found in all aerobic organisms, facilitates the transformation of H_2O_2 into water and oxygen under environmental stress. [59]
HMPREF0798_01742	Pyridine nucleotide-disulfide oxidoreductase family protein	Controls mitochondrial function. [60]
HMPREF0798_00999	Pyridine nucleotide-disulfide oxidoreductase	
HMPREF0798_01779	Dihydrolipoyl dehydrogenase	This oxidoreductase is a key factor in bacterial pathogenesis and is responsible for energy metabolism [61]
HMPREF0798_01752	Dihydrolipoyl dehydrogenase	
HMPREF0798_00272	Dihydrolipoyl dehydrogenase	
HMPREF0798_00629	Dihydrolipoyl dehydrogenase	
HMPREF0798_00484, infB	Translation initiation factor, IF-2	IF2 is a crucial protein that binds GTP and increases the rate of translation. [62]
HMPREF0798_01681	Cell division protein SufI	SufI is a protein responsible for cell division and is also considered a bacterial twin-arginine translocation protein [63]
HMPREF0798_01771	Mercury(II) reductase	This protein helps the organism to withstand the toxic Hg concentrations [64]

4. Conclusions

In this study, pure-grade citrate-capped CuO NP was synthesized using simple pre-cipitation and grafting methods. From two types of copper mine tailings, CuO NPs were also extracted and functionalized with citrate. The XRD spectrum confirmed the formation of monoclinic crystals of CuO NPs with space group C2/C. FESEM and EDAX revealed the morphology of CuO NPs. The average SEM diameter of CuO NPs was around 38.1 nm, which agreed fairly well with the XRD and DLS data. FTIR and UV-vis spectroscopy confirmed the surface functionalization of the CuO NPs with the citrate ligand. The syn-thesized nanohybrid was found to generate ROS under UV light exposure in the DCFH assay and showed excellent antimicrobial activity against *S. hominis* bacterial strains in a triparted study consisting of growth curve analysis, agar plate assay method, and mi-croscopic studies after staining using DAPI and PI. The C-CuO NPs were also found to possess antioxidant properties in the DPPH assay. Consequently, citrate-functionalized CuO NPs can, therefore, be used as an antibacterial agent in surface coatings, on a variety of substrates to stop microbes from adhering to them, colonizing and growing on them, and producing biofilms, such as in habitational medical equipment. This study suggests that the mechanisms of the antimicrobial response of a citrate-CuO nanohybrid in different species of bacteria should be further explored. The use of copper mine tailings as a source of raw materials for nanohybrid synthesis as a means of, simultaneously, catering low-cost antibiotics to a wider population across the globe and remediating the burden from the copper mines was also explored.

Author Contributions: A.B.: (Conceptualization: Lead; Data curation: Lead; Formal analysis: Lead; Investigation: Lead; Methodology: Lead; Project administration: Lead; Resources: Lead; Software: Lead; Validation: Lead; Visualization: Supporting; Writing—original draft: Lead; Writing—review & editing: Supporting); R.G.: (Data curation: Equal; Formal analysis: Supporting; Investigation: Supporting; Methodology: Equal; Validation: Supporting; Writing—review & editing: Supporting); T.A. (Data curation: Supporting; Formal analysis: Supporting; Validation: Supporting; Writing—review & editing: Supporting); S.M.: (Resources: Equal; Writing—review & editing: Supporting); A.C.: (Investigation: Supporting; Methodology: Supporting; Writing—review & editing: Supporting, Resources: Supporting); S.K.P.: (Conceptualization: Lead; Investigation: Supporting; Methodology: Equal; Supervision: Lead; Writing—review & editing: Supporting Formal analysis: Equal; Project administration: Lead; Resources: Equal). All authors have read and agreed to the published version of the manuscript.

Funding: This research received no external funding.

Data Availability Statement: All data are available on request to the corresponding author.

Acknowledgments: SKP wants to thank the Indian National Academy of Engineering (INAE) for the Abdul Kalam Technology Innovation National Fellowship, INAE/121/AKF.

Conflicts of Interest: The authors declare no conflict of interest.

References

1. Borkow, G.; Gabbay, J. Copper, an ancient remedy returning to fight microbial, fungal and viral infections. *Curr. Chem. Biol.* **2009**, *3*, 272–278.
2. Dollwet, H. Historic uses of copper compounds in medicine. *Trace Elem. Med.* **1985**, *2*, 80–87.
3. Council, N.R. *Copper in Drinking Water*; National Academies Press: Washington, DC, USA, 2000.
4. Giannousi, K.; Sarafidis, G.; Mourdikoudis, S.; Pantazaki, A.; Dendrinou-Samara, C. Selective synthesis of Cu_2O and Cu/Cu_2O NPs: Antifungal activity to yeast Saccharomyces cerevisiae and DNA interaction. *Inorg. Chem.* **2014**, *53*, 9657–9666. [CrossRef] [PubMed]
5. Gandhare, N.V.; Chaudhary, R.G.; Meshram, V.P.; Tanna, J.A.; Lade, S.; Gharpure, M.P.; Juneja, H.D. An efficient and one-pot synthesis of 2,4,5-trisubstituted imidazole compounds catalyzed by copper nanoparticles. *J. Chin. Adv. Mater. Soc.* **2015**, *3*, 270–279. [CrossRef]
6. Tanna, J.A.; Chaudhary, R.G.; Sonkusare, V.N.; Juneja, H.D. CuO nanoparticles: Synthesis, characterization and reusable catalyst for polyhydroquinoline derivatives under ultrasonication. *J. Chin. Adv. Mater. Soc.* **2016**, *4*, 110–122. [CrossRef]
7. Sun, H.; Kim, H.; Song, S.; Jung, W. Copper foam-derived electrodes as efficient electrocatalysts for conventional and hybrid water electrolysis. *Mater. Rep. Energy* **2022**, *2*, 100092. [CrossRef]
8. Han, X.; Liu, P.; Ran, R.; Wang, W.; Zhou, W.; Shao, Z. Non-metal fluorine doping in Ruddlesden–Popper perovskite oxide enables high-efficiency photocatalytic water splitting for hydrogen production. *Mater. Today Energy* **2022**, *23*, 100896. [CrossRef]
9. Kuang, P.; Ni, Z.; Yu, J.; Low, J. New progress on MXenes-based nanocomposite photocatalysts. *Mater. Rep. Energy* **2022**, *2*, 100081. [CrossRef]
10. He, J.; Liu, P.; Ran, R.; Wang, W.; Zhou, W.; Shao, Z. Single-atom catalysts for high-efficiency photocatalytic and photoelectrochemical water splitting: Distinctive roles, unique fabrication methods and specific design strategies. *J. Mater. Chem. A* **2022**, *10*, 6835–6871. [CrossRef]
11. Ssekatawa, K.; Byarugaba, D.K.; Angwe, M.K.; Wampande, E.M.; Ejobi, F.; Nxumalo, E.; Kirabira, J.B. Phyto-Mediated Copper Oxide Nanoparticles for Antibacterial, Antioxidant and Photocatalytic Performances. *Front. Bioeng. Biotechnol.* **2022**, *10*, 820218. [CrossRef]
12. Palmer, A.; Kishony, R. Understanding, predicting and manipulating the genotypic evolution of antibiotic resistance. *Nat. Rev. Genet.* **2013**, *14*, 243–248. [CrossRef]
13. Chaudhary, R.G.; Sonkusare, V.N.; Bhusari, G.S.; Mondal, A.; Shaik, D.P.; Juneja, H.D. Microwave-mediated synthesis of spinel CuAl2O4 nanocomposites for enhanced electrochemical and catalytic performance. *Res. Chem. Intermed.* **2018**, *44*, 2039–2060. [CrossRef]
14. Kaweeteerawat, C.; Chang, C.H.; Roy, K.R.; Liu, R.; Li, R.; Toso, D.; Fischer, H.; Ivask, A.; Ji, Z.; Zink, J.I.; et al. Cu Nanoparticles Have Different Impacts in *Escherichia coli* and *Lactobacillus brevis* than Their Microsized and Ionic Analogues. *ACS Nano* **2015**, *9*, 7215–7225. [CrossRef]
15. Bogdanović, U.; Vodnik, V.; Mitrić, M.; Dimitrijević, S.; Škapin, S.D.; Žunič, V.; Budimir, M.; Stoiljković, M. Nanomaterial with High Antimicrobial Efficacy—Copper/Polyaniline Nanocomposite. *ACS Appl. Mater. Interfaces* **2015**, *7*, 1955–1966. [CrossRef] [PubMed]
16. Hsueh, Y.-H.; Tsai, P.-H.; Lin, K.-S. pH-dependent antimicrobial properties of copper oxide nanoparticles in Staphylococcus aureus. *Int. J. Mol. Sci.* **2017**, *18*, 793. [CrossRef] [PubMed]
17. Nishino, F.; Jeem, M.; Zhang, L.; Okamoto, K.; Okabe, S.; Watanabe, S. Formation of CuO nano-flowered surfaces via submerged photo-synthesis of crystallites and their antimicrobial activity. *Sci. Rep.* **2017**, *7*, 1–11. [CrossRef] [PubMed]
18. Weaver, L.; Noyce, J.; Michels, H.; Keevil, C. Potential action of copper surfaces on meticillin-resistant Staphylococcus aureus. *J. Appl. Microbiol.* **2010**, *109*, 2200–2205. [CrossRef]
19. Jadhav, S.; Gaikwad, S.; Nimse, M.; Rajbhoj, A. Copper oxide nanoparticles: Synthesis, characterization and their antibacterial activity. *J. Clust. Sci.* **2011**, *22*, 121–129. [CrossRef]
20. Pelgrift, R.Y.; Friedman, A.J. Nanotechnology as a therapeutic tool to combat microbial resistance. *Adv. Drug Deliv. Rev.* **2013**, *65*, 1803–1815. [CrossRef]
21. Zhu, Z.; Wan, S.; Zhao, Y.; Gu, Y.; Wang, Y.; Qin, Y.; Bu, Y. Recent advances in bismuth-based multimetal oxide photocatalysts for hydrogen production from water splitting: Competitiveness, challenges, and future perspectives. *Mater. Rep. Energy* **2021**, *1*, 100019. [CrossRef]

22. Xiao, H.; Liu, P.; Wang, W.; Ran, R.; Zhou, W.; Shao, Z. Enhancing the photocatalytic activity of Ruddlesden-Popper Sr2TiO4 for hydrogen evolution through synergistic silver doping and moderate reducing pretreatment. *Mater. Today Energy* **2022**, *23*, 100899. [CrossRef]

23. McDonnell, G.; Russell, A.D. Antiseptics and disinfectants: Activity, action, and resistance. *Clin. Microbiol. Rev.* **1999**, *12*, 147–179. [CrossRef]

24. Pang, H.; Gao, F.; Lu, Q. Morphology effect on antibacterial activity of cuprous oxide. *Chem. Commun.* **2009**, *9*, 1076–1078. [CrossRef] [PubMed]

25. Ren, G.; Hu, D.; Cheng, E.W.; Vargas-Reus, M.A.; Reip, P.; Allaker, R.P. Characterisation of copper oxide nanoparticles for antimicrobial applications. *Int. J. Antimicrob. Agents* **2009**, *33*, 587–590. [CrossRef] [PubMed]

26. Woźniak-Budych, M.J.; Przysiecka, Ł.; Maciejewska, B.M.; Wieczorek, D.; Staszak, K.; Jarek, M.; Jesionowski, T.; Jurga, S. Facile Synthesis of Sulfobetaine-Stabilized Cu_2O Nanoparticles and Their Biomedical Potential. *ACS Biomater. Sci. Eng.* **2017**, *3*, 3183–3194. [CrossRef] [PubMed]

27. Zhou, J.; Xiang, H.; Zabihi, F.; Yu, S.; Sun, B.; Zhu, M. Intriguing anti-superbug Cu_2O@ ZrP hybrid nanosheet with enhanced antibacterial performance and weak cytotoxicity. *Nano Res.* **2019**, *12*, 1453–1460. [CrossRef]

28. Hotze, E.M.; Phenrat, T.; Lowry, G.V. Nanoparticle Aggregation: Challenges to Understanding Transport and Reactivity in the Environment. *J. Environ. Qual.* **2010**, *39*, 1909–1924. [CrossRef]

29. Mondal, S.; Ghosh, R.; Adhikari, A.; Pal, U.; Mukherjee, D.; Biswas, P.; Darbar, S.; Singh, S.; Bose, S.; Saha-Dasgupta, T.; et al. In vitro and Microbiological Assay of Functionalized Hybrid Nanomaterials To Validate Their Efficacy in Nanotheranostics: A Combined Spectroscopic and Computational Study. *Chemmedchem* **2021**, *16*, 3739–3749. [CrossRef]

30. Zhang, L.; Thomas, J.C.; Miragaia, M.; Bouchami, O.; Chaves, F.; d'Azevedo, P.A.; Robinson, D.A. Multilocus sequence typing and further genetic characterization of the enigmatic pathogen, Staphylococcus hominis. *PLoS ONE* **2013**, *8*, e66496. [CrossRef]

31. Kim, S.-D.; McDonald, L.C.; Jarvis, W.R.; McAllister, S.K.; Jerris, R.; Carson, L.A.; Miller, J.M. Determining the Significance of Coagulase-Negative Staphylococci Isolated From Blood Cultures at a Community Hospital A Role for Species and Strain Identification. *Infect. Control. Hosp. Epidemiology* **2000**, *21*, 213–217. [CrossRef]

32. Iyer, M.N.; Wirostko, W.J.; Kim, S.H.; Simons, K.B. Staphylococcus hominis Endophthalmitis Associated With a Capsular Hypopyon. *Am. J. Ophthalmol.* **2005**, *139*, 930–932. [CrossRef]

33. Lam, T.H.; Verzotto, D.; Brahma, P.; Ng, A.H.Q.; Hu, P.; Schnell, D.; Tiesman, J.; Kong, R.; Ton, T.M.U.; Li, J.; et al. Understanding the microbial basis of body odor in pre-pubescent children and teenagers. *Microbiome* **2018**, *6*, 1–14. [CrossRef] [PubMed]

34. Zhu, J.; Li, D.; Chen, H.; Yang, X.; Lu, L.; Wang, X. Highly dispersed CuO nanoparticles prepared by a novel quick-precipitation method. *Mater. Lett.* **2004**, *58*, 3324–3327. [CrossRef]

35. Mallakpour, S.; Dinari, M.; Azadi, E. Grafting of Citric Acid as a Green Coupling Agent on the Surface of CuO Nanoparticle and its Application for Synthesis and Characterization of Novel Nanocomposites Based on Poly(amide-imide) Containing*N*-trimellitylimido-L-valine Linkage. *Polym. Technol. Eng.* **2015**, *54*, 594–602. [CrossRef]

36. Nawaz, A.; Goudarzi, S.; Asghari, M.A.; Pichiah, S.; Selopal, G.S.; Rosei, F.; Wang, Z.M.; Zarrin, H. Review of Hybrid 1D/2D Photocatalysts for Light-Harvesting Applications. *ACS Appl. Nano Mater.* **2021**, *4*, 11323–11352. [CrossRef]

37. Ahmed, S.A.; Hasan, N.; Bagchi, D.; Altass, H.M.; Morad, M.; Althagafi, I.I.; Hameed, A.M.; Sayqal, A.; Khder, A.E.R.S.; Asghar, B.H.; et al. Nano-MOFs as targeted drug delivery agents to combat antibiotic-resistant bacterial infections. *R. Soc. Open Sci.* **2020**, *7*, 200959. [CrossRef] [PubMed]

38. Tinajero-Díaz, E.; Salado-Leza, D.; Gonzalez, C.; Martínez Velázquez, M.; López, Z.; Bravo-Madrigal, J.; Hernández-Gutiérrez, R. Green metallic nanoparticles for cancer therapy: Evaluation models and cancer applications. *Pharmaceutics* **2021**, *13*, 1719. [CrossRef] [PubMed]

39. Mondal, S.; Adhikari, A.; Ghosh, R.; Singh, M.; Das, M.; Darbar, S.; Bhattacharya, S.S.; Pal, D.; Pal, S.K. Synthesis and spectroscopic characterization of a target-specific nanohybrid for redox buffering in cellular milieu. *MRS Adv.* **2021**, *6*, 427–433. [CrossRef]

40. Polley, N.; Saha, S.; Adhikari, A.; Banerjee, S.; Darbar, S.; Das, S.; Pal, S.K. Safe and symptomatic medicinal use of surface-functionalized Mn_3O_4 nanoparticles for hyperbilirubinemia treatment in mice. *Nanomedicine* **2015**, *10*, 2349–2363. [CrossRef]

41. Bera, A.; Hasan, N.; Pal, U.; Bagchi, D.; Maji, T.K.; Saha-Dasgupta, T.; Das, R.; Pal, S.K. Fabrication of nanohybrids toward improving therapeutic potential of a NIR photo-sensitizer: An optical spectroscopic and computational study. *J. Photochem. Photobiol. A Chem.* **2022**, *424*, 113610. [CrossRef]

42. Hasan, N.; Bera, A.; Maji, T.K.; Mukherjee, D.; Pan, N.; Karmakar, D.; Pal, S.K. Functionalized nano-MOF for NIR induced bacterial remediation: A combined spectroscopic and computational study. *Inorganica Chim. Acta* **2022**, *532*, 120733. [CrossRef]

43. Szklarczyk, D.; Santos, A.; Von Mering, C.; Jensen, L.J.; Bork, P.; Kuhn, M. STITCH 5: Augmenting protein–chemical interaction networks with tissue and affinity data. *Nucleic Acids Res.* **2016**, *44*, D380–D384. [CrossRef] [PubMed]

44. Kuhn, M.; Szklarczyk, D.; Pletscher-Frankild, S.; Blicher, T.H.; von Mering, C.; Jensen, L.J.; Bork, P. STITCH 4: Integration of protein–chemical interactions with user data. *Nucleic Acids Res.* **2013**, *42*, D401–D407. [CrossRef] [PubMed]

45. Wang, B.; Wu, X.-L.; Shu, C.-Y.; Guo, Y.-G.; Wang, C.-R. Synthesis of CuO/graphene nanocomposite as a high-performance anode material for lithium-ion batteries. *J. Mater. Chem.* **2010**, *20*, 10661–10664. [CrossRef]

46. Asbrink, S.; Waskowska, A. CuO: X-ray single-crystal structure determination at 196 K and room temperature. *J. Physics Condens. Matter* **1991**, *3*, 8173–8180. [CrossRef]

47. Patterson, A.L. The Scherrer Formula for X-Ray Particle Size Determination. *Phys. Rev.* **1939**, *56*, 978–982. [CrossRef]

48. Ahamed, M.; Alhadlaq, H.A.; Khan, M.A.M.; Karuppiah, P.; Al-Dhabi, N.A. Synthesis, Characterization, and Antimicrobial Activity of Copper Oxide Nanoparticles. *J. Nanomater.* **2014**, *2014*, 1–4. [CrossRef]
49. Moniri, S.; Ghoranneviss, M.; Hantehzadeh, M.R.; Asadabad, M.A. Synthesis and optical characterization of copper nanoparticles prepared by laser ablation. *Bull. Mater. Sci.* **2017**, *40*, 37–43. [CrossRef]
50. Sen, S.; Sarkar, K. Effective Biocidal and Wound Healing Cogency of Biocompatible Glutathione: Citrate-Capped Copper Oxide Nanoparticles Against Multidrug-Resistant Pathogenic Enterobacteria. *Microb. Drug Resist.* **2021**, *27*, 616–627. [CrossRef]
51. Cantu, J.M.; Ye, Y.; Valdes, C.; Cota-Ruiz, K.; Hernandez-Viezcas, J.A.; Gardea-Torresdey, J.L. Citric Acid-Functionalized CuO Nanoparticles Alter Biochemical Responses in Candyland Red Tomato (*Solanum lycopersicum*). *ACS Agric. Sci. Technol.* **2022**, *2*, 359–370. [CrossRef]
52. Pabisch, S.; Feichtenschlager, B.; Kickelbick, G.; Peterlik, H. Effect of interparticle interactions on size determination of zirconia and silica based systems–A comparison of SAXS, DLS, BET, XRD and TEM. *Chem. Phys. Lett.* **2012**, *521*, 91–97. [CrossRef] [PubMed]
53. Abboud, Y.; Saffaj, T.; Chagraoui, A.; El Bouari, A.; Brouzi, K.; Tanane, O.; Ihssane, B. Biosynthesis, characterization and antimicrobial activity of copper oxide nanoparticles (CONPs) produced using brown alga extract (Bifurcaria bifurcata). *Appl. Nanosci.* **2014**, *4*, 571–576. [CrossRef]
54. Zhang, Y.; Li, N.; Xiang, Y.; Wang, D.; Zhang, P.; Wang, Y.; Lu, S.; Xu, R.; Zhao, J. A flexible non-enzymatic glucose sensor based on copper nanoparticles anchored on laser-induced graphene. *Carbon* **2020**, *156*, 506–513. [CrossRef]
55. Adhikari, A.; Bhutani, V.K.; Mondal, S.; Das, M.; Darbar, S.; Ghosh, R.; Polley, N.; Das, A.K.; Bhattacharya, S.S.; Pal, D.; et al. Chemoprevention of bilirubin encephalopathy with a nanoceutical agent. *Pediatr. Res.* **2022**, 1–11. [CrossRef]
56. Pirilä, M.; Saouabe, M.; Ojala, S.; Rathnayake, B.; Drault, F.; Valtanen, A.; Huuhtanen, M.; Brahmi, R.; Keiski, R.L. Photocatalytic Degradation of Organic Pollutants in Wastewater. *Top. Catal.* **2015**, *58*, 1085–1099. [CrossRef]
57. Ajmal, A.; Majeed, I.; Malik, R.N.; Idriss, H.; Nadeem, M.A. Principles and mechanisms of photocatalytic dye degradation on TiO$_2$ based photocatalysts: A comparative overview. *RSC Adv.* **2014**, *4*, 37003–37026. [CrossRef]
58. Sibhatu, A.K.; Weldegebrieal, G.K.; Sagadevan, S.; Tran, N.N.; Hessel, V. Photocatalytic activity of CuO nanoparticles for organic and inorganic pollutants removal in wastewater remediation. *Chemosphere* **2022**, *300*, 134623. [CrossRef]
59. Sharma, I.; Ahmad, P. Catalase: A versatile antioxidant in plants. In *Oxidative Damage to Plants*; Elsevier: Amsterdam, The Netherlands, 2014; pp. 131–148.
60. Wang, T.; Xie, X.; Liu, H.; Chen, F.; Du, J.; Wang, X.; Jiang, X.; Yu, F.; Fan, H. Pyridine nucleotide-disulphide oxidoreductase domain 2 (PYROXD2): Role in mitochondrial function. *Mitochondrion* **2019**, *47*, 114–124. [CrossRef]
61. Moran, J.F.; Sun, Z.; Sarath, G.; Arredondo-Peter, R.; James, E.K.; Becana, M.; Klucas, R.V. Molecular cloning, functional characterization, and subcellular localization of soybean nodule dihydrolipoamide reductase. *Plant Physiol.* **2002**, *128*, 300–313. [CrossRef]
62. Marzi, S.; Knight, W.; Brandi, L.; Caserta, E.; Soboleva, N.; Hill, W.E.; Gualerzi, C.O.; Lodmell, J.S. Ribosomal localization of translation initiation factor IF2. *Rna* **2003**, *9*, 958–969. [CrossRef] [PubMed]
63. Tarry, M.; Arends, S.R.; Roversi, P.; Piette, E.; Sargent, F.; Berks, B.C.; Weiss, D.S.; Lea, S.M. The Escherichia coli Cell Division Protein and Model Tat Substrate SufI (FtsP) Localizes to the Septal Ring and Has a Multicopper Oxidase-Like Structure. *J. Mol. Biol.* **2009**, *386*, 504–519. [CrossRef] [PubMed]
64. Freedman, Z.; Zhu, C.; Barkay, T. Mercury Resistance and Mercuric Reductase Activities and Expression among Chemotrophic Thermophilic Aquificae. *Appl. Environ. Microbiol.* **2012**, *78*, 6568–6575. [CrossRef] [PubMed]

A Comparison Study between Wood Flour and Its Derived Biochar for the Enhancement of the Peroxydisulfate Activation Capability of Fe₃O₄

Yu Han and Lijie Xu *

College of Biology and the Environment, Nanjing Forestry University, Nanjing 210037, China
* Correspondence: xulijie@njfu.edu.cn

Abstract: In this study, both wood flour (WF) and wood flour-derived biochar (WFB) were used as supports for Fe_3O_4 to activate peroxydisulfate (PDS). The role of different carriers was investigated emphatically from the aspects of catalyst properties, the degradation kinetics of bisphenol A (BPA), the effects of important parameters, and the generation of reactive oxygen species (ROS). Results showed that both WF and WFB could serve as good support for Fe_3O_4, which could control the release of iron into solution and increase the specific surface areas (SSAs). The WFB/Fe_3O_4 had stronger PDS activation capability than WF/Fe_3O_4 mainly due to the larger SSA of WFB/Fe_3O_4 and the PDS activation ability of WFB. Both radical species ($\bullet OH$ and $SO_4^{\bullet-}$) and non-radical pathways, including $^1O_2$ and high-valent iron-oxo species, contributed to the degradation of BPA in the WFB/Fe_3O_4–PDS process. Moreover, the WFB/Fe_3O_4 catalyst also showed stronger ability to control the iron release, better reusability, and higher BPA mineralization efficiency than WF/Fe_3O_4.

Keywords: wood flour; biochar; peroxydisulfate; bisphenol A; Fe_3O_4; activation

1. Introduction

Citation: Han, Y.; Xu, L. A Comparison Study between Wood Flour and Its Derived Biochar for the Enhancement of the Peroxydisulfate Activation Capability of Fe_3O_4. *Catalysts* **2023**, *13*, 323. https://doi.org/10.3390/catal13020323

Academic Editors: Gassan Hodaifa, Rafael Borja and Mha Albqmi

Received: 29 December 2022
Revised: 29 January 2023
Accepted: 30 January 2023
Published: 1 February 2023

The pollution of endocrine disrupting chemicals (EDCs) has attracted extensive attention due to the serious danger to the exposed environments. EDCs, also known as environmental hormones, are contaminants that may cause endocrine disorders [1]. The ingestion of EDCs affects the normal function of endogenous hormones, which can further cause adverse health effects on the human body [2]. EDCs accumulate in sewage treatment plant effluents and surface water at levels as high as μg/L, which can pose potential hazard to the safety of drinking water [3].

To decrease the concentration of EDCs in wastewater, sulfate radical ($SO_4^{\bullet-}$)-based advanced oxidation technologies (Sr-AOTs) have been applied as effective methods [4,5]. The high oxidation potential ($E^0 = 2.5\sim3.1$ V vs. NHE), wide adaptable pH range (2~8), and relatively long half-life (30–40 μs) make $SO_4^{\bullet-}$ radicals own a strong oxidizing ability to decompose recalcitrant contaminants [6,7]. $SO_4^{\bullet-}$ can be produced by the activation of persulfate, including peroxymonosulfate (PMS) and peroxydisulfate (PDS), during which processes the unstable peroxide (-O-O-) bonds are apt to be attacked by electrons or energy for cleavage [8]. PDS was selected as the radical precursor in this study instead of PMS because of its cost effectiveness, high solubility, and chemical stability [9].

Among various PDS activation methods, transition metal possesses the advantages of low energy demand, high efficiency, and mild reaction conditions [10]. Since Fe was regarded as the most efficient, economical, and environmentally friendly metal for PDS activation [11,12], varieties of Fe-based catalysts were developed (e.g., Fe_3O_4/SBA-15 [13], Fe_3C/porous carbon [14]). Fe-based heterogeneous catalysts can well improve some drawbacks of the homogeneous Fe^{2+}/PDS process. The application of Fe_3O_4 can not only avoid Fe precipitation at neutral and alkaline conditions, but it can also decrease the self-quenching of precious radicals, and good separation is able to be achieved [15,16]. However,

Fe_3O_4 particles easily aggregate during the reaction due to the high surface energy and the interaction induced by magnetic force, which can influence the catalytic capability. Therefore, Fe_3O_4 particles are often immobilized onto various carriers to prevent agglomeration. Several studies have proven that Fe_3O_4 loaded on carriers could improve the catalytic activity and stability. Fe_3O_4 particles could anchor on sepiolite with good dispersion, which was efficient for PDS activation even at alkaline conditions [17]. Fe_3O_4@carbon nanotube composites also exhibited high PDS catalytic activity and a stronger charge transfer ability and oxidation ability [18]. Moreover, carbon material, such as reduced graphene oxide (rGO), has also been found to be a promising support. The rGO-Fe_3O_4 nanoparticle was prepared as a PDS activator for the efficient degradation of trichloroethylene, which could well prevent Fe_3O_4 agglomeration [19]. Although most of these carriers could demonstrate strong stability, they are normally costly and have complex synthetic methods. Therefore, it is necessary to develop cheap and easily prepared carriers to support Fe_3O_4 as an efficient PDS activator.

Biochar has been verified as a promising catalyst carrier due to its low price, high specific surface area, and excellent chemical and biological stability. In addition, biochar alone also has the ability to activate PDS due to the abundant functional groups on the surface [20–22]. Moreover, both radical pathways (e.g., hydroxyl radicals ($\bullet$OH), surface-bound radicals) and nonradical pathways (e.g., singlet oxygen ($^1O_2$) and direct electron transfer) were identified as the dominant mechanisms [23,24]. The annual yield of waste wood materials from both agriculture and forestry is very high, which is in urgent need for recycling. The wood materials have strong mechanical strength and stability [14], and the wood-derived biochar with rich oxygen-containing functional groups and π-electron density demonstrated excellent performance in PDS activation [25,26]. Although both the wood-based biomass (WBS) and wood-based biochar (WBC) can be potentially good carriers for the Fe_3O_4 catalyst, so far, hardly any studies have been conducted comparing the influence and the role of WBS and WBC on the catalytic performance of Fe_3O_4 for PDS activation.

In this study, both wood flour (WF) and wood flour-derived biochar (WFB) were used as the carriers to support Fe_3O_4 for the activation of PDS. Bisphenol A (BPA) was selected as a typical EDC due to the toxicity and its recalcitrant structure [27,28]. The role of different carriers was investigated emphatically from the aspects of catalyst properties, BPA degradation kinetics, and the generation of reactive oxygen species (ROS). The effects of important parameters were also evaluated. Finally, the BPA degradation mechanisms were proposed.

2. Results and Discussion

2.1. Characterization of the Catalysts

2.1.1. SEM and FESEM

The surface morphologies of WF_x/Fe_3O_4 and WFB_x/Fe_3O_4 were characterized by SEM and FESEM. As shown in Figure S1, the WF had a sheet-like appearance, while after the pyrolysis treatment, the WFB generally retained the sheet-like structure interspersed with some fibers, which is the typical structure of biochar (Figure 1a). Moreover, the size of WFB became smaller than WF, and the surface of WFB became rougher and more irregular. This was probably because high pyrolysis temperature removed lots of the functional groups and destroyed the polymer framework of the pristine WF [29]. Figure 1b shows that the shape of Fe_3O_4 was almost spherical, and the average sizes of Fe_3O_4 nps were found to be ~30 nm. The FESEM images of WF_{20}/Fe_3O_4 and WFB_{20}/Fe_3O_4 shown in Figure 1c–f further demonstrate the successful synthesis of the composites, in which large numbers of Fe_3O_4 nps appeared on the surface of WF_{20}/Fe_3O_4 and WFB_{20}/Fe_3O_4, and the sheet-like structures could not be observed clearly.

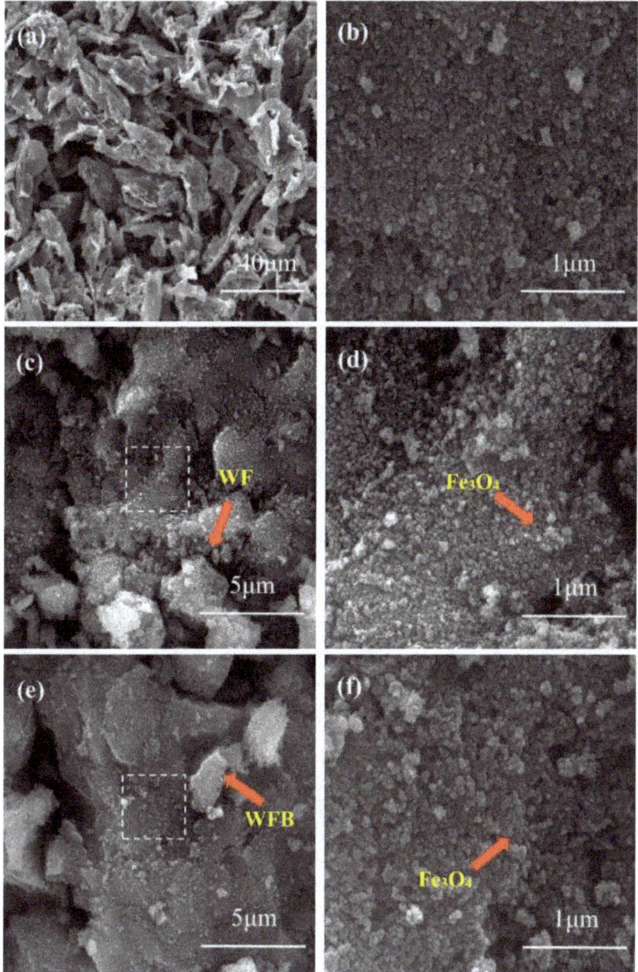

Figure 1. SEM images of (**a**) WFB, and FESEM images of (**b**) Fe$_3$O$_4$, (**c**,**d**) WF$_{20}$/Fe$_3$O$_4$, and (**e**,**f**) WFB$_{20}$/Fe$_3$O$_4$ ((**d**,**f**) are the enlarged view of the dotted boxes shown in (**c**,**e**), respectively).

2.1.2. XRD

The XRD patterns of Fe$_3$O$_4$, WF/Fe$_3$O$_4$, and WFB/Fe$_3$O$_4$ are shown in Figure 2a,b. The peaks located at 30.2° (220), 35.6° (311), 43.2° (400), 53.8° (422), 57.1° (511), and 62.8° (440) were consistent with the characteristic peaks of Fe$_3$O$_4$ (JCPDS No.19-0629) [30]. It was observed that the characteristic peaks were not very sharp, which may be ascribed to the moderate crystallinity of the prepared Fe$_3$O$_4$ [31]. After the incorporation of WF or WFB, the obtained WF/Fe$_3$O$_4$ and WFB/Fe$_3$O$_4$ with different WF or WFB amounts retained the characteristic peaks of Fe$_3$O$_4$, which indicated that the prepared WF/Fe$_3$O$_4$ and WFB/Fe$_3$O$_4$ composites kept the crystal phase of Fe$_3$O$_4$. Additionally, the broad diffraction peaks were observed near 24.5° for all WF/Fe$_3$O$_4$ and WFB/Fe$_3$O$_4$ composites, which could be assigned to the (002) crystal planes of the graphite structure (JCPDS No.75-1621) [32]. It could be obviously seen that the broad diffraction peaks of WFB/Fe$_3$O$_4$ were stronger than those of WF/Fe$_3$O$_4$, which indicated that the addition of WFB brought a more graphitized structure than that of WF.

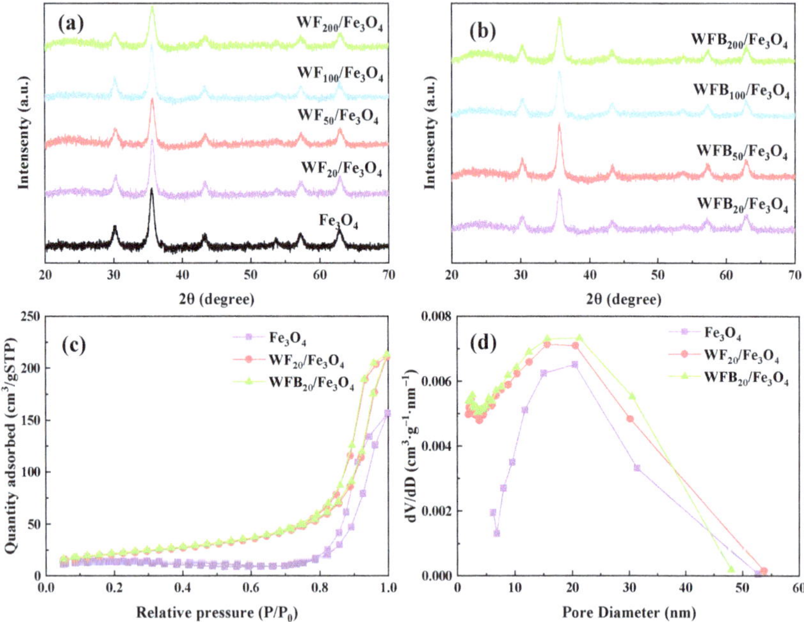

Figure 2. XRD patterns of (**a**) Fe_3O_4, WF/Fe_3O_4, and (**b**) WFB/Fe_3O_4; (**c**) the N_2 adsorption-desorption isotherms and (**d**) pore size distributions of Fe_3O_4, WF_{20}/Fe_3O_4, and WFB_{20}/Fe_3O_4.

The average crystallite size of the Fe_3O_4 nps was calculated with the Debye–Scherrer formula (Equation (1)) [33]. The results are shown in Table S1, and the average crystallite sizes were in the range of 13~16 nm:

$$D = K\lambda/(B\cos\theta) \tag{1}$$

where D is the crystallite size, K is the constant (0.89), λ is the X-ray wavelength (0.1541841 nm), θ is the Bragg diffraction angle, and B is the full width at half maximum.

Differences in the average grain size of Fe_3O_4 nps were obtained based on FESEM and XRD results, which was probably because the particles observed from FESEM normally consist of several crystal cells and the average crystallite size that could be obtained from XRD should be smaller than that obtained from FESEM [34].

2.1.3. N_2 Adsorption-Desorption

The N_2 adsorption-desorption isotherms and pore size distributions of catalysts are shown in Figure 2c,d. According to the isotherm classification from the International Union of Pure and Applied Chemistry (IUPAC), the isotherms of Fe_3O_4, WF_{20}/Fe_3O_4, and WFB_{20}/Fe_3O_4 belong to type IV with an H1-type hysteresis loop, indicating the existence of typical mesoporous structures of the catalysts [35]. This is consistent with the pore size distribution results. Moreover, compared to Fe_3O_4 (41.32 m^2/g), WF_{20}/Fe_3O_4 (76.25 m^2/g) and WFB_{20}/Fe_3O_4 (78.65 m^2/g) had larger specific surface areas (SSAs) and more micropores, which might provide abundant active sites for catalytic reactions and facilitate the adsorption of the pollutant.

2.1.4. VSM

The magnetic properties of Fe_3O_4, WF_{20}/Fe_3O_4, and WFB_{20}/Fe_3O_4 were assessed by VSM at room temperature (Figure 3). As illustrated, the saturation magnetization (M_s) of Fe_3O_4 was calculated to be 71.58 emu g^{-1}. Although the M_s values of WF_{20}/Fe_3O_4

(67.54 emu g^{-1}) and WFB$_{20}$/Fe$_3$O$_4$ (62.37 emu g^{-1}) were both lower than that of Fe$_3$O$_4$ due to the existence of non-magnetic WF or WFB, the two composites still exhibited strong magnetic responses, which ensured their easy recycle via magnetic force without introducing secondary pollution in practical use.

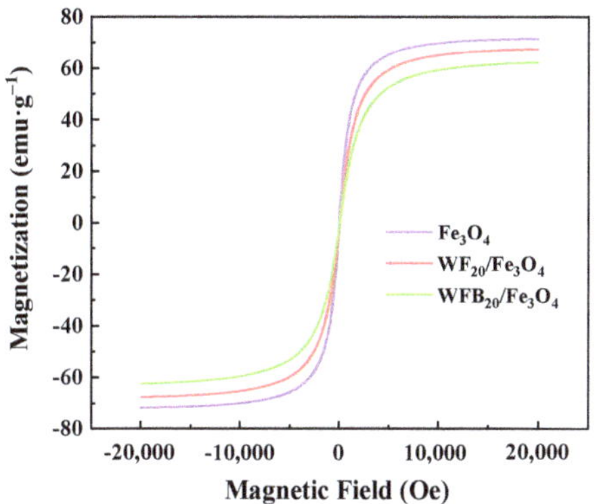

Figure 3. The magnetization curves of Fe$_3$O$_4$, WF$_{20}$/Fe$_3$O$_4$, and WFB$_{20}$/Fe$_3$O$_4$.

2.1.5. FTIR

FTIR spectra were used to further validate the successful composition of Fe$_3$O$_4$ nps with WF or WFB (Figure S2). All the catalysts had characteristic peaks at 3426.87 cm^{-1} and 1628.81 cm^{-1}, corresponding to the -OH stretching and bending vibration of water, respectively [36]. Meanwhile, the peak intensities of C-H (~2920 cm^{-1}), C=O (1736.27 cm^{-1}), and C-O-C (1052.56 cm^{-1}) vibrations in WFB and WFB$_{20}$/Fe$_3$O$_4$ were weaker than those of WF and WF$_{20}$/Fe$_3$O$_4$ samples, indicating that the oxygen-containing groups in the lignocellulose structures were decomposed and removed during the pyrolysis process [29,37]. In the spectra of Fe$_3$O$_4$, WF$_{20}$/Fe$_3$O$_4$, and WFB$_{20}$/Fe$_3$O$_4$, the vibrations of Fe^{2+}-O^{2-} (585.33 cm^{-1}) and Fe^{3+}-O^{2-} (445.16 cm^{-1}) appeared, implying the formation of Fe$_3$O$_4$, which was also consistent with the results of XRD patterns [38]. After the addition of WF or WFB, the band intensities at 2924.59 cm^{-1} (asymmetric stretching for aliphatic functional groups) and 2856.07 cm^{-1} (symmetric stretching for aliphatic functional groups) slightly increased [39]. These changes of FTIR peaks proved that the composites of WF/Fe$_3$O$_4$ and WFB/Fe$_3$O$_4$ were successfully synthesized.

2.2. Evaluation of Catalytic Performance

The degradation of BPA in different processes was evaluated (Figure 4). Before adding PDS, 30 min pre-adsorption was performed to enable the equilibrium of BPA adsorption. As shown in Figure 4a, WF and WFB showed moderate adsorption capacity for BPA, with removal rates of 19.9% and 15.2% within 30 min, while Fe$_3$O$_4$ could only adsorb 4.7% of BPA. After compositing with WF or WFB, the adsorption capacities of WF/Fe$_3$O$_4$ and WFB/Fe$_3$O$_4$ were slightly improved compared to that of Fe$_3$O$_4$, which was consistent with the results of SSA. Control experiments were conducted to preclude the direct oxidation of BPA by PDS (Figure 4a). In the WF-PDS process, hardly any BPA could be degraded, which demonstrated that pure WF could not activate PDS to generate effective ROS. In contrast, WFB showed the effective activation of PDS leading to the complete degradation of BPA within 90 min, which might be due to the formation of active sites such as graphitized carbon, oxygen-containing groups, and so on after high-temperature carbonization.

As seen in Figure 4b, 75.5% of BPA was degraded in the Fe$_3$O$_4$-PDS system. For catalysts with different amounts of WF, WF$_{20}$/Fe$_3$O$_4$ exhibited excellent catalytic efficiency with a removal rate of 89.7% in 90 min. With more than 20 mg of WF loading, the BPA removal rate decreased, and the BPA removal rate in the WF$_{200}$/Fe$_3$O$_4$-PDS process was even lower than that in the Fe$_3$O$_4$-PDS process. These results showed that excessive

WF loading could decrease the catalytic efficiency of WF/Fe$_3$O$_4$. However, the catalytic performance of all the WFB/Fe$_3$O$_4$ catalysts was better than that of Fe$_3$O$_4$. Although the catalytic efficiency of WFB/Fe$_3$O$_4$ slightly decreased as the amount of WFB increased, almost 100% of BPA in these systems was degraded within 50 min. Table 1 compared the degradation efficiencies of BPA in different Sr-AOPs. It can be seen that the WFB$_{20}$/Fe$_3$O$_4$-PDS process of the present study could have comparable efficiency toward BPA degradation with other Sr-AOPs. It should also be noted that the same dose of WF and WFB showed different influences on the catalytic performance of Fe$_3$O$_4$. To further reveal the roles of WF and WFB in the composites for the activation of PDS, WF$_{20}$/Fe$_3$O$_4$ and WFB$_{20}$/Fe$_3$O$_4$ were selected for the following study.

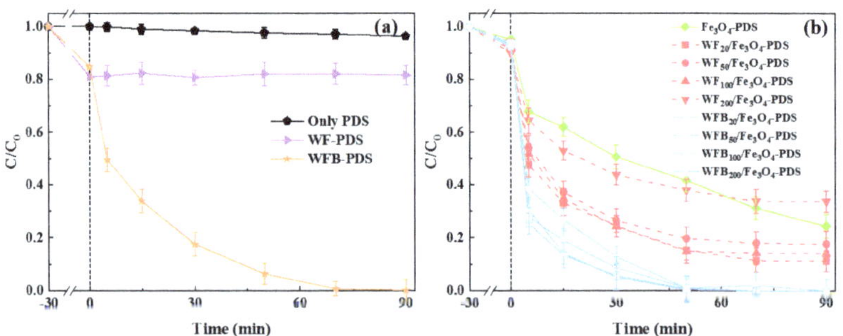

Figure 4. (**a**) The BPA removal performance in control processes; (**b**) the BPA removal performance in different Fe$_3$O$_4$-containing processes (conditions: [BPA]$_0$ = 0.02 mM, catalyst 1.0 g/L, [PDS]$_0$ = 5.0 mM, pH$_0$ = 3.00 ± 0.1).

Table 1. Comparison of the BPA degradation efficiency in different Sr-AOPs.

Sr-AOPs	Conditions	Removal Rate	Main Reactive Species	Ref.
ZnFe$_2$O$_4$-PDS	pH$_0$ = 6, [BPA]$_0$ = 0.1 mM, catalyst 0.5 g/L, [PDS]$_0$ = 5.0 mM	96.5%, 120 min	h$^+$, •OH and SO$_4$•$^-$	[40]
ZnO@AC@FeO-PMS-UV	pH$_0$ = 7, [BPA]$_0$ = 30 mg/L, catalyst 0.4 g/L, [PMS]$_0$ = 4 mM	98.3%, 120 min	^{1}O$_2$, O$_2$•$^-$, •OH and SO$_4$•$^-$	[41]
UVA-LED/CFO-rGO-PMS	pH$_0$ = 7, [BPA]$_0$ = 20 mg/L, catalyst 0.4 g/L, [PMS]$_0$ = 150 mg/L	99.5%, 30 min	•OH and SO$_4$•$^-$	[42]
Fe^{2+}/g-C$_3$N$_4$/LED-PMS	pH$_0$ = 3.5 ± 0.1, [BPA]$_0$ = 0.01 mM, g-C$_3$N$_4$ dosage 0.5 g/L, [Fe^{2+}]$_0$ = 0.01 mM, [PMS]$_0$ = 0.1 mM,	100%, 90 min	O$_2$•$^-$, ^{1}O$_2$, Fe(IV), •OH	[43]
Fe$_3$C/C1000-PMS	pH$_0$ = 3.1, [BPA]$_0$ = 0.1 mM, catalyst 0.1 g/L, [PMS]$_0$ = 2 mM	100%, <30 min	^{1}O$_2$, transferring electrons, •OH and SO$_4$•$^-$	[14]

2.3. Effects of Reaction Parameters on BPA Degradation

2.3.1. Effect of Initial Solution pH

It is well known that solution pH has significant influence on the performance of various AOTs. Herein, the effects of the initial solution pH (pH$_0$) on BPA degradation in Fe$_3$O$_4$-PDS, WF$_{20}$/Fe$_3$O$_4$-PDS, and WFB$_{20}$/Fe$_3$O$_4$-PDS processes were explored (Figures S3 and 5a,b). The BPA removal rate in Fe$_3$O$_4$-PDS and WF$_{20}$/Fe$_3$O$_4$-PDS processes decreased from 75.5% to 3.8% and from 88.7% to 5.1%, respectively, as the pH$_0$ increased from 3.0 to 7.0. By contrast, the WFB$_{20}$/Fe$_3$O$_4$-PDS process showed stronger pH tolerance, and the BPA removal rate decreased from 100% to 11.5% as the pH$_0$ increased from 3.0 to 7.0. These results indicated that these catalytic processes preferred more acidic conditions for BPA degradation. Under acidic conditions, ferrous ions could more easily enter into the solution from the catalyst surface, and the dissolved Fe^{2+} could activate PDS to accelerate BPA decomposition [44].

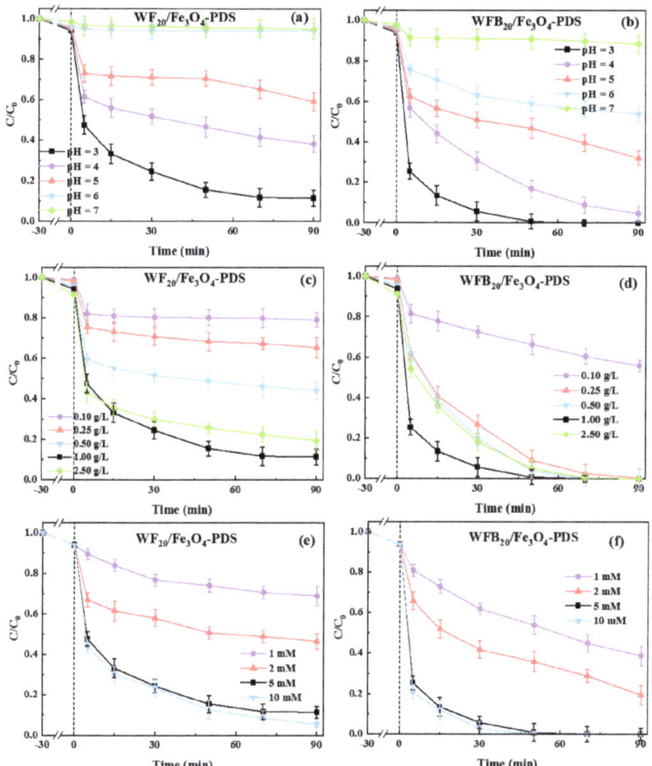

Figure 5. Effects of (**a**,**b**) the initial solution pH, (**c**,**d**) catalyst dosage, and (**e**,**f**) PDS concentration on the degradation of BPA in WF$_{20}$/Fe$_3$O$_4$-PDS and WFB$_{20}$/Fe$_3$O$_4$-PDS processes (conditions: catalyst 1.0 g/L, [PDS]$_0$ = 5.0 mM, [BPA]$_0$ = 0.02 mM, pH$_0$ = 3.00 ± 0.1).

2.3.2. Effect of Catalyst Dosage

The effects of WF$_{20}$/Fe$_3$O$_4$ and WFB$_{20}$/Fe$_3$O$_4$ dosages on BPA degradation were investigated (Figure 5c,d). It was found that the increase of WF$_{20}$/Fe$_3$O$_4$ and WFB$_{20}$/Fe$_3$O$_4$ dosages from 0.1 g/L to 1.0 g/L resulted in the significant and continuous promotion of BPA degradation, with the removal rate increasing from 20.9% to 88.7% and from 44.1% to 100%, respectively. This could be explained by the increase of active sites with higher dosages of catalysts to promote the production of ROS. However, when the dosages of WF$_{20}$/Fe$_3$O$_4$ or WFB$_{20}$/Fe$_3$O$_4$ were further increased to 2.5 g/L, the removal rate of BPA decreased slightly. This probably resulted from the self-quenching effects of explosive ROS and the quenching reactions between ROS and PDS when the catalysts were excessive (Equations (2) and (3)) [45]. Moreover, the Fe(II) on the surface of Fe$_3$O$_4$ nps also had a strong scavenging effect towards ROS (Equation (4)) [46]. Additionally, excess dosage of catalyst may also increase the mutual magnetic attraction and agglomeration of the catalysts, interfering with their uniform dispersion in solution, which could reduce the surface area of the catalysts and the corresponding catalytic efficiency [47].

$$SO_4^{\bullet-} + SO_4^{\bullet-} \rightarrow S_2O_8^{2-}, \, k = 5 \times 10^8 \, M^{-1} \, s^{-1} \tag{2}$$

$$SO_4^{\bullet-} + S_2O_8^{2-} \rightarrow SO_4^{2-} + S_2O_8^{\bullet-}, \, k = 5.5 \times 10^5 \, M^{-1} \, s^{-1} \tag{3}$$

$$SO_4^{\bullet-} + Fe(II) \rightarrow Fe(III) + SO_4^{2-} \tag{4}$$

2.3.3. Effect of PDS Concentration

Generally, the amount of generated ROS directly relates to the PDS concentration. Herein, the effects of PDS concentration on BPA degradation in WF_{20}/Fe_3O_4-PDS and WFB_{20}/Fe_3O_4-PDS processes were examined (Figure 5e,f). It was found that, with the PDS concentration increasing from 1.0 to 5.0 mM, the BPA removal rate increased from 31.2% to 88.7% in the WF_{20}/Fe_3O_4-PDS process and from 61.1% to 100.0% in the WFB_{20}/Fe_3O_4-PDS process, respectively. However, when the PDS concentration further increased to 10.0 mM, the BPA degradation was only improved slightly. The results demonstrated that the excess amount of PDS did not benefit the degradation of target pollutants, mainly because PDS could also act as the ROS quencher competing for the ROS with target pollutants (Equations (2) and (3)).

2.4. Identification of ROS

2.4.1. Quenching Experiments

In order to further understand the different roles of WF and WFB in the evolution of ROS during PDS activation, quenching experiments were conducted in different systems using various quenchers. It is widely accepted that EtOH can rapidly quench both $\bullet$OH and $SO_4^{\bullet-}$ ($k_{\bullet OH} = 1.6{\sim}7.7 \times 10^7$ M^{-1} s^{-1}, $k_{SO4\bullet-} = 1.2{\sim}2.8 \times 10^8$ M^{-1} s^{-1}), while TBA is usually used as an effective quencher for $\bullet$OH ($k_{\bullet OH} = 3.8{\sim}7.6 \times 10^8$ M^{-1} s^{-1}) but is inert towards $SO_4^{\bullet-}$ ($k_{SO4\bullet-} = 4.0{\sim}9.1 \times 10^5$ M^{-1} s^{-1}) [48,49]. As shown in Figures S4 and 6a,b, the inhibiting effect of EtOH on BPA degradation in both systems was significantly stronger than that of TBA. Therefore, both $\bullet$OH and $SO_4^{\bullet-}$ should be involved and contribute to BPA degradation, and $SO_4^{\bullet-}$ may play a more important role than $\bullet$OH. In addition, $^1O_2$ is often the essential ROS in Sr-AOTs, especially for the processes involving biochar as the catalyst. Therefore, FFA was used as the scavenger for $^1O_2$ ($k_{1O2} = 1.2 \times 10^8$ M^{-1} s^{-1}) to examine the possible contribution of $^1O_2$ [50]. It can be seen that FFA showed strong inhibition on BPA degradation in both systems, which implied that $^1O_2$ possibly contributed to BPA degradation. However, since FFA also reacts with $\bullet$OH ($k_{\bullet OH} = 1.5 \times 10^{10}$ M^{-1} s^{-1}) [51], the generation of $^1O_2$ needs to be further verified by ESR.

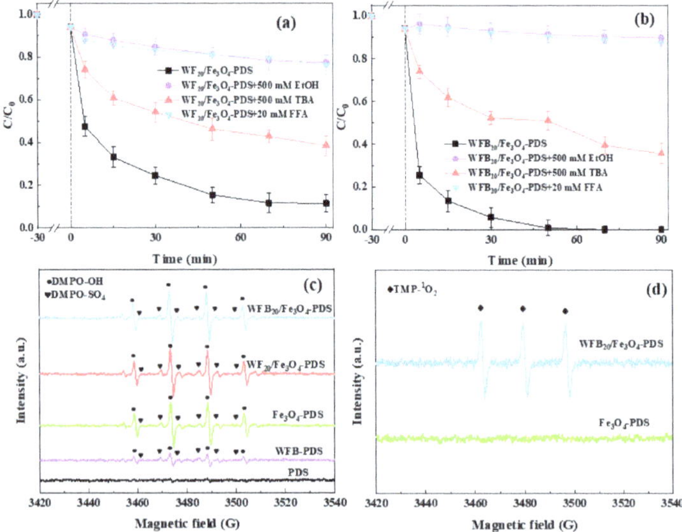

Figure 6. Effects of different radical scavengers on BPA degradation in (a) WF_{20}/Fe_3O_4-PDS and (b) WFB_{20}/Fe_3O_4-PDS systems; (c) DMPO-OH and DMPO-SO$_4$ spectra in different processes; (d) TMP-$^1O_2$ spectra in different processes (conditions: catalyst 1.0 g/L, $[PDS]_0$ = 5.0 mM, $[BPA]_0$ = 0.02 mM, $pH_0 = 3.00 \pm 0.1$).

2.4.2. ESR Analysis

ESR was performed to confirm the presence of possible ROS in different processes. DMPO was used as the spin trapping agent to identify •OH and $SO_4^{•-}$. As shown in Figure 6c, hardly any signals could be detected in the sole-PDS process, indicating that PDS could hardly self-decompose to generate ROS. This phenomenon was consistent with the results of the catalytic degradation experiment (Figure 4a). When WFB and PDS were added together, DMPO-OH ($\alpha_N = \alpha_H = 14.9$ G) and DMPO-SO$_4$ ($\alpha_N = 13.2$ G, $\alpha_H = 9.6$ G, $\alpha_H = 1.48$ G and $\alpha_H = 0.78$ G) signals could be detected, which proved the formation of •OH and $SO_4^{•-}$. The above phenomenon might be due to the activation of PDS by the active sites on WFB (Equations (5) and (6)) [52,53]. It was observed that much stronger signals were detected in the Fe_3O_4-PDS, WF_{20}/Fe_3O_4-PDS, and WFB_{20}/Fe_3O_4-PDS systems, which was consistent with the quenching results (Figures S4 and 6a,b). This might be because Fe(II), with high catalytic activity on Fe_3O_4 nps, could have reacted with PDS to generate $SO_4^{•-}$, and then the generated $SO_4^{•-}$ further reacted with H_2O to form •OH (Equations (6) and (7)). Furthermore, TMP was used as the spin trapping agent to trap $^1O_2$ (Figure 6d). The TMP-$^1O_2$ ($\alpha_N = 16.9$ G) signal could be detected in the WFB_{20}/Fe_3O_4-PDS process, but hardly any signal could be detected in the Fe_3O_4-PDS process, which was probably due to the formation of $^1O_2$ between WFB and PDS. Therefore, $^1O_2$ should also be the contributor to BPA degradation in the WFB_{20}/Fe_3O_4-PDS process.

$$S_2O_8^{2-} + \text{active sites}_{WFB} \rightarrow SO_4^{•-} + SO_4^{2-} + \text{active sites}_{WFB}^* \tag{5}$$

$$SO_4^{•-} + H_2O \rightarrow H^+ + SO_4^{2-} + •OH, \ k = 2 \times 10^{-3} \ M^{-1} s^{-1} \tag{6}$$

$$S_2O_8^{2-} + Fe(II) \rightarrow Fe(III) + SO_4^{•-} + SO_4^{2-} \tag{7}$$

2.4.3. Identification of High-Valent Iron-Oxo Species

Some previous studies have shown that, during the activation of PDS by Fe(II) or Fe(III), corresponding high-valent iron-oxo species (Fe(IV)=O and Fe(V)=O) can form via double-electron transfer, which can further promote the degradation of pollutants [54,55]. Since Fe_3O_4 involves both Fe(II) and Fe(III), both Fe(IV)=O and Fe(V)=O are likely to be generated during the activation of PDS. Moreover, according to some previous results of density functional theory calculation, the oxidation capacity of high-valent iron-oxo species under acidic conditions is stronger than that under alkaline conditions [56], which agrees well with the degradation performance of BPA in WF_{20}/Fe_3O_4-PDS and WFB_{20}/Fe_3O_4-PDS processes, as discussed in Section 2.3.1. Therefore, it is necessary to study the existence of high-valent iron-oxo species in this study.

It is known that both Fe(IV)=O and Fe(V)=O can oxidize PMSO to $PMSO_2$ through the oxygen atom transfer pathway (Equations (8) and (9)), differing markedly from radical-mediated routes [57]. The degradation of PMSO and the formation of $PMSO_2$ in different processes are depicted in Figure 7. It can be seen in Figure 7a that PMSO (0.5 mM) could not be oxidized by PDS alone. It was found in Figure 7b–d that 0.057 mM, 0.218 mM, and 0.219 mM of PMSO was degraded within 90 min in Fe_3O_4-PDS, WF_{20}/Fe_3O_4-PDS, and WFB_{20}/Fe_3O_4-PDS processes, respectively. Correspondingly, about 0.047 mM, 0.140 mM, and 0.145 mM of $PMSO_2$ was generated within 90 min in respective processes, which could well indicate the formation of high-valent iron-oxo species in these three processes. It was also observed that the WFB_{20}/Fe_3O_4-PDS process showed a stronger capability to form high-valent iron-oxo species than the WF_{20}/Fe_3O_4-PDS process. Then, the yield of $PMSO_2$ ($\eta = \Delta[PMSO_2]/\Delta[PMSO]$) was quantified to evaluate the contribution of high-valent iron-oxo species to PMSO degradation. Figure S5 shows that η values gradually decreased as the reaction proceeded in the Fe_3O_4-PDS process, and it was about 80% at 90 min. Comparatively, the η values kept stable at approximately 65% during the reaction in both WF_{20}/Fe_3O_4-PDS and WFB_{20}/Fe_3O_4-PDS processes. Since the PMSO can also be oxidized by •OH and $SO_4^{•-}$ via different pathways rather than form $PMSO_2$ [54], the η value below 100% indicated the presence of •OH and/or $SO_4^{•-}$, which was consistent with the results of ESR. Moreover, the majority of PMSO was oxidized to $PMSO_2$, which also indicated that the high-valent iron-oxo species was the dominant contributor to PMSO degradation.

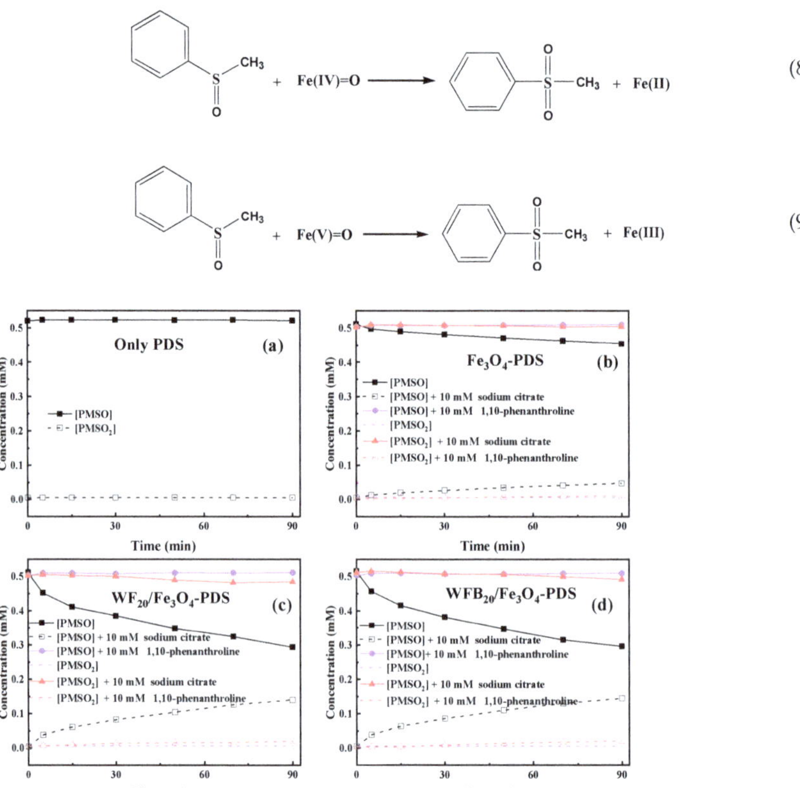

(8)

(9)

Figure 7. PMSO degradation and $PMSO_2$ production in the (a) PDS, (b) Fe_3O_4-PDS, (c) WF_{20}/Fe_3O_4-PDS, and (d) WFB_{20}/Fe_3O_4-PDS systems (conditions: catalyst 1.0 g/L, $[PDS]_0$ = 5.0 mM, $[PMSO]_0$ = 0.5 mM, pH_0 = 3.00 ± 0.1).

Further investigations were conducted to elucidate the formation mechanisms of high-valent iron-oxo species in Fe_3O_4-PDS, WF_{20}/Fe_3O_4-PDS, and WFB_{20}/Fe_3O_4-PDS processes. As shown in Figure 8a, the dissolved Fe^{2+} and dissolved total irons in Fe_3O_4-PDS, WF_{20}/Fe_3O_4-PDS, and WFB_{20}/Fe_3O_4-PDS processes could be detected, which indicated the presence of homogeneous Fe^{2+} and Fe^{3+} during the catalytic reaction. In addition, with the addition of WF or WFB, the total iron dissolution decreased significantly, and the use of WFB as the support could make more significant control of iron dissolution than that of WF. Previous studies showed that Fe^{2+} could react with PDS to generate Fe(IV)=O under acidic conditions (Equation (10)) [54], but Fe^{3+} could not [58]. Additionally, the surface ≡Fe(II) and ≡Fe(III) may also play important roles in the generation of high-valent iron-oxo species [55]. Some studies assumed that •OH, $SO_4^{•-}$ and ≡Fe(IV)=O could come from the reaction between surface ≡Fe(II) and PDS (Equations (6), (7) and (11)), while ≡Fe(V)=O could come from the reaction between ≡Fe(III) and PDS (Equation (12)) [59]. To further prove the formation mechanisms of high-valent iron-oxo species, chelating agents (sodium citrate and 1,10-phenanthroline) were added to examine the important roles of surface ≡Fe(II) and ≡Fe(III) of Fe_3O_4 nps. Some studies have reported that ligand exchange could happen between ligands and specific sites of Fe (such as ≡Fe-OH) [55], which could then suppress the capability of catalysts for PDS activation. In other words, 1,10-phenanthroline and sodium citrate could coordinate with Fe(II) and Fe(III) [60], respectively, leading to the inactivation of surface ≡Fe(II) and ≡Fe(III). From Figure 7b–d, it was clearly seen that PMSO degradation and PMSO$_2$ production were obviously suppressed by the addition of 1,10-phenanthroline and sodium citrate. The results manifested that chelating agents indeed suppressed the capability of surface ≡Fe(II) and ≡Fe(III), leading to a decrease of ≡Fe(IV)=O and ≡Fe(V)=O.

$$Fe_{aq}^{2+} + S_2O_8^{2-} + H_2O \ \rightarrow \ Fe_{aq}^{IV}O^{2+} + 2SO_4^{2-} + 2H^+ \tag{10}$$

$$\equiv Fe(II) + S_2O_8^{2-} \ \rightarrow \ \equiv Fe(IV) = O + 2SO_4^{2-} \tag{11}$$

$$\equiv Fe(III) + S_2O_8^{2-} \ \rightarrow \ \equiv Fe(V) = O + 2SO_4^{2-} \tag{12}$$

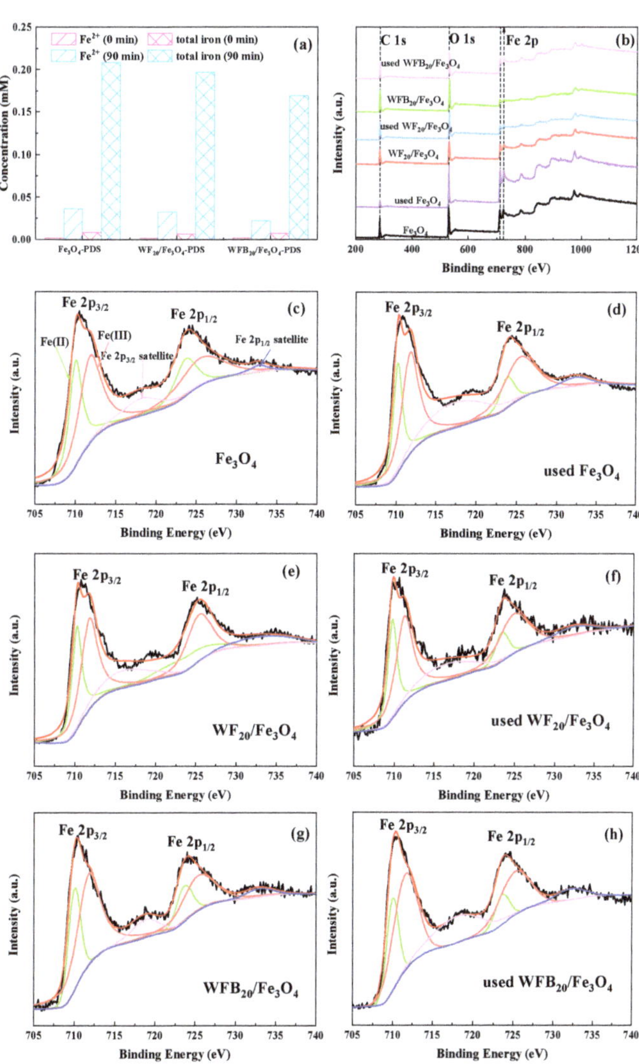

Figure 8. (**a**) Concentration of dissolved Fe^{2+} and total irons in Fe$_3$O$_4$-PDS, WF$_{20}$/Fe$_3$O$_4$-PDS, and WFB$_{20}$/Fe$_3$O$_4$-PDS processes; the XPS of (**b**) full survey spectra and (**c–h**) the Fe 2p spectra of different catalysts (conditions: catalyst 1.0 g/L, [PDS]$_0$ = 5.0 mM, [BPA]$_0$ = 0.02 mM, pH$_0$ = 3.00 ± 0.1).

To sum up, the high-valent iron-oxo species should be mainly produced by two sources: (i) the reaction of dissolved Fe^{2+} with PDS to form $Fe(IV)=O$; (ii) the reaction of surface $\equiv Fe(II)$ and $\equiv Fe(III)$ with PDS to form $\equiv Fe(IV)=O$ and $\equiv Fe(V)=O$, respectively.

2.4.4. Active Sites Analysis

The valence state of the Fe element in catalysts was investigated by XPS to further examine the contribution of surface $\equiv Fe(II)$ and $\equiv Fe(III)$ during the catalytic process. Figure 8b–h showed the XPS full survey spectra and the spectra of Fe 2p of different catalysts. As presented in Figure 8b, C (285.6 eV), O (530.2 eV), and Fe (711.8 eV, 724.3 eV) could be detected in these catalysts [61,62]. The Fe 2p spectrum could be deconvolved into six peaks, with 709.86 eV and 723.46 eV corresponding to Fe $2p^{3/2}$, 711.62 eV and 725.22 eV corresponding to Fe $2p_{1/2}$ [63], and the peaks at 718.16 eV and 732.14 eV being the satellite peaks of Fe $2p_{3/2}$ to Fe $2p_{1/2}$ [64], respectively. According to peak intensity, the content of Fe(II) and Fe(III) could be calculated (Table S2). Fe(II) and Fe(III) accounted for 41.56% and 58.44% of total Fe species in Fe_3O_4, 40.00% and 60.00% in WF_{20}/Fe_3O_4, and 37.21% and 62.79% in WFB_{20}/Fe_3O_4, respectively. After the catalytic reaction, the content of Fe(II) decreased to 36.41%, 32.32%, and 32.12%, respectively, and the content ratio of Fe(III) increased in the three catalysts, which indicated that the electrons transferred from Fe(II) to PDS, and thus, the fraction of Fe(III) increased accordingly.

To further study the catalytic mechanism of WFB in WFB_{20}/Fe_3O_4, FT-IR spectra of fresh WFB and the used WFB were compared (Figure S6). After the catalytic reaction, the C=O groups (1736.27 cm^{-1}) almost disappeared, indicating that the C=O groups on WFB were consumed during the catalytic reaction [65]. Thus, it was assumed that the C=O groups were mainly the functional groups contributing to the catalytic activity of WFB, which was consistent with the findings of previous studies that carbon-based materials containing C=O groups, such as quinone or ketone, could activate PDS to produce $^1O_2$ (Equations (13)–(15)) [66,67].

$$WFB{=}O + S_2O_8^{2-} + OH^- \rightarrow WFB\text{-}OH/SO_3\text{-}O\text{-}O^- + SO_4^{2-} \tag{13}$$

$$WFB\text{-}OH/SO_3\text{-}O\text{-}O^- + OH^- \rightarrow WFB\text{-}O^-/SO_3\text{-}O\text{-}O^- + H_2O \tag{14}$$

$$WFB\text{-}O^-/SO_3\text{-}O\text{-}O^- + S_2O_8^{2-} + 2OH^- \rightarrow WFB{=}O + 3SO_4^{2-} + {}^1O_2 + H_2O \tag{15}$$

2.5. Stability and Reusability of Catalysts

In order to test the stability and reusability of different catalysts, the used catalysts were separated from the reaction solution by a magnet to conduct consecutive catalytic reactions under the same conditions. The separated catalysts were washed with ultrapure water five times, dried in the oven, and directly used for the next cycles. As shown in Figure 9a, the BPA removal rate in the Fe_3O_4-PDS process decreased from 74% in the first cycle to 51% in the 4th cycle. The decrease of the BPA degradation efficiency was probably due to the agglomeration of Fe_3O_4 nps and the reduced content of Fe(II) in it. Compared to Fe_3O_4, both WF_{20}/Fe_3O_4 and WFB_{20}/Fe_3O_4 showed better reusability (Figure 9b,c). Specifically, WFB_{20}/Fe_3O_4 showed the best reusability with almost all BPA being degraded within 90 min after four cycles. These results were closely related to the following synergistic effects between supports (WF or WFB) and the Fe_3O_4 nps: (i) the prevention of agglomeration of Fe_3O_4 nps; (ii) the prevention of leakage of Fe(II). In addition to acting as a carrier, WFB could also react with PDS to form $^1O_2$, thus promoting the degradation of BPA.

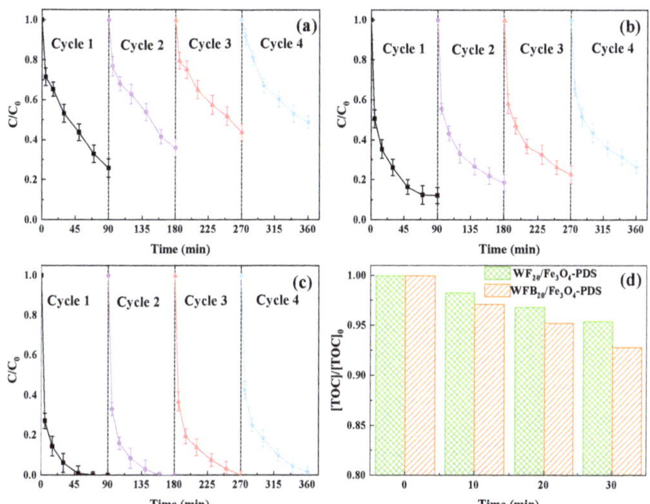

Figure 9. The removal performance of BPA in four consecutive runs in (**a**) Fe_3O_4-PDS, (**b**) $WF_{20}/Fe_3O_4/$-PDS, and (**c**) WF_{20}/Fe_3O_4-PDS processes; (**d**) the TOC removal performance of BPA in WF_{20}/Fe_3O_4-PDS and WF_{20}/Fe_3O_4-PDS processes (conditions: catalyst 1.0 g/L, $[PDS]_0$ = 5.0 mM, $[BPA]_0$ = 0.02 mM, pH_0 = 3.00 ± 0.1).

2.6. Mineralization Performance

The degree of BPA mineralization is also a critical factor to evaluate the catalytic performance of different catalysts. The changes of TOC in WF_{20}/Fe_3O_4-PDS and WFB_{20}/Fe_3O_4-PDS processes are shown in Figure 9d. The TOC removal rate reached 4.6% and 7.2% within 30 min in WF_{20}/Fe_3O_4-PDS and WFB_{20}/Fe_3O_4-PDS processes, respectively, indicating that BPA was more efficiently mineralized in the WFB_{20}/Fe_3O_4-PDS process.

3. Materials and Methods

3.1. Chemicals

BPA (≥99.0%), PDS ($K_2S_2O_8$, ≥99.0%), $FeCl_3·6H_2O$ (≥99.0%), $FeSO_4·7H_2O$ (≥99.0%), *p*-benzoquinone (*p*-BQ, ≥98.0%), furfuryl alcohol (FFA, 98%), and tert-butyl alcohol (TBA, ≥99.0%) were purchased from Sigma-Aldrich, St. Louis, MI, USA. Methanol (MeOH, ≥99.5%), ethanol (EtOH, ≥99.7%), *L*-histidine, *L*-ascorbic acid (≥99.7%), CH_3COOH (≥99.5%), CH_3COONa (≥99.0%), sodium citrate (98%), $NH_3·H_2O$ (75%), and 1,10-phenanthroline (≥99.0%) were all purchased from Sinopharm Chemical, Shanghai, China. The methyl phenyl sulfoxide (PMSO, 98.0%), methyl phenyl sulfone ($PMSO_2$, 98.0%) and 5-dimethyl-1-pyrroline N-oxide (DMPO, 97.0%) were obtained from Macklin, Shanghai, China. The 2, 2, 6, 6-tetramethylpiperidine (TMP, ≥98.0%) was purchased from Shanghai Aladdin Biochemical, Shanghai, China. The dimethyl sulfoxide (DMSO, 99.7%) was obtained from J&K, China. Ultrapure water was used for all experiments and prepared by the HHitech water purification machine.

3.2. Synthesis of the Catalysts

3.2.1. Preparation of WFB

The WFB was prepared by the pyrolysis method. Poplar wood flour with 280–300 mesh size was used as the precursor of WFB, which was obtained from Yixing Wood Flour Factory (Linyi, Shandong Province, China). The wood flour was put into a tubular furnace and heated at 600 °C for 3 h in the N_2 atmosphere with a raping rate of 5 °C/min. After cooling down to room temperature, the black product was obtained as the WFB, which was ground into powders for further use.

3.2.2. Preparation of Fe_3O_4

The coprecipitation method was used to prepare Fe_3O_4 nanoparticles (nps) [68,69]. Briefly, 2.2992 g $FeCl_3 \cdot 6H_2O$ and 1.1831 g $FeSO_4 \cdot 7H_2O$ (2:1) were dissolved in 50 mL ultrapure water with N_2 purging in advance, which was heated to 80 °C and maintained for 30 min under mild mechanical agitation. Then, 25 mL of 25% (w/w) $NH_3 \cdot H_2O$ was added into the above mixture dropwise until the pH increased to around 10. The mixture was maintained at 80 °C with continuous stirring for 1 h, and N_2 was purged throughout to coprecipitate. The solid phase was separated by the external magnetic field, which was then washed repeatedly with ultrapure water until the liquid supernatant was neutral. Finally, the solid mixture was freeze-dried overnight to obtain the Fe_3O_4 powder.

3.2.3. Preparation of WF/Fe_3O_4 and WFB/Fe_3O_4

WF/Fe_3O_4 and WFB/Fe_3O_4 composites were also prepared by the coprecipitation method. Different amounts (20, 50, 100, and 200 mg) of WF or WFB were dispersed in 50 mL of ultrapure water with the aid of sonication. After full dispersion, N_2 was purged into the mixture to blow out the dissolved oxygen (DO). Then, 2.2992 g $FeCl_3 \cdot 6H_2O$ and 1.1831 g $FeSO_4 \cdot 7H_2O$ (2:1) were dissolved in the solution, which was heated to 80 °C and maintained for 30 min under mechanical agitation. The remaining steps were the same as the preparation of Fe_3O_4. The obtained samples with different amounts of WF or WFB were denoted as WF_x/Fe_3O_4 and WFB_x/Fe_3O_4 (x = 20, 50, 100 and 200), respectively.

3.3. Characterization

The morphology of the obtained catalysts was examined by scanning electron microscopy (SEM, FEI Quanta 200, Portland, OR, USA) and field emission scanning electron microscopy (FESEM, Hitachi SU8220, Tokyo, Japan). X-ray diffraction (XRD) patterns of the samples were collected by a Rigaku Ultima IV X-ray diffractometer (Tokyo, Japan) with a Cu-Kα radiation source (1.541841 Å). Scanning rate and 2θ collection range were set at $10°/min$ and $20\sim70°$, respectively. The functional groups on the surface were identified using a Fourier transforms infrared spectrophotometer (FT-IR, Bruker VERTEX 80 V, Billerica, MA, USA) with the wavelength range of $400\sim4000$ cm^{-1}. Brunauer-Emmett-Teller (BET) specific surface area and pore size distribution were investigated by the nitrogen (N_2) adsorption-desorption method on a Micromeritics ASAP 2020 HD88 instrument (Norcross, GA, USA). The magnetic properties of catalysts were assessed by a vibrating sample magnetometer (VSM, Quantum Design PPMS-9T, San Diego, CA, USA). X-ray photoelectron spectroscopy (XPS, Kratos Axis Ultra DLD, Manchester, UK) was used to analyze the surface chemical composition and element valence states.

3.4. Experimental Conditions

Unless otherwise illustrated, all experiments were carried out in a 50 mL beaker at room temperature under mechanical agitation. In a typical experiment, the catalyst with a final concentration of 1.0 g/L was added and dispersed well in 50 mL of BPA solution, and the reaction was initiated by adding PDS to obtain the final concentration of 5.0 mM. At certain time intervals, 1.0 mL of reaction solution was withdrawn, which was immediately mixed with 0.1 mL of MeOH to terminate the reaction and then filtered through a 0.22 µm membrane for further measurement.

For the experiments of PMSO oxidation, the catalyst was dispersed well into the beaker containing a certain amount of PMSO solution, and the reaction was initiated by adding PDS stock solutions. Samples (1.0 mL) were collected at predetermined time intervals and quickly quenched by DMSO (0.1 mL) and filtered through the 0.22 µm membranes into vials for further measurement.

The initial pH was adjusted with 1.0 M H_2SO_4 or 1.0 M NaOH solution. The batch experiments were conducted in duplicates at least until the errors were below 5%, and the average values obtained were used for plotting.

3.5. Analytical Methods

The BPA concentration was analyzed by High Performance Liquid Chromatography (HPLC; Dionex Ultimate 3000, Sunnyvale, CA, USA) equipped with a reverse-phase C18 column (250 mm $\times$ 4.6 mm $\times$ 5.0 μm) and an ultraviolet and visible (UV-Vis) spectrophotometry detector, with the detection wavelength set at 225 nm. The mobile phase was a mixture of 70/30% (v/v) methanol and 0.1% phosphoric acid water solution at a flow rate of 1.0 mL/min. The concentrations of PMSO and $PMSO_2$ were detected by an HPLC equipped with a C18-A column (250 mm $\times$ 3.0 mm $\times$ 3.0 μm) and a UV-Vis detector at wavelengths of 215 nm. The mobile phase was a mixture of 20/80% (v/v) methanol and 0.1% phosphoric acid water solution at a flow rate of 0.5 mL/min. All of the column temperatures were set at 35 $°$C. The electron spin resonance (ESR) spectra were obtained by a Bruker EMX-10/12 device with X-band field scanning. The applied instrumental conditions were set as a central magnetic field of 3480 G, resonance frequency of 9.74 GHz, microwave power of 20.00 mW, and sweep time of 30.00 s. The total organic carbon (TOC) concentration was quantified by Analytic Jena multi N/C 3100 TOC. The concentration of total dissolved iron and dissolved Fe^{2+} in the solution was determined by a spectrophotometric method at 510 nm via forming the complex with 1, 10-phenanthroline [43].

4. Conclusions

In this work, WF and WFB were compared as the supports of Fe_3O_4 to enhance the PDS activation performance for the degradation of BPA. Results showed that WFB had more significant control of iron dissolution from Fe_3O_4 than WF. Moreover, WFB/Fe_3O_4 had a stronger PDS activation capability than WF/Fe_3O_4, which was likely due to the larger SSA of WFB/Fe_3O_4 and the PDS activation ability of WFB. Both radicals ($\bullet$OH and $SO_4^{\bullet-}$) and the non-radical pathways, including $^1O_2$ and high-valent iron-oxo species, contributed to the degradation of BPA in the WFB/Fe_3O_4-PDS process. In addition, the WFB/Fe_3O_4 also demonstrated better reusability and stronger BPA mineralization performance during the activation of PDS than WF/Fe_3O_4. The use of WFB as the support for Fe_3O_4 may offer a simple and cost-effective option to enhance the PDS activation performance of Fe_3O_4, which can be applied in organic wastewater treatment.

Supplementary Materials: The following supporting information can be downloaded at: https://www.mdpi.com/article/10.3390/catal13020323/s1, Figure S1: The SEM image of WF; Figure S2: FTIR spectra of different catalysts; Figure S3: Effects of initial solution pH on the degradation of BPA in Fe_3O_4-PDS process (conditions: catalyst 1.0 g/L, $[PDS]_0$ = 5.0 mM, $[BPA]_0$ = 0.02 mM, pH_0 = 3.00 $\pm$ 0.1); Figure S4: Effects of different radical scavengers on BPA degradation in Fe_3O_4-PDS system (conditions: catalyst 1.0 g/L, $[PDS]_0$ = 5.0 mM, $[BPA]_0$ = 0.02 mM, pH_0 = 3.00 $\pm$ 0.1); Figure S5: The calculated η values (η = $\Delta[PMSO_2]/\Delta[PMSO]$) in Fe_3O_4-PDS, WF_{20}/Fe_3O_4-PDS and WFB_{20}/Fe_3O_4-PDS processes (conditions: catalyst 1.0 g/L, $[PDS]_0$ = 5.0 mM, $[PMSO]_0$ = 0.5 mM, pH_0 = 3.00 $\pm$ 0.1); Figure S6: FTIR spectra of WFB and used WFB; Table S1: XRD spectral data of Fe_3O_4, WF_{20}/Fe_3O_4 and WFB_{20}/Fe_3O_4 catalysts; Table S2: The content of Fe(II) and Fe(III) in different catalysts.

Author Contributions: Conceptualization, resources, and funding acquisition: L.X.; methodology, formal analysis, investigation: Y.H. All authors have read and agreed to the published version of the manuscript.

Funding: This research was funded by the Qing Lan Project of Jiangsu Province (2020).

Data Availability Statement: Data are available on request.

Acknowledgments: The Advanced Analysis and Testing Center of Nanjing Forestry University is acknowledged.

Conflicts of Interest: The authors declare no conflict of interest.

References

1. Anipsitakis, G.P.; Dionysiou, D.D. Radical generation by the interaction of transition metals with common oxidants. *Environ. Sci. Technol.* **2004**, *38*, 3705–3712. [CrossRef] [PubMed]
2. Zhou, X.; Yang, Z.; Luo, Z.; Li, H.; Chen, G. Endocrine disrupting chemicals in wild freshwater fishes: Species, tissues, sizes and human health risks. *Environ. Pollut.* **2019**, *244*, 462–468. [CrossRef]
3. Lei, K.; Lin, C.-Y.; Zhu, Y.; Chen, W.; Pan, H.-Y.; Sun, Z.; Sweetman, A.; Zhang, Q.; He, M.-C. Estrogens in municipal wastewater and receiving waters in the Beijing-Tianjin-Hebei region, China: Occurrence and risk assessment of mixtures. *J. Hazard. Mater.* **2020**, *389*, 121891. [CrossRef] [PubMed]
4. Cai, S.; Zhang, Q.; Wang, Z.; Hua, S.; Ding, D.; Cai, T.; Zhang, R. Pyrrolic N-rich biochar without exogenous nitrogen doping as a functional material for bisphenol A removal: Performance and mechanism. *Appl. Catal. B-Environ.* **2021**, *291*, 120093. [CrossRef]
5. Wang, F.; Lai, Y.; Fang, Q.; Li, Z.; Ou, P.; Wu, P.; Duan, Y.; Chen, Z.; Li, S.; Zhang, Y. Facile fabricate of novel Co(OH)F@MXenes catalysts and their catalytic activity on bisphenol A by peroxymonosulfate activation: The reaction kinetics and mechanism. *Appl. Catal. B-Environ.* **2020**, *262*, 118099. [CrossRef]
6. Bu, Z.; Hou, M.; Li, Z.; Dong, Z.; Zeng, L.; Zhang, P.; Wu, G.; Li, X.; Zhang, Y.; Pan, Y. Fe^{3+}/Fe^{2+} cycle promoted peroxymonosulfate activation with addition of boron for sulfamethazine degradation: Efficiency and the role of boron. *Sep. Purif. Technol.* **2022**, *298*, 121596. [CrossRef]
7. Qi, L.; Lu, W.; Tian, G.; Sun, Y.; Han, J.; Xu, L. Enhancement of Sono-Fenton by P25-Mediated Visible Light Photocatalysis: Analysis of Synergistic Effect and Influence of Emerging Contaminant Properties. *Catalysts* **2020**, *10*, 1297. [CrossRef]
8. Oh, W.-D.; Lua, S.-K.; Dong, Z.; Lim, T.-T. A novel three-dimensional spherical $CuBi_2O_4$ consisting of nanocolumn arrays with persulfate and peroxymonosulfate activation functionalities for 1H-benzotriazole removal. *Nanoscale* **2015**, *7*, 8149–8158. [CrossRef]
9. Wang, J.; Wang, S. Activation of persulfate (PS) and peroxymonosulfate (PMS) and application for the degradation of emerging contaminants. *Chem. Eng. J.* **2018**, *334*, 1502–1517. [CrossRef]
10. Cai, Z.; Luo, Y.; Gan, L. ZIF-67(Co)-Loaded Filter Paper for In Situ Catalytic Degradation of Bisphenol A in Water. *Separations* **2022**, *9*, 340. [CrossRef]
11. Bu, Z.; Li, X.; Xue, Y.; Ye, J.; Zhang, J.; Pan, Y. Hydroxylamine enhanced treatment of highly salty wastewater in $Fe-0/H_2O_2$ system: Efficiency and mechanism study. *Sep. Purif. Technol.* **2021**, *271*, 118847. [CrossRef]
12. You, W.; Liu, L.; Xu, J.; Jin, T.; Fu, L.; Pan, Y. Effect of Anions and Cations on Tartrazine Removal by the Zero-Valent Iron/Peroxymonosulfate Process: Efficiency and Major Radicals. *Catalysts* **2022**, *12*, 1114. [CrossRef]
13. Huang, L.; Zhang, H.; Zeng, T.; Chen, J.; Song, S. Synergistically enhanced heterogeneous activation of persulfate for aqueous carbamazepine degradation using Fe_3O_4@SBA-15. *Sci. Total Environ.* **2021**, *760*, 144027. [CrossRef]
14. Gan, L.; Wang, L.; Xu, L.; Fang, X.; Pei, C.; Wu, Y.; Lu, H.; Han, S.; Cui, J.; Shi, J.; et al. Fe_3C-porous carbon derived from Fe_2O_3 loaded MOF-74(Zn) for the removal of high concentration BPA: The integrations of adsorptive/catalytic synergies and radical/non-radical mechanisms. *J. Hazard. Mater.* **2021**, *413*, 125305. [CrossRef]
15. Ayoub, G.; Ghauch, A. Assessment of bimetallic and trimetallic iron-based systems for persulfate activation: Application to sulfamethoxazole degradation. *Chem. Eng. J.* **2014**, *256*, 280–292. [CrossRef]
16. Wang, Z.-Y.; Ju, C.-J.; Zhang, R.; Hua, J.-Q.; Chen, R.-P.; Liu, G.-X.; Yin, K.; Yu, L. Acceleration of the bio-reduction of methyl orange by a magnetic and extracellular polymeric substance nanocomposite. *J. Hazard. Mater.* **2021**, *420*, 126576. [CrossRef] [PubMed]
17. Xu, X.; Chen, W.; Zong, S.; Ren, X.; Liu, D. Atrazine degradation using Fe_3O_4-sepiolite catalyzed persulfate: Reactivity, mechanism and stability. *J. Hazard. Mater.* **2019**, *377*, 62–69. [CrossRef]
18. Liu, B.; Song, W.; Zhang, W.; Zhang, X.; Pan, S.; Wu, H.; Sun, Y.; Xu, Y. Fe_3O_4@CNT as a high-effective and steady chainmail catalyst for tetracycline degradation with peroxydisulfate activation: Performance and mechanism. *Sep. Purif. Technol.* **2021**, *273*, 118705. [CrossRef]
19. Yan, J.; Gao, W.; Dong, M.; Han, L.; Qian, L.; Nathanail, C.P.; Chen, M. Degradation of trichloroethylene by activated persulfate using a reduced graphene oxide supported magnetite nanoparticle. *Chem. Eng. J.* **2016**, *295*, 309–316. [CrossRef]
20. Zhong, Q.; Lin, Q.; He, W.; Fu, H.; Huang, Z.; Wang, Y.; Wu, L. Study on the nonradical pathways of nitrogen-doped biochar activating persulfate for tetracycline degradation. *Sep. Purif. Technol.* **2021**, *276*, 119354. [CrossRef]
21. Lu, H.; Gan, L. Catalytic Degradation of Bisphenol A in Water by Poplar Wood Powder Waste Derived Biochar via Peroxymonosulfate Activation. *Catalysts* **2022**, *12*, 1164. [CrossRef]
22. Lu, H.; Xu, G.; Gan, L. N Doped Activated Biochar from Pyrolyzing Wood Powder for Prompt BPA Removal via Peroxymonosulfate Activation. *Catalysts* **2022**, *12*, 1449. [CrossRef]
23. Zhang, R.; Li, Y.; Wang, Z.; Tong, Y.; Sun, P. Biochar-activated peroxydisulfate as an effective process to eliminate pharmaceutical and metabolite in hydrolyzed urine. *Water Res.* **2020**, *177*, 115809. [CrossRef] [PubMed]
24. Wang, J.; Liao, Z.; Ifthikar, J.; Shi, L.; Du, Y.; Zhu, J.; Xi, S.; Chen, Z.; Chen, Z. Treatment of refractory contaminants by sludge-derived biochar/persulfate system via both adsorption and advanced oxidation process. *Chemosphere* **2017**, *185*, 754–763. [CrossRef]
25. Zhu, K.; Wang, X.; Chen, D.; Ren, W.; Lin, H.; Zhang, H. Wood-based biochar as an excellent activator of peroxydisulfate for Acid Orange 7 decolorization. *Chemosphere* **2019**, *231*, 32–40. [CrossRef]

26. Du, L.; Xu, W.; Liu, S.; Li, X.; Huang, D.; Tan, X.; Liu, Y. Activation of persulfate by graphitized biochar for sulfamethoxazole removal: The roles of graphitic carbon structure and carbonyl group. *J. Colloid Interface Sci.* **2020**, *577*, 419–430. [CrossRef]
27. Hassani, A.; Scaria, J.; Ghanbari, F.; Nidheesh, P.V. Sulfate radicals-based advanced oxidation processes for the degradation of pharmaceuticals and personal care products: A review on relevant activation mechanisms, performance, and perspectives. *Environ. Res.* **2023**, *217*, 114789. [CrossRef]
28. Yaghoot-Nezhad, A.; Waclawek, S.; Madihi-Bidgoli, S.; Hassani, A.; Lin, K.-Y.A.; Ghanbari, F. Heterogeneous photocatalytic activation of electrogenerated chlorine for the production of reactive oxygen and chlorine species: A new approach for Bisphenol A degradation in saline wastewater. *J. Hazard. Mater.* **2022**, *445*, 130626. [CrossRef]
29. Geng, A.; Xu, L.; Gan, L.; Mei, C.; Wang, L.; Fang, X.; Li, M.; Pan, M.; Han, S.; Cui, J. Using wood flour waste to produce biochar as the support to enhance the visible-light photocatalytic performance of BiOBr for organic and inorganic contaminants removal. *Chemosphere* **2020**, *250*, 126291. [CrossRef]
30. Gurav, R.; Bhatia, S.K.; Choi, T.-R.; Park, Y.-L.; Park, J.Y.; Han, Y.-H.; Vyavahare, G.; Jadhav, J.; Song, H.-S.; Yang, P.; et al. Treatment of furazolidone contaminated water using banana pseudostem biochar engineered with facile synthesized magnetic nanocomposites. *Bioresour. Technol.* **2020**, *297*, 122472. [CrossRef]
31. Lu, J.-D. The effect of two ferromagnetic metal stripes on valley polarization of electrons in a graphene. *Phys. Lett. A* **2020**, *384*, 126402. [CrossRef]
32. Aslam, S.; Bokhari, T.H.; Anwar, T.; Khan, U.; Nairan, A.; Khan, K. Graphene oxide coated graphene foam based chemical sensor. *Mater. Lett.* **2019**, *235*, 66–70. [CrossRef]
33. Jaafarzadeh, N.; Ghanbari, F.; Ahmadi, M. Catalytic degradation of 2,4-dichlorophenoxyacetic acid (2,4-D) by nano-Fe_2O_3 activated peroxymonosulfate: Influential factors and mechanism determination. *Chemosphere* **2017**, *169*, 568–576. [CrossRef]
34. Iranizad, E.S.; Dehghani, Z.; Nadafan, M. Nonlinear optical properties of nematic liquid crystal doped with different compositional percentage of synthesis of Fe_3O_4 nanoparticles. *J. Mol. Liq.* **2014**, *190*, 6–9. [CrossRef]
35. Lin, T.; Yu, L.; Sun, M.; Cheng, G.; Lan, B.; Fu, Z. Mesoporous alpha-MnO_2 microspheres with high specific surface area: Controlled synthesis and catalytic activities. *Chem. Eng. J.* **2016**, *286*, 114–121. [CrossRef]
36. Zhang, Y.; Yang, M.; Dou, X.M.; He, H.; Wang, D.S. Arsenate adsorption on an Fe-Ce bimetal oxide adsorbent: Role of surface properties. *Environ. Sci. Technol.* **2005**, *39*, 7246–7253. [CrossRef]
37. Wei, L.; McDonald, A.G.; Freitag, C.; Morrell, J.J. Effects of wood fiber esterification on properties, weatherability and biodurability of wood plastic composites. *Polym. Degrad. Stab.* **2013**, *98*, 1348–1361. [CrossRef]
38. Ahmad, S.; Riaz, U.; Kaushik, A.; Alam, J. Soft Template Synthesis of Super Paramagnetic Fe_3O_4 Nanoparticles a Novel Technique. *J. Inorg. Organomet. Polym. Mater.* **2009**, *19*, 355–360. [CrossRef]
39. Zhu, X.; Liu, Y.; Zhou, C.; Luo, G.; Zhang, S.; Chen, J. A novel porous carbon derived from hydrothermal carbon for efficient adsorption of tetracycline. *Carbon* **2014**, *77*, 627–636. [CrossRef]
40. Tang, H.; Li, R.; Fan, X.; Xu, Y.; Lin, H.; Zhang, H. A novel S-scheme heterojunction in spent battery-derived $ZnFe_2O_4/g$-C_3N_4 photocatalyst for enhancing peroxymonosulfate activation and visible light degradation of organic pollutant. *J. Environ. Chem. Eng.* **2022**, *10*, 107797. [CrossRef]
41. Hayati, F.; Moradi, S.; Saei, S.F.; Madani, Z.; Giannakis, S.; Isari, A.A.; Kakavandi, B. A novel, Z-scheme ZnO@AC@FeO photocatalyst, suitable for the intensification of photo-mediated peroxymonosulfate activation: Performance, reactivity and bisphenol A degradation pathways. *J. Environ. Manag.* **2022**, *321*, 115851. [CrossRef]
42. Hassani, A.; Eghbali, P.; Mahdipour, F.; Waclawek, S.; Lin, K.-Y.A.; Ghanbari, F. Insights into the synergistic role of photocatalytic activation of peroxymonosulfate by UVA-LED irradiation over $CoFe_2O_4$-rGO nanocomposite towards effective Bisphenol A degradation: Performance, mineralization, and activation mechanism. *Chem. Eng. J.* **2023**, *453*, 139556. [CrossRef]
43. Xu, L.; Qi, L.; Han, Y.; Lu, W.; Han, J.; Qiao, W.; Mei, X.; Pan, Y.; Song, K.; Ling, C.; et al. Improvement of Fe^{2+}/peroxymonosulfate oxidation of organic pollutants by promoting Fe^{2+} regeneration with visible light driven g-C_3N_4 photocatalysis. *Chem. Eng. J.* **2022**, *430*, 132828. [CrossRef]
44. Li, R.; Kong, J.; Liu, H.; Chen, P.; Liu, G.; Li, F.; Lv, W. A sulfate radical based ferrous-peroxydisulfate oxidative system for indomethacin degradation in aqueous solutions. *RSC Adv.* **2017**, *7*, 22802–22809. [CrossRef]
45. Liu, S.; Lai, C.; Li, B.; Zhang, C.; Zhang, M.; Huang, D.; Qin, L.; Yi, H.; Liu, X.; Huang, F.; et al. Role of radical and non-radical pathway in activating persulfate for degradation of p-nitrophenol by sulfur-doped ordered mesoporous carbon. *Chem. Eng. J.* **2020**, *384*, 123304. [CrossRef]
46. Liu, C.; Lai, L.; Yang, X. Sewage sludge conditioning by Fe(II)-activated persulphate oxidation combined with skeleton builders for enhancing dewaterability. *Water Environ. J.* **2016**, *30*, 96–101. [CrossRef]
47. Yao, Y.; Cai, Y.; Lu, F.; Wei, F.; Wang, X.; Wang, S. Magnetic recoverable $MnFe_2O_4$ and $MnFe_2O_4$-graphene hybrid as heterogeneous catalysts of peroxymonosulfate activation for efficient degradation of aqueous organic pollutants. *J. Hazard. Mater.* **2014**, *270*, 61–70. [CrossRef]
48. Fang, G.-D.; Dionysiou, D.D.; Al-Abed, S.R.; Zhou, D.-M. Superoxide radical driving the activation of persulfate by magnetite nanoparticles: Implications for the degradation of PCBs. *Appl. Catal. B-Environ.* **2013**, *129*, 325–332. [CrossRef]
49. Yu, F.; Zhang, Y.; Zhang, Y.; Gao, Y.; Pan, Y. Promotion of the degradation perfluorooctanoic acid by electro-Fenton under the bifunctional electrodes: Focusing active reaction region by Fe/N co-doped graphene modified cathode. *Chem. Eng. J.* **2023**, *457*, 141320. [CrossRef]

50. Mostafa, S.; Rosario-Ortiz, F.L. Singlet Oxygen Formation from Wastewater Organic Matter. *Environ. Sci. Technol.* **2013**, *47*, 8179–8186. [CrossRef]
51. Yang, Y.; Banerjee, G.; Brudvig, G.W.; Kim, J.-H.; Pignatello, J.J. Oxidation of Organic Compounds in Water by Unactivated Peroxymonosulfate. *Environ. Sci. Technol.* **2018**, *52*, 5911–5919. [CrossRef] [PubMed]
52. Yu, J.; Tang, L.; Pang, Y.; Zeng, G.; Wang, J.; Deng, Y.; Liu, Y.; Feng, H.; Chen, S.; Ren, X. Magnetic nitrogen-doped sludge-derived biochar catalysts for persulfate activation: Internal electron transfer mechanism. *Chem. Eng. J.* **2019**, *364*, 146–159. [CrossRef]
53. Zhou, X.; Zeng, Z.; Zeng, G.; Lai, C.; Xiao, R.; Liu, S.; Huang, D.; Qin, L.; Liu, X.; Li, B.; et al. Insight into the mechanism of persulfate activated by bone char: Unraveling the role of functional structure of biochar. *Chem. Eng. J.* **2020**, *401*, 126127. [CrossRef]
54. Wang, Z.; Jiang, J.; Pang, S.; Zhou, Y.; Guan, C.; Gao, Y.; Li, J.; Yang, Y.; Qu, W.; Jiang, C. Is Sulfate Radical Really Generated from Peroxydisulfate Activated by Iron(II) for Environmental Decontamination? *Environ. Sci. Technol.* **2018**, *52*, 11276–11284. [CrossRef]
55. Li, H.; Shan, C.; Pan, B. Fe(III)-Doped g-C3N4 Mediated Peroxymonosulfate Activation for Selective Degradation of Phenolic Compounds via High-Valent Iron-Oxo Species. *Environ. Sci. Technol.* **2018**, *52*, 2197–2205. [CrossRef]
56. Yang, T.; Wang, L.; Liu, Y.; Huang, Z.; He, H.; Wang, X.; Jiang, J.; Gao, D.; Ma, J. Comparative study on ferrate oxidation of BPS and BPAF: Kinetics, reaction mechanism, and the improvement on their biodegradability. *Water Res.* **2019**, *148*, 115–125. [CrossRef]
57. Pang, S.-Y.; Jiang, J.; Ma, J. Oxidation of Sulfoxides and Arsenic(III) in Corrosion of Nanoscale Zero Valent Iron by Oxygen: Evidence against Ferryl Ions (Fe(IV)) as Active Intermediates in Fenton Reaction. *Environ. Sci. Technol.* **2011**, *45*, 307–312. [CrossRef]
58. Meng, S.; Zhou, P.; Sun, Y.; Zhang, P.; Zhou, C.; Xiong, Z.; Zhang, H.; Liang, J.; Lai, B. Reducing agents enhanced Fenton-like oxidation (Fe(III)/Peroxydisulfate): Substrate specific reactivity of reactive oxygen species. *Water Res.* **2022**, *218*, 118412. [CrossRef]
59. Lai, L.; Zhou, H.; Zhang, H.; Ao, Z.; Pan, Z.; Chen, Q.; Xiong, Z.; Yao, G.; Lai, B. Activation of peroxydisulfate by natural titanomagnetite for atrazine removal via free radicals and high-valent iron-oxo species. *Chem. Eng. J.* **2020**, *387*, 124165. [CrossRef]
60. Ai, Z.; Gao, Z.; Zhang, L.; He, W.; Yin, J.J. Core-Shell Structure Dependent Reactivity of Fe@Fe2O3 Nanowires on Aerobic Degradation of 4-Chlorophenol. *Environ. Sci. Technol.* **2013**, *47*, 5344–5352. [CrossRef]
61. Huang, H.; Guo, T.; Wang, K.; Li, Y.; Zhang, G. Efficient activation of persulfate by a magnetic recyclable rape straw biochar catalyst for the degradation of tetracycline hydrochloride in water. *Sci. Total Environ.* **2021**, *758*, 143957. [CrossRef]
62. Feng, Y.; Chen, G.; Zhang, Y.; Li, D.; Ling, C.; Wang, Q.; Liu, G. Superhigh co-adsorption of tetracycline and copper by the ultrathin g-C$_3$N$_4$ modified graphene oxide hydrogels. *J. Hazard. Mater.* **2022**, *424*, 127362. [CrossRef] [PubMed]
63. Ul Ain, Q.; Rasheed, U.; Yaseen, M.; Zhang, H.; Tong, Z. Superior dye degradation and adsorption capability of polydopamine modified Fe$_3$O$_4$-pillared bentonite composite. *J. Hazard. Mater.* **2020**, *397*, 122758. [CrossRef] [PubMed]
64. Du, J.; Bao, J.; Liu, Y.; Kim, S.H.; Dionysiou, D.D. Facile preparation of porous Mn/Fe3O4 cubes as peroxymonosulfate activating catalyst for effective bisphenol A degradation. *Chem. Eng. J.* **2019**, *376*, 119193. [CrossRef]
65. Zhang, R.; Chang, Z.-Y.; Wang, L.-L.; Cheng, W.-X.; Chen, R.-P.; Yu, L.; Qiu, X.-H.; Han, J.-G. Solid-liquid separation of real cellulose-containing wastewaters by extracellular polymeric substances: Mechanism and cost evaluation. *Sep. Purif. Technol.* **2021**, *279*, 119665. [CrossRef]
66. Pei, X.; Peng, X.; Jia, X.; Wong, P.K. N-doped biochar from sewage sludge for catalytic peroxydisulfate activation toward sulfadiazine: Efficiency, mechanism, and stability. *J. Hazard. Mater.* **2021**, *419*, 126446. [CrossRef]
67. Zhou, Y.; Jiang, J.; Gao, Y.; Ma, J.; Pang, S.-Y.; Li, J.; Lu, X.-T.; Yuan, L.-P. Activation of Peroxymonosulfate by Benzoquinone: A Novel Nonradical Oxidation Process. *Environ. Sci. Technol.* **2015**, *49*, 12941–12950. [CrossRef]
68. Wu, D.; Zheng, P.; Chang, P.R.; Ma, X. Preparation and characterization of magnetic rectorite/iron oxide nanocomposites and its application for the removal of the dyes. *Chem. Eng. J.* **2011**, *174*, 489–494. [CrossRef]
69. Rajput, S.; Pittman, C.U., Jr.; Mohan, D. Magnetic magnetite (Fe$_3$O$_4$) nanoparticle synthesis and applications for lead (Pb^{2+}) and chromium (Cr^{6+}) removal from water. *J. Colloid Interface Sci.* **2016**, *468*, 334–346. [CrossRef] [PubMed]

Article

Differently Prepared PbO$_2$/Graphitic Carbon Nitride Composites for Efficient Electrochemical Removal of Reactive Black 5 Dye

Aleksandar Marković [1], Slađana Savić [1], Andrej Kukuruzar [1], Zoltan Konya [2,3], Dragan Manojlović [1,4], Miloš Ognjanović [5] and Dalibor M. Stanković [1,5,*]

1 Faculty of Chemistry, University of Belgrade, Studentski trg 12-16, 11000 Belgrade, Serbia
2 Interdisciplinary Excellence Centre, Department of Applied and Environmental Chemistry, University of Szeged, Rerrich Béla tér 1, H-6720 Szeged, Hungary
3 MTA-SZTE Reaction Kinetics and Surface Chemistry Research Group, Rerrich Béla tér 1, H-6720 Szeged, Hungary
4 South Ural State University, Lenin Prospekt 76, 454080 Chelyabinsk, Russia
5 Department of Theoretical Physics and Condensed Matter Physics, VINČA Institute of Nuclear Sciences, National Institute of the Republic of Serbia, University of Belgrade, Mike Petrovića Alasa 12-14, 11000 Belgrade, Serbia
* Correspondence: dalibors@chem.bg.ac.rs

Abstract: In this paper, electrochemical degradation of Reactive Black 5 (RB5) textile azo dye was examined in regard to different synthesis procedures for making PbO$_2$–graphitic carbon nitride (g-C$_3$N$_4$) electrode. The reaction of $Pb(OH)_3^-$ with ClO^- in the presence of different surfactants, i.e., cetyltrimethylammonium bromide (CTAB) and tetrabutylammonium phosphate (TBAP), under conventional conditions, resulted in the formation of PbO$_2$ with varying morphology. The obtained materials were combined with g-C$_3$N$_4$ for the preparation of the final composite materials, which were then characterized morphologically and electrochemically. After optimizing the degradation method, it was shown that an anode comprising a steel electrode coated with the composite of PbO$_2$ synthesized using CTAB as template and g-C$_3$N$_4$, and using 0.15 M Na$_2$SO$_4$ as the supporting electrolyte, gave the best performance for RB5 dye removal from a 35 mg/L solution. The treatment duration was 60 min, applying a current of 0.17 A (electrode surface 4 cm^2, current density of 42.5 mA/cm^2), while the initial pH of the testing solution was 2. The reusability and longevity of the electrode surface (which showed no significant change in activity throughout the study) may suggest that this approach is a promising candidate for wastewater treatment and pollutant removal.

Keywords: reactive azo dye; surfactant-assisted synthesis; electrode morphology; advanced oxidation processes; lead dioxide; energy efficiency

Citation: Marković, A.; Savić, S.; Kukuruzar, A.; Konya, Z.; Manojlović, D.; Ognjanović, M.; Stanković, D.M. Differently Prepared PbO$_2$/Graphitic Carbon Nitride Composites for Efficient Electrochemical Removal of Reactive Black 5 Dye. *Catalysts* **2023**, *13*, 328. https://doi.org/10.3390/catal13020328

Academic Editors: Rafael Borja, Gassan Hodaifa and Mha Albqmi

Received: 25 December 2022
Revised: 18 January 2023
Accepted: 26 January 2023
Published: 1 February 2023

1. Introduction

Reactive Black 5 (RB5) dye, or Remazol Black B (Scheme 1), belongs to the vinyl sulphone type of azo dyes. Azo dyes, aromatic compounds with one or more –N=N– groups, constitute the largest class of synthetic dyes in commercial application [1,2]. RB5 is most commonly used to dye cotton and other cellulosic fibers, wool, and nylon [3,4]. This type of dye is highly soluble in water and has reactive groups which can form covalent bonds with fibers [4].

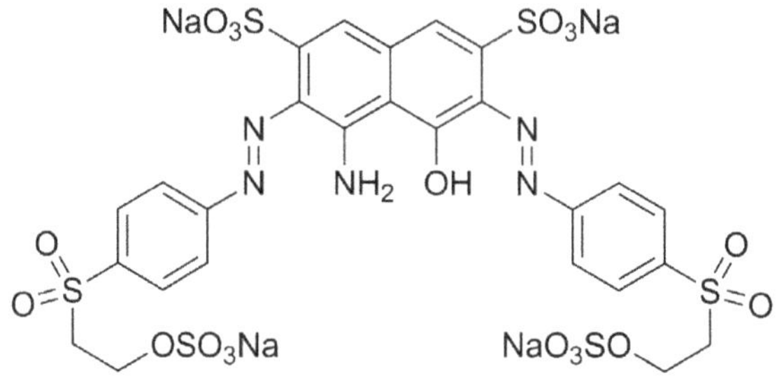

Scheme 1. Chemical structure of Reactive Black 5 azo dye.

Industries such as textile, leather, paper, pulp, printing, and dyeing are major consumers of synthetic dyes. Such industries produce large quantities of colored wastewater which cause significant adverse effects on the environment. The release of colored wastewaters is not only aesthetically unpleasant, but also obstructs penetration of light, which affects biological processes. It can also potentially generate toxicity for aquatic organisms and finally, via food chain or exposure, in humans [5–9].

Different methods for degradation and removal of RB5 can be found in scientific literature, such as biodegradation and biocatalysis [1,10,11], photocatalysis [12–15], electrocoagulation [2], Fenton or Fenton-like reaction [6,16], adsorption [9], or by activation of peroxymonosulfate (PMS) [17,18].

Each of these methods has its positive and negative sides. For instance, electrocoagulation, membrane separation processes, adsorption, and precipitation only change the phase of pollutants. Photo- and chemical oxidation require additional chemicals and oxidation agents that are considered highly toxic and can produce additional hazardous waste. Photochemical oxidation is one of the most commonly used methods for the degradation of waterborne pollutants [19]. TiO_2-based materials are most widely used because of their unprecedented photocatalytic activity, but other metallic and nonmetallic catalysts are giving promising results [20–22]. Biodegradation can yield very good results but it can also be less effective than other methods, because dyes can be toxic for bacteria and can thus inhibit their activity.

Electrochemical oxidation processes present an effective and logical choice for the development of green, time-effective, and, at the same time, potent methods for the removal of various pollutants, including reactive textile dyes [23–25]. Electrocatalytic processes can be considered green primarily due to mild and environmentally acceptable working conditions (ambient pressure and temperature, work in aqueous media, no additional toxic chemicals). These methods may also be coupled with renewable energy sources [26], they do not require systems for temperature or pressure control nor expensive gases and give the possibility of avoiding expensive catalysts based on precious metals, so their potential for large-scale application is enormous.

Therefore, electrochemical methods, through the rational design of the electrocatalytic setup, enable simple and practical systems for water purification and pollutant removal. Our research group proposed several approaches for the efficient removal of organic pollutants using electrooxidation methods, such as for the removal of Reactive Blue 52 [27,28], triketone herbicides [29], or ibuprofen [30].

For electrochemical oxidative processes, anodes with high oxygen evolution potentials (non-active anodes) achieve complete mineralization of organics and are thus favored for wastewater remediation. The most commonly used non-active anode is the boron-doped diamond electrode, but cheaper alternatives include various metal oxides such as TiO_2, SnO_2, and PbO_2 [31]. Metal oxides exhibit exceptional properties valuable to electrochemical processes, while featuring reduced costs, availability and environmental compatibility, and have thus found diverse applications. The properties of metal oxides can be fine-tuned by controlling their morphology, particle size, and crystallinity, which can be achieved using template synthesis.

Lead dioxide is an inexpensive and widely available material, readily incorporated in composite materials with outstanding properties [32]. Lead oxide-based anodes gave promising performances for electrochemical degradation of organic pollutants [33], as they can possess unique catalytically active surfaces [34], good stability, and a long electrode life [35].

Graphitic carbon nitride (g-C_3N_4) has found numerous applications owing to its properties, such as high surface area, low cost, and biocompatibility. The use of g-C_3N_4 in composites proved to enhance its photo- and electrocatalytic properties. Good results for the photocatalytic degradation of organic pollutants were achieved when using nanocomposites made from metal oxides and g-C_3N_4 [36–39]. On the other hand, reports of g-C_3N_4 being used for water remediation in a purely electrochemical setting are scarce.

The main objective of this paper was to investigate the effects of different surfactant templates in the synthesis of PbO_2, to prepare of composite PbO_2/g-C_3N_4, and to optimize working conditions for electrocatalytic degradation of RB5 using a coated stainless-steel anode. The conditions in question include the solution pH value, current density (applied voltage) during the reaction, and concentrations of both the supporting electrolyte and RB5. The electrocatalytic properties of PbO_2 synthesized with cetyltrimethylammonium bromide (CTAB) as template (hereafter abbreviated to PbO_2-CTAB) vs. PbO_2 synthesized with tetrabutylammonium phosphate (TBAP) as template (hereafter abbreviated to PbO_2-TBAP), as well as their composites with graphitic carbon nitride (hereafter abbreviated to PbO_2-CTAB/g-C_3N_4 and PbO_2-TBAP/g-C_3N_4) were investigated. Moreover, the morphological and electrocatalytic properties of these materials were studied. Finally, the best combination of conditions was used for the real-world water sample treatment, artificially polluted with RB5.

2. Results and Discussions

The present research was designed to determine whether the stainless steel (SS) electrodes coated with composites made from PbO_2 and g-C_3N_4 could be applied for RB5 electrochemical decoloration.

The electrode materials were based on PbO_2 nanoparticles, synthesized with CTAB and TBAP surfactants as templates, later combined with g-C_3N_4 to obtain composites tested in the removal of the selected textile dye. To optimize the electrochemical procedure, the initial pH value, the supporting electrolyte concentration, the current density, and the RB5 concentration were varied. Finally, the composites were compared in terms of efficiency of RB5 removal, under previously optimized conditions. Additionally, the stability, longevity, and energy consumption were also reported and set side by side with the literature data.

2.1. Morphological and Electrochemical Properties of Materials

Microstructure characterization and phase analysis were performed by powder X-ray diffraction (PXRD) (Figure 1A). The PbO_2-CTAB/g-C_3N_4 crystallizes in group P42/mnm, which can be ascribed to the β-PbO_2 (JCPDS #89-2805) crystal phase. The main diffraction peaks appeared at 2θ = 25.4°, 32.0°, 36.1°, and 49.1° which can be associated with the (110), (101), (200), and (211) standard diffraction peaks of tetragonal β-PbO_2 [40]. The remaining small intensity reflections can be assigned to α-PbO_2, as a small part of the sample crystallizes in the Pbcn space group [41]. On the other hand, PbO_2-TBAP/g-C_3N_4 (red line) crystallizes both in alpha and beta lead oxides. From the PXRD patterns, it is also clear that the intensity of diffraction peaks is weaker than for PbO_2-CTAB, indicating lower crystallinity of PbO_2-TBAP.

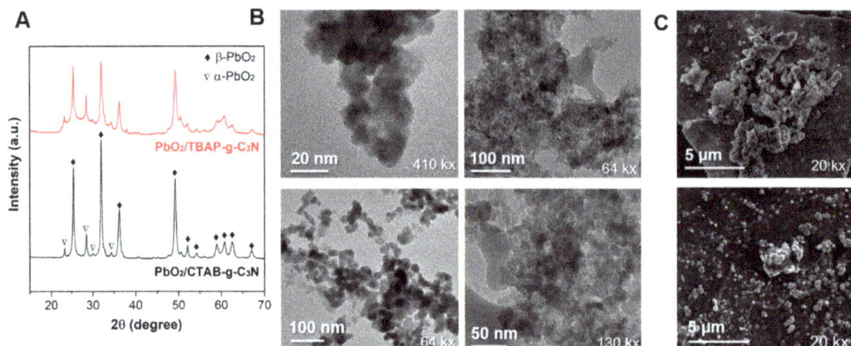

Figure 1. Morphological and microstructural analysis. (**A**) PXRD profiles of PbO_2-CTAB/g-C$_3$N$_4$ and PbO_2-TBAP/g-C$_3$N$_4$, (**B**) TEM micrographs of PbO_2-CTAB (upper left), PbO_2-CTAB/g-C$_3$N$_4$ (upper right), PbO_2-TBAP (lower left), PbO_2-TBAP/g-C$_3$N$_4$ (lower right) and (**C**) FE-SEM micrographs of PbO_2-CTAB/g-C$_3$N$_4$ (up) and PbO_2-TBAP/g-C$_3$N$_4$ (down).

The morphology and size analysis of the prepared catalysts was performed using electron microscopy. The transmission electron microscopy (TEM) images of the composite materials are displayed in Figure 1B. As can be seen, particles of both PbO_2-CTAB and PbO_2-TBAP are of elongated spherical to irregular shape, partially agglomerated, with not-so-distinctive boundaries between particles. The average particle size of PbO_2-CTAB is about 10 nm, while the PbO_2-TBAP particles are twice as large (~20 nm). The micrographs of composites with g-C$_3$N$_4$ reveal that smaller particles are interconnected with large sheets of graphitic carbon nitride that enable a larger specific surface. For further analysis of composite materials, field emission scanning electron microscopy (FE-SEM) measurements of PbO_2-CTAB/g-C$_3$N$_4$ and PbO_2-TBAP/g-C$_3$N$_4$ are shown in Figure 1C. In these micrographs, it can be seen that the small PbO_2 nanoparticles are scattered all over the sheet of g-C$_3$N$_4$. The oxide particles in the PbO_2-TBAP/g-C$_3$N$_4$ sample are better dispersed on the carbon nitride surface, while the PbO_2-CTAB/g-C$_3$N$_4$ sample particles are more agglomerated on top of the g-C$_3$N$_4$ sheet.

To confirm potential applicability of the materials for the removal of organic pollutants from wastewaters using electrochemical advanced oxidation processes, electrocatalytic characterization of the materials was conducted. Electrochemical properties of the materials were scrutinized by employing cyclic voltammetry (CV) and electrochemical impedance spectroscopy (EIS) in $Fe^{2+/3+}$ solution in 0.1 M KCl. Results are depicted in Figure 2. As can be seen, PbO_2-CTAB provides well defined and oval shaped redox peaks of the $Fe^{2+/3+}$ couple. Using the composite material PbO_2-CTAB/g-C$_3$N$_4$ resulted in a significant increase in the redox couple current, with negligible or no changes in peak potentials. This behavior indicates that the prepared composite significantly improves electrocatalytic properties of the electrode regarding diffusion and mass transfer. In the case of PbO_2-TBAP, it is noticeable that additional peaks are present, which can be attributed to the metal oxide modifier. In the case of the PbO_2-TBAP/g-C$_3$N$_4$ composite, its successful formation can be confirmed by the absence of additional peaks. However, a significant difference in peak potentials can be assigned to the lower mass transfer ability. Similar to these results, by employing impedance spectroscopy, we can conclude that the involvement of graphitic carbon nitride in the composite structure of PbO_2-CTAB/g-C$_3$N$_4$ results in increased diffusion capabilities at the electrode/solution interface, based on the linear part of the spectra, while the decrease in the R_{ct} value in the semicircle region is negligible.

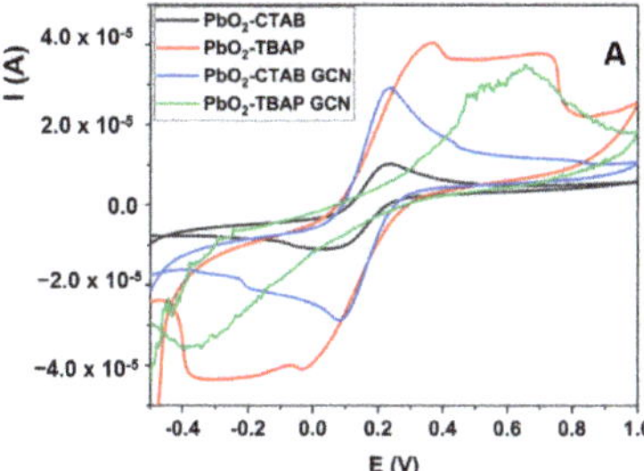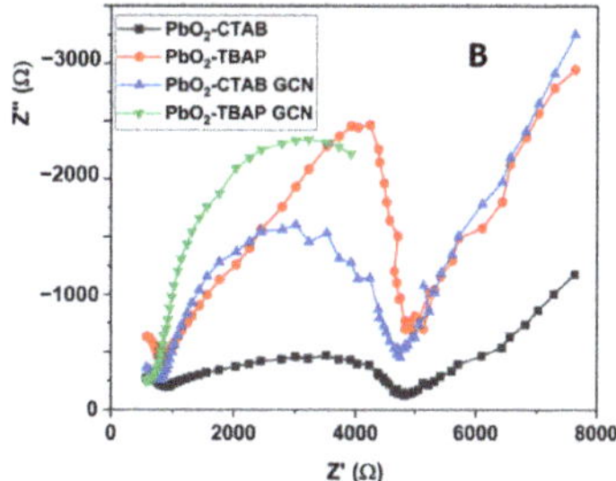

Figure 2. (**A**) CV voltammograms of modified electrodes in $Fe^{2+/3+}$ redox couple at the scan rate of 50 mV/s; (**B**) EIS spectra of modified electrodes in $Fe^{2+/3+}$ redox couple. Working potential 0.1 V.

2.2. Application of Composites for the Electrochemical Removal of RB5

The initial phase of this study was selecting the optimal operating parameters to validate the applicability of the PbO_2/g-C_3N_4 composite on SS electrode in a new electrochemical degradation method for real water treatment, to ultimately exploit all its advantages and possibilities. In this regard, we optimized several important parameters: initial pH, supporting electrolyte (Na_2SO_4) concentration, current density-applied voltage, and the initial concentration of the selected pollutant.

2.2.1. pH Optimization

Starting pH of the supporting solution represents one of the most important factors for the performance of electrochemical processes. In order to determine the optimal initial pH for electrochemical degradation, RB5 was dissolved in 0.1 M Na_2SO_4 to make a 70 mg/L solution with native pH 5.63 (measured using pH meter). The λ_{max} of this solution obtained by the spectrometer was 598 nm. Three initial pH values were tested—2, 4, and 6. The native pH of the stock solution was adjusted to these values using 0.1 M NaOH and 0.1 M H_2SO_4. The electrochemical reaction was allowed to continue almost to the point when the absorbance on the UV/Vis spectrum reached a plateau. For the reaction, one stainless steel (SS) electrode (4 cm^2 of surface area) modified with PbO_2-CTAB/g-C_3N_4 was used as the anode while an unmodified SS electrode was used as the cathode. The experiments were performed under constant a voltage of 5 V and the results are shown in Figure 3.

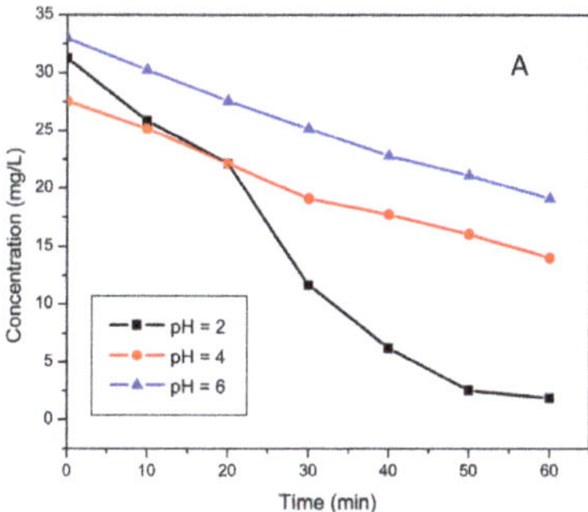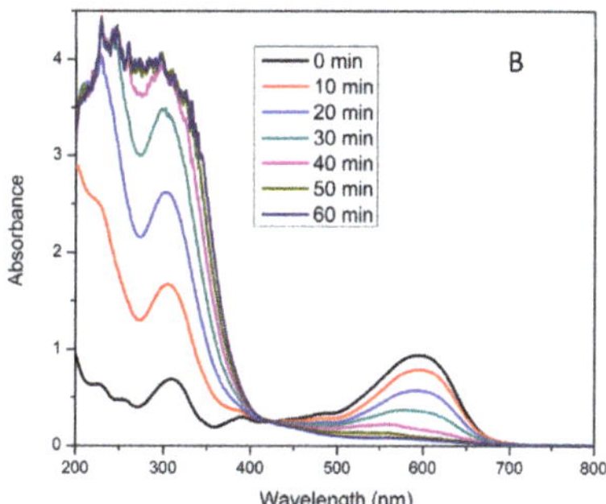

Figure 3. (**A**) RB5 removal under pH 2, 4, and 6; (**B**) UV–Vis spectra of RB5 solution in the range from 200 to 800 nm during 60 min of treatment (pH 2, 0.1 M Na_2SO_4, 5V) at different time intervals.

As can be seen in Figure 3A, the best degradation rate was observed at initial pH 2. After 60 min of treatment using this starting pH, 98% of the color was removed. This can be attributed to the cleavage of the –N=N– bonds and aromatic rings, resulting in the decrease of optical density of the dye solution (Figure 3B).

Moreover, in the acidic medium, OH• radicals are produced by the anodic discharge in water in the indirect electrochemical oxidation of organic dyes at the anode. These OH•

radicals adsorb onto the anode surface and oxidize the organic material. At higher pH, it could be expected that a larger amount of hydroxyl radicals would be generated, which would result in greater efficiency of the system. However, it was determined that the degree of ionization of the cationic dye is highly dependent on the initial pH value of the solution, which most often leads to the formation of a precipitated photochromic compound and a decrease in resistance on mass transfer at higher pH. Therefore, an acidic environment leads to better dissolution and degree of ionization. In a basic environment, an elevated consumption of electrolytes occurs dominantly, which directly affects the conductivity of the solution. Lower pH values were not tested as recent studies showed that low pH values lead to a higher dissolution rate of lead dioxide [42] and that these values are not appropriate for work with lead dioxide films, as they can cause the formation of lead sulfate [43].

The occurrence of new peaks In the spectrum confirms the formation of smaller molecules. The increase in absorbance for these peaks was followed by the decrease of the main RB5 peak, suggesting a relationship between these processes and successful decomposition of RB5. Thus, based on the conducted study, all further experiments were done using starting pH 2 for the dye solutions.

2.2.1. Optimization of Supporting Electrolyte Concentration

As mentioned earlier, Na_2SO_4 was used as the supporting electrolyte. This choice was made based on the fact that this salt is of low cost, is often found in textile industry wastewater In high concentrations, has excellent conductive properties, and, most importantly, is inert under oxidative conditions, so that the degradation efficiency can entirely be attributed to the developed method. The function of the supporting electrolyte is to improve the electrical conductivity and current transfer and presents one of the most important factors in electrochemical processes. In this study, we investigated the effect of different concentrations of supporting electrolytes, testing the following concentrations: 0.01 M, 0.05 M, 0.1 M, 0.15 M, and 0.2 M (Figure 4). Other applied conditions were pH 2, constant voltage (5 V), and RB5 concentration of 35 mg/L.

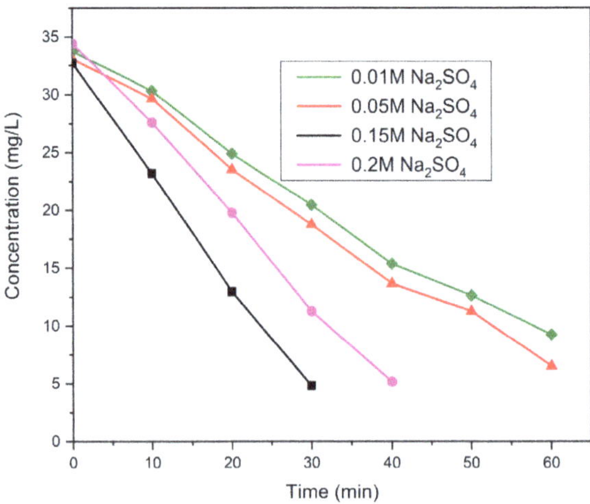

Figure 4. Degradation of RB5 dye under pH 2 and different concentrations of Na_2SO_4.

As expected, the increase in electrolyte concentration was followed by a more rapid removal rate. Up to 0.15 M of sodium sulfate, these changes were more noticeable, while when the concentration was increased to 0.2 M, the removal rate slightly diminished. Belal and coworkers studied the effect of supporting electrolytes and their concentrations on dye removal efficiency and found a similar trend [44]. The decrease in removal rate can be connected with the generation of fewer oxidants at higher electrolyte concentrations, due to increased production of $S_2O_8^{2-}$ from sulfate ions. It should be noted that in all the probes using the selected concentrations, the decolorization of RB5 was carried out up to nearly 100%. Therefore, to further extend the optimization of the experimental conditions we selected the supporting electrolyte concentration of 0.15 M, at the previously optimized initial pH of 2.

2.2.2. Optimization of Current Density-Applied Voltage

In previous optimization experiments, the voltage was held constant (at 5 V) and the current was constantly changing, depending on the duration of the experiment. In this study, probes were made under the opposite conditions—the current was held constant while the voltage was allowed to vary, but it was still monitored. The currents used for the probes were 0.08 A, 0.10 A, 0.13 A, 0.17 A, and 0.20 A (Figure 5). These currents were applied on an electrode with a working area of 4 cm^2, therefore the calculated current densities were as follows: 20 mA/cm^2, 25 mA/cm^2, 32.5 mA/cm^2, 42.5 mA/cm^2, and 50 mA/cm^2. The rest of the conditions were pH 2, 0.15 M Na$_2$SO$_4$, and 35 mg/L of RB5.

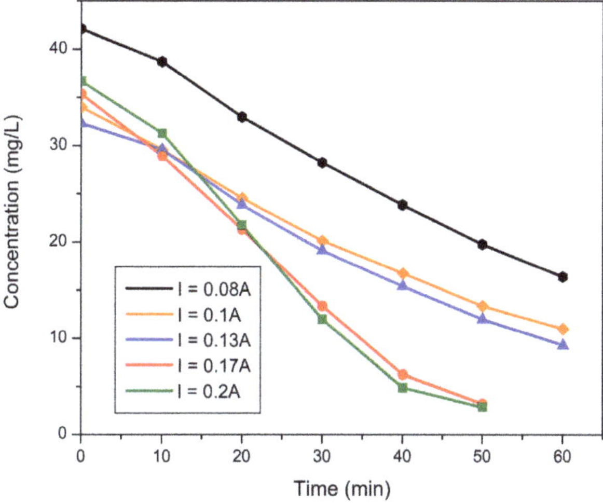

Figure 5. Degradation of RB5 dye under pH 2, 0.15 M of Na$_2$SO$_4$, and different applied currents.

During the experiments, the voltage was monitored because this information can be important for the possible construction of pilot reactors and further improvement of the method. Results pertaining to the relationship between current densities and the observed voltages are summarized in Table 1.

Table 1. Corresponding values of voltage when current was held constant.

I (A)	U (V)
0.08	~3.63
0.10	~3.65
0.13	~4.2
0.17	~4.8–4.9
0.20	~5.1–5.3

An applied voltage ranging from 3.6 V to 5.3 V, with an output current that is less than or equal to 200 mA, meets the conditions required for possible future technology transfer. Additionally, the low values of applied voltage and current density do not cause significant temperature changes of the test solution. Namely, after the treatment, the increase in solution temperature was lower than 3 °C for all optimization experiments. The results were as expected, since increasing the current density accelerated the decolorization of RB5. This occurs up to the current density of 42.5 mA/cm^2 (applied current 0.17 A → potential 4.8–4.9 V). Increasing the current to 0.2 A does not cause a significant improvement in efficiency. This can be explained by side reactions, where sulfate ions migrate to the electrode surface, occupy active sites and reduce the effective surface area. This ultimately results in surface inactivation and reduced contaminant removal capacity.

2.2.3. Optimization of RB5 Dye Concentration

It is known that industrial wastewater usually contains different concentrations of textile dyes, depending on its nature and reactivity. Moreover, it is known that RB5 is distinguishable in water at 1 mg/L concentration with the naked eye while its concentrations in the effluents of industries are usually in the range of 10–100 mg/L [45]. From this, it can be concluded that evaluating the initial concentration of RB5 in the removal process is a very important aspect. Based on that, we investigated the performance of our method in the removal of different amounts of RB5 in the working solution. Concentrations of RB5 that were examined were 20 mg/L, 35 mg/L, 70 mg/L, and 100 mg/L. The rest of the conditions were as previously optimized: pH 2, 0.15 M Na$_2$SO$_4$, and applied current 0.17 A. The results of these measurements are presented in Figure 6. It is noticeable that increasing the concentration of RB5 causes a decrease in efficiency for this method. When comparing the performance in lower concentrations, it can be concluded that for an increase in the amount of dye from 20 to 35 mg/L, treatment time only had to be slightly prolonged (in order to achieve the same degradation efficiency), while further increasing the dye concentration required extended treatment times. Based on the performed study, the dye concentration of 35 mg/L could be considered as optimal for further experiments.

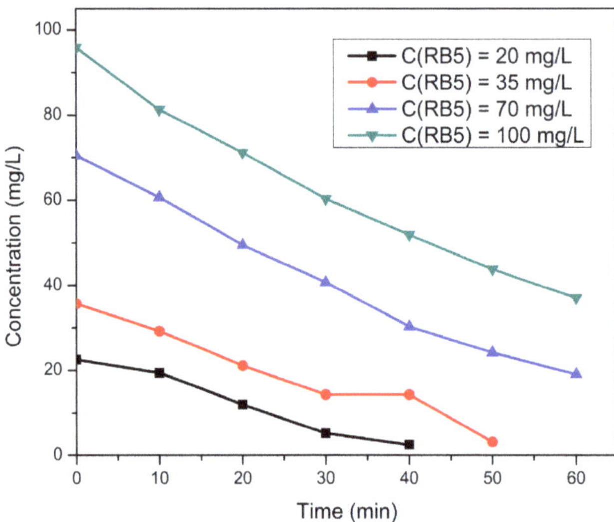

Figure 6. Degradation of RB5 dye under pH 2, 0.15 M Na$_2$SO$_4$, I = 0.17 A, and different initial concentrations of RB5 ranging from 20 mg/L to 100 mg/L.

2.3. Performance Comparison of the Synthesized Materials and Composites

In order to determine the electrocatalytic properties of the prepared materials and anodes, we tested their performances in a 35 mg/L solution of RB5 under fully optimized conditions: initial pH 2, current density 42.5 mA/cm^2, and supporting electrolyte concentration of 0.15 M. The results are summarized in Figure 7. As can be seen, the introduction of graphitic carbon nitride in the materials and formation of composites strongly influenced the results of the developed method. This can be assigned to the previously reported medium bandgap value for g-C_3N_4 and its characteristic catalytic performances [46–52]. However, it is obvious that the morphology of lead oxide is strongly correlated with the catalytic properties of g-C_3N_4 [53–55]. In the studied group of materials, the nanocomposite prepared from lead dioxide synthesized with assistance of linear surfactant cetyltrimethylammonium bromide as template and graphitic carbon nitride showed superior properties for electrocatalytic application, since using this electrode material led to almost 90% removal of RB5 in an hour.

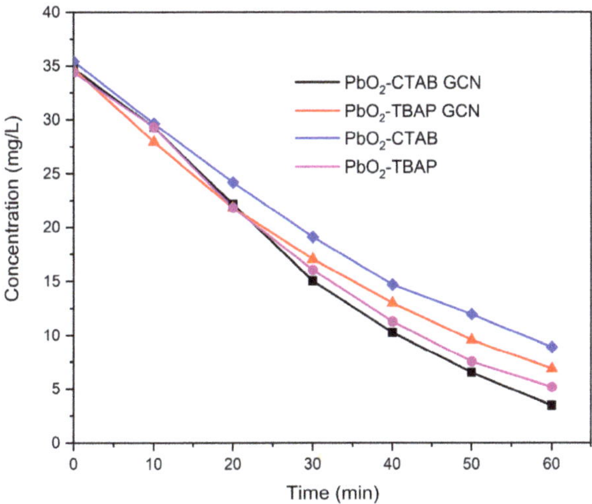

Figure 7. Degradation of RB5 dye under pH 2, Na_2SO_4 concentration of 0.15 M, I = 0.17 A, RB5 concentration of 35 mg/L, using different anodes.

Table 2 provides an overview of different electrochemical systems employed to remove RB5, with focus on results with Na_2SO_4 as supporting electrolyte. All selected studies reported faster RB5 removal when using NaCl or KBr [4,56–58], but the use of these non-inert salts is discouraged in electrochemistry, because they lead to the formation of chlorinated and brominated degradation products, which are often more toxic compared to the parent compound [57,59,60].

Table 2. The RB5 degradation efficiency comparison with the literature data.

Material	Supporting Electrolyte	% of RB5 Removal and Time	Energy Consumption	Ref
$RuO_2/IrO_2/TiO_2$@DSA®	0.008 M NaCl	~100% in 15 min	Not reported	[4]
Ti/CoO_x–RuO_2–SnO_2–Sb_2O_5	0.07 M Na_2SO_4	~40% in 2 h	34.5 kWh/kg [a]	[56]
Ti/SnO_2-Sb_2O_5-IrO_2	0.1 M Na_2SO_4	~60% in 2 h	Not reported	[58]
Graphite	0.1 M Na_2SO_4	~100% in 3 h	Not reported	[57]
PbO_2-CTAB/g-C_3N_4 on SS	0.15 M Na_2SO_4	~90% in 1 h	0.4374 kWh/g	This study

[a] calculated as general current efficiency, based on COD removal [61].

When excluding the results with NaCl, our study gave the fastest (under 1 h) and the highest RB5 removal (nearly 90%). The system with Ti/CoO_x–RuO_2–SnO_2–Sb_2O_5 removed only 40% of the selected dye, during 2 h of treatment in the presence of Na_2SO_4, but increased to over 95% when NaCl was used [56]. The next study with Ti/SnO_2-Sb_2O_5-IrO_2 was slightly improved in the same time interval without the addition of NaCl [58]. Graphite electrode led to complete RB5 degradation and quite efficient COD removal, but this experimental setup lasted for three hours [57].

2.4. Specific Energy Consumption

The specific energy consumption of an advanced oxidation process is an important indicator of economic efficiency, taking into account that a major part of operating costs for the proposed procedure derives from electric energy consumption. For this reason, the average energy consumption for a 60 min treatment was calculated when using each of the four prepared electrodes. This study was performed under all the previously optimized parameters. The results are listed in Table 3. As can be seen, PbO_2-CTAB in synergy with graphitic carbon nitride showed the lowest energy consumption during treatment of RB5 and can potentially be used for designing electrochemical systems for the removal of organic pollutants.

Table 3. Specific energy consumption (SEC) during RB5 treatment, a comparison of different electrode combinations.

T (min)	0	10	20	30	40	50	60	
electrode				RB5 degradation rate (%)				SEC (kWh/g)
I	0.00	16.45	37.43	57.69	70.19	81.01	88.53	**0.4374**
II	0.00	15.81	31.31	45.49	58.15	67.64	74.89	**0.5173**
III	0.00	19.94	37.52	50.93	62.79	72.14	80.24	**0.4895**
IV	0.00	14.83	36.57	53.38	67.21	78.09	85.01	**0.4698**

I—PbO_2-CTAB/g-C_3N_4 on SS; **II**—PbO_2-CTAB on SS; **III**—PbO_2-TBAP/g-C_3N_4 on SS; **IV**—PbO_2-TBAP on SS.

2.5. Stability and Longevity of the Method

To scrutinize stability and longevity of the method we monitored electrode efficiency during 7 days of testing. During that period, we performed 5 to 7 experiments per day. Each experiment lasted for 60 min (Figure 8). The relative standard deviation between absorbances was lower than 7% indicating that the proposed electrode possesses excellent stability and durability during this time period.

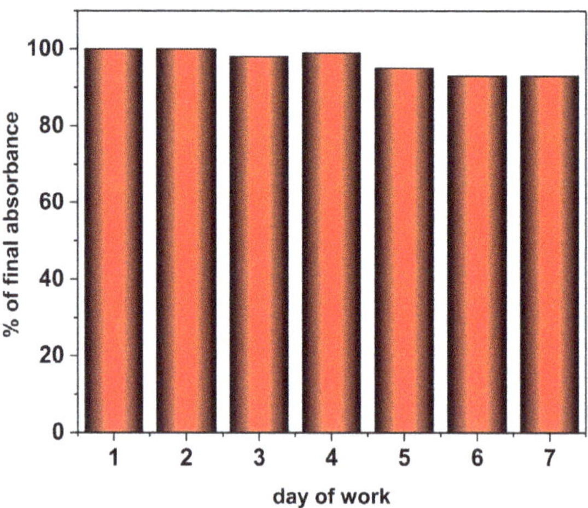

Figure 8. Changes in the final absorbance during seven days of work with the same electrode.

3. Materials and Methods

3.1. Chemicals

Reactive Black 5 dye (M-991.82, CAS number 17095-24-8, chemical purity grade, dye content ~50%) was purchased from Sigma-Aldrich (St. Louis, MO, USA). Cetyltrimethylammonium bromide (CTAB), tetrabutylammonium phosphate (TBAP), sodium hydroxide, hydrochloric acid, sulfuric acid, ethanol (96%), acetone, urea, sodium hypochlorite, lead nitrate, dimethylformamide, and sodium sulfate were p.a. grade and purchased from Sigma-Aldrich. All chemicals were used as supplied without any purification. All the solutions were prepared in double distilled water. Working solutions were prepared before every experiment and used accordingly. Calibration solution for UV/Vis spectrometry was prepared using the supporting electrolyte.

3.2. Material Synthesis

Graphitic carbon nitride (g-C_3N_4) was synthesized from urea. Briefly, 25 g of urea was heated in air in a ceramic crucible with a lid, starting from room temperature up to 550 °C, and this temperature was maintained for 4 h. PbO_2 was synthesized using the procedure proposed by Cao et al. [62]. In summary, CTAB or TBAP were added to a 0.015 M solution of $Pb(OH)_3^-$. After the resulting solution was stirred for 30 min at 50 °C, which ensured the complete dissolution of the surfactants, 1 mL of 1.5 M (10%) aqueous NaClO solution was added under constant stirring. This gave a homogeneous solution. After heating the solution at 85 °C for 3 h, the resulting black-brown precipitate was collected, washed several times with absolute ethanol and distilled water, centrifuged, and dried under vacuum at room temperature for 5 h [62]. The materials thus obtained were labeled PbO_2-CTAB and PbO_2-TBAP. The composite materials were prepared using the above-mentioned metal oxide materials and graphitic carbon nitride by ultrasonification—5 mg of PbO_2-CTAB or PbO_2-TBAP and 50 mg of g-C_3N_4 were suspended in 1 mL of dimethylformamide and ultrasonicated for 90 min. The prepared materials were labeled PbO_2-CTAB/g-C_3N_4 and PbO_2-TBAP/g-C_3N_4.

3.3. Electrode Preparation

To clean up and roughen the surface of the (2 cm × 5 cm) raw stainless-steel mesh, it was first washed ultrasonically in 1 M H_2SO_4 for 60 min and then in acetone for 40 min, fully rinsed with distilled water, and dried at 80 °C in air. A measurement of 20 µL of prepared nanomaterial suspensions (5 mg of PbO_2/CTAB or PbO_2/TBAP and 50 mg of g-C_3N_4 in 1 mL of dimethylformamide) was added to the electrode surface to cover its entire surface area. An electrode prepared in such a manner was allowed to dry overnight and the same procedure was applied on the opposite side of the electrode. The coated electrodes were used as anodes while the uncoated electrodes served as cathodes.

3.4. Electrochemical and Morphological Characterization

Microstructural and morphological characterization was performed using transition electron microscopy (TEM), scanning electron microscopy (SEM), and powder X-ray diffraction (PXRD). PXRD measurements were performed on Rigaku's high-resolution SmartLab® diffractometer equipped with a CuKα radiation source, an accelerating voltage of 40 kV, and a current of 30 mA. The diffraction patterns were recorded in the 10–70° 2θ range with a measurement speed of 1°/min and a step of 0.05°. The morphology of prepared PbO_2-CTAB/g-C_3N_4 and PbO_2-TBAP/g-C_3N_4 were examined by high-resolution transmission electron microscopy (HR-TEM, FEI Technai G2, Hillsboro, Thermo Fisher Scientific, Waltham, MA, USA) applying an accelerating voltage of 200 kV. The samples were prepared by doping diluted aqueous suspensions into copper grids (300 mesh carbon, Ted Pella Inc., Redding, CA, USA) and left to dry at RT. The composite materials were also analyzed with scanning electron microscopy (Hitachi S-4700) operating at 10 kV acceleration voltage, equipped with the X-ray spectroscopy (EDX) accessory (Röntec QX2 spectrometer). A few nm of gold was condensed on the sample surface to prevent them from becoming charged.

Electrochemical properties of the materials were examined using cyclic voltammetry (CV) and electrochemical impedance spectroscopy (EIS) measurements (done on CHInstruments model CHI760b, Austin, TX, USA) in the solution of 5 mM $Fe^{2+/3+}$ redox couple in 0.1 M KCl. Measurements were done in a three-electrode system, where material properties were tested using carbon paste electrodes (CPE) modified with 10% of the synthesized nanomaterials and composites. As a reference electrode, a Ag/AgCl (3M KCl) electrode was used while a platinum wire was employed as an auxiliary electrode. EIS measurements were done using the same assembly in the range from 0.01 to 10^5 Hz at the potential of 0.1 V.

3.5. Experimental Setup

Solutions of RB5 with Na_2SO_4 were made in concentrations ranging from 20 mg/L up to 100 mg/L. The blank probe for UV–Vis was made with 0.1 M Na_2SO_4. For pH adjustments, 0.1 M NaOH and 0.1 M H_2SO_4 were used.

The optimization of RB5 degradation experiments was held in a vessel mounted on a magnetic stirrer, with a constant starting volume of the treated solutions (60 cm^3 of RB5 dissolved in Na_2SO_4 solution). The electrodes were immersed in the solution and connected to the voltage source PeakTech 1525. For RB5 concentration monitoring, samples were withdrawn in 10-min intervals, up to 90 min of electrochemical oxidation, and analyzed by Evolution™ One/One Plus UV–Vis Spectrophotometer, Waltham, MA, USA.

The first parameter to be optimized was starting pH value (adjusted using H_2SO_4 and NaOH), followed by the optimization concentration of the supporting electrolyte, with a fixed pH value. When these two parameters were set right, in terms of the fastest RB5 degradation, the current density was varied to improve the degradation rate. The last criterion was the RB5 concentration and with the optimized previously stated values, the synthesized electrode materials were compared to distinguish the most excellent conditions for RB5 degradation. Data were processed using MS Excel 2016 and OriginPro 8.

3.6. Specific Energy Consumption

To deduce the most efficient conditions for degradation of RB5, the specific energy consumption (SEC; kWh/g) was calculated:

$$SEC_{RB5}\left(\frac{kWh}{g}\right) = \frac{I\,(A) \times E_{cell}(V) \times T\,(h)}{V\left(dm^3\right) \times \Delta C_{RB5}\left(\frac{g}{dm^3}\right) \times 1000}$$

where I (A) was the applied electrical current; E_{cell} (V) the cell voltage, T (h) the treatment time, V (dm^3) the treated solution volume, ΔC_{RB5} (g/dm^3) the difference between the starting and final RB5 concentration, while a conversion factor of 1000 was needed to convert from Wh/g to kWh/g. This equation was adopted from elsewhere [61].

4. Conclusions

In this paper, an electrochemical degradation/removal method for Reactive Black 5 azo dye was created, based on PbO_2-CTAB/g-C_3N_4 anode material as an electrocatalyst. The parameters that were optimized include initial pH, concentration of Na_2SO_4 as supporting electrolyte, electric current, and concentration of RB5 dye. PbO_2 probes synthesized using different surfactants as templates were used separately in pristine form and as composite materials with graphitic carbon nitride. Morphological and electrochemical properties of the materials were evaluated using PXRD, TEM, SEM, CV, and EIS techniques. The efficiency of the electrolytic reaction vastly depended on procedure parameters. The results showed that PbO_2-CTAB/g-C_3N_4 exhibits the best electrocatalytic properties and the highest removal rate for RB5, with 90% of the dye removed after 60 min of treatment. The specific energy consumption for this process was 0.4374 kWh/g. Finally, the prepared material showed good stability and enabled the treatment of highly polluted wastewaters with excellent results, but also, due to its electrocatalytic properties, could be applied in various fields of electrochemistry, such as sensing and biosensing probes. With regard to efficiency and longevity, the obtained results suggest that the proposed method can offer a low-cost, effective, and green approach in the field of environmental control, and due to its simplicity, this method has high potential for technology transfer.

Author Contributions: Conceptualization, A.M., M.O., A.K., S.S. and D.M.S.; methodology, investigation, A.M., S.S., A.K., M.O. and D.M.; resources, M.O. and D.M.S.; writing—original draft preparation, A.M., S.S., M.O. and D.M.S.; writing—review and editing, A.M., S.S., M.O. and D.M.S.; supervision, D.M.S.; project administration, Z.K., D.M. and D.M.S.; funding acquisition, D.M.S. All authors have read and agreed to the published version of the manuscript.

Funding: This work was supported by Ministry of Education, Science and Technological Development of Republic of Serbia Contract number: 451-03-68/2022-14/200168, EUREKA project E!13303 and Ministry of Science and Higher Education of the Russian Federation (agreement No. 075-15-2022-1135) and South Ural State University.

Data Availability Statement: The study did not report any data.

Conflicts of Interest: The authors declare that they have no known competing financial interest or personal relationship that could have appeared to influence the work reported in this paper.

References

1. El Bouraie, M.; El Din, W.S. Biodegradation of Reactive Black 5 by Aeromonas Hydrophila Strain Isolated from Dye-Contaminated Textile Wastewater. *Sustain. Environ. Res.* **2016**, *26*, 209–216. [CrossRef]
2. Kothari, M.S.; Vegad, K.G.; Shah, K.A.; Aly Hassan, A. An Artificial Neural Network Combined with Response Surface Methodology Approach for Modelling and Optimization of the Electro-Coagulation for Cationic Dye. *Heliyon* **2022**, *8*, e08749. [CrossRef] [PubMed]
3. Droguett, T.; Mora-Gómez, J.; García-Gabaldón, M.; Ortega, E.; Mestre, S.; Cifuentes, G.; Pérez-Herranz, V. Electrochemical Degradation of Reactive Black 5 Using Two-Different Reactor Configuration. *Sci. Rep.* **2020**, *10*, 4482. [CrossRef] [PubMed]
4. Jager, D.; Kupka, D.; Vaclavikova, M.; Ivanicova, L.; Gallios, G. Degradation of Reactive Black 5 by Electrochemical Oxidation. *Chemosphere* **2018**, *190*, 405–416. [CrossRef]
5. Brillas, E.; Martínez-Huitle, C.A. Decontamination of Wastewaters Containing Synthetic Organic Dyes by Electrochemical Methods. An Updated Review. *Appl. Catal. B Environ.* **2015**, *166–167*, 603–643. [CrossRef]
6. Yu, J.; Shu, S.; Wang, Q.; Gao, N.; Zhu, Y. Evaluation of Fe^{2+}/Peracetic Acid to Degrade Three Typical Refractory Pollutants of Textile Wastewater. *Catalysts* **2022**, *12*, 684. [CrossRef]
7. Emadi, Z.; Sadeghi, R.; Forouzandeh, S.; Mohammadi-Moghadam, F.; Sadeghi, R.; Sadeghi, M. Simultaneous Anaerobic Decolorization/Degradation of Reactive Black-5 Azo Dye and Chromium(VI) Removal by Bacillus Cereus Strain MS038EH Followed by UV-C/H2O2 Post-Treatment for Detoxification of Biotransformed Products. *Arch. Microbiol.* **2021**, *203*, 4993–5009. [CrossRef]
8. Saroyan, H.; Ntagiou, D.; Rekos, K.; Deliyanni, E. Reactive Black 5 Degradation on Manganese Oxides Supported on Sodium Hydroxide Modified Graphene Oxide. *Appl. Sci.* **2019**, *9*, 2167. [CrossRef]
9. De Luca, P.; Nagy, B.J. Treatment of Water Contaminated with Reactive Black-5 Dye by Carbon Nanotubes. *Materials* **2020**, *13*, 5508. [CrossRef]
10. Al-Tohamy, R.; Sun, J.; Fareed, M.F.; Kenawy, E.-R.; Ali, S.S. Ecofriendly Biodegradation of Reactive Black 5 by Newly Isolated Sterigmatomyces Halophilus SSA1575, Valued for Textile Azo Dye Wastewater Processing and Detoxification. *Sci. Rep.* **2020**, *10*, 12370. [CrossRef]
11. Mandic, M.; Djokic, L.; Nikolaivits, E.; Prodanovic, R.; O'Connor, K.; Jeremic, S.; Topakas, E.; Nikodinovic-Runic, J. Identification and Characterization of New Laccase Biocatalysts from Pseudomonas Species Suitable for Degradation of Synthetic Textile Dyes. *Catalysts* **2019**, *9*, 629. [CrossRef]
12. Khalik, W.F.; Ho, L.-N.; Ong, S.-A.; Voon, C.-H.; Wong, Y.-S.; Yusoff, N.; Lee, S.-L.; Yusuf, S.Y. Optimization of Degradation of Reactive Black 5 (RB5) and Electricity Generation in Solar Photocatalytic Fuel Cell System. *Chemosphere* **2017**, *184*, 112–119. [CrossRef] [PubMed]
13. Liu, F.; Wang, X.; Liu, Z.; Miao, F.; Xu, Y.; Zhang, H. Peroxymonosulfate Enhanced Photocatalytic Degradation of Reactive Black 5 by ZnO-GAC: Key Influencing Factors, Stability and Response Surface Approach. *Sep. Purif. Technol.* **2021**, *279*, 119754. [CrossRef]
14. Rosli, N.I.M.; Lam, S.-M.; Sin, J.-C.; Putri, L.K.; Mohamed, A.R. Comparative Study of G-C3N4/Ag-Based Metals (V, Mo, and Fe) Composites for Degradation of Reactive Black 5 (RB5) under Simulated Solar Light Irradiation. *J. Environ. Chem. Eng.* **2022**, *10*, 107308. [CrossRef]
15. Rao, M.P.; Ponnusamy, V.K.; Wu, J.J.; Asiri, A.M.; Anandan, S. Hierarchical CuO Microstructures Synthesis for Visible Light Driven Photocatalytic Degradation of Reactive Black-5 Dye. *J. Environ. Chem. Eng.* **2018**, *6*, 6059–6068. [CrossRef]
16. Liu, X.; Qiu, M.; Huang, C. Degradation of the Reactive Black 5 by Fenton and Fenton-like System. *Procedia Eng.* **2011**, *15*, 4835–4840. [CrossRef]
17. Mengelizadeh, N.; Mohseni, E.; Dehghani, M.H. Heterogeneous Activation of Peroxymonosulfate by GO-CoFe2O4 for Degradation of Reactive Black 5 from Aqueous Solutions: Optimization, Mechanism, Degradation Intermediates and Toxicity. *J. Mol. Liq.* **2021**, *327*, 114838. [CrossRef]

18. Fadaei, S.; Noorisepehr, M.; Pourzamani, H.; Salari, M.; Moradnia, M.; Darvishmotevalli, M.; Mengelizadeh, N. Heterogeneous Activation of Peroxymonosulfate with Fe_3O_4 Magnetic Nanoparticles for Degradation of Reactive Black 5: Batch and Column Study. *J. Environ. Chem. Eng.* **2021**, *9*, 105414. [CrossRef]
19. López-Ramón, M.V.; Rivera-Utrilla, J.; Sánchez-Polo, M. Photocatalytic Degradation of Organic Wastes in Water. *Catalysts* **2022**, *12*, 114. [CrossRef]
20. Ramírez, J.I.D.L.; Villegas, V.A.R.; Sicairos, S.P.; Guevara, E.H.; Brito Perea, M.D.C.; Sánchez, B.L. Synthesis and Characterization of Zinc Peroxide Nanoparticles for the Photodegradation of Nitrobenzene Assisted by UV-Light. *Catalysts* **2020**, *10*, 1041. [CrossRef]
21. Fernández-Perales, M.; Rozalen, M.; Sánchez-Polo, M.; Rivera-Utrilla, J.; López-Ramón, M.V.; Álvarez, M.A. Solar Degradation of Sulfamethazine Using RGO/Bi Composite Photocatalysts. *Catalysts* **2020**, *10*, 573. [CrossRef]
22. Amin Marsooli, M.; Rahimi Nasrabadi, M.; Fasihi-Ramandi, M.; Adib, K.; Pourmasoud, S.; Ahmadi, F.; Eghbali, M.; Sobhani Nasab, A.; Tomczykowa, M.; Plonska-Brzezinska, M.E. Synthesis of Magnetic $Fe_3O_4/ZnWO_4$ and $Fe_3O_4/ZnWO_4/CeVO_4$ Nanoparticles: The Photocatalytic Effects on Organic Pollutants upon Irradiation with UV-Vis Light. *Catalysts* **2020**, *10*, 494. [CrossRef]
23. Jović, M.; Stanković, D.; Manojlović, D.; Anđelković, I.; Milić, A.; Dojčinović, B.; Roglić, G. Study of the Electrochemical Oxidation of Reactive Textile Dyes Using Platinum Electrode. *Int. J. Electrochem. Sci.* **2013**, *8*, 16.
24. Savić, B.G.; Stanković, D.M.; Živković, S.M.; Ognjanović, M.R.; Tasić, G.S.; Mihajlović, I.J.; Brdarić, T.P. Electrochemical Oxidation of a Complex Mixture of Phenolic Compounds in the Base Media Using PbO2-GNRs Anodes. *Appl. Surf. Sci.* **2020**, *529*, 147120. [CrossRef]
25. Žunić, M.J.; Milutinović-Nikolić, A.D.; Stanković, D.M.; Manojlović, D.D.; Jović-Jovičić, N.P.; Banković, P.T.; Mojović, Z.D.; Jovanović, D.M. Electrooxidation of P-Nitrophenol Using a Composite Organo-Smectite Clay Glassy Carbon Electrode. *Appl. Surf. Sci.* **2014**, *313*, 440–448. [CrossRef]
26. Ganiyu, S.O.; Martínez-Huitle, C.A.; Rodrigo, M.A. Renewable Energies Driven Electrochemical Wastewater/Soil Decontamination Technologies: A Critical Review of Fundamental Concepts and Applications. *Appl. Catal. B Environ.* **2020**, *270*, 118857. [CrossRef]
27. Manojlović, D.; Lelek, K.; Roglić, G.; Zherebtsov, D.; Avdin, V.; Buskina, K.; Sakthidharan, C.; Sapozhnikov, S.; Samodurova, M.; Zakirov, R.; et al. Efficiency of Homely Synthesized Magnetite: Carbon Composite Anode toward Decolorization of Reactive Textile Dyes. *Int. J. Environ. Sci. Technol.* **2020**, *17*, 2455–2462. [CrossRef]
28. Stanković, D.M.; Ognjanović, M.; Espinosa, A.; del Puerto Morales, M.; Bessais, L.; Zehani, K.; Antić, B.; Dojcinović, B. Iron Oxide Nanoflower–Based Screen Print Electrode for Enhancement Removal of Organic Dye Using Electrochemical Approach. *Electrocatalysis* **2019**, *10*, 663–671. [CrossRef]
29. Jović, M.; Manojlović, D.; Stanković, D.; Dojčinović, B.; Obradović, B.; Gašić, U.; Roglić, G. Degradation of Triketone Herbicides, Mesotrione and Sulcotrione, Using Advanced Oxidation Processes. *J. Hazard. Mater.* **2013**, *260*, 1092–1099. [CrossRef]
30. Marković, M.; Jović, M.; Stanković, D.; Kovačević, V.; Roglić, G.; Gojgić-Cvijović, G.; Manojlović, D. Application of Non-Thermal Plasma Reactor and Fenton Reaction for Degradation of Ibuprofen. *Sci. Total Environ.* **2015**, *505*, 1148–1155. [CrossRef]
31. Jiang, Y.; Zhao, H.; Liang, J.; Yue, L.; Li, T.; Luo, Y.; Liu, Q.; Lu, S.; Asiri, A.M.; Gong, Z.; et al. Anodic Oxidation for the Degradation of Organic Pollutants: Anode Materials, Operating Conditions and Mechanisms. A Mini Review. *Electrochem. Commun.* **2021**, *123*, 106912. [CrossRef]
32. Duan, P.; Qian, C.; Wang, X.; Jia, X.; Jiao, L.; Chen, Y. Fabrication and Characterization of Ti/Polyaniline-Co/PbO2–Co for Efficient Electrochemical Degradation of Cephalexin in Secondary Effluents. *Environ. Res.* **2022**, *214*, 113842. [CrossRef] [PubMed]
33. Xia, Y.; Wang, G.; Guo, L.; Dai, Q.; Ma, X. Electrochemical Oxidation of Acid Orange 7 Azo Dye Using a PbO2 Electrode: Parameter Optimization, Reaction Mechanism and Toxicity Evaluation. *Chemosphere* **2020**, *241*, 125010. [CrossRef] [PubMed]
34. Shmychkova, O.; Luk'yanenko, T.; Dmitrikova, L.; Velichenko, A. Modified Lead Dioxide for Organic Wastewater Treatment: Physicochemical Properties and Electrocatalytic Activity. *J. Serb. Chem. Soc.* **2019**, *84*, 187–198. [CrossRef]
35. Zhou, M.; Dai, Q.; Lei, L.; Ma, C.; Wang, D. Long Life Modified Lead Dioxide Anode for Organic Wastewater Treatment: Electrochemical Characteristics and Degradation Mechanism. *Environ. Sci. Technol.* **2005**, *39*, 363–370. [CrossRef]
36. Kumar, A.; Kumari, A.; Sharma, G.; Du, B.; Naushad, M.; Stadler, F.J. Carbon Quantum Dots and Reduced Graphene Oxide Modified Self-Assembled S@C3N4/B@C3N4 Metal-Free Nano-Photocatalyst for High Performance Degradation of Chloramphenicol. *J. Mol. Liq.* **2020**, *300*, 112356. [CrossRef]
37. Hu, J.; Zhang, P.; An, W.; Liu, L.; Liang, Y.; Cui, W. In-Situ Fe-Doped g-C3N4 Heterogeneous Catalyst via Photocatalysis-Fenton Reaction with Enriched Photocatalytic Performance for Removal of Complex Wastewater. *Appl. Catal. B Environ.* **2019**, *245*, 130–142. [CrossRef]
38. Mohammad, A.; Ahmad, K.; Qureshi, A.; Tauqeer, M.; Mobin, S.M. Zinc Oxide-Graphitic Carbon Nitride Nanohybrid as an Efficient Electrochemical Sensor and Photocatalyst. *Sens. Actuators B Chem.* **2018**, *277*, 467–476. [CrossRef]
39. Ma, S.; Xue, J.; Zhou, Y.; Zhang, Z.; Cai, Z.; Zhu, D.; Liang, S. Facile Fabrication of a Mpg-C3N4/TiO2 Heterojunction Photocatalyst with Enhanced Visible Light Photoactivity toward Organic Pollutant Degradation. *RSC Adv.* **2015**, *5*, 64976–64982. [CrossRef]
40. Duan, X.; Sui, X.; Wang, Q.; Wang, W.; Li, N.; Chang, L. Electrocatalytic Oxidation of PCP-Na by a Novel Nano-PbO_2 Anode: Degradation Mechanism and Toxicity Assessment. *Environ. Sci. Pollut. Res.* **2020**, *27*, 43656–43669. [CrossRef]
41. Velichenko, A.; Luk'yanenko, T.; Nikolenko, N.; Shmychkova, O.; Demchenko, P.; Gladyshevskii, R. Composite Electrodes PbO2-Nafion®. *J. Electrochem. Soc.* **2020**, *167*, 063501. [CrossRef]

42. Xie, Y.; Wang, Y.; Singhal, V.; Giammar, D.E. Effects of PH and Carbonate Concentration on Dissolution Rates of the Lead Corrosion Product PbO2. *Environ. Sci. Technol.* **2010**, *44*, 1093–1099. [CrossRef] [PubMed]

43. Broda, B.; Inzelt, G. Investigation of the Electrochemical Behaviour of Lead Dioxide in Aqueous Sulfuric Acid Solutions by Using the in Situ EQCM Technique. *J. Solid State Electrochem.* **2020**, *24*, 1–10. [CrossRef]

44. Belal, R.M.; Zayed, M.A.; El-Sherif, R.M.; Abdel Ghany, N.A. Advanced Electrochemical Degradation of Basic Yellow 28 Textile Dye Using IrO$_2$/Ti Meshed Electrode in Different Supporting Electrolytes. *J. Electroanal. Chem.* **2021**, *882*, 114979. [CrossRef]

45. Jalali Sarvestani, M.R.; Doroudi, Z. Removal of Reactive Black 5 from Waste Waters by Adsorption: A Comprehensive Review. *J. Water Environ. Nanotechnol.* **2020**, *5*, 180–190. [CrossRef]

46. Ejeta, S.Y.; Imae, T. Cobalt Incorporated Graphitic Carbon Nitride as a Bifunctional Catalyst for Electrochemical Water-Splitting Reactions in Acidic Media. *Molecules* **2022**, *27*, 6445. [CrossRef]

47. Iravani, S.; Varma, R.S. MXene-Based Photocatalysts in Degradation of Organic and Pharmaceutical Pollutants. *Molecules* **2022**, *27*, 6939. [CrossRef]

48. Isahak, W.N.R.W.; Kamaruddin, M.N.; Ramli, Z.A.C.; Ahmad, K.N.; Al-Azzawi, W.K.; Al-Amiery, A. Decomposition of Formic Acid and Acetic Acid into Hydrogen Using Graphitic Carbon Nitride Supported Single Metal Catalyst. *Sustainability* **2022**, *14*, 3156. [CrossRef]

49. Liu, J.; Guo, H.; Yin, H.; Nie, Q.; Zou, S. Accelerated Photodegradation of Organic Pollutants over BiOBr/Protonated g-C3N4. *Catalysts* **2022**, *12*, 1109. [CrossRef]

50. Khan, H.; Kang, S.; Lee, C.S. Evaluation of Efficient and Noble-Metal-Free NiTiO3 Nanofibers Sensitized with Porous GC3N4 Sheets for Photocatalytic Applications. *Catalysts* **2021**, *11*, 385. [CrossRef]

51. Li, C.; Wu, X.; Shan, J.; Liu, J.; Huang, X. Preparation, Characterization of Graphitic Carbon Nitride Photo-Catalytic Nanocomposites and Their Application in Wastewater Remediation: A Review. *Crystals* **2021**, *11*, 723. [CrossRef]

52. Shen, Y.; Dos santos-Garcia, A.J.; Martín de Vidales, M.J. Graphitic Carbon Nitride-Based Composite in Advanced Oxidation Processes for Aqueous Organic Pollutants Removal: A Review. *Processes* **2021**, *9*, 66. [CrossRef]

53. Tang, J.; Cheng, Z.; Li, H.; Xiang, L. Electro-Chemical Degradation of Norfloxacin Using a PbO2-NF Anode Prepared by the Electrodeposition of PbO$_2$ onto the Substrate of Nickel Foam. *Catalysts* **2022**, *12*, 1297. [CrossRef]

54. Zheng, T.; Wei, C.; Chen, H.; Xu, J.; Wu, Y.; Xing, X. Fabrication of PbO$_2$ Electrodes with Different Doses of Er Doping for Sulfonamides Degradation. *Int. J. Environ. Res. Public. Health* **2022**, *19*, 3503. [CrossRef] [PubMed]

55. Zhang, Y.; Ni, Z.; Yao, J. Enhancement of the Activity of Electrochemical Oxidation of BPS by Nd-Doped PbO$_2$ Electrodes: Performance and Mechanism. *Water* **2020**, *12*, 1317. [CrossRef]

56. Saxena, P.; Ruparelia, J. Influence of Supporting Electrolytes on Electrochemical Treatability of Reactive Black 5 Using Dimensionally Stable Anode. *J. Inst. Eng. India Ser. A* **2019**, *100*, 299–310. [CrossRef]

57. Rivera, M.; Pazos, M.; Sanromán, M.Á. Development of an Electrochemical Cell for the Removal of Reactive Black 5. *Desalination* **2011**, *274*, 39–43. [CrossRef]

58. Eguiluz, K.I.B.; Hernandez-Sanchez, N.K.; Dória, A.R.; Santos, G.O.S.; Salazar-Banda, G.R.; Ponce de Leon, C. Template-Made Tailored Mesoporous Ti/SnO$_2$-Sb$_2$O$_5$-IrO$_2$ Anodes with Enhanced Activity towards Dye Removal. *J. Electroanal. Chem.* **2022**, *910*, 116153. [CrossRef]

59. Zambrano, J.; Min, B. Comparison on Efficiency of Electrochemical Phenol Oxidation in Two Different Supporting Electrolytes (NaCl and Na2SO4) Using Pt/Ti Electrode. *Environ. Technol. Innov.* **2019**, *15*, 100382. [CrossRef]

60. Carneiro, J.F.; Aquino, J.M.; Silva, A.J.; Barreiro, J.C.; Cass, Q.B.; Rocha-Filho, R.C. The Effect of the Supporting Electrolyte on the Electrooxidation of Enrofloxacin Using a Flow Cell with a BDD Anode: Kinetics and Follow-up of Oxidation Intermediates and Antimicrobial Activity. *Chemosphere* **2018**, *206*, 674–681. [CrossRef]

61. Xu, L.; Guo, Z.; Du, L.; He, J. Decolourization and Degradation of C.I. Acid Red 73 by Anodic Oxidation and the Synergy Technology of Anodic Oxidation Coupling Nanofiltration. *Electrochim. Acta* **2013**, *97*, 150–159. [CrossRef]

62. Cao, M.; Hu, C.; Peng, G.; Qi, Y.; Wang, E. Selected-Control Synthesis of PbO2 and Pb3O4 Single-Crystalline Nanorods. *J. Am. Chem. Soc.* **2003**, *125*, 4982–4983. [CrossRef]

Review

State of Art and Perspectives in Catalytic Ozonation for Removal of Organic Pollutants in Water: Influence of Process and Operational Parameters

Naghmeh Fallah [1,*], Ermelinda Bloise [1], Domenico Santoro [2] and Giuseppe Mele [1]

[1] Department of Engineering for Innovation, University of Salento, Via Monteroni, 73100 Lecce, Italy
[2] Department of Chemical and Biochemical Engineering, University of Western Ontario, London, ON N6A 5B9, Canada
* Correspondence: naghmeh.fallah@unisalento.it

Abstract: The number of organic pollutants detected in water and wastewater is continuously increasing thus causing additional concerns about their impact on public and environmental health. Therefore, catalytic processes have gained interest as they can produce radicals able to degrade recalcitrant micropollutants. Specifically, catalytic ozonation has received considerable attention due to its ability to achieve advanced treatment performances at reduced ozone doses. This study surveys and summarizes the application of catalytic ozonation in water and wastewater treatment, paying attention to both homogeneous and heterogeneous catalysts. This review integrates bibliometric analysis using VOS viewer with systematic paper reviews, to obtain detailed summary tables where process and operational parameters relevant to catalytic ozonation are reported. New insights emerging from heterogeneous and homogenous catalytic ozonation applied to water and wastewater treatment for the removal of organic pollutants in water have emerged and are discussed in this paper. Finally, the activities of a variety of heterogeneous catalysts have been assessed using their chemical–physical parameters such as point of zero charge (PZC), pKa, and pH, which can determine the effect of the catalysts (positive or negative) on catalytic ozonation processes.

Keywords: catalytic ozonation; homogenous catalysts; heterogeneous catalysts; water treatment; VOSviewer; reaction mechanism

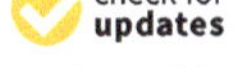
check for **updates**

Citation: Fallah, N.; Bloise, E.; Santoro, D.; Mele, G. State of Art and Perspectives in Catalytic Ozonation for Removal of Organic Pollutants in Water: Influence of Process and Operational Parameters. *Catalysts* **2023**, *13*, 324. https://doi.org/10.3390/catal13020324

Academic Editors: Gassan Hodaifa, Rafael Borja and Mha Albqmi

Received: 23 December 2022
Revised: 24 January 2023
Accepted: 28 January 2023
Published: 1 February 2023

1. Introduction

Industrial wastewater has been under extensive research due to its hazardous effect on the aquatic, air, and soil environment, as well as human and animal health. Developing wastewater treatment technologies that are simple, safe, and efficient for the environment is becoming the twenty-first century's primary goal.

Advanced oxidation processes (AOPs) have shown great potential for the degradation and mineralization of recalcitrant and toxic organic pollutants compared to conventional treatment processes. This process is classified into two main general categories. The first category utilizes light energy such as ultraviolet (UV) light in conjunction with other chemical additives. There are processes under this category that associate other agents with UV, such as hydrogen peroxide (UV/H_2O_2), ozone (UV/O_3), titanium oxide (UV/TiO_2), and Fenton reagents (UV/Fenton). When no light source is used, the technology can be termed a dark oxidative process. Processes in this category include ozonation, Fenton's reagent, ultrasound, and microwaves. These processes are simultaneously based on the in situ generations of highly reactive transitory species (H_2O_2, $HO^{\bullet}$, O^{2-}, O_3) for the mineralization of refractory organic compounds and the inactivation of waterborne pathogens. Due to rapid oxidation reactions, AOPs are characterized by high reaction rates and short treatment times, which make them promising in wastewater treatment [1].

Ozone is a powerful oxidant with several advantages that make it an excellent material in the AOPs. There are many noticeable benefits of using O_3 in the wastewater treatment process. Benefits such as it rapidly reacting with bacteria, viruses, and protozoa, being efficient for organics degradation and inorganics removal, and removing color, taste, and odor. Although the ozonation process is a practical system, O_3 has low solubility and stability in water and a high production cost. To solve the mentioned disadvantages, some solutions have been explored, such as using fixed beds of porous glass or metals, solid catalysts, stirring, line mixers, contact towers, and an increase in retention time by large bubble columns or diffusers [1]. The combination of ozonation with other techniques is suggested as an intelligent solution. In this regard, various O_3-based AOPs, such as OH^-/O_3, O_3/H_2O_2, O_3/UV, $O_3/H_2O_2/UV$, $O_3/S_2O_8^{2-}$, O_3/biological treatment, and catalytic and photocatalytic ozonation, were introduced to the industry [2–7]. Each of these processes has its specific features and conditions.

In the OH^-/O_3 process, the pH value of the water matrices has a significant influence on both the direct ozonation efficiency and the generation of $HO^\bullet$ (indirect ozonation). At significantly high pH (pH > 8), the abundance of OH^- can improve $HO^\bullet$ generation, which will enhance the ozonation of pollutants. However, high pH might cause the precipitation of calcium carbonate or other problems, which should be considered. In addition, the pH adjustment will increase the operational cost. In the so-called peroxone technique, O_3 and H_2O_2 would be combined [8]. The critical effect of combining O_3 and H_2O_2 is increasing oxidation efficiency. This occurs by converting O_3 to $HO^\bullet$ and improving O_3 transfer from the gas to the liquid phase [8]. The chemistry of the main reactions described above is shown in Equations (1)–(3).

$$H_2O_2 + O_3 \rightarrow HO^\bullet + HO_2^\bullet + O_2 \tag{1}$$

$$HO^\bullet + O_3 \rightarrow O_2 + HO_2^\bullet \tag{2}$$

$$O_3 + HO_2^\bullet \rightarrow 2O_2 + HO^\bullet \tag{3}$$

Another ozonation technique is the usage of ultraviolet light in combination with O_3 in an aqueous medium. This combination causes the increase in the $HO^\bullet$ formation and its concentration, consequently increasing the degradation efficiency. Equations (4) and (5) show that the formation of H_2O_2 as a by-product is possible, which will be degraded by the exact mechanism of H_2O_2/UV [8], also increasing the treatment efficiency.

$$O_3 + H_2O + h\upsilon \rightarrow O_2 + 2HO^\bullet \tag{4}$$

$$2HO^\bullet \rightarrow H_2O_2 \tag{5}$$

The introduction of UV with H_2O_2/O_3 makes the previously mentioned techniques more efficient. This combination enhances $HO^\bullet$ generation and more efficiently allows the transformation of H_2O_2 to $HO^\bullet$ (Equation (6)) [9], consequently increasing the degradation rate.

$$O_3 + H_2O_2 + h\upsilon \rightarrow 2HO^\bullet + 3O_2 \tag{6}$$

Persulfate $S_2O_8^{2-}$ as a practicable material for water treatment would be combined with O_3. It is assumed that O_3 decomposes with the formation of $HO^\bullet$, which can then activate persulfate to generate $SO_4^{\bullet-}$ (Equation (7)) [10]. In turn, $SO_4^{\bullet-}$ can increase the formation of $HO^\bullet$, which leads to a multiradical system (Equations (8) and (9))

$$HO^\bullet + S_2O_8^{2-} \rightarrow SO_4^{\bullet-} + HSO_4^- + 1/2O_2 \tag{7}$$

$$SO_4^{\bullet-} + H_2O \rightarrow SO_4^{2-} + HO^\bullet + H^+ \tag{8}$$

$$SO_4^{\bullet-} + OH^- \rightarrow SO_4^{2-} + HO^\bullet \tag{9}$$

A popular ozonation process is an ozonation/biological treatment technology. This process can be divided into two types: (1) ozonation is used as pre-treatment, such as an O_3-biological activated carbon process, ozonation/batch aerobic biological system, and ozonation/aerated biological filter; (2) ozonation is used as post-treatment, such as membrane bioreactor/ozonation, activated-sludge biological treatment/ozonation, and sequencing batch biofilm reactor/ozonation. Since the intermediate products formed by ozonation and O_3-based processes are generally more biodegradable than their precursors, these intermediates can be much more easily removed by biological treatment processes. Therefore, if the water containing many inhibiting compounds is toxic to the biological cultures, in such cases, a biological treatment followed by pre-treatment ozonation is suitable for the application. On the other hand, if there are many biodegradable compounds, the pre-oxidation step obviously will only lead to the unnecessary consumption of chemicals. In this case, a biological pre-treatment followed by ozonation (removing non-biodegradable and toxic components with less oxidant consumption) may be more suitable [11].

In the ozonation process, the addition of some catalysts can promote the decomposition of the oxidant (O_3) to generate active free radicals, such as $HO^{\bullet}$. Compared with other O_3-based treatment methods, catalytic ozonation can reduce operational costs since it does not need additional energy costs such as for UV or for pH adjustment due to its effectiveness in a wide range of pH values. Moreover, the catalytic ozonation systems have shown exemplary performance in water treatment, with several advantages compared to ozonation alone. Several pieces of evidence based on published articles show that the catalytic ozonation process achieved higher mineralization of various organic compounds than the sole ozonation system [12]. All of these reasons make catalytic ozonation an interesting water treatment process and one of the main AOPs processes that received significant attention from scientists. Figure 1 shows the gradual increase in scientific publication since 2000 based on approximately 600 published articles found in the Web of Science collection database.

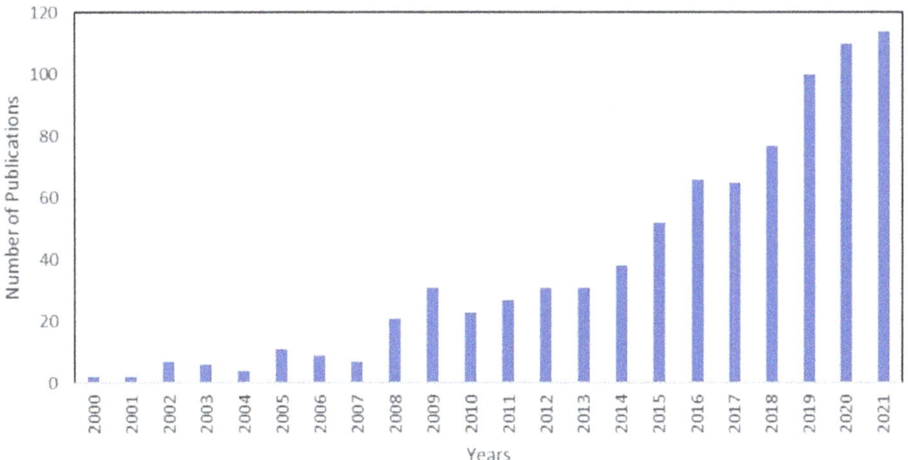

Figure 1. Published articles related to catalytic ozonation since 2000.

During these years, several catalysts have been proven to be effective in the enhancement of ozonation efficiency. Generally, the catalytic ozonation process can be divided into two types:

(1) Homogeneous catalytic ozonation, in which transition metal ions used as catalysts influence the rate of reaction, the selectivity of O_3 oxidation, and the efficiency of O_3 utilization. Two major mechanisms of homogeneous catalytic ozonation can be found: the O_3 decomposition by metal ions which generates free radicals; and the

complex formation between the metal ions and organic molecule following oxidation of the complex.

(2) Heterogeneous catalytic ozonation, which is based on the activation of O_3 to improve the ozonation of pollutants in the presence of a solid catalyst. Obviously, the key point in this process is to find the most appropriate catalyst, which is a solid material that in combination with O_3 shows a greater removal of a pollutant at a given pH value, compared to the separate processes of adsorption or ozonation alone. Among the most widely used catalysts in heterogeneous catalytic ozonation are metal/bimetal/polymetal oxides, metal/metal oxides on supports, carbon-based materials, and the emerging category of multifunctional porous materials as metal–organic frameworks. The role of the catalyst in this process is to provide reaction sites for adsorption and catalysis. So, based on the interaction of catalysts with O_3 and micropollutants, three general major mechanisms can be found. (a) Adsorption of O_3 on the catalyst surface following O_3 decomposition to generate free radicals; (b) adsorption of micropollutants on the catalyst surface, then attacking by O_3 molecule; (c) adsorption of both O_3 and micropollutants on the catalyst and their reaction together.

Many factors may have an impact on the performance of the catalysts. The PZC value of the material, the acid/basic sites of the surface, the oxidative potential of the metals contained in the solid structure, the cation exchange capacity, the oxygen vacancies, etc., are examples of these factors [13–15]. An important research aspect is the specific role played by the point of zero charge (PZC) in the overall efficiency of the catalytic ozonation process. Several researchers measured the PZC value of their synthesized catalyst and explained their surface characteristics based on this factor [16–21]. On the other hand, there are studies that examined the role of oxygen vacancies in catalytic ozonation [14,22], and the effect of the lewis acid/basic sites of the surface [23–27]. Although there are some studies that prove that these two factors can also influence O_3 decomposition, there are a variety of papers in which authors do not report, nor discuss, the impact of these factors.

In the terms of process efficiency of catalytic ozonation, operational parameters and reaction mechanism in the process have a big impact. According to operational parameters, initial solution pH, O_3 dose, initial pollutant concentration, pressure, catalyst dosage, and temperature have major effects on efficiency. Our attention has been placed on the governing mechanisms of the process based on the chemical properties of catalysts, physical properties of catalysts, natural properties of target pollutants (pKa), and pH of the solution as described in this article.

2. A Bibliometric Analysis Using VOSviewer

The scientific articles about catalytic ozonation published between 2000 and 2021 were scanned in the Web of Science (WOS) collection database. The words "Catalytic Ozonation" were used as the keywords to achieve the relevant publications. VOS viewer was applied to perform the bibliometric analysis of these articles. In this respect, 600 publications on the topic of catalytic ozonation were identified in the WOS core database.

Bibliometric analysis of the keywords in publications was studied. In this respect, all provided keywords in the articles related to catalytic ozonation that occurred more than 20 times in the WOS core database were enrolled in the final analysis. Based on the catalytic ozonation articles in the English language, of the 1441 keywords accrued, 26 of these keywords had appeared 20 times more frequently than the others. The keyword "catalytic ozonation" was the most frequently occurring one, with an occurrence of 197 and total link strength of 177. Following the previously mentioned keyword, "ozone", "degradation", and "oxidation" occurred 169, 154, and 129 times, respectively. Figure 2 illustrated the bibliometric analysis and VOSviewer visualization of the keywords in articles related to catalytic ozonation.

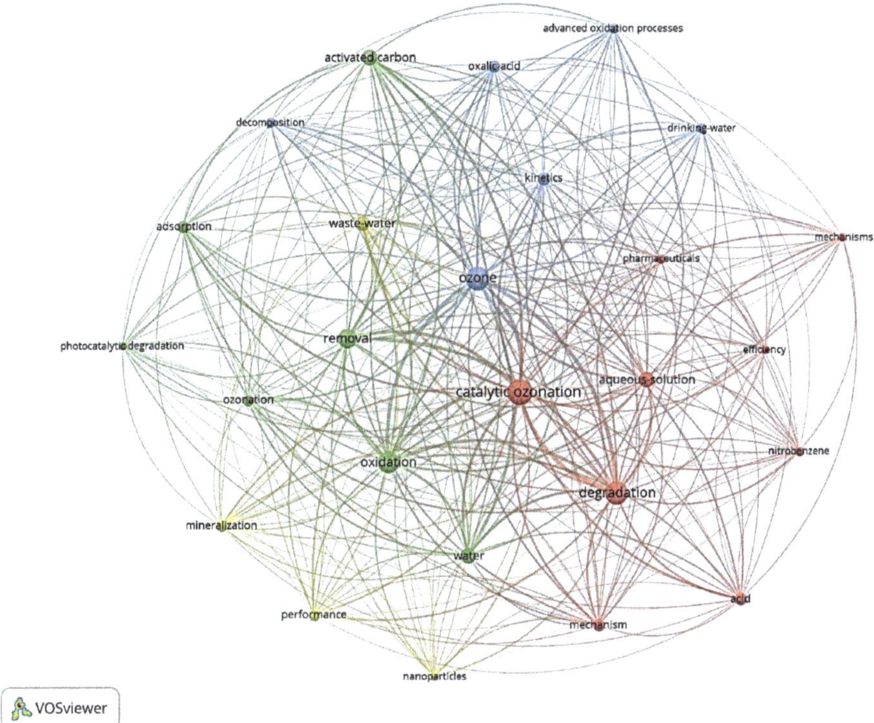

Figure 2. Co-occurrence of the keywords in articles related to catalytic ozonation visualized by VOSviewer software.

Based on the WOS collection database results in the case of 'catalytic ozonation', a bibliometric analysis of the co-authorship between countries was studied. All countries that published more than five articles in the WOS core database related to catalytic ozonation were enrolled in the final analysis.

In this respect, the top ten most active countries in the field of catalytic ozonation based on the number of citations, publications, and total link strength, are listed in Table 1. Of the 47 countries that worked on catalytic ozonation topics, 14 countries had more than 5 published articles. Based on the information gathered, the country most active in the field of catalytic ozonation is "China", with 185 publications, 422 citations, and a total link strength of 20. Following that, Iran and Canada take second and third places, respectively.

All these bibliometric analyses show the importance of catalytic ozonation among scientists all around the world. In the absence of a comprehensive and precise survey about the developments in catalytic ozonation processes, it would not be easy to propose novel investigations to optimize water treatment performance in terms of mineralization, industrial application, and economic capability. In light of these considerations, this study aims to recognize the popular and leading articles in catalytic ozonation, summarizing new visions on the evaluation of both heterogeneous and homogenous processes for the degradation and mineralization of various toxic organic pollutants in water.

Table 1. The top 10 active countries in the field of catalytic ozonation.

Rank	Country	Number of Citations	Number of Documents	Total Link Strength
1	China	422	185	20
2	Iran	82	37	4
3	Canada	38	14	5
4	USA	41	13	8
5	Pakistan	28	12	8
6	Brazil	22	8	1
7	Turkey	25	8	1
8	France	23	8	0
9	Australia	10	7	5
10	Greece	14	7	1

3. Homogeneous Catalytic Ozonation

Transition metals such as Fe(II), Fe(III), Mn(II), Ni(II), Co(II), Pb(II), and Zn(II) used as homogenous catalysts in the aqueous solution are commonly involved in two mechanistic steps of the ozonation processes. initiation of the O_3 decomposition reaction followed by the generation of the hydroxyl radicals, and oxidation reaction between catalyst and organic pollutants. Some homogeneous catalytic systems selected among the most influencing articles reported in the current literature are listed in Table 2. Details regarding the kind of used catalysts, target pollutants, operating conditions, and removal results are included in this table.

Table 2. Literature reports on different homogenous catalysts in the ozonation process.

Target Pollutants	Catalysts	Operating Conditions	Removal Results	Ref.
Aniline	Fe(II)	O_3 dose: 0.5 g/h; Catalyst dose: 1 mmol/dm^3; pH: 3.3; T: 25 °C; t: 15 min	132 TOC removal	[28]
4-chlorophenol	Fe(II)	O_3 dose: 0.5 g/h; Catalyst dose: 1 mmol/dm^3; pH: 3.3; T: 25 °C; t: 15 min	144 TOC removal	[28]
Oxalic acid (OA)	Fe(III)	O_3 dose: 8.2 mg/L; Catalyst dose: 1 mg/L; pH: 2; T: 20 °C; t: 3 h	7% OA removal	[29]
1,3,6-naphthalenetrisulfonic acid (NTS)	Fe(II)	O_3 dose: 1.04×10^{-4} mol/dm^{-3}; Catalyst dose: 1.25×10^{-4} mol/dm^{-3}; pH: 2; T: 25 °C; t: 30 min	79% NTS degradation	[30]
Lipid	Fe(II)	O_3 dose: 0.6 g/L; Catalyst dose: 7 mg/L; pH: 6.75; T: 25 °C; t: 60 min	96.7% lipid degradation	[31]
Chlorobenzenes	Fe(II)	O_3 dose: 1.5 g O_3/TOC; Catalyst dose: 6×10^{-5} mol/L; pH: 7; t: 20 min	55% COD removal	[32]
Chlorobenzenes	Fe(III)	O_3 dose: 1.5 g O_3/TOC; Catalyst dose: 6×10^{-5} mol/L; pH: 7; t: 20 min	12% COD removal	[32]
Aniline aerofloat (AAF)	Fe(II)	O_3 dose: 2.08 mg/min. L; Catalyst dose: 10 mg/L; pH: 8; T: 25 °C; t: 180 min	80% COD removal	[33]
AAF	Fe(III)	O_3 dose: 2.08 mg/min. L; Catalyst dose: 10 mg/L; pH: 8; T: 25 °C; t: 180 min	76% COD removal	[33]
C.I. Reactive Red 2 (RR2)	Fe(III)	O_3 dose: 200 mL/min; Catalyst dose: 0.6 mM; pH: 2; T:NR; t: 6 min	1.278 of decolorization rates (1/min)	[34]
RR2	Fe(II)	O_3 dose: 200 mL/min; Catalyst dose: 0.6 mM; pH: 2; T:NR; t: 6 min	1.299 of decolorization rates (1/min)	[34]
p-Chlorobenzoic acid (p-CBA)	Fe(II)	O_3 dose: 2 mg/L; Catalyst dose: 1 mg/L; pH: 7; T: 23 °C; t: 15 min	92.5% p-CBA degradation	[35]
RR2	Zn(II)	O_3 dose: 200 mL/min; Catalyst dose: 0.6 mM; pH: 2; T:NR; t: 6 min	1.015 of decolorization rates (1/min)	[34]

Table 2. *Cont.*

Target Pollutants	Catalysts	Operating Conditions	Removal Results	Ref.
p-CBA	Co(II)	O_3 dose: 2 mg/L; Catalyst dose: 1 mg/L; pH: 7; T: 23 °C; t: 15 min	95.5% p-CBA degradation	[35]
RR2	Co(II)	O_3 dose: 200 mL/min; Catalyst dose: 0.6 mM; pH: 2; T:NR; t: 6 min	0.843 of decolorization rates (1/min)	[34]
OA	Co(II)	O_3 dose: 30 mg/L; Catalyst dose: 0.8 mg/L; pH: 2.5; T:NR; t: 90 min	70% OA removal	[36]
OA	Co(II)	O_3 dose: 5×10^{-3} mol/L; Catalyst dose: 4 mg/L; pH: 2.5; T: 25 °C; t: 30 min	99.3% TOC removal	[37]
Formic acid	Co(II)	O_3 dose: 5×10^{-3} mol/L; Catalyst dose: 4 mg/L; pH: 2.5; T: 25 °C; t: 35 min	60.2% TOC removal	[37]
Carboxylic acids	Cu(II)	O_3 dose: 71 mg/L; Catalyst dose: 20 μg/L; pH: natural; T:NR; t:NR	75% TOC reduction	[38]
AAF	Cu(II)	O_3 dose: 2.08 mg/min. L; Catalyst dose: 10 mg/L; pH: 8; T: 25 °C; t: 180 min	75% COD removal	[33]
N-dimethylpropyl-2-pyrrolidone (NDPP)	Pd(II)	O_3 dose: 250 mL/min. L; Catalyst dose: 9.4×10^{-5} M; pH: 2; T: 25 °C; t: 30 min	73% NDPP removal	[39]
RR2	Ni(II)	O_3 dose: 200 mL/min; Catalyst dose: 0.6 mM; pH: 2; T:NR; t: 6 min	0.822 of decolorization rates (1/min)	[34]
RR2	Mn(II)	O_3 dose: 200 mL/min; Catalyst dose: 0.6 mM; pH: 2; T:NR; t: 6 min	3.295 of decolorization rates (1/min)	[34]
Pyruvic acid	Mn(IV)	O_3 dose: 0.5 g/h; Catalyst dose: 200 mg; pH: 3; T: 25 °C; t: 60 min		[40]
Atrazine (ATZ)	Mn(II)	O_3 dose: 2.54 mg/L; Catalyst dose: 0.3 mg/L; pH: 7; T: 21 °C; t: 4 min	96% ATZ removal	[41]
ATZ	Mn(II)	O_3 dose: 2.28 mg/L; Catalyst dose: 1 mg/L; pH: 7; T: 23 °C; t: 5 min	70% ATZ removal	[42]
ATZ	Mn(IV)	O_3 dose: 2.28 mg/L; Catalyst dose: 1 mg/L; pH: 7; T: 23 °C; t: 5 min	37% ATZ removal	[42]
2,4-dinitrotoluene (DNT)	Mn(II)	O_3 dose: 5.6 mg/L; Catalyst dose: 0.2 mg/L Mn^{2+} and 4 mg/L OA; pH: 5.5; T: 25 °C; t: 15 min	65% DNT removal	[43]
2,4-dichlorophenol	Mn(II)	O_3 dose: 8.4 mg/L; Catalyst dose: 0.5 mg/L; pH: 5.5; T: 25 °C; t: 30 min	80% TOC removal	[44]
NTS	Mn(II)	O_3 dose: 1.04×10^{-4} mol/dm^{-3}; Catalyst dose: 1.25×10^{-4} mol/dm^{-3}; pH: 2; T: 25 °C; t: 30 min	72% NTS degradation rate	[30]
OA	Mn(II)	O_3 dose: 5×10^{-3} mol/L; Catalyst dose: 4 mg/L; pH: 2.5; T: 25 °C; t: 30 min	82% TOC removal	[37]
Formic acid	Mn(II)	O_3 dose: 5×10^{-3} mol/L; Catalyst dose: 4 mg/L; pH: 2.5; T: 25 °C; t: 35 min	61% TOC removal	[37]
Chlorobenzenes	Mn(II)	O_3 dose: 1.5 g O_3/TOC; Catalyst dose: 6×10^{-5} mol/L; pH: 7; T:NR; t: 20 min	66% COD removal	[32]
Lipid	Mn(II)	O_3 dose: 0.6 g/L; Catalyst dose: 3 mg/L; pH: 6.75; T: 25 °C; t: 60 min	93% lipid degradation	[31]
Simazine	Mn(II)	O_3 dose: 9.5 mg/L; Catalyst dose: 0.2 mg/L; pH: 7; T: 25 °C; t: 30 min	90% Simazine conversion	[45]

NR-value not reported, TOC total organic carbon, COD chemical oxygen demand.

Different kinds of transitional metals are used as homogeneous catalysts however, Fe(II) and Mn(II) have been reported to be the most efficient catalysts for water purification purposes. This trend in the research has underlying scientific logic. By focusing on the general features of these transition metals' chemistry, their wide range of oxidation states, and complex ion formation, this popularity for using them as a catalyst makes sense. Mn with an atomic number of 25 has the highest number of unpaired electrons in the d-subshell, and it shows variable oxidation states in its compounds such as (II) in Mn^{2+}, (III) in Mn_2O_3, (IV) in MnO_2, (VI) in MnO_4^{2-}, and (VII) in MnO^{4-}. Fe has two standard oxidation states, Fe^{2+} and Fe^{3+}, and a less common (VI) oxidation state in FeO_4^{2-}. Existing unpaired electrons and vacant orbitals in these transition metal structures enable them to accept electrons from other ions of molecules to form complex compounds. So this ability causes the adsorption of other substances onto their surface and activates them in the process, which is the objective function of a catalyst. The application of these two important transitional metals is explained in the following publications as examples.

Ramos et al. [31] evaluated the efficiency of the Fe(II) catalyst in the ozonation process for lipid degradation. Milk was chosen as the lipid source. It is observed that under neutral conditions, low catalyst dosages are enough to cause the almost complete degradation of lipids (96.7%). Fu et al. [33] investigated the homogeneous catalytic ozonation of AAF collector by coexisting transition metallic ions (Fe(II), Fe(III), Cu(II), Pb(II), and Zn(II)) in flotation wastewaters. Based on this research, the following order of the degradation rate was achieved: $O_3/Fe(II) > O_3/Fe(III) > O_3/Cu(II) > O_3/Pb(II) > O_3/Zn(II) \approx O_3$-alone. The best catalytic activity gained by Fe(II) had a 31.15% growth of degradation rate and achieved an increase of 42.26% for the AAF mineralization compared to O_3-alone. Xiao et al. [44] studied the mineralization of DCP in the ozonation process with Mn(II) as a catalyst. This study suggested that in the optimal condition of 0.5 (mg/L) catalyst dose and pH: 5.5, Mn(II) catalytic ozonation had a strong ability to degrade DCP and had 80% TOC removal in water solution.

In terms of the popularity of using this process, it is worth noting that most of the studies related to homogenous catalytic ozonation belong to the first decade of the 20th century. As can be observed in the table, the most significant disadvantage is that this catalytic process is mainly carried out in acidic pH values and not near the natural pH. At the same time, micropollutants, mostly emergent organic pollutants, usually exist in wastewater at the pH range of 6–8. Noting that although homogeneous catalytic ozonation processes can effectively improve the removal of organic contaminants in water in some cases, the addition of metal ions might result in secondary pollution, which causes limiting of their application.

However, in previous years, some scientists showed interest in using transitional metals for the catalytic ozonation process, working at the natural pH and real wastewater. Furthermore, some scientists solved the drawback of introducing these harmful metal ions in the aqueous environment by presenting the idea that some of these transition metallic ions usually coexisted in real wastewater. Several studies confirmed that Fe^{2+}, Cu^{2+}, Zn^{2+}, Pb^{2+}, and Co^{2+}, usually coexist with flotation reagents in the flotation pulp because of the dissolution of minerals or in the bastnaesite flotation pulp, some transition metal ions such as Fe^{2+}, Cu^{2+}, Zn^{2+}, and Pb^{2+} were determined [46,47]. Finally, Fu and colleagues [33] showed that these coexisting transition metallic ions can be used as in situ catalysts. So, it can be deduced that this process can have pleasing prospects by considering some improvements in the future.

4. Heterogeneous Catalytic Ozonation

Catalysts in solid form with high stability and efficiency were widely studied in catalytic ozonation systems. In this section, the activity and efficiency of heterogeneous catalysts in the ozonation process have been evaluated by focusing our attention on essential parameters such as:

- Chemical properties of catalysts: crystallographic and morphological, chemical stability.
- Physical properties of catalysts: point of zero charge (PZC), mechanical strength, surface area, pore volume, and porosity.
- Natural properties of target pollutants (pKa) and pH of the solution.

In each category of catalysts, these parameters are different to achieve the optimum efficiency in the water treatment process. Previous review articles [1,15,48] mentioned that the mechanism of catalytic ozonation is too complicated due to the contradictory catalytic mechanisms proposed by different research groups. In this article, by scrutinizing the governed mechanism of the process based on their conditions, this lack of understanding would have new interpretations.

Generally, the interaction between catalyst, pollutant, and O_3 determines the governing mechanism of this process. Moreover, each of these active components' behaviors depends on other factors and some of these have more influence than others. In the following, possible conditions based on the most influencing factors for each active component are expressed, and the governing mechanism in each specific condition is discussed.

Considering that the PZC is generally described as the pH value at which the net charge of the catalyst's surface is equal to zero.

The positively charged surface catalyst could be found under three different conditions which are proposed in Figure 3.

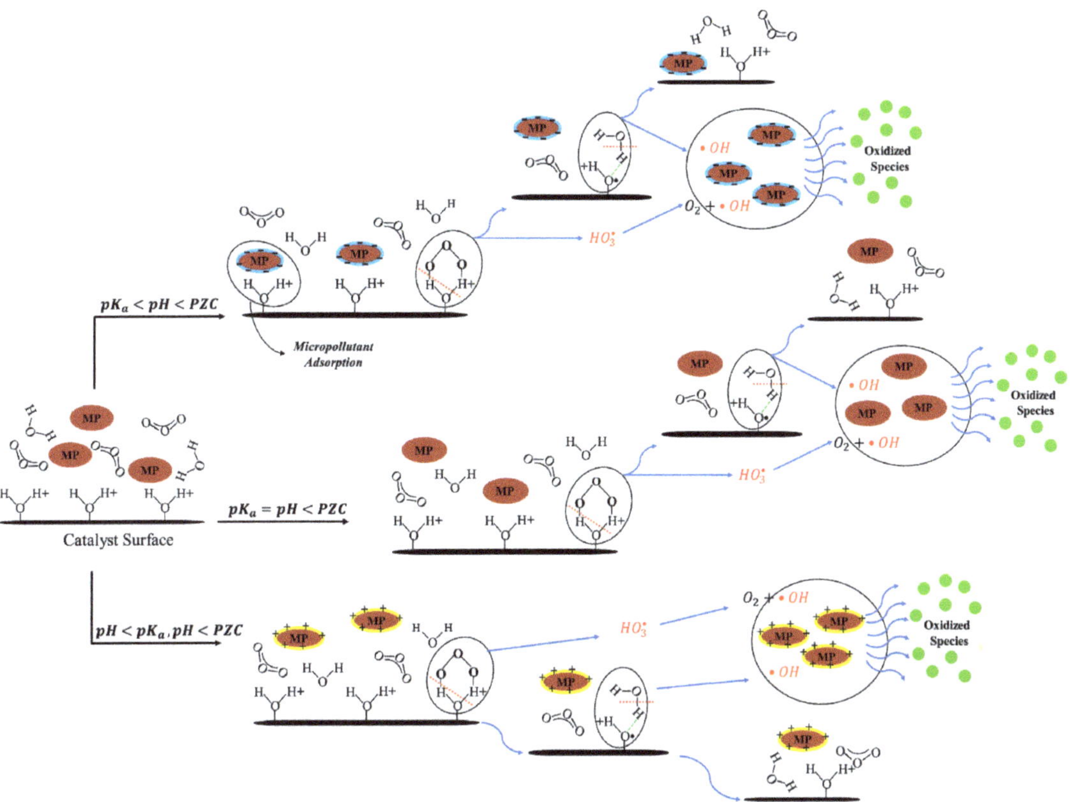

Figure 3. Proposed three influencing mechanisms using positively charged catalysts.

$pK_a < pH < PZC$: Catalyst is positively charged, pollutant is negatively charged.

In this condition, the negatively charged micropollutants can be adsorbed on the positively charged catalyst surface. Thus, the contaminants would be close to the area where the generation of HO$^\bullet$ radicals happens, which means HO$^\bullet$ can quickly oxidize them in the environment. On the other hand, by adsorbing the micropollutants on the catalyst surface, the active area for adsorption of O_3 will be limited, which has a negative effect on the HO$^\bullet$ generation in the environment.

$pH < PZC$, $pH = pK_a$: Catalyst is positively charged, pollutant is uncharged.

There is no effective interaction between the catalyst and micropollutant in this condition, so the generation of HO$^\bullet$ by O_3 decomposition is the only effective parameter here.

$pH < PZC$, $pH < pK_a$: Catalyst and pollutant are positively charged.

In these conditions, the HO$^\bullet$ radicals, which are highly useful in the oxidation of micropollutants, would be generated in a short time with the adsorption of O_3 on the catalyst surface. Adsorption of O_3 molecules by hydroxyl radical on the surface causes a generation of intermediate species ($OH_3^\bullet$) and another radical on the catalyst surface. The produced intermediate ($OH_3^\bullet$) turns into reactive $HO^\bullet$ radicals and O_2 in an in situ reaction. In parallel, the radical species on the catalyst surface would adsorb water molecules and produce reactive $HO^\bullet$ radicals. Due to the non-selectively behavior of $HO^\bullet$, it can oxidize almost all organic contaminants, which causes the high removal efficiency of micropollutants. Furthermore, based on the charge of micropollutants (pK_a), pulling of micropollutants or expelling of micropollutants on the catalyst surface might happen, and each of these conditions can affect removal efficiency. In this condition, the desorption of micropollutants from the catalyst would happen due to the repulsive electrostatic forces, which means the contaminants do not occupy the active surface sites, and pore blocking and associated fouling on the surface would be limited.

There are conditions that the catalyst is uncharged. Uncharged catalysts usually have hydroxyl radicals on the surface. Although these radicals are uncharged, they can be the starter part for O_3 decomposition. In the O_3 decomposition reaction chain, after the adsorption of O_3 on the surface, chemical bond stretching and breaking can happen in several ways. In one case, after bond breaking, $HO^\bullet$ and O_2 would be generated directly. In another case, some intermediate species such as $OH_3^\bullet$ and O_3^- would be produced. The produced intermediate ($OH_3^\bullet$) turns into reactive $HO^\bullet$ radicals and O_2 in an in situ reaction. In parallel, O_3^- causes other chain reactions, which finally produce $HO^\bullet$. Figure 4 proposed influencing mechanisms when the catalyst is uncharged.

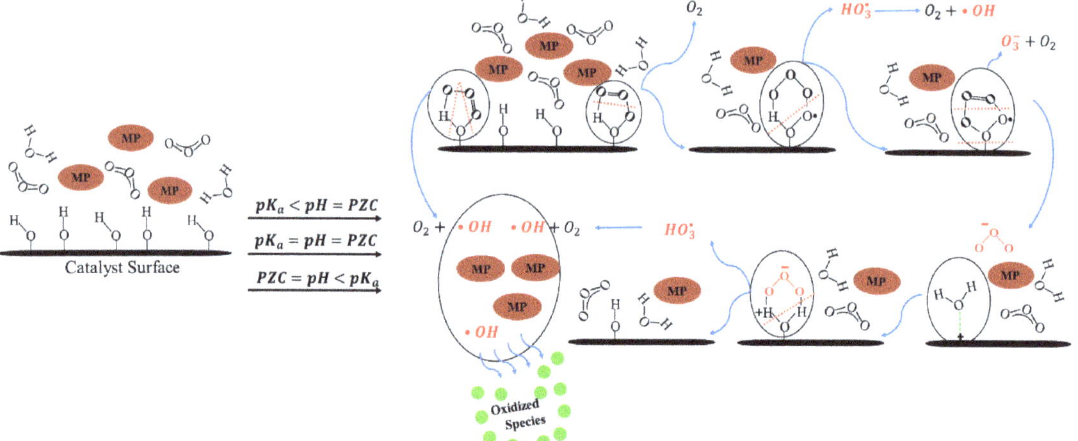

Figure 4. Proposed three influencing mechanisms using uncharged catalysts.

$pK_a < pH$, $PZC = pH$: Catalyst is uncharged, Pollutant is negatively charged.
$pH < pK_a$, $PZC = pH$: Catalyst is uncharged, Pollutant is positively charged.
$PZC = pH = pK_a$: Catalyst is uncharged, Pollutant is uncharged.

Due to the uncharged catalyst surface, there is no considerable difference between the three conditions in this subcategory.

Figure 5 summarizes three conditions related to the negatively charged surface of the catalyst. This situation seems to be the most favored for the adsorption and subsequent decomposition of O_3.

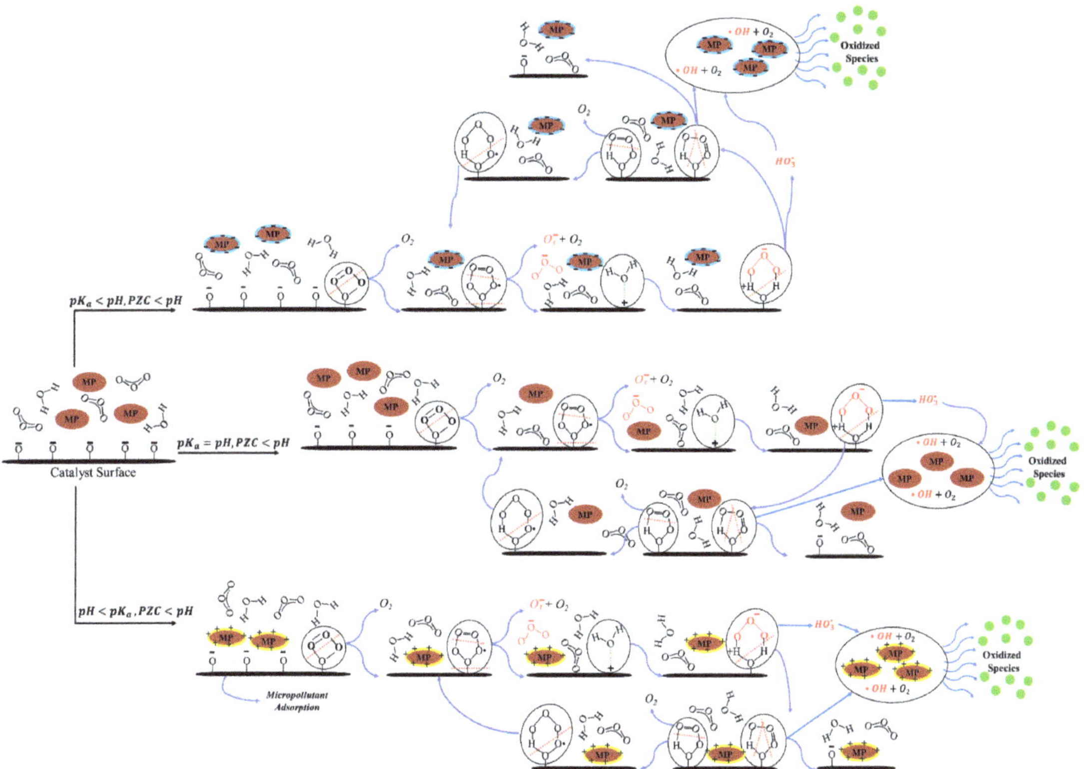

Figure 5. Proposed three influencing mechanisms using negatively charged catalysts.

In fact, the adsorption of O_3 on a surface would generate intermediate species such as O_3^- and $OH_3^\bullet$, considered as precursors of the high oxidant species $HO^\bullet$.

$PZC < pH < pK_a$: Catalyst is negatively charged, pollutant is positively charged.

Although in this condition, adsorption of O_3 is favored, adsorption of the pollutant on the catalyst surface can occur too. Thus, the pore blocking and associated fouling on the surface and reducing the active surface sites might have a negative effect on the $HO^\bullet$ generation in the environment. On the other side, closing to the area of $HO^\bullet$ radicals generation might cause quickly oxidize of micropollutants.

$PZC < pH,\ pH = pK_a$: Catalyst is negatively charged, pollutant is uncharged.

There is no effective interaction between the catalyst and micropollutants in this condition, so the generation of $HO^\bullet$ by O_3 decomposition is the only effective parameter here.

$PZC < pH$, $pK_a < pH$: Catalyst and pollutant negatively charged.

In this condition, the negatively charged micropollutants would be repulsed from the negatively charged catalyst surface due to the electrostatic forces. Thus, the micropollutants do not occupy the active surface sites.

The values of pKa for different pollutants reported in Table 3 have been examined in order to compare and better evaluate the fundamental mechanisms proposed in this section.

Table 3. Values of pKa for different pollutants.

Target Pollutants	pKa	Ref.	Target Pollutants	pKa	Ref.
Isoniazid	1.82	[49]	Ciprofloxacin (CPF)	6.38	[50]
Oxalic acid (OA)	pKa1 = 1.14; pKa2 = 3.64	[51]	4-nitrophenol	7.15	[52]
Phenacetin	2.2	[53]	Fluoxetine	8.7	[50]
Amoxicillin	2.4	[50]	Atenolol	9.16	[50]
Humic acids	2.5	[54]	Acetaminophen	9.38	[55]
Sulfamethazine	2.65	[50]	4-Chloro phenol	9.41	[52]
Salicylic acid	3.5	[50]	Paracetamol (PCT)	9.39–9.5	[50,56]
Methylene blue	3.8	[57]	4-Chloro-2-methyl	9.71	[52]
Reactive black-5	3.8	[58]	Phenol (PH)	9.98	[59]
Furosemide	3.9	[50]	m-cresol	10.1	[60]
Diclofenac	4.15	[50]	Bisphenol-A (BPA)	10.29	[61]
Naproxen	4.2	[50]	2,4-dimethylphenol	10.4	[52]
Ibuprofen	4.51	[50]	Orange (II)	11.4	[62]
Acetic acid	4.76	[63]	RR189	11.7	[64]
(SMX)	5.6–5.8	[65,66]	Carbamazepine	13.9	[50]
Naphthenic acid	5–6	[67]			

The PZC values for selected catalysts referring to the reviewed articles are shown in Table 4. Based on the type of catalysts, the pathway of the catalytic ozonation can also be varied. Although some catalysts exhibit PZC values in narrow ranges due to various impurity contents, synthesis routes, or thermal histories, this difference appears to be of little relevance to their catalytic properties.

Table 4. Range of PZC for selected catalysts.

Catalysts	PZC	Ref.	Catalysts	PZC	Ref.
SiO_2	2.6	[68,69]	$CoFe_2O_4$	7.31	[70]
MnO_2	3–5	[1,71]	Fe_3O_4 nanoparticles	7.4	[72]
MWCNTs	4.2	[68]	CeO_2	8.1	[73]
α-Al_2O_3	4.2	[74]	γ-Al_2O_3	8.3–8.9	[75,76]
AC	4.9	[68]	NiO	8.45	[77]
$NiCo_2O_4$	5	[78]	ZnO	9	[1]
FeOOH	5.9	[79]	α-Al_2O_3	9.4	[75]
Ce_3O_4	5–8	[69,80]	$MgFe_2O_4$	9.8	[81]
$CuFe_2O_4$	6–7	[53]	CuO	10	[82]
TiO_2	6.2–6.6	[1]	MgO	12–13	[83]
Fe_2O_3	6–9	[84]			

The main catalyst types applied in heterogeneous catalytic ozonation are metal/bimetal/polymetal oxides, metals or metal oxides on supports, and carbon-based materials. In each of these categories, the most popular catalysts and published works in recent years are summarized in this review.

4.1. Metal/Bimetal/Polymetal Oxides

Several metal oxides have been introduced to promote the heterogeneous catalytic ozonation processes. Some of these metal oxides are more popular than others in the catalytic ozonation process, such as Al_2O_3, MgO, CeO_2, MnO_2, NiO, ZnO, etc. Furthermore, some bimetal/polymetal oxides were widely applied due to their high stability as well as high catalytic activity, such as $CuFe_2O_4$, Mn-Ce-O, $ZnAl_2O_4$, etc. Table 5 compiles the literature results employing metal/bimetal/polymetal oxides for the degradation of pollutants in the catalytic ozonation process. As can be seen, operating conditions (pH, O_3 dosage, catalyst dose, etc.), the kind of used catalysts, target pollutants, and removal results are reported. The articles for summarizing in this part were chosen according to their high citations, and newness.

Table 5. Literature reports on different metal/bimetal/polymetal oxides as catalysts in the ozonation process (see Figures 3–5, respectively).

Catalysts	Target Pollutants	Operating Conditions		Removal Results	Year	Ref.
		pH < PZC	pH > pK_a, pH ≈ pK_a, pH < pK_a			
γ-Al$_2$O$_3$ PZC = 8.9 +	Ibuprofen pKa = 4.51 −	O$_3$ dose: 0.5 mg/min; Catalyst dose: 5 g; pH: 7.2; T: 20 °C; t: 30 min		83% ibuprofen removal	2015	[76]
CoFe$_2$O$_4$ PZC = 7.31 +	OA pKa1 = 1.14; pKa2 = 3.64	O$_3$ dose: 14 ± 1 mg/L; Catalyst dose: 1 g/L; pH: 2.3; T: NR; t: 120 min		68.3% TOC removal	2017	[70]
γ-Al$_2$O$_3$ PZC = 8.9 +	Cumene	O$_3$ dose: 0.5 mg/min; Catalyst dose: 5 g; pH: 7.2; T: 20 °C; t: 30 min		58% cumene removal	2015	[76]
MgO PZC = 12–13 +	Methylene blue pKa = 3.8 −	O$_3$ dose 5 mg/L; Catalyst dose: NR; pH: 9; T: NR; t: 60 min		50% COD removal	2016	[85]
γ-Al$_2$O$_3$ PZC = 8.9 +	1,2-dichlorobenzene	O$_3$ dose: 0.5 mg/min; Catalyst dose: 5 g; pH: 7.2; T: 20 °C; t: 30 min		45% 1,2 dichlorobenzene removal	2015	[76]
γ-Al$_2$O$_3$ PZC = 8.9 +	Acetic acid pKa = 4.76 −	O$_3$ dose: 0.5 mg/min; Catalyst dose: 5 g; pH: 7.2; T: 20 °C; t: 30 min		19% acetic acid removal	2015	[76]
Ce-O PZC = 8.5 +	CI Reactive Blue 5 (RB5)	O$_3$ dose: 50 g/Nm^3; Catalyst dose: 350 mg; pH: 5.6; T: 25 °C; t: 3 h		85% TOC removal	2009	[86]
γ-Al$_2$O$_3$ PZC = 8.3 +	Petroleum refinery wastewater	O$_3$ dose: 5 mg/min; Catalyst dose: 0.5 g; pH: 8.15; T: 30 °C; t: 40 min		45.9% COD removal	2017	[87]
MgO PZC = 12–13 +	Acetaminophen pKa = 9.38 +	O$_3$ dose: 50 mg/L; Catalyst dose: 2 g/L; pH: 5.4; T: NR; t: 10 min		100% ACT degradation	2017	[16]
MgO PZC = 12–13 +	RR198 pKa = 11.7 +	O$_3$ dose: 0.2 g/h; Catalyst dose: 5 g/L; pH: 8; T: 23 °C; t: 9 min		100% RR198 removal	2009	[88]
MgO PZC = 12–13 +	4-Chloro phenol pKa = 9.41 +	O$_3$ dose: 2.5 mg/min; Catalyst dose: 1.0 g/L; pH: 6.2; T: NR; t: NR		99.5% removal efficiency	2015	[17]
β-FeOOH PZC = 5.9 +	4- Chloro phenol pKa = 9.41 +	O$_3$ dose: 28.24 mg/L; Catalyst dose: 1 g/L; pH: 3.5; T: NR; t: 40 min		99% removal efficiency	2015	[89]
γ-Al$_2$O$_3$ PZC = 8.3–8.9 +	PCT pKa = 9.39–9.5 +	O$_3$ dose: 3 mg/min; Catalyst dose: 5 mg/L; pH: 7; T:NR; t: 9 min		98% PCT removal	2018	[90]
MgO PZC = 12–13 +	PH pKa = 9.98 +	O$_3$ dose: 0.25 g/h; Catalyst dose: 4 g/L; pH: 7; T: 25 °C; t: 80 min		96% PH removal, 70% COD removal	2010	[91]
γ-Al$_2$O$_3$ PZC = 8.3–8.9 +	Fluoxetine pKa = 8.7 +	O$_3$ dose: 30 mg/L; Catalyst dose: 1 g/L; pH: 7; T: 25 °C; t: 17 min		86% Fluoxtenine removal	2019	[92]
MgFe$_2$O$_4$ PZC = 9.8 +	Acid Orange II pKa = 11.4 +	O$_3$ dose: 5 mg/L; Catalyst dose: 0.5 mmol/L; pH: 4.6–9.6; T: 25 °C; t:NR		90% degradation efficiency	2016	[81]
NiO PZC = 8.45 +	Carbamazepine pKa = 13.9 +	O$_3$ dose: 5.5 g/m^3; Catalyst dose: 500 mg/L; pH: 3,4; T: 25 °C; t: 5 min		79.2% TOC removal	2020	[93]
CeO$_2$ PZC = 8.1 +	SMX pKa = 5.6–5.8 +	O$_3$ dose: 50 g/Nm^3; Catalyst dose: 100 mg; pH: 4.8; T: NR; t: 3 h		61% TOC removal	2013	[94]
γ-Al$_2$O$_3$ PZC = 8.3 +	2,4 dimethylphenol pKa = 10.4 +	O$_3$ dose: 2 g/Nm^3; Catalyst dose: 5 g/L; pH: 4.5; T: 25 °C; t: 300 min		57% TOC removal	2015	[95]
γ-Al$_2$O$_3$ PZC = 8.3–8.9 +	Landfill leachate	O$_3$ dose: 22 mg/min; Catalyst dose: 50 g/L; pH: 7.3; T: NR; t: 30 min		70% COD removal	2018	[96]
Al$_2$O$_3$ PZC = 7.2–9.2 +	Textile wastewater	O$_3$ dose: 0.9 mmol/L; Catalyst dose: 300 g; pH: 4; T: NR; t: NR		25.83% COD removal	2015	[97]

Table 5. *Cont.*

Catalysts	Target Pollutants	Operating Conditions		Removal Results	Year	Ref.
		pH ≈ PZC	pH > pK$_a$, pH ≈ pK$_a$, pH < pK$_a$			
NiCo$_2$O$_4$ PZC = 5 N	Sulfamethazine pKa = 2.65 −	O_3 dose: 4.5 mg/min; Catalyst dose: 0.05 g/L; pH: 5.2; T: NR; t: 60 min		34.1% TOC removal	2021	[78]
α-MnO$_2$ PZC = 3–5 N	4-Nitrophenol pKa = 7.15 +	O_3 dose: 20 mg/L; Catalyst dose: 100 mg, pH 3.5–5.9; T: NR; t: NR		96.7% degradation 79.5% TOC removal	2015	[98]
		pH > PZC	pH > pK$_a$, pH ≈ pK$_a$, pH < pK$_a$			
α-Al$_2$O$_3$ PZC = 4.2 −	Humic acids pKa = 2.5 −	O_3 dose: 0.063 m^3/h; Catalyst dose: 0.5 g/L; pH: 5.5; T: 25 °C; t: 1 h		100% Humic acid removal	2020	[74]
γ-Al$_2$O$_3$ PZC = 8.3–8.9 −	CPF pKa = 6.38 −	O_3 dose: 1.4 mg/L.min; Catalyst dose: 0.55 g/L; pH: 9.5; t: 60 min		93% removal efficiency	2019	[99]
Ca$_2$Fe$_2$O$_5$ PZC = 9.5 −	Quinoline	O_3 dose: 17 mg/L; Catalyst dose: 1 g; pH: 10.5; T: 25 °C; t: 120 min		92% COD removal	2022	[100]
CeO$_2$–MnO$_2$ PZC = 10.13 −	Ammonium pKa = 9.25 −	O_3 dose: 12 mg/min; Catalyst dose: 1 g/L; pH: 11; T: NR; t: 60 min		88.14% Ammonium removal	2022	[101]
γ-Al$_2$O$_3$ PZC = 8.3–8.9 −	Naphthenic acid pKa = 5–6 −	O_3 dose: NR; Catalyst dose: 1 g/L; pH: 8.5; T: 25 °C; t: 50 min		88% Naphthenic acids removal	2019	[102]
CuFe$_2$O$_4$ PZC = 6–7 −	Phenacetin pKa = 2.2 −	O_3 dose: 0.36 mg/L; Catalyst dose: 2.0 g/L; pH: 7.72; T: NR; t: 30 min		95% of degradation	2015	[53]
α-MnO$_2$ PZC = 3–5 −	BPA pKa = 10.29 +	O_3 dose: 4.47 mmol/min; Catalyst 0.1 mg/L; pH: 6.25; T: 20 °C; t: NR		93.5% removal efficiency	2015	[103]

NR—value not reported, TOC—total organic carbon, COD—chemical oxygen demand.

In Table 5, the PZC values of the metal oxide catalysts with the pKa of the pollutants in different operating conditions are correlated in order to identify the most favorable combinations for their removal.

From this comparison, it can be seen that the best removal efficiency trend is provided for experimental pH values in which the catalyst surface is charged. The electrostatic repulsion between pollutant and catalyst allows rapid O_3 decomposition, as well as the formation of radical species that occurs in the proximity of the pollutant adsorbed on the surface of the catalyst, seem to be the most effective mechanisms.

Al$_2$O$_3$ is one of the most popular materials in the catalytic ozonation process, among which several studies performed at pH close to neutral confirm it as a catalyst with excellent removal yields. The PZC value of Al$_2$O$_3$ can be different due to the catalyst's various impurities content, synthesis route, or thermal history, but the approximate range of PZC value is 7.2–9.4 [75,90,97]. Scrutinizing the study of Ziylan-Yavaş et al. [90] that had more than 95% removal of PCT (pKa ≈ 9.5 (Table 3)) by using γ-Al$_2$O$_3$ (PZC = 8.3–8.9 (Table 4)) as a catalyst verified that the optimal condition was when pH of the solution was lower than pKa and PZC. In these conditions, both catalyst and pollutant are positively charged. So, the governing mechanism was the repulsive electrostatic forces resulting in the desorption of the micropollutant from the catalyst which does not occupy the active surface sites, favoring the O_3 adsorption and its decomposition. Nemati Sani et al. [99] studied the catalytic efficiency of Al$_2$O$_3$ for CPF degradation in the catalytic ozonation process. Based on their work, the highest removal efficiency was at pH = 9.5. In this condition PZC < pH, pK$_a$ < pH, which means both catalyst and pollutant are negatively charged. So, as previously mentioned, in this condition there is not any effective adsorption of pollutants on the surface of the catalyst, and the primary mechanism is related to O_3 decomposition. In explaining their work, they clearly mentioned that ozonation is responsible for 88% of the CPF removal efficiency, which is another proof of the accurate understanding of govern mechanism categories in this review.

MgO is another efficient metal oxide that had excellent results in micropollutants degradation. The PZC value of MgO is in the range of 12–13 approximately [83]. With this PZC in a wide range of solution pH, MgO is positively charged. So oxidation of micropollutants due to the generation of HO$^\bullet$ radicals in the solution is the governing mechanism for catalytic ozonation by using this metal oxide. Based on using MgO as a catalyst, several works with considerable target pollutants removal are reported. Mashayekh-Salehi et al. [16] achieved complete degradation of acetaminophen in only 10 min of reaction time in the catalytic ozonation process. Based on our categorization, this work is in pH < PZC, pH < pK$_a$ condition. So, as previously mentioned, there is no effective adsorption of pollutants on the catalyst's surface due to the similar charges (positive charges). The primary mechanism is related to the oxidation of micropollutants due to the generation of HO$^\bullet$ radicals in the solution. It is worth mentioning, the authors confirmed that the reaction with HO$^\bullet$ radical was the leading cause of ACT oxidation using the MgO/O$_3$ process. The same condition (pH < PZC, pH < pK$_a$) was applied by Moussavi et al. [88], Kong et al. [17], and Mousavi et al. [91]. The exciting results of comparing these works was that they all achieved more than 95% degradation of their target pollutant, which is almost the top result for the catalytic ozonation process. This observation may prove that MgO is one of the best catalysts for the ozonation process, and this condition (pH < PZC, pH < pK$_a$) may be one of the best catalytic ozonation conditions, especially for the removal of very weak acid pollutants.

Furthermore, some scientists indicated that the catalytic mechanism and removal efficiency highly depend on the catalyst's characteristics. In this regard, introducing bimetal/polymetal oxides and using different synthesis routes with better chemical and physical properties was another way of improving this field. The proposed catalytic ozonation mechanisms are also rationalized for these kinds of catalysts. Oputu et al. [89] studied catalytic ozonation activity using β-FeOOH as a catalyst and 4-chloro phenol as the target pollutant. The pH of the solution was 3.5, so based on reported PZC (Table 4) and pKa (Table 3), both the pollutant and catalyst surface was a positive charge. According to the governing mechanism, the O$_3$ adsorption then its decomposition are favored, and there is no adsorption of pollutants on the surface due to the same positive charges. The interesting results based on this work were that in the presence of O$_3$ and catalyst, the removal efficiency was 99% (almost complete removal); however, in the absence of O$_3$, using only catalyst, the removal efficiency was 3%. This thought-provoking achievement shows that O$_3$ decomposition is the main effecting parameter in this condition; furthermore, the effect of pollutant adsorption on the catalyst surface is negligible.

Other metal oxide catalysts need to be mentioned, such as magnetic Fe$_3$O$_4$ nanoparticles [104] and δ-MnO$_2$ [105]. These metal oxides have shown good removal efficiencies but have not been reported in Table 5 as their pH values are not reported in the articles so their unclear process conditions do not allow to frame them in the proposed mechanism.

After perusing a lot of published articles from previous years, it is important to highlight that due to a large number of studies on these materials and the development of our understanding and knowledge about them, these materials can now be used for real wastewater plants. By scrutinizing the trend of published articles, changes in applications from laboratory scale to real wastewater plant can be observed.

4.2. Metal/Metal Oxides on Supports

This kind of catalyst was prepared by loading metal or metal oxides on supporters with unique surface properties due to increasing the catalytic activity of metal/metal oxides in the ozonation process. By applying this kind of material, both the surface area and the active sites of the materials would be increased. By the combination of metals/metal oxides and supporters, the PZC value of the prepared catalyst shifts to more acidic/basic, resulting in a new PZC value. This PZC change is attributed to the complexation of loaded metal/metal oxides onto the support surface. So, the kind of loading materials is important in this case. The following will clarify the governing mechanisms by scrutinizing several

Catalysts 2023, 13, 324

related articles in this field. Table 6 systemizes the literature that employed metal/metal oxides as supports for the degradation of pollutants in the catalytic ozonation process. Operating conditions (pH, O_3 dosage, catalyst dose, etc.), the catalysts used, target pollutants, and removal results are reported.

Table 6. Literature reports on different metals/metal oxides on support as catalysts in the ozonation process (see Figures 3–5, respectively).

Catalysts	Target Pollutants	Operating Conditions	Removal Results	Year	Ref.
		pH < PZC pH > pK_a, pH ≈ pK_a, pH < pK_a			
$Ce_{1.0}$ $Fe_{0.9}$ OOH PZC = 7.8 +	Sulphamethazine pKa = 2.65 −	O_3 dose: 15 mg/L; Catalyst dose: 0.2 g/L; pH 7.3; T: NR; t: 15 min	41.2% mineralization efficiency	2016	[106]
$LaTi_{0.15}Cu_{0.85}O_3$ PZC = 9.8 +	SMX pKa = 5.6–5.8 −	O_3 dose: 25 mg/L; Catalyst dose: NR; pH: 7; T: 20 °C; t: 2 h	85% TOC removal	2009	[107]
Ce deposited magnetic pyrite cinder PZC = 7.63 +	Reactive black-5 pKa = 3.8 −	O_3 dose 5.6 mg/min; Catalyst dose: 2.5 g/L; pH: 5.5; T: NR; t: 2 h	83.32% TOC removal efficiency	2016	[69]
$Ca-C/Al_2O_3$ PZC = 9.53 +	High-salt organic wastewater	O_3 dose: 12 mg/L; Catalyst dose: 20 g; pH: 8.36; T: NR; t: 40 min	64.4% COD removal efficiency	2022	[108]
$Mn-CeOx@\gamma-Al_2O_3$ PZC = 9.37 +	CPF pKa = 6.38 −	O_3 dose: 13.969 ± 0.434 mg/L; Catalyst dose: 80 g/L; pH: 8.5; T: NR; t: 60 min	100% CPF removal	2022	[109]
Mg-doped ZnO PZC = 11–11.2 +	Isoniazid pKa = 1.82 −	O_3 dose: 10–25 mg/L; Catalyst dose: 0.1 g/L; pH: 7.2; T: NR; t: 9 min	76.3% removal efficiency	2020	[49]
$MnOx/SBA-15$ PZC = 4.27–6.35 +	Clofibric acid pKa = 3.2 −	O_3 dose: 1 mg/L; Catalyst dose: 0.2 g/L; pH: 3.85; T: 293 K; t: 15 min	43.8% TOC removal	2015	[20]
Fe-SBA-15 PZC = 4.0 +	OA pKa1 = 1.14; pKa2 = 3.64 N/−	O_3 dose: 100 mg/h; Catalyst dose: 0.24 g; pH: 3; T: NR; t: 60 min	86.6% removal efficiency	2016	[21]
Fe-MCM-41 PZC = 4.95 +	OA pKa1 = 1.25; pKa2 = 3.81 N/−	O_3 dose: 21.8 mg/L; Catalyst dose: 0.4 g; pH: 3.6; T: NR; t: 30 min	94% degradation 6% TOC/TOC_0 reduction	2017	[110]
$SnOx-MnOx@Al_2O_3$ PZC = 8.7 +	PH pKa = 9.98 +	O_3 dose: 6 mg/L.min; Catalyst dose: 40 g/L; pH: 7; T: 20 ± 5 °C; t: 240 min	93.8% COD removal efficiency	2022	[111]
		pH ≈ PZC pH > pK_a, pH ≈ pK_a, pH < pK_a			
$4\%Mn/\gamma-Al_2O_3$ PZC = 6.75 N	PH pKa = 9.98 +	O_3 dose: 8 mg/min; Catalyst dose: 0.20 g; pH: 6.5; T:15 °C; t: NR	82.67% degradation efficiency	2016	[18]
$MnOx-0.013/KCC-1$ PZC = 4.0 N	OA pKa1 = 1.14; pKa2 = 3.64 −	O_3 dose: 20 mg/L; Catalyst dose: 0.25 g/L; pH: 3.8; T: 25 °C; t: NR	86% TOC removal	2016	[112]
$\gamma-Ti-Al_2O_3$ PZC = 7.3 N	SMX pKa = 5.6–5.8 −	O_3 dose: 30 mg/Nm^3; Catalyst dose: 1.5 g; pH: 7; T: NR; t: 1 h	92% TOC removal	2017	[113]
		pH > PZC pH > pK_a, pH ≈ pK_a, pH < pK_a			
$Cu–O–Mn/\gamma-Al_2O_3$ (CMA) PZC = 7.9 −	Polyvinyl alcohol pKa = 5–6.5 −	O_3 dose: 5.52 mg/L.min; Catalyst dose: 150 mg/L; pH: 10, T:25 °C; t: 10 min	99.3% PVA removal	2020	[19]
Fe-MCM-41 PZC = 4.85 −	Diclofenac pKa = 4.15 −	O_3 dose: 100 mg/h; Catalyst dose: 1 g/L; pH: 7; T: NR; t: 30 min	76.3% mineralization 70% TOC reduction	2016	[114]
Cu/Al_2O_3	Herbicide Alachlor	O_3 dose: 12.2 mg/L.min; Catalyst dose: 0.27 g/L; pH: 4.4; T: 20 °C; t:NR	75% TOC removal	2013	[115]
FeMn-MCM-41	methyl orange	O_3 dose: 35 mg/L; Catalyst dose: 0.2 g; pH: 7; T: 25 °C; t:NR	78% TOC removal	2021	[116]
$MnOx/SBA-15$ PZC = 5.33–6.06	Norfloxacin	O_3 dose: 100 mg/h; Catalyst dose: 0.1 g/L, Catalyst loading 2%; pH: 5; T: 298 K; t: 60 min	54% mineralization efficiency	2017	[117]

Table 6. *Cont.*

Catalysts	Target Pollutants	Operating Conditions	Removal Results	Year	Ref.
MnO$_2$/Al$_2$O$_3$	Quinoline	O$_3$ dose: 135 mg/L; Catalyst dose: 70 g/L, 8% MnO$_2$ loading; pH: NR; T: NR; t: 90 min	95% quinoline removal 65% TOC removal	2021	[118]
Fe silicate-loaded pumice	Diclofenac pKa = 4.15	O$_3$ dose: 5 g/L; Catalyst dose: 8 g/L; pH 5; t: 30 min	73.3% mineralization 21.17% TOC removal	2017	[119]
MOF-derived Co$_3$O$_4$–C composite	Norfloxacin	O$_3$ dose: 15 mg/L; Catalyst dose: 0.05 g/L; pH: 6.7; T: NR; t: NR	48% TOC removal	2021	[25]
Fe silicate doped FeOOH	p-Chloronitro benzene	O$_3$ dose: 0.6 mg/L; Catalyst dose: 500 g/L; pH: 7; T: NR; t: 15 min	61.3% TOC removal	2017	[120]
Ni/NHPC	linear alkylbenzene sulfonate	O$_3$ dose: NR; Catalyst dose: 0.3 g; pH: 10; T: NR; t: 30 min	98.1% LAS removal	2021	[121]
CuO/SiO$_2$	Oxalate	O$_3$ dose: 4 mg/L; Catalyst dose: 0.5 g, 4.5% metal loading; pH: 7; T: NR; t: NR	95% oxalate removal	2019	[122]
Pt-Al$_2$O$_3$	PCT pKa = 9.4–9.5	O$_3$ dose: 6 mg/L; Catalyst dose: 5 mg/L; pH: 7; T: NR; t: 5 min	100% PCT removal	2018	[90]
MnO$_2$/CeO$_2$	Sulfosalicylic acid	O$_3$ dose: 4 mg/min; Catalyst dose: 0.1 g; pH: 6.5; T: NR; t: 30 min	97% TOC removal	2016	[123]
Fe$_3$O$_4$/Co$_3$O$_4$	SMX pKa = 5.6–5.8	O$_3$ dose: 6 mg/L; Catalyst dose: 0.1 g/L; pH: 5.1; T: 25 °C; t: NR	60% TOC	2019	[124]
Mn-Ce-O	SMX pKa = 5.6–5.8	O$_3$ dose: 120 mg/h; Catalyst dose: 1 g/L; pH: 6.9; T: NR; t:120 min	69% COD removal	2015	[125]
CuO/Al$_2$O$_3$-EPC	SMX pKa = 5.6–5.8	O$_3$ dose: 4.75 µM; Catalyst dose: 0.5 g/L; pH: 6.2; T: 20 °C; t: 15 min	87% SMX removal 21.2% TOC removal	2019	[126]
Fe^{2+}-Montmorillonite	SMX pKa = 5.6–5.8	O$_3$ dose: 5 mg/min; Catalyst dose: 1 g/L; pH: 2.88; T: NR; t: 20 min	97% COD removal	2015	[127]

NR—value not reported, TOC—total organic carbon, COD—chemical oxygen demand.

Wang et al. [18] studied the enhancement of PH removal in the catalytic ozonation process by introducing an Mn-doped Al$_2$O$_3$ nanocatalyst. Based on the Mn weight ratios (2 wt.% Mn/γ-Al$_2$O$_3$, 4 wt.% Mn/γ-Al$_2$O$_3$, 8 wt.% Mn/γ-Al$_2$O$_3$), the PZC values were measured. The study showed that by increasing the amount of loaded Mn, the PZC values of the catalyst decreased from 7.31 to 6.75 to 5.54, respectively. Studying the effect of Mn loaded, they observed that at the natural pH (pH = 6.5), 4 wt.% Mn/γ-Al$_2$O$_3$ showed the best efficiency on PH (pKa $\approx$ 9.98 (Table 3)) removal. It means O$_3$ decomposition is happening in the environment according to our suggested mechanism. Yan et al. [19] used γ-Al$_2$O$_3$ support but with another modification of the surface. In this study, Cu–O–Mn/γ-Al$_2$O$_3$ was used as a catalyst in the ozonation system for the degradation of polyvinyl alcohol. Based on their investigation related to PZC measurement, the PZC of γ-Al$_2$O$_3$ decreased from 8.4 to 7.9 after loading Cu and Mn. They mentioned that the optimal condition is when the PZC of the catalyst and pH of the solution are the same or the pH is more than PZC. In these conditions, the governing mechanism is related to the existing hydroxyl groups on the surface of the catalyst, the decomposition of O$_3$, and the generation of reactive oxidation species. In another work, Bing et al. [113] studied the mineralization of pharmaceuticals over γ-Ti-Al$_2$O$_3$ catalyst. The exciting part related to this article was scanning the surface reaction mechanism. Using in situ Raman spectroscopy, they characterized intermediate species formed on the γ-Ti-Al$_2$O$_3$ surface with an aqueous O$_3$ solution. They observed the appearance of two new peaks at 880 and 930 cm^{-1} in γ-Ti-Al$_2$O$_3$ suspension with O$_3$, which were related to surface peroxide (O$_2$$^{\bullet}$) and surface atomic oxygen species (O$^{\bullet}$), respectively. As mentioned before in the governing mechanism explanation, these species would generate when the PZC of the catalyst is the same as the pH of the solution. They reported that the PZC value for this synthesized catalyst was 7.3 and the pH of the solution was 7. This work is an excellent observation for reliability even the details on intermediate species of our proposed govern mechanism.

One of the prominent supporting materials for catalysts is silica-based materials (such as SiO$_2$, SBA-15, MCM-41, etc.) due to their large surface area, good flexibility, stability, adjustable structure, and biocompatibility. The following examples comprehensively illustrate the diversity of using these materials in catalytic ozonation.

Jeirani et al. [110] worked on a modified mesoporous Fe-MCM-41 catalyst to remove OA as a target pollutant. Their evaluation of the adsorption and ozonation process and the determination of PZC and pKa was thought-provoking. They specifically describe the dissociation of OA in water. OA (target pollutant) was ionized to hydrogen-oxalate anion ($C_2O_4H^-$), having a negative charge on the surface. On the other hand, the PZC value for Fe-MCM-41 and Mn, Ce/Fe-MCM-41 were measured (4.95 and 6, respectively); accordingly, the catalyst's surface was positively charged (pH < PZC). Then by comparing the adsorption and ozonation efficiency of the catalyst, they observed that adsorption was the governing mechanism for this treatment system, which is in line with our previous explanation of the mechanism.

Yan et al. [21] applied another silica-based material for OA removal. They found Fe-doped SBA-15 (PZC = 4) as a potential catalyst for the catalytic ozonation process. They studied the influence of initial pH in the range of 1 to 9 on the removal process. Based on their results and the known information about the pKa value of OA and PZC value of Fe-SBA-15, the best pH was 3 with 97.4% OA removal efficiency. This achievement affirms that the negatively charged micropollutants adsorbed on the positively charged catalyst surface. Thus, the contaminants were near where the generation of $^•$OH radicals happens, which means $^•$OH can quickly oxidize them in the environment. This mechanism is similar to the previous example of OA removal from wastewater.

Shen et al. [49] studied a new Mg-doped ZnO catalyst. By modifying the catalyst surface, the value of PZC was changed from 8 for ZnO support to 11.2 for the Mg-ZnO catalyst. Furthermore, by increasing the doping amount of Mg, the PZC value was increased. Interestingly, by changing the pH value from 3 to 9, the efficiency of the catalyst did not noticeably change because all of those pH values were less than the PZC of the catalyst (pH < PZC), so the reaction mechanism for all of those conditions were the same. Furthermore, their explanations related to the catalytic mechanism, surface charge, and kind of active radicals were another validation of our suggested governing mechanism.

For many of these supported catalysts, the lack of studies on the new PZC values makes it difficult to predict their mechanism of action, although they show high removal efficiency as reported in Table 6.

4.3. Carbon-Based Materials

Applying carbon-based materials has received significant popularity in the catalytic ozonation process during past decades. The variety of these materials, their good catalytic performance, and their environmentally friendly properties make them very attractive for these studies. Bulky carbons (such as activated carbon (AC)), nano carbons (such as carbon nanotubes (CNTs) and graphene), and carbon-based nanocomposites are the three most popular materials in catalytic ozonation. Table 7 compiles the literature results employing carbon-based materials for the degradation of organics. As can be seen, the systems have been investigated over a wide range of operating conditions (pH, O_3 dosage, the mass ratio between solid and organic matter load, etc.), the kind of used catalysts, target pollutants, and removal results. The articles for summarizing this part were chosen according to their high citations.

Table 7. Literature reports on different carbon-based materials as catalysts in the ozonation process (see Figures 3–5, respectively).

Catalysts	Target Pollutants	Operating Conditions	Removal Results	Year	Ref.
		pH < PZC $\quad$ pH > pK$_a$, pH ≈ pK$_a$, pH < pK$_a$			
Fe/AC PZC = 7.95 +	Crystal violet dye pKa1 = 1.15 pKa2 = 1.8 −	O$_3$ dose: 4.44 mg/min; Catalyst dose: 2.5 g/L; pH: 7; T: NR; t: NR	>96% decolorization	2015	[128]
AC PZC = 8.5 +	SMX pKa = 5.6–5.8 +	O$_3$ dose: 50 g/Nm^3; Catalyst dose: 100 mg; pH: 4.8; T: NR; t: 3 h	45% TOC removal	2011	[129]
AC PZC = 8.5 +	SMX pKa = 5.6–5.8 +	O$_3$ dose: 48 mg/L; Catalyst dose: 2 g/L; pH: 5; T: 26 °C; t: 20 min	78% TOC removal	2011	[130]
Treated Commercial MWCNT-HNO$_3$-N$_2$-900 PZC = 7.3 +	SMX pKa = 5.6–5.8 +	O$_3$ dose: 50 g/Nm^3; Catalyst dose: 100 mg; pH: 4.8; T: NR; t: 3 h	45% TOC removal	2013	[131]
Treated Commercial MWCNT-O$_2$ PZC = 5.2 +	SMX pKa = 5.6–5.8 +	O$_3$ dose: 50 g/Nm^3; Catalyst dose: 100 mg; pH: 4.8; T: NR; t: 3 h	41% TOC removal	2013	[131]
MWCNT PZC = 7 +	SMX pKa = 5.6–5.8 +	O$_3$ dose: 50 g/Nm^3; Catalyst dose: 100 mg; pH: 4.8; T: NR; t: 3 h	35% TOC removal	2011	[130]
AC$_0$ PZC = 8.5 +	RB5	O$_3$ dose: 50 g/Nm^3; Catalyst dose: 350 mg; pH: 5.6; T: 25 °C; t: 2 h	70% TOC removal	2009	[86]
AC$_0$-Ce-O composite PZC = 8.5 +	RB5	O$_3$ dose: 50 g/Nm^3; Catalyst dose: 350 mg; pH: 5.6; T: 25 °C; t: 2 h	100% TOC removal	2009	[86]
		pH ≈ PZC $\quad$ pH > pK$_a$, pH ≈ pK$_a$, pH < pK$_a$			
MnOx/sewage sludge-derived AC PZC = 3.5 N	OA pKa1 = 1.14; pKa2 = 3.64 N/−	O$_3$ dose: 5 mg/L; Catalyst dose: 100 mg/L, Catalyst loading 30%; pH: 3.5; T: NR; t: 60 min	92.2% removal efficiency	2017	[132]
Fe-MnO$_X$/AC PZC = 6.1 N	PH pKa = 9.9 +	O$_3$ dose: 60 mg/L; Catalyst dose: 1 g/L; pH: 6; T: NR; t: 20 min	90.75% TOC removal	2022	[133]
		pH > PZC $\quad$ pH > pK$_a$, pH ≈ pK$_a$, pH < pK$_a$			
AC Darco 12–20 PZC = 6.4 −	SMX pKa = 5.6–5.8 −	O$_3$ dose: 25 mg/L; Catalyst dose: NR; pH: 7; T: 20 °C; t: 2 h	92% TOC removal	2012	[107]
AC/nano-Fe$_3$O$_4$ PZC = 6.08–7.7 −	PH pKa = 9.9 +	O$_3$ dose: 33 mg/L.min; Catalyst dose: 2 g/L; pH: 8; T: NR; t: 60 min	98.5% PH removal 69.8% COD removal	2014	[134]
CeO$_2$/MWCNT	SMX	O$_3$ dose: 50 g/Nm^3; Catalyst dose: 100 mg; pH: 4.8; T: NR; t: 3 h	56% TOC removal	2013	[94]
Fe$_2$O$_3$/CeO$_2$ loaded AC (MOPAC)	SMX	O$_3$ dose: 48 mg/L; Catalyst dose: 2 g/L; pH: 5; T: 26 °C; t: 20 min	86% TOC removal	2011	[130]
rGO	p-Hydroxylbenzoic Acid (PHBA)	O$_3$ dose: 20 mg/L; Catalyst dose: 0.1 g/L mg; pH: 3.5; T: 25 °C; t: 30 min	100% PHBA removal	2016	[135]
α-MnO$_2$/RGO	BPA	O$_3$ dose: 4.47 mmol/min; Catalyst dose: 0.1 mg/L; pH: 6.25; T: 20 °C; t: 60 min	90.5% BPA removal	2015	[103]
GO/Fe$_3$O$_4$	ρ-chlorobenzoic acid (pCBA)	O$_3$ dose: 4 mg/L; Catalyst dose: 20 mg/L; pH: 7; T: NR; t: 5 min	51% TOC removal	2018	[136]
Heteroatom doped graphene oxide PGO	SMX	O$_3$ dose: 2 g/h; Catalyst dose: 1 g/L; pH: 9; T: 25 °C; t: 5 min	99% SMX removal	2017	[137]

NR—value not reported, TOC—total organic carbon, COD—chemical oxygen demand.

As it is apparent in Table 7, AC or carbon black is one of the most used catalysts for the ozonation process. High porosity, surface functionalities, its low cost are the reasons for its extensive utilization. The PZC value of AC can be different due to the catalyst's various impurities content, synthesis route, or thermal history, and the method for investigation of PZC, while the reported range of PZC value for AC is between 4.9–11.9 [68,138]. On the other hand, commercially available AC can be modified by minerals such as alkali metals

(Ca, Na, K, Li, Mg) or multivalent metals (Al, Fe, Ti, Si) and metal oxides. The presence of impurities on the surface of AC would significantly affect the PZC values. In most articles, the authors reported this value for the AC used in their work.

Shahamat et al. [134] studied a new carbon-based catalyst called AC/nano-Fe_3O_4 to remove PH. During this study, they calculated the PZC of the catalyst and illustrated that when the pH of the solution is between PZC and pKa, the negative catalytic charge and positive charge of the pollutant can attract each other on the surface of the catalyst. O_3 decomposition is the primary reaction mechanism when the catalyst has a negative charge. These conditions were responsible for achieving the optimal efficiency for PH removal. In another work, Huang et al. [132] synthesized MnOx/sewage sludge-derived AC (MnOx/SAC) to improve the catalytic efficiency of OA degradation in ozonation. Based on the report, the best organic contaminant removal was at the pH equal to the PZC of the catalyst (PZC = 3.5), and the governing mechanism was related to existing hydroxyl radicals on the catalyst's surface which is the starter part for O_3 decomposition. Synthesized Fe-loaded AC for dibutyl phthalate removal was another work by Huang et al. [139], which had the same result that an uncharged surface with hydroxyl radicals on the surface was more active than the charged surface.

CNTs and MWCNTs are used frequently in the catalytic ozonation process as the mixed mesoporous structured nanocarbons. Acceleration of reaction kinetics, rapid mass diffusion, large surface area, and facile modification of surface are the main advantages of this kind of catalyst. Several techniques were used to promote the catalytic activity of this material, such as substituting carbon atoms with metal free heteroatoms (e.g., N, S, and F).

Gonçalves et al. [131] studied the effect of MWCNTs on the catalytic ozonation of SMX (pKa ≈ 5.6–5.8 (Table 3)). A set of modified MWCNTs with different levels of acidity/basicity was prepared and tested. The PZC value of the original MWCNT was 7; however, by modification of the original catalyst, the amount of PZC was changed to more acidic and basic. Based on the results, all those catalysts illustrated excellent efficiency for SMX removal, but one of the modified MWCNTs, called MWCNT-HNO$_3$_N$_2$_900 with PZC 7.3, illustrated better catalytic efficiency than the others. Based on our categorization and the observation in this work, at pH < PZC, pH < pKa condition, there is no effective adsorption of pollutants on the catalyst's surface due to the similar charges (positive charges). So the primary mechanism is related to the oxidation of micropollutants due to the generation of HO• radicals in the solution.

Graphene oxide (GO) and reduced graphene oxide (rGO) are other prominent catalysts that have been extensively employed to accelerate the degradation of various contaminants by O_3. Scrutinizing the study of Wang et al. [135] that had complete PHBA removal (pKa ≈ 4.85) by using rGO (PZC = 4.7) as a catalyst verified that the optimal condition was at the pH of 3.5. As mentioned before, in the condition that pH < PZC, pH < pKa, both catalyst and pollutant are positively charged, which leads to no adsorption on the system so that the primary mechanism would be related to the generation of HO• by O_3 decomposition.

4.4. Metal–Organic Frameworks (MOFs)

As a rapidly emerging category of porous materials, metal–organic frameworks (MOFs) are widely used in different research fields due to their unique topology, adjustable features, large surface area, ultrahigh porosity, and ease of access to numerous functional groups [27,140–143]. The presence of hydroxyl groups on the surface of MOFs and the open metal sites of MOFs are two powerful catalytic active sites for the ozonation process. Their presence plays an important role in the adsorption and decomposition of O_3. The catalytic efficiency of MOF highly depends on the type of metal incorporated in the MOF. Therefore, there are several studies on designing and synthesizing MOFs to produce an appropriate catalyst to be used in the catalytic ozonation process [26,121,143–149]. In recent years, several studies have been reported on the applicability of MOFs in the catalytic ozonation process, including Co/Ni-MOF [150], Ce-doped MIL-88A(Fe) [26], and Fe-based

MOFs [151], and this emerging category of materials could be one of high-potential materials by considering some improvements in the future.

5. Conclusions

This review focused on a bibliometric study of catalytic ozonation as one of the popular AOPs methods, conducted from 2000–2021. Nearly 600 articles published during this period, identified by the Web of Science (WOS) database and a bibliometric analysis using VOS viewer software, have been carefully examined and evaluated in terms of future development and practicability. The most impacting articles have been scrutinized and discussed in terms of the interpretation of mechanism and a new vision outlined on the evaluation of both heterogeneously and homogenously catalyzed ozonation processes for the degradation and mineralization of various toxic organic pollutants in water.

Particular attention has been devoted to describing the activities and efficiency of heterogeneous catalysts in the ozonation process related to the chemical properties of catalysts such as crystallographic and morphological, chemical stability as well as the opportune combination of their PZC values with pKa of the target pollutant and pH of the solution.

Examining the results related to the catalytic activity of the metal oxide catalysts, it can be emphasized that the best performance can be obtained when the PZC and pKa values produce positively charged catalyst surfaces and target pollutants, respectively. Despite the small number of citing works, the negatively charged pollutant and the catalyst surface seem a favorable combination for obtaining a good removal efficiency. At least for this type of heterogeneous catalyst, it can be assumed that the repulsion between the pollutants and catalysts promotes the formation of $HO^\bullet$ as the species responsible for the enhancement of the removal processes of the target pollutant. Carbon-based catalysts do not seem to follow this trend; for this reason, deeper investigations could be expected for this class of materials in the future.

Finally, we believe that this study may be of help to authors aiming to improve knowledge in this field.

Author Contributions: Conceptualization, N.F. and E.B.; methodology, N.F.; software, N.F.; validation, N.F., E.B., D.S. and G.M.; formal analysis, N.F. and E.B.; investigation, N.F. and E.B.; resources, N.F. and E.B.; data curation, N.F. and E.B.; writing—original draft preparation, N.F.; writing—review and editing, N.F., E.B. and G.M.; visualization, D.S.; supervision, D.S. and G.M; project administration, G.M.; funding acquisition, G.M. All authors have read and agreed to the published version of the manuscript.

Funding: PANIWATER project funded jointly by the European Union's Horizon 2020 research and innovation programme: 820718; Programma Operativo Nazionale Ricerca e Innovazione 2014–2020: F85F20000290007.

Data Availability Statement: Not applicable.

Acknowledgments: All individuals included in this section have consented to the acknowledgment.

Conflicts of Interest: The authors declare no conflict of interest.

References

1. Boczkaj, G.; Fernandes, A. Wastewater treatment by means of advanced oxidation processes at basic pH conditions: A review. *Chem. Eng. J.* **2017**, *320*, 608–633. [CrossRef]
2. Méndez-Arriaga, F.; Otsu, T.; Oyama, T.; Gimenez, J.; Esplugas, S.; Hidaka, H.; Serpone, N. Photooxidation of the antidepressant drug Fluoxetine (Prozac®) in aqueous media by hybrid catalytic/ozonation processes. *Water Res.* **2011**, *45*, 2782–2794. [CrossRef] [PubMed]
3. Kusic, H.; Koprivanac, N.; Bozic, A.L. Minimization of organic pollutant content in aqueous solution by means of AOPs: UV- and ozone-based technologies. *Chem. Eng. J.* **2006**, *123*, 127–137. [CrossRef]
4. Zangeneh, H.; Zinatizadeh, A.A.L.; Feizy, M. A comparative study on the performance of different advanced oxidation processes ($UV/O_3/H_2O_2$) treating linear alkyl benzene (LAB) production plant's wastewater. *J. Ind. Eng. Chem.* **2014**, *20*, 1453–1461. [CrossRef]

5. Lucas, M.; Peres, J.; Puma, G.L. Treatment of winery wastewater by ozone-based advanced oxidation processes (O_3, O_3/UV and O_3/UV/H_2O_2) in a pilot-scale bubble column reactor and process economics. *Sep. Purify. Technol.* **2010**, *72*, 235–241. [CrossRef]
6. Gimeno, O.; Rivas, F.J.; Beltrán, F.J.; Carbajo, M. Photocatalytic Ozonation of Winery Wastewaters. *J. Agric. Food Chem.* **2007**, *55*, 9944–9950. [CrossRef] [PubMed]
7. Adil, S.; Maryam, B.; Kim, E.-J.; Dulova, N. Individual and simultaneous degradation of sulfamethoxazole and trimethoprim by ozone, ozone/hydrogen peroxide and ozone/persulfate processes: A comparative study. *Environ. Res.* **2020**, *189*, 109889. [CrossRef]
8. Urbina-Suarez, N.A.; Machuca-Martínez, F.; Barajas-Solano, A.F. Advanced Oxidation Processes and Biotechnological Alternatives for the Treatment of Tannery Wastewater. *Molecules* **2021**, *26*, 3222. [CrossRef]
9. Jiménez, S.; Andreozzi, M.; Micó, M.M.; Álvarez, M.G.; Contreras, S. Produced water treatment by advanced oxidation processes. *Sci. Total Environ.* **2019**, *666*, 12–21. [CrossRef]
10. Yang, Y.; Guo, H.; Zhang, Y.; Deng, Q.; Zhang, J. Degradation of Bisphenol A Using Ozone/Persulfate Process: Kinetics and Mechanism. *Water Air Soil Pollut.* **2016**, *227*, 53. [CrossRef]
11. Ding, P.; Chu, L.; Wang, J. Advanced treatment of petrochemical wastewater by combined ozonation and biological aerated filter. *Environ. Sci. Pollut. Res.* **2018**, *25*, 9673–9682. [CrossRef] [PubMed]
12. Wang, J.; Chen, H. Catalytic ozonation for water and wastewater treatment: Recent advances and perspective. *Sci. Total Environ.* **2020**, *704*, 135249. [CrossRef] [PubMed]
13. Azzouz, A.; Kotbi, A.; Niquette, P.; Sajin, T.; Ursu, A.V.; Rami, A.; Monette, F.; Hausler, R. Ozonation of oxalic acid catalyzed by ion-exchanged montmorillonite in moderately acidic media. *React. Kinet. Mech. Catal.* **2010**, *99*, 289–302. [CrossRef]
14. Afzal, S.; Quan, X.; Lu, S. Catalytic performance and an insight into the mechanism of CeO_2 nanocrystals with different exposed facets in catalytic ozonation of p-nitrophenol. *Appl. Catal. B Environ.* **2019**, *248*, 526–537. [CrossRef]
15. Nawrocki, J.; Kasprzyk-Hordern, B. The efficiency and mechanisms of catalytic ozonation. *Appl. Catal. B Environ.* **2010**, *99*, 27–42. [CrossRef]
16. Mashayekh-Salehi, A.; Moussavi, G.; Yaghmaeian, K. Preparation, characterization and catalytic activity of a novel mesoporous nanocrystalline MgO nanoparticle for ozonation of acetaminophen as an emerging water contaminant. *Chem. Eng. J.* **2017**, *310*, 157–169. [CrossRef]
17. Chen, J.; Tian, S.; Lu, J.; Xiong, Y. Catalytic performance of MgO with different exposed crystal facets towards the ozonation of 4-chlorophenol. *Appl. Catal. A Gen.* **2015**, *506*, 118–125. [CrossRef]
18. Wang, Y.; Yang, W.; Yin, X.; Liu, Y. The role of Mn-doping for catalytic ozonation of phenol using Mn/γ-Al_2O_3 nanocatalyst: Performance and mechanism. *J. Environ. Chem. Eng.* **2016**, *4*, 3415–3425. [CrossRef]
19. Yan, Z.; Zhu, J.; Hua, X.; Liang, D.; Dong, D.; Guo, Z.; Zheng, N.; Zhang, L. Catalytic ozonation for the degradation of polyvinyl alcohol in aqueous solution using catalyst based on copper and manganese. *J. Clean. Prod.* **2020**, *272*, 122856. [CrossRef]
20. Sun, Q.; Wang, Y.; Li, L.; Bing, J.; Wang, Y.; Yan, H. Mechanism for enhanced degradation of clofibric acid in aqueous by catalytic ozonation over MnO_x/SBA-15. *J. Hazard. Mater.* **2015**, *286*, 276–284. [CrossRef]
21. Yan, H.; Chen, W.; Liao, G.; Li, X.; Ma, S.; Li, L. Activity assessment of direct synthesized Fe-SBA-15 for catalytic ozonation of oxalic acid. *Sep. Purify. Technol.* **2016**, *159*, 1–6. [CrossRef]
22. Hong, W.; Zhu, T.; Sun, Y.; Wang, H.; Li, X.; Shen, F. Enhancing Oxygen Vacancies by Introducing Na+ into OMS-2 Tunnels To Promote Catalytic Ozone Decomposition. *Environ. Sci. Technol.* **2019**, *53*, 13332–13343. [CrossRef] [PubMed]
23. Yang, L.; Hu, C.; Nie, Y. Surface acidity and reactivity of β-FeOOH/Al_2O_3 for pharmaceuticals degradation with ozone: In situ ATR-FTIR studies. *Appl. Catal. B Environ.* **2010**, *97*, 340–346. [CrossRef]
24. Wei, K.; Cao, X.; Gu, W.; Liang, P.; Huang, X.; Zhang, X. Ni-Induced C-Al_2O_3-Framework (NiCAF) Supported Core–Multishell Catalysts for Efficient Catalytic Ozonation: A Structure-to-Performance Study. *Environ. Sci. Technol.* **2019**, *53*, 6917–6926. [CrossRef]
25. Chen, H.; Zhang, Z.; Hu, D.; Chen, C.; Zhang, Y.; He, S.; Wang, J. Catalytic ozonation of norfloxacin using Co_3O_4/C composite derived from ZIF-67 as catalyst. *Chemosphere* **2021**, *265*, 129047. [CrossRef]
26. Yu, D.; Wang, L.; Yang, T.; Yang, G.; Wang, D.; Ni, H.; Wu, M. Tuning Lewis acidity of iron-based metal-organic frameworks for enhanced catalytic ozonation. *Chem. Eng. J.* **2021**, *404*, 127075. [CrossRef]
27. Yu, D.; Li, L.; Wu, M.; Crittenden, J.C. Enhanced photocatalytic ozonation of organic pollutants using an iron-based metal-organic framework. *Appl. Catal. B Environ.* **2019**, *251*, 66–75. [CrossRef]
28. Sauleda, R.; Brillas, E. Mineralization of aniline and 4-chlorophenol in acidic solution by ozonation catalyzed with Fe^{2+} and UVA light. *Appl. Catal. B Environ.* **2001**, *29*, 135–145. [CrossRef]
29. Beltrán, F.J.; Rivas, F.J.; Montero-de-Espinosa, R. Iron type catalysts for the ozonation of oxalic acid in water. *Water Res.* **2005**, *39*, 3553–3564. [CrossRef]
30. Sánchez-Polo, M.; Rivera-Utrilla, J. Ozonation of 1,3,6-naphthalenetrisulfonic acid in presence of heavy metals. *J. Chem. Technol. Biotechnol.* **2004**, *79*, 902–909. [CrossRef]
31. Ramos, R.M.B.; Biz, A.P.; Tavares, D.A.; Kolicheski, M.B.; Dantas, T.L.P. Homogeneous Catalytic Ozonation of Lipids Involving Fe^{2+} and Mn^{2+}: An Experimental and Modeling Study. *CLEAN—Soil Air Water* **2020**, *48*, 1900430. [CrossRef]
32. Cortés, S.; Sarasa, J.; Ormad, P.; Gracia, R.; Ovelleiro, J.L. Comparative Efficiency of the Systems O_3/High pH And $O3$/catalyst for the Oxidation of Chlorobenzenes in Water. *Ozone Sci. Eng.* **2000**, *22*, 415–426. [CrossRef]

33. Fu, P.; Wang, L.; Li, G.; Hou, Z.; Ma, Y. Homogenous catalytic ozonation of aniline aerofloat collector by coexisted transition metallic ions in flotation wastewaters. *J. Environ. Chem. Eng.* **2020**, *8*, 103714. [CrossRef]

34. Wu, C.-H.; Kuo, C.-Y.; Chang, C.-L. Homogeneous catalytic ozonation of C.I. Reactive Red 2 by metallic ions in a bubble column reactor. *J. Hazard. Mater.* **2008**, *154*, 748–755. [CrossRef]

35. Psaltou, S.; Karapatis, A.; Mitrakas, M.; Zouboulis, A. The role of metal ions on p-CBA degradation by catalytic ozonation. *J. Environ. Chem. Eng.* **2019**, *7*, 103324. [CrossRef]

36. Beltrán, F.J.; Rivas, F.J.; Montero-de-Espinosa, R. Ozone-Enhanced Oxidation of Oxalic Acid in Water with Cobalt Catalysts. 1. Homogeneous Catalytic Ozonation. *Ind. Eng. Chem. Res.* **2003**, *42*, 3210–3217. [CrossRef]

37. El-Raady, A.A.A.; Nakajima, T. Decomposition of Carboxylic Acids in Water by O_3, O_3/H_2O_2, and O_3/Catalyst. *Ozone Sci. Eng.* **2005**, *27*, 11–18. [CrossRef]

38. Petre, A.L.; Carbajo, J.B.; Rosal, R.; García-Calvo, E.; Letón, P.; Perdigón-Melón, J.A. Influence of water matrix on copper-catalysed continuous ozonation and related ecotoxicity. *Appl. Catal. B Environ.* **2015**, *163*, 233–240. [CrossRef]

39. Tachibana, Y.; Nogami, M.; Sugiyama, Y.; Ikeda, Y. Effect of Pd(II) Species on Decomposition Reactions of Pyrrolidone Derivatives by Ozone. *Ozone Sci. Eng.* **2012**, *34*, 359–369. [CrossRef]

40. Andreozzi, R.; Caprio, V.; Insola, A.; Marotta, R.; Tufano, V. The ozonation of pyruvic acid in aqueous solutions catalyzed by suspended and dissolved manganese. *Water Res.* **1998**, *32*, 1492–1496. [CrossRef]

41. Ma, J.; Graham, N.J.D. Degradation of atrazine by manganese-catalysed ozonation influence of radical scavengers. *Water Res.* **2000**, *34*, 3822–3828. [CrossRef]

42. Ma, J.; Graham, N.J.D. Preliminary Investigation of Manganese-Catalyzed Ozonation for the Destruction of Atrazine. *Ozone Sci. Eng.* **1997**, *19*, 227–240. [CrossRef]

43. Xiao, H.; Liu, R.; Zhao, X.; Qu, J. Enhanced degradation of 2,4-dinitrotoluene by ozonation in the presence of manganese(II) and oxalic acid. *J. Mol. Catal. A Chem.* **2008**, *286*, 149–155. [CrossRef]

44. Xiao, H.; Liu, R.; Zhao, X.; Qu, J. Effect of manganese ion on the mineralization of 2,4-dichlorophenol by ozone. *Chemosphere* **2008**, *72*, 1006–1012. [CrossRef] [PubMed]

45. Rivas, J.; Rodríguez, E.; Beltrán, F.J.; García-Araya, J.F.; Alvarez, P. Homogeneous Catalyzed Ozonation of Simazine. Effect of Mn(II) and Fe(II). *J. Environ. Sci. Health B* **2001**, *36*, 317–330. [CrossRef] [PubMed]

46. Xian, Y.; Wen, S.; Liu, J.; Deng, J.; Bai, S. Discovery of a new source of unavoidable ions in pyrite aqueous solutions. *Min. Metall. Explor.* **2013**, *30*, 117–121. [CrossRef]

47. Cao, S.; Cao, Y.; Ma, Z.; Liao, Y. Metal Ion Release in Bastnaesite Flotation System and Implications for Flotation. *Minerals* **2018**, *8*, 203. [CrossRef]

48. Kasprzyk-Hordern, B.M.; Ziółek; Nawrocki, J. Catalytic ozonation and methods of enhancing molecular ozone reactions in water treatment. *Appl. Catal. B Environ.* **2003**, *46*, 639–669. [CrossRef]

49. Shen, T.; Zhang, X.; Lin, K.-Y.A.; Tong, S. Solid base Mg-doped ZnO for heterogeneous catalytic ozonation of isoniazid: Performance and mechanism. *Sci. Total Environ.* **2020**, *703*, 134983. [CrossRef]

50. Bones, J.; Thomas, K.; Nesterenko, N.; Paull, B. On-line preconcentration of pharmaceutical residues from large volume water samples using short reversed-phase monolithic cartridges coupled to LC-UV-ESI-MS. *Talanta* **2006**, *70*, 1117–1128. [CrossRef]

51. Vasca, E.; Ferri, D.; Manfredi, C.; Torello, L.; Fontanella, C.; Caruso, T.; Orrù, S. Complex formation equilibria in the binary Zn^{2+}–oxalate and In^{3+}–oxalate systems. *Dalton Trans.* **2003**, *13*, 2698–2703. [CrossRef]

52. Svobodová Vařeková, R.; Geidl, S.; Ionescu, C.-M.; Skřehota, O.; Kudera, M.; Sehnal, D.; Bouchal, T.; Abagyan, R.; Huber, H.J.; Koča, J. Predicting pKa Values of Substituted Phenols from Atomic Charges: Comparison of Different Quantum Mechanical Methods and Charge Distribution Schemes. *J. Chem. Inf. Model.* **2011**, *51*, 1795–1806. [CrossRef] [PubMed]

53. Qi, F.; Chu, W.; Xu, B. Ozonation of phenacetin in associated with a magnetic catalyst $CuFe_2O_4$, The reaction and transformation. *Chem. Eng. J.* **2015**, *262*, 552–562. [CrossRef]

54. Paul, S.; Sharma, T.R.; Saikia, D.; Saikia, P.; Borah, D.; Baruah, M.K. Evaluation of pKa Values of Soil Humic Acids and their Complexation Properties. *Int. J. Plant Soil Sci.* **2015**, *6*, 218–228. [CrossRef] [PubMed]

55. Lim, E.-B.; Vy, T.A.; Lee, S.-W. Comparative release kinetics of small drugs (ibuprofen and acetaminophen) from multifunctional mesoporous silica nanoparticles. *J. Mater. Chem. B* **2020**, *8*, 2096–2106. [CrossRef] [PubMed]

56. Bernal, V.; Erto, A.; Giraldo, L.; Moreno-Piraján, J.C. Effect of Solution pH on the Adsorption of Paracetamol on Chemically Modified Activated Carbons. *Molecules* **2017**, *22*, 1032. [CrossRef]

57. Sousa, H.R.; Silva, L.S.; Sousa, A.A.; Sousa, R.R.M.; Fonseca, M.G.; Osajima, J.A.; Silva-Filho, E.C. Evaluation of methylene blue removal by plasma activated palygorskites. *J. Mater. Res. Technol.* **2019**, *8*, 5432–5442. [CrossRef]

58. Karatay, S.E.; Aksu, Z.; Özeren, İ.; Dönmez, G. Potentiality of newly isolated Aspergillus tubingensis in biosorption of textile dyes: Equilibrium and kinetic modeling. *Biomass Convers. Biorefinery* **2021**, *155*, 1–8. [CrossRef]

59. Liptak, M.D.; Gross, K.C.; Seybold, G.; Feldgus, S.; Shields, G.C. Absolute pKa Determinations for Substituted Phenols. *J. Am. Chem. Soc.* **2002**, *124*, 6421–6427. [CrossRef]

60. Shiu, W.-Y.; Ma, K.-C.; Varhaníčková, D.; Mackay, D. Chlorophenols and alkylphenols: A review and correlation of environmentally relevant properties and fate in an evaluative environment. *Chemosphere* **1994**, *29*, 1155–1224. [CrossRef]

61. Corrales, J.; Kristofco, L.A.; Steele, W.B.; Yates, B.S.; Breed, C.S.; Williams, E.S.; Brooks, B.W. Global Assessment of Bisphenol A in the Environment:Review and Analysis of Its Occurrence and Bioaccumulation. *Dose-Response* **2015**, *13*, 1559325815598308. [CrossRef] [PubMed]

62. Dükkancı, M.; Vinatoru, M.; Mason, T.J. The sonochemical decolourisation of textile azo dye Orange II: Effects of Fenton type reagents and UV light. *Ultrason. Sonochem.* **2014**, *21*, 846–853. [CrossRef] [PubMed]

63. Goldberg, R.N.; Kishore, N.; Lennen, R.M. Thermodynamic Quantities for the Ionization Reactions of Buffers. *J. Phys. Chem. Ref. Data* **2002**, *31*, 231–370. [CrossRef]

64. Kornmüller, A.; Karcher, S.; Jekel, M. Adsorption of reactive dyes to granulated iron hydroxide and its oxidative regeneration. *Water Sci. Technol.* **2002**, *46*, 43–50. [CrossRef]

65. Nguyen Dang Giang, C.; Sebesvari, Z.; Renaud, F.; Rosendahl, I.; Hoang Minh, Q.; Amelung, W. Occurrence and Dissipation of the Antibiotics Sulfamethoxazole, Sulfadiazine, Trimethoprim, and Enrofloxacin in the Mekong Delta, Vietnam. *PLoS ONE* **2015**, *10*, e01318552015. [CrossRef]

66. Esteki, M.; Dashtaki, E.; Heyden, Y.V.; Simal-Gandara, J. Application of Rank Annihilation Factor Analysis for Antibacterial Drugs Determination by Means of pH Gradual Change-UV Spectral Data. *Antibiotics* **2020**, *9*, 383. [CrossRef]

67. Headley, J.V.; McMartin, D.W. A Review of the Occurrence and Fate of Naphthenic Acids in Aquatic Environments. *J. Environ. Sci. Health A* **2004**, *39*, 1989–2010. [CrossRef]

68. Psaltou, S.; Kaprara, E.; Triantafyllidis, K.; Mitrakas, M.; Zouboulis, A. Heterogeneous catalytic ozonation: The significant contribution of PZC value and wettability of the catalysts. *J. Environ. Chem. Eng.* **2021**, *9*, 106173. [CrossRef]

69. Wu, D.; Liu, Y.; He, H.; Zhang, Y. Magnetic pyrite cinder as an efficient heterogeneous ozonation catalyst and synergetic effect of deposited Ce. *Chemosphere* **2016**, *155*, 127–134. [CrossRef]

70. Zhang, F.; Wei, C.; Wu, K.; Zhou, H.; Hu, Y.; Preis, S. Mechanistic evaluation of ferrite AFe2O4 (A=Co, Ni, Cu, and Zn) catalytic performance in oxalic acid ozonation. *Appl. Catal. A Gen.* **2017**, *547*, 60–68. [CrossRef]

71. Morgan, J.J.; Stumm, W. Colloid-chemical properties of manganese dioxide. *J. Colloid Sci.* **1964**, *19*, 347–359. [CrossRef]

72. Rajput, S.; Pittman, C.U.; Mohan, D. Magnetic magnetite (Fe_3O_4) nanoparticle synthesis and applications for lead (Pb^{2+}) and chromium (Cr^{6+}) removal from water. *J. Colloid Interface Sci.* **2016**, *468*, 334–346. [CrossRef] [PubMed]

73. Farooq, M.; Ramli, A.; Subbarao, D. Physiochemical Properties of γ-Al_2O_3–MgO and γ-Al_2O_3–CeO_2 Composite Oxides. *J. Chem. Eng. Data* **2012**, *57*, 26–32. [CrossRef]

74. Salla, J.S.; Padoin, N.; Amorim, S.M.; Li Puma, G.; Moreira, R.F.M. Humic acids adsorption and decomposition on Mn_2O_3 and α-Al_2O_3 nanoparticles in aqueous suspensions in the presence of ozone. *J. Environ. Chem. Eng.* **2020**, *8*, 102780. [CrossRef]

75. Qi, F.; Xu, B.; Chen, Z.; Ma, J.; Sun, D.; Zhang, L. Influence of aluminum oxides surface properties on catalyzed ozonation of 2,4,6-trichloroanisole. *Sep. Purify. Technol.* **2009**, *66*, 405–410. [CrossRef]

76. Ikhlaq, A.; Brown, D.R.; Kasprzyk-Hordern, B. Catalytic ozonation for the removal of organic contaminants in water on alumina. *Appl. Catal. B Environ.* **2015**, *165*, 408–418. [CrossRef]

77. Naeem, A.; Saddique, M.T.; Mustafa, S.; Kim, Y.; Dilara, B. Cation exchange removal of Pb from aqueous solution by sorption onto NiO. *J. Hazard. Mater.* **2009**, *168*, 364–368. [CrossRef] [PubMed]

78. Chen, H.; Wang, J. Catalytic ozonation for degradation of sulfamethazine using $NiCo_2O_4$ as catalyst. *Chemosphere* **2021**, *268*, 128840. [CrossRef]

79. Li, F.; Geng, D.; Cao, Q. Adsorption of As(V) on aluminum-, iron-, and manganese-(oxyhydr)oxides: Equilibrium and kinetics. *Desalin. Water Treat.* **2015**, *56*, 1829–1838. [CrossRef]

80. Pirovano, C.; Trasatti, S. The point of zero charge of Co_3O_4, Effect of the preparation procedure. *J. Electroanal. Chem. Interf. Electrochem.* **1984**, *180*, 171–184. [CrossRef]

81. Lu, J.; Wei, X.; Chang, Y.; Tian, S.; Xiong, Y. Role of Mg in mesoporous $MgFe_2O_4$ for efficient catalytic ozonation of Acid Orange II. *J. Chem. Technol. Biotechnol.* **2016**, *91*, 985–993. [CrossRef]

82. Sousa, V.S.; Teixeira, M.R. Aggregation kinetics and surface charge of CuO nanoparticles: The influence of pH, ionic strength and humic acids. *Environ. Chem.* **2013**, *10*, 313–322. [CrossRef]

83. Tengvall, P. 4.6 Protein Interactions with Biomaterials ☆. In *Comprehensive Biomaterials II*; Ducheyne, Ed.; Elsevier: Oxford, UK, 2017; pp. 70–84.

84. Parks, G.A.; Bruyn, L.D. The Zero Point of Charge of Oxides1. *J. Phys. Chem.* **1962**, *66*, 967–973. [CrossRef]

85. Kong, L.; Diao, Z.; Chang, X.; Xiong, Y.; Chen, D. Synthesis of recoverable and reusable granular MgO-SCCA-Zn hybrid ozonation catalyst for degradation of methylene blue. *J. Environ. Chem. Eng.* **2016**, *4*, 4385–4391. [CrossRef]

86. Faria, C.C.; Órfão, J.J.M.; Pereira, M.F.R. Activated carbon and ceria catalysts applied to the catalytic ozonation of dyes and textile effluents. *Appl. Catal. B Environ.* **2009**, *88*, 341–350. [CrossRef]

87. Chen, C.; Chen, Y.; Yoza, B.A.; Du, Y.; Wang, Y.; Li, Q.X.; Yi, L.; Guo, S.; Wang, Q. Comparison of Efficiencies and Mechanisms of Catalytic Ozonation of Recalcitrant Petroleum Refinery Wastewater by Ce, Mg, and Ce-Mg Oxides Loaded Al_2O_3. *Catalysts* **2017**, *7*, 72. [CrossRef]

88. Moussavi, G.; Mahmoudi, M. Degradation and biodegradability improvement of the reactive red 198 azo dye using catalytic ozonation with MgO nanocrystals. *Chem. Eng. J.* **2009**, *152*, 1–7. [CrossRef]

89. Oputu, O.; Chowdhury, M.; Nyamayaro, K.; Fatoki, O.; Fester, V. Catalytic activities of ultra-small β-FeOOH nanorods in ozonation of 4-chlorophenol. *J. Environ. Sci.* **2015**, *35*, 83–90. [CrossRef]

90. Ziylan-Yavaş, A.; Ince, N.H. Catalytic ozonation of paracetamol using commercial and Pt-supported nanocomposites of Al_2O_3, The impact of ultrasound. *Ultrason. Sonochem.* **2018**, *40*, 175–182. [CrossRef]

91. Moussavi, G.; Khavanin, A.; Alizadeh, R. The integration of ozonation catalyzed with MgO nanocrystals and the biodegradation for the removal of phenol from saline wastewater. *Appl. Catal. B Environ.* **2010**, *97*, 160–167. [CrossRef]

92. Aghaeinejad-Meybodi, A.; Ebadi, A.; Shafiei, S.; Khataee, A.; Kiadehi, A.D. Degradation of Fluoxetine using catalytic ozonation in aqueous media in the presence of nano-γ-alumina catalyst: Experimental, modeling and optimization study. *Sep. Purify. Technol.* **2019**, *211*, 551–563. [CrossRef]

93. Aguilar, C.M.; Vazquez-Arenas, J.; Castillo-Araiza, O.O.; Rodríguez, J.L.; Chairez, I.; Salinas, E.; Poznyak, T. Improving ozonation to remove carbamazepine through ozone-assisted catalysis using different NiO concentrations. *Environ. Sci. Pollut. Res.* **2020**, *27*, 22184–22194. [CrossRef] [PubMed]

94. Gonçalves, A.G.; Órfão, J.J.M.; Pereira, M.F.R. Ceria dispersed on carbon materials for the catalytic ozonation of sulfamethoxazole. *J. Environ. Chem. Eng.* **2013**, *1*, 260–269. [CrossRef]

95. Vittenet, J.; Aboussaoud, W.; Mendret, J.; Pic, J.-S.; Debellefontaine, H.; Lesage, N.; Faucher, K.; Manero, M.-H.; Thibault-Starzyk, F.; Leclerc, H.; et al. Catalytic ozonation with γ-Al_2O_3 to enhance the degradation of refractory organics in water. *Appl. Catal. A Gen.* **2015**, *504*, 519–532. [CrossRef]

96. He, Y.; Zhang, H.; Li, J.J.; Zhang, Y.; Lai, B.; Pan, Z. Treatment of Landfill Leachate Reverse Osmosis Concentrate from by Catalytic Ozonation with γ-Al_2O_3. *Environ. Eng. Sci.* **2018**, *35*, 501–511. [CrossRef]

97. Polat, D.; Balcı, İ.; Özbelge, T.A. Catalytic ozonation of an industrial textile wastewater in a heterogeneous continuous reactor. *J. Environ. Chem. Eng.* **2015**, *3*, 1860–1871. [CrossRef]

98. Nawaz, F.; Xie, Y.; Cao, H.; Xiao, J.; Wang, Y.; Zhang, X.; Li, M.; Duan, F. Catalytic ozonation of 4-nitrophenol over an mesoporous α-MnO_2 with resistance to leaching. *Catal. Today* **2015**, *258*, 595–601. [CrossRef]

99. Nemati Sani, O.; Navaei fezabady, A.A.; Yazdani, M.; Taghavi, M. Catalytic ozonation of ciprofloxacin using γ-Al_2O_3 nanoparticles in synthetic and real wastewaters. *J. Water Process. Eng.* **2019**, *32*, 100894. [CrossRef]

100. Wang, S.; Zhou, L.; Zheng, M.; Han, J.; Liu, R.; Yun, J. Catalytic Ozonation over $Ca_2Fe_2O_5$ for the Degradation of Quinoline in an Aqueous Solution. *Ind. Eng. Chem. Res.* **2022**, *61*, 6343–6353. [CrossRef]

101. He, C.; Chen, Y.; Guo, L.; Yin, R.; Qiu, T. Catalytic ozonation of NH^{4+}-N in wastewater over composite metal oxide catalyst. *J. Rare Earths* **2022**, *40*, 73–84. [CrossRef]

102. Al jibouri, A.K.H.; Wu, J.; Upreti, S.R. Heterogeneous catalytic ozonation of naphthenic acids in water. *Can. J. Chem. Eng.* **2019**, *97*, 67–73. [CrossRef]

103. Li, G.; Lu, Y.; Lu, C.; Zhu, M.; Zhai, C.; Du, Y.; Yang, P. Efficient catalytic ozonation of bisphenol-A over reduced graphene oxide modified sea urchin-like α-MnO_2 architectures. *J. Hazard. Mater.* **2015**, *294*, 201–208. [CrossRef] [PubMed]

104. Yin, R.; Guo, W.; Zhou, X.; Zheng, H.; Du, J.; Wu, Q.; Chang, J.; Ren, N. Enhanced sulfamethoxazole ozonation by noble metal-free catalysis based on magnetic Fe_3O_4 nanoparticles: Catalytic performance and degradation mechanism. *RSC Adv.* **2016**, *6*, 19265–19270. [CrossRef]

105. Luo, K.; Zhao, S.-X.; Wang, Y.-F.; Zhao, S.-J.; Zhang, X.-H. Synthesis of petal-like δ-MnO_2 and its catalytic ozonation performance. *New J. Chem.* **2018**, *42*, 6770–6777. [CrossRef]

106. Bai, Z.; Yang, Q.; Wang, J. Catalytic ozonation of sulfamethazine using $Ce_{0.1}Fe_{0.9}OOH$ as catalyst: Mineralization and catalytic mechanisms. *Chem. Eng. J.* **2016**, *300*, 169–176. [CrossRef]

107. Beltrán, F.J.; Pocostales, P.; Álvarez, M.; López-Piñeiro, F. Catalysts to improve the abatement of sulfamethoxazole and the resulting organic carbon in water during ozonation. *Appl. Catal. B Environ.* **2009**, *92*, 262–270. [CrossRef]

108. Chen, J.; Tu, Y.; Shao, G.; Zhang, F.; Zhou, Z.; Tian, S.; Ren, Z. Catalytic ozonation performance of calcium-loaded catalyst (Ca-C/Al_2O_3) for effective treatment of high salt organic wastewater. *Sep. Purify. Technol.* **2022**, *301*, 121937. [CrossRef]

109. Liu, H.; Gao, Y.; Wang, J.; Pan, J.; Gao, B.; Yue, Q. Catalytic ozonation performance and mechanism of Mn-CeOx@γ-Al_2O_3/O_3 in the treatment of sulfate-containing hypersaline antibiotic wastewater. *Sci. Total Environ.* **2022**, *807*, 150867. [CrossRef]

110. Jeirani, Z.; Soltan, J. Improved formulation of Fe-MCM-41 for catalytic ozonation of aqueous oxalic acid. *Chem. Eng. J.* **2017**, *307*, 756–765. [CrossRef]

111. Guo, H.; Li, X.; Li, G.; Liu, Y.; Rao, P. Preparation of SnO_x-MnO_x@Al_2O_3 for Catalytic Ozonation of Phenol in Hypersaline Wastewater. *Ozone Sci. Eng.* **2022**, 1–14. [CrossRef]

112. Afzal, S.; Quan, X.; Chen, S.; Wang, J.; Muhammad, D. Synthesis of manganese incorporated hierarchical mesoporous silica nanosphere with fibrous morphology by facile one-pot approach for efficient catalytic ozonation. *J. Hazard. Mater.* **2016**, *318*, 308–318. [CrossRef] [PubMed]

113. Bing, J.; Hu, C.; Zhang, L. Enhanced mineralization of pharmaceuticals by surface oxidation over mesoporous γ-Ti-Al_2O_3 suspension with ozone. *Appl. Catal. B Environ.* **2017**, *202*, 118–126. [CrossRef]

114. Chen, W.; Li, X.; Pan, Z.; Ma, S.; Li, L. Effective mineralization of Diclofenac by catalytic ozonation using Fe-MCM-41 catalyst. *Chem. Eng. J.* **2016**, *304*, 594–601. [CrossRef]

115. Li, H.; Huang, Y.; Cui, S. Removal of alachlor from water by catalyzed ozonation on Cu/Al_2O_3honeycomb. *Chem. Cent. J.* **2013**, *7*, 143. [CrossRef]

116. Feng, C.; Zhao, J.; Qin, G.; Diao, P. Construction of the Fe^{3+}-O-$Mn^{3+/2+}$ hybrid bonds on the surface of porous silica as active centers for efficient heterogeneous catalytic ozonation. *J. Solid State Chem.* **2021**, *300*, 122266. [CrossRef]

117. Chen, W.; Li, X.; Pan, Z.; Ma, S.; Li, L. Synthesis of MnO_x/SBA-15 for Norfloxacin degradation by catalytic ozonation. *Sep. Purify. Technol.* **2017**, *173*, 99–104. [CrossRef]

118. Liu, X.; Wang, S.; Yang, H.; Liu, Z.; Wang, Y.; Meng, F.; Ma, J.; Izosimova, O.S. Characterization of a Doped $MnO_2Al_2O_3$ Catalyst and its Application Inmicrobubble Ozonation for Quinoline Degradation. *Ozone Sci. Eng.* **2021**, *43*, 173–184. [CrossRef]

119. Gao, G.; Shen, J.; Chu, W.; Chen, Z.; Yuan, L. Mechanism of enhanced diclofenac mineralization by catalytic ozonation over iron silicate-loaded pumice. *Sep. Purify. Technol.* **2017**, *173*, 55–62. [CrossRef]

120. Liu, Y.; Wang, S.; Gong, W.; Chen, Z.; Liu, H.; Bu, Y.; Zhang, Y. Heterogeneous catalytic ozonation of p-chloronitrobenzene (pCNB) in water with iron silicate doped hydroxylation iron as catalyst. *Catal. Common.* **2017**, *89*, 81–85. [CrossRef]

121. Mohammadi, V.; Tabatabaee, M.; Fadaei, A.; Mirhoseini, S.A. Study of Nickel Nanoparticles in Highly Porous Nickel Metal–Organic Framework for Efficient Heterogeneous Catalytic Ozonation of Linear Alkyl Benzene Sulfonate in Water. *Ozone Sci. Eng.* **2021**, *43*, 239–253. [CrossRef]

122. Yang, W.; Wu, T. Investigation of Matrix Effects in Laboratory Studies of Catalytic Ozonation Processes. *Ind. Eng. Chem. Res.* **2019**, *58*, 3468–3477. [CrossRef]

123. Xing, S.; Lu, X.; Liu, J.; Zhu, L.; Ma, Z.; Wu, Y. Catalytic ozonation of sulfosalicylic acid over manganese oxide supported on mesoporous ceria. *Chemosphere* **2016**, *144*, 7–12. [CrossRef]

124. Chen, H.; Wang, J. Catalytic ozonation of sulfamethoxazole over Fe_3O_4/Co_3O_4 composites. *Chemosphere* **2019**, *234*, 14–24. [CrossRef]

125. Martins, R.C.; Cardoso, M.; Dantas, R.F.; Sans, C.; Esplugas, S.; Quinta-Ferreira, R.M. Catalytic studies for the abatement of emerging contaminants by ozonation. *J. Chem. Technol. Biotechnol.* **2015**, *90*, 1611–1618. [CrossRef]

126. Yan, J.; Lai, L.; Ji, F.; Zhang, Y.; Li, Y.; Lai, B. Catalytic Ozonation of Sulfamethoxazole in the Presence of CuO/Al_2O_3-EPC: Optimization, Degradation Pathways, and Toxicity Evaluation. *Environ. Eng. Sci.* **2019**, *36*, 912–921. [CrossRef]

127. Shahidi, D.; Moheb, A.; Abbas, R.; Larouk, S.; Roy, R.; Azzouz, A. Total mineralization of sulfamethoxazole and aromatic pollutants through Fe^{2+}-montmorillonite catalyzed ozonation. *J. Hazard. Mater.* **2015**, *298*, 338–350. [CrossRef] [PubMed]

128. Wu, J.; Gao, H.; Yao, S.; Chen, L.; Gao, Y.; Zhang, H. Degradation of Crystal Violet by catalytic ozonation using Fe/activated carbon catalyst. *Sep. Purify. Technol.* **2015**, *147*, 179–185. [CrossRef]

129. Gonçalves, A.G.; Órfão, J.J.M.; Pereira, M.F.R. Catalytic ozonation of sulphamethoxazole in the presence of carbon materials: Catalytic performance and reaction pathways. *J. Hazard. Mater.* **2012**, *239*, 167–174. [CrossRef]

130. Akhtar, J.; Amin, N.S.; Aris, A. Combined adsorption and catalytic ozonation for removal of sulfamethoxazole using Fe_2O_3/CeO_2 loaded activated carbon. *Chem. Eng. J.* **2011**, *170*, 136–144. [CrossRef]

131. Gonçalves, A.G.; Órfão, J.J.M.; Pereira, M.F.R. Ozonation of sulfamethoxazole promoted by MWCNT. *Catal. Common.* **2013**, *35*, 82–87. [CrossRef]

132. Huang, Y.; Sun, Y.; Xu, Z.; Luo, M.; Zhu, C.; Li, L. Removal of aqueous oxalic acid by heterogeneous catalytic ozonation with MnOx/sewage sludge-derived activated carbon as catalysts. *Sci. Total Environ.* **2017**, *575*, 50–57. [CrossRef] [PubMed]

133. Zhang, J.; Guo, Q.; Wu, W.; Shao, S.; Li, Z.; Liu, Y.; Jiao, W. Preparation of Fe-MnO_X/AC by high gravity method for heterogeneous catalytic ozonation of phenolic wastewater. *Chem. Eng. Sci.* **2022**, *255*, 117667. [CrossRef]

134. Shahamat, Y.D.; Farzadkia, M.; Nasseri, S.; Mahvi, A.H.; Gholami, M.; Esrafili, A. Magnetic heterogeneous catalytic ozonation: A new removal method for phenol in industrial wastewater. *J. Environ. Health Sci. Eng.* **2014**, *12*, 50. [CrossRef] [PubMed]

135. Wang, Y.; Xie, Y.; Sun, H.; Xiao, J.; Cao, H.; Wang, S. Efficient Catalytic Ozonation over Reduced Graphene Oxide for p-Hydroxylbenzoic Acid (PHBA) Destruction: Active Site and Mechanism. *ACS Appl. Mater. Interfaces* **2016**, *8*, 9710–9720. [CrossRef]

136. Jothinathan, L.; Hu, J. Kinetic evaluation of graphene oxide based heterogenous catalytic ozonation for the removal of ibuprofen. *Water Res.* **2018**, *134*, 63–73. [CrossRef]

137. Yin, R.; Guo, W.; Du, J.; Zhou, X.; Zheng, H.; Wu, Q.; Chang, J.; Ren, N. Heteroatoms doped graphene for catalytic ozonation of sulfamethoxazole by metal-free catalysis: Performances and mechanisms. *Chem. Eng. J.* **2017**, *317*, 632–639. [CrossRef]

138. Kosmulski, M. pH-dependent surface charging and points of zero charge. IV. Update and new approach. *J. Colloid Interface Sci.* **2009**, *337*, 439–448. [CrossRef]

139. Huang, Y.; Cui, C.; Zhang, D.; Li, L.; Pan, D. Heterogeneous catalytic ozonation of dibutyl phthalate in aqueous solution in the presence of iron-loaded activated carbon. *Chemosphere* **2015**, *119*, 295–301. [CrossRef]

140. Karimi, M.; Sadeghi, S.; Mohebali, H.; Azarkhosh, Z.; Safarifard, V.; Mahjoub, A.; Heydari, A. Fluorinated solvent-assisted photocatalytic aerobic oxidative amidation of alcohols via visible-light-mediated HKUST-1/Cs-POMoW catalysis. *New J. Chem.* **2021**, *45*, 14024–14035. [CrossRef]

141. Karimi, M.; Mohebali, H.; Sadeghi, S.; Safarifard, V.; Mahjoub, A.; Heydari, A. Additive-free aerobic C-H oxidation through a defect-engineered Ce-MOF catalytic system. *Microporous Mesoporous Mater.* **2021**, *322*, 111054. [CrossRef]

142. Rozveh, Z.S.; Kazemi, S.; Karimi, M.; Ali, G.A.M.; Safarifard, V. Effect of functionalization of metal-organic frameworks on anion sensing. *Polyhedron* **2020**, *183*, 114514. [CrossRef]

143. Liao, X.; Wang, F.; Wang, Y.; Cai, Y.; Liu, H.; Wang, X.; Zhu, Y.; Chen, L.; Yao, Y.; Hao, Q. Constructing Fe-based bi-MOFs for photo-catalytic ozonation of organic pollutants in Fischer-Tropsch waste water. *Appl. Surf. Sci.* **2020**, *509*, 145378. [CrossRef]

144. Chávez, A.M.; Rey, A.; López, J.; Álvarez, M.; Beltrán, F.J. Critical aspects of the stability and catalytic activity of MIL-100(Fe) in different advanced oxidation processes. *Sep. Purify. Technol.* **2021**, *255*, 117660. [CrossRef]

145. Ye, G.; Luo, P.; Zhao, Y.; Qiu, G.; Hu, Y.; Preis, S.; Wei, C. Three-dimensional Co/Ni bimetallic organic frameworks for high-efficient catalytic ozonation of atrazine: Mechanism, effect parameters, and degradation pathways analysis. *Chemosphere* **2020**, *253*, 126767. [CrossRef] [PubMed]
146. Zheng, H.; Hou, Y.; Li, S.; Ma, J.; Nan, J.; Li, T. Recent advances in the application of metal organic frameworks using in advanced oxidation progresses for pollutants degradation. *Chin. Chem. Lett.* **2022**, *33*, 5013–5022. [CrossRef]
147. Chen, H.; Wang, J. MOF-derived Co_3O_4-C@FeOOH as an efficient catalyst for catalytic ozonation of norfloxacin. *J. Hazard. Mater.* **2021**, *403*, 123697. [CrossRef]
148. Hu, Q.; Zhang, M.; Xu, L.; Wang, S.; Yang, T.; Wu, M.; Lu, W.; Li, Y.; Yu, D. Unraveling timescale-dependent Fe-MOFs crystal evolution for catalytic ozonation reactivity modulation. *J. Hazard. Mater.* **2022**, *431*, 128575. [CrossRef]
149. Hu, Q.; Xu, L.; Fu, K.; Zhu, F.; Yang, T.; Yang, T.; Luo, J.; Wu, M.; Yu, D. Ultrastable MOF-based foams for versatile applications. *Nano Res.* **2022**, *15*, 2961–2970. [CrossRef]
150. Pang, Z.; Luo, P.; Wei, C.; Qin, Z.; Wei, T.; Hu, Y.; Wu, H.; Wei, C. In-situ growth of Co/Ni bimetallic organic frameworks on carbon spheres with catalytic ozonation performance for removal of bio-treated coking wastewater. *Chemosphere* **2022**, *291*, 132874. [CrossRef]
151. Yu, D.; Wu, M.; Hu, Q.; Wang, L.; Lv, C.; Zhang, L. Iron-based metal-organic frameworks as novel platforms for catalytic ozonation of organic pollutant: Efficiency and mechanism. *J. Hazard. Mater.* **2019**, *367*, 456–464. [CrossRef]

 **catalysts**

MDPI

Article

Enhanced Photodegradation of Organic Pollutants by Novel Samarium-Doped Zinc Aluminium Spinel Ferrites

Ionela Grecu [1], Petrisor Samoila [1,*], Petronela Pascariu [1], Corneliu Cojocaru [1,*], Maria Ignat [1,2], Ioan-Andrei Dascalu [1] and Valeria Harabagiu [1]

[1] "Petru Poni" Institute of Macromolecular Chemistry, 41A Grigore Ghica Voda Alley, 700487 Iasi, Romania
[2] Department of Chemistry, "Alexandru Ioan Cuza" University, 11 Carol I Blvd, Iasi 700506, Romania
* Correspondence: samoila.petrisor@icmpp.ro (P.S.); cojocaru.corneliu@icmpp.ro (C.C.);
  Tel.: +40-332-880-050 (P.S.)

Abstract: $ZnAlFe_{1-x}Sm_xO_4$ (x = 0, 0.02, 0.04, 0.06, 0.08) spinel ferrites were successfully obtained for the first time via a sol–gel autocombustion technique using citric acid as the combustion/chelating agent. These materials were then employed as photocatalysts for the degradation of Evans Blue, considered herein as a model organic pollutant. The XRD and FTIR analysis confirmed the achievement of pure spinel ferrite structures for all the materials. TEM analysis showed that the average particle sizes decline from about 27 for the undoped material to 17 nm for samarium-doped materials, and the magnetic characterization at room temperature indicated the paramagnetic conduct for the studied samples. All the photocatalysts were active in Evans Blue photodegradation. The best photocatalytic performances were observed for the $ZnAlFe_{0.94}Sm_{0.06}O_4$ formulation and explained by the smallest values calculated for lattice parameter, interplanar distance, and particle-size values. By adding H_2O_2 and applying the modelling and optimization of the photocatalytic process for the best material, the half-life of the pollutant decreased significantly from 115 min to about 7 min (about 16-times), and the colour-removal efficiency was almost 100%.

Keywords: aluminium-substituted zinc ferrite; samarium-doped spinel ferrite; H_2O_2/UV-Vis; wastewater treatment; modelling and optimization of photocatalytic process

Citation: Grecu, I.; Samoila, P.; Pascariu, P.; Cojocaru, C.; Ignat, M.; Dascalu, I.-A.; Harabagiu, V. Enhanced Photodegradation of Organic Pollutants by Novel Samarium-Doped Zinc Aluminium Spinel Ferrites. *Catalysts* **2023**, *13*, 266. https://doi.org/10.3390/catal13020266

Academic Editors: Gassan Hodaifa, Rafael Borja and Mha Albqmi

Received: 28 December 2022
Revised: 18 January 2023
Accepted: 20 January 2023
Published: 24 January 2023

1. Introduction

Spinel ferrites (MFe_2O_4—where M is a divalent cation) are a special category of mixed iron oxides of great interest to the research community because of their unique properties, from magnetic, electrical, electronic, dielectric, mechanical, optical, as well as (photo)catalytic points of view [1,2]. Their properties are making spinel ferrites strong candidates for a wide range of applications from microwave and electronic devices [2] to hyperthermia, contrast agents for MRI applications, and drug delivery systems [3] or catalytic [4], adsorption [5], and photocatalytic applications for wastewater treatment [6].

The characteristics of spinel ferrites strongly depend on their chemical composition, phase purity, grain and/or crystallite size, or specific surface area. In order to fine-tune the spinel ferrite properties for a specific application, several strategies have been implemented to date: judicious selection of the synthesis method considering the targeted application [1], variation in the sintering temperature and/or time [7], the substitution of divalent or iron ions [8,9], doping [10], etc.

Spinel ferrites with zinc as a divalent cation are known as zinc spinel ferrites. $ZnFe_2O_4$ is one of the most representative compounds of the spinel ferrite class of materials, along with $CoFe_2O_4$ and $NiFe_2O_4$ [11]. Because of its relatively narrow band gap (around 1.9 eV) [12,13], zinc ferrite was often proposed as a highly active photocatalyst for wastewater purification by advanced oxidation of organic pollutants [13]. Thus, in their exhaustive review, Sonu et al. [13] observed that the number of publications on $ZnFe_2O_4$ was constantly growing from 2010 to 2020, and more than half of the studies proposed this material as an efficient photocatalyst for oxidative water-purification applications compared to only 24% for inductors, 14% for adsorption processes, and 9% for electronic devices, respectively. Despite the encouraging photocatalytic behaviour of the pristine zinc ferrite, different strategies were developed to enhance the performances of these materials, such as [9,14,15] substitution, composite formation, and doping. In a previous work, we demonstrated that the substitution of Fe cations with Al cations into Zinc ferrite clearly improved the photodegradation efficiency of Orange I azo dye [9]. On the other hand, we previously proved that the properties of Cobalt and Nickel ferrites can be significantly changed by doping with rare-earth cations [4,16]. Thus, in this work, we propose, for the first time, to simultaneously study the influence of substituting Fe cations with Al and of doping the as-obtained $ZnAlFeO_4$ with a representative rare-earth element: samarium. Therefore, in the present work, for the first time, ternary spinel ferrites with the general formula $ZnAlFe_{1-x}Sm_xO_4$ (x = 0, 0.02, 0.04, 0.06, 0.08), as highly efficient photocatalysts for Evans Blue dye removal from synthetic wastewaters under UV-Vis light, are studied.

2. Results

2.1. Characterization of Photocatalysts

2.1.1. X-Ray Diffraction Analysis

XRD analysis was performed in order to disclose the structural features of the obtained catalysts. The recorded XRD patterns for the powders are depicted in Figure 1, and the chemical formulas, sample codes, and calculated data are given in Table 1. Close inspection of the patterns shown in Figure 1 suggests that each of the observed peaks for each of the studied materials closely resembles the typical patterns for zinc ferrite samples according to JCPDS card No. 22-1012 [17,18]. As a result, the exclusive presence of the planes (220), (311), (222), (400), (422), (511), (440), (620), and (533) in the diffraction patterns reveals that all of the samples formed a pure cubic spinel structure free of any observable impurities. This fact demonstrates that the selected preparation technique was successfully achieved for, on one hand, the insertion of the aluminium cations in the zinc ferrite spinel matrix and, on the other hand, the incorporation of Sm cations into the spinel structure of the aluminium-substituted zinc ferrite.

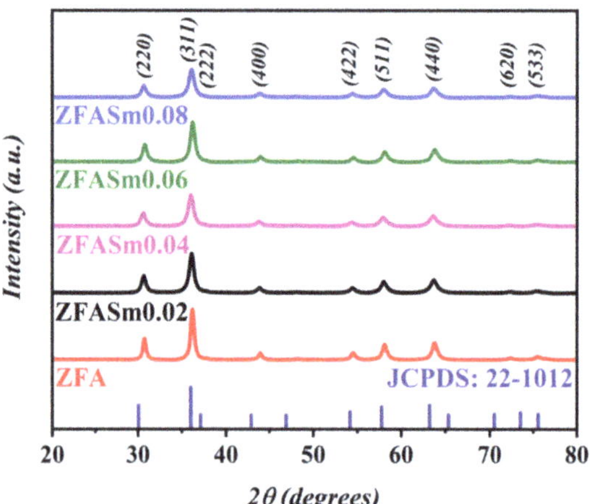

Figure 1. XRD patterns of $ZnAlFe_{1-x}Sm_xO_4$ (x = 0, 0.02, 0.04, 0.06, 0.08) photocatalysts compared to the standard XRD pattern of $ZnFe_2O_4$ (JCPDS card No. 22-1012).

Table 1. Calculated data for $ZnAlFe_{1-x}Sm_xO_4$ (x = 0, 0.02, 0.04, 0.06, 0.08) photocatalysts.

Photocatalyst Chemical Formula	Code	D (nm) [1]	a (Å) [1]	d_{311} (Å) [1]	$\bar{d}$ (nm) [2]	S_{BET} [3] (m^2/g)	V_{tot} [3] (cc/g)	D_{pore} [3] (nm)
$ZnAlFeO_4$	ZFA	13.00	8.2403	1.8423	26.6	15.076	6.82×10^{-2}	31.7
$ZnAlFe_{0.98}Sm_{0.02}O_4$	ZFASm0.02	11.17	8.2501	1.8446	22.9	19.625	9.49×10^{-2}	21.5
$ZnAlFe_{0.96}Sm_{0.04}O_4$	ZFASm0.04	10.34	8.2457	1.8434	20.9	26.556	9.49×10^{-2}	15.3
$ZnAlFe_{0.94}Sm_{0.06}O_4$	ZFASm0.06	10.26	8.2359	1.8410	17.3	28.239	9.20×10^{-2}	14.7
$ZnAlFe_{0.92}Sm_{0.08}O_4$	ZFASm0.08	10.04	8.2525	1.8453	19.4	34.617	9.59×10^{-2}	11.8

[1] Data from XRD; [2] Data from TEM; [3] Data from N_2 sorption.

Despite the greater ionic radius of samarium cations (0.958 Å) compared to Fe^{3+} (0.645 Å), the similarity between the diffraction patterns of Sm-doped materials and those corresponding to the zinc–aluminium ferrite one suggested that Sm^{3+} may have replaced Fe^{3+} ions in the host spinel structure [19]. However, as a consequence of the inclusion of the larger samarium cations in the spinel structure, the diffraction peaks in doped samples are widened with increasing amounts of lanthanide. These findings are in close agreement with our previous work on the rare-earth doping of spinel ferrites [4,16,20].

Deeper exploitation of the registered XRD patterns was achieved by determining the crystallite sizes (D) using the Debye–Scherrer formula, the lattice parameter (a), and the interplanar distance (d), according to Laue and Bragg's equations for cubic lattices using the relations presented in previous works [4,20]. These data are given in Table 1. One may observe that the crystallite sizes decrease regularly with the samarium content increasing; doping the ZFA sample with samarium led to a decrease. This fact indicates that the replacement of Fe^{3+} with larger Sm^{3+} cations obstructs the growth of crystallites. Such behaviour is often reported on doping spinels with rare-earth cations [4,16,20,21]. Both lattice parameters and interplanar distances are higher for Sm-containing materials, except for the ZFASm0.06 sample, but no direct correlation between rare-earth concentration and these values can be observed. This fact suggests that increasing samarium concentration leads to the redistribution of the Zn, Fe, and Al cations between the tetrahedral and octahedral sites [21,22].

2.1.2. Fourier-Transform Infrared Spectroscopy Analysis

FTIR spectroscopy was employed to identify the metal–oxygen bonds once the ferrite structure was obtained in order to confirm the XRD findings. Further, this technique was used to estimate the presence of combustion by-products in the samples. In this respect, the spectra were registered in a wavenumber range of 400–4000 cm^{-1}, as shown in Figure 2.

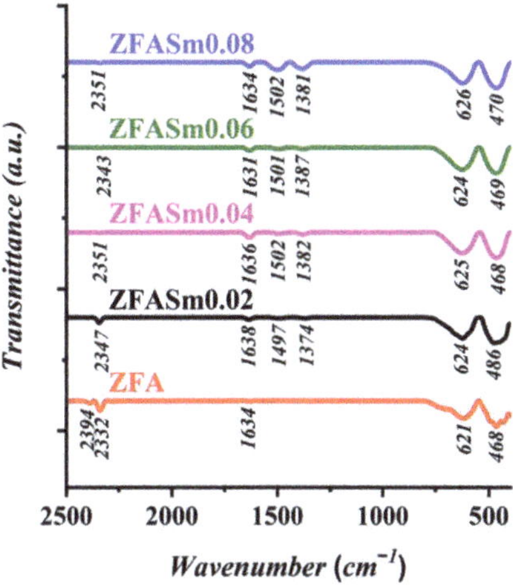

Figure 2. FTIR spectra of $ZnAlFe_{1-x}Sm_xO_4$ (x = 0, 0.02, 0.04, 0.06, 0.08) photocatalysts.

First, the fingerprint bands for metal oxides are clearly observed in a range of 700–400 cm^{-1}. Thus, the band observed in the 621–626 cm^{-1} range is usually attributed to the stretching vibration of the metal–oxygen bond in the tetrahedral positions of the spinel structure, whereas the band observed at 468–486 cm^{-1} is assigned to the stretching vibration of the metal–oxygen bond in the octahedral sites of the spinel matrix [23,24]. It is important to note that all the samples containing samarium cations show a slight increase in the wavenumber value for both tetrahedral and octahedral vibrations compared to the ZFA sample. These observations are consistent with previous works on rare-earth-doped ferrites [25,26] and indicate perturbations occurring in the Fe–O, Zn–O and Al–O bonds, respectively, as well as cation migration between the two types of sites, due to the replacement of small amounts of iron cations by bulkier samarium cations [27]. It is worth noting that these observations are in very close agreement with the XRD discussion.

Second, the weak bands in the 2931–1374 cm^{-1} range are due to the presence of very small amounts of combustion by-products and can be assigned as follows [28,29]: the bands in the 2931–2862 cm^{-1} range are ascribed to antisymmetric and symmetric –CH$_2$– group vibrations resulting from the citric acid combustion, the bands in the 2394–1497 cm^{-1} range are most probably due to the water and CO$_2$ absorbed by the samples from the atmosphere, and the bands around 1380 cm^{-1} indicate the presence of nitrate ions from the metallic salts used during synthesis.

2.1.3. Transmission Electron Microscopy

Representative TEM micrographs and grain size distributions for the materials under study are shown in Figure 3. Table 1 lists the average particle sizes, as measured by analysing more than 100 grains from at least five different TEM images, for each photocatalyst. For the five materials under study, the TEM micrographs demonstrate the nanometric nature of grains, a narrow grain-size distribution, and a low degree of agglomeration. The average particle sizes fall from about 27, for the undoped material, to 17 nm with an increase in samarium content. Nevertheless, it is important to note that the average grain sizes are not directly correlated with the rare-earth concentration, the ZFASm0.06 presenting the smallest particles. It should also be noted that the distributions of the particle sizes are narrowed when Sm ions replace Fe ions, except for the material most loaded in samarium.

2.1.4. Magnetic Properties

The magnetic properties of the studied materials were evaluated at room temperature by registering the hysteresis cycles (Figure 4).

The shapes of the loops from Figure 4 are in close agreement with previous findings on zinc ferrite materials studied at room temperature and indicate the paramagnetic comportment with zero remanent magnetization and zero intrinsic coercivity (see insert in Figure 4) for all the photocatalysts [30,31]. According to the literature, the paramagnetic nature of zinc ferrite materials is usually explained by the arrangement of the cations within the tetrahedral and octahedral sites close to the ideal normal spinel structure [31]. One may observe that the replacement of half of the iron cations with non-magnetic aluminium cations and paramagnetic samarium cations does not interfere with the paramagnetic behaviour of the original zinc ferrite. Nevertheless, a slight decrease in magnetization is observed with increasing samarium content.

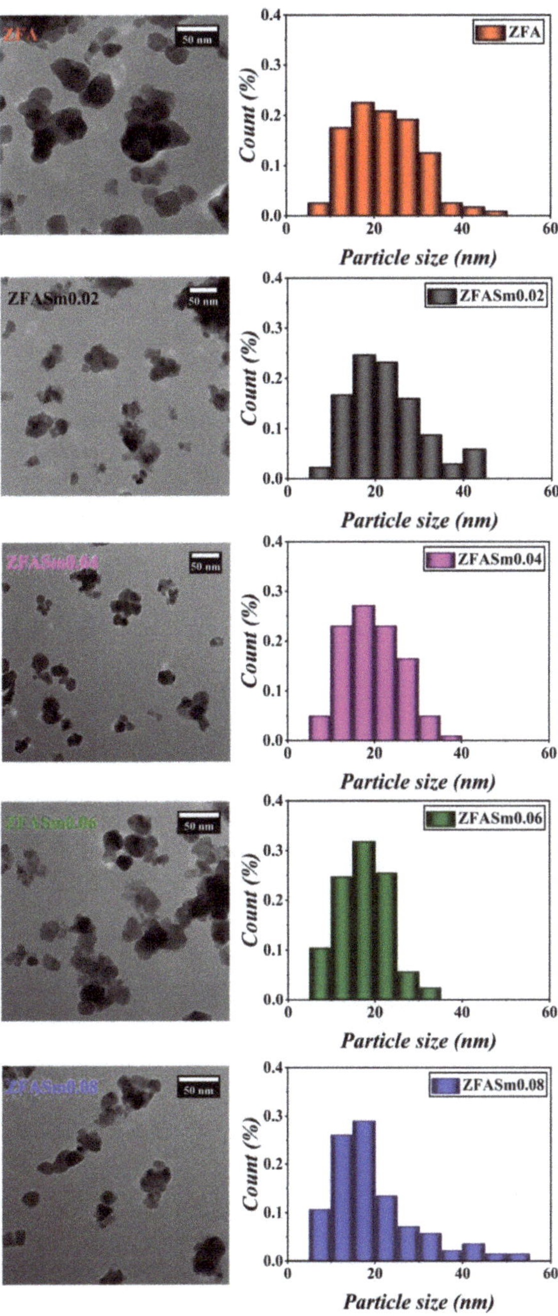

Figure 3. Representative TEM micrographs (left) and particle-size distributions (right) of $ZnAlFe_{1-x}Sm_xO_4$ (x = 0, 0.02, 0.04, 0.06, 0.08) photocatalysts.

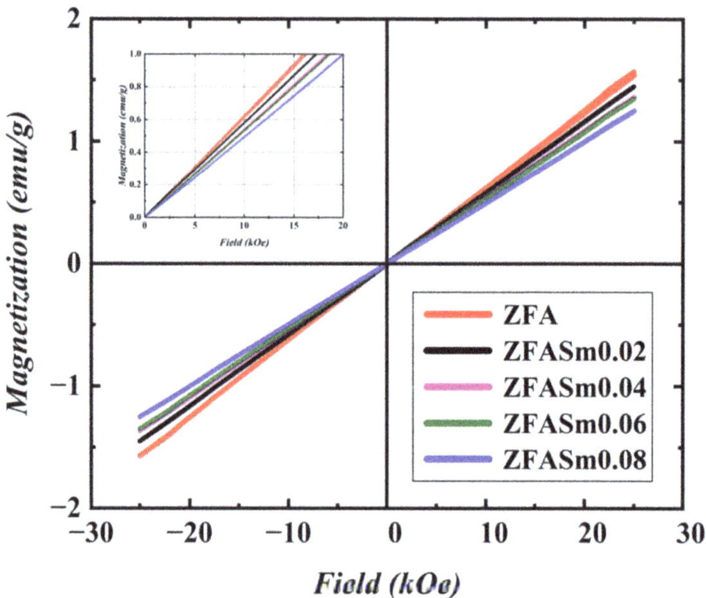

Figure 4. Hysteresis loops recorded at room temperature for $ZnAlFe_{1-x}Sm_xO_4$ (x = 0, 0.02, 0.04, 0.06, 0.08) photocatalysts.

2.1.5. Textural Properties

The nitrogen adsorption–desorption isotherms and the corresponding pore-size distribution are depicted in Figure 5, and the calculated data are provided in Table 1.

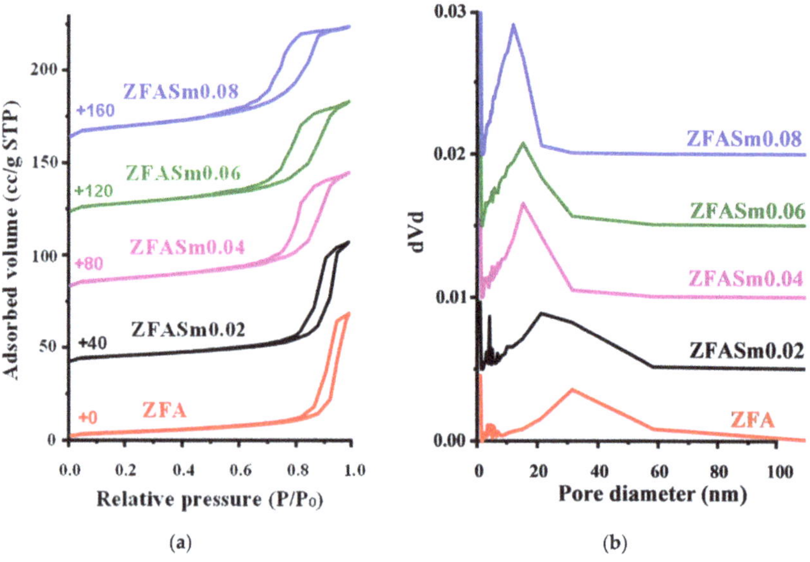

(a)

(b)

Figure 5. N_2-sorption isotherms (**a**) and corresponding pore-size distributions (**b**) for $ZnAlFe_{1-x}Sm_xO_4$ (x = 0, 0.02, 0.04, 0.06, 0.08) photocatalysts.

According to IUPAC classification [32], the registered nitrogen adsorption–desorption isotherms are of type IVa, characterizing mesoporous adsorbent materials. All isotherms are accompanied by a H1-type hysteresis loop, meaning that the adsorption occurs in pores larger than 4 nm and the characterized material exhibits a narrow range of uniform ink-bottle-like mesopores [33]. As can be observed, the condensation in mesopores occurs at high relative pressures ($P/P_0 = 0.8$–1), indicating that the mesopores are large in size.

It should be mentioned that the capillary condensation shifts to lower relative pressures as the Sm quantity increases. Thus, by applying the BJH (Barret–Joyner–Halenda) theory, pore-size distributions were found and drawn, allowing for estimation of the mean pore diameter for each characterized sample. As expected, the ZFA pore diameters decrease as the Sm quantity increases in doped ferrite samples, from 31.7 nm to 11.8 nm, respectively. Additionally, an inverse trend for the BET (Brunauer–Emmett–Teller) specific surface area was observed starting from 15.076 m^2/g for the ZFA sample, increasing up to 34.617 m^2/g for ZFASm0.08, respectively. This finding could be explained by the larger Sm ion diameter compared to the other containing elements of ZFA. The Sm doping led to an increase in ZFA pore volume, but no trend was observed with the increasing Sm dosage.

2.1.6. Optical Properties

By drawing the Tauc plots resulting from the UV–Vis DR spectra, the optical band gap values were obtained using the Kubelka-Munk function (see Section 3.2. Photocatalyst characterization). As observed from Figure 6a,b the direct band gap energies were found to be about 2.1–2.2 eV, while the indirect band gap energies of 1.8–1.85 eV. As observed, the band gap energies for the Sm-doped samples did not change in comparison with the ZFA sample (undoped), meaning that all samples will be photoactivated at the same wavelength.

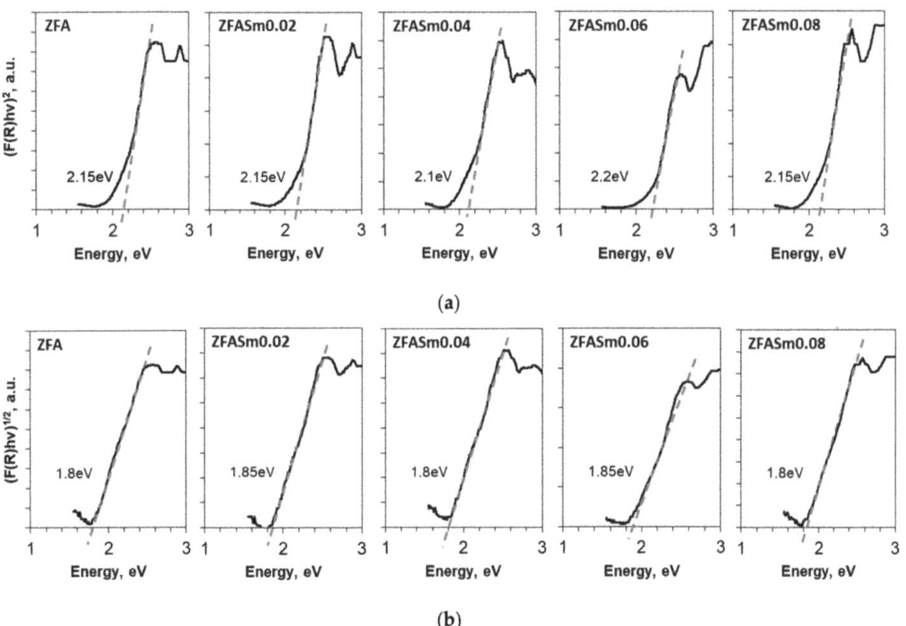

Figure 6. Direct band gap transition (**a**) and indirect band gap transition (**b**) in ZnAlFe$_{1-x}$Sm$_x$O$_4$ (x = 0, 0.02, 0.04, 0.06, 0.08) photocatalysts.

2.2. Photocatalytic Properties

2.2.1. Photocatalytic Activity under UV-Vis Light

The photocatalytic activity of $ZnAlFe_{1-x}Sm_xO_4$ (x = 0, 0.02, 0.04, 0.06, 0.08) materials was evaluated in photodegradation processes of Evans Blue (EB) dye under UV-Vis light at room temperature. Evans blue ($C_{34}H_{24}N_6Na_4O_{14}S_4$) was selected as the model organic pollutant, on one hand, due to its wide range of applications from the diagnostic tools in biomedicine to colouring agent for cotton and silk fibres in the textile industry and, on the other hand, owing to its high toxicity for lung, liver, intestine, and kidney function [34,35].

Figure 7a–e show the modifications of the UV-Vis spectra of EB in the presence of the studied photocatalysts at a regular time interval from 0 to 240 min. Several observations can be seen by close inspection of Figure 7a–e. First, one may observe that, in the 300–800 nm range, EB registered UV-Vis spectra are characterized by two adsorption peaks: one broad peak located at 606 nm and another peak around 320 nm, in close agreement with the literature [36]. Second, with reaction time, both EB peaks became weaker, which shows that all of the investigated photocatalysts are attacking the pollutant molecule. Third, after 4 h of the photocatalytic process, the colour removal was (almost) complete for some of the samarium-containing samples, as compared to the ZFA sample. For comparison purposes, Figure 7f shows the colour-removal efficiency registered after 1 h of the photocatalytic process for all samples. One may note all materials doped with samarium ions are significantly more active as photocatalysts than the undoped zinc aluminium ferrite sample. Among the materials with Sm, the ZFASm0.06 is the most active by far. The photocatalytic behaviours of the samples can be easily correlated with lattice parameters, interplanar distance, and particle-size values (see Table 1). Furthermore, XRD and FTIR analysis indicated that the insertion of samarium into the spinel structure led to the redistribution of the cations between tetrahedral and octahedral sites. The most suitable cation arrangement from a photocatalytic performance standpoint was, most probably, for x = 0.06.

2.2.2. Kinetics of the Photodegradation Process under UV-Vis Light

Based on the registered UV-Vis spectra, kinetics studies of EB photodegradation under UV-Vis light were performed. Thus, Figure 8 illustrates the kinetics data related to the photodegradation of EB dye in aqueous solutions under UV-Vis light, in the absence of a catalyst (photolysis) and in the presence of catalysts ($ZnAlFe_{1-x}Sm_xO_4$ (x = 0, 0.02, 0.04, 0.06, 0.08)). For the experiments with photocatalytic materials, one should mention that before irradiation, the suspension was magnetically stirred for 30 min to ensure adsorption–desorption equilibrium. Thus, it is important to observe the adsorption effect of EB onto the surface of the catalyst for all studied materials. The adsorption capacity of the materials seems very sensitive to the samarium presence in the samples and strongly influences the dye-removal performances. Throughout the reaction time, the photocatalytic activity of the zinc–aluminium ferrite was enhanced by doping with samarium. Nevertheless, the best photocatalyst was proven to be ZFASm0.06. From Figure 8, it is also important to note that without a catalyst, the photodegradation of the EB dye is practically negligible.

By means of nonlinear regression techniques, the experimental data were fitted to the pseudo-first-order (PFO) kinetic model. The goodness of fit was ascertained by the chi-square test (χ^2-value). The PFO equation can be expressed as:

$$C = C_0\, e^{-kt} \tag{1}$$

where C_0 designates the initial EB dye concentration (~10 mg/L), k —pseudo-first-order reaction rate constant (min^{-1}), and t —irradiation time (min). The fitted parameters of the PFO model are listed in Table 2.

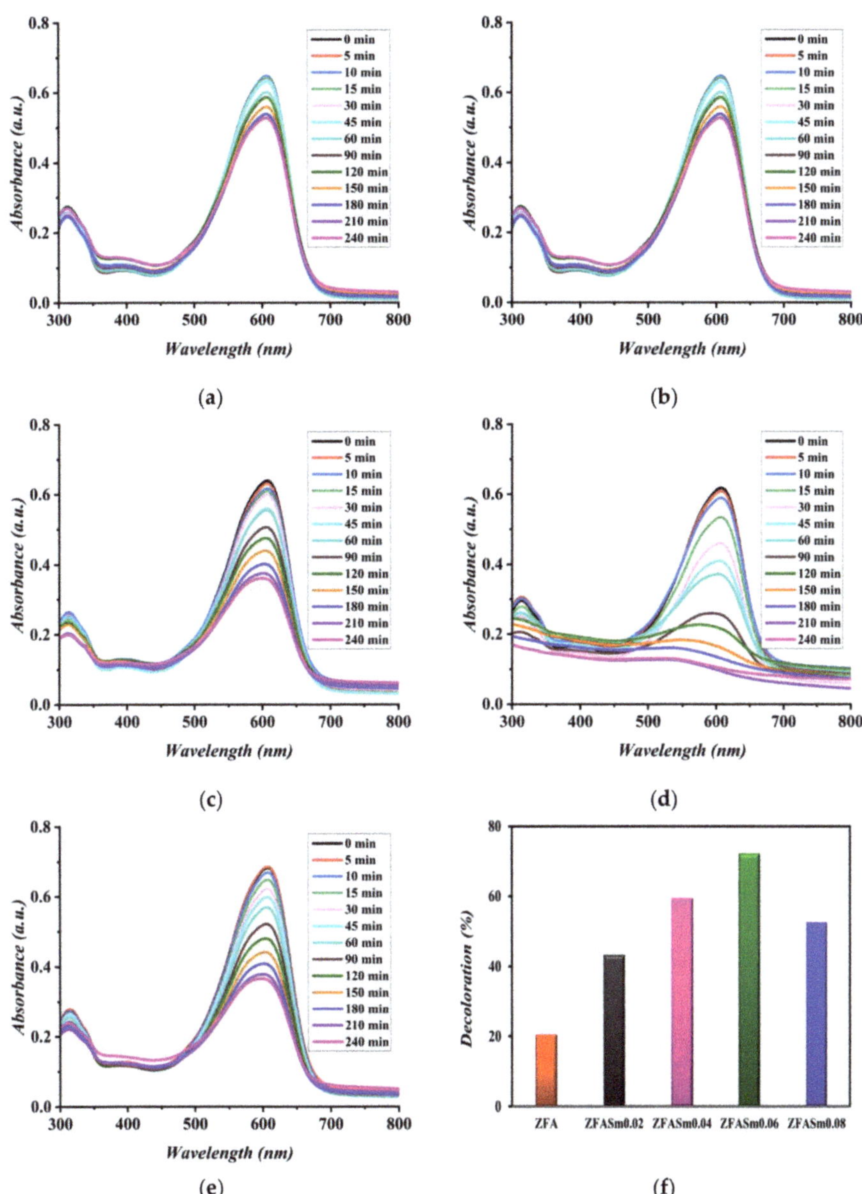

Figure 7. Spectral response of EB during photodegradation in the presence of samples (**a**) ZnAlFeO$_4$ (ZFA), (**b**) ZnAlFe$_{0.98}$Sm$_{0.02}$O$_4$ (ZFASm0.02), (**c**) ZnAlFe$_{0.96}$Sm$_{0.04}$O$_4$ (ZFASm0.04), (**d**) ZnAlFe$_{0.94}$Sm$_{0.06}$O$_4$ (ZFASm0.06), (**e**) ZnAlFe$_{0.92}$Sm$_{0.08}$O$_4$ (ZFASm0.08), and (**f**) colour-removal efficiency after 1 h of photocatalytic reaction, in the presence of ZnAlFe$_{1-x}$Sm$_x$O$_4$ (x = 0, 0.02, 0.04, 0.06, 0.08) photocatalysts.

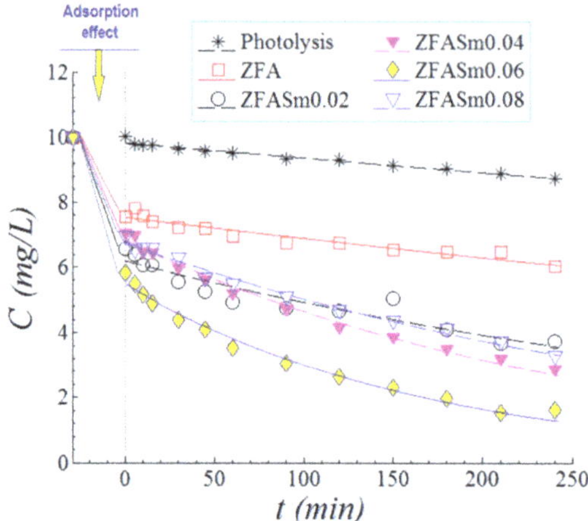

Figure 8. Kinetics of EB dye concentration decay versus irradiation time recorded during photolysis and photocatalysis processes under UV-Vis light; $ZnAlFe_{1-x}Sm_xO_4$ (x = 0, 0.02, 0.04, 0.06, 0.08) were used as photocatalysts; solid and dashed lines provide predictions according to PFO kinetic model; experimental conditions: catalyst dosage = 0.75 g/L; T = 25 ± 1 °C.

Table 2. Kinetic parameters for EB dye photodegradation under UV-Vis light in the presence of catalytic materials $ZnAlFe_{1-x}Sm_xO_4$ (x = 0, 0.02, 0.04, 0.06, 0.08).

Photocatalyst Code	Pseudo-First-Order Reaction Rate Constant, k (min^{-1})	χ^2-Test Value
na (photolysis)	4.886×10^{-4}	0.005
ZFA	8.983×10^{-4}	0.041
ZFASm0.02	2.275×10^{-3}	0.234
ZFASm0.04	3.842×10^{-3}	0.058
ZFASm0.06	6.014×10^{-3}	0.154
ZFASm0.08	2.945×10^{-3}	0.039

Consequently, analysing the pseudo-first-order reaction rate constants reported in Table 2, one can see that for the experiments with photocatalytic material, the k values were at least two-times higher as compared to the experiment without catalyst (photolysis). More importantly, the k values for the samarium-doped photocatalysts were clearly higher than for the undoped ZFA sample, demonstrating the positive impact of the Sm^{3+} presence in the spinel structure on the photocatalysts. On the other hand, among the samarium-doped materials, as previously observed for the colour-removal efficiency after 1 h of photocatalytic reaction, the most performant catalyst was the ZFASm0.06. This fact indicates that, for our series of photocatalysts, lattice parameters, interplanar distances, and particle sizes are playing a key role in the photocatalytic behaviour of the obtained materials, as compared to the crystallite size.

2.2.3. Effect of Hydrogen Peroxide Addition and Catalyst Dose: Process Optimization

The intensification of the photocatalytic process (under UV-Vis light irradiation) was carried out by adding hydrogen peroxide (H_2O_2) to the system and by varying the photocatalyst dose. In this regard, the best photocatalyst sample, ZFASm0.06 (with formulation $ZnAlFe_{0.94}Sm_{0.06}O_4$), was used in these experiments for EB photodegradation. Thus, herein, we investigated the synergistic effect of two operating parameters (factors), i.e., the initial concentration of the hydrogen peroxide ((H_2O_2), M) in the system and the catalyst dose (CatDose, g/L). For this purpose, we applied the design of experiments (DoE) and response surface methodology (RSM) as the modelling-optimization tools. More details regarding these methodologies (DoE and RSM) can be found in the following references [37,38].

Hence, the simultaneous influence of both factors ((H_2O_2) and CatDose) was investigated in a systematic fashion by using DoE and RSM. The purpose of this study was to intensify the photocatalytic process by maximization of the colour-removal efficiency, Y (the process response). In all experiments, the aqueous solutions of 10 mg/L EB dye (initial concentration) were subjected to photodegradation at a temperature level of $25 \pm 1\,°C$.

For each experimental trial, the process response Y (colour-removal efficiency) was determined for an irradiation time equal to 60 min. The aim of the optimization was to maximize the value of the colour-removal efficiency $Y(\%)$, which may be expressed as:

$$Y = \left(1 - \frac{C}{C_0}\right) \times 100 \tag{2}$$

where C_0 is the initial concentration of EB dye (10 mg/L) and C is the remaining concentration of EB dye determined after $t = 60$ min irradiation time.

For modelling purposes, both operating parameters ((CatDose) and (H_2O_2)) were scaled into the coded variables x_1 and x_2 in order to compare their influence in the same dimensionless scale. The mathematical relations used for coding factors are given elsewhere [37,38].

In this section, a central composite experimental design of the face-centred type was adopted to conduct experiments (Table 3). According to Table 3, the operating parameters (factors) are reported in actual values ((CatDose) and (H_2O_2)) as well as in coded values (x_1 and x_2). The experimental design matrix (Table 3) included 11 experimental trials, where both factors were changed simultaneously. As a result, the process response (Y, %) was determined for each run (set of conditions). Note that the central assays (runs no.9 to 11) were performed to estimate the reproducibility of the experiment.

Table 3. Central composite design of face-centred type employed for experimentation.

Run	Catalyst Dose (g/L)		Concentration of Hydrogen Peroxide, (M = mol/L)		Colour-Removal Efficiency (Response), Determined after 60 min Irradiation Time
	Coded x_1	Actual CatDose, g/L	Coded x_2	Actual [H_2O_2], M	Y (%)
1	−1	0.50	−1	0.00	29.55
2	+1	1.00	−1	0.00	50.79
3	−1	0.50	+1	0.10	91.44
4	+1	1.00	+1	0.10	92.41
5	−1	0.50	0	0.05	90.98
6	+1	1.00	0	0.05	88.99
7	0	0.75	−1	0.00	49.69
8	0	0.75	+1	0.10	95.44
9	0	0.75	0	0.05	92.81
10	0	0.75	0	0.05	93.75
11	0	0.75	0	0.05	93.01

Based on the data shown in Table 3, a mathematical model was built using multiple regression techniques [37,38]. Finally, the multiple-regression model for estimated response ($\hat{Y}$) can be written in terms of coded variables (x_1 and x_2) as follows:

$$\hat{Y} = 93.89 + 3.37x_1 + 24.88x_2 - 5.07x_1x_2 - 4.94x_1^2 - 22.37x_2^2 \text{ subjected to}: \quad -1 \leq x_j \leq +1; \ (j = 1, 2) \tag{3}$$

The obtained multiple-regression model (Equation (3)) was validated from a statistical point of view by using analysis of variance (ANOVA) [37]. The statistical estimators calculated by the ANOVA method are given in Table 4.

Table 4. Analysis of variance (ANOVA) for the multiple-regression model $\hat{Y}(x_1, x_2)$.

Source	DF [1]	SS [2]	MS [3]	F-value [4]	P-value [5]	R2 [6]	R_{adj}^2 [7]
Model	5476.64	5	1095.33	71.50	0.0001	0.986	0.972
Residual	76.59	5	15.32				
Total	5553.23	10					

[1] degree of freedom; [2] sum of squares; [3] mean square; [4] ratio between mean squares; [5] probability of randomness; [6] coefficient of determination; [7] adjusted coefficient of determination.

According to ANOVA outcomes (Table 4), the F-value of 71.50 and a very small P-value (0.0001) indicated a significant model from a statistical standpoint. In other words, the model is adequate to estimate the process response in the valid region (region of experimentation). Moreover, the value of the determination coefficient R^2 points out that the model can explain more than 98% of data variation. Further, the adjusted coefficient R^2_{adj} is close to R^2, suggesting that the multiple-regression model implying main, interaction, and quadratic effects offers good predictions. Figure 9 illustrates the goodness of fit between the experimental data and model predictions. The data scattered around the bisector (45° straight line) indicate a good accordance between the model and observations (Figure 9a). In addition, Figure 9b highlights the normal plot of residuals. In fact, the residual designates the difference between the experimental response and the model prediction ($Y_i - \hat{Y}_i$). Hence, this type of graph illustrates the departure of the residual errors from the normal distribution. From Figure 9b, one can see that the residuals are located in the vicinity of the straight line underlining a normal distribution.

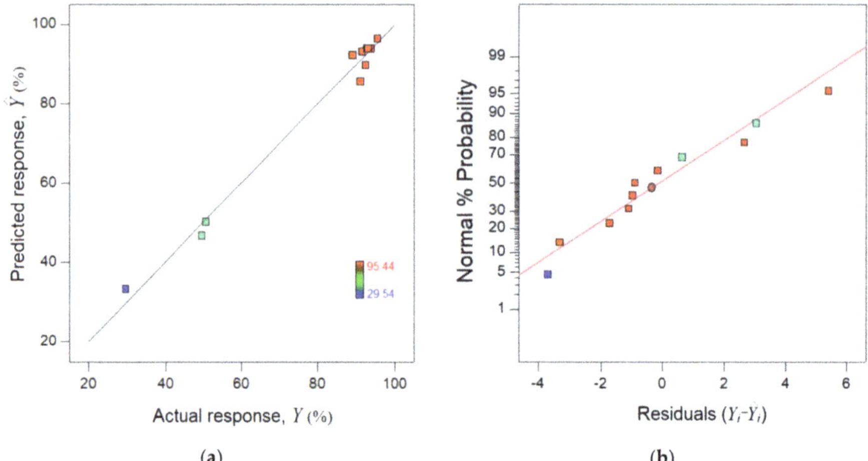

Figure 9. Agreement between experimental data and model predictions: (a) parity plot and (b) normal plot of residuals.

By using the substitution technique, the final empirical model in terms of actual factors was developed, which can be expressed as:

$$\hat{Y} = -23.17 + 152.4 \times CatDose + 1696.5 \times [H_2O_2] - 405.5 \times CatDose \times [H_2O_2] - 79.1 \times CatDose^2 - 8948.5 \times [H_2O_2]^2 \text{ subjected to :}$$
$$0.5 \leq CatDose \leq 1.0 \text{ (g/L)}; \quad 0 \leq [H_2O_2] \leq 0.10 \text{ M} \tag{4}$$

This empirical model (Equation (4)) was applied for simulation to detail the 3D response surface diagram and 2D contour-lines map in order to reveal the mutual effect of factors on the process response (Figure 10).

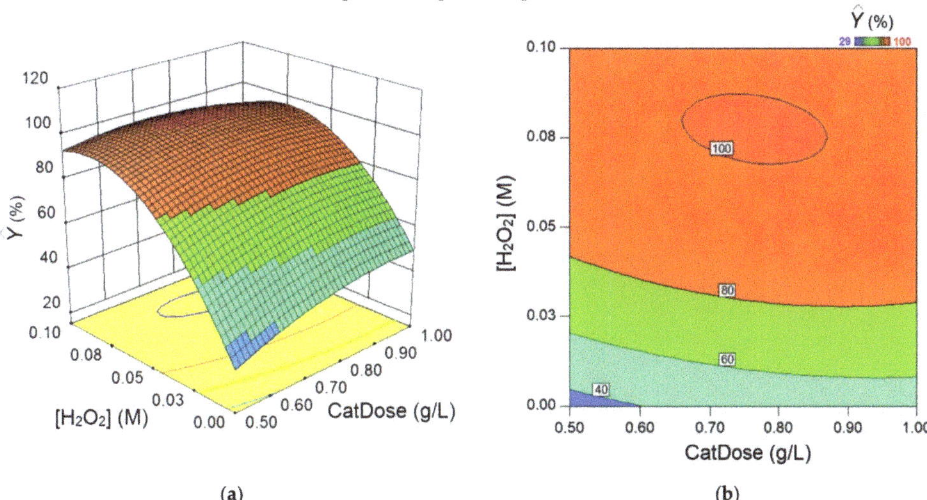

(a)   (b)

Figure 10. Response surface 3D plot (**a**) and contour-lines 2D map (**b**) showing the mutual effect of CatDose and (H_2O_2) factors on the estimated process response ($\hat{Y}$, %).

As one can see from Figure 10, the main effects of CatDose and (H_2O_2) operating parameters are positive with respect to the estimated process response ($\hat{Y}$, %). In other words, the greater the factors (CatDose and (H_2O_2)), the greater the estimated colour-removal efficiency ($\hat{Y}$, %). However, it should be noted that the main effect of (H_2O_2) is stronger than the main effect of the CatDose factor. The quadratic effects of both factors induce curvature features on the response surface, pointing out an optimal zone (maximum) on the plateau. In addition, there is a discernible interaction effect between both operating parameters CatDose and (H_2O_2). In accordance with this interaction effect, the influence of the CatDose factor on the estimated response is more evident at low concentrations of hydrogen peroxide.

The graphical analysis of the response surface (Figure 10) indicated an extreme point (a local maximum) inside the valid region. To determine exactly the coordinates of the optimal point, we used the classical approach [39]. According to this, we located, firstly, the stationary point ($x_S = [x_1\ x_2]^T$) by vanishing the first derivatives of the objective function ($\hat{Y}$) in relation to the decision variables (x_1 and x_2—coded values). Consequently, a system of two algebraic equations resulted, which was solved analytically to establish the stationary point, that is,

$$\begin{cases} \dfrac{\partial \hat{Y}}{\partial x_1} = 3.37 - 5.07x_2 - 9.88x_1 = 0 \\[2mm] \dfrac{\partial \hat{Y}}{\partial x_2} = 24.88 - 5.07x_1 - 44.74x_2 = 0 \end{cases} \tag{5}$$

Hence, the established stationary point is equal to $x_S = [0.059\ 0.549]^T$ (in terms of coded variables). In order to confirm that the stationary point (x_S) is a point of maximum, we performed the second partial derivative test based on the Hessian matrix (H) [39]. Generally, if det(H) > 0 and ($\partial^2 \hat{Y}/\partial x_1^2$) < 0, then the stationary point is a point of maximum. If det(H) > 0 and ($\partial^2 \hat{Y}/\partial x_1^2$) > 0, then the stationary point is a point of minimum. Instead, if det(H) < 0, then the stationary point is a saddle point (i.e., a point of inflection or minimax). Otherwise, for det(H) = 0, the second derivative test is inconclusive [39]. In our particular case, the Hessian matrix (H) can be written as:

$$(H)_{i,j} = \frac{\partial^2 \hat{Y}}{\partial x_i \partial x_j} = \begin{vmatrix} -9.88 & -5.07 \\ -5.07 & -44.74 \end{vmatrix} \tag{6}$$

where det(H) = 416.3; hence, the optimal point (x_S = [0.059 0.549]^T) is a point of maximum since det(H) > 0 and ($\partial^2 \hat{Y}/\partial x_1{}^2$) < 0. By converting the coded variables into the actual factors, the optimal point (x_S = [0.059 0.549]^T) can be written as CatDose = 0.765 g/L and (H_2O_2) = 0.077 M. Subsequently, the experimental validation (confirmation run) was carried out under these established optimal conditions (CatDose = 0.765 g/L and (H_2O_2) = 0.077 M). Under these optimal conditions, the colour-removal efficiency (determined at 60 min irradiation time) was equal to $\hat{Y}$ = 100.82 (estimated value by model) and Y = 98.83% (experimental value). The difference ($\hat{Y}$-Y) of 1.99% was assigned to the residual error (between the model and experiment).

In addition, under the determined optimal conditions, we recorded the full kinetics of photo-degradation of EB dye (Figure 11). Results revealed a significant PFO rate constant (k = 1.020 × 10^{-1} min^{-1}) for these optimal conditions if compared with a similar system (k = 6.014 × 10^{-3} min^{-1}) that did not contain hydrogen peroxide as a booster.

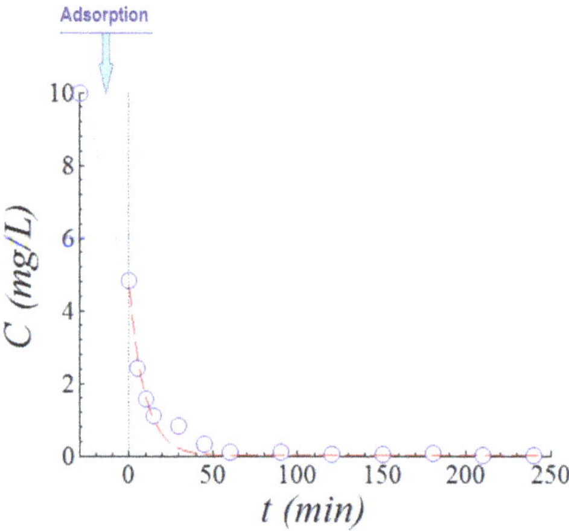

Figure 11. Kinetics of photocatalytic degradation (UV-Vis light) of EB dye recorded under optimal conditions in the presence of ZnAlFe$_{0.94}$Sm$_{0.06}$O$_4$ (sample ZFASm0.06); experimental conditions: CatDose = 0.765 g/L, (H_2O_2) = 0.077 M, and T = 25 ± 1 °C.

By using the term, the half-life of reaction (τ = *ln2/k*), the process was evidently intensified after optimization. Thus, the half-life of the reaction decreased significantly from 115 min to about 7 min, which corresponds to a process intensification of about 16-fold. It should be mentioned here that after an irradiation time of 240 min (UV-Vis light), the final colour-removal efficiency was equal to 99.9% (Figure 11).

For modelling and optimization purposes, the scientific computations were carried out by using the Design-Expert and Matlab software packages.

3. Materials and Methods

3.1. Photocatalyst Synthesis

Zinc ferrite nanoparticles substituted with aluminium and doped with samarium cations with the formula ZnAlFe$_{1-x}$Sm$_x$O$_4$ (x = 0.02, 0.04, 0.06, 0.08) were obtained via the sol–gel autocombustion technique using citric acid as fuel. Likewise, the undoped ZnAlFeO$_4$ (x = 0) sample was prepared for comparison. Analytical reagent-graded zinc nitrate (Zn(NO$_3$)$_2$·6H$_2$O), aluminium nitrate (Al(NO$_3$)$_3$·9H$_2$O), iron nitrate (Fe(NO$_3$)$_3$·9H$_2$O), and samarium nitrate (Sm(NO$_3$)$_3$·6H$_2$O) and citric acid (C$_6$H$_8$O$_7$·H$_2$O) were used, as purchased from Sigma-Aldrich, to obtain the ferrite samples. First, stoichiometric mixtures of the metal nitrate solutions were prepared. Second, each sample of the metal nitrate combination was mixed with a solution of citric acid in a 1:1 molar ratio of combustion agent to metallic cations. The mixtures were heated at 80 °C in a water bath to produce viscous gels. In order to provoke self-ignition, the gels were progressively heated up to 350 °C in a sand bath. The resulting powders underwent two stages of thermal treatment, including pre-sinterization at 500 °C for five hours and sinterization at 700 °C for five hours.

The obtained samples were denoted as ZFA, ZFASm0.02, ZFASm0.04, ZFASm0.06, and ZFASm0.08, respectively, after the cation initials present in the system and the samarium content. For a better understanding, a synthesis flowchart is provided in Scheme 1.

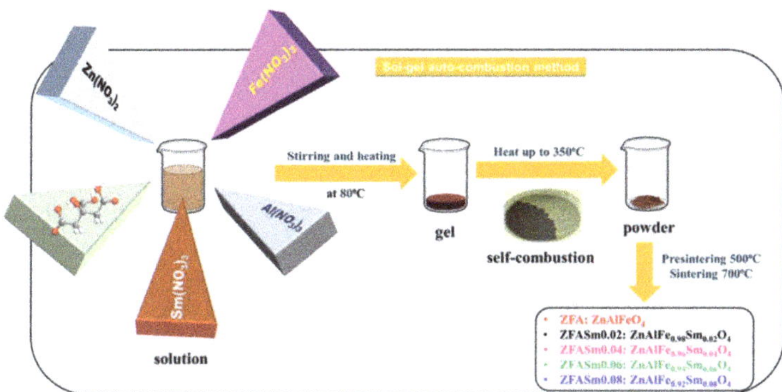

Scheme 1. A synthesis flowchart of $ZnAlFe_{1-x}Sm_xO_4$ (x = 0, 0.02, 0.04, 0.06, 0.08) photocatalysts.

3.2. Photocatalyst Characterization

Structural characterization of the ferrites heated at 7000 C was performed by recording the powder X-ray diffraction (XRD) patterns on a Rigaku Miniflex 600 diffractometer (Rigaku Corporation, Tokio, Japan) using CuKα emission in a 2θ range 20–80°, with a scanning step of 0.01° and a recording rate of 1°/min.

The formation of spinel structure at 7000 C was observed by infrared spectroscopy using a Bruker Vertex 70 FTIR spectrometer (Ettlingen, Germany) using the KBr pellets technique, at room temperature, with a resolution of 2 cm^{-1}. The spectra were recorded in a range of 4000–400 cm^{-1}.

The morphology and microstructure of photocatalysts were investigated using a Transmission Electron Microscope (TEM) Hitachi High-Tech HT7700 (Hitachi High Technologies Corporation, Tokyo, Japan), operated in high-contrast mode at 100 kV accelerating voltage. Samples were prepared by dispersion in acetone, ultrasonication for 30 min, drop casting on 300 mesh, Ted Pella carbon-coated copper grids, and drying in a vacuum at 60 °C.

Magnetic measurements were performed on a (vibrating sample magnetometer) VSM LakeShore 8607 system (Lake Shore Cryotronics, Westerville, OH, USA) at room temperature. The samples were demagnetized in the alternating field before the experiment.

Textural characteristics of the synthesized materials were determined by registering nitrogen adsorption–desorption isotherms on an NOVA 2200e Quantachrome instrument (Quantachrome instruments, Boynton Beach, FL, USA), working at N_2 liquefaction temperature of −196 °C. Before measurements, each sample was outgassed at room temperature for two hours, for the purpose of emptying the pores from physically adsorbed molecules. The obtained data allowed us to calculate the specific surface area by applying BET theory (Brunauer–Emmet–Teller) for the isotherm region of P/P_0 = 0.05–0.35. Moreover, the BJH (Barrett–Joyner–Halenda) equation was used to draw the pore-size distributions, and the total pore volume was calculated by considering the volume of nitrogen adsorbed on the material's surface at a relative pressure of 0.95 P/P_0.

The optical behaviour of ZFA and Sm-doped ZFA samples was investigated through DRS technique (Diffuse Reflectance Spectroscopy) by recording UV-Vis spectra on a Schimadzu UV-2450 spectrophotometer (Kyoto, Japan) equipped with an integrating sphere to measure the reflectance of the solid sample. All samples were compacted into pellets using MgO as white standard. The remission function of Kubelka–Munk was applied to registered optical absorption spectra, and direct and indirect band gap energies were determined using Equation (7):

$$F(R) = \frac{(1-R)^2}{2R} = \frac{k}{S} = \frac{Ac}{S} \tag{7}$$

where k is the Kubelka–Munk absorption coefficient, S is the scattering coefficient, R is the reflectance, c is the concentration of the absorbing species, and A is the absorbance.

3.3. Photocatalytic Activity Evaluation

The photodegradation under UV-Vis light of Evans Blue (T-1824) dye, as provided by Sigma-Aldrich, was studied in this work.

A medium-pressure Hg lamp was placed in the centre of a cylindrical photoreactor and used for the photodegradation of Evans Blue. In order to maintain the temperature at 25 °C during the reaction, the lamp was placed inside a recirculating water jacket. For a typical photocatalytic run, 600 mL of 10 mg L^{-1} Evans Blue solution was placed in the cylindrical photoreactor and then 0.45 g of spinel ferrite photocatalyst, in powder form, was added under continuous stirring. After 30 min of magnetic stirring in the dark, to achieve adsorption–desorption equilibrium at a constant temperature, the suspension was exposed to irradiation. This moment was considered as zero time for all photodegradation experiments. Stirring was continued throughout the irradiation process to keep the mixture suspended. At time intervals of 5, 10, 15, 30, 45, 60, 90, 120, 150, 180, 210, and 240 min, analytical samples were extracted from the photoreactor in order to evaluate the change in the concentration of each irradiated solution by measuring the absorbance in a range of 300–800 nm for Evans blue, using a Hitachi U-2910 Spectrophotometer (Hitachi High Technologies Corporation, Tokyo, Japan). The estimated absorbance of Evans Blue solutions at 606 nm was used to calculate the organic pollutant concentration. For comparison purposes, an experiment without a photocatalyst was performed respecting the previous protocol.

4. Conclusions

In this study, the efficiency of zinc aluminium ferrite and samarium-doped zinc aluminium ferrite, with the chemical formula ZnAlFe$_{1-x}$Sm$_x$O$_4$ (x = 0, 0.02, 0.04, 0.06, 0.08), as photocatalysts for the degradation of a dangerous organic pollutant such as Evans Blue was examined. The sol–gel autocombustion preparation technique successfully allowed for the insertion of aluminium cations into the zinc ferrite spinel structure, as well as the incorporation of Sm cations into the spinel unit cell of the aluminium-substituted zinc ferrite, as proven by XRD and FTIR analysis. TEM analysis showed a decrease in particle sizes with samarium doping, but no strict correlation between samarium content and grain sizes was found. Magnetic characterization of the materials indicated the paramagnetic behaviour, for all studied samples, with zero intrinsic coercivity and zero remanent magnetization.

The photodegradation of Evans Blue was successfully achieved by using the studied photocatalysts. The samarium-doped photocatalysts were clearly more active than the undoped material, demonstrating the positive impact of Sm^{3+} presence in the spinel structure on the photocatalysts. The photocatalytic performances followed the same trend as the lattice parameter, interplanar distance, and particle-size values. Thus, the sample with the chemical formula ZnAlFe$_{0.94}$Sm$_{0.06}$O$_4$ was the most active by far. The intensification of the photocatalytic process was realized by adding hydrogen (H$_2$O$_2$) peroxide to the system and by varying the best catalyst dose in the frame of the optimization procedure based on a central composite experimental design of the face-centred type. The obtained results revealed an increase in pseudo-first-order reaction rate constant of about 16-times using hydrogen peroxide as a photodegradation reaction booster.

Author Contributions: Conceptualization, P.S. and C.C.; methodology, P.S., C.C., I.G., M.I. and P.P; software, C.C. and I.-A.D.; validation, P.S., C.C. and V.H.; investigation, I.G., P.S., C.C., P.P., M.I. and I.-A.D.; resources, P.S. and V.H.; writing—original draft preparation, P.S., C.C., I.G., M.I. and P.P.; writing—review and editing, P.S., C.C. and V.H.; supervision, V.H.; project administration, P.S and C.C.; funding acquisition, P.S. All authors have read and agreed to the published version of the manuscript.

Funding: This work was supported by a grant from the Ministry of Research, Innovation and Digitization, CNCS-UEFISCDI, project number PN-III-P1-1.1-TE-2021-0030, within PNCDI III.

Data Availability Statement: Not applicable.

Conflicts of Interest: The authors declare no conflict of interest.

References

1. Soufi, A.; Hajjaoui, H.; Elmoubarki, R.; Abdennouri, M.; Qourzal, S.; Barka, N. Spinel ferrites nanoparticles: Synthesis methods and application in heterogeneous Fenton oxidation of organic pollutants—A review. *Appl. Surf. Sci. Adv.* **2021**, *6*, 100145. [CrossRef]
2. Gong, L.; Chen, G.; Lv, J.; Lu, M.; Zhang, J.; Wu, X.; Wang, J. Phase transition-enabled $MnFe_2O_4$ nanoparticles modulated by high-pressure with enhanced electrical transport properties. *Appl. Surf. Sci.* **2021**, *565*, 150532. [CrossRef]
3. Anila, I.; Mathew, M.J. Study on the physico-chemical properties, magnetic phase resolution and cytotoxicity behavior of chitosan-coated cobalt ferrite nanocubes. *Appl. Surf. Sci.* **2021**, *556*, 149791. [CrossRef]
4. Samoila, P.; Cojocaru, C.; Mahu, E.; Ignat, M.; Harabagiu, V. Boosting catalytic wet-peroxide-oxidation performances of cobalt ferrite by doping with lanthanides for organic pollutants degradation. *J. Environ. Chem. Eng.* **2021**, *9*, 104961. [CrossRef]
5. Samoila, P.; Cojocaru, C.; Cretescu, I.; Stan, C.D.; Nica, V.; Sacarescu, L.; Harabagiu, V. Nanosized spinel ferrites synthesized by sol-gel autocombustion for optimized removal of azo dye from aqueous solution. *J. Nanomater.* **2015**, *2015*, 713802. [CrossRef]
6. Kefeni, K.K.; Mamba, B.B. Photocatalytic application of spinel ferrite nanoparticles and nanocomposites in wastewater treatment: Review. *Sustain. Mater. Technol.* **2019**, *23*, e00140. [CrossRef]
7. Hcini, F.; Hcini, S.; Alzahrani, B.; Zemni, S.; Bouazizi, M.L. Effects of sintering temperature on structural, infrared, magnetic and electrical properties of $Cd_{0.5}Zn_{0.5}FeCrO_4$ ferrites prepared by sol–gel route. *J. Mater. Sci. Mater. Electron.* **2020**, *31*, 14986–14997. [CrossRef]
8. Sharma, A.; Harmanpreet, H. Influence of different substitution metal ion on magnetic properties of Mn-Zn ferrite. *Mater. Today Proc.* **2020**, *37*, 3058–3060. [CrossRef]
9. Borhan, A.I.; Samoila, P.; Hulea, V.; Iordan, A.R.; Palamaru, M.N. Effect of Al^{3+} substituted zinc ferrite on photocatalytic degradation of Orange I azo dye. *J. Photochem. Photobiol. A* **2014**, *279*, 17–23. [CrossRef]
10. Pham, T.N.; Huy, T.Q.; Le, A.-T. Spinel ferrite (AFe_2O_4)-based heterostructured designs for lithium-ion battery, environmental monitoring, and biomedical applications. *RSC Adv.* **2020**, *10*, 31622–31661. [CrossRef]
11. Amiri, M.; Salavati-Niasari, M.; Akbari, A. Magnetic nanocarriers: Evolution of spinel ferrites for medical applications. *Adv. Colloid Interface Sci.* **2019**, *265*, 29–44. [CrossRef]
12. Wu, K.; Li, J.; Zhang, C. Zinc ferrite based gas sensors: A review. *Ceram. Int.* **2019**, *45*, 11143–11157. [CrossRef]
13. Sharma, S.S.; Dutta, V.; Raizada, P.; Hosseini-Bandegharaei, A.; Thakur, V.; Nguyen, V.-H.; Vanle, Q.; Singh, P. An overview of heterojunctioned $ZnFe_2O_4$ photocatalyst for enhanced oxidative water purification. *J. Environ. Chem. Eng.* **2021**, *9*, 105812. [CrossRef]
14. Katrapally, V.K.; Bhavani, S.D. Impact of calcination temperature on structural and optical properties of erbium-doped zinc ferrite nanoparticles. *Phase Transit.* **2022**, *95*, 770–785. [CrossRef]
15. Qin, M.; Shuai, Q.; Wu, G.; Zheng, B.; Wang, Z.; Wu, H. Zinc ferrite composite material with controllable morphology and its applications. *Mater. Sci. Eng. B* **2017**, *224*, 125–138. [CrossRef]
16. Samoila, P.; Cojocaru, C.; Sacarescu, L.; Dorneanu Pascariu, P.; Domocos, A.A.; Rotaru, A. Remarkable catalytic properties of rare-earth doped nickel ferrites synthesized by sol-gel auto-combustion with maleic acid as fuel for CWPO of dyes. *Appl. Catal. B Environ.* **2017**, *202*, 21–32. [CrossRef]
17. Zhang, J.; Song, J.-M.; Niu, H.-L.; Mao, C.-J.; Zhang, S.-Y.; Shen, Y.-H. ZnFe2O4 nanoparticles: Synthesis, characterization, and enhanced gas sensing property for acetone. *Sens. Actuators B Chem.* **2015**, *221*, 55–62. [CrossRef]
18. Ignat, M.; Samoila, P.; Cojocaru, C.; Sacarescu, L.; Harabagiu, V. Novel Synthesis Route for Chitosan-Coated Zinc Ferrite Nanoparticles as Potential Sorbents for Wastewater Treatment (Chitosan-ZnFe2O4 Sorbent for Wastewater Treatment). *Chem. Eng. Commun.* **2016**, *203*, 1591–1599. [CrossRef]
19. Shannon, R.D. Revised effective ionic radii and systematic studies of interatomic distances in halides and chalcogenides. *Acta Cryst.* **1976**, *A32*, 751–767. [CrossRef]
20. Samoila, P.; Sacarescu, L.; Borhan, A.I.; Timpu, D.; Grigoras, M.; Lupu, N.; Zaltariov, M.; Harabagiu, V. Magnetic properties of nanosized Gd doped Ni-Mn-Cr ferrites prepared using the sol-gel autocombustion technique. *J. Magn. Magn. Mater.* **2015**, *378*, 92–97. [CrossRef]

21. Slimani, Y.; Almessiere, M.A.; Guner, S.; Aktas, B.; Shirsath, S.E.; Silibin, M.V.; Trukhanov, A.V.; Baykal, A. Impact of Sm^{3+} and Er^{3+} cations on the structural, optical, and magnetic traits of spinel cobalt ferrite nanoparticles: Comparison investigation. *ACS Omega* **2022**, *7*, 6292–6301. [CrossRef] [PubMed]
22. Xavier, S.; Thankachan, S.; Jacob, B.P.; Mohammed, E.M. Effect of samarium substitution on the structural and magnetic properties of nanocrystalline cobalt ferrite. *J. Nanosci.* **2013**, *524380*, 524380. [CrossRef]
23. Borhan, A.I.; Iordan, A.R.; Palamaru, M.N. Correlation between structural, magnetic and electrical properties of nanocrystalline Al^{3+} substituted zinc ferrite. *Mater. Res. Bull.* **2013**, *48*, 2549–2556. [CrossRef]
24. Hajdu, V.; Muránszky, G.; Nagy, M.; Kopcsik, E.; Kristály, F.; Fiser, B.; Viskolcz, B.; Vanyorek, L. Development of high-efficiency, magnetically separable palladium-decorated manganese-ferrite catalyst for nitrobenzene hydrogenation. *Int. J. Mol. Sci.* **2022**, *23*, 6535. [CrossRef]
25. Kumar, N.H.; Ravinder, D.; Edukondalu, A. Effect of Ce^{3+} ion doped Ni-Zn Ferrites: Structural, Optical and Low temperature Magnetic Properties. *Chin. J. Phys.* **2022**, *81*, 171–180. [CrossRef]
26. Naik, P.P.; Tangsali, R.B.; Meena, S.S.; Yusuf, S.M. Influence of rare earth (Nd^{3+}) doping on structural and magnetic properties of nanocrystalline manganese-zinc ferrite. *Mater. Chem. Phys.* **2017**, *191*, 215–224. [CrossRef]
27. Amri, N.; Massoudi, J.; Nouri, K.; Triki, M.; Dhahri, E.; Bessais, L. Influence of neodymium substitution on structural, magnetic and spectroscopic properties of Ni–Zn– Al nano-ferrites. *RSC Adv.* **2021**, *11*, 13256–13268. [CrossRef]
28. Kanagesan, S.; Hashim, M.; Aziz, S.A.B.; Ismail, I.; Tamilselvan, S.; Alitheen, N.B.; Swamy, M.K.; Rao, B.P.C. Evaluation of Antioxidant and Cytotoxicity Activities of Copper Ferrite (CuFe2O4) and Zinc Ferrite (ZnFe2O4) Nanoparticles Synthesized by Sol Gel Self-Combustion Method. *Appl. Sci.* **2016**, *6*, 184. [CrossRef]
29. Singh, J.P.; Dixit, G.; Srivastava, R.C.; Negi, P.; Agrawal, H.M.; Kumar, R. HRTEM and FTIR investigation of nanosized zinc ferrite irradiated with 100 MeV oxygen ions. *Spectrochim. Acta A Mol. Biomol. Spectrosc.* **2013**, *107*, 326–333. [CrossRef]
30. Ochmann, M.; Vrba, V.; Kopp, J.; Ingr, T.; Malina, O.; Machala, L. Microwave-enhanced crystalline properties of zinc ferrite nanoparticles. *Nanomaterials* **2022**, *12*, 2987. [CrossRef]
31. Xiang, Q.-Y.; Wu, D.; Bai, Y.; Yan, K.; Yao, W.-Q.; Zhang, L.; Zhang, J.; Cao, J.-L. Atomic hydrogenation-induced paramagnetic-ferromagnetic transition in zinc ferrite. *Ceram. Int.* **2016**, *42*, 16882–16887. [CrossRef]
32. Thommes, M.; Kaneko, K.; Neimark, A.V.; Olivier, J.P.; Rodriguez-Reinoso, F.; Rouquerol, J.; Sing, K.S.W. Physisorption of gases, with special reference to the evaluation of surface area and pore size distribution (IUPAC Technical Report). *Pure Appl. Chem.* **2015**, *87*, 1051–1069. [CrossRef]
33. Thommes, M.; Cychosz, K.A. Physical adsorption characterization of nanoporous materials: Progress and challenges. *Adsorption* **2014**, *20*, 233–250. [CrossRef]
34. Adarsha, J.R.; Ravishankar, T.N.; Manjunatha, C.R.; Ramakrishnappa, T. Green synthesis of nanostructured calcium ferrite particles and its application to photocatalytic degradation of Evans blue dye. *Mater. Today Proc.* **2022**, *49*, 777–788. [CrossRef]
35. Jinendra, U.; Bilehal, D.; Nagabhushana, B.M.; Jithendra Kumara, K.M.; Kollur, S.P. Nano-catalytic behavior of highly efficient and regenerable mussel-inspired Fe_3O_4@CFR@GO and Fe_3O_4@CFR@TiO_2 magnetic nanospheres in the reduction of Evans blue dye. *Heliyon* **2021**, *7*, e06070. [CrossRef]
36. Perniss, A.; Wolf, A.; Wichmann, L.; Schönberger, M.; Althaus, M. Evans Blue is not a suitable inhibitor of the epithelial sodium channel d-subunit. *Biochem. Biophys. Res. Commun.* **2015**, *466*, 468–474. [CrossRef]
37. Bezerra, M.A.; Santelli, R.E.; Oliveira, E.P.; Villar, L.S.; Escaleira, L.A. Response surface methodology (RSM) as a tool for optimization in analytical chemistry. *Talanta* **2008**, *76*, 965–977. [CrossRef]
38. Mäkelä, M. Experimental design and response surface methodology in energy applications: A tutorial review. *Energy Convers. Manage.* **2017**, *151*, 630–640. [CrossRef]
39. Rao, S.S. *Engineering Optimization: Theory and Practice*, 4th ed.; John Wiley and Sons: Hoboken, NJ, USA, 2009; pp. 63–75.

Review

Recent Advances in the Development of Novel Iron–Copper Bimetallic Photo Fenton Catalysts

Gabriela N. Bosio [1], Fernando S. García Einschlag [1], Luciano Carlos [2,*] and Daniel O. Mártire [1,*]

[1] Instituto de Investigaciones Fisicoquímicas Teóricas y Aplicadas (INIFTA), Departamento de Química, Facultad de Ciencias Exactas, Universidad Nacional de la Plata, Consejo Nacional de Investigaciones Científicas y Técnicas (CONICET), La Plata 1900, Argentina

[2] Instituto de Investigación y Desarrollo en Ingeniería de Procesos, Biotecnología y Energías Alternativas, PROBIEN (CONICET-UNCo), Universidad Nacional del Comahue, Neuquén 8300, Argentina

* Correspondence: luciano.carlos@probien.gob.ar (L.C.); dmartire@inifta.unlp.edu.ar (D.O.M.)

Abstract: Advanced oxidation processes (AOPs) have been postulated as viable, innovative, and efficient technologies for the removal of pollutants from water bodies. Among AOPs, photo-Fenton processes have been shown to be effective for the degradation of various types of organic compounds in industrial wastewater. Monometallic iron catalysts are limited in practical applications due to their low catalytic activity, poor stability, and recyclability. On the other hand, the development of catalysts based on copper oxides has become a current research topic due to their advantages such as strong light absorption, high mobility of charge carriers, low environmental toxicity, long-term stability, and low production cost. For these reasons, great efforts have been made to improve the practical applications of heterogeneous catalysts, and the bimetallic iron–copper materials have become a focus of research. In this context, this review focuses on the compilation of the most relevant studies on the recent progress in the application of bimetallic iron–copper materials in heterogeneous photo–Fenton-like reactions for the degradation of pollutants in wastewater. Special attention is paid to the removal efficiencies obtained and the reaction mechanisms involved in the photo–Fenton treatments with the different catalysts.

Keywords: bimetallic catalysts; copper; heterogeneous catalysis; iron; photo-Fenton

Citation: Bosio, G.N.; García Einschlag, F.S.; Carlos, L.; Mártire, D.O. Recent Advances in the Development of Novel Iron–Copper Bimetallic Photo Fenton Catalysts. *Catalysts* **2023**, *13*, 159. https://doi.org/10.3390/catal13010159

Academic Editors: Gassan Hodaifa, Rafael Borja and Mha Albqmi

Received: 23 November 2022
Revised: 21 December 2022
Accepted: 4 January 2023
Published: 10 January 2023

1. Introduction

In recent years, the rapid development of the industrial sector has resulted in numerous effluents that contain toxic substances and affect the environment in a dangerous way [1]. Some of these pollutants, due to their chemical nature, are resistant to the conventional physical and biological processes commonly used in wastewater treatment plants [2]. For this reason, advanced oxidation processes (AOPs) have been postulated as viable, innovative, and efficient technologies for the removal of these compounds from water bodies [3]. Among the AOPs, the heterogeneous Fenton and photo–Fenton processes have been shown to be effective for the degradation of various types of organic compounds in industrial wastewater [4,5].

Equations (1)–(5) are the main pathways involved in the Fenton processes. Reaction 1 is a key step because it generates the strongly oxidizing radical $HO^\bullet$, while regeneration of Fe^{2+} is the rate–determining step under dark conditions and in the absence of alternative Fe^{3+}–reducing species (Equation (2)) [6,7].

$$Fe^{2+} + H_2O_2 \rightarrow Fe^{3+} + HO^- + HO^\bullet \; (k = 40\text{–}80 \, M^{-1}s^{-1}) \tag{1}$$

$$Fe^{3+} + H_2O_2 \rightarrow Fe^{2+} + HO_2^\bullet + H^+ \; (k = 9.1 \times 10^{-7} \, M^{-1}s^{-1}) \tag{2}$$

$$Fe^{2+} + HO^\bullet \rightarrow Fe^{3+} + HO^- \qquad (k = 2.5\text{–}5 \times 10^8 \, M^{-1}s^{-1}) \tag{3}$$

$$Fe^{3+} + HO_2^\bullet \rightarrow Fe^{2+} + H^+ + O_2 \quad (k = 0.33\text{--}2.1 \times 10^6 \text{ M}^{-1}\text{s}^{-1}) \tag{4}$$

$$Fe^{2+} + HO_2^\bullet \rightarrow Fe^{3+} + HO_2^- \quad (k = 0.72\text{--}1.5 \times 10^6 \text{ M}^{-1}\text{s}^{-1}) \tag{5}$$

Equations (1)–(5) are the most important steps in the dark Fenton chemistry because H_2O_2 is consumed and Fe^{2+} is regenerated from Fe^{3+} by these reactions. The $HO^\bullet$ and $HO_2{}^\bullet$ radicals are the main species involved in the oxidation of pollutants. In addition to $Fe(II)$, other transition metal ions can also promote similar processes, which are then referred to as Fenton–like or Fenton–type. For example, both oxidation states of copper can react with H_2O_2 to form $HO_2{}^\bullet$ and $HO^\bullet$ radicals (Equations (6) and (7)) [8].

$$Cu^{2+} + H_2O_2 \rightarrow Cu^+ + HO_2^\bullet + H^+ \quad (k = 1.15 \times 10^{-6} \text{ M}^{-1}\text{s}^{-1}) \tag{6}$$

$$Cu^+ + H_2O_2 \rightarrow Cu^{2+} + HO^- + HO^\bullet \quad (k = 1.0 \times 10^4 \text{ M}^{-1}\text{s}^{-1}) \tag{7}$$

Under suitable conditions, copper can act as a better catalyst compared to iron, as it is able to form temporary complexes with oxidation products and rapidly convert Cu^+ to Cu^{2+} and vice versa [9–11]. Since the oxidation products do not form stable complexes with copper, reactive coordination sites remain available for continuous catalytic cycling. Therefore, copper not only provides a better redox cycle but is also active in the mineralization of organic matter. However, a major disadvantage of copper–based catalysts is the high excess of H_2O_2 required to maintain catalytic activity, which increases treatment costs [12]. This disadvantage has often been mitigated by the use of bimetallic composites of copper and iron [9,13–15].

Since Reaction 1 proceeds much faster than Reaction 2, the Fe^{2+} ions are consumed faster than generated, so that a large amount of ferric hydroxide is formed as sludge during the process, which causes additional problems in separation and disposal [6,16,17]. Fenton (H_2O_2/Fe^{2+}) and Fenton–like (H_2O_2/Fe^{3+}) processes can be significantly enhanced under ultraviolet (UV)/visible irradiation ($\lambda < 600$ nm). The UV region of the electromagnetic spectrum extends from about 100 nm to 400 nm, while the visible region extends from about 400 nm to 760 nm [18]. Photoirradiation prevents the accumulation of Fe^{3+} ions in the system since Fe^{2+} ions are regenerated from Fe^{3+} by photoreduction [17,19–21]. The improvement in pollutant removal rates achieved with the photo–Fenton process compared to the Fenton process can be explained by the following contributions [6,22–24]: (i) the generation of $HO^\bullet$ and Fe^{2+} via the photolysis of iron (III) hydroxo complexes ($\lambda < 580$ nm) (Equations (8) and (9)) [25,26], followed by the participation of Fe^{2+} in Equation (1) to generate further $HO^\bullet$ radicals [27], (ii) the photolysis of H_2O_2 when $\lambda < 310$ nm is used, and (iii) the photolysis of $Fe(III)$ chelates (Fe_3L) formed between Fe^{3+} and the organic substrate, its degradation intermediates or other ligands present in the reaction medium (Equation (10)) [6]. However, the photo–Fenton process still has some disadvantages, such as the narrow pH range (2.8–3.5), poor switching between Fe^{2+} and Fe^{3+}, and sludge formation. Therefore, many efforts have been made to develop catalysts that can operate efficiently at circumneutral pH conditions [28–31].

$$[Fe(OH)]^{2+} + h\nu \rightarrow Fe^{2+} + HO^\bullet \tag{8}$$

$$[Fe(H_2O)]^{3+} + h\nu \rightarrow Fe^{2+} + HO^\bullet + H^+ \tag{9}$$

$$[Fe^{3+}L] + h\nu \rightarrow [Fe^{3+}L]^\bullet + L^\bullet \tag{10}$$

In the heterogeneous Fenton process, solid iron catalysts (such as the iron minerals magnetite (Fe_3O_4) or hematite (α–Fe_2O_3)) are used instead of aqueous Fe^{2+}, to catalyze the generation of hydroxyl radicals in acidic or near–neutral media [32,33]. The mechanism of this process starts with the adsorption of H_2O_2 on the surface of Fe–based catalysts [34,35]. Then, the formation of a surface complex ($\equiv Fe^{2+}$–H_2O_2 and/or $\equiv Fe^{3+}$–H_2O_2) precursor occurs (Equations (11) and (12)). Subsequently, the Fe^{2+}–H_2O_2 complex generates $\equiv Fe^{3+}$ and the surface–bound $HO^\bullet$ radical by intramolecular electron transfer (Reaction 13) [36].

The $\equiv Fe^{3+}$–H_2O_2 complex can be converted to $\equiv Fe^{2+}$ and $^{\bullet}OOH$ (Equation (14)) [36], and the $HOO^{\bullet}$ can further regenerate $\equiv Fe^{2+}$ (Equation (15)) [37]. The $HO^{\bullet}$ can either attack the adsorbed organic compounds in the vicinity or oxidize the non–adsorbed organic compounds.

$$\equiv Fe^{2+} + H_2O_2 \rightarrow \equiv Fe^{2+} - H_2O_2 \tag{11}$$

$$\equiv Fe^{3+} + H_2O_2 \rightarrow \equiv Fe^{3+} - H_2O_2 \tag{12}$$

$$\equiv Fe^{2+} - H_2O_2 \rightarrow \equiv Fe^{3+} + HO^{\bullet} + HO^{-} \tag{13}$$

$$\equiv Fe^{3+} - H_2O_2 \rightarrow \equiv Fe^{2+} + HOO^{\bullet} + H^{+} \tag{14}$$

$$\equiv Fe^{3+} + HOO^{\bullet} \rightarrow \equiv Fe^{2+} + O_2 + H^{+} \tag{15}$$

Monometallic iron catalysts (ZVI, Fe_2O_3, Fe_3O_4, and $Fe(OOH)$) are limited in practical applications due to their low catalytic activity, poor stability, and recyclability [38]. On the other hand, copper–based oxides have become the priorities of research into novel photocatalysts, due to their advantages such as strong light absorption, high carrier mobility, non–toxicity, environmental friendliness, long–term stability, and low production cost. In particular, copper has the property of a Lewis acid, which can create a localized acidic microenvironment by interacting with iron species. The effect of expanding the pH range of the system is achieved by the localized acidic microenvironment when bimetallic Fe–Cu photocatalysts are used [39]. In this context, great efforts have been made to improve the practical applications of heterogeneous catalysts, and bimetallic oxides (composite oxides of iron and another metallic element) are an active research area. Similarly, bimetallic catalysts composed of iron and copper are advantageous for the reduction of Fe^{3+} in Fenton–like reactions. Previous reports have suggested that the cooperation between the redox pairs of iron (Fe^{3+}/Fe^{2+}) and copper (Cu^{2+}/Cu^{+}) [40,41] can accelerate electron transfer at the interface to promote the rapid reduction of Fe^{3+}. Han et al. [42] explained the synergistic effect due to the reduced ΔE for the Fe^{3+}/Fe^{2+} redox cycle obtained for the bimetallic Fe–Cu catalyst compared to the monometallic Fe catalyst. Sun et al. [43] suggested that the Fe^{2+} species of the Fe–Cu bimetallic catalyst is mainly regenerated by the reaction between Fe^{3+} and Cu^{+} (Equation (16)) rather than by the reduction of Fe^{3+} by H_2O_2 (Equation (2)). Thus, the coexistence of Fe and Cu on the surface of a catalyst can accelerate the process of electron transfer in the reaction environment and create a suitable condition for the generation of reactive radical species.

$$Cu^{+} + Fe^{3+} \rightarrow Cu^{2+} + Fe^{2+} \quad \Delta E^0 = 0.6 \text{ V} \tag{16}$$

In summary, for the above reasons, bimetallic iron–copper materials are likely to perform better as heterogeneous Fenton and photo–Fenton catalysts compared to monometallic iron or copper materials. In this context, this review focuses on the compilation of the most relevant studies on the recent advances in the application of bimetallic iron– copper materials as Fe and Cu sources in heterogeneous photo–Fenton–like reactions for the degradation of pollutants in wastewater. Although the differences in the model pollutants and the experimental conditions make an evaluation difficult, we used as central parameters for the comparisons between the different materials, on the one hand, the combinations "achieved efficiency/treatment time" (which can be found in the tables) and, on the other hand, the dominant mechanisms in each study. For clarity, the results in the next sections are classified into the following groups according to the chemical composition of the catalysts (Figure 1): (i) $CuFe_2O_4$, (ii) mixed ferrites containing Fe and Cu, (iii) $CuFeO_2$ and $CuFeS_2$ materials, (iv) Fe–Cu oxide composites, (v) Metal–Organic Frameworks (MOF) based on Fe and Cu.

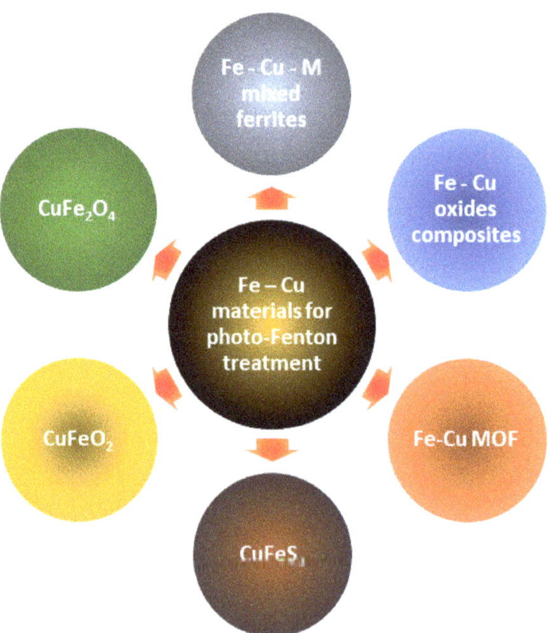

Figure 1. Classification of the different *Fe–Cu* based materials applied as catalysts in photo–Fenton processes.

2. CuFe$_2$O$_4$

Ferrites are a broad class of iron–bearing oxides that include spinel ferrites, perovskites, ilmenite (*FeTiO$_3$*), and hexagonal ferrites (or hexaferrites). In general, spinel ferrites can be described as ferrites with the formula *MFe$_2$O$_4$*, where *M* is a divalent metal cation (such as *Fe, Co, Zn, Ni, Cu*, or others) [44]. Due to their thermal stability [45], unique structural [46], optical [47], magnetic [48], electrical, and dielectric properties [49], spinel ferrites have broad potential technological applications in photoluminescence [50], photocatalysis [51], biosensor development [52], magnetic drug delivery [53], corrosion protection [54], antimicrobial agents [55], and biomedicine (hyperthermia) [56]. In particular, *CuFe$_2$O$_4$* has an inverse spinel structure with 8 *Cu^{2+}* ions in octahedral sites and 16 Fe^{3+} ions evenly distributed between the tetrahedral (A) and octahedral (B) sites [57].

The Tauc's equation was used to determine the optical band gap (*E$_g$*) of copper ferrite [58]:

$$\alpha h\nu = B\left(h\nu - E_g\right)^m$$

where α is the energy–dependent absorption coefficient, $h\nu$ is the photon energy, B is an energy independent constant, and the factor m depends on the nature of the electron transition and is equal to 2 or 1/2 for the direct and indirect transition band gaps, respectively. Assuming a direct semiconductor, E_g can be obtained from the extrapolation of the linear section of Tauc's plot of $(\alpha h\nu)^{0.5}$ vs. $h\nu$ to the energy axis. In comparison with other spinel ferrites such as *CoFe$_2$O$_4$* (*E$_g$* = 2.7 eV) and *NiFe$_2$O$_4$* NPs (*E$_g$* = 2.2 eV) [59], *CuFe$_2$O$_4$* NPs have a relatively small band gap of (1.7–1.9 eV) which allows them to use freely available energy in the form of sunlight to degrade pollutants in water [59]. Since *CuFe$_2$O$_4$* is a narrow band gap material, it can be successfully used in type II or Z–scheme heterojunction photocatalysts [60].

Wei et al. [39] prepared magnetic *Fe–Cu* materials by the sol–gel method. Solutions containing 0–40% *Cu^{2+}* were prepared separately to obtain catalysts with different iron and copper ratios. The molar ratio of iron and copper had significant effects on the surface and structural properties of the materials. These materials were used for the removal of pyridine under UV irradiation in a Fenton–like system. The effects of various parameters, such as the molar ratio of *Fe* and *Cu*, catalyst and H_2O_2 dosage, initial pH, and initial pyridine concentration were studied. The best degradation performance was obtained for the M–40% material prepared from the 40% Cu^{2+} solution, which consisted of Fe_2O_3 and $CuFe_2O_4$. This co–doped *Fe–Cu* catalyst showed high activity, good stability, and easy recovery for the degradation of pyridine when the catalyst loading was 0.90 g L^{-1}, the H_2O_2 dosage was 832.50 mg L^{-1}, the initial pH was 7, and the initial pyridine concentration was 100.00 mg L^{-1}. The degradation efficiency of pyridine was more than 99% within 30 min and the TOC removal efficiency was 97.4% within 50 min. Therefore, the M–40% catalyst could be a potential material for wastewater treatment, which could extend the active pH range to facilitate recycling and reuse. The proposed reaction mechanism involves the reduction of Cu^{2+} to Cu^+ at the surface of the photocatalyst (Equation (6)), and the reaction of Cu^+ with H_2O_2 to form hydroxyl radicals and Cu^{2+} (Equation (7)). In addition, $HO^\bullet$ is obtained from Cu^{2+} and water under UV irradiation (Equation (17)). Part of Cu^{2+} is reduced to Cu^+ by $O_2^{\bullet-}$ according to Equation (18). Furthermore, the monovalent copper ions can also be oxidized by H_2O_2 to produce Cu^{3+}, which reacts with water molecules under UV light to form hydroxyl radicals (Equations (19) and (20)). Direct evidence for the occurrence of Equation (19) was obtained by EPR spectroscopy using the spin–trapping approach [61]. It is very likely that this reaction occurs at the surface of the photocatalyst. The complete mechanism is shown in Figure 2.

$$\equiv Cu^{2+} + H_2O + h\nu \rightarrow \equiv Cu^+ + HO^\bullet + H^+ \tag{17}$$

$$Cu^{2+} + O_2^{\bullet-} \rightarrow Cu^+ + O_2 \tag{18}$$

$$Cu^+ + H_2O_2 \rightarrow Cu^{3+} + 2\,HO^- \tag{19}$$

$$\equiv Cu^{3+} + H_2O + h\nu \rightarrow \equiv Cu^{2+} + HO^\bullet + H^+ \tag{20}$$

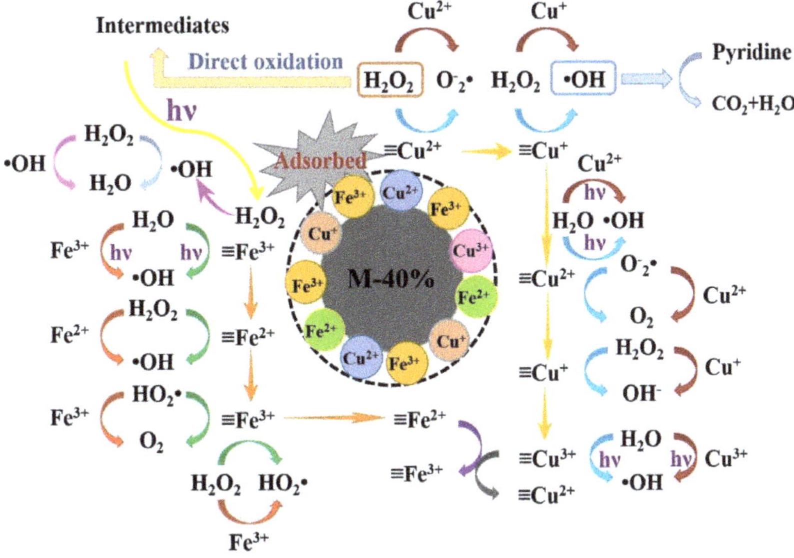

Figure 2. Reaction mechanism proposed for the degradation of pyridine by the UV/M–40%/H_2O_2 system. Reprinted with permission from Ref. [39]. Copyright 2020, Elsevier.

Silva et al. [62] prepared mixed iron and copper oxides by the modified Pechini's method. The samples were calcined at different temperatures and used for the removal of Methylene Blue dye (MB) by solar photo–Fenton catalysis. The results of X–ray diffraction analysis showed that hematite and copper ferrite phases were mainly responsible for the high efficiency of the photochemical reaction. MB removal from the aqueous solution was carried out in the presence of H_2O_2 (300 mg L^{-1}), with a 1.0 g L^{-1} catalyst, at a neutral pH, and under solar irradiation. One of the catalysts was more efficient than pure iron (Fe_2O_3) and copper oxides (CuO), indicating the synergistic effect produced by combining the two metals on the same material. The authors proposed a mechanism for the regeneration of the active sites that satisfactorily explained the experimental results.

Cao et al. [63] prepared $CuFe_2O_4$ composites via combustion solution synthesis. The authors studied the effect of Cu content on the synthesized composites. The composite with 18 wt.% Cu had the best photo–Fenton activity. This nanocomposite, which was composed of $CuFe_2O_4$ and CuO, could degrade 40 mg L^{-1} MB, 20 mg L^{-1} Rhodamine B (RhB), and 20 mg L^{-1} methyl orange (MO) in 40, 30, and 50 min, respectively. In addition, the composite showed superparamagnetic behavior and could be recycled.

Guo et al. [40] synthesized self–assembled hollow nanospheres of $CuFe_2O_4$ using the solvothermal method. The catalytic activity of the nanospheres was evaluated by the degradation of MB. In addition, these authors compared the properties of $CuFe_2O_4$ particles prepared by different methods. It was found that the good performance of the solid catalyst depends on both the large specific surface area and the degree of the optical response. The increase in photoelectric response and conductivity are beneficial for the improvement of catalytic performance. Transient photo–current experiments showed that the samples obtained by the solvothermal method exhibited a better optical response in the on–off cycle under light conditions.

Leichtweis et al. [64] prepared a novel composite catalyst by doping $CuFe_2O_4$ nanoparticles in the malt bagasse biochar. Malt bagasse (the malted barley residue) is the main residue obtained in the manufacture of beer. Composites with different ratios of malt biochar and $CuFe_2O_4$ were produced. The composites had lower band gap energy than $CuFe_2O_4$, which increased the photocatalytic activity. At 60 min of heterogeneous photo–Fenton treatment (pH 3), visible light tests showed that pure $CuFe_2O_4$ removed only 39% of the RhB color, while the composites removed up to 88% of the dye. A total of 100% color removal (pH 3) was achieved at 10- and 20-min reaction times for 10 and 50 mg L^{-1} of dye under solar irradiation. Tests performed in the presence of specific radical scavengers showed that $HO^\bullet$, $O_2^{\bullet-}$, and h^+ were the predominant reactive species involved in the degradation of the dye.

Jiang et al. [65] prepared a magnetic $Bi_2WO_6/CuFe_2O_4$ catalyst to be used in the removal of the antibiotic tetracycline hydrochloride (TCH). The obtained catalyst achieved 92.1% TCH (20 mg L^{-1}) degradation efficiency in a photo–Fenton–like system and a mineralization performance of 50.7% and 35.1% for TCH and a raw secondary effluent from a wastewater treatment plant, respectively. The excellent performance was attributed to the fact that photogenerated electrons accelerated the conversion $Fe(III)/Fe(II)$ and $Cu(II)/Cu(I)$, which increased the reaction rates of $Fe(II)/Cu(I)$ with H_2O_2 and generated abundant $HO^\bullet$ radicals for pollutant oxidation. EPR assays confirmed that $O_2^{\bullet-}$ and $HO^\bullet$ were mainly responsible for THC degradation in dark and photo–Fenton–like systems, respectively.

The low efficiency of photo–Fenton processes at a neutral pH is mainly due to the precipitation of iron and can therefore be prevented by the proper addition of iron complexing agents, such as tartaric acid [66]. Guo et al. [67] prepared $CuFe_2O_4$ particles by the sol–gel method and investigated the effects of tartaric acid (TA) on the degradation of MB in the presence of $CuFe_2O_4$ and H_2O_2 under light irradiation. The results showed that the introduction of TA increased the decolorization rate of MB decolorization from 52.0% to 92.1% within 80 min. The contribution of $O_2^{\bullet-}$ was only 10%, whereas that of $HO^\bullet$ was about 88%. The enhancement of MB degradation in the presence of TA was explained by the complexation of Fe^{3+} and Cu^{2+} with TA (Equation (21)), followed by a photo–induced ligand to metal charge transfer process (Equation (22)). In the literature, similar equations with iron species and other organic radicals have been proposed to explain the photo–Fenton mechanism [66]. The photo–induced intramolecular charge transfer of the copper (II) complex of tartaric acid has already been described in the mechanism of the catalytic reduction of Cr(VI) by tartaric acid under the irradiation of simulated sunlight [68]. The excited species $Fe^{2+}/Cu^{+}-(TA)^{\bullet}$ further produced Fe^{2+}/Cu^{+} and $TA^{\bullet}$ radicals according to Equation (23). Moreover, the presence of molecular oxygen also accelerated the transformation of the chemical state of the metal ions (see Equation (24)).

$$Fe^{3+}/Cu^{2+} + TA \rightarrow Fe^{3+}(TA)/Cu^{2+}(TA) \tag{21}$$

$$Fe^{3+}(TA)/Cu^{2+}(TA) + h\nu \rightarrow Fe^{2+}(TA)^{\bullet}/Cu^{+}(TA)^{\bullet} \tag{22}$$

$$Fe^{2+}(TA)^{\bullet}/Cu^{+}(TA)^{\bullet} \rightarrow Fe^{2+}/Cu^{+} + (TA)^{\bullet} \tag{23}$$

$$Fe^{2+}(TA)^{\bullet}/Cu^{+}(TA)^{\bullet} + O_2 \rightarrow Fe^{3+}(TA)^{\bullet}/Cu^{+}(TA)^{\bullet} + O_2^{\bullet -} \tag{24}$$

Rocha et al. [69] prepared $CuFe_2O_4$ from copper recycled from spent lithium–ion batteries. The photocatalytic properties were analyzed by monitoring the decolorization of MB in a heterogeneous photo–Fenton process in the presence of solar radiation. The decolorization efficiency was 96.1% in 45 min of reaction Formic and acetic acids were detected as degradation products of MB by ion chromatography.

Lin and Lu [70] developed $CuFe_2O_4$ nanoparticles decorated on partially reduced graphene oxides ($CuFe_2O_4@rGO$), to selectively and efficiently cleave lignin model compounds into value–added aromatic chemicals via a sunlight–assisted heterogeneous Fenton process. Controlled oxidative cleavage of the lignin model compound enabled the production of high–value aromatic chemicals, guaiacol and 2–methoxy–4–propylphenol, in high yields, instead of complete mineralization of the lignin model compound. The partially reduced graphene oxides serve as the large size support to accommodate and immobilize the $CuFe_2O_4$ nanoparticles for easy recycling of the catalyst, to attract the lignin model compounds through π–π stacking and hydrogen bonding for efficient cleavage reaction, and to accelerate the transport of photo–induced electrons for better charge separation and thus higher photocatalytic activities.

Membrane separation offers many advantages in wastewater treatment, including high efficiency, low energy consumption, and ease of operation [71]. However, membrane fouling always lowers the separation efficiency and shortens the membrane life, which greatly hinders the application of membrane technology. For this reason, Wang et al. [72] combined photo–Fenton and membrane processes for water treatment (Figure 3). These authors synthesized $CuFe_2O_4$ particles and doped them into the $PVDF@CuFe_2O_4$ membranes. The photo–Fenton process can degrade various foulants through the generation of hydroxyl radicals, which improves the filtration performance of the membranes.

The $PVDF@CuFe_2O_4$ membrane (1.0% $CuFe_2O_4$) showed significant improvement in both permeability and separation, with a tripling of flux and doubling of rejection compared to the values obtained by the membrane filtration process itself. In addition, the $PVDF@CuFe_2O_4$ nanofiltration membrane showed excellent stability and reusability after repeated tests over fifteen cycles.

Recently reported studies on the use of $CuFe_2O_4$ as photo–Fenton photocatalysts are summarized in Tables below.

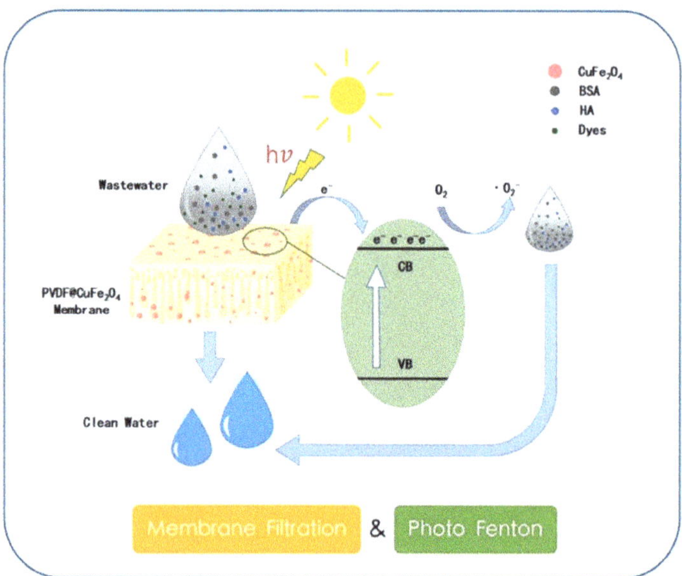

Figure 3. Schematic diagram illustrating the integration system of membrane filtration and the photo–Fenton process. BSA: bovine serum albumin; HA: humic acid. Reprinted with permission from Ref. [72]. Copyright 2019, Elsevier.

3. Mixed Ferrites Containing Fe and Cu

As already mentioned, among the iron–based materials used as heterogeneous photo–Fenton catalysts, ferrites have attracted much attention because they are chemically and thermally stable magnetic materials [73,74]. Most importantly, ferrites have a narrow band gap, which enables them to efficiently utilize the visible region of solar energy in photocatalysis [75]. $MnFe_2O_4$ has a large specific surface area, good biocompatibility, and excellent magnetic properties [76,77]. However, mixed ferrites have better catalytic behavior compared to single ferrites [78]. In particular, doping with Cu^{2+} could improve the optical properties of $MnFe_2O_4$, which is due to the lattice defects in the spinel structure generated as a consequence of the smaller ionic radius of Cu^{2+}. In addition, doping with other metals can also introduce oxygen vacancies, leading to a significant improvement in catalytic performance due to structural distortions [79]. Meena et al. [80] prepared Ce–doped $MnFe_2O_4$ by a low–temperature solution combustion synthesis using Oxalyl Dihydrazine (ODH) as fuel and reported that the band gap decreased after Ce–doping compared to that of pure $MnFe_2O_4$. Moreover, copper, manganese, and iron have been shown to have synergistic effects [81]. Thus, Cu^{2+}–doped $MnFe_2O_4$ could further improve the catalytic activity because the different radii of metal ions in the spinel structure may lead to some defects and distortions, which result in the production of oxygen vacancies that are beneficial for the formation of reactive oxygen species (ROS) [82].

Yang et al. [83] synthesized porous $Cu_{0.5}Mn_{0.5}Fe_2O_4$ nanoparticles by a co–precipitation process and a subsequent high–temperature annealing treatment method. $Cu_{0.5}Mn_{0.5}Fe_2O_4$ nanoparticles exhibited much higher catalytic activity towards the degradation of bisphenol A (BPA) by the activation of H_2O_2 under UV light irradiation compared with $CuFe_2O_4$ and $MnFe_2O_4$ nanoparticles. The authors proved, through radical scavenger and EPR/DMPO experiments, that the hydroxyl radical is involved and plays a critical role in the presence of H_2O_2. They also proposed a possible BPA degradation pathway.

Sun et al. [82] employed $Cu_{0.8}Mn_{0.2}Fe_2O_4$ in the presence of H_2O_2 and achieved a great improvement in the degradation efficiency of TCH, which was due to the fact that H_2O_2 was activated by the *Fe*, *Mn*, and *Cu* ions on the $Cu_{0.8}Mn_{0.2}Fe_2O_4$ surface to generate $HO^\bullet$ radicals. Two possible mechanisms for H_2O_2 activation by $Cu_{0.8}Mn_{0.2}Fe_2O_4$ at pH = 3 and pH = 11 were proposed. Firstly, $Cu_{0.8}Mn_{0.2}Fe_2O_4$ could be excited by visible light to generate electrons (e^-) and holes (h^+), then H_2O_2 could react with the generated e^- to produce $HO^\bullet$ (Equation (25)). Photoelectrons migrating to the surface of the catalyst immediately react with the metal ions of the catalyst, which not only promotes the circulation of *Fe*, *Mn*, and *Cu* components but also strengthens the synergistic effect between them (Equations (16), (26) and (27)). When H_2O_2 came into contact with *Fe*, *Mn*, and *Cu* ions, the Fenton reaction was triggered immediately and TCH was mineralized by these reactive species into small molecule compounds.

$$H_2O_2 + e^- \rightarrow HO^\bullet + HO^- \tag{25}$$

$$Mn^{3+} + Fe^{2+} \rightarrow Mn^{2+} + Fe^{3+} \tag{26}$$

$$Mn^{3+} + Cu^+ \rightarrow Mn^{2+} + Cu^{2+} \tag{27}$$

Wang et al. [84] used a p–n heterostructured nano–photocatalyst based on the p–type Cu_2ZnSnS_4 (CZTS) nanosheets, which were successfully assembled on the surface of $ZnFe_2O_4$ (ZFO) nanospheres, forming CZTS/ZFO p–n heterostructures. These p–n heterostructures could not only efficiently expand the spectral response and promote photo–induced charge separation, but also increase the specific surface areas for photocatalytic and photo–Fenton reactions. All these factors resulted in the p–n hetero–structured CZTS/ZFO nano–photocatalyst with significantly enhanced photocatalytic activity for the degradation of MO in the presence of H_2O_2 with visible light irradiation, compared to pure ZFO. This behavior was due to the synergistic enhancement effects of the CZTS/ZFO p–n heterostructure combined with the photo–Fenton mechanism. (Figure 4).

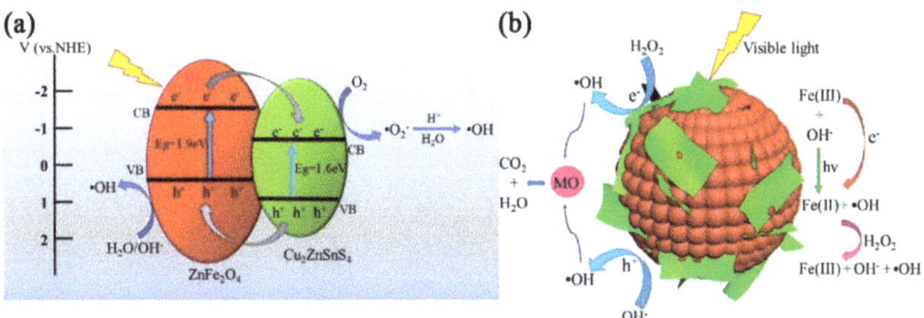

Figure 4. Schematic diagram for the photocatalytic degradation of MO by CZTS/ZFO p–n heterostructure plus H_2O_2: (**a**) the energy band structure, and the photo–induced carrier separation and transfer and (**b**) possible mechanism of photo–Fenton reaction in CZTS/ZFO + H_2O_2 photocatalytic system. Reprinted with permission from Ref. [84]. Copyright 2020, Elsevier.

Shi et al. [85] fabricated a multi–functional honeycomb ceramic plate by coating a layer of $CuFeMnO_4$ on the surface of a cordierite (magnesium iron aluminum cyclosilicate) material. The honeycomb structure was beneficial for light trapping and energy recycling and thus improved the solar–to–water evaporation efficiency. The $CuFeMnO_4$ coating layer acted as both the photothermal material for the solar–driven water evaporation process and the catalyst for the removal of volatile organic compounds (VOCs) via the heterogeneous photo–Fenton reaction. With the integration of the photo–Fenton reaction into the solar distillation process, clean distilled water was produced with efficient removal of the potential VOCs from the contaminated water sources.

All the results described in this section are summarized in Table 1.

Table 1. Summary of recently used *Fe–Cu* ferrites and mixed ferrites as photo–Fenton photocatalysts in wastewater treatment.

Material [a]	Conditions	Contaminant	Degradation Efficiency	Reference
$Fe_2O_3/CuFe_2O_4$ [b]	UV light, pH 7, $[H_2O_2]$ = 832.50 mg L^{-1}	Pyridine (100 mg L^{-1})	>99% within 30 min; TOC removal of 97% within 50 min	[39]
$\alpha–Fe_2O_3/CuFe_2O_4$ [c]	Natural solar light, pH 7, $[H_2O_2]$ = 300 mg L^{-1}	MB (MB, 100 mg L^{-1})	100% removal of the dye in 180 min	[62]
$CuFe_2O_4/CuO$ [d]	Halogen lamp, pH not indicated, $[H_2O_2]$ not indicated	MB (40 mg L^{-1}), Rhodamine B (20 mg L^{-1}), Methyl Orange (20 mg L^{-1})	100% dye degradation within 50 min	[63]
Hollow $CuFe_2O_4$ nanospheres [e]	300-W UV curing lamp (λ > 400 nm), pH not indicated, $[H_2O_2]$ = 0.02 M	MB (30 mg L^{-1})	96.4% in 60 min	[40]
$CuFe_2O_4$/biochar nanocomposites [f]	Fluorescent lamp (395–580 nm)/Sunlight, pH 3–7, $[H_2O_2]$ from 2.5 to 10 µM	Rhodamine B	100% color removal was obtained at 10 and 20 min of reaction for 10 and 50 mg L^{-1} of dye with solar radiation	[64]
$Bi_2WO_6/CuFe_2O_4$ [e]	Visible light, pH 2.6–6.3, $[H_2O_2]$ = 10 mM	Tetracycline hydrochloride (TCH)	After 30 min, 92.1% TCH (20 mg L^{-1}) degradation (pH 2.6) efficiency and 50.7% and 35.1% mineralization performance.	[65]
$CuFe_2O_4$/tartaric acid (*TA*) [b]	UV–curing lamp (365–450 nm), pH 5.0, $[H_2O_2]$ = 0.02 M	MB (50 mg L^{-1})	Introducing TA enhanced MB decolorization rate from 52.0% to 92.1% within 80 min.	[67]
$CuFe_2O_4$ [f]	Sunlight, pH 7.0, $[H_2O_2]$ = 0.3 M	MB (10 mg L^{-1})	Decolorization efficiency was 96.1% in 45 min of reaction.	[69]
$CuFe_2O_4@rGO$ [g]	Simulated sunlight, pH not indicated	Guaiacylglycerol–β–guaiacyl ether (lignin model compound)	Yields of 72.6% and 52.5% were achieved for guaiacol and 2–methoxy–4–propylpheno, respectively, in 60 min.	[70]
$CuFe_2O_4$ nanoparticles doped in polyvinylidene fluoride (PVDF) membranes [h]	Xe arc lamp, pH 3.0–11.0. The H_2O_2 (30 wt.%) dosage was in the range 50–1200 µL in 50 mL MB solution	MB (100 mg L^{-1})	MB was thoroughly degraded in 30 min when the pH is 3.0, while there is still 15.6% of the MB left at solution pH of 11.0.	[72]
$Cu_{0.5}Mn_{0.5}Fe_2O_4$ [i]	$Cu_{0.5}Mn_{0.5}Fe_2O_4$ 0.08 g L^{-1} pH: 4.2 $[H_2O_2]$ = 10 mM	BPA (10 mg L^{-1})	100% degradation and 47.6% mineralization 5 min	[83]
$Cu_{0.8}Mn_{0.2}Fe_2O_4$ [e]	Catalyst 0.100 g L^{-1} pH 3 and 11 300 W Xe lamp with a 420 nm UV–cut off filter	TC–HCL 100 mL 80 mg L^{-1}	99% 30 min	[82]
CZTS/ZFO p–n heterostructures [e]	Catalyst 0.5 g L^{-1} pH 3 to 9 $[H_2O_2]$ = 10 mM. 500 W Xe high intensity discharge lamp > 450 nm	MO 10 mg L^{-1}	91% 120 min At optimum pH: 6	[84]
$CuFeMnO_4$ on the surface of a honeycomb ceramic substrate [j]	0.05g of catalyst pH = 6.71. $[H_2O_2]$ from 0 to 0.1 M Solar light	Phenol as a VOC model 10, 20,50 and 100 mg L^{-1} MB 10 mg L^{-1}	~99.18% COD and MB was removed 20 min	[85]

[a] The synthesis method is shown in the footnotes. [b] Sol–gel method. [c] Modified Pechini method. [d] Solution combustion synthesis method. [e] Solvothermal method. [f] Co–precipitation. [g] Solvent–assisted interfacial reaction process. [h] Non–solvent induced phase separation method. [i] Chemical co–precipitation method followed by high–temperature annealing treatment. [j] See original papers for detail.

4. $CuFeO_2$ and $CuFeS_2$ Materials

Bimetallic $Cu(II)$ and $Fe(II)$ oxides and sulfides capable of activating H_2O_2 for the degradation of organic pollutants have become a focus of research. Among these catalysts, $CuFeO_2$ has gained wide concern due to its simple synthesis, low cost, interesting catalytic activity, and high chemical stability. In addition, due to its narrow band gap and cooperative impact of Fe^{3+}/Fe^{2+} and Cu^{2+}/Cu^+ redox cycles, $CuFeO_2$ materials usually exhibit significant visible–light absorption and a good performance for H_2O_2 activation.

Schmachtenberg et al. [86] obtained delafossite–type powders by conventional hydrothermal ($CuFeO_2$) and microwave–assisted hydrothermal ($CuFeO_2$–MW) routes. Both materials were tested as potential catalysts in the photo–Fenton reaction under visible light. An experimental design was used for optimizing the degradation efficiency of the dye Reactive Red 141 and assessing the effects of the operating variables pH, catalyst loading, and oxidant concentration. Both a substantial reduction in the treatment times and a significant efficiency improvement were observed for the catalyst prepared by microwave irradiation (i.e., about 98% of dye degradation at 30 min) in comparison with the material obtained by the conventional method (i.e., 84% at 150 min). The latter difference was ascribed to the fact that $CuFeO_2$–MW samples presented higher values of surface area and pore volume, as well as smaller band–gap energy in comparison with conventional $CuFeO_2$. In addition, under optimal conditions (i.e., pH = 3.0, $[CuFeO_2$–MW] = 0.25 g L^{-1}, and $[H_2O_2]$ = 8 mmol L^{-1}) 80% of TOC removal was achieved at 180 min. Finally, reusability tests conducted with the $CuFeO_2$–MW catalyst showed only a marginal loss of decolorization efficiency after four cycles (from 98% to 93%) and total concentrations of leached Fe and Cu ions of less than 1.0 wt.% of the catalyst content.

Liu et al. [87] prepared stoichiometric $CuFeO_2$ microcrystals with high crystallinity and single crystal phase via the optimized hydrothermal process. The authors presented a detailed characterization including the analysis of crystal structure, optical properties, and photoelectrochemical behavior. TCH was used as a model compound to evaluate the activity of $CuFeO_2$ under different conditions. Since the microcrystals showed a nonnegligible photocatalytic activity in the absence of H_2O_2, the authors proposed that the photo–generated electrons in the CB of $CuFeO_2$ could interact with dissolved oxygen and the photo–generated holes in the valence band react with H_2O, thus forming active radicals such as $O_2^{\bullet-}$ and $HO^\bullet$, respectively. On the other hand, in the presence of H_2O_2, the material showed some catalytic activity for the Fenton reaction under dark conditions, which was enhanced more than four times upon illumination. The heterogeneous photo–Fenton–like treatment of TCH, carried out at pH 8 and using a Xe–Lamp in a quartz cell, showed conversion degrees of about 80 and 90% within the first 60 and 120 min, respectively. Photo–Fenton assays performed in the presence of different radical scavengers indicated that $HO^\bullet$ is not directly responsible for the degradation, but that the photogenerated holes and $O_2^{\bullet-}$ are the predominant reactive species.

Da Silveira Salla et al. [88] evaluated the catalytic properties of $CuFeS_2$–MW through BPA degradation tests conducted at a near–neutral pH and using simulated visible light. Results showed a remarkable enhancement of the catalytic efficiency (i.e., about 90% in the first 15 min) for the catalyst obtained in the presence of citrate, with a rate about 10 times faster than that of $CuFeS_2$ prepared without citrate. This behavior suggested that the presence of citric acid accelerates the photo–conversion of Fe^{3+} to Fe^{2+}, thus increasing the generation of $HO^\bullet$ and the efficiency of the overall process. Reusability assays showed that, after the fourth cycle, the degradation rate remained higher than 95% and the TOC decrease was 76.6% after 60 min of reaction. In addition, concentrations of Fe and Cu leached into the solution were lower than 0.5% of the total amounts of Fe and Cu present in the catalyst at the beginning of the reaction. To assess the main reactive species, tests were conducted in the presence of different ROS scavengers (i.e., t–butanol for $HO^\bullet$ and p–benzoquinone for $O_2^{\bullet-}$). Results showed that BPA degradation may be ascribed to the generation of $HO^\bullet$. As expected, due to the low selectivity of the $HO^\bullet$ radicals, the degradation efficiency was affected by the presence of other chemical species in the real effluent such as carbonates/bicarbonates, nitrates/nitrites, sulfates, phosphates, and chlorides. Using LC–MS analysis, the authors identified several hydroxylated derivates resulting from the initial reaction stages. These primary intermediates are likely to undergo ring–opening reactions to form typical aliphatic acids as degradation by–products, which may further undergo a series of oxidation steps that finally lead to complete mineralization.

More recently, Da Silveira Salla et al. [89] prepared a chalcopyrite powder by a microwave–assisted method ($CuFeS_2$–MW), which exhibited higher catalytic activity than that obtained with the material prepared by the conventional method ($CuFeS_2$). Both materials were synthesized in the presence of citric acid, which is capable of enhancing the photoreduction of $Fe(III)$. Tartrazine dye was used as a model compound for the degradation by the photo–Fenton reaction under visible irradiation at pH 3.0. $CuFeS_2$–MW reached 99.1% of tartrazine decolorization at 40 min and 87.3% of mineralization at 150 min, the decolorization rate being twice as fast as that obtained with $CuFeS_2$. The latter difference was attributed to the synergy between higher crystallinity and increased amount of Fe^{2+} on the $CuFeS_2$–MW surface when compared to $CuFeS_2$. Reusability tests performed on both $CuFeS_2$–MW and $CuFeS_2$ showed that, after five cycles, the decolorization efficiencies remained at 93.6 and 69.6%, respectively. Moreover, the highest concentrations of leached iron were less than 1% of the total amount of iron present in the catalysts. The addition of t–butanol decreased 20–fold the tartrazine decolorization rate, while p–benzoquinone did not inhibit tartrazine decolorization. These results show that $HO^{\bullet}$ radicals are the key species for tartrazine removal and that $O_2^{\bullet-}$ radicals do not play a significant role in the degradation mechanism.

Cai et al. [90] anchored $CuFeO_2$ on a Manganese residue (MR) through mechanical activation (MA) for obtaining $Fe–Cu@SiO_2$/starch–derived carbon composites. The material was tested as a heterogeneous catalyst for the photo–Fenton treatment of TC using H_2O_2 and visible light. Under optimal conditions (i.e., 15 mM of H_2O_2 and pH 7.0) 100% of TC conversion (50 mg L^{-1}) was achieved within 40 min. Moreover, the observed decay constants were 4.00, 2.77, and 2.14 times higher than those of Cu/SC, $MAMR–Fe_3O_4@SiO_2/SC$, and $MR–Fe–Cu@SiO_2/SC$, respectively. Reutilization tests showed good material stability since the photocatalytic performance of $MAMR–Fe–Cu@SiO_2/SC$ changed from 99.2% to 96.3% after five cycles and metal leaching was below 0.1 mg L^{-1} for Cu, Fe, and Mn. Based on XPS analyses before and after material usage, the authors proposed that the efficiency of the prepared catalyst is closely related to the interaction of Cu^{2+}/Cu^+, Fe^{3+}/Fe^{2+}, and Mn^{3+}/Mn^{2+} couples. The study of reactive species through the use of scavengers and ESR showed that both $HO^{\bullet}$ and $O_2^{\bullet-}$ contribute to TC degradation in the $MAMR–Fe–Cu@SiO_2/SC + H_2O_2$ + visible light system, while surface photogeneration of electrons and holes seems to play a negligible role.

Xin et al. [38] synthesized $CuFeO_2$/biochar composites for heterogeneous photo–Fenton–like processes (HPF–like) via the hydrothermal method. Biochar prevented agglomeration and avoided the usage of added reductants. Moreover, the introduction of biochar had several advantages, such as enhancing visible–light absorption, narrowing the bandgap of $CuFeO_2$, and partially suppressing the recombination between photoelectron and hole pairs. By applying the design of experiments and the surface response methodology the authors studied the effects of the operating conditions on TC degradation efficiency. The results showed that the optimum parameters were 220 mg L^{-1} of catalysts, 22 mM of H_2O_2, and pH 6.4, with a pH dependence relatively gentle in the range of 4.0 to 8.0. Under optimal conditions, the efficiency of TC degradation in HPF–like systems was 96.7% after 120 min of treatment. Furthermore, the catalyst exhibited excellent stability since the activity only decreased by 3.9% after five utilization cycles and the total dissolved concentrations of both iron and copper ions at 120 min were below 0.02 mg L^{-1}. Active species scavenging experiments were carried out by using p–benzoquinone, silver nitrate, methyl alcohol, and ammonium oxalate as the scavengers of $O_2^{\bullet-}$, photoelectrons, $HO^{\bullet}$, and photogenerated holes, respectively. Interestingly, ammonium oxalate did not decrease the catalytic performance but accelerated the TC degradation rate. The latter effect was attributed to a lower photoelectron quenching rate, which could promote Fe^{3+}/Fe^{2+} and Cu^{2+}/Cu^+ cycles, thus significantly improving H_2O_2 activation. In addition, ESR experiments allowed the identification of $HO^\bullet$ as the predominant active species, whereas photoelectrons and $O_2^{\bullet-}$, were auxiliary species for TC degradation. The mechanistic findings are schematically summarized in Figure 5.

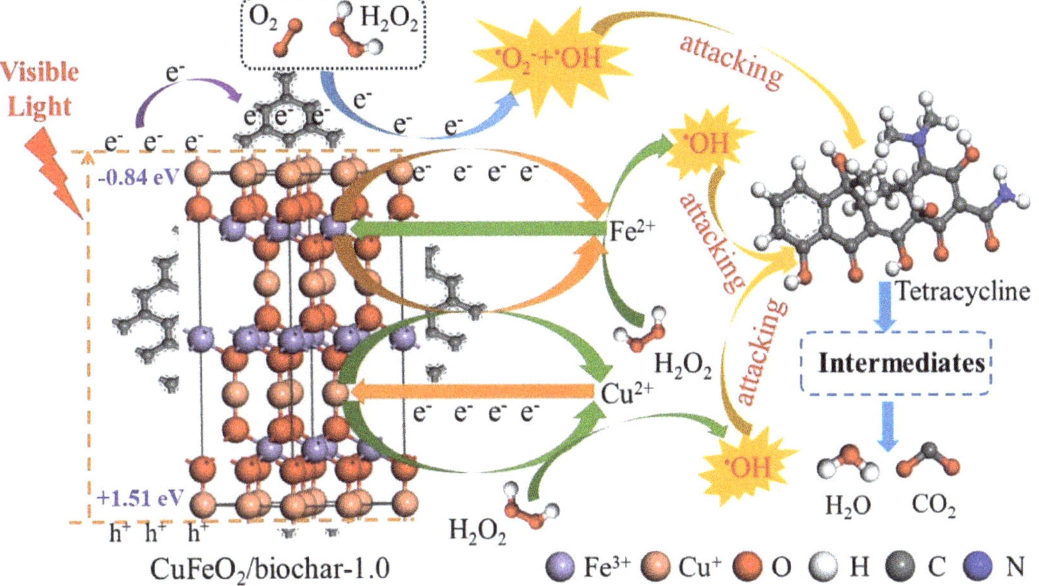

Figure 5. Possible mechanism of tetracycline degradation by $CuFeO_2$/biochar in the heterogeneous photo–Fenton system. Reprinted with permission from Ref. [38]. Copyright 2020, Elsevier.

It is worth mentioning that the authors also explored the TC degradation intermediate products by HPLC–MS and proposed plausible transformation pathways. On the other hand, Xin et al. [91] also studied the heterogeneous visible–light Photo–electro–Fenton (H–VL–PEF) treatment of TC using an undivided photoelectrochemical cell in the presence of $CuFeO_2$/biochar particles. A nitrogen/oxygen self–doped biomass porous carbon cathode was used as a gas diffusion electrode (GDE) to efficiently produce H_2O_2, while a Xe–Lamp (filter > 420 nm) was used as a visible irradiation source for $CuFeO_2$/biochar particles. The performances of different treatments (including electro–catalysis, photo–catalysis, photo–electro–catalysis, electro–Fenton, and H–VL–PEF) were evaluated for comparable experimental setups. The H–VL–PEF showed excellent performance, achieving the highest TC degradation rate and the best TOC removal with the lowest energy consumption. The authors analyzed the effect of several operational parameters and found that the full decomposition of TC (20 to 200 mg L^{-1}) was attained within 60–70 min under optimal conditions (i.e., 100 mg L^{-1} $CuFeO_2$/biochar, 80 mA cm^{-2} of current density, and pH 5.0). Reutilization tests showed negligible metal leaching and small efficiency losses after 10 cycles. Experiments performed in the presence of selective scavengers of active species as well as ESR measurements indicated that $HO^\bullet$ radicals are the main species responsible for TC degradation, whereas $O_2^{\bullet-}$ radicals play a subsidiary role. The schematic representation of the involved processes is presented in Figure 6. In addition, based on the intermediates detected by HPLC–MS analyses, the authors proposed general pathways for TC transformation and used a software tool for predicting, through QSAR techniques, the evolution of sample toxicity with the treatment time.

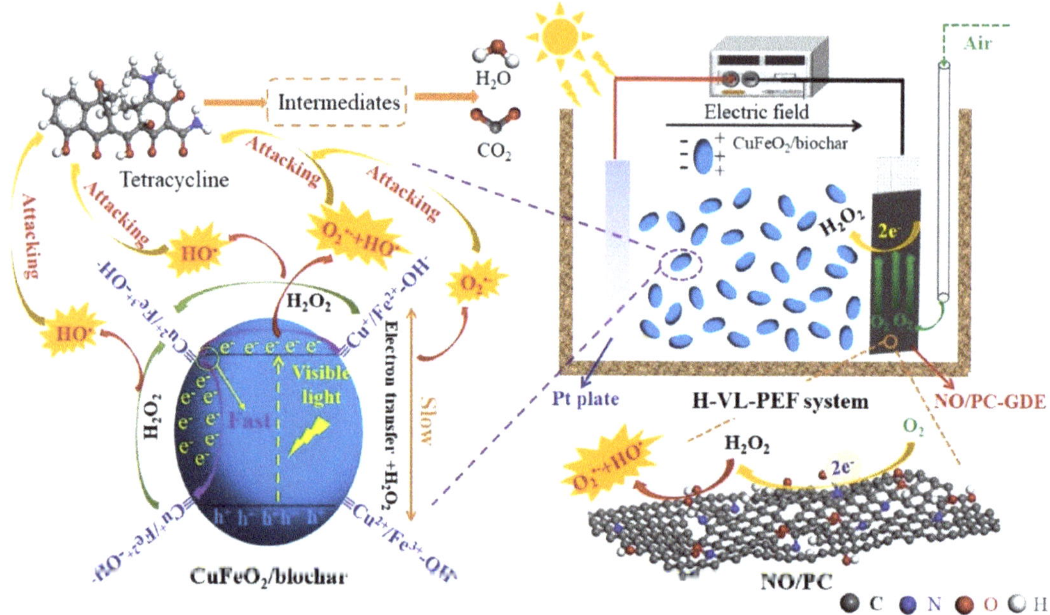

Figure 6. Possible mechanism of tetracycline degradation in the H–VL–PEF system. Reprinted with permission from Ref. [91]. Copyright 2020, Elsevier.

More recently, Xin et al. [92] achieved a further enhancement in the performance of photo–electro–Fenton (PEF) degradation of TC by introducing modifications to the latter treatment strategy. Three GDE composites were fabricated by mixing different weight ratios of a nitrogen/oxygen self–doped porous biochar (NO/PBC) cathode for H_2O_2 electrogeneration with a NO/PBC–supported $CuFeO_2$ ($CuFeO_2$–NO/PBC) catalyst for H_2O_2 activation. The NO/PBC was prepared by the pyrolysis method, whereas $CuFeO_2$–NO/PBC was prepared by the hydrothermal method without additional chemical reductants. The PEF treatment of TC was carried out in an undivided quartz reactor with a Pt anode and using visible light (Xe–Lamp, cutoff > 420 nm). The authors analyzed the effects of the NO/PBC to $CuFeO_2$–NO/PBC ratio, the current density, and initial pH on system performance, which depends on the compromise between H_2O_2 formation, parallel reactions such as 4–electron oxygen reduction reaction (ORR) and H_2 evolution, and H_2O_2 decomposition. Using a NO/PBC to $CuFeO_2$–NO/PBC ratio of 1:1, a current density of 80 mA cm^{-2}, and pH 5, an efficiency of 96.1% was achieved for TC (20 mg L^{-1}) degradation at 180 min in the PEF treatment. The latter value is twice the rate observed in EF and an order of magnitude higher than that recorded under anodic oxidation. The degradation efficiency at 180 min of PEF treatment steadily declined from 96.1% at pH 5.0 to 46.8% at pH 11.0. After five cycles of material use, the PEF treatment efficiency was higher than 80%, the amounts of leached Fe and Cu were negligible, and surface hydrophobicity remained practically constant. Assays in the presence of radical scavengers showed that $HO^\bullet$ and $O_2^{\bullet-}$ played the primary and auxiliary roles for tetracycline degradation, respectively. The authors proposed that despite the photoinduced hole could not effectively participate in TC degradation, the photogenerated electrons reacted with H_2O_2, O_2, and Fe^{3+}/Cu^{2+} to form $HO^\bullet$, $O_2^{\bullet-}$, and Fe^{2+}/Cu^+, respectively, thus promoting an efficient TC transformation. Finally, from the reaction intermediates detected by HPLC–MS, QSAR tools, and experiments of *E. coli* growth inhibition, a general picture of TC oxidation pathways and toxicity evolution was outlined.

All the results described in this section are summarized in Table 2.

Table 2. Summary of recently used $CuFeO_2$ and $CuFeS_2$ as photo–Fenton photocatalysts in wastewater treatment.

Material [a]	Conditions	Contaminant	Degradation Efficiency	Reference
Delafossite–type $CuFeO_2$ [b,c]	visible light, pH from 2.4 to 3.6, H_2O_2 from 3 to 13 mM, catalyst from 0.13 to 0.33 gL^{-1}	Reactive Red 141 dye, 50 mg L^{-1}	about 98% at 30 min	[86]
3R–delafossite $CuFeO_2$ microcrystals [b]	200 mL reactor, 20 mM of H_2O_2, 1 g L^{-1} catalyst, initial pH 8	Tetracycline hydrochloride 20 mg L^{-1}	96.1% in 180 min	[87]
$CuFeS_2$ [c]	visible light, 20 mM of H_2O_2, 0.2 g L^{-1} catalyst, pH 6	bisphenol A (BPA) 20 mg L^{-1}	97% in 60 min	[88]
$CuFeS_2$ chalcogenide powders [b,c]	visible light, pH 3.0, 8.33 mM of H_2O_2, 0.2 g L^{-1} of catalyst	tartrazine, 100 mg L^{-1}	99.1% of tartrazine decolorization after 40 min and 87.3% of mineralization after 150 min	[89]
MAMR–Fe–$Cu@SiO_2$/SC, a core–shell structure of $CuFeO_2@SiO_2$/starch–derived carbon anchored on a Manganese Residue [d]	Xe–lamp with UV cut-off, pH from 2.5 to 9.5, 15 mM of H_2O_2, 0.7 g L^{-1} of catalyst	Tetracycline 50 mg L^{-1}	100% in 40 min	[90]
$CuFeO_2$/biochar [b]	Xe Lamp with UV cutoff, pH 4 to 8, 20 mM of H_2O_2, 0.2 g L^{-1} of catalyst	Tetracycline 20 mg L^{-1}	97.6% in 120 min	[38]
$CuFeO_2$/biochar [b]	photo–electro–Fenton, Xe–Lamp with UV cutoff, pH from 3 to 11, H_2O_2 generated by a NO–doped/porous carbon cathode	Tetracycline (20 to 200 mg L^{-1})	100% in 60–70 min	[91]
Nitrogen/oxygen self–doped porous biochar (NO/PBC) and NO/PBC–supported $CuFeO_2$ ($CuFeO_2$–NO/PBC) [b]	undivided quartz reactor, visible light, pH from 3 to 11, H_2O_2 by a gas diffusion electrode	Tetracycline 20 mg L^{-1}	98% at 30 min	[92]

[a] The synthesis method is shown in the footnotes. [b] Hydrothermal method. [c] Microwave–assisted hydrothermal method. [d] Reductive roasting and mechanical activation.

5. Fe–Cu Oxide Composites

In this section, we present the results obtained with other types of heterogeneous catalysts consisting of mixtures of iron and copper oxides and mixed oxides of defined composition. Heterojunctions formed by different copper and iron oxides and/or sulfides decrease the recombination rate of photogenerated e^- and h^+ [93,94].

Mansoori et al. [95] prepared an iron–copper oxide impregnated $NaOH$–activated biochar (Fe–Cu/ABC) through the pyrolysis of activated biochar, followed by the impregnation method. The catalytic activity of the bimetallic catalyst was tested for ciprofloxacin (CIP) degradation through a heterogeneous photo–electro–Fenton process) at a natural pH. The heterogeneous catalyst exhibited remarkable catalytic activity and showed great stability and structural integrity for five cycles. Furthermore, from a practical point of view, the catalyst exhibited an acceptable performance by oxidizing CIP dissolved in various water matrices such as tap water, river water, and a real sample of wastewater. The intermediates by–products of CIP were determined, and a plausible degradation mechanism was proposed. The adsorption of as–generated H_2O_2 onto the well–developed surface of the mesoporous bimetallic catalyst facilitated the reduction of Fe^{3+} to Fe^{2+} and Cu^{2+} to Cu^+. The co–existence of Fe and Cu on the surface of activated biochar accelerates the process of electron transfer in the reaction and enhances the production of reactive species.

Iron oxide NPs supported on the surface of hydroxylated diamond exhibited photocatalytic activity and stability for the visible light–assisted Fenton reaction even working at pH values around 6 and demonstrated superiority as supports compared to other alternative solids such as activated carbon, graphite, carbon nanotubes, and even the benchmark semiconductor TiO_2 [96]. In particular, Manickam–Periyaraman, et al. [97] prepared bimetallic $Fe_{20}Cu_{80}$ NPs supported on various surface hydroxylated diamond NPs (D) and compared them to analogous catalysts based on graphite. The assays were conducted at pH 6 for the heterogeneous Fenton degradation of phenol (Ph) assisted by natural or simulated sunlight irradiation, achieving a mineralization degree of 90% at an H_2O_2 to Ph molar ratio of 6. The authors explained this result based on the high activity of reduced copper to form hydroxyl radicals and the favorable redox process of Fe^{2+} maintaining a pool of reduced Cu^+ species. Thus, a heterogeneous catalyst based on abundant iron or copper metals allowed the promotion of H_2O_2 activation to yield hydroxyl radicals under visible light irradiation according to Equations (28)–(36).

$$Fe_{20}Cu_{80}/D_3 \,+\, h\nu \,\rightarrow\, e^- + h^+ \tag{28}$$

$$H_2O_2 + e^- \,\rightarrow\, HO^\bullet + HO^- \tag{29}$$

$$Fe^{3+} + e^- \,\rightarrow\, Fe^{2+} \tag{30}$$

$$Cu^{2+} + e^- \,\rightarrow\, Cu^+ \tag{31}$$

$$Cu^+ + e^- \,\rightarrow\, Cu^0 \tag{32}$$

$$h^+ + \, H_2O_2 \,\rightarrow\, HO_2^\bullet + H^+ \tag{33}$$

$$h^+ + \, Fe^{2+} \,\rightarrow\, Fe^{3+} \tag{34}$$

$$h^+ + \, Cu^+ \,\rightarrow\, Cu^{2+} \tag{35}$$

$$h^+ + \, organic\ compunds \,\rightarrow\, oxidized\ products \tag{36}$$

Khan et al. [98] compared the photocatalytic activity of novel peculiar–shaped CuO, Fe_2O_3, and FeO nanoparticles (NPs) to that of the iron(II)–doped copper ferrite, $Cu^{II}_{0.4}Fe^{II}_{0.6}Fe^{III}_2O_4$, through the degradation of MB and RhB. The catalysts were synthesized via the simple co–precipitation and calcination technique. The highest degradation efficiency was achieved by CuO for RhB and by $Cu^{II}_{0.4}Fe^{II}_{0.6}Fe^{III}_2O_4$ for MB. The $CuO/FeO/Fe_2O_3$ composite proved to be the second–best catalyst in both cases, with excellent reusability.

Asenath–Smith et al. [99] used iron oxide (α–Fe_2O_3, hematite) colloids synthesized under hydrothermal conditions as catalysts for the photodegradation of MO. To enhance the photocatalytic performance, Fe_2O_3 was combined with other transition–metal oxide (TMO) colloids (e.g., CuO and ZnO), which are sensitive to different regions of the solar spectrum (visible and UV, respectively), using a ternary blending approach for compositional mixtures. For a variety of $ZnO/Fe_2O_3/CuO$ mole ratios, the pseudo–first–order rate constant for MO degradation was at least twice the sum of the individual Fe_2O_3 and CuO rate constants, indicating that an underlying synergy governs the photocatalytic reactions for these combinations of TMOs. The increased photo–catalytic performance of Fe_2O_3 in the presence of CuO was associated with the hydroxyl radical, consistent with heterogeneous photo–Fenton mechanisms, which are not accessible by ZnO. Then, the authors proposed a mechanism where CuO plays a supportive role in Fe_2O_3 photocatalysis by decomposing H_2O_2 to generate $HO^\bullet$ radicals. With additional ROS available, the reaction kinetics are accelerated. The bandgap energies of the TMOs used in this study further corroborated the obtained results. Smaller bandgap materials, such as Fe_2O_3 and CuO, have valence–band edges above the oxidation potential of H_2O. As a result, it is energetically favorable for a photogenerated h^+ on Fe_2O_3 or CuO to participate in the generation of $HO^\bullet$ radicals by H_2O oxidation. Larger band gap materials, such as ZnO, have valence–band edges that are below the H_2O oxidation potential for $HO^\bullet$ formation, but their conduction band edges are below the reduction potential of O_2.

Lu et al. [100] used a magnesium aluminum silicate known as attapulgite (ATP) supported *Fe–Mn–Cu* polymetallic oxide as a catalyst with UV irradiation in the photocatalytic oxidation (photo–Fenton) treatment of a synthetic pharmaceutical wastewater. *Fe–Mn–Cu*@ATP had good catalytic potential and a significant synergistic effect since it removed almost all heterocyclic compounds, as well as humus–like and fulvic acid. The degradation efficiency of the nanocomposite only decreased by 5.8% after repeated use for six cycles (COD removal reached 64.9%, and the BOD $_5$/COD increased from 0.179 to 0.387 after 180 min of reaction time). The proposed mechanism involves: (i) the direct reaction between the catalyst and hydrogen peroxide, which generates $HO^\bullet$ radicals and oxidized ions of the three metals, (ii) the UV photoreductions of these metal ions, and (iii) indirect reactions between photogenerated free radicals and the catalyst. In this mechanism, the consumed metal ions are effectively regenerated under UV light, and $HO^\bullet$ finally reacts with the organic matter.

Davarnejad et al. [101] synthesized alginate–based hydrogel–coated bimetallic iron–copper nanocomposite beads through a green method and used them as heterogeneous catalysts for metronidazole elimination from wastewater. These authors employed the response surface methodology (RSM) based on the Box–Behnken design (BBD) to assess both the individual and interaction effects of five main variables involving catalyst loading, initial pH, reaction time, metronidazole, and H_2O_2 concentrations. The data obtained from the model were in good agreement with the experimental ones. The optimum conditions (for 95.3%, based on model, and for 95.0%, based on experiments) were found at a catalyst (Fe_2O_3–*CuO*@*Ca–Alg*) loading of 44.7 mg L^{-1}, an initial pH of 3.5, a metronidazole concentration of 10 mg L^{-1}, a H_2O_2 concentration of 33.17 mmol L^{-1}, and a reaction time of 85 min.

Zhang et al. [102] used Fe_3O_4@Cu_2O/carbon quantum dots (CQDs)/nitrogen–doped carbon quantum dots (N–CQDs) (FCCN) for the degradation of azo compounds, such as MO, acid orange II, and mordant yellow 10, even at a neutral and alkaline pH (pH: 7–12) with a shortest time for complete degradation of 15 min. There is a cooperative interaction between each component, so the synergism from all the components of FCCN enhances the photocatalytic properties. The authors proposed a mechanism where in the first place, the up–conversion characteristics of CQDs and N–CQDs improve the light utilization of FCCN. Secondly, the load of CQDs and N–CQDs can separate and transfer photo–generated charges between Fe_3O_4 and Cu_2O more efficiently, accelerating the circulation of Fe^{3+}/Fe^{2+} for the decomposition of H_2O_2 to yield the active species: $HO^\bullet$ radical. Thirdly, the light–induced protons of CQDs change the partial pH, so that the FCCN system is able to work in alkaline solutions. Finally, the insoluble layers of CQDs and N–CQDs are beneficial to enhance the stability of the composite (see in Figure 7 the schematic diagram of the catalytic mechanism proposed by the authors). The magnetism of Fe_3O_4 made this catalyst easily separable and the insoluble layers of CQDs and N–CQDs allowed it to be repeatedly used without activity change even after 10 cycles. The degradation rate constant of MO in the FCCN/H_2O_2/light system was 5.4 times higher than that of the Fe_3O_4@Cu_2O/H_2O_2/light system, indicating that the loaded quantum dots greatly enhanced the photocatalytic activity. The degradation rate constant in the FCCN/H_2O_2/light system was 2.5 times higher than that of the simple mixture (Fe_3O_4@Cu_2O + CQDs + N–CQDs)/H_2O_2/light system, which suggests that these excellent catalytic performances should come from the synergism effect of all components in FCCN.

All the results described in this section are summarized in Table 3.

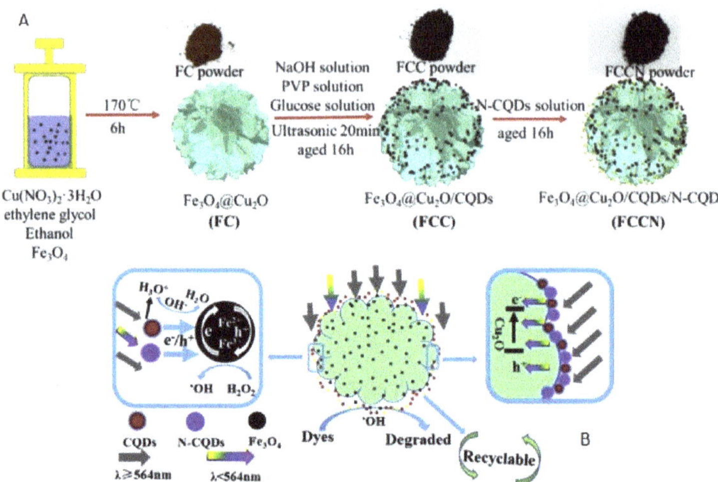

Figure 7. (**A**) Schematic illustration of the synthesis of FCCN. (**B**) Schematic diagram of the catalytic mechanism of FCCN (see text). Reprinted with permission from Ref. [91]. Copyright 2020, Elsevier.

Table 3. Summary of recently used *Fe–Cu* oxides composites as photo–Fenton photocatalysts in wastewater treatment.

Material [a]	Conditions	Contaminant	Degradation Efficiency	Reference
Iron–copper oxide impregnated *NaOH*–activated biochar (*FeCu/ABC* catalyst) [b]	Heterogeneous PEF process, pH = 5.8 catalyst dosage of 1 g L⁻¹, electrical current of 200 mA	CIP (45 mg L⁻¹)	100% removal 2 h	[95]
$Fe_{20}Cu_{80}$(0.2 wt.%)/D3 [c]	catalyst ~200 mg L⁻¹, [H_2O_2] (200 mg L⁻¹; 5.88 mM), 20 °C, simulated sunlight. pH = 4	Phenol (100 mg L⁻¹)	90% removal 2 h	[103]
NP–3 ($Cu^{II}0.4Fe^{II}0.6Fe^{III}_2O_4$) [d]	NP–3 = 400 mg L⁻¹, [H_2O_2] = 1.76 × 10⁻¹ mol L⁻¹, pH = 7.5 Optonica SP1275 LED lamp (GU10, 7 W, 400 Lm, 6000 K, Optonica LED, Sofia, Bulgaria)	MB (1.5 × 10⁻⁵ mol L⁻¹) RhB (1.75 × 10⁻⁵ mol L⁻¹)	100% 140 min	[98]
Iron oxide (α–Fe₂O₃, hematite) colloids combined with other transition–metal oxide (TMO) colloids (e.g., *CuO* and *ZnO*) [e]	750 mg L⁻¹ catalyst, [H_2O_2] = 0.025 mol L⁻¹. Tungsten halogen lamps	MO (25 μM)	7 to 78% 60 min	[99]
Fe–Mn–Cu@ATP [f]	500 mL of *Fe–Mn–Cu@ATP* dosage (1–12 g L⁻¹), pH = 3 [H_2O_2] (0.1–0.6 mol L⁻¹) UV 40 W UV lamp	pharmaceutical wastewater	COD removal: 64.9% 180 min	[100]
Fe_2O_3–*CuO@Ca*–Alg hydrogel [g]	4.7 mg L⁻¹ of catalyst, pH = 3.5 [H_2O_2] = 33.17 mmol L⁻¹ UV light	[MNZ]₀ 10 mg L⁻¹	Removal = 95% 85 min	[101]
FCCN [h]	Catalyst 1 g L⁻¹ pH: 7–12 [H_2O_2] = 15 mM 500W Xe lamp 564 nm cut–off filter	MO, acid orange II and mordant yellow 10–20 mg L⁻¹	100% 15 min	[102]

[a] The synthesis method is shown in the footnotes. [b] Pyrolysis of activated biochar followed by impregnation. [c] Deposition of the NPs (*Fe, Cu,* or *Fe–Cu*) onto the surface of commercial carbonaceous supports [d] Co–precipitation. [e] Hydrothermal method. [f] Aeration–coprecipitation method. [g] Green method using walnut green shells. [h] See Figure 7A.

6. Metal–Organic Frameworks Based on Fe and Cu

Metal–organic frameworks (MOFs) are inorganic–organic porous polymers composed of metal ions or metal clusters linked to each other by bridging organic ligands to form three–dimensional structures with a high specific surface area. In particular, *Fe*–based MOFs, made by polydentate organic ligands as linkers and inorganic *Fe* ions or *Fe*–oxo clusters as nodes, have recently attracted great attention in various applications such as gas adsorption, catalysis, and sensor development [104]. Within the applications, *Fe*–based MOFs have been considered promising catalysts in photo–Fenton processes due to their interfacial electron transfer properties and the cycling of the *Fe(III)/Fe(II)* redox pair. Additionally, the band gaps of *Fe*–based MOFs are suitable for visible light photoactivation and thus, *Fe–O* clusters could be directly excited to generate electron hole pairs ($e^- - h^+$), which subsequently degrade organic pollutants or lead to the generation of reactive oxygen species [105]. Likewise, it has been shown that the separation efficiency of photogenerated carriers in the *Fe–O* cluster remains low due to fast electron–hole recombination [105,106]. Taking this into account, different modifications in the structure of MOFs have been proposed to delay the recombination rate and improve pollutant degradation efficiencies [106,107]. In this sense, the combination of MOFs with metals or metal oxide nanoparticles may form a new interface or heterojunction that efficiently couples the catalytic process of photo–induced electrons and radicals' generation [108,109]. Moreover, the addition of a second metal ion into the nodes of frameworks could significantly improve the catalytic properties of MOFs [41]. Recently, different reports have incorporated *Cu* into *Fe*–based MOFs structure yielding *Fe–Cu* bimetallic MOFs in order to improve the performance of these materials in photo–Fenton processes.

Do et al. [110] reported the preparation of *Fe*–doped *Cu* 1,4–benzenedicarboxylate MOFs (*Fe–CuBDC*) and its application as a heterogeneous photo–Fenton catalyst for the degradation of MB in aqueous solution under visible light irradiation. In this work, the authors compared the efficiency of the bimetallic *Fe–CuBDC* with each single metal MOF. The degradation efficiency of MB with *Fe–CuBDC* was much higher than those obtained with the others, which indicates that the partial substitution of *Cu* by *Fe* significantly improves the photocatalytic activity of the material. Complete removal of MB was achieved after 70 min under light irradiation with 1.0 g L^{-1} of *Fe–CuBDC* (Table 4). The degradation performance of the *Fe–CuBDC* catalyst was relatively constant with a slight reduction in removal efficiency from 99.9% to 97.2% after five reuse cycles.

Shi et al. [111] prepared a binary bimetallic heterojunction, which consisted of a core–shell magnetic $CuFe_2O_4$@MIL–100(Fe, Cu) metal–organic framework, via an in–situ derivation strategy. The synthesis methods, which involved the in situ surface pyrolysis of $CuFe_2O_4$ nanoparticles and complexation with trimesic acid to derive MIL–100(Fe, Cu), allowed the formation of a bimetallic core–shell $CuFe_2O_4$@MIL–100(Fe, Cu) heterojunction (*MCuFe* MOF). A schematic diagram of the synthesis process of *MCuFe* MOF is shown in Figure 8. These materials showed notable catalytic performance in a photo–Fenton process towards the degradation of various organic pollutants by increasing H_2O_2 activation efficiency and decreasing the required dosage of *MCuFe* MOF (0.05 g L^{-1}) over a wide pH range (4–9) (Table 4). Furthermore, the *MCuFe* MOF showed a high stability for the degradation of organic contaminants, with almost no decrease in activity and negligible metal leaching after five successive cycles. According to the proposed mechanism, the combination of the photothermal conversion effect with the formed heterojunction cannot only accelerate the generation and separation of photogenerated e^- and h^+ but also improve the continuous and efficient transformation of $\equiv Fe(III)/Fe(II)$ and $\equiv Cu(II)/Cu(I)$ redox couples, leading to enhanced photo–Fenton efficiency (see Figure 9).

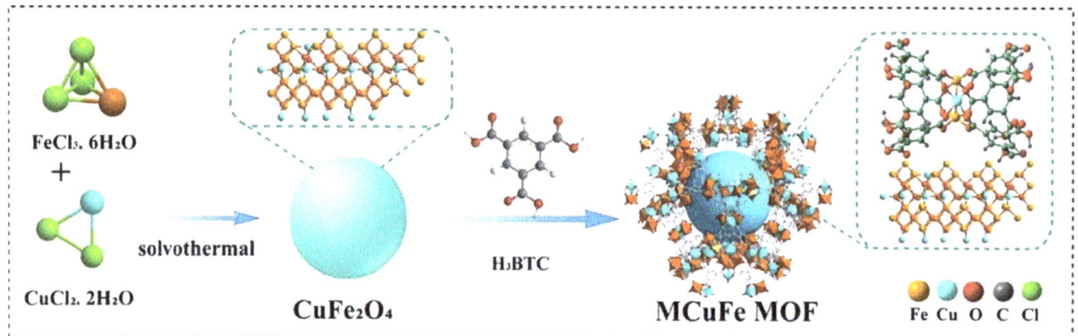

Figure 8. Schematic diagram of the synthesis process of *MCuFe* MOF. Reprinted with permission from Ref. [111]. Copyright 2020, Elsevier.

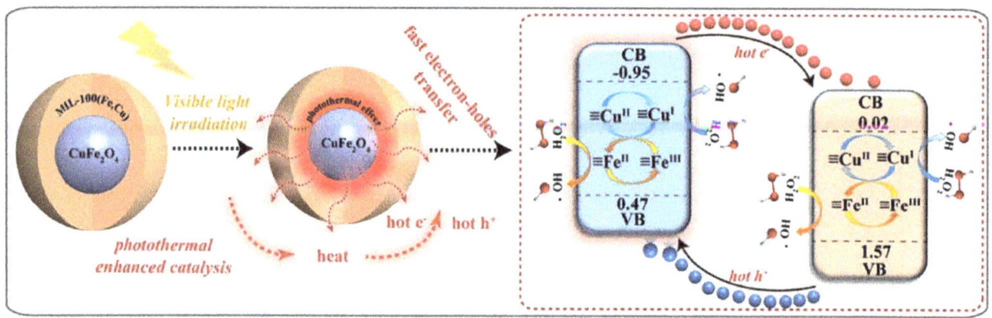

Figure 9. Photo–Fenton reaction mechanism of the charge transfer for hydroxyl radical generation over the MCuFe MOF under visible light irradiation. Reprinted with permission from Ref. [111]. Copyright 2020, Elsevier.

Table 4. Summary of recently used *Fe–Cu* bimetallic MOFs as photo–Fenton photocatalysts in wastewater treatment.

Material [a]	Conditions	Contaminant	Degradation Efficiency	Reference
Fe–CuBDC [b]	Simulated sunlight, pH 6, $[H_2O_2]$ = 50 mM, [Catalyst] = 1 g L^{-1}	Methyl Blue (50 mg L^{-1})	100% removal of the dye in 70 min	[110]
MCuFe MOF [b,c]	300 W xenon lamp equipped with a UV cut–off filter (λ > 400 nm), pH 4–9, $[H_2O_2]$ = 5 mM, [Catalyst] = 0.05 g L^{-1}	Methyl Blue (50 mg L^{-1})	100% removal of the dye in 40 min	[111]
Cu₂O/MIL(Fe/Cu) [b]	500 W xenon lamp, pH 7.47, $[H_2O_2]$ = 49 mM, [Catalyst] = 0.5 g L^{-1}	Thiacloprid (80 mg L^{-1})	100% removal of TCL in 20 min; 82% TOC removal in 80 min	[112]
L–MIL–53 *(Fe, Cu)* [b]	300 W xenon lamp equipped with a UV cut–off filter (λ > 420 nm), pH 7, $[H_2O_2]$ = 5 mM, [Catalyst] = 0.1 g L^{-1}	Ciprofloxacin (20 mg L^{-1})	60% removal of CIP in 30 min	[113]

[a] The synthesis method is shown in the footnotes. [b] solvothermal method. [c] Hydrothermal method.

Zhong et al. [112] performed Cu_2O growth on the surface of Cu–doped MIL–100 (*Fe*) to obtain a $Cu_2O/MIL(Fe/Cu)$ composite, which showed an enhanced interfacial synergistic effect and was successfully used as photo–Fenton catalysts for thiacloprid (TCL) degradation. The authors reported that Cu doping into $MIL(Fe)$ led to the reduction in the band gap, and a boost of the redox cycle Fe^{2+}/Fe^{3+}. Likewise, the growth of Cu_2O extended the light absorption range of $MIL(Fe/Cu)$ from UV to the visible region. A proposed TCL degradation mechanism is shown in Figure 10. $Cu_2O/MIL(Fe/Cu)$ composite exhibited a TCL degradation rate nearly 10 times faster than those of Cu_2O and $MIL–100(Fe)$, separately, achieving a complete TCL degradation within 20 min of reaction. Moreover, TOC removal reached 82.3% within 80 min under neutral conditions, which highlights the good performance of the $Cu_2O/MIL(Fe/Cu)$ composite. Catalyst reuse tests showed no significant efficiency loss in reaction rates even after the tenth cycle, reaching complete TCL degradation within 20 min in each cycle.

Low–crystalline MOFs, with long–range disorder but local crystallinity, allow the availability of more active sites and defects than highly crystalline materials. These materials could improve the activation capacity of Fe–based MOFs towards H_2O_2 by increasing the number of metallic coordinately unsaturated active sites (CUS) within the frameworks. Wu et al. [113] prepared low–crystalline bimetallic MOFs of $MIL–53(Fe, M)$ (*M: Mn* or *Cu*), via a one–pot solvothermal method, as photo–Fenton catalysts for the degradation of CIP. The results showed a significant improvement in photo–Fenton performance for the low–crystallinity catalyst compared to the crystalline counterparts, which was mainly attributed to the enhancement of the synergism between the hetero–metal nodes. In particular, low crystallinity $MIL–53(Fe, Cu)$ exhibited a much higher removal efficiency and a faster reaction rate than that of crystalline $MIL–53(Fe, Cu)$ within 30 min (Table 4). Besides the increased metal CUSs in the low–crystalline state, both Cu and Mn could increase the specific surface area and promote the visible–light absorption and separation/transportation of carriers in the low–crystalline state, thus leading to the acceleration of $Fe(II)/Fe(III)$ and $M(red)/M(ox)$ cycles.

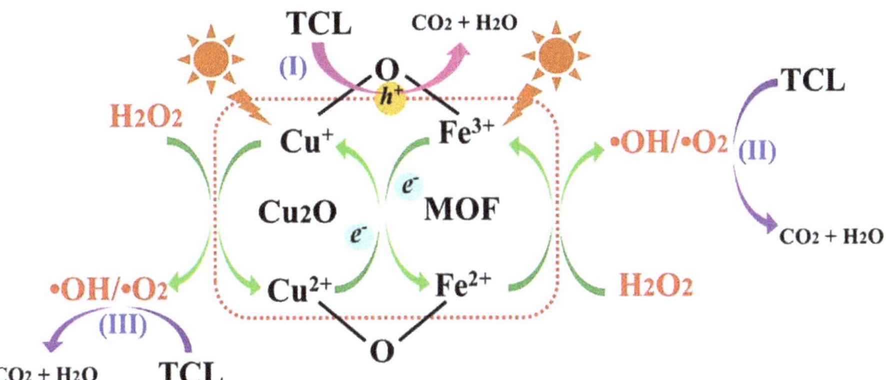

Figure 10. Proposed reaction mechanism for the photo–Fenton degradation of TCL over $Cu_2O/MIL(Fe/Cu)$. Reprinted with permission from Ref. [112]. Copyright 2020, Elsevier.

All the results described in this section are summarized in Table 4.

7. Conclusions and Future Perspective

This review has shown the present trends in the development of bimetallic iron–copper materials and their application as catalysts in heterogeneous photo–Fenton–like reactions for the degradation of wastewater pollutants. Bimetallic $Fe–Cu$ catalysts have shown better performance than monometallic Fe catalysts.

We have compared many materials with very different chemical compositions and physicochemical properties. In general, bimetallic *Fe–Cu* catalysts have shown better performance than monometallic Fe catalysts. In the works summarized here, it is difficult to find a common explanation of the synergistic effect of iron and copper on the catalytic activity of the materials. Some works attribute the improvement to changes in the band–gap, others to the decrease in the electron–hole recombination rates, and even to the reduction of H_2O_2 by electrons, or the beneficial effect of the binary redox couples of *Fe(II)/Fe(III)* and *Cu(I)/Cu(II)* on the decomposition efficiency of H_2O_2. However, the common factor in all the materials seems to be that the coexistence of both metals evidently favors the redox cycles of *Fe* and *Cu*, resulting in higher catalytic activity. The co–existence of *Fe* and *Cu* on the surface of a catalyst accelerates the process of electron–transfer in the reaction environment, providing an appropriate condition for the activation of hydrogen peroxide and the generation of reactive radical species. This hypothesis is supported by the excellent degradation and mineralization results obtained when aqueous solutions of the contaminant are treated with these catalysts. In particular, lower degradation times and catalyst dosages are required to achieve complete pollutant degradation. Most of the bimetallic *Fe–Cu* catalysts have shown very good results at a circumneutral pH, which represents one of the major advantages of these materials since the development of an efficient and sustainable photo–Fenton treatment at neutral pH values remains a challenge for the scientific community working in the field. In addition, many of the works summarized here report the stability of the bimetallic catalysts even after having been used in four or five cycles.

From the analysis of the collected research, we can conclude that further efforts should focus on the preparation of new catalysts by green synthesis routes, that is, using bioactive agents such as plant materials, microorganisms, or biological waste. This approach would help to reduce the overall production cost and to limit environmental impact. Only a few of the compiled articles employed renewable energies such as natural solar light as the photoirradiation source. Most of them used either Xe lamps with or without cut–off filters for simulating sunlight or other light sources, such as UV lamps or visible light from lamps or LEDs. In this context, further studies devoted to addressing the effect of using solar light as the photoirradiation source on the performance of the catalysts are certainly encouraged.

Although excellent degradation and mineralization results have been obtained by applying this technique at the laboratory scale, its effectiveness needs to be evaluated in continuous flow systems for real industrial effluents. Most of the published research focused on the degradation of single pollutants (mainly dyes and antibiotics), disregarding the effects of the presence of humic–like dissolved organic matter and other contaminants. Moreover, to increase the efficiency and decrease the cost of the treatment of recalcitrant contaminants in real systems, the integration of photo–Fenton methods with biological technologies should be explored.

Author Contributions: Conceptualization, investigation, writing—original draft preparation, writing—review and editing: G.N.B., F.S.G.E., L.C. and D.O.M. contributed equally. All authors have read and agreed to the published version of the manuscript.

Funding: This project received funding from ANPCyT, Argentina (PICT 2017–1628, and PICT 2019–03140).

Data Availability Statement: Not applicable.

Acknowledgments: G.N. Bosio, F. García Einschlag, and L. Carlos and are staff researchers of CONICET (Argentina). D.O. Mártire is staff researcher of Comisión de Investigaciones Científicas (CIC, Buenos Aires, Argentina).

Conflicts of Interest: There are no conflicts of competing interest to declare.

References

1. Afrad, M.S.I.; Monir, M.B.; Haque, M.E.; Barau, A.A.; Haque, M.M. Impact of Industrial Effluent on Water, Soil and Rice Production in Bangladesh: A Case of Turag River Bank. *J. Environ. Health Sci. Eng.* **2020**, *18*, 825–834. [CrossRef] [PubMed]
2. Amor, C.; Marchão, L.; Lucas, M.S.; Peres, J.A. Application of Advanced Oxidation Processes for the Treatment of Recalcitrant Agro–Industrial Wastewater: A Review. *Water* **2019**, *11*, 205. [CrossRef]
3. Lama, G.; Meijide, J.; Sanromán, A.; Pazos, M. Heterogeneous Advanced Oxidation Processes: Current Approaches for Wastewater Treatment. *Catalysts* **2022**, *12*, 344. [CrossRef]
4. Shokri, A.; Fard, M.S. A Critical Review in Fenton–like Approach for the Removal of Pollutants in the Aqueous Environment. *Environ. Chall.* **2022**, *7*, 100534. [CrossRef]
5. Aparicio, F.; Escalada, J.P.; De Gerónimo, E.; Aparicio, V.C.; Einschlag, F.S.G.; Magnacca, G.; Carlos, L.; Mártire, D.O. Carbamazepine Degradation Mediated by Light in the Presence of Humic Substances–Coated Magnetite Nanoparticles. *Nanomaterials* **2019**, *9*, 1379. [CrossRef] [PubMed]
6. Litter, M.I.; Slodowicz, M. An Overview on Heterogeneous Fenton and PhotoFenton Reactions Using Zerovalent Iron Materials. *J. Adv. Oxid. Technol.* **2017**, *20*, 20160164. [CrossRef]
7. Babuponnusami, A.; Muthukumar, K. A Review on Fenton and Improvements to the Fenton Process for Wastewater Treatment. *J. Environ. Chem. Eng.* **2014**, *2*, 557–572. [CrossRef]
8. Bokare, A.D.; Choi, W. Review of Iron–Free Fenton–like Systems for Activating H_2O_2 in Advanced Oxidation Processes. *J. Hazard. Mater.* **2014**, *275*, 121–135. [CrossRef]
9. Hussain, S.; Aneggi, E.; Goi, D. Catalytic Activity of Metals in Heterogeneous Fenton–like Oxidation of Wastewater Contaminants: A Review. *Environ. Chem. Lett.* **2021**, *19*, 2405–2424. [CrossRef]
10. Nichela, D.A.; Berkovic, A.M.; Costante, M.R.; Juliarena, M.P.; García Einschlag, F.S. Nitrobenzene Degradation in Fenton–like Systems Using Cu(II) as Catalyst. Comparison between Cu(II)– and Fe(III)–Based Systems. *Chem. Eng. J.* **2013**, *228*, 1148–1157. [CrossRef]
11. Berkovic, A.M.; Costante, M.R.; García Einschlag, F.S. Combining Multivariate Curve Resolution and Lumped Kinetic Modelling for the Analysis of Lignin Degradation by Copper–Catalyzed Fenton–like Systems. *React. Chem. Eng.* **2022**, *7*, 1954–1967. [CrossRef]
12. Bali, U.; Karagözoğlu, B. Performance Comparison of Fenton Process, Ferric Coagulation and H_2O_2/Pyridine/Cu(II) System for Decolorization of Remazol Turquoise Blue G–133. *Dye Pigment.* **2007**, *74*, 73–80. [CrossRef]
13. Tian, X.; Jin, H.; Nie, Y.; Zhou, Z.; Yang, C.; Li, Y.; Wang, Y. Heterogeneous Fenton–like Degradation of Ofloxacin over a Wide PH Range of 3.6–10.0 over Modified Mesoporous Iron Oxide. *Chem. Eng. J.* **2017**, *328*, 397–405. [CrossRef]
14. Do, Q.C.; Kim, D.G.; Ko, S.O. Catalytic Activity Enhancement of a Fe_3O_4@SiO_2 Yolk–Shell Structure for Oxidative Degradation of Acetaminophen by Decoration with Copper. *J. Clean. Prod.* **2018**, *172*, 1243–1253. [CrossRef]
15. Salem, I.A. Kinetics of the Oxidative Color Removal and Degradation of Bromophenol Blue with Hydrogen Peroxide Catalyzed by Copper(II)–Supported Alumina and Zirconia. *Appl. Catal. B Environ.* **2000**, *28*, 153–162. [CrossRef]
16. Pignatello, J.; Oliveros, E.; MacKay, A. Advanced oxidation processes for organic contaminant destruction based on the fenton reaction and related chemistry. *Crit. Rev. Environ. Sci. Technol.* **2006**, *36*, 1–84. [CrossRef]
17. Zhang, M.H.; Dong, H.; Zhao, L.; Wang, D.-X.; Meng, D. A Review on Fenton Process for Organic Wastewater Treatment Based on Optimization Perspective. *Sci. Total Environ.* **2019**, *670*, 110–121. [CrossRef]
18. McNaught, A.D.; Wilkinson, A. *The IUPAC Compendium of Chemical Terminology*; International Union of Pure and Applied Chemistry (IUPAC): Durham, NC, USA, 2019; ISBN 0-9678550-9-8.
19. Sun, C.; Chen, C.; Ma, W.; Zhao, J. Photodegradation of Organic Pollutants Catalyzed by Iron Species under Visible Light Irradiation. *Phys. Chem. Chem. Phys.* **2011**, *13*, 1957–1969. [CrossRef] [PubMed]
20. Kalal, S.; Singh Chauhan, N.P.; Ameta, N.; Ameta, R.; Kumar, S.; Punjabi, P.B. Role of Copper Pyrovanadate as Heterogeneous Photo–Fenton like Catalyst for the Degradation of Neutral Red and Azure–B: An Eco–Friendly Approach. *Korean J. Chem. Eng.* **2014**, *31*, 2183–2191. [CrossRef]
21. Lee, H.J.; Lee, H.; Lee, C. Degradation of Diclofenac and Carbamazepine by the Copper(II)–Catalyzed Dark and Photo–Assisted Fenton–like Systems. *Chem. Eng. J.* **2014**, *245*, 258–264. [CrossRef]
22. Faust, B.C.; Hoigné, J. Photolysis of Fe (III)–Hydroxy Complexes as Sources of OH Radicals in Clouds, Fog and Rain. *Atmos. Environ. Part A Gen. Top.* **1990**, *24*, 79–89. [CrossRef]
23. Malato, S.; Fernández–Ibáñez, P.; Maldonado, M.I.; Blanco, J.; Gernjak, W. Decontamination and Disinfection of Water by Solar Photocatalysis: Recent Overview and Trends. *Catal. Today* **2009**, *147*, 1–59. [CrossRef]
24. García Einschlag, F.S.; Braun, A.M.; Oliveros, E. Fundamentals and Applications of the Photo–Fenton Process to Water Treatment. In *Handbook of Environmental Chemistry*; Springer: Berlin/Heidelberg, Germany, 2015; Volume 35, pp. 301–342.
25. Bauer, R.; Fallmann, H. The Photo–Fenton Oxidation—A Cheap and Efficient Wastewater Treatment Method. *Res. Chem. Intermed.* **1997**, *23*, 341–354. [CrossRef]
26. Pliego, G.; Zazo, J.A.; Garcia–Muñoz, P.; Munoz, M.; Casas, J.A.; Rodriguez, J.J. Trends in the Intensification of the Fenton Process for Wastewater Treatment: An Overview. *Crit. Rev. Environ. Sci. Technol.* **2015**, *45*, 2611–2692. [CrossRef]
27. Rahim Pouran, S.; Abdul Aziz, A.R.; Wan Daud, W.M.A. Review on the Main Advances in Photo–Fenton Oxidation System for Recalcitrant Wastewaters. *J. Ind. Eng. Chem.* **2015**, *21*, 53–69. [CrossRef]

28. Miralles–Cuevas, S.; Oller, I.; Pérez, J.A.S.; Malato, S. Application of Solar Photo–Fenton at Circumneutral PH to Nanofiltration Concentrates for Removal of Pharmaceuticals in MWTP Effluents. *Environ. Sci. Pollut. Res.* **2015**, *22*, 846–855. [CrossRef]
29. Maniakova, G.; Salmerón, I.; Aliste, M.; Inmaculada Polo–López, M.; Oller, I.; Malato, S.; Rizzo, L. Solar Photo–Fenton at Circumneutral PH Using Fe(III)–EDDS Compared to Ozonation for Tertiary Treatment of Urban Wastewater: Contaminants of Emerging Concern Removal and Toxicity Assessment. *Chem. Eng. J.* **2022**, *431*, 133474. [CrossRef]
30. López–Vinent, N.; Cruz–Alcalde, A.; Lai, C.; Giménez, J.; Esplugas, S.; Sans, C. Role of Sunlight and Oxygen on the Performance of Photo–Fenton Process at near Neutral PH Using Organic Fertilizers as Iron Chelates. *Sci. Total Environ.* **2022**, *803*, 149873. [CrossRef]
31. Ahile, U.J.; Wuana, R.A.; Itodo, A.U.; Sha'Ato, R.; Dantas, R.F. A Review on the Use of Chelating Agents as an Alternative to Promote Photo–Fenton at Neutral PH: Current Trends, Knowledge Gap and Future Studies. *Sci. Total Environ.* **2020**, *710*, 134872. [CrossRef]
32. Sanabria, P.; Wilde, M.L.; Ruiz–Padillo, A.; Sirtori, C. Trends in Fenton and Photo–Fenton Processes for Degradation of Antineoplastic Agents in Water Matrices: Current Knowledge and Future Challenges Evaluation Using a Bibliometric and Systematic Analysis. *Environ. Sci. Pollut. Res.* **2022**, *29*, 42168–42184. [CrossRef]
33. Matta, R.; Hanna, K.; Kone, T.; Chiron, S. Oxidation of 2,4,6-trinitrotoluene in the presence of different iron-bearing minerals at neutral pH. *Chem. Eng. J.* **2008**, *144*, 453–458. [CrossRef]
34. He, J.; Yang, X.; Men, B.; Wang, D. Interfacial Mechanisms of Heterogeneous Fenton Reactions Catalyzed by Iron–Based Materials: A Review. *J. Environ. Sci.* **2016**, *39*, 97–109. [CrossRef] [PubMed]
35. Zuo, S.; Jin, X.; Wang, X.; Lu, Y.; Zhu, Q.; Wang, J.; Liu, W.; Du, Y.; Wang, J. Sandwich Structure Stabilized Atomic Fe Catalyst for Highly Efficient Fenton–like Reaction at All PH Values. *Appl. Catal. B Environ.* **2021**, *282*, 119551. [CrossRef]
36. Xu, L.; Wang, J. Degradation of 2,4,6–Trichlorophenol Using Magnetic Nanoscaled Fe_3O_4/CeO_2 Composite as a Heterogeneous Fenton–like Catalyst. *Sep. Purif. Technol.* **2015**, *149*, 255–264. [CrossRef]
37. Wang, J.; Tang, J. Fe–Based Fenton–like Catalysts for Water Treatment: Catalytic Mechanisms and Applications. *J. Mol. Liq.* **2021**, *332*, 115755. [CrossRef]
38. Xin, S.; Ma, B.; Liu, G.; Ma, X.; Zhang, C.; Ma, X.; Gao, M.; Xin, Y. Enhanced Heterogeneous Photo–Fenton–like Degradation of Tetracycline over $CuFeO_2$/Biochar Catalyst through Accelerating Electron Transfer under Visible Light. *J. Environ. Manag.* **2021**, *285*, 112093. [CrossRef]
39. Wei, Y.; Wang, C.; Liu, D.; Jiang, L.; Chen, X.; Li, H.; Zhang, F. Photo–Catalytic Oxidation for Pyridine in Circumneutral Aqueous Solution by Magnetic Fe–Cu Materials Activated H_2O_2. *Chem. Eng. Res. Des.* **2020**, *163*, 1–11. [CrossRef]
40. Guo, X.; Xu, Y.; Wang, K.; Zha, F.; Tang, X.; Tian, H. Synthesis of Magnetic CuFe2O4 Self–Assembled Hollow Nanospheres and Its Application for Degrading Methylene Blue. *Res. Chem. Intermed.* **2020**, *46*, 853–869. [CrossRef]
41. Tang, J.; Wang, J. Iron–Copper Bimetallic Metal–Organic Frameworks for Efficient Fenton–like Degradation of Sulfamethoxazole under Mild Conditions. *Chemosphere* **2020**, *241*, 125002. [CrossRef]
42. Han, Z.; Dong, Y.; Dong, S. Copper–Iron Bimetal Modified PAN Fiber Complexes as Novel Heterogeneous Fenton Catalysts for Degradation of Organic Dye under Visible Light Irradiation. *J. Hazard. Mater.* **2011**, *189*, 241–248. [CrossRef]
43. Sun, Y.; Yang, Z.; Tian, P.; Sheng, Y.; Xu, J.; Han, Y.F. Oxidative Degradation of Nitrobenzene by a Fenton–like Reaction with Fe–Cu Bimetallic Catalysts. *Appl. Catal. B Environ.* **2019**, *244*, 1–10. [CrossRef]
44. Diodati, S.; Walton, R.I.; Mascotto, S.; Gross, S. Low–Temperature Wet Chemistry Synthetic Approaches towards Ferrites. *Inorg. Chem. Front.* **2020**, *7*, 3282–3314. [CrossRef]
45. Bastianello, M.; Gross, S.; Elm, M.T. Thermal Stability, Electrochemical and Structural Characterization of Hydrothermally Synthesised Cobalt Ferrite ($CoFe_2O_4$). *RSC Adv.* **2019**, *9*, 33282–33289. [CrossRef]
46. Ranga, R.; Kumar, A.; Kumari, P.; Singh, P.; Madaan, V.; Kumar, K. Ferrite Application as an Electrochemical Sensor: A Review. *Mater. Charact.* **2021**, *178*, 111269. [CrossRef]
47. Rana, G.; Dhiman, P.; Kumar, A.; Vo, D.V.N.; Sharma, G.; Sharma, S.; Naushad, M. Recent Advances on Nickel Nano–Ferrite: A Review on Processing Techniques, Properties and Diverse Applications. *Chem. Eng. Res. Des.* **2021**, *175*, 182–208. [CrossRef]
48. Kaur, B.; Kaushal, G.; Rana, S.; Kumar, P.; Khanra, P.; Dhiman, M. Magnetic Ferrites: A Brief Review About Substation on Electric and Magnetic Properties. *ECS Trans.* **2022**, *107*, 9093–9101. [CrossRef]
49. Almessiere, M.A.; Slimani, Y.; Trukhanov, A.V.; Sadaqat, A.; Korkmaz, A.D.; Algarou, N.A.; Aydın, H.; Baykal, A.; Toprak, M.S. Review on Functional Bi–Component Nanocomposites Based on Hard/Soft Ferrites: Structural, Magnetic, Electrical and Microwave Absorption Properties. *Nano-Struct. Nano-Objects* **2021**, *26*, 100728. [CrossRef]
50. Chand, P.; Vaish, S.; Kumar, P. Structural, Optical and Dielectric Properties of Transition Metal (MFe_2O_4; M = Co, Ni and Zn) Nanoferrites. *Phys. B Condens. Matter* **2017**, *524*, 53–63. [CrossRef]
51. Kefeni, K.K.; Mamba, B.B. Photocatalytic Application of Spinel Ferrite Nanoparticles and Nanocomposites in Wastewater Treatment: Review. *Sustain. Mater. Technol.* **2020**, *23*, e00140. [CrossRef]
52. Shobana, M.K. Nanoferrites in Biosensors—A Review. *Mater. Sci. Eng. B Solid-State Mater. Adv. Technol.* **2021**, *272*, 115344. [CrossRef]
53. Amiri, M.; Salavati–Niasari, M.; Akbari, A. Magnetic Nanocarriers: Evolution of Spinel Ferrites for Medical Applications. *Adv. Colloid Interface Sci.* **2019**, *265*, 29–44. [CrossRef] [PubMed]

54. Alao, A.O.; Popoola, A.P.; Sanni, O. The Influence of Nanoparticle Inhibitors on the Corrosion Protection of Some Industrial Metals: A Review. *J. Bio- Tribo-Corrosion* **2022**, *8*, 68. [CrossRef]
55. Mmelesi, O.K.; Masunga, N.; Kuvarega, A.; Nkambule, T.T.; Mamba, B.B.; Kefeni, K.K. Cobalt Ferrite Nanoparticles and Nanocomposites: Photocatalytic, Antimicrobial Activity and Toxicity in Water Treatment. *Mater. Sci. Semicond. Process.* **2021**, *123*, 105523. [CrossRef]
56. Dippong, T.; Levei, E.A.; Cadar, O. Recent Advances in Synthesis and Applications of MFe_2O_4 (M = Co, Cu, Mn, Ni, Zn) Nanoparticles. *Nanomaterials* **2021**, *11*, 1560. [CrossRef]
57. Iqbal, M.J.; Yaqub, N.; Sepiol, B.; Ismail, B. A Study of the Influence of Crystallite Size on the Electrical and Magnetic Properties of $CuFe_2O_4$. *Mater. Res. Bull.* **2011**, *46*, 1837–1842. [CrossRef]
58. Bagade, A.; Nagwade, P.; Nagawade, A.; Thopate, S.; Pandit, V.; Pund, S. Impact of Mg^{2+} Substitution on Structural, Magnetic and Optical Properties of Cu–Cd Ferrites. *Mater. Today Proc.* **2022**, *53*, 144–152. [CrossRef]
59. Holinsworth, B.S.; Mazumdar, D.; Sims, H.; Sun, Q.C.; Yurtisigi, M.K.; Sarker, S.K.; Gupta, A.; Butler, W.H.; Musfeldt, J.L. Chemical Tuning of the Optical Band Gap in Spinel Ferrites: $CoFe_2O_4$ vs. $NiFe_2O_4$. *Appl. Phys. Lett.* **2013**, *103*, 082406. [CrossRef]
60. Lai, Y.J.; Lee, D.J. Solid Mediator Z–Scheme Heterojunction Photocatalysis for Pollutant Oxidation in Water: Principles and Synthesis Perspectives. *J. Taiwan Inst. Chem. Eng.* **2021**, *125*, 88–114. [CrossRef]
61. Feng, Y.; Lee, P.H.; Wu, D.; Zhou, Z.; Li, H.; Shih, K. Degradation of Contaminants by Cu+–Activated Molecular Oxygen in Aqueous Solutions: Evidence for Cupryl Species (Cu^{3+}). *J. Hazard. Mater.* **2017**, *331*, 81–87. [CrossRef]
62. Silva, E.D.N.; Brasileiro, I.L.O.; Madeira, V.S.; De Farias, B.A.; Ramalho, M.L.A.; Rodríguez–Aguado, E.; Rodríguez–Castellón, E. Reusable $CuFe_2O_4$–Fe_2O_3 catalyst Synthesis and Application for the Heterogeneous Photo–Fenton Degradation of Methylene Blue in Visible Light. *J. Environ. Chem. Eng.* **2020**, *8*, 104132. [CrossRef]
63. Cao, Z.; Zuo, C.; Wu, H. One Step for Synthesis of Magnetic $CuFe_2O_4$ Composites as Photo–Fenton Catalyst for Degradation Organics. *Mater. Chem. Phys.* **2019**, *237*, 121842. [CrossRef]
64. Leichtweis, J.; Silvestri, S.; Welter, N.; Vieira, Y.; Zaragoza–Sánchez, P.I.; Chávez–Mejía, A.C.; Carissimi, E. Wastewater Containing Emerging Contaminants Treated by Residues from the Brewing Industry Based on Biochar as a New $CuFe_2O_4$/Biochar Photocatalyst. *Process Saf. Environ. Prot.* **2021**, *150*, 497–509. [CrossRef]
65. Jiang, J.; Gao, J.; Niu, S.; Wang, X.; Li, T.; Liu, S.; Lin, Y.; Xie, T.; Dong, S. Comparing Dark– and Photo–Fenton–like Degradation of Emerging Pollutant over Photo–Switchable Bi2WO6/CuFe2O4: Investigation on Dominant Reactive Oxidation Species. *J. Environ. Sci.* **2021**, *106*, 147–160. [CrossRef]
66. Clarizia, L.; Russo, D.; Di Somma, I.; Marotta, R.; Andreozzi, R. Homogeneous Photo–Fenton Processes at near Neutral PH: A Review. *Appl. Catal. B Environ.* **2017**, *209*, 358–371. [CrossRef]
67. Guo, X.; Wang, K.; Xu, Y. Tartaric Acid Enhanced $CuFe_2O_4$–Catalyzed Heterogeneous Photo–Fenton–like Degradation of Methylene Blue. *Mater. Sci. Eng. B Solid-State Mater. Adv. Technol.* **2019**, *245*, 75–84. [CrossRef]
68. Li, Y.; Qin, C.; Zhang, J.; Lan, Y.; Zhou, L. Cu(II) Catalytic Reduction of Cr(VI) by Tartaric Acid under the Irradiation of Simulated Solar Light. *Environ. Eng. Sci.* **2014**, *31*, 447–452. [CrossRef] [PubMed]
69. Rocha, A.K.S.; Magnago, L.B.; Santos, J.J.; Leal, V.M.; Marins, A.A.L.; Pegoretti, V.C.B.; Ferreira, S.A.D.; Lelis, M.F.F.; Freitas, M.B.J.G. Copper Ferrite Synthesis from Spent Li–Ion Batteries for Multifunctional Application as Catalyst in Photo Fenton Process and as Electrochemical Pseudocapacitor. *Mater. Res. Bull.* **2019**, *113*, 231–240. [CrossRef]
70. Lin, Y.Y.; Lu, S.Y. Selective and Efficient Cleavage of Lignin Model Compound into Value–Added Aromatic Chemicals with $CuFe_2O_4$ Nanoparticles Decorated on Partially Reduced Graphene Oxides via Sunlight–Assisted Heterogeneous Fenton Processes. *J. Taiwan Inst. Chem. Eng.* **2019**, *97*, 264–271. [CrossRef]
71. Ayala, L.I.M.; Paquet, M.; Janowska, K.; Jamard, P.; Quist–Jensen, C.A.; Bosio, G.N.; Mártire, D.O.; Fabbri, D.; Boffa, V. Water Defluoridation: Nanofiltration vs. Membrane Distillation. *Ind. Eng. Chem. Res.* **2018**, *57*, 14740–14748. [CrossRef]
72. Wang, T.; Wang, Z.; Wang, P.; Tang, Y. An Integration of Photo–Fenton and Membrane Process for Water Treatment by a PVDF@CuFe2O4 Catalytic Membrane. *J. Membr. Sci.* **2019**, *572*, 419–427. [CrossRef]
73. Guin, D.; Baruwati, B.; Manorama, S.V. A Simple Chemical Synthesis of Nanocrystalline AFe_2O_4 (A = Fe, Ni, Zn): An Efficient Catalyst for Selective Oxidation of Styrene. *J. Mol. Catal. A Chem.* **2005**, *242*, 26–31. [CrossRef]
74. Zhenmin, L.; Xiaoyong, L.; Hong, W.; Dan, M.; Chaojian, X.; Dan, W. General Synthesis of Homogeneous Hollow Core–Shell Ferrite Microspheres. *J. Phys. Chem. C* **2009**, *113*, 2792–2797. [CrossRef]
75. Sharma, R.; Bansal, S.; Singhal, S. Tailoring the Photo–Fenton Activity of Spinel Ferrites (MFe_2O_4) by Incorporating Different Cations (M = Cu, Zn, Ni and Co) in the Structure. *RSC Adv.* **2015**, *5*, 6006–6018. [CrossRef]
76. Li, N.; Fu, F.; Lu, J.; Ding, Z.; Tang, B.; Pang, J. Facile Preparation of Magnetic Mesoporous $MnFe_2O_4$@SiO_2−CTAB Composites for Cr(VI) Adsorption and Reduction. *Environ. Pollut.* **2017**, *220*, 1376–1385. [CrossRef]
77. Deng, J.; Xu, M.; Qiu, C.; Chen, Y.; Ma, X.; Gao, N.; Li, X. Magnetic $MnFe_2O_4$ Activated Peroxymonosulfate Processes for Degradation of Bisphenol A: Performance, Mechanism and Application Feasibility. *Appl. Surf. Sci.* **2018**, *459*, 138–147. [CrossRef]
78. Velinov, N.; Petrova, T.; Genova, I.; Ivanov, I.; Tsoncheva, T.; Idakiev, V.; Kunev, B.; Mitov, I. Synthesis and Mössbauer Spectroscopic Investigation of Copper–Manganese Ferrite Catalysts for Water–Gas Shift Reaction and Methanol Decomposition. *Mater. Res. Bull.* **2017**, *95*, 556–562. [CrossRef]
79. Fu, W.; Yi, J.; Cheng, M.; Liu, Y.; Zhang, G.; Li, L.; Du, L.; Li, B.; Wang, G.; Yang, X. When Bimetallic Oxides and Their Complexes Meet Fenton–like Process. *J. Hazard. Mater.* **2022**, *424*, 127419. [CrossRef] [PubMed]

80. Meena, S.; Anantharaju, K.S.; Vidya, Y.S.; Renuka, L.; Uma, B.; Sharma, S.C.; Prasad, B.D.; More, S.S. Enhanced Sunlight Driven Photocatalytic Activity and Electrochemical Sensing Properties of Ce–Doped $MnFe_2O_4$ Nano Magnetic Ferrites. *Ceram. Int.* **2021**, *47*, 14760–14774. [CrossRef]

81. Niu, J.; Qian, H.; Liu, J.; Liu, H.; Zhang, P.; Duan, E. Process and Mechanism of Toluene Oxidation Using $Cu_{1-Y}Mn_2Ce_yO_x$/Sepiolite Prepared by the Co–Precipitation Method. *J. Hazard. Mater.* **2018**, *357*, 332–340. [CrossRef] [PubMed]

82. Sun, Y.; Zhou, J.; Liu, D.; Li, X.; Liang, H. Enhanced Catalytic Performance of Cu–Doped $MnFe_2O_4$ Magnetic Ferrites: Tetracycline Hydrochloride Attacked by Superoxide Radicals Efficiently in a Strong Alkaline Environment. *Chemosphere* **2022**, *297*, 134154. [CrossRef]

83. Yang, J.; Zhang, Y.; Zeng, D.; Zhang, B.; Hassan, M.; Li, P.; Qi, C.; He, Y. Enhanced Catalytic Activation of Photo–Fenton Process by $Cu_{0.5}Mn_{0.5}Fe_2O_4$ for Effective Removal of Organic Contaminants. *Chemosphere* **2020**, *247*, 125780. [CrossRef] [PubMed]

84. Wang, X.T.; Li, Y.; Zhang, X.Q.; Li, J.F.; Luo, Y.N.; Wang, C.W. Fabrication of a Magnetically Separable Cu_2ZnSnS_4/$ZnFe_2O_4$ p–n Heterostructured Nano–Photocatalyst for Synergistic Enhancement of Photocatalytic Activity Combining with Photo–Fenton Reaction. *Appl. Surf. Sci.* **2019**, *479*, 86–95. [CrossRef]

85. Shi, L.; Shi, Y.; Zhuo, S.; Zhang, C.; Aldrees, Y.; Aleid, S.; Wang, P. Multi–Functional 3D Honeycomb Ceramic Plate for Clean Water Production by Heterogeneous Photo–Fenton Reaction and Solar–Driven Water Evaporation. *Nano Energy* **2019**, *60*, 222–230. [CrossRef]

86. Schmachtenberg, N.; Silvestri, S.; Da Silveira Salla, J.; Dotto, G.L.; Hotza, D.; Jahn, S.L.; Foletto, E.L. Preparation of Delafossite– Type $CuFeO_2$ Powders by Conventional and Microwave–Assisted Hydrothermal Routes for Use as Photo–Fenton Catalysts. *J. Environ. Chem. Eng.* **2019**, *7*, 102954. [CrossRef]

87. Liu, Q.L.; Zhao, Z.Y.; Zhao, R.D.; Yi, J.H. Fundamental Properties of Delafossite $CuFeO_2$ as Photocatalyst for Solar Energy Conversion. *J. Alloys Compd.* **2020**, *819*, 153032. [CrossRef]

88. da Silveira Salla, J.; da Boit Martinello, K.; Dotto, G.L.; García–Díaz, E.; Javed, H.; Alvarez, P.J.J.; Foletto, E.L. Synthesis of Citrate–Modified $CuFeS_2$ Catalyst with Significant Effect on the Photo–Fenton Degradation Efficiency of Bisphenol a under Visible Light and near–Neutral PH. *Colloids Surfaces A Physicochem. Eng. Asp.* **2020**, *595*, 124679. [CrossRef]

89. Da Silveira Salla, J.; Dotto, G.L.; Hotza, D.; Landers, R.; Da Boit Martinello, K.; Foletto, E.L. Enhanced Catalytic Performance of $CuFeS_2$ chalcogenide Prepared by Microwave–Assisted Route for Photo–Fenton Oxidation of Emerging Pollutant in Water. *J. Environ. Chem. Eng.* **2020**, *8*, 104077. [CrossRef]

90. Cai, X.; Huang, Q.; Hong, Z.; Zhang, Y.; Hu, H.; Huang, Z.; Liang, J.; Qin, Y. Cu Anchored on Manganese Residue through Mechanical Activation to Prepare a Fe–Cu@SiO_2/Starch–Derived Carbon Composites with Highly Stable and Active Visible Light Photocatalytic Performance. *J. Environ. Chem. Eng.* **2021**, *9*, 104710. [CrossRef]

91. Xin, S.; Huo, S.; Zhang, C.; Ma, X.; Liu, W.; Xin, Y.; Gao, M. Coupling Nitrogen/Oxygen Self–Doped Biomass Porous Carbon Cathode Catalyst with $CuFeO_2$/Biochar Particle Catalyst for the Heterogeneous Visible–Light Driven Photo–Electro–Fenton Degradation of Tetracycline. *Appl. Catal. B Environ.* **2022**, *305*, 121024. [CrossRef]

92. Xin, S.; Huo, S.; Xin, Y.; Gao, M.; Wang, Y.; Liu, W.; Zhang, C.; Ma, X. Heterogeneous Photo–Electro–Fenton Degradation of Tetracycline through Nitrogen/Oxygen Self–Doped Porous Biochar Supported $CuFeO_2$ Multifunctional Cathode Catalyst under Visible Light. *Appl. Catal. B Environ.* **2022**, *312*, 121442. [CrossRef]

93. Aparicio, F.; Mizrahi, M.; Ramallo–López, J.M.; Laurenti, E.; Magnacca, G.; Carlos, L.; Mártire, D.O. Novel Bimetallic Magnetic Nanocomposites Obtained from Waste–Sourced Bio–Based Substances as Sustainable Photocatalysts. *Mater. Res. Bull.* **2022**, *152*, 111846. [CrossRef]

94. Ayala, L.I.M.; Aparicio, F.; Boffa, V.; Magnacca, G.; Carlos, L.; Bosio, G.N.; Mártire, D.O. Removal of As(III) via Adsorption and Photocatalytic Oxidation with Magnetic Fe–Cu Nanocomposites. *Photochem. Photobiol. Sci.* **2022**, *1*, 1–10. [CrossRef]

95. Mansoori, S.; Davarnejad, R.; Ozumchelouei, E.J.; Ismail, A.F. Activated Biochar Supported Iron–Copper Oxide Bimetallic Catalyst for Degradation of Ciprofloxacin via Photo–Assisted Electro–Fenton Process: A Mild PH Condition. *J. Water Process Eng.* **2021**, *39*, 101888. [CrossRef]

96. Espinosa, J.C.; Catalá, C.; Navalón, S.; Ferrer, B.; Álvaro, M.; García, H. Iron Oxide Nanoparticles Supported on Diamond Nanoparticles as Efficient and Stable Catalyst for the Visible Light Assisted Fenton Reaction. *Appl. Catal. B Environ.* **2018**, *226*, 242–251. [CrossRef]

97. Manickam–Periyaraman, P.; Espinosa, J.C.; Ferrer, B.; Subramanian, S.; Álvaro, M.; García, H.; Navalón, S. Bimetallic Iron–Copper Oxide Nanoparticles Supported on Nanometric Diamond as Efficient and Stable Sunlight–Assisted Fenton Photocatalyst. *Chem. Eng. J.* **2020**, *393*, 124770. [CrossRef]

98. Khan, A.; Valicsek, Z.; Horváth, O. Comparing the Degradation Potential of Copper(II), Iron(II), Iron(III) Oxides, and Their Composite Nanoparticles in a Heterogeneous Photo–Fenton System. *Nanomaterials* **2021**, *11*, 225. [CrossRef]

99. Asenath–Smith, E.; Ambrogi, E.K.; Barnes, E.; Brame, J.A. CuO Enhances the Photocatalytic Activity of Fe2O3 through Synergistic Reactive Oxygen Species Interactions. *Colloids Surfaces A Physicochem. Eng. Asp.* **2020**, *603*, 125179. [CrossRef]

100. Lu, M.; Wang, J.; Wang, Y.; He, Z. Heterogeneous Photo–Fenton Catalytic Degradation of Practical Pharmaceutical Wastewater by Modified Attapulgite Supported Multi–Metal Oxides. *Water* **2021**, *13*, 156. [CrossRef]

101. Davarnejad, R.; Hassanvand, Z.R.; Mansoori, S.; Kennedy, J.F. Metronidazole Elimination from Wastewater through Photo–Fenton Process Using Green–Synthesized Alginate–Based Hydrogel Coated Bimetallic Iron-copper Nanocomposite Beads as a Reusable Heterogeneous Catalyst. *Bioresour. Technol. Rep.* **2022**, *18*, 101068. [CrossRef]

102. Zhang, Y.; Guo, P.; Jin, M.; Gao, G.; Xi, Q.; Zhou, H.; Xu, G.; Xu, J. Promoting the Photo–Fenton Catalytic Activity with Carbon Dots: Broadening Light Absorption, Higher Applicable PH and Better Reuse Performance. *Mol. Catal.* **2020**, *481*, 110254. [CrossRef]

103. Zhang, B.; Hou, Y.; Yu, Z.; Liu, Y.; Huang, J.; Qian, L.; Xiong, J. Three–Dimensional Electro–Fenton Degradation of Rhodamine B with Efficient Fe–Cu/Kaolin Particle Electrodes: Electrodes Optimization, Kinetics, Influencing Factors and Mechanism. *Sep. Purif. Technol.* **2019**, *210*, 60–68. [CrossRef]

104. Joseph, J.; Iftekhar, S.; Srivastava, V.; Fallah, Z.; Zare, E.N.; Sillanpää, M. Iron–Based Metal–Organic Framework: Synthesis, Structure and Current Technologies for Water Reclamation with Deep Insight into Framework Integrity. *Chemosphere* **2021**, *284*, 131171. [CrossRef] [PubMed]

105. Wang, D.; Wang, M.; Li, Z. Fe–Based Metal–Organic Frameworks for Highly Selective Photocatalytic Benzene Hydroxylation to Phenol. *ACS Catal.* **2015**, *5*, 6852–6857. [CrossRef]

106. Ahmad, M.; Chen, S.; Ye, F.; Quan, X.; Afzal, S.; Yu, H.; Zhao, X. Efficient Photo–Fenton Activity in Mesoporous MIL–100(Fe) Decorated with ZnO Nanosphere for Pollutants Degradation. *Appl. Catal. B Environ.* **2019**, *245*, 428–438. [CrossRef]

107. He, X.; Fang, H.; Gosztola, D.J.; Jiang, Z.; Jena, P.; Wang, W.N. Mechanistic Insight into Photocatalytic Pathways of MIL–100(Fe)/TiO$_2$ Composites. *ACS Appl. Mater. Interfaces* **2019**, *11*, 12516–12524. [CrossRef] [PubMed]

108. Oladipo, A.A. MIL–53 (Fe)–Based Photo–Sensitive Composite for Degradation of Organochlorinated Herbicide and Enhanced Reduction of Cr(VI). *Process Saf. Environ. Prot.* **2018**, *116*, 413–423. [CrossRef]

109. Wang, Q.; Gao, Q.; Al–Enizi, A.M.; Nafady, A.; Ma, S. Recent Advances in MOF–Based Photocatalysis: Environmental Remediation under Visible Light. *Inorg. Chem. Front.* **2020**, *7*, 300–339. [CrossRef]

110. Do, T.L.; Ho, T.M.T.; Doan, V.D.; Le, V.T.; Hoai Thuong, N. Iron–Doped Copper 1,4–Benzenedicarboxylate as Photo–Fenton Catalyst for Degradation of Methylene Blue. *Toxicol. Environ. Chem.* **2019**, *101*, 13–25. [CrossRef]

111. Shi, S.; Han, X.; Liu, J.; Lan, X.; Feng, J.; Li, Y.; Zhang, W.; Wang, J. Photothermal–Boosted Effect of Binary Cu—Fe Bimetallic Magnetic MOF Heterojunction for High–Performance Photo–Fenton Degradation of Organic Pollutants. *Sci. Total Environ.* **2021**, *795*, 148883. [CrossRef]

112. Zhong, Z.; Li, M.; Fu, J.; Wang, Y.; Muhammad, Y.; Li, S.; Wang, J.; Zhao, Z.; Zhao, Z. Construction of Cu–Bridged Cu$_2$O/MIL(Fe/Cu) Catalyst with Enhanced Interfacial Contact for the Synergistic Photo–Fenton Degradation of Thiacloprid. *Chem. Eng. J.* **2020**, *395*, 125184. [CrossRef]

113. Wu, Q.; Siddique, M.S.; Guo, Y.; Wu, M.; Yang, Y.; Yang, H. Low–Crystalline Bimetallic Metal–Organic Frameworks as an Excellent Platform for Photo–Fenton Degradation of Organic Contaminants: Intensified Synergism between Hetero–Metal Nodes. *Appl. Catal. B Environ.* **2021**, *286*, 119950. [CrossRef]

catalysts

Article

Fenton-like Remediation for Industrial Oily Wastewater Using Fe$_{78}$Si$_9$B$_{13}$ Metallic Glasses

Yulong Liu [1,2], Bowen Zhao [1,3,4], Guofeng Ma [2], Shiming Zhang [5], Haifeng Zhang [1,4] and Zhengwang Zhu [1,4,*]

[1] Shi-Changxu Innovation Center for Advanced Materials, Institute of Metal Research, Chinese Academy of Sciences, Shenyang 110016, China
[2] College of Mechanical Engineering, Shenyang University, Shenyang 110044, China
[3] School of Materials Science and Engineering, University of Science and Technology of China, Shenyang 110016, China
[4] CAS Key Laboratory of Nuclear Materials and Safety Assessment, Institute of Metal Research, Chinese Academy of Sciences, Shenyang 110016, China
[5] Qingdao Yunlu Advanced Materials Technology Co., Ltd., Qingdao 266232, China
* Correspondence: zwzhu@imr.ac.cn

Abstract: Metallic glasses (MGs) with a unique atomic structure have been widely used in the catalytic degradation of organic pollutants in the recent years. Fe$_{78}$Si$_9$B$_{13}$ MGs exhibited excellent catalytic performance for the degradation of oily wastewater in a Fenton-like system for the first time. The oil removal and chemical oxygen demand (COD) removal from the oily wastewater were 72.67% and 70.18% within 60 min, respectively. Quenching experiments were performed to verify the production of active hydroxyl radicals ($\cdot$OH) by activating hydrogen peroxide (H$_2$O$_2$). The formation of $\cdot$OH species can significantly contribute to the degradation reaction of oily wastewater. Fe$_{78}$Si$_9$B$_{13}$ MG ribbons were highly efficient materials that exhibited superior reactivity towards H$_2$O$_2$ activation in oily wastewater treatment. The study revealed the catalytic capability of metallic glasses, presenting extensive prospects of their applications in oily wastewater treatment.

Keywords: metallic glasses; oily wastewater; activation; catalytic degradation

Citation: Liu, Y.; Zhao, B.; Ma, G.; Zhang, S.; Zhang, H.; Zhu, Z. Fenton-like Remediation for Industrial Oily Wastewater Using Fe$_{78}$Si$_9$B$_{13}$ Metallic Glasses. *Catalysts* **2022**, *12*, 1038. https://doi.org/10.3390/catal12091038

Academic Editors: Gassan Hodaifa, Rafael Borja and Mha Albqmi

Received: 10 August 2022
Accepted: 5 September 2022
Published: 12 September 2022

Publisher's Note: MDPI stays neutral with regard to jurisdictional claims in published maps and institutional affiliations.

1. Introduction

Waste sludge containing large amounts of mineral oil is produced in the metal manufacturing process, which is classified as a high concentration hazardous waste [1–3]. About 10~20 tons of oily wastewater will be produced from cleaning 1 ton of waste sludge. This leads to a huge discharge of oily wastewater. The organic pollutants in oily wastewater have certain characteristics, such as high concentration, complex composition, biological toxicity and refractory [4–7]. At present, a large number of advanced treatment technologies for oily wastewater are systematically implemented [8–11]. Compared to conventional techniques, advanced oxidation processes (AOPs) have been extensively studied as a promising technique due to their superior degradation and mineralization efficiency of pollutants in wastewater [12,13]. Very recently, owing to the advantage of abundant natural resources, low cost and environmental friendliness, Fe-based catalytic materials have been extensively used for the degradation of organic contaminants in AOPs system. However, the catalysts like magnetite (Fe$_3$O$_4$) [14] and zero valent iron (ZVI) [15] still have certain disadvantages, such as low efficiency, poor reusability, fast decay and secondary pollution [16]. To overcome the aforementioned limitations, Fe-based MGs are utilized, which could effectively enhance the catalytic activity by improving the internal atomic arrangement [17] and tuning the chemical composition [18,19].

Metallic glasses with short range ordered and long range disordered atomic structure were usually employed as structural materials for different industrial applications due to improved mechanical properties, soft ferromagnetism and corrosion resistance [20–22].

Additionally, a great number of studies have been showed that the metallic glasses with various alloy systems exhibited superior catalytic degradation behaviors of organic pollutants owing to high reactivity [23–25]. For instance, the $Cu_{46}Zr_{42}Al_7Y_5$ MG ribbons showed excellent catalytic performance for acid orange II degradation with a degradation efficiency of 96.05% and a COD removal of 51.73% [26]. A maximum degradation efficiency of 98% for congo red is achieved using $Mg_{60}Zn_{35}Ca_5$ MG powders [27]. Moreover, the direct blue 2B solutions can be completely degraded by Al-based MG ribbons under wide pH conditions [19]. Similarly, the unexpected organic pollutant degradation performance of Fe-based MGs was revealed. Amorphous $Fe_{84}B_{16}$ exhibited degradation of direct blue 6 dye 1.8 and 89 times faster compared to the Fe-B crystalline alloy and commercial iron powder, respectively [28]. Furthermore, $Fe_{78}Si_9B_{13}$ MGs exhibited excellent production rate of active radical species among various Fe-based catalysts [29]. As a result, $Fe_{78}Si_9B_{13}$ MGs have been widely used for wastewater treatment [30–32], providing an environmentally functional material for the catalytic degradation of organic contaminants.

In this work, oily wastewater degradation and mineralization were investigated using $Fe_{78}Si_9B_{13}$ MGs. These catalytic materials serve as an alternative Fe^{2+} releasing source that can produce ·OH radicals by activating H_2O_2 (Fenton-like system). $Fe_{78}Si_9B_{13}$ MGs were found to be highly efficient catalysts that showed improved degradation performance of oily wastewater in terms of oil removal and COD removal. Moreover, the stability and durability of $Fe_{78}Si_9B_{13}$ MGs were discussed based on recycling experiments. The catalytic mechanism was proposed based on the structural and morphological variations of catalysts combined with free radical quenching experiments.

2. Materials and Methods

2.1. Materials

The alloy ingots with nominal composition of $Fe_{78}Si_9B_{13}$ were prepared by arc melting of a mixture of Fe, Si and B with greater purity than 99.9 wt% under a Ti-gettered Ar atmosphere. The master alloy ingot was melted by induction heating in a quartz crucible. The molten master alloy ingot was ejected onto a chilled copper roll surface to prepare the as-melt MG ribbons of $Fe_{78}Si_9B_{13}$. The waste oily sludge was supplied by Shenyang General Magnetic Co., Ltd., Shenyang, China. The oily wastewater was prepared by stirring for 30 min and centrifuging sludge mixtures with an oil–water ratio of 1:20, 1:40, 1:60 and 1:80 (wt%).

2.2. Characterization

The structural features of the as-melt and reacted $Fe_{78}Si_9B_{13}$ ribbons were characterized by X-ray diffraction (XRD, Rigaku D/max-2500PC, Tokyo, Japan) with Co-Kα radiation. The surface morphologies of the ribbons before and after the catalytic organic pollutant degradation were characterized using a scanning electron microscope (SEM, Zeiss SUPPA 55, Oberkochen, Germany). The crystallization behavior of the melt and reacted ribbons was characterized by differential scanning calorimetry (Netzsch DSC 404C, Selbu, Germany).

2.3. Analytical Methods

All oily wastewater removal experiments were conducted in a 250 mL glass beaker with a stirring speed of 300 rpm by a mechanical stirrer (Changzhou Jaboson Instrument JJ-1, Changzhou, China). The oily wastewater of 200 mL in a glass beaker was placed in a thermostat water bath at the desired temperature. The solution pH values were adjusted by diluted HCl solution (1 mol·L^{-1}) and diluted NaOH (0.1 mol·L^{-1}). At every 10 min interval, about 10 mL aliquot of the reaction mixture was collected and centrifuged. The supernatant liquid was tested by infrared spectrophotometer (Tianjin Tianguang TJ270-30A, Tianjin, China) and chemical oxygen demand detector (COD, Beijing Lianhua 5B-3(B), Beijing, China). The ion concentration of the oily wastewater samples before and after the reactions was measured by inductively coupled plasma-optical emission spectrometer (ICP-OES 720, Agilent Technologies Inc., Palo Alto, America). The active radical species were analyzed using an electron paramagnetic resonance spectrometer (EPR, Bruker A300, Karlsruhe, Germany) with 5,5-dimethyl-1-pyrroline N-oxide (DMPO) as the spin-trapping agent.

3. Results and Discussion

3.1. Catalytic Capability

3.1.1. Effect of pH Value

Figure 1a shows the effect of pH value (from 2 to 7) on oil removal. It is observed that the oil removal sharply reduces with increasing of pH value. Oil removal of 72.67% can be achieved within 60 min at pH = 3. In contrast, only 52.14% oil removal was observed at pH = 7. The enhanced degradation performance at lower pH value may be due to the influence of pH on stability of H_2O_2 [33,34]. As shown in Figure 1b, the COD removal of 70.18% was achieved within 60 min at pH = 3, indicating the favorable mineralization efficiency of oily wastewater. Under a high pH system, some side effects lead to reducing the degradation capability due to formation of sediment, following Equations (1) and (2) [35,36].

$$Fe^{2+} + 2OH^- \rightarrow Fe(OH)_2 \tag{1}$$

$$4Fe(OH)_2 + 2H_2O + O_2 \rightarrow 4Fe(OH)_3 \tag{2}$$

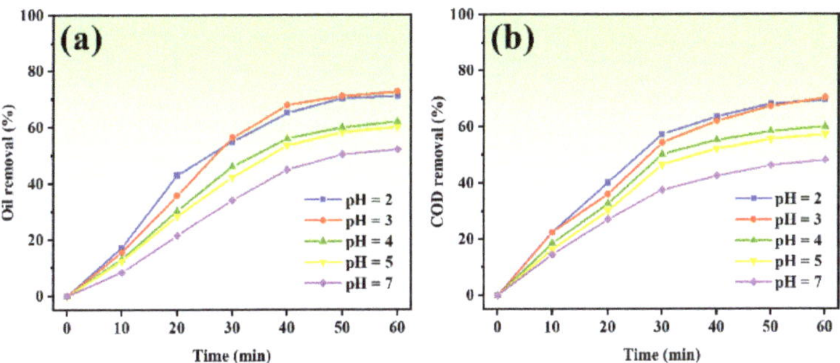

Figure 1. Effect of pH value on (**a**) oil removal and (**b**) COD removal. (H_2O_2 concentration: 0.12 mol·L^{-1}, catalyst dosage: 1.5 g·L^{-1}, oil–water mass ratio: 1:60, temperature: 40 °C).

3.1.2. Effect of H_2O_2 Concentration

As shown in Figure 2a, the oil removal is 55.10% when the H_2O_2 concentration is 0.08 mol·L^{-1}. Apparently, the oil removal increased to 63.34% and 70.97% within 60 min at H_2O_2 concentrations of 0.10 and 0.14 mol·L^{-1}, respectively. As the H_2O_2 concentration increased further to 0.16 mol·L^{-1}, the oil removal decreased. COD removal displayed the similar results as shown in Figure 2b. The degradation capability of COD removal increases and then decreases with H_2O_2 concentration. The reason may be that the quantity of ·OH radicals play a significant role in the oily wastewater degradation [37,38]. Moreover, the excessive H_2O_2 concentration results in ·OH radicals self-quenching to form the weak oxidizing hydroperoxy radicals (·O_2H radicals) via Equations (3) and (4).

$$H_2O_2 + \cdot OH \rightarrow \cdot O_2H + H_2O \tag{3}$$

$$HO_2\cdot + HO_2\cdot \rightarrow H_2O_2 + O_2 \tag{4}$$

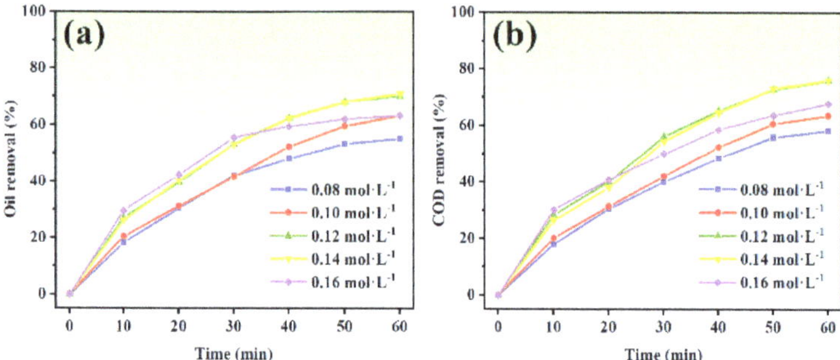

Figure 2. Effect of H_2O_2 concentration on (**a**) oil removal and (**b**) COD removal. (pH = 3, catalyst dosage: 1.5 g·L^{-1}, oil–water mass ratio: 1:60, temperature: 40 °C).

3.1.3. Effect of Catalyst Dosage

As seen from Figure 3a, the oil removal significantly enhanced from 53.35% to 72.23% within 60 min when the catalyst dosage increased from 0.5 g·L^{-1} to 1.5 g·L^{-1}. The oil removal slightly increased with a further increase in catalyst loading to 2.0 g·L^{-1}. COD removal increased with an increase in the catalyst loading as shown in Figure 3b. When the MG ribbon dosage is 2.0 g·L^{-1}, the COD removal stabilized at around 72%. Normally, the organic molecule decomposition takes place via direct surface reaction on active iron species [39]. Increasing the amount of catalyst provides more active sites, thereby achieving faster production rate of ·OH radical species. However, excessive addition of catalyst would provide more Fe^{3+}, which would work against catalytic efficiency [40,41].

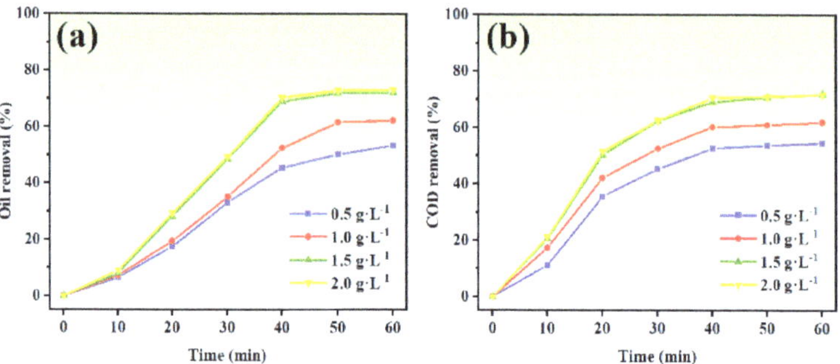

Figure 3. Effect of catalyst dosage on (**a**) oil removal and (**b**) COD removal. (H_2O_2 concentration: 0.12 mol·L^{-1}, pH = 3, oil–water mass ratio: 1:60, temperature: 40 °C).

3.1.4. Effect of Oil–Water Mass Ratio

As the oil–water mass ratio increases, the initial oil concentration increases. As seen in Figure 4a, the faster oil removal is observed for the lower the oil–water mass ratio in the first 10 min of reaction. Figure 4b shows the effect of the oil–water mass ratio on COD removal with an oil to water in the ratio (by mass) 1:80 to 1:20. With the increase of the oil–water mass ratio, the COD removal is obviously reduced. This is because more organic molecules exist in the system of larger oil–water mass ratio as the reaction progresses, thereby leading to the weak degradation behaviors in the presence of the same amount of ·OH radicals.

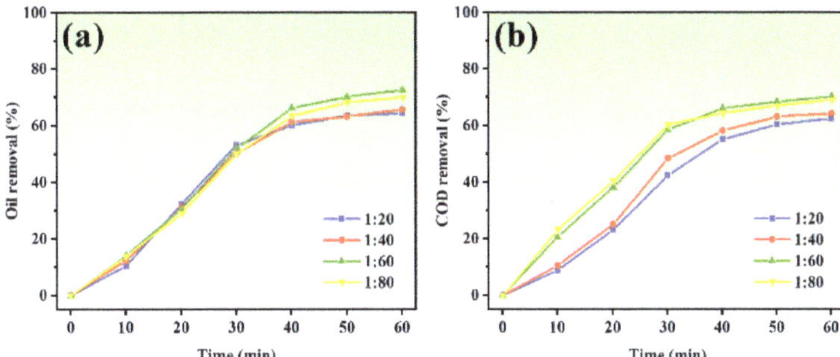

Figure 4. Effect of oil–water mass ratio on (**a**) oil removal and (**b**) COD removal (H_2O_2 concentration: 0.12 mol·L^{-1}, pH = 3, catalyst dosage: 1.5 g·L^{-1}, temperature: 40 °C).

3.1.5. Effect of Reaction Temperature

Regarding chemical reactions, the reaction temperature can be always considered as an important experimental parameter. According to the principle of reaction kinetics, the molecules colliding with each other in solution will be accelerated and energized with the increase of reaction temperature, thus speeding up the reaction. The evaluation of the degradation behaviors in temperature ranging from 25 to 45 °C as shown in Figure 5a,b. The oil removal and COD removal remarkably increase along with the increase of reaction temperature. When the temperature increases to 45 °C, the degradation behavior is not significantly improved.

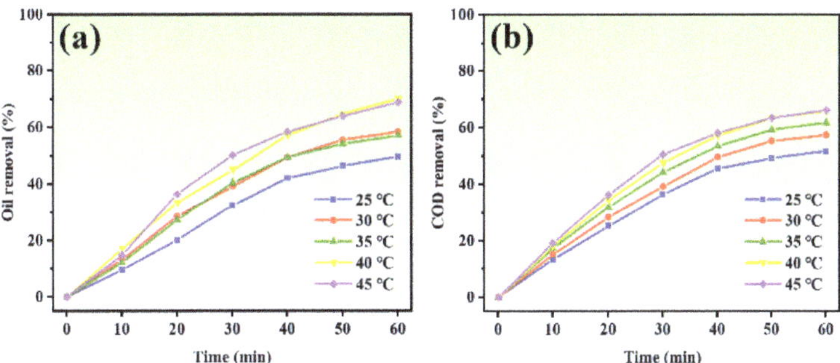

Figure 5. Effect of reaction temperature on (**a**) oil removal and (**b**) COD removal. (H_2O_2 concentration: 0.12 mol·L^{-1}, pH = 3, catalyst dosage: 1.5 g·L^{-1}, oil–water mass ratio: 1:60).

3.1.6. Stability and Reusability

The stability and reusability of metallic glassy catalysts are the extremely important capabilities for degradation of organic pollutants [42]. Figure 6 shows the catalytic reusability using $Fe_{78}Si_9B_{13}$ MGs from the 1st to 3rd run for degrading oily wastewater. Clearly, the COD removal still maintains as high as 68% after 60 min for three cycles, suggesting the excellent reuse life. Notably, the COD removal for the 2nd run is little better than that for the 1st run due to surface activation. The slight decay is observed on the ribbon surface after the 2nd run, forming SiO_2 layers during the degradation of oily wastewater [8]. Abundant Fe^{2+} irons can be supplied by falling the oxide layers for activating H_2O_2, thereby further improving the catalytic reusability [43,44]. For the 3rd run, the COD removal slightly decreases due to the side effect of oxide deposition.

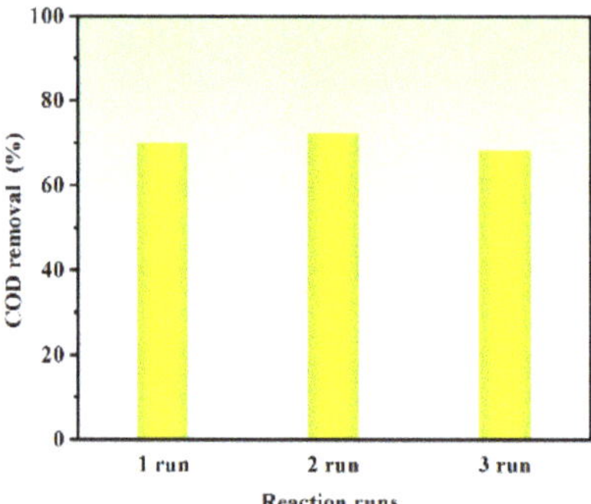

Figure 6. Reaction runs for oily wastewater removal by $Fe_{78}Si_9B_{13}/H_2O_2$ system (H_2O_2 concentration: 0.12 mol·L^{-1}, pH = 3, catalyst dosage: 1.5 g·L^{-1}, oil–water mass ratio: 1:60, temperature: 40 °C).

3.2. Structures and Surface Morphology

In order to further investigate stability, the structures and surface morphology of the ribbons were characterized, respectively. Figure 7a displays the XRD patterns of the as-received and reused 3rd $Fe_{78}Si_9B_{13}$ metallic glass ribbons. The XRD patterns of as-received and reused ribbons present a broad diffraction peak at $2\theta = 40{\sim}50°$, indicating that all the ribbons are mainly in the amorphous state [45]. However, the 3rd run recycled $Fe_{78}Si_9B_{13}$ metallic glass ribbons have a crystallization peak, indicating crystalline precipitated phase of α-Fe on the surface of the ribbons [46,47]. As shown in Figure 7b, each of DSC curves clearly displays the exothermic peak, further obtaining thermodynamic parameters of T_{p1} and T_{p2}. The amorphous nature was also verified by DSC measurements. According to the existence of two exothermic peaks in DSC curves and the intensity of T_{p1} and T_{p2} is raised, the crystallization process can be roughly divided into two steps: one is the precipitation process of primary α-Fe phase and the other is the precipitation process of boride [48]. According to the Dubois model [49], Fe-Si-B series alloys are composed of Fe-B and Fe-Si regions. The precipitation of α-Fe may accelerate the decomposition of Fe-B region due to heteronucleation [50].

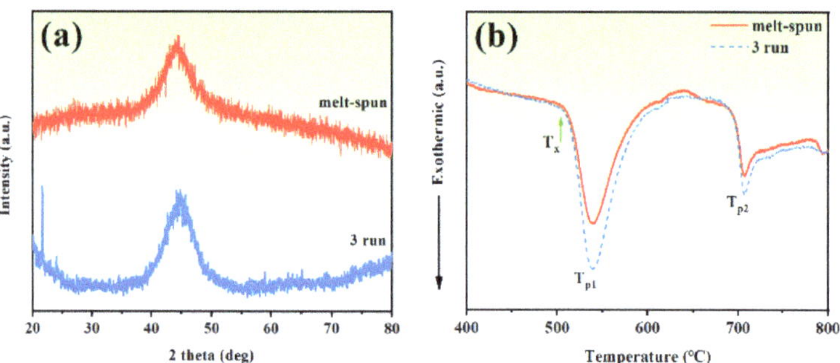

Figure 7. (**a**) XRD patterns and (**b**) DSC curves of the melt-spun and 3rd reused $Fe_{78}Si_9B_{13}$ metallic glass ribbons.

Figure 8 shows the SEM images of the melt-spun, 1st and 3rd reused $Fe_{78}Si_9B_{13}$ ribbons. It can be observed that the surface morphology of melt-spun ribbons is very smooth without obvious surface defects as shown in Figure 8a. Although the surface of the ribbons presents a slight decay with several corrosion areas in Figure 8b, most of the surface remains relative smooth. As seen from Figure 8c, some corrosion products are precipitated and accumulated on the surface of the ribbons for area B, except for the small regions of smooth surface for area A. These products are covered with active substances, further leading to the reduction of catalytic degradation reaction.

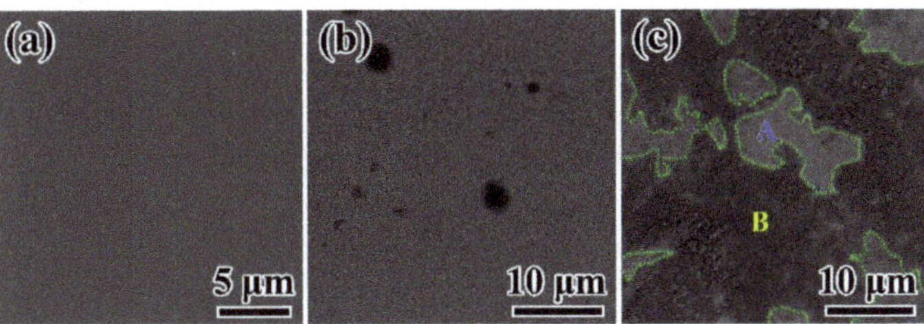

Figure 8. SEM images of the (**a**) as-received, (**b**) 1st and (**c**) 3rd reused $Fe_{78}Si_9B_{13}$ metallic glass ribbons.

3.3. Reaction Mechanism

It has been recognized that the mechanism for degradation of organic contaminants is attributed to the catalytic oxidation of activated ·OH radicals in the Fenton process. The o-Phenylenediamine (OPDA) as typical catcher of ·OH radicals was used to verify the catalytic contribution of ·OH radicals. The OPDA will react with ·OH radicals to form 2,3-diaminophenazine (DAPN) stabilizing in solution for a long time. Figure 9a shows the effect of OPDA concentration on COD removal from 0 to 20 mmol·L^{-1}. It is observed that the COD removal sharply decreases with increasing of OPDA concentration. Meanwhile, the reaction rate rapidly decreases in Figure 9b. The reason may be that the active ·OH radicals preferentially combine with OPDA to produce DAPN rather than degrading organic pollutants.

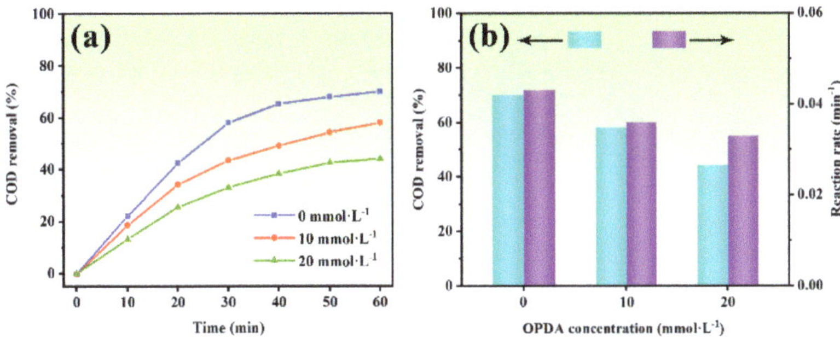

Figure 9. Effect of OPDA concentration on (**a**) the normalized as a function of reaction time, (**b**) COD removal and reaction rate. (H_2O_2 concentration: 0.12 mol·L^{-1}, pH = 3, catalyst dosage: 1.5 g·L^{-1}, oil–water mass ratio: 1:60, temperature: 40 °C).

The active radical species were observed by EPR technique with DMPO as spin-trapping agent. As shown in Figure 10, variation for aliquots of samples collected at 10 min and 20 min of reaction interval in intensity of ·OH radicals is obvious between 3480 and 3540 of magnetic field. The relative intensity of ·OH radicals increased with the prolongation of reaction time. To combine with the above results, ·OH radicals play an important role in the whole Fenton-like degradation process of oily wastewater.

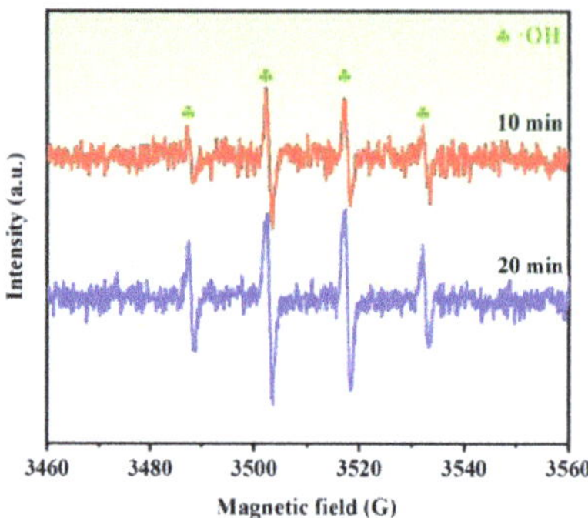

Figure 10. Intensity of ·OH radicals in oily wastewater during the Fenton-like process.

Various ion concentrations of solution system before and after the reactions are analyzed in Table 1. The variation on majority of ions concentration is slight, whereas iron ions in solution have a great increased. A mass of Fe^{2+} and Fe^{3+} flow into the solution because of corrosion action and Fe^0 plays an important role is confirmed at the same time. Furthermore, Figure 11 shows the forming process of ·OH radicals in the Fenton reaction. It is well known that Fe^{2+} acts as the main active source to activate H_2O_2 producing active species for ·OH radicals (Equation (5)) [16]. However, the formation of Fe^{2+} is more likely by direct reaction between zero-valent iron in the $Fe_{78}Si_9B_{13}$ ribbons and a small amount of H_2O_2 molecules (Equation (6)) [23]. In addition, the electrons of amorphous Fe atom on $4s^2$ orbital are extraordinarily unstable and active, thereby leading to forming Fe^{2+} by losing electrons (Equation (7)) [11] Besides this, there are reciprocal transitions between the iron ions for the Fe^{2+} and Fe^{3+} under certain conditions as the reaction continues (Equations (8)–(10)). According to the continuous reaction process, the Fe^{2+} will react with H_2O_2 to produce moderate activated ·OH radicals, contributing to enhancing the catalytic degradation reaction [9]. Therefore, the major reaction equations are as the following Equations (5)–(10):

$$Fe^{2+} + H_2O_2 \rightarrow Fe^{3+} + OH^- + \cdot OH \tag{5}$$

$$Fe^0 + H_2O_2 \rightarrow Fe^{2+} + 2OH^- \tag{6}$$

$$Fe^0 \rightarrow Fe^{2+} + 2e^- \tag{7}$$

$$Fe^{3+} + H_2O_2 \rightarrow Fe^{2+} + O_2H + H^+ \tag{8}$$

$$Fe^{3+} + \cdot O_2H \rightarrow Fe^{2+} + O_2 + H^+ \tag{9}$$

$$Fe^0 + 2Fe^{3+} \rightarrow 3Fe^{2+} \tag{10}$$

Table 1. Comparative variations of ion concentration before and after the reaction.

Ion	Ion Concentration (mg·L^{-1})	
	Before Reaction	**After Reaction**
Li^+	<0.2	<0.2
K^+	6.4	8.1
Mg^+	5.4	7.1
Fe^{2+}/Fe^{3+}	2.5	251.3
Ca^{2+}	29.9	40.3
Si^{4+}	4.0	10.7
B^{3+}	5.8	11.7

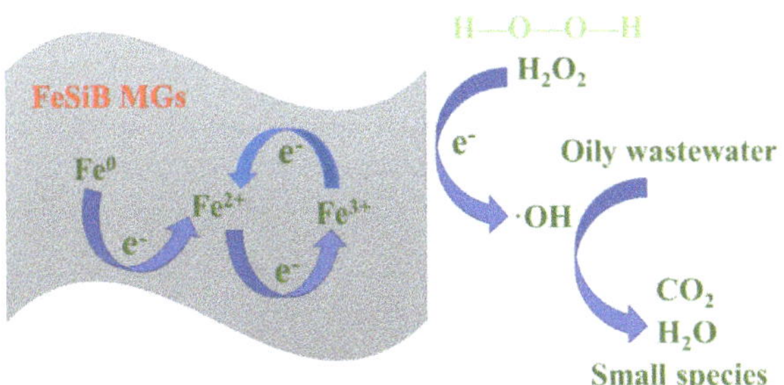

Figure 11. Schematic illustration for the Fenton-like degradation of oily wastewater using Fe$_{78}$Si$_9$B$_{13}$ metallic glassy ribbons.

4. Conclusions

In this work, the Fe$_{78}$Si$_9$B$_{13}$ MGs as efficient catalysts demonstrate excellent catalytic degradation behavior towards oily wastewater treatment in H$_2$O$_2$ activation system. The conclusions are as follows:

(1) The oil removal and COD removal of oily wastewater are achieved as high as 72.67% and 70.18% within 60 min using Fe$_{78}$Si$_9$B$_{13}$ MG ribbons under the optimum conditions, respectively.

(2) The Fe$_{78}$Si$_9$B$_{13}$ MGs present the superior stability and reusability for 3 times with high COD removal during oily wastewater degradation.

(3) The enhanced degradation performance may be mainly attributed to activate H$_2$O$_2$ molecules to generate ·OH radicals in the EPR analysis and quenching experiments. The Fe$_{78}$Si$_9$B$_{13}$ MGs provide a potential strategy for activating abundant ·OH radicals during oily wastewater degradation.

Author Contributions: Conceptualization, Y.L. and Z.Z.; methodology, B.Z. and Z.Z.; validation, Y.L., B.Z. and Z.Z.; writing—original draft preparation, Y.L.; writing—review and editing, Y.L., B.Z., G.M., S.Z., H.Z. and Z.Z.; supervision, G.M. and Z.Z.; funding acquisition, H.Z. and Z.Z.; All authors have read and agreed to the published version of the manuscript.

Funding: The work was supported by the National Natural Science Foundation of China (51801209 and 52074257), the Fund of Qingdao (19-9-2-1-wz).

Data Availability Statement: All relevant data are included in the paper.

Conflicts of Interest: The authors declare no conflict of interest.

References

1. Zhang, B.; Huang, K.; Wang, Q.; Li, G.; Wu, T.; Li, Y. Highly efficient treatment of oily wastewater using magnetic carbon nanotubes/layered double hydroxides composites. *Colloids Surf. A* **2020**, *586*, 124187. [CrossRef]
2. Khalifa, O.; Banat, F.; Srinivasakannan, C.; AlMarzooqi, F.; Hasan, S.W. Ozonation-assisted electro-membrane hybrid reactor for oily wastewater treatment: A methodological approach and synergy effects. *J. Clean. Prod.* **2021**, *289*, 125764. [CrossRef]
3. Zhong, L.; Sun, C.; Yang, F.; Dong, Y. Superhydrophilic spinel ceramic membranes for oily emulsion wastewater treatment. *J. Water Process Eng.* **2021**, *42*, 102161. [CrossRef]
4. Ahmad, T.; Guria, C.; Mandal, A. A review of oily wastewater treatment using ultrafiltration membrane: A parametric study to enhance the membrane performance. *J. Water Process Eng.* **2020**, *36*, 101289. [CrossRef]
5. Behroozi, A.H.; Ataabadi, M.R. Improvement in microfiltration process of oily wastewater: A comprehensive review over two decades. *J. Environ. Chem. Eng.* **2020**, *9*, 104981. [CrossRef]
6. Zhao, C.L.; Zhou, J.Y.; Yan, Y.; Yang, L.W.; Xing, G.H.; Li, H.Y.; Wu, P.; Wang, M.Y.; Zheng, H.L. Application of coagulation/flocculation in oily wastewater treatment: A review. *Sci. Total Environ.* **2021**, *765*, 142795. [CrossRef]
7. Liang, H.; Esmaeili, H. Application of nanomaterials for demulsification of oily wastewater: A review study. *Environ. Technol. Innov.* **2021**, *22*, 101498. [CrossRef]
8. Ikhsan, S.N.W.; Yusof, N.; Aziz, F.; Ismail, A.F.; Jaafar, J.; Salleh, W.N.W.; Misdan, N. Superwetting materials for hydrophilic-oleophobic membrane in oily wastewater treatment. *J. Environ. Manag.* **2021**, *290*, 112565. [CrossRef]

9.	Ismail, N.H.; Salleh, W.N.W.; Ismail, A.F.; Hasbullah, H.; Yusof, N.; Aziz, F.; Jaafar, J. Hydrophilic polymer-based membrane for oily wastewater treatment: A review. *Sep. Purif. Technol.* **2020**, *233*, 116007. [CrossRef]

10.	Liu, C.H.; Xia, J.Y.; Gu, J.C.; Wang, W.Q.; Liu, Q.Q.; Yan, L.K.; Chen, T. Multifunctional CNTs-PAA/MIL101(Fe)@P composite membrane for high-throughput oily wastewater remediation. *J. Hazard. Mater.* **2021**, *403*, 123547. [CrossRef]

11.	Yaacob, N.; Sean, G.P.; Nazri, N.A.M.; Ismail, A.F.; Abidin, M.N.Z.; Subramaniam, M.N. Simultaneous oily wastewater adsorption and photodegradation by ZrO_2-TiO_2 heterojunction photocatalysts. *J. Water Process Eng.* **2021**, *39*, 101644. [CrossRef]

12.	Jia, Z.; Liang, S.X.; Zhang, W.C.; Wang, W.M.; Yang, C.; Zhang, L.C. Heterogeneous photo Fenton-like degradation of the cibacron brilliant red 3B-A dye using amorphous $Fe_{78}Si_9B_{13}$ and $Fe_{73.5}Si_{13.5}B_9Cu_1Nb_3$ alloys: The influence of adsorption. *J. Taiwan Inst. Chem. Eng.* **2017**, *71*, 128–136. [CrossRef]

13.	Jia, Z.; Miao, J.; Lu, H.; Habibi, D.; Zhang, W. Photocatalytic degradation and absorption kinetics of cibacron brilliant yellow 3G-P by nanosized ZnO catalyst under simulated solar light. *J. Taiwan Inst. Chem. Eng.* **2015**, *60*, 267–274. [CrossRef]

14.	Usman, M.; Faure, P.; Hanna, K.; Abdelmoula, M.; Ruby, C. Application of magnetite catalyzed chemical oxidation (Fenton-like and persulfate) for the remediation of oil hydrocarbon contamination. *Fuel* **2012**, *96*, 270–276. [CrossRef]

15.	Grčić, I.; Papić, S.; Žižek, K.; Koprivanac, N. Zero-valent iron (ZVI) Fenton oxidation of reactive dye wastewater under UV-C and solar irradiation. *Chem. Eng. J.* **2012**, *195*, 77–90. [CrossRef]

16.	Jia, Z.; Zhang, W.; Wang, W.; Habibi, D. Amorphous Fe78Si9B13 alloy: An efficient and reusable photo-enhanced Fenton-like catalyst in degradation of cibacron brilliant red 3B-A dye under UV–vis light. *Appl. Catal. B Environ.* **2016**, *192*, 46–56. [CrossRef]

17.	Song, D.S.; Kim, J.H.; Fleury, E.; Kim, W.T.; Kim, D.H. Synthesis of ferromagnetic Fe-based bulk glassy alloys in the Fe-Nb-B-Y system. *J. Alloys Compd.* **2005**, *389*, 159–164. [CrossRef]

18.	Liu, P.; Zhang, J.L.; Zha, M.Q.; Shek, C.H. Synthesis of an Fe Rich Amorphous Structure with a Catalytic Effect to Rapidly Decolorize Azo Dye at Room Temperature. *ACS Appl. Mater. Interfaces* **2014**, *6*, 5500–5505. [CrossRef]

19.	Wang, P.; Wang, J.-Q.; Li, H.; Yang, H.; Huo, J.; Wang, J.; Chang, C.; Wang, X.; Li, R.-W.; Wang, G. Fast decolorization of azo dyes in both alkaline and acidic solutions by Al-based metallic glasses. *J. Alloys Compd.* **2017**, *701*, 759–767. [CrossRef]

20.	Wang, W.H. Bulk metallic glasses with functional physical properties. *Adv. Mater.* **2009**, *21*, 1524–1544. [CrossRef]

21.	Li, H.; Pang, S.; Liu, Y.; Sun, L.; Liaw, P.K.; Zhang, T. Biodegradable Mg–Zn–Ca–Sr bulk metallic glasses with enhanced corrosion performance for biomedical applications. *Mater. Des.* **2015**, *67*, 9–19. [CrossRef]

22.	Baron-Wiechec, A.; Szewieczek, D.; Nawrat, G. Corrosion of amorphous and nanocrystalline Fe-based alloys and its influence on their magnetic behavior. *Electrochim. Acta* **2007**, *52*, 5690–5695. [CrossRef]

23.	Wang, X.; Pan, Y.; Zhu, Z.; Wu, J. Efficient degradation of rhodamine B using Fe-based metallic glass catalyst by Fenton-like process. *Chemosphere* **2014**, *117*, 638–643. [CrossRef]

24.	Zhao, B.W.; Zhu, Z.W.; Qin, X.D.; Li, Z.K.; Zhang, H.F. Highly efficient and stable CuZr-based metallic glassy catalysts for azo dye degradation. *J. Mater. Sci. Technol.* **2020**, *46*, 88–97. [CrossRef]

25.	Wang, J.Q.; Liu, Y.H.; Chen, M.W.; Luzgin, D.V.L.; Inoue, A.; Perepezko, J.H. Excellent capability in degrading azo dyes by MgZn-based metallic glass powders. *Sci. Rep.* **2012**, *2*, 418. [CrossRef]

26.	Yang, X.; Xu, X.C.; Xiang, Q.C.; Qu, Y.D.; Ren, Y.L.; Qiu, K.Q. The catalytic performance of $Cu_{46}Zr_{47-x}Al_7Y_x$ amorphous ribbons in the degradation of AOII dye wastewater. *Environ. Sci. Pollut. R.* **2021**, *28*, 48038–48052. [CrossRef]

27.	Ramya, M.; Karthika, M.; Selvakumar, R.; Raj, B.; Ravi, K. A facile and efficient single step ball milling process for synthesis of partially amorphous Mg-Zn-Ca alloy powders for dye degradation. *J. Alloys Compd.* **2017**, *696*, 185–192. [CrossRef]

28.	Tang, Y.; Shao, Y.; Chen, N.; Yao, K.-F. Rapid decomposition of Direct Blue 6 in neutral solution by Fe–B amorphous alloys. *RSC Adv.* **2015**, *5*, 6215–6221. [CrossRef]

29.	Jia, Z.; Kang, J.; Zhang, W.C.; Wang, W.M.; Yang, C.; Sun, H.; Habibi, D.; Zhang, L.C. Surface aging behavior of Fe-based amorphous alloys as catalysts during heterogeneous photo Fenton-like process for water treatment. *Appl. Catal. B Environ.* **2017**, *204*, 537–547. [CrossRef]

30.	Ghorbani, M.; Vakili, M.H.; Ameri, E. Fabrication and evaluation of a biopolymer-based nanocomposite membrane for oily wastewater treatment. *Mater. Today Commun.* **2021**, *28*, 102560. [CrossRef]

31.	Liang, S.X.; Jia, Z.; Zhang, W.C.; Li, X.; Wang, W.M.; Lin, H.C.; Zhang, L.C. Ultrafast activation efficiency of three peroxides by $Fe_{78}Si_9B_{13}$ metallic glass under photo-enhanced catalytic oxidation: A comparative study. *Appl. Catal. B Environ.* **2018**, *221*, 108–118. [CrossRef]

32.	Zhang, C.; Wang, Y.; Zhang, X.; Guo, H. Fe78Si9B13 amorphous alloys for the decolorization process of azo dye aqueous solutions with different initial concentrations at different reaction temperatures. *Vacuum* **2020**, *176*, 109301. [CrossRef]

33.	Lai, B.; Zhang, Y.; Chen, Z.; Yang, P.; Zhou, Y.; Wang, J. Removal of p-nitrophenol (PNP) in aqueous solution by the micron-scale iron–copper (Fe/Cu) bimetallic particles. *Appl. Catal. B Environ.* **2014**, *144*, 816–830. [CrossRef]

34.	Weng, C.-H.; Tao, H. Highly efficient persulfate oxidation process activated with Fe0 aggregate for decolorization of reactive azo dye Remazol Golden Yellow. *Arab. J. Chem.* **2018**, *11*, 1292–1300. [CrossRef]

35.	Yamaguchi, R.; Kurosu, S.; Suzuki, M.; Kawase, Y. Hydroxyl radical generation by zero-valent iron/Cu (ZVI/Cu) bimetallic catalyst in wastewater treatment: Heterogeneous Fenton/Fenton-like reactions by Fenton reagents formed in-situ under oxic conditions. *Chem. Eng. J.* **2018**, *334*, 1537–1549. [CrossRef]

36. Xiong, Z.; Lai, B.; Yang, P.; Zhou, Y.; Wang, J.; Fang, S. Comparative study on the reactivity of Fe/Cu bimetallic particles and zero valent iron (ZVI) under different conditions of N_2, air or without aeration. *J. Hazard. Mater.* **2015**, *297*, 261–268. [CrossRef] [PubMed]

37. Chen, Q.; Yan, Z.C.; Zhang, H.; Kim, K.; Wang, W.M. Role of nanocrystallites of Al-based glasses and H2O2 in degradation azo dyes. *Materials* **2021**, *14*, 39. [CrossRef]

38. Qin, X.; Xu, J.; Zhu, Z.; Li, Z.; Fang, D.; Fu, H.; Zhang, S.; Zhang, H. Co78Si8B14 metallic glass: A highly efficient and ultra-sustainable Fenton-like catalyst in degrading wastewater under universal pH conditions. *J. Mater. Sci. Technol.* **2022**, *113*, 105–116. [CrossRef]

39. Weber, E.J. Iron-Mediated Reductive Transformations: Investigation of Reaction Mechanism. *Environ. Sci. Technol.* **1996**, *30*, 716–719. [CrossRef]

40. SLiang, X.; Jia, Z.; Zhang, W.C.; Wang, W.M.; Zhang, L.C. Rapid malachite green degradation using $Fe_{73.5}Si_{13.5}B_9Cu_1Nb_3$ metallic glass for activation of persulfate under UV-vis light. *Mater. Des.* **2017**, *119*, 244–253.

41. Qin, X.D.; Li, Z.K.; Zhu, Z.W.; Fu, H.M.; Li, H.; Wang, A.M.; Zhang, H.W.; Zhang, H.F. Mechanism and kinetics of treatment of acid orange II by aged Fe-Si-B metallic glass powders. *J. Mater. Sci. Technol.* **2017**, *33*, 1147–1152. [CrossRef]

42. Zhu, S.; Xiang, Q.; Ma, C.; Ren, Y.; Qiu, K. Continuous electrocoagulation degradation of oily wastewater with $Fe_{78}Si_9B_{13}$ amorphous ribbons. *Environ. Sci. Pollut. Res.* **2020**, *27*, 40101–40108. [CrossRef] [PubMed]

43. Liang, C.; Guo, Y.-Y. Mass Transfer and Chemical Oxidation of Naphthalene Particles with Zerovalent Iron Activated Persulfate. *Environ. Sci. Technol.* **2010**, *44*, 8203–8208. [CrossRef] [PubMed]

44. Wang, L.; Yao, Y.; Zhang, Z.; Sun, L.; Lu, W.; Chen, W.; Chen, H. Activated carbon fibers as an excellent partner of Fenton catalyst for dyes decolorization by combination of adsorption and oxidation. *Chem. Eng. J.* **2014**, *251*, 348–354. [CrossRef]

45. Yurdakal, S.; Loddo, V.; Augugliaro, V.; Berber, H.; Palmisano, G. Photodegradation of pharmaceutical drugs in aqueous TiO_2 suspensions: Mechanism and kinetics. *Catal. Today* **2007**, *129*, 9–15. [CrossRef]

46. Huang, C.-M.; Pan, G.-T.; Li, Y.-C.M.; Li, M.-H.; Yang, T.C.-K. Crystalline phases and photocatalytic activities of hydrothermal synthesis Ag_3VO_4 and $Ag_4V_2O_7$ under visible light irradiation. *Appl. Catal. A Gen.* **2009**, *358*, 164–172. [CrossRef]

47. Ma, Y.; Rheingans, B.; Liu, F.; Mittemeijer, E.J. Isochronal crystallization kinetics of $Fe_{40}Ni_{40}B_{20}$ amorphous alloy. *J. Mater. Sci.* **2013**, *48*, 5596–5606. [CrossRef]

48. Niu, Y.; Bian, X.; Wang, W. Origin of ductile–brittle transition of amorphous $Fe_{78}Si_9B_{13}$ ribbon during low temperature annealing. *J. Non-Cryst. Solids* **2004**, *341*, 40–45. [CrossRef]

49. Mariano, N.; Souza, C.; May, J.; Kuri, S. Influence of Nb content on the corrosion resistance and saturation magnetic density of FeCuNbSiB alloys. *Mater. Sci. Eng. A* **2003**, *354*, 1–5. [CrossRef]

50. Lin, B.; Bian, X.; Wang, P.; Luo, G. Application of Fe-based metallic glasses in wastewater treatment. *Mater. Sci. Eng. B* **2012**, *177*, 92–95. [CrossRef]

Article

Real-Time Degradation of Indoor Formaldehyde Released from Actual Particle Board by Heterostructured g-C$_3$N$_4$/TiO$_2$ Photocatalysts under Visible Light

Qing Jin [1], Youlin Xiang [1] and Lu Gan [1,2,*]

[1] College of Materials Science and Engineering, Nanjing Forestry University, Nanjing 210037, China
[2] Jiangsu Co-Innovation Center of Efficient Processing and Utilization of Forest Resources, International Innovation Center for Forest Chemicals and Materials, College of Materials Science and Engineering, Nanjing Forestry University, Nanjing 210037, China
* Correspondence: ganlu@njfu.edu.cn

Abstract: Indoor formaldehyde pollution causes a serious threat to human health since it is uninterruptedly released from wooden furniture. Herein, we prepared a g-C$_3$N$_4$-modified TiO$_2$ composite photocatalyst and coated it on the surface of a commercial artificial particle board with the assistance of melamine formaldehyde adhesive. The g-C$_3$N$_4$/ TiO$_2$ coating was then used to degrade formaldehyde which was released in real-time from the particle board under the irradiation of visible light. The results showed that compared with pure TiO$_2$, the g-C$_3$N$_4$/ TiO$_2$ composite with a heterojunction structure had a lower band gap energy (~2.6 eV), which could effectively capture luminous energy from the visible light region. Under continuous irradiation, the g-C$_3$N$_4$/ TiO$_2$ photocatalytic coating was capable of degrading more than 50% of formaldehyde constantly released from the particle board. In the meantime, the photocatalytic coating also exhibited promising catalytic stability towards various formaldehyde release speeds, air flow velocities and environmental humidities. The hydroxyl radical and superoxide radical were found to be the predominant active species which triggered formaldehyde degradation. This study provides a feasible and practical approach for the improvement in indoor air quality through photocatalyst surface engineering.

Keywords: photocatalyst; TiO$_2$; g-C$_3$N$_4$; heterostructure; formaldehyde degradation

Citation: Jin, Q.; Xiang, Y.; Gan, L. Real-Time Degradation of Indoor Formaldehyde Released from Actual Particle Board by Heterostructured g-C$_3$N$_4$/TiO$_2$ Photocatalysts under Visible Light. *Catalysts* **2023**, *13*, 238. https://doi.org/10.3390/catal13020238

Academic Editors: Gassan Hodaifa, Rafael Borja and Mha Albqmi

Received: 29 December 2022
Revised: 13 January 2023
Accepted: 17 January 2023
Published: 19 January 2023

1. Introduction

Indoor volatile organic compounds (VOCs) released from wooden furniture have received tremendous attention, and of these formaldehyde (HCHO) is regarded as one of the most frequently released VOCs [1,2]. If HCHO cannot be completely removed, it may cause severe threats to human health, such as nasal tumors and skin cancer [3]. Although many techniques such as ventilation, adsorption, plasma and thermal catalytic oxidation have been conducted to remove HCHO, it is still relatively difficult to find an all-weather approach which can detect the released HCHO in real-time and promptly remove it, since HCHO is continuously volatilized from indoor wooden building materials during daily life [4].

More recently, photocatalytic technology has emerged as a promising means to quickly remove HCHO from indoor air [5]. Under the irradiation of light, HCHO molecules can be effectively oxidized to inorganic H$_2$O and CO$_2$ by semiconductor-based photocatalysts [6]. Amongst all applicable photocatalysts, titanium oxide (TiO$_2$) is the most frequently selected due to high stability, low toxicity and low cost [7]. However, the wide application of TiO$_2$ still encounters many limitations since it is a UV light-responsive photocatalyst which cannot effectively utilize the energy from the visible light region [8]. The hybridization of other semiconductors, such as ZnO, CdS, SnO2, etc., with TiO$_2$ can readily improve the photocatalytic activity of TiO$_2$ through constructing a heterojunction structure, which can lead to higher charge collection and separation efficiency [9]. Graphitic carbon nitride (g-C$_3$N$_4$) is a two-dimensional metal-free n-type semiconductor which has been intensively studied recently as a visible light-responsive photocatalyst. Compared with other metal compound-based photocatalysts, g-C$_3$N$_4$ has the merits of a proper band gap energy (~2.7 eV), low toxicity and ease of accessibility [10]. Many studies have shown that g-C$_3$N$_4$ can be introduced as a second phase to hybridize with other semiconductors and build heterostructure photocatalysts with highly elevated catalytic capabilities [11]. Therefore, it is possible to prepare a heterostructure composite photocatalyst with high HCHO removal efficiency through integrating g-C$_3$N$_4$ with TiO$_2$.

Another concern when utilizing photocatalysts for indoor HCHO degradation in real living spaces is that most photocatalysts have a powdery form, which cannot be easily attached to wooden materials. Furthermore, most previous studies focusing on the photocatalytic degradation of HCHO mainly investigated HCHO degradation with a fixed concentration of HCHO in air [12]. However, in actual conditions, HCHO is continuously released from wooden furniture and the concentration of HCHO in the air fluctuates [13]. Thus, in this study, g-C$_3$N$_4$/TiO$_2$ composite photocatalysts were prepared and used for real-time degradation of the HCHO released from commercial artificial particle board. The prepared photocatalyst was adhered onto the surface of the board as a photocatalytic coating with the assistance of melamine formaldehyde adhesive. The stability for long-term use of the g-C$_3$N$_4$/TiO$_2$ composite towards continuous HCHO release from the artificial board was studied in detail.

2. Results and Discussion

The morphology of the prepared photocatalysts were observed first in terms of SEM, with the results shown in Figure 1. It was seen from Figure 1a that pure TiO$_2$ had a nanoparticle appearance. Based on the Scherrer formula, the particle size of TiO$_2$ was calculated to be ~10 nm. Meanwhile, g-C$_3$N$_4$ exhibited a layered structure (Figure 1b). When two components were hybridized, it could be observed that TiO$_2$ nanoparticles were uniformly anchored on the surface of g-C$_3$N$_4$ (Figure 1c). With the increase in g-C$_3$N$_4$ content in the composite, a lower amount of TiO$_2$ nanoparticle aggregations could be seen (Figure 1d) since g-C$_3$N$_4$ could provide more surface area.

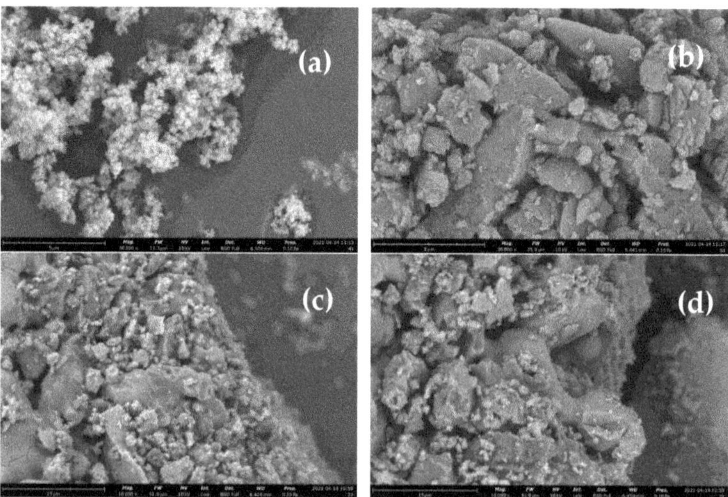

Figure 1. SEM images of (**a**) TiO$_2$, (**b**) 10-g-C$_3$N$_4$/TiO$_2$, (**c**) 20-g-C$_3$N$_4$/TiO$_2$, (**d**) 30-g-C$_3$N$_4$/TiO$_2$.

The structure of g-C$_3$N$_4$/TiO$_2$ composites was further investigated. Figure 2a shows the XRD patterns of the prepared samples. As illustrated, the prepared TiO$_2$ exhibited a typical anatase structure (JCPDS 04-0477) which had characteristic peaks located at 25° (101), 38° (004), 48° (200), 54 (105) and 55° (211) [14]. Based on the Bragg formula, the d-spacing for the (101) lattice plane was calculated to be ~0.35 nm. It was also seen that g-C_3N_4 had two distinct peaks at 13° and 27°, which were indexed to the (100) and (002) lattice planes of its hexagonal graphitic structure (JCPDS 87-1526) [15]. According to the Bragg equation, the d spacing for the (002) lattice plane was calculated to be 0.33 nm, which was close to the interlayer spacing of the graphite crystal. The g-C_3N_4/TiO_2 composites showed integrated patterns in which both characteristic peaks from TiO_2 and g-C_3N_4 could be observed. With the increase in g-C_3N_4 content, the (100) peak of g-C_3N_4 became more apparent, and increased incorporation of g-C_3N_4 did not break the crystalline structure of TiO_2.

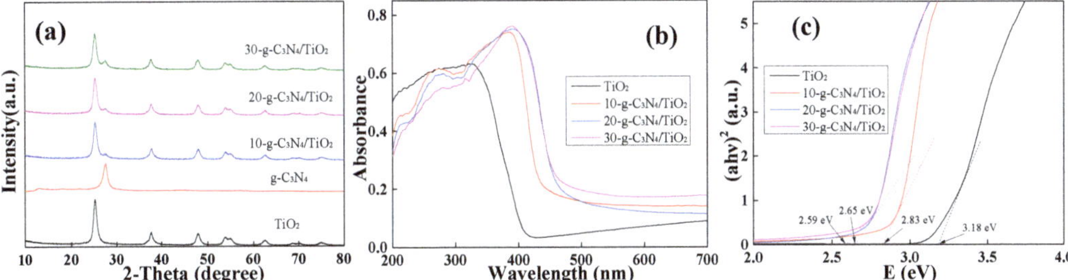

Figure 2. (**a**) XRD patterns, (**b**) DRS spectra and (**c**) corresponding band gap energy of TiO_2 and g-C_3N_4/TiO_2 composites.

The absorbance properties of the prepared photocatalysts were investigated through DRS spectra, with the results shown in Figure 2b. For pristine TiO_2, the absorption edge occurred at a wavelength of lower than 400 nm, indicating that the anatase phase TiO_2 was only responsive to UV light [16]. When g-C_3N_4 was incorporated, the absorption edges of the resulting composites all shifted to the visible light region, indicating that heterostructured g-C_3N_4/TiO_2 could capture visible light to initiate electron/hole separation. The band gap energy (E_g) of the prepared photocatalysts could be further calculated via Tauc plots, with the results shown in Figure 2c. As expected, pristine TiO_2 had an E_g of 3.18 eV, which was in accordance with many previous studies [17]. In the meantime, all the composite photocatalysts had smaller E_g values, in which a higher g-C_3N_4 incorporation amount could lead to narrower E_g. It was noted that the E_g values for 20-g-C_3N_4/TiO_2 (2.65 eV) and 30-g-C_3N_4/TiO_2 (2.59 eV) were both lower than that of pristine g-C_3N_4 (~2.7 eV) [18], indicating that g-C_3N_4 and TiO_2 formed a heterojunction structure which could effectively enhance the light absorption ability of the composite photocatalysts.

The photocatalytic performance of the prepared g-C_3N_4/TiO_2 towards indoor HCHO removal (continuously released from artificial particle board) was then investigated. Figure 3 shows the HCHO degradation capability of the prepared samples. As illustrated in Figure 3a, HCHO was quickly released from the particle board and the adsorption/desorption equilibrium of ~0.64 g/m^3 was reached within 30 min. When the light was off, all the tested boards, coated with different photocatalysts, exhibited a similar HCHO release curve, indicating that each photocatalyst coating barely adsorbed the HCHO in the air. When the UV lamp was on (Figure 3b), the equilibrium concentration of HCHO quickly rose to ~0.78 g/m^3, since UV light could introduce more thermal energy into the reaction chamber which facilitated HCHO release. It was observed that under the irradiation of UV light, HCHO concentration in the air was significantly reduced to a different extent, which was caused by the photocatalytic degradation of HCHO on the coating surface. It was seen that all the g-C_3N_4/TiO_2 composites showed better HCHO degradation performance than pure TiO_2, e.g., the real-time HCHO concentration could be reduced to lower than 0.2 g/m^3 by both 20-g-C_3N_4/TiO_2 and 30-g-C_3N_4/TiO_2. In the meantime, the catalyst had very promising long-term use stability, as it could unceasingly degrade perpetually released HCHO from the particle boards for more than 24 h. This signified that the high-performance photocatalyst coating on the wooden furniture was a potential and applicable way to lower the concentration of HCHO in the air when the release of HCHO from the wooden furniture theoretically could not be inhibited.

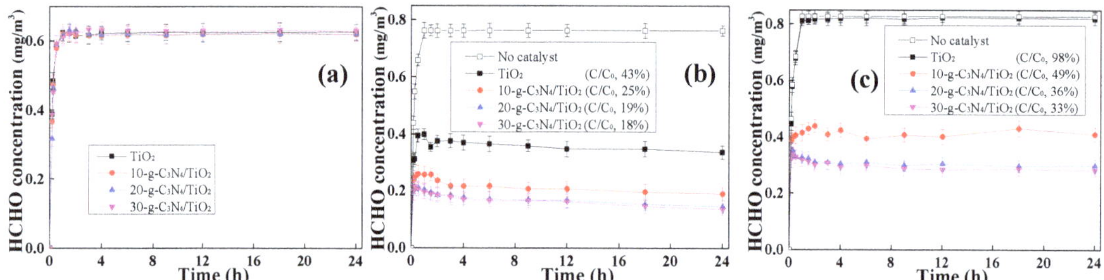

Figure 3. (**a**) Adsorptive removal of HCHO and photocatalytic degradation of HCHO under (**b**) UV light and (**c**) visible light by TiO_2 and $g\text{-}C_3N_4/TiO_2$ composites ($m_{(photocatalyst)}$ = 1 g).

When the light source was changed to a Xe lamp (Figure 3c), the HCHO equilibrium concentration in the reaction box was increased to even higher than 0.80 g/m³ due to the thermal radiation source nature of the Xe lamp. As anticipated, TiO_2 showed negligible HCHO removal capability since the UV responsible photocatalyst could not effectively generate energy from visible light irradiation [19]. On the contrary, all the composite photocatalysts still exhibited excellent HCHO degradation performance under visible light, with $20\text{-}g\text{-}C_3N_4/TiO_2$ and $30\text{-}g\text{-}C_3N_4/TiO_2$ also presented the optimal efficiency. Specifically, more than 60% of the HCHO could be degraded by these two photocatalyst-based coatings at any time of release. The results demonstrated that compared with TiO_2, $g\text{-}C_3N_4/TiO_2$ was more suitable to be coated on wooden furniture to reduce indoor HCHO concentration since visible light is more than 50% of natural light, whereas UV light is only 5% [20].

The influence of the coating parameters, including the catalyst dosage, of the coating and HCHO concentration on the performance of the composite catalyst was then investigated, with the results shown in Figure 4. In these cases, $20\text{-}g\text{-}C_3N_4/TiO_2$ was selected as the test catalyst due to its high performance. A Xe lamp was used as the light source. As demonstrated in Figure 4a, the performance of $20\text{-}g\text{-}C_3N_4/TiO_2$ did not fluctuate much with changes in its initial dosage in the coating solution, which might because a magnified amount of the catalyst particles could not increase the exposed active sites on the coating surface, and the contact between HCHO and the active sites of $20\text{-}g\text{-}C_3N_4/TiO_2$ was not enhanced as a result, since only surface photocatalyst particles could be irradiated by the incident light and generate active species for HCHO degradation. When the particle board was changed to those with different HCHO releasing rates, it was shown from Figure 4b that the photocatalyst still exhibited high HCHO real-time degradation performance towards all selected panels. Notably, the $20\text{-}g\text{-}C_3N_4/TiO_2$ coating gave the highest HCHO removal efficiency to the board with the highest HCHO releasing rate, which might because more HCHO molecules were in contact with the $g\text{-}C_3N_4/TiO_2$ particles in the fixed volume chamber.

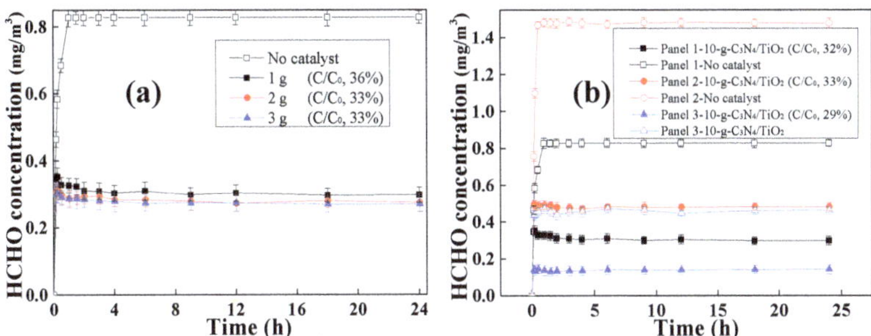

Figure 4. Impact of (**a**) photocatalyst coating amount and (**b**) HCHO release velocity on the photocatalytic performance of 20-g-C$_3$N$_4$/TiO$_2$ composites.

Figure 5 shows the impact of the environmental parameters on the performance of the photocatalyst. The results shown in Figure 5a indicated that by reducing the air flow velocity, the HCHO degradation efficiency was effectively enhanced due to higher contact between the HCHO and 20-g-C$_3$N$_4$/TiO$_2$ at lower air flow velocities. Furthermore, the alteration of environmental air humidity also influenced the HCHO removal rate. It was inferred from Figure 5b that increased humidity could impede the photocatalytic degradation of HCHO. This was because when the water vapor content in the air increased, more water molecules adhered to the 20-g-C$_3$N$_4$/TiO$_2$ coating surface, which hindered the contact rate between the catalyst molecules and pollutant molecules [21]. It was also observed from Figure 5b that a lower humidity also suppressed the performance of the photocatalyst. This indicated that the existence of H$_2$O in the system was the key for photocatalytic degradation of HCHO. It is generally known that in many cases, when incident light triggers the electron/hole separation in the photocatalyst, the hole in the valence band (VB) needs to react with an ambient hydroxyl ion (OH$^-$) to form OH to degrade the organic pollutant [22]. Thus, sufficient H$_2$O molecules existing in the reaction system is necessary to guarantee the conversion from hole to OH [23], which was why a relatively low environmental humidity led to a negative impact to the performance of the composite photocatalyst. However, excessive H$_2$O molecules could also occupy the active sites on photocatalyst surface when further increasing the humidity to 80%, which also led to a significant decline in the HCHO degradation efficiency.

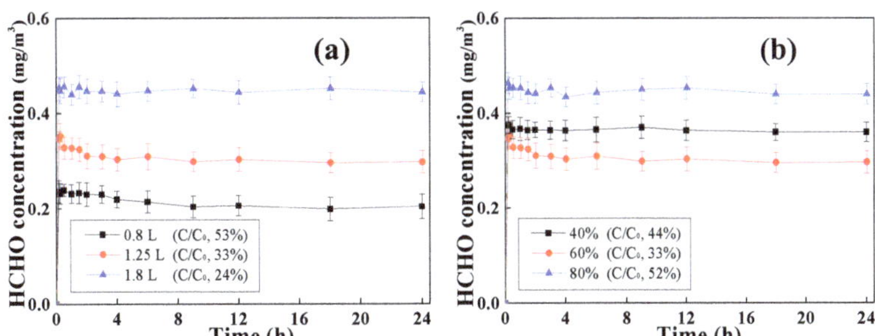

Figure 5. Impact of (**a**) air flow velocity and (**b**) environmental humidity on the photocatalytic performance of 20-g-C$_3$N$_4$/TiO$_2$ composites.

The predominant active species which initiated HCHO degradation was then investigated, with the results shown in Figure 6a. It was clearly seen that both MeOH and t-BQ could inhibit the HCHO removal rate, whereas EDTA could not hinder HCHO degradation. This meant the radicals, including OH and O$_2{}^-$, were the main species responsible for the degradation of HCHO [24], where OH and O$_2{}^-$ were produced from the reaction of holes in the VB with OH^- and electrons in the conduction band (CB) with ambient oxygen (O_2), respectively. This result could also explain why the performance of 20-g-C_3N_4/TiO_2 was slowed down at low humidity. Figure 6b shows the XRD pattern of used 20-g-C_3N_4/TiO_2 after 24 h of service. It was seen that the used photocatalyst had the same XRD pattern as the freshly made one, indicating a promising long-term use stability.

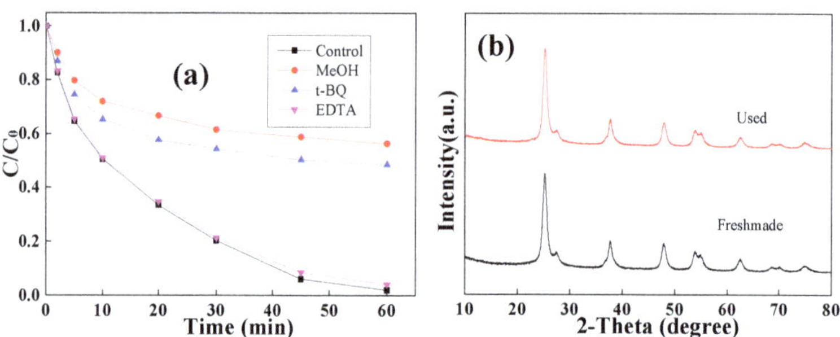

Figure 6. (a) HCHO degradation in water with the existence of different scavengers, (b) XRD pattern of used 20-g-C_3N_4/TiO_2.

Based on the above overall results, the real-time HCHO degradation process by 20-g-C_3N_4/TiO_2 coating was proposed in Figure 7. In practical conditions, HCHO is incessantly released from indoor wooden furniture, such as the particle board used in this work. If no valid inhibitory approach is conducted, HCHO will quickly occupy indoor spaces. When the photocatalyst coating, which is g-C_3N_4/TiO_2 in this study, is adhered on the furniture surface, the coating can unceasingly degrade HCHO released in real-time under the irradiation of light. Specifically, when a very small amount of g-C_3N_4 was hybridized with TiO_2, a heterojunction-structured composite photocatalyst could be formed. Under such conditions, g-C_3N_4/TiO_2 had a narrower band gap energy with an expanded light adsorption region compared with TiO_2. Therefore, the composite photocatalyst can adsorb visible light for HCHO degradation, where g-C_3N_4 accelerates the separation and suppresses the recombination of the electron/hole pairs. After the electron/hole pairs were produced by visible light irradiation and stabilized by the g-C_3N_4/TiO_2 heterojunction, the excited electrons in the CB quickly reacted with the O_2 in the reaction system to produce O_2^-, and the remaining holes in the VB also reacted with the nearby OH^- to generate OH, both of which caused HCHO degradation in the reaction chamber. As long as the visible light source was on, the constantly released HCHO from the particle board could be removed in real-time with high efficiency by the photocatalyst coating prepared in this study.

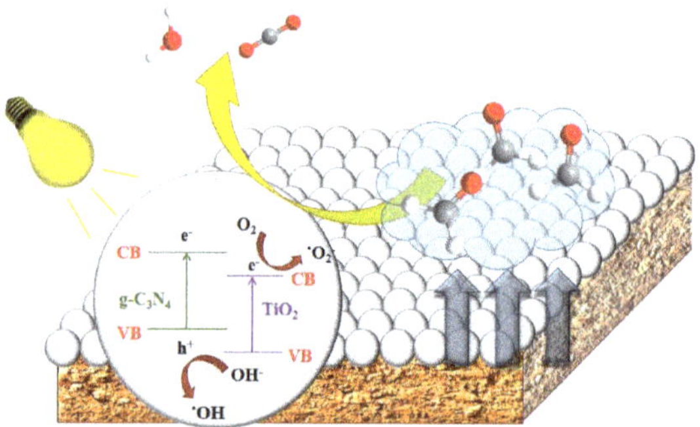

Figure 7. Schematic illustration of real-time HCHO degradation by g-C_3N_4/TiO_2 (Red dot: Oxygen, gray dot: carbon, white dot: hydrogen).

3. Materials and Methods

3.1. Chemical Reagents

Titanium butoxide (TBT, $C_{16}H_{36}O_4Ti$, >99%), melamine ($C_3H_6N_6$, 99%), hexadecyl trimethyl ammonium bromide (CTAB, $C_{19}H_{42}BrN$, 99%) and tert-benzoquinone ($C_6H_4O_2$, 99%) were purchased from Aladdin Co., Ltd. (Shanghai, China). Edetate disodium ($C_{10}H_{14}N_2Na_2O_8 \cdot 2H_2O$, 99–101%) was purchased from Sigma-Aldrich (St. Louis, MI, USA). The artificial particle boards that released different HCHO amounts were provided by Linyi wood factory (Linyi, China). Other chemicals and reagents were of analytical grade and used without further purification.

3.2. Preparation of g-C₃N₄/TiO₂

g-C_3N_4 was prepared by pyrolyzing melamine at 550 °C for 3 h in a muffle furnace. The preparation of g-C_3N_4/TiO_2 composites with different g-C_3N_4 contents was conducted as follows. Firstly, a certain amount of g-C_3N_4 and CTAB (20 mg) were added into 100 mL of ethanol, and the mixture was sonicated for 1 h to obtain a g-C_3N_4 dispersion. Then, TBT (0.5 g) was added and dissolved in the above dispersion, and the mixture transferred into a Teflon-lined stainless-steel autoclave. The autoclave was afterwards heated to 120 °C for 24 h. After the temperature was cooled to room temperature, the photocatalysts named as x-g-C_3N_4/TiO_2 were obtained after the light brown precipitate was separated via centrifuging, washing with H_2O and ethanol twice and drying at 60 °C for 24 h. In the formula, x represents for the mass dosage of g-C_3N_4 in the composites. For comparison, pure TiO_2 was also prepared following a similar procedure without the introduction of g-C_3N_4.

3.3. Characterization and Analytical Methods

The morphology of the prepared samples was determined by scanning electron microscopy (SEM, JEOL SEM 6490, Tokyo, Japan). The X-ray diffraction spectra (XRD) were recorded by a Rigaku Smartlab XRD instrument. The diffuse reflectance spectra (DRS) were conducted by a PerkinElmer Lambda 950 UV/Vis/NIR spectrophotometer.

3.4. Real-Time HCHO Degradation Process

The photocatalytic system for HCHO degradation used in this study is shown schematically in Figure 8. A $50 \times 50 \times 30$ m³ glass box, equipped with an air inlet pump and a HCHO sensor, was used as the reaction chamber. The glass (with an area of 20×20 cm²) on the top of the chamber was changed to quartz glass for the passage of light. UV light and visible light were both introduced as the light source, in which a 250 W UV lamp was used as the UV light source and a 300 W Xe lamp equipped with a 420 nm cut-off filter was used as the visible light source. All particle boards, with a thickness of 1.5 cm, were cut into 10×10 cm² rectangular shapes. The particle board was first brushed with a layer of melamine formaldehyde adhesive. Afterwards, 10 mL, 20 mL or 30 mL of the photocatalyst suspension with a concentration of 100 g/L was coated onto the surface of the particle board using a spin-coater at 2000 rpm. After being air-dried for 24 h, the photocatalyst-coated particle board was placed on a lifting table in the glass box, with the distance between the particle board and light source fixed at 18 cm. At the same time, the lamp was turned on to initiate the photocatalytic process, and the HCHO sensor (Dart 2-FE5, Exeter EX4 3AZ, UK) recorded the real-time HCHO concentration in the chamber. The humidity of the reaction chamber was tuned by a commercial mini humidifier. The humidity of the reaction chamber for photocatalytic tests was set at ~60% if not otherwise stated.

The predominant active species generated by g-C_3N_4/TiO_2 for HCHO degradation was identified in aqueous solution by the introduction of different scavengers, in which a Xe lamp was introduced as the light source. Typically, 50 mg of the photocatalyst was added into a 50 mL HCHO solution (10 mg/L) containing 10 mM of methanol (MeOH), tert-benzoquinone (t-BQ) or edetate disodium (EDTA), which were used to trap hydroxyl radicals (OH), superoxide radicals (O_2^-) and holes (h), respectively. After the Xe lamp was turned on, approximately 2 mL of solution was withdrawn from the reaction solution at predetermined intervals and centrifuged to separate the solid. The concentration of HCHO was quantified using gas chromatography (Agilent 7890A, Agilent, Santa Clara, CA, USA).

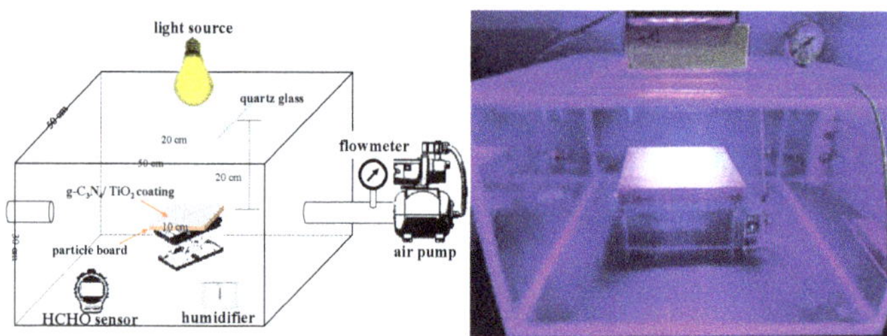

Figure 8. Photocatalytic system for real-time degradation of HCHO released from artificial particle board.

4. Conclusions

To conclude, g-C$_3$N$_4$ was modified with TiO$_2$ to construct composite photocatalysts with a heterojunction structure. The prepared g-C$_3$N$_4$/TiO$_2$ was coated onto the surface of artificial particle board and used for HCHO degradation as it was released in real-time from the particle board under the irradiation of visible light. The prepared g-C$_3$N$_4$/TiO$_2$ exhibited high visible light energy adsorption efficiency since the two components formed an effective heterojunction structure. The g-C$_3$N$_4$/TiO$_2$ coating could unceasingly degrade HCHO which was continuously released from the particle board. The photocatalyst coating also exhibited promising stability and adaptability. The heterostructured g-C$_3$N$_4$/TiO$_2$ prepared in this study can be used for practical indoor air purification.

Author Contributions: Investigation, writing—original draft preparation, Q.J.; methodology, investigation, data curation, Y.X.; conceptualization, writing—review and editing, funding acquisition, L.G. All authors have read and agreed to the published version of the manuscript.

Funding: This research was funded by the Natural Science Foundation of Jiangsu Province, China (BK20201385).

Data Availability Statement: Data are available upon request.

Conflicts of Interest: The authors declare no conflict of interest.

References

1. Chen, J.; Guo, X.; Lang, L.; Yin, X.; Wang, A.; Rui, Z. Multifunctional z-scheme Cu$_x$O/Ag/SrTiO$_3$ heterojunction for photothermo-catalytic vocs degradation and antibiosis. *Appl. Surf. Sci.* **2022**, 153275. [CrossRef]
2. Zhao, Z.Y.; Sun, S.J.; Wu, D.; Zhang, M.; Huang, C.X.; Umemura, K.; Yong, Q. Synthesis and characterization of sucrose and ammonium dihydrogen phosphate (sadp) adhesive for plywood. *Polymers* **2019**, *11*, 16. [CrossRef] [PubMed]
3. Zhang, G.; Sun, Z.; Duan, Y.; Ma, R.; Zheng, S. Synthesis of nano-TiO$_2$/diatomite composite and its photocatalytic degradation of gaseous formaldehyde. *Appl. Surf. Sci.* **2017**, *412*, 105–112. [CrossRef]
4. Salthammer, T. Formaldehyde sources, formaldehyde concentrations and air exchange rates in european housings. *Build. Environ.* **2019**, *150*, 219–232. [CrossRef]
5. Wang, X.; Hong, S.; Lian, H.; Zhan, X.; Cheng, M.; Huang, Z.; Manzo, M.; Cai, L.; Nadda, A.; Le, Q.V.; et al. Photocatalytic degradation of surface-coated tourmaline-titanium dioxide for self-cleaning of formaldehyde emitted from furniture. *J. Hazard. Mater.* **2021**, *420*, 126565. [CrossRef] [PubMed]
6. Qin, Y.; Wang, Z.; Jiang, J.; Xing, L.; Wu, K. One-step fabrication of TiO$_2$/Ti foil annular photoreactor for photocatalytic degradation of formaldehyde. *Chem. Eng. J.* **2020**, *394*, 124917. [CrossRef]
7. Li, X.; Li, H.; Huang, Y.; Cao, J.; Huang, T.; Li, R.; Zhang, Q.; Lee, S.-c.; Ho, W. Exploring the photocatalytic conversion mechanism of gaseous formaldehyde degradation on TiO$_2$–x-Ov surface. *J. Hazard. Mater.* **2022**, *424*, 127217. [CrossRef]
8. Xu, L.; Meng, L.; Zhang, X.; Mei, X.; Guo, X.; Li, W.; Wang, P.; Gan, L. Promoting Fe^{3+}/Fe^{2+} cycling under visible light by synergistic interactions between P25 and small amount of fenton reagents. *J. Hazard. Mater.* **2019**, *379*, 120795. [CrossRef]
9. Li, D.; Li, R.; Zeng, F.; Yan, W.; Deng, M.; Cai, S. The photoexcited electron transfer and photocatalytic mechanism of g-C$_3$N$_4$/TiO$_2$ heterojunctions: Time-domain ab initio analysis. *Appl. Surf. Sci.* **2023**, *614*, 156104. [CrossRef]

10. Xu, L.; Qi, L.; Han, Y.; Lu, W.; Han, J.; Qiao, W.; Mei, X.; Pan, Y.; Song, K.; Ling, C.; et al. Improvement of Fe^{2+}/peroxymonosulfate oxidation of organic pollutants by promoting Fe^{2+} regeneration with visible light driven g-C_3N_4 photocatalysis. *Chem. Eng. J.* **2022**, *430*, 132828. [CrossRef]

11. Li, Y.; Zhou, M.; Cheng, B.; Shao, Y. Recent advances in g-C_3N_4-based heterojunction photocatalysts. *J. Mater. Sci. Technol.* **2020**, *56*, 1–17. [CrossRef]

12. Wu, C. Facile one-step synthesis of n-doped ZnO micropolyhedrons for efficient photocatalytic degradation of formaldehyde under visible-light irradiation. *Appl. Surf. Sci.* **2014**, *319*, 237–243. [CrossRef]

13. He, Z.; Xiong, J.; Kumagai, K.; Chen, W. An improved mechanism-based model for predicting the long-term formaldehyde emissions from composite wood products with exposed edges and seams. *Environ. Int.* **2019**, *132*, 105086. [CrossRef] [PubMed]

14. Manisha; Kumar, V.; Sharma, D.K. Fabrication of dimensional hydrophilic TiO_2 nanostructured surfaces by hydrothermal method. *Mater. Today Proc.* **2021**, *46*, 2171–2174.

15. Jiang, M.; Zou, Y.; Xu, F.; Sun, L.; Hu, Z.; Yu, S.; Zhang, J.; Xiang, C. Synthesis of g-C_3N_4/Fe_3O_4/MoS_2 composites for efficient hydrogen evolution reaction. *J. Alloys Compd.* **2022**, *906*, 164265. [CrossRef]

16. Janus, M.; Choina, J.; Morawski, A.W. Azo dyes decomposition on new nitrogen-modified anatase TiO_2 with high adsorptivity. *J. Hazard. Mater.* **2009**, *166*, 1–5. [CrossRef]

17. Bouzourâa, M.B.; Battie, Y.; En Naciri, A.; Araiedh, F.; Ducos, F.; Chaoui, N. $N2^+$ ion bombardment effect on the band gap of anatase TiO_2 ultrathin films. *Opt. Mater.* **2019**, *88*, 282–288. [CrossRef]

18. Che, H.; Li, C.; Zhou, P.; Liu, C.; Dong, H.; Li, C. Band structure engineering and efficient injection rich-π-electrons into ultrathin g-C_3N_4 for boosting photocatalytic H_2-production. *Appl. Surf. Sci.* **2020**, *505*, 144564. [CrossRef]

19. Khan, T.T.; Bari, G.A.R.; Kang, H.-J.; Lee, T.-G.; Park, J.-W.; Hwang, H.J.; Hossain, S.M.; Mun, J.S.; Suzuki, N.; Fujishima, A. Synthesis of n-doped TiO_2 for efficient photocatalytic degradation of atmospheric NOx. *Catalysts* **2021**, *11*, 109. [CrossRef]

20. Kumar, S.G.; Devi, L.G. Review on modified TiO_2 photocatalysis under UV/visible light: Selected results and related mechanisms on interfacial charge carrier transfer dynamics. *J. Phys. Chem. A* **2011**, *115*, 13211–13241. [CrossRef]

21. Li, Q.; Zhang, S.; Xia, W.; Jiang, X.; Huang, Z.; Wu, X.; Zhao, H.; Yuan, C.; Shen, H.; Jing, G. Surface design of g-C_3N_4 quantum dot-decorated TiO_2(001) to enhance the photodegradation of indoor formaldehyde by experimental and theoretical investigation. *Ecotox. Environ. Saf.* **2022**, *234*, 113411. [CrossRef] [PubMed]

22. Li, M.; Song, C.; Wu, Y.; Wang, M.; Pan, Z.; Sun, Y.; Meng, L.; Han, S.; Xu, L.; Gan, L. Novel z-scheme visible-light photocatalyst based on $CoFe_2O_4$/$BiOBr$/graphene composites for organic dye degradation and Cr(VI) reduction. *Appl. Surf. Sci.* **2019**, *478*, 744–753. [CrossRef]

23. Li, X.; Jia, Y.; Zhou, M.; Ding, L.; Su, X.; Sun, J. Degradation of diclofenac sodium by pre-magnetization Fe^0/persulfate system: Efficiency and degradation pathway study. *Water Air Soil Pollut.* **2020**, *231*, 311. [CrossRef]

24. Li, X.; Qin, Y.; Song, H.; Zou, W.; Cao, Z.; Ding, L.; Pan, Y.; Zhou, M. Efficient removal of bisphenol a by a novel biochar-based Fe/C granule via persulfate activation: Performance, mechanism, and toxicity assessment. *Process Saf. Environ.* **2023**, *169*, 48–60. [CrossRef]

 *catalysts*

Review

A Critical Review of Photo-Based Advanced Oxidation Processes to Pharmaceutical Degradation

Isabelle M. D. Gonzaga , Caio V. S. Almeida and Lucia H. Mascaro *

LIEC-Interdisciplinary Laboratory of Electrochemistry and Ceramics, Federal University of São Carlos, São Carlos, SP 13565-905, Brazil
* Correspondence: lmascaro@ufscar.br

Abstract: Currently, the production and consumption of pharmaceuticals is growing exponentially, making them emerging contaminants that cause hazards to the ecological environment and human health. These drugs have been detected in surface water and drinking water around the world. This indicates that the conventional treatments used are ineffective for the removal of these compounds from the water, since they are very complex, with high stability and have high persistence in aquatic environments. Considering this problem, several types of alternative treatments, such as advanced oxidative processes, have been studied. Of these, AOPs using irradiation have received increasing interest due to their fast reaction rate and the ability to generate oxidizing species, which leads to an efficient degradation and mineralization of organic compounds, thus improving the quality of water and allowing its reuse. Therefore, in this review, we focus on the advances made in the last five years of irradiated AOPs in the degradation of different classes of pharmaceutical compounds. The articles address different study parameters, such as the method of the synthesis of materials, oxidants used, treatment time, type of light used and toxicity of effluents. This review highlights the success of irradiated AOPs in the removal of pharmaceuticals and hopes to help the readers to better understand these processes and their limitations for removing drugs from the environment. It also sheds light on some paths that future research must follow so that the technology can be fully applied.

Keywords: degradation; photoelectrochemical; toxicity; environment

Citation: Gonzaga, I.M.D.; Almeida, C.V.S.; Mascaro, L.H. A Critical Review of Photo-Based Advanced Oxidation Processes to Pharmaceutical Degradation. *Catalysts* **2023**, *13*, 221. https://doi.org/10.3390/catal13020221

Academic Editors: Gassan Hodaifa, Rafael Borja and Mha Albqmi

Received: 20 December 2022
Revised: 12 January 2023
Accepted: 16 January 2023
Published: 18 January 2023

1. Introduction

Water pollution occurs when dangerous substances enter the oceans, rivers and lakes, among others, and can be found suspended or dissolved in these water bodies [1]. It is reported that areas of higher population density have a greater amount of water bodies polluted with different types of contaminants [2]. So, with the increase in human life expectancy, millions of medicines have been developed to maintain good health and quality of life [3]. In the last decades, pharmaceutical residues have been discovered in almost all different aquatic matrices on all continents [4]. Due to the high solubility and low biodegradability of the drugs, these compounds are extremely difficult to eliminate from water using conventional methods of treatment [5].

Existing wastewater treatment plants around the world are equipped with conventional treatment methods for the removal of simple organic substances, such as filtration, sedimentation, flocculation, adsorption and/or biological treatment [6]. However, these methods do not completely destroy organic pharmaceutical pollutants. Therefore, many researchers are in a relentless search for unconventional methods for pharmaceutical remediation in wastewater. Advanced oxidation processes (AOPs) are becoming attractive alternative techniques due to the promising chance of complete removal of drugs with production of non-toxic by-products [7].

AOPs are based on the generation of strong oxidants capable of oxidizing a wide range of organic compounds with one or several double bonds [8], such as process thermal, electrochemical, sonochemical, transition metal-based and irradiated. Of these, irradiated AOPs have been highlighted (Figure 1). Irradiated AOPs are green processes that do not involve secondary pollution, have mild reaction conditions, are easy to operate and have high pollutant degradation efficiency [9,10].

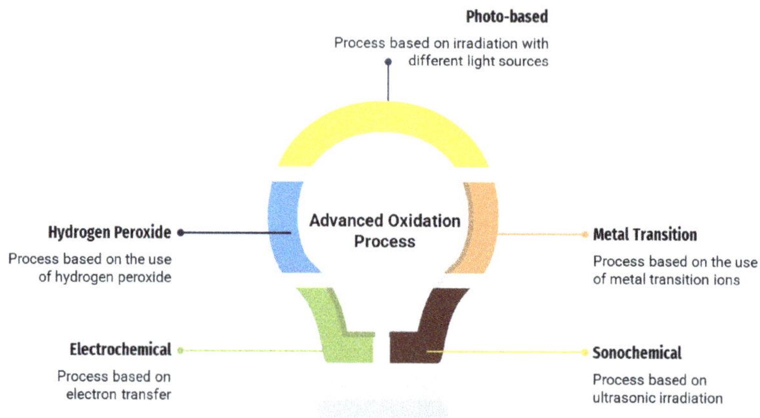

Figure 1. Different advanced oxidative process for the removal of pharmaceuticals.

This review aims to provide a discussion of recent literature that uses irradiated AOPs to potentiate radicals formation to completely remove drugs and/or transform by-products into less toxic by-products. Research works that produce new materials, study different techniques for removing drugs and/or evaluate the toxicity of effluents after treatment are discussed. Additionally, the treatment of water contaminated with various drugs is examined, which is classified according to the functional group. Thus, six different classes are approached, namely the nitrogenated, chlorinated, fluorinated, sulfured, phosphorated and oxygenated. Finally, a critical discussion of these approaches is also presented for a better understanding of the gaps opened in this topic in the literature.

2. Status of Pharmaceutical Pollution

Water is one of the most important compounds on the planet, as it is directly related to the existence of living beings. With the growth of the world population and the development of technology, water pollution has become a very important issue. Water contamination can be the result of several factors; one of the most polluting is the rapid development of industries and, consequently, the disposal of industrial waste [11]. Among the existing hazardous effluents, we can see that pharmaceutical products often play a vital role in the emission of toxic substances into the environment [12]. Drugs, which are organic and/or inorganic chemical compounds, can be classified according to their therapeutic actions as antibiotics, analgesics, beta-blockers, anesthetics, radiographic contrast agents and antidepressants [13].

Drugs have received enormous attention regarding detection and degradation due to their potential adverse effects on humans and the ecosystem [14]. Thus, the non-treatment of waste with antibiotics can give rise to antibiotic-resistant bacteria. Another class of drugs can cause various other problems, such as the development of reproductive disorders, birth defects, severe bleeding, cancer, organ damage, endocrine disorders and toxic effects on aquatic organisms, which could have immeasurable consequences for humanity [15]. In Table 1, it is possible to observe widely used drugs, which have been covered in the literature.

Table 1. Drugs reported in the literature as objects of study for removal in wastewater.

Class	Pharmaceuticals	Abbreviation	Chemical Formula	Use
NITROGENATED	Acetaminophen	ACT	$C_8H_9NO_2$	Analgesic and antipyretic for mild pain and fever
	Tramadol	TRA	$C_{16}H_{25}NO_2$	Analgesic for rigorous pain
	Azithromycin	AZT	$C_{38}H_{72}N_2O_{12}$	Antibiotic used to treat respiratory, ear, skin or sexually transmitted infections
	Doxorubicin	DOX	$C_{27}H_{29}NO_4$	Antibiotic used for the therapy of lymphoma, leukemia and sarcoma
	Metronidazole	MTZ	$C_6H_9N_3O_3$	Antiparasitics and antiprotozoa
	Trimethoprim	TRIM	$C_{14}H_{18}N_4O_3$	Antibiotic used for mild-to-moderate bacterial infections
	Caffeine	CAF	$C_8H_{10}N_4O_2$	Used for mild pain and is mostly consumed in the form of coffee
	Propranolol	PROP	$C_{16}H_{21}NO_2$	Used for hypertension, cardiac arrhythmias, angina and hyperthyroidism
CHLORINATED	Lamotrigine	LAM	$C_9H_7Cl_2N$	Used to fight seizures
	Chloramphenicol	CAP	$C_{11}H_{12}Cl_2N_2O_5$	Antibiotics used for severe or life-threatening infections
	Diclofenac	DIC	$C_{14}H_{11}Cl_2NO_2$	Used for the therapy of mild-to-moderate acute pain
	Indomethacin	IDM	$C_{19}H_{16}ClNO$	Used for inflammatory arthritis
	Chlorhexidine	CLH	$C_{22}H_{30}Cl_2N_{10}$	Used topical antiseptic and dental practice for the treatment of dental inflammation
	Clorazepate	CLZ	$C_{16}H_{11}ClN_2O_3$	Anticonvulsant for therapy in epilepsy and as an anxiolytic
	Hydroxyzine	HDZ	$C_{21}H_{27}ClN_2O_2$	Used largely for symptoms of allergy, nausea and anxiety
	Indapamide	IDP	$C_{16}H_{16}ClNO_3$	Antihypertensive and a diuretic
	Bezafibrate	BZF	$C_{19}H_{20}ClNO_4$	Used for hyperlipidemia
FLUORATED	Ciprofloxacin	CIP	$C_{17}H_{18}FN_3O_3$	Used in the therapy of mild-to-moderate urinary or respiratory infections
	Ofloxacin	OFLO	$C_{18}H_{20}FN_3O_4$	Used for treatment rare instances of hepatocellular injury
	Enrofloxacin	ENR	$C_{19}H_{22}FN_3O_3$	Antibiotic used to treat infections of the urinary system
	Levofloxacin	LEV	$C_{18}H_{20}FN_3O_4$	Antibiotic used to treat infections, such as pneumonia and sinusitis
	Etofinamate	ETO	$C_{18}H_{18}F_3NO_4$	Used in the treatment of muscles and joints
	Fluoxetine	FLU	$C_{17}H_{18}F_3NO$	Used to treat major depressive, moderate-to-severe bulimia and obsessive-compulsive disorder
SULFURATED	Eprosartan	EPS	$C_{23}H_{24}N_2O_4S$	Used for the treatment of high blood pressure
	Captopril	CPT	$C_9H_{15}NO_3S$	Used in the therapy of hypertension
	Sulpiride	SUL	$C_{15}H_{23}N_3O_4S$	Used therapy as an antidepressant
	Sulfamonomethoxine	SFX	$C_{11}H_{12}N_4O_3S$	Long acting antibacterial agent
	Amoxicillin	AMX	$C_{16}H_{19}N_3O_5S$	Used in the treatment of mild-to-moderate bacterial infections
PHOSPORATED	Ifosfamide	IFD	$C_7H_{15}Cl_2N_2O_2P$	Chemotherapy used to treat certain types of cancer, such as cancer of the ovary and cervix
	Cyclophosphamide	CFD	$C_7H_{15}Cl_2N_2O_2P$	Cyclophosphamide is used to treat cancer of the ovaries, breast, blood and lymph systems
OXIGENATED	Aspirin	AAS	$C_9H_8O_4$	Used as a medicine to treat pain, fever and inflammation
	Ketoprofen	KET	$C_{16}H_{14}O_3$	Anti-inflammatory, with analgesic, anti-inflammatory and antirheumatic effects
	Naproxen	NAP	$C_{14}H_{14}O_3$	Analgesic for pain
	Ibuprofen	IBU	$C_{13}H_{18}O_2$	Anti-inflammatory used to treat pain and fever

These compounds are recalcitrant towards conventional treatment methods commonly used (sedimentation, flocculation, adsorption and ultrafiltration, for example), which are efficient for the removal of simple substances and materials on a macro/micro scale, while the advanced oxidation processes utilize the high reactivity of oxidants to degrade organic compounds to harmless products in water progressively have shown better efficiency in degrading these drugs. The combination of two or more AOPs is increasingly promising since, in some cases, only one AOPs is not enough to mineralize the effluent. Thus, combining technologies is a promising alternative for complete drug removal.

3. Photocatalysis as a Tool for the Degradation of Different Pharmaceuticals

3.1. Nitrogenous-Based Compounds

Pharmaceuticals that have heterocyclic compounds with nitrogen, such as pyridine, are widely used in living beings. Some examples of commonly used nitrogenous drugs are acetaminophen, metronidazole, tramadol and azithromycin, among others [16]. Heterocyclic nitrogenous compounds have high toxicity and recalcitrance and there are reports of frequent detection occurring in the aquatic environment, and thus the existing treatment methods are not efficient [17]. Thus, it is necessary to seek efficient systems for the complete mineralization of these compounds during water treatment processes.

In this sense, Li et al. [18] studied the photodegradation of acetaminophen (ACT) by combining UV light with different oxidants, such as chloramine, hydrogen peroxide and persulfate. The researchers showed that the drug can be effectively degraded by the combined processes when compared to the exclusive use of UV irradiation and oxidants. When comparing the efficiency of the three hybrid methods, they found that the efficiency was given in the following order: UV-LED/Persulfate > UV-LED/Hydrogen peroxide > UV-LED/Chloramines. Furthermore, it was observed that the kinetics of acetominophen removal increased when the oxidant dosage was increased in all cases. Finally, acute toxicity tests carried out using *Vibrio fisheri* and UV-LED/Persulfate were more efficient. In this view, persulfate was found as the most promising oxidant for the combination with UV-LED in the degradation of acetaminophen. On the other hand, Cai and co-workers carried out work investigating the efficiency of ACT removal with thermal treatment/peroxymonosulfate (T/PMS). They achieved efficient removal of the compound. Comparing the works that use T/PMS [19–21] with photo-assisted works [18,22], it was observed that the light-based processes more efficiently remove the toxicity of the studied effluents.

More recently, Qiangwei Li and co-workers [22] investigated the use of the Co_3O_4/TiO_2 heterojunction to degrade ACT through the photocatalytic activation of the sulfite. They discovered through X-ray photoelectron spectroscopy analysis and theoretical calculations that electron transfer from Co_3O_4 to TiO_2 directly contributes to the catalytic activation of the sulfite. Therefore, due to the unique configuration, 96% of ACMPH was removed in just 10 min. Showing that sulfite radicals activation is extremely promising.

Another widely used nitrogenated compound, azithromycin, is a common antibiotic, and during the COVID-19 pandemic, its use increased even more. Therefore, it is extremely important to search for new technologies to remove this antibiotic. Thus, Sayadi et al. [23] investigated the synthesis of a new material with iron, zinc and tin oxide to apply it to the photocatalytic removal of azithromycin. The effect of pH, contact time, catalyst content and the initial concentration of azithromycin was analyzed, and under optimal conditions (pH = 3, 120 min with 1 g/L of catalyst), obtained 90.06% of azithromycin degraded under UVC irradiation.

More recently, Tenzi and co-workers [24] studied a new material for photodegradation with UV-LED from AZT, $SrTiO_3$. In its optimal condition, with 40 mg of the material and 4 h of treatment, 99% of drug removal occurred. The authors justified its excellent performance through two important factors: (i) the nanostructured and irregular morphology and (ii) the hole action and hydroxyl formation.

Another drug in this important class is metronidazole. In this sense, Martins et al. [25] studied the synthesis of ginite with different alkaline hydroxides (Na, Li and K) and subsequently applied it to remove metronidazole using photo-Fenton technology. The authors reported that cations directly influence the morphology of the material, resulting in different surface areas. Na@giniita showed the largest surface area, 10.9 m^2 g^{-1}, when compared to the others. When applying them in the degradation, Na@giniita obtained 91% degradation, so the authors concluded that the greater the surface area, the more efficient the removal of MTZ.

Still, Neghi and collegues [26] carried out the removal of the MTZ, but unlike the previous works cited, they carried out the degradation on a laboratory scale (Figure 2a) and on a pilot scale (Figure 2b). They used different methodologies, such as Persulfate (PS), TiO_2 activated by UV-C (UV/TiO_2), PS activated by UV-C (UV/PS) and UV/PS with TiO_2 (UV/PS/TiO_2). The great difficulty was optimizing the parameters, since the scales of the systems were very different. In the optimized condition, which was the combination of the three technologies, regardless of the system used, 90% of the MTZ was removed. In addition, toxicity studies using Vibrio Fisheri were performed and, in addition to antibiotic removal, effluent toxicity was removed.

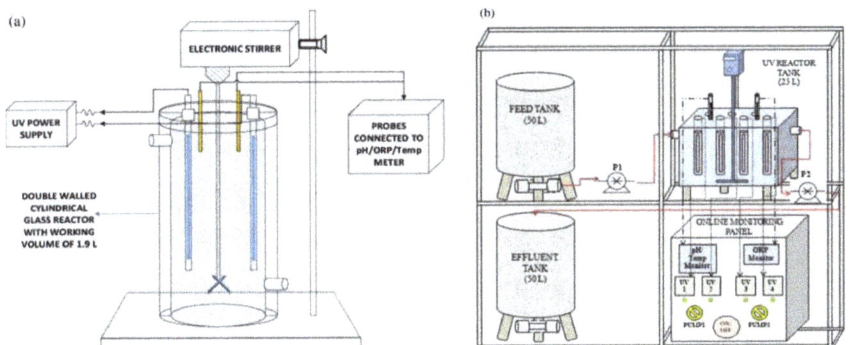

Figure 2. Comparison of photodegradation systems used with UV, PS or TiO_2, single or coupled. (**a**) laboratory scale and (**b**) pilot scale. Reprinted with permission from [26]. Copyright © 2020 Elsevier Ltd.

3.2. Chlorinated-Based Compounds

Chloramphenicol (CAP), Diclofenac (DC), Hydroxyzine (HDZ) and Indomethacin (IMT), among others, are drugs widely used worldwide. In this context, Leeladevi et al. [27] proposed the $SmVO_4$/g-C_3N_4 synthesis using a simple hydrothermal method. Due to the excellent photocatalytic activity, when the material was applied for CAP removal it achieved 94.35% efficiency. The excellent performance was justified by its unique composition and an excellent metal-free decorated photocatalyst alternative. Additionally, Hu and co-workers [28] studied a new strategy, photocatalysis coupled with microbial fuel cell (photo-MFC) on a Ni/Ti_2O_3 photocathode to increase the degradation efficiency of CAP. It was observed that the best degradation efficiency of CAP can reach 82.62% removal. Finally, the evaluation of the ecotoxicity of the degradation products showed that the degradation of CAP in the photo-MFC system modified with Ni/Ti_2O_3 presented a remarkable trend towards low toxicity levels. Therefore, it represents a very promising system for industrial waste degradation.

Kumar et al. [29] reported the synthesis of a Fe_3O_4@$SrTiO_3$/$Bi_4O_5I_2$ heterojunction through a hydrothermal process. The heterojunction was used for DC removal under simulated sunlight irradiation. Current results reveal that the optimized heterojunction exhibited a 98.4% removal of diclofenac within 90 min. EPR analyzes show that ·OH and ·O_2 radicals are present in DC degradation. The improved performance was justified by the proposal of a Z scheme (Figure 3), since the charge separation was obviously improved and facilitated by Fe_3O_4 and Ti^{4+}/Ti^{3+}, the involved redox couples. Thus, the authors demonstrate a promising material for rational design and fabrication of semiconductor heterojunctions for environmental applications.

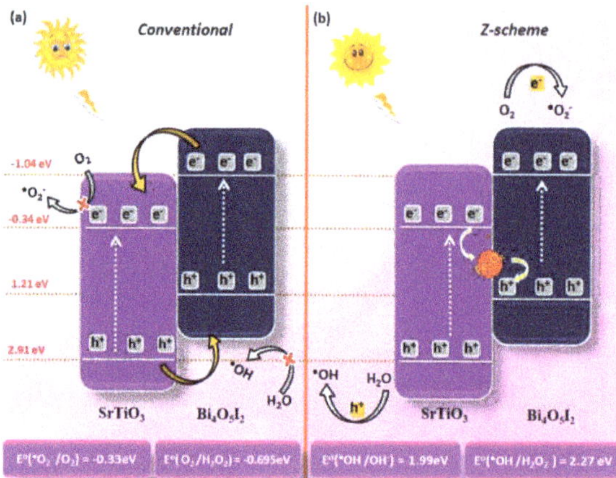

Figure 3. (**a,b**) Photocatalytic mechanism Z proposed by Kumar et al. Reprinted with permission from [29]. Copyright © 2020 Elsevier Ltd.

Examining Table 2, it is possible to conclude that the degradation of nitrogenous and chlorinated base compounds was successfully degraded via photo-assisted and photocatalytic AOPs systems. New materials and technologies were proposed to improve the removal rate and mineralization of the systems. Still, few works carry out their tests on a pilot scale.

Table 2. Selected recent studies of AOPs photo-irradiated processes for nitrogenated and chlorated base compound removal.

Pharmaceutical	Process	Experimental Conditions	Results	Ref.
DC	Photocatalysis	pH = 3; t = 150 min; [Catalyst] = 0.1 g L^{-1}; [DC] = 60 mg L^{-1}	99.4% DC and 87.8% TOC removal	[30]
CAP	Electrochemical and photo-assisted	pH = 6.8; t = 8h; [CAP] = 50 mg L^{-1}	100% CAP removal and complete removal of antibiotic activity	[31]
IMT	Photocatalysis	pH = 6.8; t = 45 min; [Catalyst] = 0.03 g mL^{-1}; [IMT] = 20 mg L^{-1}	97.3% IMT removal	[32]
IMT	UV–vis/peroxydisulfate	pH = 7; t = 45 min; [PDS] = 20 μM [IMT] = 20 μM	~80% IMT and 30% TOC removal	[33]
MET	Photocatalysis	pH = 2; t = 15 min; [Catalyst] = 1.2 g L^{-1} [MET] = 10 mg L^{-1}	94.3% MET removal	[34]
DC	US/photo-Fenton	pH = 3; t = 90 min; [Fe^{2+}] = 0.20 mg L^{-1}; [H$_2$O$_2$] = 1 mg L^{-1}; [DC] = 1 mg L^{-1}	94.4% DC and 17% TOC removal	[35]
AZT	Photocatalysis	t = 8h; [Catalyst] = 2 mg mL^{-1} [AZT] = 20 mg L^{-1}	90% AZT removal and complete removal of toxicity	[36]
AZT	UV/H$_2$O$_2$	pH = 9; t = 120 min; [H$_2$O$_2$] = 482 mg L^{-1}; [DC] = 1 mg L^{-1}	53% AZT and 100% TOC removal	[37]
CAF	Fenton, photo-Fenton, UV/H$_2$O$_2$ and UV/Fe^{3+}	pH = 5.8; t = 120 min; [H$_2$O$_2$] = 0.5 mL L^{-1}; [CAF] = 20 mg L^{-1}	99% CAF removal	[38]
LEV	Photo-electrocatalyst	pH = 5.9; t = 4 h; [LEV] = 20 mg L^{-1}	99% LEV and 100% TOC removal	[39]

3.3. Fluorinated-Based Compounds

Fluoroquinolones (FQ) are a family of fluorinated-based antibiotics used for both human and animal extensively. Most common FQs used include CIP, ENR, NOR and others [40–42]. They all present a bicyclic core structure related to the substance 4-quinolone with a fluorine substituted on the C6 position. FQs are incompletely metabolized in bodies and excreted into wastewaters. Because of the incomplete removal in conventional wastewater treatment facilities, since these treatment plants were never designed to deal with antibiotics, FQs are released to the environment via wastewater discharge on a continuous basis [42]. In recent years, these substances were detected in other aqueous environments around the world, such as lakes, rivers, seas, ground water and even in tap water [41–45].

Thus, it is crucial to develop efficient and economical treatment technologies to eliminate FQs from natural waters and wastewaters due to the potential threats of FQs to aquatic ecosystems and human beings. Photodegradation was found to be an important mechanism responsible for their removal in surface waters. The quinolone chromophore group enables the molecules to absorb UV-A light efficiently [46].

In this sense, Geng and co-workers [40] investigated the photo-degradation of three FQs (NOR, CIP, LEV) by VUV/UV and UV irradiation. They evaluated several effects such as flow rate (30, 60 and 90 L h^{-1}), FQs initial concentration (15, 30 and 60 μmol L^{-1}), temperature (5, 15, 25, 35 and 45 °C), initial pH (ranging 3 to 11) and typical anions in degradation. FQs was degraded more efficiently with high temperatures and low initial concentration (Figure 4a). The effect of the flow rate was small and exhibited a maximum value at around 60 L h^{-1} (Figure 4a). In the pH range of 3–9, the photo-degradation rate constants of NOR and CIP increased with the pH but decreased at pH 11, which was related to different dissociation forms in the water at different pH. However, the photo-degradation kinetic constant of LEV was slightly affected by pH in the VUV/UV system (Figure 4c). In the VUV/UV system, carbonate and phosphate species enhanced the removal rate of FQs. However, sulfate, chloride and nitrate restrained the degradation process of FQs (Figure 4b,d).

The photo-degradation mechanisms of FQs in the VUV/UV system involved the direct photolysis, O_2 and ·OH oxidation. The degradation pathways of FQs were defluorination, decarboxylation and piperazine ring oxidation, and demethylation was a particular degradation pathway for LEV. In addition, antibacterial activities significantly decreased in the VUV/UV process due to defluorination and decarboxylation induced by VUV and ·OH. Particularly, the comparison of the energy cost showed that the VUV/UV process was more energy efficient than UV process, indicating that the VUV/UV system was a promising treatment technology for removing FQs from water or wastewater [40].

In another work, Mondal, Saha, and Sinha [47] studied the degradation of ciprofloxacin by using various AOPs, such as UV, H_2O_2, UV/H_2O_2, modified Fenton (nanoscale zerovalent iron-nZVI/H_2O_2) and modified photo-Fenton at near neutral pH. The degradation of CIP by photolysis was performed at the initial pH of 7 and with the initial concentration of 10 mg L^{-1}. Approximately 60% of the antibiotic was removed within 120 min, confirming ciprofloxacin as a photosensitive compound. However, only 4% of the total organic carbon (TOC) removal was achieved, indicating that the degradation of CIP by UV irradiation may not prove to be feasible in 120 min. Later, hydrogen peroxide (H_2O_2) with several concentrations was added to ciprofloxacin solution (10 mg L^{-1}) in presence of UV light at initial pH 7. The TOC % removal increased from 10.47% for only H_2O_2 to 35.41% for UV/H_2O_2 combination due to a higher amount of hydroxyl radicals (·OH) generation due to photolysis.

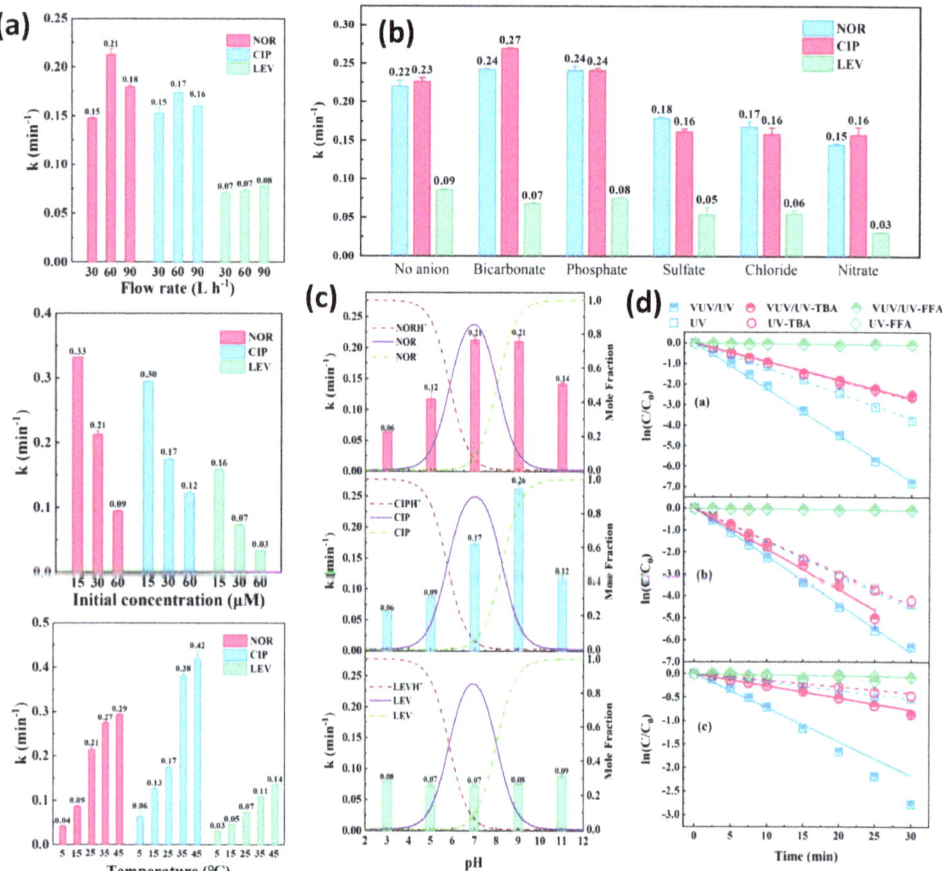

Figure 4. Comparison of the first-order rate constants of three FQs by the VUV/UV irradiation processes in pure water with various factors. (**a**) Flow rate; initial concentration of FQs and temperature; (**b**) effect of anion on the photodegradation of FQs; (**c**) initial pH and chemical specifications of the FQs; (**d**) degradation of the three antibiotics by UV and VUV/UV. Conditions: temperature = 25 °C, pH = 7.0 ± 0.05; [FQs] = 30 μmol L^{-1}; flow rate = 60 L h^{-1}; [Anion] = 300 μmol L^{-1}; [TBA] = 100 mmol L^{-1}. Reprinted with permission from [40]. Copyright © 2020 Elsevier Ltd.

The UV/H$_2$O$_2$ combination was highly effective in achieving 100% removal efficiency of CIP in 40 min for higher doses of H$_2$O$_2$ (100 mmol L^{-1}), while the modified photo-Fenton oxidation process using 100 mmol L^{-1} of H$_2$O$_2$ and nZVI dose of 5 mmol L^{-1} was capable of completely removing ciprofloxacin in 30 min. Additionally, this combination reached the highest percentage of mineralization, removing 60% of the initial TOC. According to the authors, the degradation of ciprofloxacin may occur through: (i) the reaction at the piperazinyl ring; (ii) the oxidation of quinolone moiety, thereby leading to defluorination and hydroxyl substitution reaction; and (iii) the oxidation of the cyclopropyl group, resulting in ring cleavage.

In summary, the use of photolysis was not sufficient to completely remove CIP as it requires a high reaction time and has a very low mineralization rate. On the other hand, the UV/H$_2$O$_2$ method could be suited for complete removal of ciprofloxacin in comparison to the modified Fenton and modified photo-Fenton oxidation process as there is no sludge formation; however, it may require a high dose of H$_2$O$_2$ [47].

In an interesting work, Liu et al. [48] studied the degradation behavior of OFLO and Levofloxacin LEV using UV/H_2O_2 and UV/PS. The effects of oxidant dose (50 to 300 $\mu mol \cdot L^{-1}$), solution pH (3 to 11) and coexisting substances (such as Cl^-, NO_3^-, SO_4^{2-}, Ca^{2+}, Mg^{2+} and NOMs (nature organic matters)) were investigated. The addition of H_2O_2 or PS significantly improves the degradation efficiency of OFLO and LEV (5 mg L^{-1}), while UV/PS achieves a better degradation effect than UV/H_2O_2 under the same oxidant concentration. The removal efficiencies of OFLO and LEV were 96.40% and 99.23% with PS concentration of 150 $\mu mol \cdot L^{-1}$, where the removal efficiencies were 22.52 and 13.11 times as high as that of UV process, respectively (pH = 3 and t = 30 min). The degradation of both OFLO and LEV revealed pronounced pH dependence in UV/H_2O_2 and UV/PS processes, where the impact on LEV was greater than that on OFLO, indicating that LEV might be more easily degraded due to its shorter half-life. The TOC removal efficiencies of OFLO and LEV were 36.24% and 38.36%, respectively, for the UV/H_2O_2 system. By contrast, the UV/PS system exhibited the highest activity for the mineralization of OFL and LEV, with TOC removal efficiency of about 46.43% and 49.74%, respectively, indicating that $SO_4^{\cdot-}$ generated in the UV/PS process yields a greater mineralization of antibiotics than $\cdot OH$ in the UV/H_2O_2 process.

The coexisting substances (Cl^-, NO_3^- and NOM) exhibited more inhibition regarding the degradation of OFLO and LEV in the UV/H_2O_2 than the UV/PS system. SO_4^{2-} has negligible influence on the UV/H_2O_2 processes, while it showed positive effect on UV/PS degradation. The results of acute toxicity assay demonstrated that OFLO exhibited higher acute toxicity than LEV, while the triphenyl tetrazolium chloride (TTC) dehydrogenase activity of the intermediate products of OFLO was lower than LEV [48].

Recently, Wang and co-workers [49] reported the fast removal, after 4 min, and high mineralization (63.3% at 8 min) of 45 μmol L^{-1} norfloxacin at neutral pH by the $VUV/Fe^{2+}/H_2O_2$ system (90 μmol L^{-1} Fe^{2+} and 3 $mmol$ L^{-1} H_2O_2). Compared with other UV-based and VUV-based systems (UV, UV/H_2O_2, $UV/Fe^{2+}/H_2O_2$, VUV, UV/Fe^{2+}, VUV/H_2O_2, $VUV/Fe^{2+}/H_2O_2$), the $VUV/Fe^{2+}/H_2O_2$ system increased the pseudo-first-order reaction rate constant of NOR removal by 2.3–14.9 times; increased the mineralization by 20.4–59.4%; and reduced the residual ratio of H_2O_2 by 19.9−70.1%. Moreover, $HO\cdot$, $O^{\cdot-2}$ and singlet oxygen ($^1O_2$) were the main ROS during NOR removal at neutral pH. The degradation pathways of NOR included defluorination, attack by $HO\cdot$, decarboxylation and piperazine ring conversion by O_2^-. Moreover, adding Fe^{2+} to the VUV/H_2O_2 system reduced cost by 36.8%, 36.2% and 36.2% in ultrapure water, tap water and secondary wastewater, respectively. As a result, the $VUV/Fe^{2+}/H_2O_2$ process also achieved the rapid removal of NOR in real waters at neutral pH, while saving considerable cost and manifesting the feasibility of $VUV/Fe^{2+}/H_2O_2$ system in real waters.

Some researchers have studied the feasibility of O_3 and UV treatment (independent or together) to remove antibiotic wastewaters [50–53]. Liu et al. [50] investigated the degradation and mineralization of CIP in high-salinity wastewater through ozonation coupled with UV. Compared with independent O_3 (37.5%), the dissolved organic carbon (DOC) reduction was significantly increased by the introduction of UV irradiation (91.4%), which was attributed to the process of UV catalyzing O_3 to produce a reactive oxygen species (ROS), including $\cdot OH$, O_2 and $^1O_2$. The existence of salinity (3.5%, *w/v*) accelerated O_3 mass transfer at the gas–liquid interface, so CIP degradation was boosted by 17.7% and 2.0% in O_3/SO_4^{2-} and O_3/Cl^- system, respectively, within 15 min. The pH had little impact on the salt-free and O_3/containing Na_2SO_4 system. On the contrary, for an O_3/containing NaCl system, CIP and DOC removal was promoted with the increase of pH from 3 to 11 due to O_3 mass-transfer rate enhancement. The results of LC-MS demonstrated that the dominant reaction site of CIP in the UV/O_3 process was the piperazine ring. The toxicity of products decreased significantly as compared with the parent pollutant, which proved that the UV/O_3 process was promising in the hyper-salinity industry.

Paucar et al. [52] examined the effectiveness of the UV/O_3 process on the removal of personal care products (PPCPs) in the secondary effluent of a municipal wastewater treatment plant (WWTP). Experiments were conducted using a pilot-scale treatment process with two flow-through reactors, equipped with three 65 W lamps (UV$_{65W}$), connected in series. Ozone dosage (1–4 and 6 mg L^{-1}) and hydraulic retention time (HRT; 5 and 10 min). Of the 38 PPCPs detected, 11 were antibiotics, with ciprofloxacin and levofloxacin highlighted. In this system, ciprofloxacin was degraded below it limit of detection (LOD) at the ozone dose of 1 mg L^{-1} in 5 min. However, levofloxacin required an ozone dose of 6 mg L^{-1} to be degraded below LOD. The authors attributed the difference in the efficiency of treatment to the chemical properties of contaminants and highlighted further comprehensive studies in the areas of the reaction kinetics, the formation of byproducts, possible pathways and toxicity of wastewater treated by the UV/O$_3$ system.

Heterogeneous photocatalysis is a promising technique capable of directly attacking pollutant molecules and generating ·OH radicals and superoxides (·O$_2$$^-$), powerful oxidants capable of destroying pollutants and transforming them into water and carbon dioxide [54]. Many studies developed photocatalysts with higher yields and easier operation [54–57].

Costa and co-workers [55] reported the ciprofloxacin photodegradation TiO$_2$/SnO$_2$ nanocomposites as heterogenous photocatalysts. The photocatalytic experiments were studied at different variables, such as catalyst dosage, drug concentration and pH solution. Titanate nanotubes (Na-TiNT) were synthesized using the hydrothermal method and then modified with Sn^{2+} using the coprecipitation method. Therefore, photolysis showed the small removal of CIP (4.63%) over 120 min of exposure time under UV-light radiation while the removal of CIP for Na-TiNT, SnO$_2$ and TiO$_2$/SnO$_2$ nanocomposites were 41.55%, 45.83% and 92.8%, respectively (Figure 5a). Increasing the catalyst dosage of TiO$_2$/SnO$_2$ nanocomposite from 25 to 50 mg (Figure 5b) in the CIP photodegradation increased the degradation rate of 27.8% to 92.8%, respectively. The authors attributed this behavior to high surface for interaction with CIP molecules and effective photon absorption. In contrast, increasing the catalyst amount to 75 mg, decreased the degradation rate for CIP, which was related to the increase of the opacity of the solution. The best photocatalytic performance was achieved at pH 7, where 97.7% of CIP molecules were oxidized (Figure 5c). In addition, the initial concentration of CIP had little influence on the removal of the compound. The study of radical scavengers indicated that the oxygen singlets, holes and superoxide radicals are the main species associated with the CIP photodegradation (Figure 5d).

The results presented in the photodegradation of ciprofloxacin from TiO$_2$/SnO$_2$ nanocomposite [55] were promising in comparison with other works in the literature [56–59]. Nguyen et al. [59] reported the optimal CIP degradation rates of 78.7% using co-doped TiO$_2$ nanomaterials (N, S-TiO$_2$) at pH 5.5, a catalyst dosage of 50 mg and CIP initial concentration of 30 mg L^{-1}. Similar results were obtained by Zhang et al. [57] using Vo-WO$_3$/Bi$_2$WO$_6$ composites under visible light. They achieved 79.5% of CIP removal within 120 min under visible light irradiation (a 300 W Xe lamp) when 40 mg of the photocatalyst was added into CIP solution (10 mg L^{-1}). According to the active species-capturing results and ESR test, h$^+$ and ·O$_2$$^-$ played crucial roles in photodegradation CIP, while ·OH played a weak role [57].

Moreover, Wu and co-workers [58], proposed an intercalated heterostructural g-C$_3$N$_4$/TiO$_2$ supported on Halloysite nanotubes (HNTs) and obtained 87% of CIP removal (15 mg L^{-1}), employing a 300W Xe lamp after 60 min of irradiation. In addition, the main active species of g-C$_3$N$_4$-TiO$_2$/HNTs heterojunction composites were ·O$_2$$^-$ and the h$^+$ in the process of photocatalytic degradation, in agreement with the results presented by Zhang et al. [57]. Despite the promising results, the authors did not evaluate the toxicity of the by-products formed nor the rates of mineralization achieved, fundamental parameters to determining the efficiency of degradation.

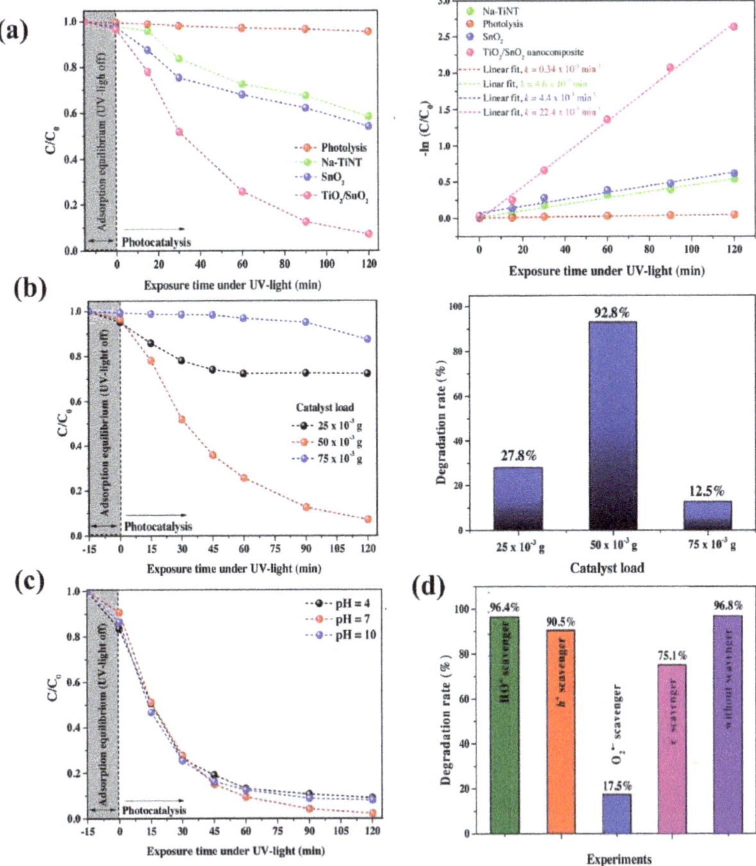

Figure 5. (**a**) Kinects study by the photocatalytic degradation curve and –ln(C/C$_0$) curve against exposure time (t) for ciprofloxacin using Na-TiNTs, SnO$_2$ and TiO$_2$/SnO$_2$ samples as photocatalyst. (**b**) Degradation rate, for different amounts of TiO$_2$/SnO$_2$ nanocomposite (25×10^{-3} g, 50×10^{-3} g and 75×10^{-3} g) against 120 min of exposure time. (**c**) Photocatalytic performance for TiO$_2$/SnO$_2$ nanocomposite at different initial pH values of ciprofloxacin solution. (**d**) Degradation rate at the end of 120 min, using different radicals scavengers. Reprinted with permission from [55]. Copyright © 2021 Elsevier Ltd.

In another research, Prabavathi et al. [59] demonstrated a novel Sm$_6$WO$_{12}$ decorated g-C$_3$N$_4$ heterojunction for the degradation of levofloxacin (10 mg L^{-1}) in the aqueous phase. TEM images confirmed the formation of heterostructure between the C$_3$N$_4$ nanosheets and Sm$_6$WO$_{12}$ nanorods. The Sm$_6$WO$_{12}$/g-C$_3$N$_4$ nanocomposite heterojunction catalyst shows higher photocatalytic efficiency towards LEV degradation (90.8%, after 70 min), which was 20.52 and 2.93 times higher than individual Sm$_6$WO$_{12}$ and g-C$_3$N$_4$, respectively. Based on the radical trapping experiments analysis, $\cdot$O$_2^-$ and $\cdot$OH were the main reactive species for the degradation of LEV. Through CG-MS, the authors identified that the intermediate steps of levofloxacin included the decarboxylation reaction of the methyl- morpholine group followed by the degradation of the N-methyl piperazine ring, de-alkylation defluorination and hydroxylation. The degradation might be continued with the breakdown of intermediates to small molecule organic acids as well as pollutant mineralization.

Several other compounds were used for the removal of levofloxacin under solar irradiation, reaching different percentages of degradation under different conditions [60–63]. For example, Zhang et al. [62] reported the removal of 89.2% of the LEV in water within 75 min under simulated sunlight using the $Bi_2O_3/P-C_3N_4$ composite. Further, through radicals capture, electron spin resonance (ESR) and the density functional theory (DFT) experiments verified the mechanism of the heterojunction degradation of LEV and revealed that holes and superoxide free radicals are the main active substances in the degradation of LEV. Finally, eleven intermediate products were identified and four possible degradation pathways were proposed. According to the authors, the entire degradation process mainly occurs in the quinolone and piperazinyl groups. The first pathway is the dehydrogenation of LEV followed by the piperazine oxidation and ring opening and then the oxidation for continuous CO removal. The third degradation pathway included the hydroxylation of LEV, oxidative decarboxylation and, furthermore, the demethylation of the methyl group on the piperazine ring. The fourth degradation path is the decarboxylation of LEV which is further oxidatively demethylated. The further degradation mechanism may be mainly manifested as the further opening of the quinolone group and oxazine ring (the degradation produces small molecules of amine compounds and carboxylic acid compounds). Finally, H_2O, CO_2 and other small molecules are generated.

Al Balushi et al. [60] achieved a 70% degradation rate of LEV (5 mg L^{-1}) after 240 min of treatment using visible light from LEDs (455 nm). The generation of hydroxy radicals was attributed as the main driver of pharmaceutical photodegradation. Upon illumination with visible light, electrons are excited from the valence band to the conduction band of the CdS microspheres. The holes created on the valence band migrate to the surface and scavenge water molecules to generate OH· radicals. The highly oxidizing radicals can degrade the pharmaceuticals. The electrons on the other hand may scavenge adsorbed molecular oxygen to form superoxide radicals ($O_2·$) that oxides water to form OH· radicals that can perform the pharmaceutical photodegradation.

While Lu et al. [61] employing $CeVO_4$-$BiVO_4$ nanocomposites, removed 95.7% of LEV after 300 min using 50 mg L^{-1} of the pollutant. According to the authors, the higher photodegradation activity for LEV degradation for the $CeVO_4$-$BiVO_4$ heterojunction, in comparison to bare photocatalyst material, is a result of the enhancement of separation and transfer efficiency of photogenerated electron–hole pairs. The trapping experiments and ESR tests identified that the contribution of active species was in the descending order of ·OH > h$^+$ > ·O_2^-. In all cases, the possible degradation of intermediate products and paths of LEV was similar to the one reported by Prabavathi et al. [59].

Fluoxetine (FLU) is another fluorinated-based antidepressant (trade name Prozac$^®$) that gained considerable attention since it was detected in surface waters and has the potential to be bioaccumulated [63]. In regards to this, several researchers investigated the removal of FLU by UV and other photo-assisted AOPs with different degradation rates [64–68]. For example, Pan and co-workers [64] investigated the toxicity of FLU and the products formed during UV photolysis by using zebrafish embryos (*Danio rerio*) as a model. The degradation rates of FLU for five days were approximately 63.6 ± 2.14%, 84.6 ± 0.99% and 97.5 ± 0.25% after 15, 30 and 60 min of UV irradiation, respectively. Using LC-MS measurements and density flooding theory (DFT) theoretical calculations, two possible degradation pathways for FLU were proposed: In pathway I, the 4O–12C in the structure of FLU was broken in the presence of UV irradiation. In pathway II, the fluorine atom passing through the parent molecule of FLU was defluorinated under UV irradiation and hydroxylated via hydroxyl substitution process. According to the results of the toxicity evaluation of the possible degradation intermediates, the byproducts retained certain biological toxicity. The authors highlighted the need to consider the toxicity of mixtures and the formation and persistence of toxicologically relevant degradation products when assessing environmental risk.

Hollman, Dominic and Achari [65] presented an evaluation of the UV/PAA (peracetic acid) process for the degradation of four pharmaceuticals venlafaxine (VEN), sulfamethoxazole (SFX), fluoxetine (FLU) and carbamazepine (CBZ) with comparison to UV/H_2O_2 process. All pharmaceuticals tested in this study were degraded below detection limits by

UV/PAA (λ = 254 nm), following pseudo-first-order kinetics. Increasing PAA and H_2O_2 dosage (between 5 and 100 mg L^{-1}) or UVC intensity (between 650 and 3500 W m^{-3}) resulted in a linear increase in pseudo-first-order rate coefficients. UV/H_2O_2 was found to be more efficient than UV/PAA for the degradation of FLU, VEN and CBZ. While UV/PAA was more effective in SFX degradation. Mass spectrum analysis revealed that the FLU mineralization pathway included hydroxylated FLU formed as a result of primary attack by ·OH, which was also reported in previous studies on the degradation of FLU using UV/H_2O_2 and UV/TiO_2 [66]. The primary attack of the ·OH on FLU could possibly target any of the open sites on both the benzene rings to form numerous monohydrated isomers, and the possibility of substitution of trifluoromethyl group also exists [66].

The use of semiconductor photocatalysis is another strategy proposed to improve the fluoxetine degradation. In this context, Sharma et al. [67] reported the construction of the hybrid heterojunction of Fe_3O_4-$BiVO_4$/$Cr_2V_4O_{13}$ (FBC) for visible and solar photo-degradation of FLU. Within 60 min of visible exposure, 99.2% of FLU was removed at pH 7. The high total organic carbon removal of 80.3% and 61.4% by FBC under visible and solar light confirmed the mineralization after 180 min of treatment. Scavenging experiments and the electron spin resonance (ESR) probe suggest ·OH and $·O_2^-$ radicals as the dominant species.

More recently, Norouzi and co-workers [68] studied FLU removal via anodic oxidation, employing different anodes Ti/RuO_2, Ti/RuO_2-IrO_2 and Ti/RuO_2-IrO_2-SnO_2 and graphite and carbon nanotubes (CNTs) as cathodes. The effect of current intensity (from 100 to 50 mA), initial pH (2, 4, 6, 8 and 10), initial FLU concentration (from 50 to 25 mg L^{-1}) and process time (40, 80, 120, 160 and 200 min) on FLU removal efficiency was investigated. Previous, all electrode combinations used (anode + cathode) were tested under the same conditions (current intensity = 300 mA, $[FLU]_0$ = 20 mg L^{-1}, time = 160 min and pH 6) to determine the optimal electrode. The Ti/RuO_2–IrO_2–SnO_2 was chosen as the optimal anode, and the CNTs were selected as the optimal cathode in the AO process for FLU removal. The results showed that at current intensity, pH, initial FLU concentration and process time of 500 mA, 6, 25 mg L^{-1} and 160 min, respectively, maximum FLU removal efficiency was observed, which was 96.25%. The TOC results demonstrated that about 81.51% of the mineralization of the FLU was achieved after 6 h under optimal experimental conditions. GC-MS results also showed that no toxic intermediates formed.

From Table 3, it is possible to state that the degradation of fluorinated-based compounds was successfully degraded via photo-combined AOPs systems. Different approaches were utilized, such as UV/H_2O_2, UV/O_3 and semiconductor materials. Despite the low rate of the removal and mineralization of the compounds using only UV radiation, the photocoupled systems showed high rates of removal and mineralization.

Table 3. Selected recent studies of photo-assisted AOPs processes for fluorinated and sulfurized base compounds removal.

Pharmaceutical Compound	Process	Experimental Conditions	Results	Ref.
NOR, CIP and LEV	VUV/UV and UV irradiation	pH = 7 Flow rate: 60 L h^{-1} [NOR] = [CIP] = [LEV] = 15 μmol L^{-1} T = 45 °C	The compounds were degraded more efficiently with high temperature and low initial concentration. The degradation pathways of FQs were defluorination, decarboxylation and piperazine ring oxidation, and demethylation was a particular degradation pathway for LEV.	[40]
CIP	UV, H_2O_2, UV/H_2O_2, modified Fenton and modified photo-Fenton	pH = 7 [CIP] = 10 mg L^{-1} $[H_2O_2]$ = 100 mmol L^{-1} [nZVI] = 5 mmol L^{-1}	Only 4% of TOC removal for UV radiation TOC removal % increased from 10.47% for only H_2O_2 to 35.41 % for UV/H_2O_2. $nZVI/H_2O_2$ removed 100% of CIP in 30 min and 60% of the initial TOC.	[47]

Table 3. *Cont.*

Pharmaceutical Compound	Process	Experimental Conditions	Results	Ref.
OFLO and LEV	UV/H_2O_2 and UV/PS	pH = 3 [LEV] = [OFLO] = 5 mg L^{-1} t = 30 min [H_2O_2] = [PS] = 150 μmol·L^{-1}	The removal efficiencies of OFLO and LEV were 96.40% and 99.23% with UV/PS. The TOC removal efficiencies of OFLO and LEV were 36.24% and 38.36%, respectively, for the UV/H_2O_2 system. By contrast, the UV/PS system exhibited mineralization of OFL and LEV, with TOC removal efficiency about 46.43% and 49.74%, respectively.	[48]
NOR	UV, UV/H_2O_2, UV/Fe^{2+}/H_2O_2, VUV, UV/Fe^{2+}, VUV/H_2O_2, VUV/Fe^{2+}/H_2O_2	pH = 7 [NOR] = 45 μmol L^{-1} [Fe^{2+}] = 90 μmol L^{-1} [H_2O_2] = 3 mmol L^{-1}	VUV/Fe^{2+}/H_2O_2 process removed 100% of NOR after 4 min, and high mineralization (63.3% at 8 min) and achieved rapid removal of NOR in real waters at neutral pH.	[49]
CIP	UV/O_3	pH = 5.6 Gas flow: 0.5 L min^{-1} [CIP] = 100.0 mg L^{-1} [O_3] = 30.0 ± 2.0 mg L^{-1} min^{-1}	Compared with single O_3 (37.5%), the DOC reduction was prompted significantly by the introduction of UV irradiation (91.4%) within 40 min. The toxicity of products decreased significantly as compared with parent pollutant.	[51]
38 PPCPs, highlighting CIP and LEV	UV/O_3	Ozone dosage (1–4 and 6 mg L^{-1}) and hydraulic retention time (5 and 10 min)	CIP was degraded below the limit of detection (LOD) at the ozone dose of 1 mg L^{-1} in 5 min. However, LEV required an O_3 dose of 6 mg L^{-1} to be degraded below LOD.	[52]
CIP	Photolysis	pH = 7 catalyst dosage = 50 mg [CIP] = 5 mg L^{-1}	Photolysis showed a small removal of CIP (4.63%) over 120 min of exposure time under UV-light radiation while the removal of CIP for Na-TiNT, SnO_2 and TiO_2/SnO_2 nanocomposite were 41.55%, 45.83% and 92.8%, respectively.	[55]
LEV	Photocatalysis	[LEV] = 10 mg L^{-1} catalyst dosage = 50 mg	The Sm_6WO_{12}/g-C_3N_4 nanocomposite heterojunction catalyst shows higher photocatalytic efficiency towards LEV degradation (90.8%, after 70 min).	[59]
FLU	UV photolysis	pH = 7 T = 20 °C [FLU] = 1 mg L^{-1} Irradiation source = 300 W	The degradation rates of FLU for five days were approximately 63.6% ± 2.14%, 84.6% ± 0.99%, and 97.5% ± 0.25% after 15, 30 and 60 min of UV irradiation, respectively.	[64]
VEN, SFX, FLU and CBZ	UV/PAA and UV/H_2O_2	pH = 7 UVC intensity = 3.50 kW m^{-3} [PAA] = [H_2O_2] = 50 mg L^{-1} [compound] = 10 mg L^{-1}	UV/H_2O_2 was found to be more efficient than UV/PAA for the degradation of FLU, VEN and CBZ. While UV/PAA was more effective in SFX degradation.	[65]
FLU	Photocatalysis	pH = 7 [FLU] = 10 mg L^{-1} Light source = 280 mW cm^{-1} catalyst dosage = 0.3 mg mL^{-1}	Within 60 min of visible exposure, 99.2% of FLU was removed at pH 7. The high total organic carbon removal of 80.3% and 61.4% by FBC under visible and solar light confirmed the mineralization after 180 min of treatment.	[67]
FLU	Ti/RuO_2, Ti/RuO_2-IrO_2 and Ti/RuO_2-IrO_2-SnO_2	pH = 6 i = 500 mA [FLU] = 25 mg L^{-1} t = 160 min	At optimized conditions, maximum FLU removal efficiency was observed, which was 96.25%. The TOC results showed that about 81.51% of the mineralization of the FLU was achieved after 6 h under optimal experimental conditions.	[68]

Table 3. *Cont.*

Pharmaceutical Compound	Process	Experimental Conditions	Results	Ref.
SMZ, SDZ and SML	UV, UV/H_2O_2 and UV/$K_2S_2O_8$	pH = 7 [SMZ]=[SDZ]=[SML] 15 mg L^{-1} [H_2O_2] = [$K_2S_2O_8$] = 4.4×10^{-4} M	UVC proved inadequate in removing these compounds. A concentration of 4.4×10^{-4} mol L^{-1} of H_2O_2 or $K_2S_2O_8$ increases SMZ degradation up to 100%. UVC/PDS requires less energy than UVC and UVC/H_2O_2 system.	[69]
SMX	UVC/VUV	pH = 7 [SMX] = 100 mg L^{-1}, Fluence = 54.9 mJ cm^{-2}	Total SMX removal was reached at a fluence of 54.9 mJ cm^{-2}, while the TOC removal of 98.5% was attained at fluence of 109.8 mJ cm^{-2}.	[70]
SMT	photo-Fenton	pH = 4 [SMT] = 1.8×10^{-2} mmol L^{-1} [H_2O_2] = 0.74 mmol L^{-1} [Fe^{3+}] = 0.25 mmol L^{-1}	The mineralization of SMT were significantly enhanced in the VUV/UV photo-Fenton process as compared to the UV and UV photo-Fenton processes after 60 min of treatment, achieving ~60% of TOC removal.	[71]
DF, SP, SMX and SIM	photo-Fenton	[H_2O_2] = 100 mmol L^{-1} [pharmaceuticals] = 20 mg L^{-1}, Fe^{2+} dosage = 1/200 mol L^{-1} of $FeSO_4$ $7H_2O$ and H_2O_2 UV intensity = 75 mW cm^{-2}	In ultrapure water, all the four pharmaceuticals were degraded by more than 95% within 4 min, the same removal rate expended about 30–60 min. Except for DF, the cytotoxicity increase during the degradation process for SP, SMX and SIM.	[72]
SMZ	Photoelectrochemical	[SMZ] = 10 mg L^{-1} pH = 3.5 Potential = 2 V	The removal efficiency of SMZ by photoelectrochemical process was 81.3%, which was approximately twice the sum of both electrochemical and photochemical processes, and over 40% of TOC was eliminated after 180 min.	[73]
SD	Photocatalysis	[SD] = 10 mg L^{-1} pH = 8	Nearly 70% of SD was degraded by the Ag_3PO_4/MoS_2/TiO_2 NTAs in 240 min, which was higher than that of MoS_2/TiO_2 (35%) and Ag_3PO_4/TiO_2 NTAs (44%). In addition, the percentage of SD removal over direct photolysis, electrochemical and photochemical was only 16, 20 and 31%, respectively.	[74]
SP	US-CN	pH = 7.5 [SP] = 0.03 mmol L^{-1} catalyst dosage = 32 mg UV intensity = 100 mW cm^{-2}	The time-dependent removal efficiency of SP over different carbon nitride samples was 90.55%, 50.77% and 26.19% of SP by US-CN, S-CN and CN after 100 min of photocatalytic reaction. The removal rates of SP at 100 min in river water (Shanghai), pharmaceutical water (Guangzhou), domestic wastewater (Guangzhou) and tap water (laboratory) reached up to 84.74, 75.66, 82.06 and 85.26%, respectively.	[75]
AMX	UV/H_2O_2 and UV/persulfate	[AMX] = 20 µmol L^{-1} [$H2O2$] = [PS] = 500 µmol L^{-1}	The direct UV photolysis system alone showed an insignificant AMX degradation. However, the addition of H_2O_2 or PS increases the degradation efficiency of AMX significantly. Despite the high percentage of AMX removal through UV/AOPs, the mineralization of AMX was insignificant. After 30 min of treatment, only 15.2% and 28.7% of TOC were removed in the UV/H_2O_2 and UV/persulfate systems, respectively.	[76]

Table 3. *Cont.*

Pharmaceutical Compound	Process	Experimental Conditions	Results	Ref.
AMX	Fenton process, photo-Fenton, solar photo-Fenton, sono-Fenton, and sono-photo-Fenton	pH = 3 [AMX] = 10 mg L^{-1} [FeSO$_4$] = 3.0 mg L^{-1} [H$_2$O$_2$] = 375 mg L^{-1} Ultrasound = 40 kHz, Light source = UV tubes 365 nm	Under the optimized conditions, Fenton's process was able to remove 100% of AMX within 12 min of reaction time. Coupling the Fenton process with UV-light illumination, solar light illumination and UV light–ultrasound treatment allowed complete antibiotic removal in 3.5, 9 and 6 min, respectively.	[77]
AMX	solar photo-Fenton	[Fe^{3+}] = 3 mg L^{-1} [H$_2$O$_2$] = 2.75 mg L^{-1} [AMX] = 1 mg L^{-1}	A total of 94 and 66% of AMX was removed after 60 and 210 min of treatment in simulated and real wastewater, respectively. In addition, the percentage of TOC removal for AMX was 19.5% in simulated wastewater. In the study carried out with real effluent, the removal rate was 6.5%.	[47]
AMX	Photolysis	[AMX] = 20 mg L^{-1} pH = 7 catalyst dosage = 0.5 g L^{-1}	Photolysis under the simulated sunlight was ineffective in the decomposition of AMX, while pure V$_2$O$_5$ and C$_3$N$_4$ reached 33.2% and 52.7% of AMX removal under 120 min illumination. The V$_2$O$_5$/C$_3$N$_4$ (1 wt% V$_2$O$_5$) nanocomposite increased AMX removal to 91.3%. Moreover, 1-V$_2$O$_5$/C$_3$N$_4$ nanocomposite attained 76.2% of TOC removal under the same conditions.	[48]
CP	solar photoelectro-Fenton	[CP] = 0.230 mmol L^{-1} pH = 3 j = 50 mA cm^{-2} [Fe^{2+}] = 0.50 mmol L^{-1}	At the best conditions, the degradation of CP by SPEF reached 100% removal in only 15 min, while using EF, a complete removal was achieved in about 20 min. The AO-H$_2$O$_2$ process was capable of removing only 36% in 30 min of treatment.	[51]

Abbreviations: CP: captopril; AMX: amoxilin; SMZ: sulfamethazine; SDZ: sulfadiazine; SML: sulfamethizole; DF: diclofenac; SP: sulpiride; SMX: sulfamethoxazole; SIM: sulfisomidine; SD: sulfadiazine; VEN: venlafaxine; SFX: sulfamethoxazole; FLU: fluoxetine; CBZ: carbamazepine.

However, most of these systems still do not use real samples or large-scale flow reactors.

3.4. Sulfurized-Based Compounds

Sulfonamides (SA) are widely used as antibiotics in human and veterinary medicine due to their high antimicrobial activity, stable chemical properties and low costs [69,78]. SA has been found in surface waters at concentrations of 148–2978 ng L^{-1} [69]. Therefore, Moradi and Moussavi [70] investigated the degradation, mineralization and mechanism of sulfamethoxazole (SMX) oxidation in a UVC/VUV reactor. Total SMX (100 mg L^{-1}) removal was reached in pH 7 and at a fluence of 54.9 mJ cm^{-2}, while the TOC removal of 98.5% was obtained at fluence of 109.8 mJ cm^{-2}. In addition, mineralization was also evaluated by measuring the concentrations of nitrate, ammonium and sulfate ions. The nitrogen and sulfur releases of 83.44% and 96.58%, respectively, during the SMX degradation confirmed high mineralization. Scavenging tests using alcoholic radicals scavengers and salicylic acid proved that ·OH was the dominant radicals specie involved in the degradation of SMX.

Acosta-Rangel et al. [69] conducted an interesting study on the degradation of three SAs, sulfamethazine (SMZ), sulfadiazine (SDZ) and sulfamethizole (SML), by UV, UV/H$_2$O$_2$ and UV/K$_2$S$_2$O$_8$. Based on the quantification of the UVC radiation, the low values of quantum yield at 60 min of treatment were observed. According to the authors, the UVC dose commonly applied for water disinfection (400 J m^{-2}) in treatment plants proved inadequate to remove these compounds, requiring higher UV radiation doses or longer exposure times. Directing the UVC photolysis of SAs is influenced by their initial concentration (5, 10 and 15 mg L^{-1}), the degradation rates were higher at pH 12 due to the SAs are in their anionic form for SMZ and SML (Figure 6a). A concentration of 4.4 × 10^{-4} mol L^{-1} of H$_2$O$_2$ or

$K_2S_2O_8$ increases SMZ degradation by up to 100%. UVC/PDS costs less energy than UVC and UVC/H_2O_2.

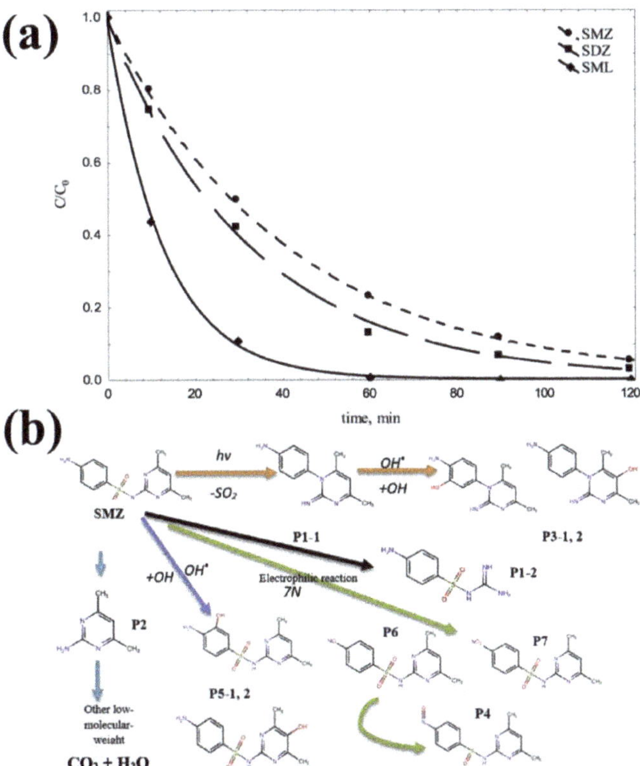

Figure 6. (**a**) Degradation kinetics of the three SAs by UVC photolysis. $[SAs]_0 = 15$ mg L^{-1}, pH = 7 and T = 298 K. The lines represent the prediction of the first-order kinetic model. (**b**) Mechanism of degradation by direct SMZ photolysis in the presence of UVC radiation. Reprinted with permission from [69]. Copyright © 2018 Elsevier Ltd.

According to Figure 6b, SO_2 removal is first produced by direct SMZ photolysis through UVC radiation, obtaining by-product P1. Next, given the oxidation of OH· radical during the process, hydroxylation is expected to be a common reaction responsible for SMZ degradation, generating by-products P3-1, P3-2, P5-1 or P5-2. A break in the $-SO_2-$ and $-NH-$ bond allows the identification of the by-product P2. By attacking nucleophilic and incision in the group $-SO_2-$, the byproduct P4 may be formed. While the adduction of seven N atoms, forming the nitroso- and nitro-substitutional SMZ through the electrophilic reaction was the dominant pathway by PDS (P6 and P7). Moreover, cytotoxicity tests revealed that the by-products formed were less toxic than the original products.

Similar behavior was observed by Lin and Wu [79] and Wen et al. [71] when investigating the degradation of sulfamethazine (SMT) in an aqueous solution by the UV/H_2O_2 and VUV/UV photo-Fenton processes, respectively. Li and Wu [79] reported that SMT (10 mg L^{-1}) was 100% degraded in half the time (15 min) after adding 10 mmol L^{-1} of H_2O_2, while the UV process degraded 79% of the SMT after 30 min. Wen et al. [71] evaluated the mineralization of SMT ($[SMT]_0 = 1.8 \times 10^{-2}$ mmol L^{-1}, $[H_2O_2]_0 = 0.74$ mmol L^{-1}, $[Fe^{3+}]_0 = 0.25$ mmol L^{-1}, $pH_0 = 4.0$) and found that it was significantly enhanced in the VUV/UV photo-Fenton process as compared to the UV and UV photo-Fenton processes after 60 min of treatment, achieving ~60% of TOC removal. The authors also reported the effect of the initial concentration of SMN on the reaction mechanism. At low concentrations of SMN ($1.8 \ \mu mol \ L^{-1}$), indirect oxidation is primarily responsible for the degradation of SMN; while at higher concentrations ($90 \ \mu mol \ L^{-1}$), both photolysis and indirect oxidation contributed to the degradation of the compound.

More recently, Hong, Wang and Lu [72] reported the degradation of a complex matrix of four common refractory pharmaceuticals, diclofenac (DF), sulpiride (SP), sulfamethoxazole (SMX) and sulfisomidine (SIM) present in a disc tubular reverse osmosis (DTRO) concentrator from the local landfill by UV-Fenton system. In the ultrapure water, all four pharmaceuticals were degraded by more than 95% within 4 min, while in the DTRO concentrates, the same removal rate expended by about $30-60$ min under the same dosage of H_2O_2 (100 times of the [pharmaceuticals]$_0$ = 20 mg L^{-1}), catalyst ($1/200$ mol L^{-1} of $FeSO_4 \cdot 7H_2O$ and H_2O_2) and UV intensity ($\lambda = 253.7, 75$ mW cm^{-2}). A total of 49 transformation products (TP) were identified by HPLC-MS, and 22 new TPs were first found and presented. Toxicity evolutions on HepG2 cells during UV–Fenton treatment revealed that, except for DF, the cytotoxicity increased during the degradation process for SP, SMX and SIM.

The degradation of sulfonamides was also performed by employing photoelectrochemical semiconductors. Jia and co-workers [73] obtained visible light-driven semiconductor-metal organic frameworks (MOFs), which were constructed by electro-anodization and the deposition growth method. ZIF-8 (zeolitic imidazolate framework) nanoparticles were deposited on the hollow TiO_2 nanotubes and N and F were added as co-doping and the electrode as nominated ZIF-8/NF-TiO_2. The removal efficiency of SMZ (10 mg L^{-1}) through the photoelectrochemical process was 81.3%, which was approximately twice the sum of both electrochemical and photochemical processes, and over 40% of TOC was eliminated after 180 min, suggesting that SMZ could be partially mineralized by ZIF-8/NF-TiO_2 under visible light irradiation. The degradation pathway of SMZ was divided into three steps: cleaving, aromatic ring opening and mineralizing.

In another study, Teng et al. [74] reported the decomposition of sulfadiazine (SD) by $Ag_3PO_4/MoS_2/TiO_2$ nanotube array (NTAs) electrode, under visible light excitation. TiO_2 NTAs were modified with MoS_2 nanosheets and Ag_3PO_4 nanoparticles through photo-assisted electrochemical deposition and chemical immersion methods. Nearly 70% of SD was degraded by the $Ag_3PO_4/MoS_2/TiO_2$ NTAs in 240 min, which was higher than that of MoS_2/TiO_2 (35%) and Ag_3PO_4/TiO_2 NTAs (44%). In addition, the percentage of SD removal over direct photolysis, electrochemical and photochemical was only 16, 20 and 31%, respectively. The results demonstrating that the Ag_3PO_4 nanoparticles increased the visible light absorption and MoS_2 nanosheet promoted the separation of photogenerated charges effectively.

Ultrathin S-doped graphitic carbon nitride nanosheets (US-CN) were synthesized and its SP removal efficiency was evaluated under various conditions via the visible-light-assisted peroxydisulfate (PDS-VL) activation method by She et al. [75]. The time-dependent removal efficiency of SP over different carbon nitride samples was 90.55%, 50.77% and 26.19% of SP by US-CN, S-CN and CN, respectively, after 100 min of photocatalytic reaction. The $^1O_2$ generated from US-CN/PDS-VL system was the major reactive oxidation species (ROS) for SP degradation. Three possible degradation pathways were proposed, combining theoretical studies and LC-MS. Pathway I included the attack on 9N atom with the C-N bond cleavage, followed by consecutives C-C bond cleavage. In pathway II, the C-N bond cleavage occurred through the attack on 12N atom, followed by SO_2 extrusion. While the attack on nine N atoms and SO_2 extrusion was ascribed as pathway III.

Later, the US-CN/PDS-VL system in SP degradation was applied in actual wastewater. The removal rates of SP at 100 min in river water, pharmaceutical water, domestic wastewater and tap water reached up to 84.74, 75.66, 82,06 and 85.26%, respectively, indicating a slight decrease compared with that in ultrapure water matrix, demonstrating the potential application of this system [75].

Amoxicillin (AMX) belongs to the β-lactams family and is often used for the treatment of bacterial infections, such as pneumonia and urinary tract infections [43]. It has been shown that over 80% of AMX is excreted through urine from the human body after 2 h of ingestion. This relatively fast egestion coupled with its extensive use has been attributed to AMX being one of the most widely reported antibiotics in wastewater, generating antibiotic-resistant bacteria and requires effective treatment methods [80–82].

Taking this into account, Zhang et al. [76] compared the reaction kinetics, degradation pathways and antibacterial activity of AMX in the UV/H_2O_2 and UV/persulfate systems. Direct UV (λ = 254 nm) photolysis system alone showed an insignificant AMX degradation (20 μmol L^{-1}), which was attributed to its low quantum yield of 9.74×10^{-3} mol E^{-1}. However, the addition of H_2O_2 or PS (500 μmol L^{-1}) increases the degradation efficiency of AMX significantly due to the generation of HO$\cdot$ and $SO_4^{-}\cdot$. Despite the high percentage of AMX removal through UV/AOPs, the mineralization of AMX was insignificant. After 30 min of treatment, only 15.2% and 28.7% of TOC were removed in the UV/H_2O_2 and UV/persulfate systems, respectively.

The by-products identified via LC/MSMS indicate three possible degradation pathways (Figure 7a), including hydroxylation, hydrolysis and decarboxylation. The intermediate with m/z of 382 indicates the mono-hydroxylation of AMX. HO$\cdot$ and $SO_4^{-}\cdot$ are expected to attack the sulfur atom on the thioether groups to yield sulfur centered radical cation by electron transfer. The radicals can be further deprotonated to generate α-thioether radicals, followed by oxidation to yield sulfoxide TP1, while HO$\cdot$ can attack the aromatic ring of the side chain to produce a hydroxyl radical, which can be further oxidized to form monohydroxy AMX TP2. Moreover, TP1 and TP2 can be hydroxylated to form the monohydroxy AMX sulfoxide TP3. The penicilloic acid, obtained by the hydrolysis by-product TP4, is formed by opening the strained four-membered beta-lactam ring of AMX. The decarboxylation product TP5 relates to the penicilloic acid derivative, generated by the subsequent release of the carboxyl group from the penicilloic acid. TP4 can also be hydroxylated at the sulfur atom and aromatic ring to generate TP6 and TP7, respectively. The products of AMX after UV/AOPs treatment possess significantly lower antibacterial activity. In addition, UV/H_2O_2 was more cost-effective than UV/PS process in the degrading AMX, according to the economic evaluation results [76].

The degradation of AMX has also been investigated using Fenton and coupled-Fenton processes. Verma and Haritash [83] studied the removal of amoxicillin using the Fenton process and hybrid Fenton-like processes, such as photo-Fenton, solar photo-Fenton, sono-Fenton and sono-photo-Fenton. Effects such as ferrous ions (Fe^{2+}), H_2O_2 and pH were evaluated. Under the optimized conditions of $[Fe^{2+}]$ = 30 mg L^{-1}, $[H_2O_2]$ = 375 mg L^{-1} and pH = 3, the Fenton process was able to remove 100% of AMX (10 mg L^{-1}) within 12 min of reaction time. Coupling the Fenton process with UV-light illumination, solar light illumination and UV light–ultrasound treatment allowed the complete antibiotic removal in 3.5, 9 and 6 min, respectively (Figure 7b). The authors concluded that the Fenton process coupled with other UV/solar light was more efficient than the stand-alone Fenton process for the degradation of AMX.

In another study, Guerra et al. [85] presented the results of the solar photo-Fenton oxidation of paracetamol (PCT) and amoxicillin in two aqueous matrices, a synthetic wastewater and real wastewater from El Ejido wastewater treatment plant effluent (Almeria). $Fe_2(SO_4)_3$ was used as the source of iron and ethylenediamine disuccinic acid (EDDS) as the iron complexing agent, employing different doses of H_2O_2. In all cases, the process was operated under conditions of natural sunlight. Amounts of 94 and 66% of AMX were removed after 60 and 210 min of treatment in simulated and real wastewater, respectively, using the optimized parameters: 3 mg L^{-1} Fe^{3+} and 2.75 mg L^{-1} H_2O_2. In addition, the percentage of TOC removal for AMX was 19.5% in simulated wastewater. In the study carried out with real effluent at concentrations of 2.75 mg L^{-1} H_2O_2, the removal rate was 6.5%, while at an H_2O_2 concentration of 5.0 mg L^{-1}, the removal rate increased to 20%. The intermediates identified suggest the hydroxylation of the aromatic ring and the opening of the four-membered β-lactam ring and subsequent formation of amoxilloic acid and amoxicilloic acid as the main transformation pathways.

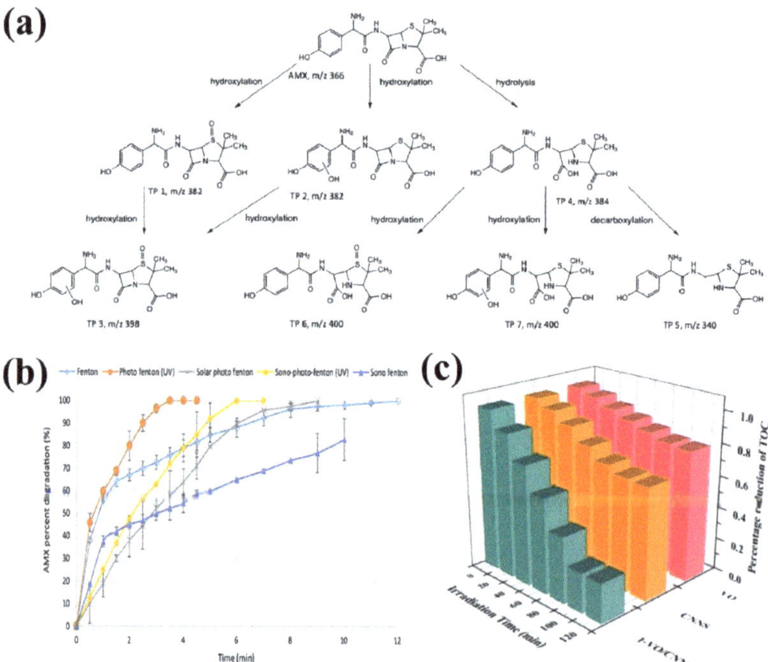

Figure 7. (**a**) Proposed pathway of AMX degradation. Reprinted with permission from [73]. Copyright © 2019 Elsevier Ltd. (**b**) Degradation of AMX at Comparison of Fenton, photo-Fenton, solar photo-Fenton, sono-Fenton and sono-photo-Fenton for complete degradation of AMX (AMX: 10 mg L^{-1}; FeSO$_4$: 3.0 mg L^{-1}; H$_2$O$_2$: 375 mg L^{-1}, pH: 3, Ultrasound: 40 kHz and light source: UV tubes (365 nm)). Reprinted with permission from [83]. Copyright © 2019 Elsevier Ltd. (**c**) TOC of AMX solution degraded under different irradiation time by as-prepared VO, CNNS and 1-VO/CNNS. Reprinted with permission from [84]. Copyright © 2022 Elsevier Ltd.

Recently, Le and co-workers [85] reported an innovative heterostructure V$_2$O$_5$/C$_3$N$_4$ nanosheets (NS) photocatalyst to degrade AMX under solar light. Photolysis under the simulated sunlight was ineffective in the decomposition of AMX, while pure V$_2$O$_5$ and C$_3$N$_4$ reached 33.2% and 52.7% of AMX removal under 120 min of illumination. The 1-V$_2$O$_5$/C$_3$N$_4$-NS (1 wt% V$_2$O$_5$) nanocomposite increased AMX removal to 91.3%. This improvement was attributed to the enlarged specific surface area, increased active sites and promoted the separation of photoinduced electron–hole pairs by the S-scheme heterojunctions. Moreover, 1-V$_2$O$_5$/C$_3$N$_4$-NS nanocomposite attained 76.2% of TOC removal under the same conditions (Figure 7c), showing the highest oxidation capacity. Furthermore, the authors confirmed that O$_2$$^{·-}$ and h$^+$ radicals are the foremost reactive species in photodegradation. Mmelesi et al. [86] also presented a high AMX removal percentage (89%) under visible light irradiation using Zn$_x$Co$_{1-x}$Fe$_2$O$_4$ (x = 0.0, 0.1, 0.2, 0.3, 0.4 and 0.5) nanoparticles. The catalysts were synthesized by a simple co-precipitation method. The Zn-doped NPs (x = 0.4) showed the highest degradation efficiency of 89% and TOC removal of 66.59% after 100 min.

Captopril is used as an antihypertensive in the treatment of heart failure, in cases of myocardial infarction, in diabetic nephropathy and as an angiotensin converting enzyme (ACE) inhibitor [87]. It is estimated that about 40 to 50% of the active ingredient is excreted from the human body and remains unchanged, while the rest is excreted as metabolites, which can be extremely dangerous for human health and must be evaluated against the contamination of surface waters due to the contamination of surface water due to incorrect disposal [88].

Dos Santos et al. [88] evaluated the electrochemical destruction of captopril in different aqueous matrices through the solar photo-electro-Fenton (SPEF) method using a solar pre-pilot flow plant with a Pt/air-diffusion cells and a planar photoreactor. The effect of j and drug content on the SPEF performance was studied, and comparative anodic oxidation (AO)-H_2O_2 and electro-Fenton (EF) assays were performed in order to confirm the superiority of SPEF process. At the best conditions (0.230 mmol L^{-1} of captopril, pH 3, j = 50 mA cm^{-2} and 0.50 mmol L^{-1} Fe^{2+}), the degradation of captopril by SPEF reached 100% of removal in only 15 min, while EF achieved complete drug removal in about 20 min. The AO-H_2O_2 process was capable of removing only 36% in 30 min of treatment. TOC was only reduced by 25% in SPEF after 300 min, suggesting that the resulting by-products were mainly oxidized by homogeneous ·OH, but contained very small amounts of Fe^{3+} complexes. The mineralization in urban wastewater (28%) was slightly accelerated as compared to those in sulfate medium because of the parallel oxidation with active chlorine. The drug removal was also feasible in urine, showing a larger mineralization of 70% [88].

Table 2 shows the main experimental conditions and results of the degradation and mineralization efficiency of the sulfurized-based drugs. In all studies presented, high degradation rates were achieved. In addition, UV alone proved ineffective in the removal of these compounds. However, it is worth noting that the authors used lamps to simulate radiation rather than sunlight to simulate light.

3.5. Phosphorus-Based Compounds

Ifosfamide (IF) and cyclophosphamide (CP) are two of the most used alkylating agents for the treatment of different cancers and autoimmune diseases that act directly on DNA, inhibiting cell division and, consequently, cell death [86–88]. Pharmacokinetic studies showed that 61% and 10−20% of the administered doses of IF and CP, respectively, are excreted unchanged via urine and feces, causing the emission of these drugs into the wastewater stream [89]. The metabolites and transformation products (TPs) residues of IF and CP have been found in the aquatic environment in amounts ranging from a few ng L^{-1} to tens of mg L^{-1}, indicating that these drugs are not readily biodegradable, making these compounds difficult to remove from wastewater by traditional methods [90].

Thus, Russo et al. [90] reported the IF and CP degradation by UV irradiation (λ = 254 nm) and investigated their acute and chronic ecotoxicity of and their commercially available human metabolites/TPs on different organisms of the aquatic trophic chain. After 48 h of treatment, IF and CP (200 mg L^{-1}) were degraded only 36.5% and 28.3%, respectively. According to the authors, this poor removal was attributed to the chemical structures of IF and CP that do not contain double bonds that could absorb photons under UV irradiation [89]. Moreover, DOC was monitored in UV irradiated samples containing IF and CP at 10 mg L^{-1} and revealed no significant decrease in DOC after 48 h of treatment, suggesting that UV degradation was not able to mineralize both compounds. Regarding the toxicological tests, the UV-irradiated compounds showed an increase in toxicity. This effect was related to a mixture of different TPs formed during treatment.

In another study, Janssens et al. [91] coupled nanofiltration and UV, UV/TiO_2 and UV/H_2O_2 processes for the removal of anti-cancer drugs, which include CP and IF, from two different matrices: secondary wastewater effluent and nanofiltration concentrate. Direct photolysis was not able to remove IF and CP (500 µg L^{-1}) in secondary effluent after 180 min of treatment (Figure 8a), and the same behavior was reported in the work of Russo et al. [53]. The inefficiency of direct photolysis explains why IF and CP showed negligible removal through UV/TiO_2 (100 mg L^{-1}) and UV/H_2O_2 (40 mg L^{-1}): <25%. The same trend was observed during the degradation from nanofiltration retentate. CP and IF showed no degradation either through UV/TiO_2 or through UV/H_2O_2.

In an effort to increase the efficiency of the UVC/H_2O_2 system for the degradation of cyclophosphamide, Graumans et al. [92] coupled thermal plasma activation with UV-C/H_2O_2 treatment. Plasma-activated water (PAW) contains highly reactive oxygen and nitrogen species because of electric gas discharges in the air over water. The oxidative degradation of CP solutions (4 ng mL^{-1}) in tap water by PAW resulted in a complete degradation within 80 min at 150 W (Figure 8a). CP was also completely degraded within 60 min of applying the UVC/H_2O_2 (10 mg L^{-1}) system. In addition, LC-MS/MS detected the reaction products 4-keto-CP, 4-hydroperoxy-CP and carboxyphosphamide. Furthermore, the authors analyzed the implications of the toxicity of the products formed and highlighted the concern regarding the 4-Hydroperoxy-CP formation, a very potent toxic compound capable of alkylating DNA strands.

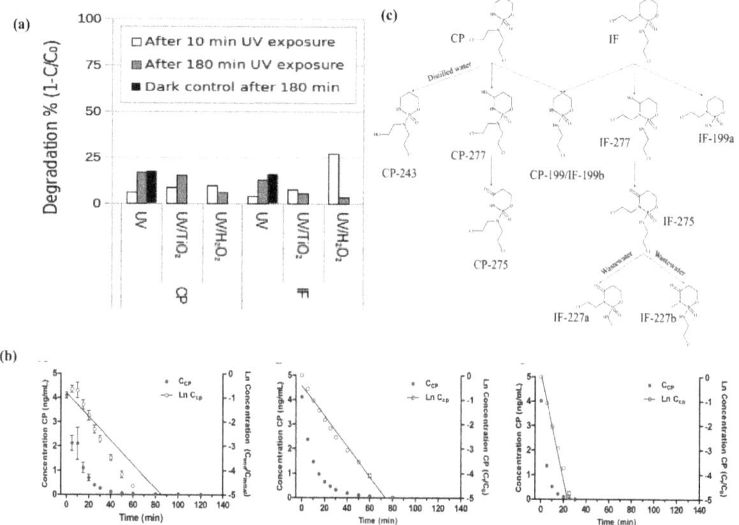

Figure 8. (**a**) Removal of cyclophosphamide (CP) and ifosfamide (IF) from a NF concentrate by UV, UV/TiO$_2$ (100 mg L^{-1}) and UV/H$_2$O$_2$ (40 mg L^{-1}). Reprinted with permission from [88]. Copyright © 2019 Elsevier Ltd. (**b**) Degradation rate of 4 ng mL^{-1} CP applied on TS1 in water at different hydrogen peroxide concentrations in combination with an UV-C source. With UV-C/H$_2$O$_2$ (970 mM) treatment, no complete CP removal was obtained within 120 min (left), in contrast with H$_2$O$_2$ concentrations 0.11 (center) or 0.22 mM (right), which showed complete and rapid CP removal rates. Reprinted with permission from [93]. Copyright © 2020 Elsevier Ltd. (**c**) Proposed photodegradation pathways for IF and CP. Reprinted with permission from [86]. Copyright © 2019 Elsevier Ltd.

Osawa et al. [89] investigated the photodegradation of CP and IF (10 mg L^{-1}) using ruthenium-doped titanate nanowires (Ru-TNW) in distilled water (DW) and wastewater (WW) from secondary wastewater treatment, under UV–Vis radiation. An improvement in the degradation of CP and If was observed using Ru-TNWas catalyst (20 mg), mainly when WW was used for solutions preparation. In addition, the results were better when the pollutants were used as single solutions. Using Ru-TNW as photocatalyst, the degradation of both compounds presented a higher removal rate independently on the matrix used and DW and WW from secondary treatment. CP and IF degradation experiments followed pseudo-first-order kinetics, and the degradation rate constant for IF was higher than the CP for both matrices.

Four CP transformation products (TPs) and six IF TPs from the photodegradation process were elucidated using high-resolution mass spectrometry (Figure 8c). The CP-277 was generated using the hydroxylation of CP while CP-275 was formed from the dehydrogenation of the hydroxy group connected with the heterocyclic ring from CP-277. According to the authors,

CP-243 was formed by replacing the chlorine atom with the hydroxy group from the CP molecule. Moreover, it was considered that CP-199 was formed by the loss of the chloroethane group from CP and along with IF-199a and IF-199b are isomers. Additionally, IF-277 degraded to IF-275 through the dehydrogenation of the hydroxy group. In the WW samples, we believe that IF227a and IF-227b were generated from IF-275. For both pollutants, in the WW, there was a higher production of TPs and two of them were detected only in this matrix, indicating that environmental matrices may produce different TPs. Finally, the ecotoxicity prediction showed that TPs had low toxic potential on aquatic organisms. However, most of the TPs resulted in positive mutagenicity [89].

According to Table 4, it was possible to state that direct photolysis was not able to remove IF and CP. However, photo-combined AOPs systems showed elevated removal rates of the compounds. In addition, further studies may focus on mineralization or the formation of less toxic compounds, the importance of degradation intermediates should not be underestimated since the by-products generated using the systems presented in this work showed a toxicity that was the same or sometimes greater than the drug of origin.

Table 4. Selected recent studies of photo-assisted AOPs processes for phosphorus-based and oxygenated compound removal.

Pharmaceutical	Process	Experimental Conditions	Results	Ref.
IF and CP	Ru-TNW	$[IF] = [CP] = 10$ mg L^{-1} catalyst dosage = 20 mg pH = 7	The degradation of both compounds presented a higher removal rate independently to the matrix used; DW and WW in secondary treatment. CP and IF degradation experiments followed the pseudo-first-order kinetics and the degradation rate constant for IF was higher than the CP for both matrices.	[89]
IF and CP	UV irradiation	$\lambda = 254$ nm $[IF] = [CP] = 10$ mg L^{-1} pH = 7	After 48 h of treatment, IF and CP were degraded only by 36.5% and 28.3%, respectively, containing IF and CP at 10 mg L^{-1} and revealed no significant decrease in DOC after 48 h of treatment, suggesting that UV degradation was not able to mineralize both compounds.	[90]
IF and CP	UV irradiation, UV/TiO$_2$ and UV/H$_2$O$_2$	$[IF] = [CP] = 500$ µg L^{-1} $[TiO_2] = 100$ mg L^{-1} $[H_2O_2] = 40$ mg L^{-1} $\lambda = 254$ nm pH = 7	Direct photolysis was not able to remove IF and CP in secondary effluent after 180 min of treatment. CP and IF did not degrade negligibly either through UV/TiO$_2$ or through UV/H$_2$O$_2$.	[91]
CP	UVC/H$_2$O$_2$/PAW	$\lambda = 254$ nm $[CP] = 4$ ng mL^{-1} $[H_2O_2] = 10$ mg L^{-1} pH = 7	The oxidative degradation of CP solutions in tap water by PAW resulted in a complete degradation within 80 min at 150 W. CP was also completely degraded within 60 min applying UVC/H$_2$O$_2$ system.	[92]
IBP, NPX and MO	MIL-53(Al)@TiO$_2$ and MIL-53(Al)/ZnO	[compound] = 6 mg L^{-1} pH = 6.8 catalyst dosage = 2 mg	MIL-53(Al)@TiO$_2$ exhibited high photodegradation efficiency (80.3%) for NPX over 4 h and was recyclable for up to three cycles with only a 13.6% decrease in photodegradability. However, for IBP degradation, MIL-53(Al)@ZnO was observed to be more efficient than MIL-53(Al)@TiO$_2$ and it was found that only 1 h of treatment was sufficient to obtain a considerable COD reduction of 58%.	[93]

Table 4. *Cont.*

Pharmaceutical	Process	Experimental Conditions	Results	Ref.
NPX	UV irradiation, UV/chorine and UV/H_2O_2	$[NPX] = 5\ \mu mol\ L^{-1}$ $[chlorine] = [H_2O_2] = 50\ \mu mol\ L^{-1}$ $\lambda = 254\ nm$ $pH = 7$	NPX was degraded by 27.3% at a UV dosage of 922 mJ cm^{-2}. The degradation of NPX by both AOPs followed pseudo-first-order kinetics, and the first-order rate constant was 4.9 times higher in UV/chlorine than that in UV/H_2O_2.	[94]
IBP, NPX and CTZ	PAN-MWCNT/TiO$_2$–NH$_2$	$[compound] = 5\ mg\ L^{-1}$ $pH = 2$ UV intensity = 40 W catalyst dosage = 15 mg	The complete degradation of IBP required 120 min of treatment, while the same degradation rate for NPX was achieved in 40 min.	[95]
KET	UV irradiation	UV intensity = 1700 $\mu W\ cm^{-2}$ $[KET] = 16\ mg\ L^{-1}$ $pH = 7$	The maximum removal rate of KET reached 99.59% within 40 min, which agrees with the pseudo-first-order kinetic equations.	[96]
KET	CuO/TiO$_2$@GCN	$[KET] = 10\ mg\ L^{-1}$ $pH = 6.2$ (GW) and 7.4 (DW) catalyst dosage = 75 mg	At the best conditions, the impregnation of 1% CuO/TiO$_2$ into GCN presented an efficiency of 94.7% in the photodegradation of KET in deionized water under simulated light. A lower removal efficiency (~50%) was achieved in ground (GW) and drinking water (DW).	[97]
IBP	Ag-Ce/TiO$_2$ (Co-DPU) and Ag-Ce/TiO$_2$ (C-IMP)	Visible light = 239 W m^{-2} $\lambda > 400$ nmcatalyst dosage = 0.1 g L^{-1} $[IBP] = 10\ mg\ L^{-1}pH = 5.3–5.6$	The obtained TOC conversion followed the decreasing order: Ag-Ce/TiO$_2$ (Co-DPU) > Ag-Ce/TiO$_2$ (C-IMP) >TiO$_2$. After 4 h, up to 98% mineralization of IBP was obtained for Ag-Ce/TiO$_2$ (Co-DPU).	[98]
IBP	UV/H_2O_2	$[IBP] = 10\ \mu mol\ L^{-1}$ $[H_2O_2] = 0.5\ mmol\ L^{-1}$ $pH = 7–7.5$ $\lambda = 254\ nm$	IBP was degraded by 8 and 3% under UV and H_2O_2 only, while the combined UV/H_2O_2 process removed 78% of IBP within 4 min.	[99]
ASA	Photo-Fenton	$[ASA] = 10\ \mu g\ L^{-1}$ $[Fe^{2+}] = 1.5\ mmol\ L^{-1}$ $[H_2O_2] = 45\ mmol\ L^{-1}$ UV intensity = 40 W $pH = 7$	Using optimized conditions, 90% of mineralization was reached in 10 min.	[100]
ASA	UV/O_3	$[ASA] = 10\ mg\ L^{-1}$ $[O_3] = 2.4\ mg\ L^{-1}$ UV intensity = 2600 $\mu W\ cm^{-2}$ $pH = 4.3$	The UV/O_3 process was able to remove 99% of ASA after 7 min of treatment. Moreover, the combined process exhibited higher mineralization rate, 47% of TOC reduction, compared with the individual ozonation process (25% of TOC) after 30 min.	[101]
ASA	UV/ZnO	UV Intensity = 6 W catalyst dosage = 375.16 mg L^{-1} $[IBP] = 33.84\ mg\ L^{-1}$ $pH = 5.05$	At the optimized conditions, the ASA removal of 83.11% was achieved. The kinetic studies showed that the pseudo-first-order model had the highest correlation with aspirin removal using the UV/ZnO photocatalytic process.	[102]
ASA	1-Ni-MnON/NG	$[ASA] = 75\ mg\ L^{-1}$ catalyst dose = 10 mg L^{-1} $pH = 3$ UV Intensity = 4.2 W	Almost complete degradation was achieved for the nanocomposites of 1-Ni-MnON/NG after 90 min of treatment.	[103]

Abbreviation: ifosfamide (IF); naproxen (NPX), methyl orange (MO); naproxen (NPX); cetirizine (CTZ); ketoprofen (KET); aspirin (ASA).

3.6. Oxygenated-Based Compounds

Ibuprofen (IBP), naproxen (NPX) and ketoprofen (KET) are non-steroidal anti-inflammatory drugs (NSAIDs), one of the pharmaceuticals groups widely detected in the environment. IBP and NPX are most used for the treatment of musculoskeletal injuries, rheumatoid arthritis and fever [93,104]. Several studies have detected these compounds in surface water and groundwater and even in drinking water sources at several concentrations due to the limited removal efficiency during the municipal sewage treatment processes [93].

In this sense, Pan et al. [104] compared the kinetics and pathways of the degradation of NPX by the UV/chorine and UV/H_2O_2 processes. NPX (5 μmol L^{-1}) was degraded by 27.3% at a UV dosage of 922 mJ cm^{-2} due to the high molar absorptivity at 254 nm of 4024 M^{-1} cm^{-1}. However, the degradation of chlorine or H_2O_2 alone was negligible. The degradation of NPX by both AOPs followed pseudo-first-order kinetics, and, at pH 7, the first-order rate constant (k$'$) was 4.9 times higher in UV/chlorine than that in UV/H_2O_2 ([chlorine]$_0$ = [H_2O_2]$_0$ = 50 μmol L^{-1}). Evaluating the relative radicals contribution, radicals chlorine species, such as Cl·, ClO· and $Cl_2{}^{·-}$, are dominant in the NPX degradation through UV/chlorine while HO· played a dominant role during the NPX removal by UV/H_2O_2. The degradation by both AOPs was associated with hydroxylation and demethylation; in particular, decarboxylation was observed in UV/H_2O_2, and chlorine substitution was observed in the UV/chlorine process. The acute toxicity of *Vibrio fischeri* in UV/chlorine was lower than that of the system using only UV radiation, besides following an increase and then decrease trend with increasing reaction times, which was related to the production of different compounds during the reaction.

In another study, Mohamed and co-workers [95] reported the photodegradation of ibuprofen, naproxen, and cetirizine in aqueous media under UV irradiation. The photocatalyst consisted of TiO_2-NH_2 nanoparticles grafted into polyacrylonitrile (PAN)/multi-walled carbon nanotube composite nanofibers (PAN−CNT). The authors investigated the effect of pharmaceutical initial concentration (5−50 mg L^{-1}), solution pH (2−9) and irradiation time on degradation efficiency. It was shown that the complete degradation was achieved at low drug concentration (5 mg L^{-1}), pH = 2, at a low power intensity of the UV lamp (40 W) and employing a dosage of 15 mg of PAN-MWCNT/TiO_2–NH_2 photocatalyst. In addition, the complete degradation of IBP required 120 min of treatment, while the same degradation rate for NPX was achieved in 40 min.

Zhen, Liu and Zhang [96] studied the removal effect of KET in UV-light and explored the influence of light turbidity, light intensity and other factors on the degradation efficiency of ketoprofen. The results show that deep ultraviolet treatment has a good degradation effect on ketoprofen. After irradiation with 1700 μW cm^{-2} UV lamps, the maximum removal rate of KET (16 mg L^{-1}) reached 99.59% within 40 min, which agrees with the pseudo-first-order kinetic equations. In addition, the irradiation strength of UV lamps showed little influence on KET degradation. Moreover, low turbidity (1-5NTU) solutions reduced the degradation rate of ketoprofen, while high turbidity (5-9NTU) solutions enhanced the removal rate of ketoprofen.

Murtaza et al. [93] reported the photodegradation studies of single and binary mixtures of naproxen, ibuprofen and methyl orange, employing photocatalysts prepared by incorporating TiO_2 and ZnO into the framework of the aluminum-based MOF (metal organic frameworks), MIL-53(Al) (MIL = Materials from the Lavoisier Institute), MIL-53(Al)@TiO_2 and MIL-53(Al)/ZnO. MIL-53(Al)@TiO_2 exhibited high photodegradation efficiency (80.3%) for NPX over 4 h and was recyclable for up to three cycles with only a 13.6% decrease in photodegradability. A ratio of 3:1 concentration of NPX to mass of photocatalyst was found to be optimum for degradation. However, for IBP degradation, MIL-53(Al)@ZnO was observed to be more efficient than MIL-53(Al)@TiO_2. Experiments conducted with scavengers showed that hydroxyl radicals played a major role in the photocatalytic process photodegradation, and it was found that only 1 h of treatment was sufficient to obtain a considerable COD reduction of 58%.

Mofokeng et al. [97] studied the formation of CuO/TiO_2@GCN (graphitic carbon nitride) and its applicability to the decomposition of KET in an aqueous environment, drinking water, and groundwater under simulated visible light. At the best conditions, the impregnation of 1% CuO/TiO_2 into GCN presented an efficiency of 94.7% in the photodegradation of KET (10 mg L^{-1}) in deionized water under simulated light. A lower removal efficiency (~50%) was achieved in ground and drinking water due to electrical current and high conductivities of the water samples. In addition, the photocatalytic degradation of KET followed the pseudo-first-order kinetics and the rate constants in ground and drinking water were slower compared to deionized water (Figure 9a). The authors justified that the ground and drinking water contains substances that could have interfered with the catalysts active sites and inhibit its catalytic activities. The electrical energy consumption (E_{EC}) utilized for KET photodegradation in ground and drinking water was 5.9 and 7.0 times higher than the E_{EC} used in deionized water. LC/MS coupled with a Q-TOF analyzer indicated that the KET degradation entailed the deprotonation of the carboxyl group, followed by decarboxylation forming KET intermediates, which included benzylic and ketyl radical structures.

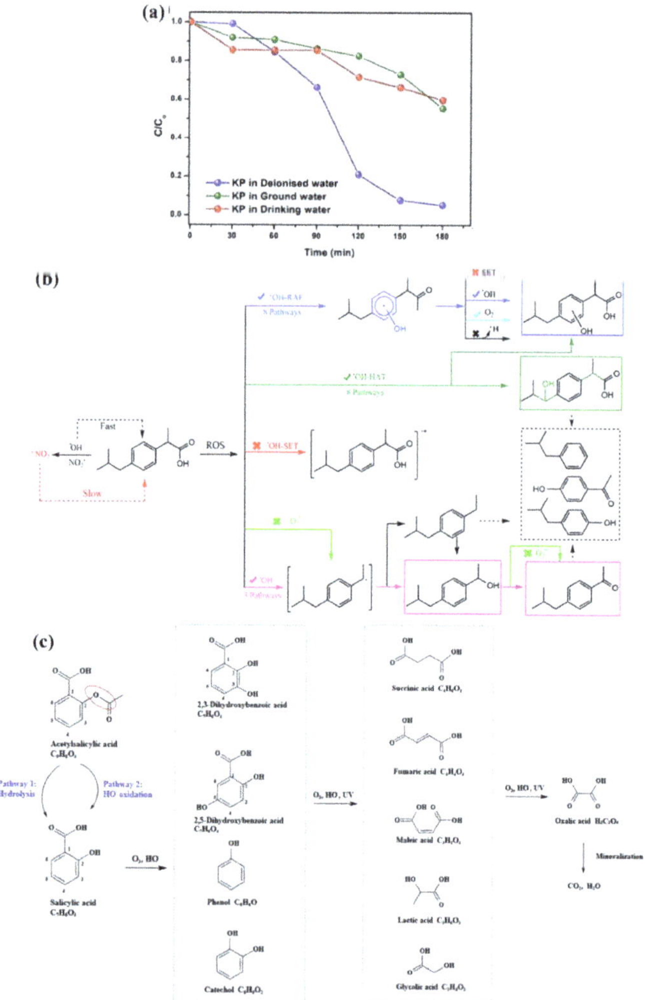

Figure 9. (**a**) Photocatalytic degradation of KP in deionized drinking and ground water using 75 mg of 1% CuO/TiO$_2$@GCN (9:1). Reprinted with permission from [97]. Copyright © 2022 John Wiley and Sons. (**b**) Proposed degradation pathways for IBP. Reprinted with permission from [99]. Copyright © 2021 Elsevier Ltd. (**c**) Reaction pathway of ASA during UV/O$_3$ process. Reprinted with permission from [101]. Copyright © 2021 Elsevier Ltd.

The degradation of IBP under visible light irradiation was also investigated by Chaker, Fourmentin, and Chérif-Aouali [98] using TiO_2 mesoporous co-doped with Ag and Ce. The co-doped photocatalysis was obtained using two different methods: co-impregnation (Co-IMP) and co-deposition precipitation with urea (Co-DPU). The obtained TOC conversion followed a decreasing order: Ag-Ce/TiO_2 (Co-DPU) > Ag-Ce/TiO_2 (C-IMP) > TiO_2. After 4 h, up to 98% mineralization of IBP was obtained for Ag-Ce/TiO_2 (Co-DPU). Additionally, the stability of Ag-Ce/TiO_2 (Co-DPU) was evaluated during three cycles of use and reuse. TOC removal remained unchanged, ranging from 97 to 98% after three cycles. Later, the authors identified the products for IBP mineralization and found that through the hydroxylation process by the attack of ·OH radicals, IBP may be transformed into hydroxy ibuprofen. The reaction follows until the ring opening for the total mineralization, leading to the formation of CO_2 and H_2O.

Wang et al. [99] investigated the degradation kinetic and transformation mechanism of IBP in UV/H_2O_2 process. Impacts of H_2O_2 dosage, pH, quenching agent and concentration of nitrite (NO_2^-) on IBP degradation were evaluated. IBP (10 μmol L^{-1}) was degraded by 8 and 3% under UV and H_2O_2 only, while the combined UV/H_2O_2 ([H_2O_2] = 0.5 mmol L^{-1}) process removed 78% of IBP within 4 min. At higher concentrations of H_2O_2, the IBP degradation became slower, which was attributed to the production of the radicals of low oxidation ability (·HO_2 or ·O_2^-). Regarding the impact of pH, the degradation of IBP decreased with the pH increases from 5.2 to 9.6. The phenomenon was explained by the formation of scavenger species and the photolyzation of H_2O_2 into water and oxygen under alkaline conditions (Figure 9b).

Moreover, the addition of different concentrations of NO_2^- had a significant inhibitory effect on the degradation of IBP. According to the authors, nitrites reacted with ·HO to produce ·NO_2, which exhibited lower reactivity to compounds with electron-withdrawing moieties [105]. Combining ultra-high-resolution mass and density functional theory calculations, the authors identified hydroxylation as the first step in IBP degradation (Figure 9b) in indirect photolysis, as observed in another papers [98], and the formation of decarboxylation products (1-(4-isobutylphenyl) ethanol and 4-isobutylacetophenone), and some of the products of subsequent degradation of 4-IBP [99].

Aspirin (acetylsalicylic acid: ASA) also belongs to the NSAIDs group and is widely used for pain, infection or inflammation. ASA as a pollutant is found with a different concentration, from 0.03 to 10 μg L^{-1}, in aqueous environment and found to be very toxic for human health and the aquatic system due to a short half-life of 2–3h, and therefore, its removal from the environment is important [100,106]. Thus, Cunha-Filho et al. [100] optimized the kinetic conditions to mineralize ASA using a photo-Fenton process with UVA radiation in a tubular photochemical reactor. Employing a statical tool-termed factorial design, a large interval of concentrations of ASA, Fe^{2+} and H_2O_2 were studied. Using the optimized conditions of H_2O_2 and Fe^{2+} (45 and 1.5 mmol L^{-1}, respectively), 90% of mineralization was reached in 10 min. Such performance was attributed to the optimized 4.5-folds excess of [H_2O_2], i.e., the ratio of the stoichiometric [H_2O_2] to the theoretic TOC for total mineralization.

In another study, Zhe and co-workers [101] investigated the removal of ASA through the UV/O_3 process. The initial reaction positions and the byproducts were also identified. The UV/O_3 process was able to remove 99% of ASA (10 mg L^{-1}) after 7 min of treatment. Moreover, the combined process exhibited higher mineralization rate, 47% of TOC reduction, compared with the individual ozonation process (25% of TOC) after 30 min. It was demonstrated, through the inhibition function of t-butanol addition, that ·HO made a significant contribution to the UV/O_3 process. The optimal values of pH were 4.3 and 10.0 for ASA removal and mineralization, respectively. Higher pH values lead to the decomposition of the ozone to produce ·HO species. These species have a stronger effect on the mineralization of organic compounds than O_3 molecules.

Frontier electron density (FED) analysis confirmed that the initial oxidation sites of HO· were located on the benzene ring of ASA, which generated salicylic acid (SA) more directly and enhanced SA transformation efficiency during the UV/O_3 process. An ASA degradation pathway was proposed and involved hydrolysis and hydroxylation reactions (Figure 9c), the opening of the benzene ring, the further oxidation of the alkyl chain byproducts and finally mineralization [101]. Pathway 1 included the hydrolysis of the ester group in ASA. In parallel, the C2 position could be attacked by $HO\cdot$, resulting in the hydroxyl substitution reaction, leading to the formation of SA (pathway 2). The ortho/para positions are more electron-rich due to the presence of hydroxyl groups (a strong electron donor). This makes the ortho and para positions of the benzene ring more susceptible to oxidant attacks with the generation of two hydroxylation byproducts, 2,3-dihydroxybenzoic acid and 2,5-dihydroxybenzoic acid (Figure 9c) Phenol was another aromatic byproduct, and its formation involved the decarboxylation of SA during $HO\cdot$ upon UV irradiation. The further hydroxylation of phenol was able to generate catechol, and the resulting aromatic byproducts were oxidized again. With the opening of the benzene ring, alkyl carboxylic acids containing two to four carbons were formed and were further degraded until the formation of oxalic acid, and after that, the mineralization occurred [101].

To optimize ASA removal from aqueous solution by the UV/ZnO photocatalytic process, Karimi, Baneshi and Malakootian [69] used response surface methodology (RSM) to study the influence of different parameters, such as ASA initial concentration ($10-100$ mg L^{-1}), ph ($3-11$), contact time ($10-120$ min) and ZnO catalyst dosage ($100-600$ mg L^{-1}). The optimized conditions included a pH solution of 5.05, after 90.5 min of treatment, employing a ZnO catalyst dosage of 375.16 mg L^{-1} and ASA initial concentration of 33.84 mg L^{-1} achieved the ASA removal of 83.11%. The effect of chloride and phosphate (20 mg L^{-1}) on the ASA degradation resulted in the increased removal efficiency (83.11% to 94.7%) and reduced the removal efficiency (from 83.11% to 56%), respectively. This behavior was attributed to the competition of the anion phosphate with the nanoparticles in ASA degradation for adsorption on the catalyst surface.

Mohan et al. [103] prepared Ni-decorated manganese oxynitride on graphene nanosheets for the degradation of ASA. Besides the effect of catalyst's composition, the authors examined the influence of other experimental parameters, such as initial concentration, catalyst dose, initial pH and additives. The best performance was achieved with the ASA initial concentration of 75 mg L^{-1}, with the catalyst dose of 10 mg L^{-1} and the initial pH 3. Almost complete degradation was achieved for the nanocomposites of 1-Ni-MnON/NG after 90 min of treatment. The detection of intermediates during photocatalysis showed that ASA undergoes hydroxylation, demethylation, aromatization, ring opening and finally complete mineralization into CO_2 and H_2O by reactive species. The catalyst remained stable even after five cycles of usage, proving its reusability. Cytotoxicity, plant toxicity and microbial toxicity studies corroborate the environmentally friendly properties of the synthesized material.

In this sense, photo-combined systems have shown promise in the removal and mineralization of oxygen-based compounds (Table 4).

4. Conclusions and Outlook

This review has showed the prospective application of photo- and photo-combined AOPs for the removal of pharmaceuticals compounds. Current research demonstrates that pharmaceuticals have been found in distinct kinds of surface waters, wastewaters and WWTP and hospital effluents. Furthermore, the presence of these compounds in water may have harmful effects on human beings as well as promote the spread of resistant bacterial strains.

Several studies have applied photo- and photo-combined AOPs to removing pharmaceuticals in water or wastewater. Overall, the works presented here discussed (I) degradation kinetics by investigating the effect of operational parameters; (II) mineralization measurements using indicators, such as TOC, DOC or COD; (III) toxicity studies; and (IV) the detection of intermediates and the proposition of degradation pathways. Although it is evident from most of the reviewed studies that several photo-combined AOP processes are efficient for the degradation of several classes of pharmaceutical compounds, the identification of intermediate products and toxicity levels are equally crucial, although they have been less explored in the literature, as these products can be more biologically active or toxic than their parent compounds, thus creating even greater hazards for the environment.

In addition, most of the literature discussed here is devoted to laboratory-scale or pilot-scale studies. The implementation of AOPs at a full scale is still quite limited. The major impediment to the application on an industrial scale is the elevated operational cost of combined AOP processes, mainly compared to conventional methods that are currently applied. Thus, if the overall cost per unit mass of pollutant removed from the unit volume of the treated wastewater is reduced, the industrial implementation of these technologies will become more appealing for companies and public administrations.

In summary, the literature of pharmaceutical compound degradation through advanced oxidation processes coupled with UV radiation has made some progress, and future research should focus on the following aspects to achieve these goals:

- More studies need to be carried out using real wastewater samples to evaluate the effectiveness of the combined advanced oxidative processes, since the matrix of real wastewater samples is complex due to the presence of organic and inorganic substances besides the variations of wastewater characteristics.
- Advanced oxidation processes need to be optimized to improve their adaptability and practicability, such as enhancing the efficiency and dosage of the photocatalysts and the utilization efficiency of O_3 or H_2O_2.
- Energy costs must also be reduced. In this context, the search for novel, affordable photocatalysts that can use a broader part of the light spectrum instead of only UV is a priority. Furthermore, the application of renewable energy sources in the treatment plants should also be investigated.
- The generation mechanism of free radicals and the degradation pathways of pollutants are not yet clear. More attention should be given to the study of mechanisms, combining experimental measurements with theoretical calculations.
- The generation of waste (e.g., sludge in the photo-Fenton process and/or exhausted or poisoned catalysts in photocatalyzed AOPs) should be minimized and possible alternatives for the valorization of such wastes should be explored.
- It is recommended that future studies should focus on the evaluation of treated water toxicity, employing ecotoxicity tests to monitor the toxicity of the by-products formed during the degradation.

Author Contributions: Conceptualization, I.M.D.G. and C.V.S.A.; methodology, I.M.D.G. and C.V.S.A.; formal analysis, I.M.D.G. and C.V.S.A.; resources, I.M.D.G., C.V.S.A. and L.H.M.; data curation, I.M.D.G., C.V.S.A. and L.H.M.; writing—original draft preparation, I.M.D.G. and C.V.S.A.; writing—review and editing, I.M.D.G., C.V.S.A. and L.H.M.; visualization, I.M.D.G., C.V.S.A. and L.H.M.; supervision, L.H.M.; project administration, L.H.M.; funding acquisition, L.H.M. All authors have read and agreed to the published version of the manuscript.

Funding: This research was funded by FAPESP (#2020/15211-0, #2021/14693-4), CEPID/FAPESP (#2013/07296-2), FAPESP/SHELL (#2017/11986-5).

Data Availability Statement: Not applicable.

Acknowledgments: The authors would like to thank FAPESP (#2020/15211-0, #2021/14693-4), CEPID/FAPESP (#2013/07296-2), FAPESP/SHELL (#2017/11986-5). The authors also thanks Shell and the strategic importance of the support given by ANP (Brazil's National Oil, Natural Gas and Biofuels Agency) through the R&D levy regulation, CNPq, CAPES finance code 001 and FINEP for the financial support.

Conflicts of Interest: The authors declare no conflict of interest.

References

1. Karimi-Maleh, H.; Karimi, F.; Fu, L.; Sanati, A.; Alizadeh, M.; Karaman, C.; Orooji, Y. Cyanazine herbicide monitoring as a hazardous substance by a DNA nanostructure biosensor. *J. Hazard. Mater.* **2022**, *423*, 127058. [CrossRef] [PubMed]
2. Chandarana, H.; Kumar, P.S.; Seenuvasan, M.; Kumar, M.A. Kinetics, equilibrium and thermodynamic investigations of methylene blue dye removal using Casuarina equisetifolia pines. *Chemosphere* **2021**, *285*, 131480. [CrossRef] [PubMed]
3. Talbot, J.; Nilsson, B. Pharmacovigilance in the pharmaceutical industry. *Br. J. Clin. Pharmacol.* **1998**, *45*, 427–431. [CrossRef] [PubMed]
4. Mansouri, F.; Chouchene, K.; Roche, N.; Ksibi, M. Removal of pharmaceuticals from water by adsorption and advanced oxidation processes: State of the art and trends. *Appl. Sci.* **2021**, *11*, 6659. [CrossRef]
5. Szabó, R.; Megyeri, C.; Illés, E.; Gajda-Schrantz, K.; Mazellier, P.; Dombi, A. Phototransformation of ibuprofen and ketoprofen in aqueous solutions. *Chemosphere* **2011**, *84*, 1658–1663. [CrossRef]
6. Majumder, A.; Gupta, B.; Gupta, A. Pharmaceutically active compounds in aqueous environment: A status, toxicity and insights of remediation. *Environ. Res.* **2019**, *176*, 108542. [CrossRef]
7. Velempini, T.; Prabakaran, E.; Pillay, K. Recent developments in the use of metal oxides for photocatalytic degradation of pharmaceutical pollutants in water—A review. *Mater. Today Chem.* **2021**, *19*, 100380. [CrossRef]
8. Brillas, E. A review on the degradation of organic pollutants in waters by UV photoelectro-fenton and solar photoelectro-fenton. *J. Braz. Chem. Soc.* **2014**, *25*, 393–417. [CrossRef]
9. Martínez-Huitle, C.; Rodrigo, M.; Sirés, I.; Scialdone, O. Single and Coupled Electrochemical Processes and Reactors for the Abatement of Organic Water Pollutants: A Critical Review. *Chem. Rev.* **2015**, *115*, 13362–13407. [CrossRef]
10. de O.S. Santos, G.; Gonzaga, I.; Dória, A.; Moratalla, A.; da Silva, R.; Eguiluz, K.; Salazar-Banda, G.; Saez, C.; Rodrigo, M. Testing and scaling-up of a novel Ti/Ru$_{0.7}$Ti$_{0.3}$O$_2$ mesh anode in a microfluidic flow-through reactor. *Chem. Eng. J.* **2020**, *398*, 125568. [CrossRef]
11. Karimi-Maleh, H.; Ranjbari, S.; Tanhaei, B.; Ayati, A.; Orooji, Y.; Alizadeh, M.; Karimi, F.; Salmanpour, S.; Rouhi, J.; Sillanpää, M.; et al. Novel 1-butyl-3-methylimidazolium bromide impregnated chitosan hydrogel beads nanostructure as an efficient nanobio-adsorbent for cationic dye removal: Kinetic study. *Environ. Res.* **2021**, *195*, 110809. [CrossRef] [PubMed]
12. Saravanan, A.; Kumar, P.; Jeevanantham, S.; Anubha, M.; Jayashree, S. Degradation of toxic agrochemicals and pharmaceutical pollutants: Effective and alternative approaches toward photocatalysis. *Environ. Pollut.* **2022**, *298*, 118844. [CrossRef] [PubMed]
13. Moratalla, Á.; Cotillas, S.; Lacasa, E.; Fernández-Marchante, C.; Ruiz, S.; Valladolid, A.; Cañizares, P.; Rodrigo, M.; Sáez, C. Occurrence and toxicity impact of pharmaceuticals in hospital effluents: Simulation based on a case of study. *Process Saf. Environ. Prot.* **2022**, *168*, 10–21. [CrossRef]
14. Jelic, A.; Cruz-Morató, C.; Marco-Urrea, E.; Sarrà, M.; Perez, S.; Vicent, T.; Petrović, M.; Barcelo, D. Degradation of carbamazepine by Trametes versicolor in an air pulsed fluidized bed bioreactor and identification of intermediates. *Water Res.* **2012**, *46*, 955–964. [CrossRef]
15. Ahuja, S. Current status of pharmaceutical contamination in water, in: Handb. *Water Purity Qual.* **2021**, 255–270. [CrossRef]
16. Li, M.; An, Z.; Huo, Y.; Jiang, J.; Zhou, Y.; Cao, H.; Jin, Z.; Xie, J.; Zhan, J.; He, M. Individual and combined degradation of N-heterocyclic compounds under sulfate radical-based advanced oxidation processes. *Chem. Eng. J.* **2022**, *442*, 136316. [CrossRef]
17. Yao, J.; Tang, Y.; Zhang, Y.; Ruan, M.; Wu, W.; Sun, J. New theoretical investigation of mechanism, kinetics, and toxicity in the degradation of dimetridazole and ornidazole by hydroxyl radicals in aqueous phase. *J. Hazard. Mater.* **2022**, *422*, 126930. [CrossRef]
18. Li, B.; Ma, X.; Deng, J.; Li, Q.; Chen, W.; Li, G.; Chen, G.; Wang, J. Comparison of acetaminophen degradation in UV-LED-based advance oxidation processes: Reaction kinetics, radicals contribution, degradation pathways and acute toxicity assessment. *Sci. Total Environ.* **2020**, *723*, 137993. [CrossRef]
19. Cai, H.; Zou, J.; Lin, J.; Li, J.; Huang, Y.; Zhang, S.; Ma, J. Sodium hydroxide-enhanced acetaminophen elimination in heat/peroxymonosulfate system: Production of singlet oxygen and hydroxyl radical. *Chem. Eng. J.* **2022**, *429*, 132438. [CrossRef]
20. Cai, H.; Zou, J.; Lin, J.; Li, Q.; Li, J.; Huang, Y.; Ma, J. Elimination of acetaminophen in sodium carbonate-enhanced thermal/peroxymonosulfate process: Performances, influencing factors and mechanism. *Chem. Eng. J.* **2022**, *449*, 137765. [CrossRef]
21. Li, J.; Zou, J.; Zhang, S.; Cai, H.; Huang, Y.; Lin, J.; Ma, J. Sodium tetraborate simultaneously enhances the degradation of acetaminophen and reduces the formation potential of chlorinated by-products with heat-activated peroxymonosulfate oxidation. *Water Res.* **2022**, *224*, 119095. [CrossRef]
22. Li, Q.; Zhang, M.; Xu, Y.; Quan, X.; Xu, Y.; Liu, W.; Wang, L. Constructing heterojunction interface of Co$_3$O$_4$/TiO$_2$ for efficiently accelerating acetaminophen degradation via photocatalytic activation of sulfite. *Chin. Chem. Lett.* **2022**, *34*, 107530. [CrossRef]
23. Sayadi, M.; Sobhani, S.; Shekari, H. Photocatalytic degradation of azithromycin using GO@Fe$_3$O$_4$/ZnO/SnO$_2$ nanocomposites. *J. Clean. Prod.* **2019**, *232*, 127–136. [CrossRef]
24. Tenzin, T.; Yashas, S.; Anilkumar, K.; Shivaraju, H. UV–LED driven photodegradation of organic dye and antibiotic using strontium titanate nanostructures. *J. Mater. Sci. Mater. Electron.* **2021**, *32*, 21093–21105. [CrossRef]
25. Martins, P.; Salazar, H.; Aoudjit, L.; Gonçalves, R.; Zioui, D.; Fidalgo-Marijuan, A.; Costa, C.; Ferdov, S.; Lanceros-Mendez, S. Crystal morphology control of synthetic giniite for enhanced photo-Fenton activity against the emerging pollutant metronidazole. *Chemosphere* **2021**, *262*, 128300. [CrossRef] [PubMed]

26. Neghi, N.; Krishnan, N.; Kumar, M. Analysis of metronidazole removal and micro-toxicity in photolytic systems: Effects of persulfate dosage, anions and reactor operation-mode. *J. Environ. Chem. Eng.* **2018**, *6*, 754–761. [CrossRef]

27. Leeladevi, K.; Kumar, J.V.; Arunpandian, M.; Thiruppathi, M.; Nagarajan, E. Investigation on photocatalytic degradation of hazardous chloramphenicol drug and amaranth dye by SmVO$_4$ decorated g-C$_3$N$_4$ nanocomposites. *Mater. Sci. Semicond. Process.* **2021**, *123*, 105563. [CrossRef]

28. Hu, X.; Qin, J.; Wang, Y.; Wang, J.; Yang, A.; Tsang, Y.; Liu, B. Synergic degradation Chloramphenicol in photo-electrocatalytic microbial fuel cell over Ni/MXene photocathode. *J. Colloid Interface Sci.* **2022**, *628*, 327–337. [CrossRef] [PubMed]

29. Kumar, A.; Sharma, S.; Kumar, A.; Sharma, G.; AlMasoud, N.; Alomar, T.; Naushad, M.; ALOthman, Z.; Stadler, F. High interfacial charge carrier separation in Fe$_3$O$_4$ modified SrTiO$_3$/Bi$_4$O$_5$I$_2$ robust magnetic nano-heterojunction for rapid photodegradation of diclofenac under simulated solar-light. *J. Clean. Prod.* **2021**, *315*, 128137. [CrossRef]

30. Li, S.; Cui, J.; Wu, X.; Zhang, X.; Hu, Q.; Hou, X. Rapid in situ microwave synthesis of Fe3O4@MIL-100(Fe) for aqueous diclofenac sodium removal through integrated adsorption and photodegradation. *J. Hazard. Mater.* **2019**, *373*, 408–416. [CrossRef]

31. Gonzaga, I.; Dória, A.; Moratalla, A.; Eguiluz, K.; Salazar-Banda, G.; Cañizares, P.; Rodrigo, M.; Saez, C. Electrochemical systems equipped with 2D and 3D microwave-made anodes for the highly efficient degradation of antibiotics in urine. *Electrochim. Acta.* **2021**, *392*, 139012. [CrossRef]

32. Kumari, A.; Kumar, A.; Sharma, G.; Iqbal, J.; Naushad, M.; Stadler, F. Constructing Z-scheme LaTiO$_2$N/g-C$_3$N$_4$@Fe$_3$O$_4$ magnetic nano heterojunctions with promoted charge separation for visible and solar removal of indomethacin. *J. Water Process Eng.* **2020**, *36*, 101391. [CrossRef]

33. Li, R.; Kong, J.; Liu, H.; Chen, P.; Su, Y.; Liu, G.; Lv, W. Removal of indomethacin using UV–vis/peroxydisulfate: Kinetics, toxicity, and transformation pathways. *Chem. Eng. J.* **2018**, *331*, 809–817. [CrossRef]

34. Cao, J.; Li, J.; Chu, W.; Cen, W. Facile synthesis of Mn-doped BiOCl for metronidazole photodegradation: Optimization, degradation pathway, and mechanism. *Chem. Eng. J.* **2020**, *400*, 125813. [CrossRef]

35. Parra-Enciso, C.; Avila, B.; Rubio-Clemente, A.; Peñuela, G. Degradation of diclofenac through ultrasonic-based advanced oxidation processes at low frequency. *J. Environ. Chem. Eng.* **2022**, *10*, 108296. [CrossRef]

36. Naraginti, S.; Yu, Y.; Fang, Z.; Yong, Y. Visible light degradation of macrolide antibiotic azithromycin by novel ZrO$_2$/Ag@TiO$_2$ nanorod composite: Transformation pathways and toxicity evaluation. *Process Saf. Environ. Prot.* **2019**, *125*, 39–49. [CrossRef]

37. Cano, P.; Jaramillo-Baquero, M.; Zúñiga-Benítez, H.; Londoño, Y.; Peñuela, G. Use of simulated sunlight radiation and hydrogen peroxide in azithromycin removal from aqueous solutions: Optimization & mineralization analysis. *Emerg. Contam.* **2020**, *6*, 53–61. [CrossRef]

38. de Almeida, L.; Josué, T.; Fidelis, M.; Abreu, E.; Bechlin, M.; Santos, O.; Lenzi, G. Process Comparison for Caffeine Degradation: Fenton, Photo-Fenton, UV/H$_2$O$_2$ and UV/Fe^{3+}. *Water. Air. Soil Pollut.* **2021**, *232*, 147. [CrossRef]

39. Goulart, L.; Moratalla, A.; Cañizares, P.; Lanza, M.; Sáez, C.; Rodrigo, M. High levofloxacin removal in the treatment of synthetic human urine using Ti/MMO/ZnO photo-electrocatalyst. *J. Environ. Chem. Eng.* **2022**, *10*, 107317. [CrossRef]

40. Geng, C.; Liang, Z.; Cui, F.; Zhao, Z.; Yuan, C.; Du, J.; Wang, C. Energy-saving photo-degradation of three fluoroquinolone antibiotics under VUV/UV irradiation: Kinetics, mechanism, and antibacterial activity reduction. *Chem. Eng. J.* **2020**, *383*, 123145. [CrossRef]

41. Jia, J.; Guan, Y.; Cheng, M.; Chen, H.; He, J.; Wang, S.; Wang, Z. Occurrence and distribution of antibiotics and antibiotic resistance genes in Ba River. *China Sci. Total Environ.* **2018**, *642*, 1136–1144. [CrossRef]

42. Wormser, G.; Tang, Y.-W. Antibiotics in Laboratory Medicine, 5th Edition Edited by Victor Lorain Philadelphia: Lippincott Williams & Wilkins, 2005 832 pp. illustrated. $199.00 (cloth). *Clin. Infect. Dis.* **2005**, *41*, 577. [CrossRef]

43. Liu, X.; Lu, S.; Guo, W.; Xi, B.; Wang, W. Antibiotics in the aquatic environments: A review of lakes. *China Sci. Total Environ.* **2018**, *627*, 1195–1208. [CrossRef]

44. He, K.; Hain, E.; Timm, A.; Tarnowski, M.; Blaney, L. Occurrence of antibiotics, estrogenic hormones, and UV-filters in water, sediment, and oyster tissue from the Chesapeake Bay. *Sci. Total Environ.* **2019**, *650*, 3101–3109. [CrossRef]

45. Boy-Roura, M.; Mas-Pla, J.; Petrovic, M.; Gros, M.; Soler, D.; Brusi, D.; Menció, A. Towards the understanding of antibiotic occurrence and transport in groundwater: Findings from the Baix Fluvià alluvial aquifer (NE Catalonia, Spain). *Sci. Total Environ.* **2018**, *612*, 1387–1406. [CrossRef]

46. Wu, M.; Que, C.; Xu, G.; Sun, Y.; Ma, J.; Xu, H.; Sun, R.; Tang, L. Occurrence, fate and interrelation of selected antibiotics in sewage treatment plants and their receiving surface water. *Ecotoxicol. Environ. Saf.* **2016**, *132*, 132–139. [CrossRef] [PubMed]

47. Mondal, S.; Saha, A.; Sinha, A. Removal of ciprofloxacin using modified advanced oxidation processes: Kinetics, pathways and process optimization. *J. Clean. Prod.* **2018**, *171*, 1203–1214. [CrossRef]

48. Liu, X.; Liu, Y.; Lu, S.; Wang, Z.; Wang, Y.; Zhang, G.; Guo, X.; Guo, W.; Zhang, T.; Xi, B. Degradation difference of ofloxacin and levofloxacin by UV/H$_2$O$_2$ and UV/PS (persulfate): Efficiency, factors and mechanism. *Chem. Eng. J.* **2019**, *385*, 123987. [CrossRef]

49. Wang, C.; Zhang, J.; Du, J.; Zhang, P.; Zhao, Z.; Shi, W.; Cui, F. Rapid degradation of norfloxacin by VUV/Fe^{2+}/H$_2$O$_2$ over a wide initial pH: Process parameters, synergistic mechanism, and influencing factors. *J. Hazard. Mater.* **2021**, *416*, 125893. [CrossRef]

50. Yao, W.; Rehman, S.U.; Wang, H.; Yang, H.; Yu, G.; Wang, Y. Pilot-scale evaluation of micropollutant abatements by conventional ozonation, UV/O$_3$, and an electro-peroxone process. *Water Res.* **2018**, *138*, 106–117. [CrossRef]

51. Liu, H.; Gao, Y.; Wang, J.; Ma, D.; Wang, Y.; Gao, B.; Yue, Q.; Xu, X. The application of UV/O$_3$ process on ciprofloxacin wastewater containing high salinity: Performance and its degradation mechanism. *Chemosphere* **2021**, *276*, 130220. [CrossRef] [PubMed]

52. Paucar, N.E.; Kim, I.; Tanaka, H.; Sato, C. Effect of O_3 dose on the O_3/UV treatment process for the removal of pharmaceuticals and personal care products in secondary effluent. *Chem. Eng.* **2019**, *3*, 53. [CrossRef]

53. Asgari, E.; Sheikhmohammadi, A.; Nourmoradi, H.; Nazari, S.; Aghanaghad, M. Degradation of ciprofloxacin by photocatalytic ozonation process under irradiation with UVA: Comparative study, performance and mechanism. *Process Saf. Environ. Prot.* **2021**, *147*, 356–366. [CrossRef]

54. Ghattavi, S.; Nezamzadeh-Ejhieh, A. A double-Z-scheme ZnO/AgI/WO$_3$ photocatalyst with high visible light activity: Experimental design and mechanism pathway in the degradation of methylene blue. *J. Mol. Liq.* **2021**, *322*, 114563. [CrossRef]

55. Costa, L.; Nobre, F.; Lobo, A.; de Matos, J. Photodegradation of ciprofloxacin using Z-scheme TiO_2/SnO_2 nanostructures as photocatalyst. *Environ. Nanotechnol. Monit. Manag.* **2021**, *16*, 100466. [CrossRef]

56. Nguyen, L.; Nguyen, H.; Pham, T.; Tran, T.; Chu, H.; Dang, H.; Nguyen, V.; Nguyen, K.; Pham, T.; Van der Bruggen, B. UV–Visible Light Driven Photocatalytic Degradation of Ciprofloxacin by N,S Co-doped TiO_2: The Effect of Operational Parameters. *Top. Catal.* **2020**, *63*, 985–995. [CrossRef]

57. Zhang, M.; Lai, C.; Li, B.; Huang, D.; Liu, S.; Qin, L.; Yi, H.; Fu, Y.; Xu, F.; Li, M.; et al. Ultrathin oxygen-vacancy abundant WO$_3$ decorated monolayer Bi$_2$WO$_6$ nanosheet: A 2D/2D heterojunction for the degradation of Ciprofloxacin under visible and NIR light irradiation. *J. Colloid Interface Sci.* **2019**, *556*, 557–567. [CrossRef]

58. Wu, D.; Li, J.; Guan, J.; Liu, C.; Zhao, X.; Zhu, Z.; Ma, C.; Huo, P.; Li, C.; Yan, Y. Improved photoelectric performance via fabricated heterojunction g-C$_3$N$_4$/TiO_2/HNTs loaded photocatalysts for photodegradation of ciprofloxacin. *J. Ind. Eng. Chem.* **2018**, *64*, 206–218. [CrossRef]

59. Prabavathi, S.; Saravanakumar, K.; Park, C.; Muthuraj, V. Photocatalytic degradation of levofloxacin by a novel Sm$_6$WO$_{12}$/g-C$_3$N$_4$ heterojunction: Performance, mechanism and degradation pathways. *Sep. Purif. Technol.* **2021**, *257*, 117985. [CrossRef]

60. Al Balushi, B.; Al Marzouqi, F.; Al Wahaibi, B.; Kuvarega, A.; Al Kindy, S.; Kim, Y.; Selvaraj, R. Hydrothermal synthesis of CdS sub-microspheres for photocatalytic degradation of pharmaceuticals. *Appl. Surf. Sci.* **2018**, *457*, 559–565. [CrossRef]

61. Lu, G.; Lun, Z.; Liang, H.; Wang, H.; Li, Z.; Ma, W. In situ fabrication of BiVO$_4$-CeVO$_4$ heterojunction for excellent visible light photocatalytic degradation of levofloxacin. *J. Alloys Compd.* **2019**, *772*, 122–131. [CrossRef]

62. Zhang, X.; Zhang, Y.; Jia, X.; Zhang, N.; Xia, R.; Zhang, X.; Wang, Z.; Yu, M. In situ fabrication of a novel S-scheme heterojunction photocatalyts Bi$_2$O$_3$/P-C$_3$N$_4$ to enhance levofloxacin removal from water. *Sep. Purif. Technol.* **2021**, *268*, 118691. [CrossRef]

63. Aghaeinejad-Meybodi, A.; Ebadi, A.; Shafiei, S.; Khataee, A.; Kiadehi, A. Degradation of Fluoxetine using catalytic ozonation in aqueous media in the presence of nano-Γ-alumina catalyst: Experimental, modeling and optimization study. *Sep. Purif. Technol.* **2019**, *211*, 551–563. [CrossRef]

64. Pan, C.; Zhu, F.; Wu, M.; Jiang, L.; Zhao, X.; Yang, M. Degradation and toxicity of the antidepressant fluoxetine in an aqueous system by UV irradiation. *Chemosphere* **2022**, *287*, 132434. [CrossRef] [PubMed]

65. Hollman, J.; Dominic, J.; Achari, G. Degradation of pharmaceutical mixtures in aqueous solutions using UV/peracetic acid process: Kinetics, degradation pathways and comparison with UV/H$_2$O$_2$. *Chemosphere* **2020**, *248*, 125911. [CrossRef]

66. Szabó, L.; Mile, V.; Kiss, D.; Kovács, K.; Földes, T.; Németh, T.; Tóth, T.; Homlok, R.; Balogh, G.; Takács, E.; et al. Applicability evaluation of advanced processes for elimination of neurophysiological activity of antidepressant fluoxetine. *Chemosphere* **2018**, *193*, 489–497. [CrossRef]

67. Sharma, S.; Kumar, A.; Sharma, G.; Naushad, M.; Vo, D.; Alam, M.; Stadler, F. Fe$_3$O$_4$ mediated Z-scheme BiVO$_4$/Cr$_2$V$_4$O$_{13}$ strongly coupled nano-heterojunction for rapid degradation of fluoxetine under visible light. *Mater. Lett.* **2020**, *281*, 128650. [CrossRef]

68. Norouzi, R.; Zarei, M.; Khataee, A.; Ebratkhahan, M.; Rostamzadeh, P. Electrochemical removal of fluoxetine via three mixed metal oxide anodes and carbonaceous cathodes from contaminated water. *Environ. Res.* **2022**, *207*, 112641. [CrossRef]

69. Acosta-Rangel, A.; Sánchez-Polo, M.; Polo, A.; Rivera-Utrilla, J.; Berber-Mendoza, M. Sulfonamides degradation assisted by UV, UV/H$_2$O$_2$ and UV/K$_2$S$_2$O$_8$: Efficiency, mechanism and byproducts cytotoxicity. *J. Environ. Manag.* **2018**, *225*, 224–231. [CrossRef]

70. Moradi, M.; Moussavi, G. Investigation of chemical-less UVC/VUV process for advanced oxidation of sulfamethoxazole in aqueous solutions: Evaluation of operational variables and degradation mechanism. *Sep. Purif. Technol.* **2018**, *190*, 90–99. [CrossRef]

71. Wen, D.; Wu, Z.; Tang, Y.; Li, M.; Qiang, Z. Accelerated degradation of sulfamethazine in water by VUV/UV photo-Fenton process: Impact of sulfamethazine concentration on reaction mechanism. *J. Hazard. Mater.* **2018**, *344*, 1181–1187. [CrossRef]

72. Hong, M.; Wang, Y.; Lu, G. UV-Fenton degradation of diclofenac, sulpiride, sulfamethoxazole and sulfisomidine: Degradation mechanisms, transformation products, toxicity evolution and effect of real water matrix. *Chemosphere* **2020**, *258*, 127351. [CrossRef] [PubMed]

73. Jia, M.; Yang, Z.; Xu, H.; Song, P.; Xiong, W.; Cao, J.; Zhang, Y.; Xiang, Y.; Hu, J.; Zhou, C.; et al. Integrating N and F co-doped TiO_2 nanotubes with ZIF-8 as photoelectrode for enhanced photo-electrocatalytic degradation of sulfamethazine. *Chem. Eng. J.* **2020**, *388*, 124388. [CrossRef]

74. Teng, W.; Xu, J.; Cui, Y.; Yu, J. Photoelectrocatalytic degradation of sulfadiazine by Ag$_3$PO$_4$/MoS$_2$/TiO_2 nanotube array electrode under visible light irradiation. *J. Electroanal. Chem.* **2020**, *868*, 114178. [CrossRef]

75. She, S.; Wang, Y.; Chen, R.; Yi, F.; Sun, C.; Hu, J.; Li, Z.; Lu, G.; Zhu, M. Ultrathin S-doped graphitic carbon nitride nanosheets for enhanced sulpiride degradation via visible-light-assisted peroxydisulfate activation: Performance and mechanism. *Chemosphere* **2021**, *266*, 128929. [CrossRef]

76. Zhang, Y.; Xiao, Y.; Zhong, Y.; Lim, T. Comparison of amoxicillin photodegradation in the UV/H$_2$O$_2$ and UV/persulfate systems: Reaction kinetics, degradation pathways, and antibacterial activity. *Chem. Eng. J.* **2019**, *372*, 420–428. [CrossRef]

77. Lu, J.; Ji, Y.; Chovelon, J.; Lu, J. Fluoroquinolone antibiotics sensitized photodegradation of isoproturon. *Water Res.* **2021**, *198*, 117136. [CrossRef]

78. Wan, Z.; Wang, J. Removal of sulfonamide antibiotics from wastewater by gamma irradiation in presence of iron ions. *Nucl. Sci. Tech.* **2016**, *27*, 104. [CrossRef]

79. Lin, C.; Wu, M. Feasibility of using UV/H$_2$O$_2$ process to degrade sulfamethazine in aqueous solutions in a large photoreactor. *J. Photochem. Photobiol. A Chem.* **2018**, *367*, 446–451. [CrossRef]

80. Aryee, A.; Han, R.; Qu, L. Occurrence, detection and removal of amoxicillin in wastewater: A review. *J. Clean. Prod.* **2022**, *368*, 133140. [CrossRef]

81. Yazidi, A.; Atrous, M.; Soetaredjo, F.E.; Sellaoui, L.; Ismadji, S.; Erto, A.; Bonilla-Petriciolet, A.; Dotto, G.L.; Lamine, A.B. Adsorption of amoxicillin and tetracycline on activated carbon prepared from durian shell in single and binary systems: Experimental study and modeling analysis. *Chem. Eng. J.* **2020**, *379*, 122320. [CrossRef]

82. Ighalo, J.; Igwegbe, C.; Aniagor, C.; Oba, S. A review of methods for the removal of penicillins from water. *J. Water Process Eng.* **2021**, *39*, 101886. [CrossRef]

83. Verma, M.; Haritash, A. Degradation of amoxicillin by Fenton and Fenton-integrated hybrid oxidation processes. *J. Environ. Chem. Eng.* **2019**, *7*, 102886. [CrossRef]

84. Le, S.; Zhu, C.; Cao, Y.; Wang, P.; Liu, Q.; Zhou, H.; Chen, C.; Wang, S.; Duan, X. V$_2$O$_5$ nanodot-decorated laminar C$_3$N$_4$ for sustainable photodegradation of amoxicillin under solar light, Appl. *Catal. B Environ.* **2022**, *303*, 120903. [CrossRef]

85. Guerra, M.H.; Alberola, I.O.; Rodriguez, S.M.; López, A.A.; Merino, A.A.; Lopera, A.E.-C.; Alonso, J.Q. Oxidation mechanisms of amoxicillin and paracetamol in the photo-Fenton solar process. *Water Res.* **2019**, *156*, 232–240. [CrossRef] [PubMed]

86. Mmelesi, O.; Patala, R.; Nkambule, T.; Mamba, B.; Kefeni, K.; Kuvarega, A. Effect of Zn doping on physico-chemical properties of cobalt ferrite for the photodegradation of amoxicillin and deactivation of *E. coli*. *Colloids Surf. A Physicochem. Eng. Asp.* **2022**, *649*, 129462. [CrossRef]

87. Freitas, J.; Quintão, F.; da Silva, J.; de Queiroz, S.; Aquino, S.; de Cássia Franco, R.J. Characterisation of captopril photolysis and photocatalysis by-products in water by direct infusion, electrospray ionisation, high-resolution mass spectrometry and the assessment of their toxicities. *Int. J. Environ. Anal. Chem.* **2017**, *97*, 42–55. [CrossRef]

88. Santos, A.; Cabot, P.; Brillas, E.; Sirés, I. A comprehensive study on the electrochemical advanced oxidation of antihypertensive captopril in different cells and aqueous matrices. *Appl. Catal. B Environ.* **2020**, *277*, 119240. [CrossRef]

89. Osawa, R.A.; Barrocas, B.T.; Monteiro, O.C.; Oliveira, M.; Florêncio, M. Photocatalytic degradation of cyclophosphamide and ifosfamide: Effects of wastewater matrix, transformation products and in silico toxicity prediction. *Sci. Total Environ.* **2019**, *692*, 503–510. [CrossRef]

90. Russo, C.; Lavorgna, M.; Česen, M.; Kosjek, T.; Heath, E.; Isidori, M. Evaluation of acute and chronic ecotoxicity of cyclophosphamide, ifosfamide, their metabolites/transformation products and UV treated samples. *Environ. Pollut.* **2018**, *233*, 356–363. [CrossRef]

91. Janssens, R.; Cristovao, M.; Bronze, M.; Crespo, J.; Pereira, V.; Luis, P. Coupling of nanofiltration and UV, UV/TiO$_2$ and UV/H$_2$O$_2$ processes for the removal of anti-cancer drugs from real secondary wastewater effluent. *J. Environ. Chem. Eng.* **2019**, *7*, 103351. [CrossRef]

92. Graumans, M.; Hoeben, W.; Russel, F.; Scheepers, P. Oxidative degradation of cyclophosphamide using thermal plasma activation and UV/H$_2$O$_2$ treatment in tap water. *Environ. Res.* **2020**, *182*, 109046. [CrossRef] [PubMed]

93. Sabouni, R.; Murtaza, S.; Ghommem, M. Facile Metal Organic Framework Composites as Photocatalysts for Lone/Simultaneous Photodegradation of Naproxen, Ibuprofen and Methyl Orange. *SSRN Electron. J.* **2022**, *27*, 102751. [CrossRef]

94. Wu, T.; Pi, M.; Zhang, D.; Chen, S. Three-dimensional porous structural MoP$_2$ nanoparticles as a novel and superior catalyst for electrochemical hydrogen evolution. *J. Power Source* **2016**, *328*, 551–557. [CrossRef]

95. Mohamed, A.; Salama, A.; Nasser, W.; Uheida, A. Photodegradation of Ibuprofen, Cetirizine, and Naproxen by PAN-MWCNT/TiO$_2$–NH$_2$ nanofiber membrane under UV light irradiation. *Environ. Sci. Eur.* **2018**, *30*, 47. [CrossRef] [PubMed]

96. Zhen, H.; Liu, J.; Zhang, T. The Effect of UV Degradation of Ketoprofen and Its Influencing Factors. *J. Phys. Conf. Ser.* **2022**, *2194*, 012049. [CrossRef]

97. Mofokeng, L.; Hlekelele, L.; Tetana, Z.; Moma, J.; Chauke, V. CuO-doped TiO$_2$ Supported on Graphitic Carbon Nitride for the Photodegradation of Ketoprofen in Drinking and Groundwater: Process Optimization and Energy Consumption evaluation. *Chem. Select.* **2022**, *7*, e202101847. [CrossRef]

98. Chaker, H.; Fourmentin, S.; Chérif-Aouali, L. Efficient Photocatalytic Degradation of Ibuprofen under Visible Light Irradiation Using Silver and Cerium Co-Doped Mesoporous TiO$_2$. *Chem. Select.* **2020**, *5*, 11787–11796. [CrossRef]

99. Wang, P.; Bu, L.; Wu, Y.; Ma, W.; Zhu, S.; Zhou, S. Mechanistic insight into the degradation of ibuprofen in UV/H$_2$O$_2$ process via a combined experimental and DFT study. *Chemosphere* **2021**, *267*, 128883. [CrossRef]

100. Cunha-Filho, F.; Mota-Lima, A.; Ratkievicius, L.; Silva, D.; Silva, D.; Chiavone-Filho, O.; Nascimento, C.O. Rapid mineralization rate of acetylsalicylic acid in a tubular photochemical reactor: The role of the optimized excess of H$_2$O$_2$. *J. Water Process Eng.* **2019**, *31*, 100856. [CrossRef]

101. Zhe, W.; Wenjuan, Z.; Haihan, W.; Zhiwei, W.; Jing, C. Oxidation of acetylsalicylic acid in water by UV/O_3 process: Removal, byproduct analysis, and investigation of degradation mechanism and pathway. *J. Environ. Chem. Eng.* **2021**, *9*, 106259. [CrossRef]
102. Karimi, P.; Baneshi, M.; Malakootian, M. Photocatalytic degradation of aspirin from aqueous solutions using the UV/ZnO process: Modelling, analysis and optimization by response surface methodology (RSM), Desalin. *Water Treat.* **2019**, *161*, 354–364. [CrossRef]
103. Mohan, H.; Yoo, S.; Thimmarayan, S.; Oh, H.; Kim, G.; Seralathan, K.; Shin, T. Nickel decorated manganese oxynitride over graphene nanosheets as highly efficient visible light driven photocatalysts for acetylsalicylic acid degradation. *Environ. Pollut.* **2021**, *289*, 117864. [CrossRef] [PubMed]
104. Pan, M.; Wu, Z.; Tang, C.; Guo, K.; Cao, Y.; Fang, J. Emerging investigators series: Comparative study of naproxen degradation by the UV/chlorine and the UV/H_2O_2 advanced oxidation processes. *Environ. Sci. Water Res. Technol.* **2018**, *4*, 1219–1230. [CrossRef]
105. Zhou, S.; Li, L.; Wu, Y.; Zhu, S.; Zhu, N.; Bu, L.; Dionysiou, D. UV365 induced elimination of contaminants of emerging concern in the presence of residual nitrite: Roles of reactive nitrogen species. *Water Res.* **2020**, *178*, 115829. [CrossRef] [PubMed]
106. Tahir, M.; Sagir, M.; Shahzad, K. Removal of acetylsalicylate and methyl-theobromine from aqueous environment using nano-photocatalyst WO$_3$-TiO$_2$ @g-C$_3$N$_4$ composite. *J. Hazard. Mater.* **2018**, *363*, 205–213. [CrossRef] [PubMed]

Article

Photocatalytic Concrete Developed by Short Seedless Hydrothermal Method for Water Purification

Marie Le Pivert [1,2] and Yamin Leprince-Wang [2,*]

[1] Université Gustave Eiffel, COSYS/ISME, F-77447 Marne-la-Vallée, France
[2] Université Gustave Eiffel, CNRS, ESYCOM, F-77454 Marne-la-Vallée, France
* Correspondence: yamin.leprince@univ-eiffel.fr

Abstract: Stormwater runoff management and treatment are significant topics for designing a sustainable city. Therefore, photocatalytic, permeable, and removable concrete is a promising solution to reduce pollution through leaching with permeable and scalable road. The objective of this work was to develop cost-effective and greener photocatalytic concretes that can be easily scaled-up, and to demonstrate their photocatalytic activities. To achieve this, seedless hydrothermal ZnO nanostructures (NSs) in 2 h were employed to functionalize a concrete surface by a soft functionalization process, avoiding overconsumption of energy and chemical products. In this work, two different concretes were studied and used for the degradation of organic dye in water. The results demonstrated the universality of the proposed functionalization process by showing similar gap values, ZnO NSs morphologies, and XRD pattern, compared to the concrete functionalized by the traditional two-step hydrothermal synthesis. The XRD results certified the presence of the ZnO Würtzite phase on the concrete surface. The synthesis feasibility was attributed to the basic pH and O^- groups' presence in concrete. Then, their photocatalytic efficiency was proved for organic dye removal in water. An almost total degradation was recorded after 5 h under artificial solar light, even after several uses, demonstrating a similar efficiency to the photocatalytic concrete functionalized by the traditional two-step synthesis.

Keywords: photocatalytic concrete; ZnO nanostructures; seedless method; water purification; photocatalysis

Citation: Le Pivert, M.; Leprince-Wang, Y. Photocatalytic Concrete Developed by Short Seedless Hydrothermal Method for Water Purification. *Catalysts* **2022**, *12*, 1620. https://doi.org/10.3390/catal12121620

Academic Editors: Gassan Hodaifa, Rafael Borja and Mha Albqmi

Received: 4 November 2022
Accepted: 4 December 2022
Published: 9 December 2022

Publisher's Note: MDPI stays neutral with regard to jurisdictional claims in published maps and institutional affiliations.

1. Introduction

Sustainable city development is facing many problems such as the development of greener urban road [1]. Indeed, most of the road infrastructures in the city are impermeable, leading to more overflow and pollutants loading of rainwater, and less groundwater recharge [2]. Porous concrete material appears as one of the most promising ways to reduce stormwater volume and pollutant concentrations [1,3]. Therefore, effort has been made to develop photocatalytic concrete able to degrade and mineralize pollutants in the presence of sunlight [1–5].

Zinc oxide (ZnO) semiconductor nanomaterials have already shown great potential as an eco-friendly photocatalyst for environmental pollution remediation due to their ability to degrade and mineralize organic toxic pollutants into CO_2, H_2O, and other light by-products by photocatalysis under UV light or solar light [4–8]. Nevertheless, these ZnO nanomaterials are often synthesized by complex and expensive routes; therefore, their integration in road infrastructures could be difficult. Thus, the development of functionalized building materials using a low-cost process with minimal chemicals and a shorter manufacturing time is an emerging field of study [9].

Recently, in our previous work [4,5,10], ZnO nanostructures' (NSs) hydrothermal direct synthesis on paving block appeared as an interesting route to produce daily life photocatalytic materials for the urban pollution remediation. In addition, it is important to highlight that the ZnO surface functionalization may not affect the hydration reactions of concrete and, consequently, its mechanical properties, contrarily to ZnO direct incorporation in concrete chemical formulation. Moreover, the hydrothermal method needs low processing temperature and short duration and could be easily scaled-up and adapted to different substrates [10–13]. Usually, ZnO NSs grow onto a substrate by two simple operating steps: (1) a seed layer deposition for creating ZnO nucleation sites on the substrate; (2) hydrothermal growth in the presence of zinc salt and hexamethylenetetramine (HMTA) to obtain ZnO NSs such as nanowires (NWs) or nanorods (NRs).

Nevertheless, this surface functionalization process for the permeable and removable concrete production needs to be improved. Indeed, depending on the ZnO seed layer deposition process employed, an annealing temperature from 300 °C to 500 °C is necessary, which could damage concrete. Moreover, even the ZnO seed layer deposition is known for allowing a better morphology and density control on ZnO NS growth; it also causes consumption of time and chemicals, leading to an additional cost [11].

The objective of this work was to determine and understand if permeable and removable concrete could be exempted from the seed layer deposition step and functionalized only by the second step to develop the cost-effective and greener photocatalytic concretes. Knowing that the ZnO NSs growth, in terms of morphology and gap value, could be influenced by the substrate textural properties (porosity, roughness, etc.) and surface chemistry (pH, surface functional groups, etc.), which, in turn, affect the photocatalytic activity (reactivity and availability of surface, light absorption, etc.) [5,8], in this work, two kinds of concrete with different porosity and chemical formulation were employed as substrates for ZnO NS growth. Indeed, for example, ZnO NSs' morphology could influence the ZnO total surface available for photocatalytic reaction. Moreover, the ZnO NSs' quality will define the gap value and, therefore, the light that could be absorbed and used in the photocatalytic process, and the surface reactivity. Their photocatalytic efficiency and durability for the degradation of two different organic dyes (Methyl Orange (MO) and Acid Red 14 (AR14)), whose degradation mechanisms have already been studied [14], under artificial solar light, were compared to their reference corresponding to concrete functionalized by two-step hydrothermal synthesis. The samples were thoroughly characterized by a scanning electron microscope (SEM) for nanostructure morphology investigation, ultraviolet–visible spectrophotometry for both ZnO bandgap measurements and the following organic dye photodegradation, and X-ray diffraction (XRD) for microstructures investigation.

2. Results and Discussion

2.1. Concrete Surface Characterization

2.1.1. Concrete Surface Observations and SEM Analysis

Figure 1 shows optical photography of both types of concrete samples before and after the ZnO NS growth by one-step seedless hydrothermal synthesis. We can note that the samples appeared more whitish after the synthesis. The same phenomenon was observed on the samples functionalized by two-step hydrothermal synthesis and was already attributed to the presence of very dense ZnO NSs on the samples surface, leading to diffuse light on the sample surface [4].

To further confirm the synthesis of the ZnO NSs and study the influence of the seedless effect, SEM analysis was used to examine concrete surfaces (Figure 2a). By comparing concrete surfaces with and without treatment, ZnO NSs were clearly observed on all sample surfaces with no significant modification whatever the concrete substrate and the hydrothermal route used. Only a very modest decline in density of ZnO NSs seemed to be observed on samples functionalized by simplified synthesis but could not be quantified. Due to the NSs' variety, no size information was precisely measured. The ZnO NS length was a size of the order of the micrometer, and the thickness was a size of the order of the nanometer.

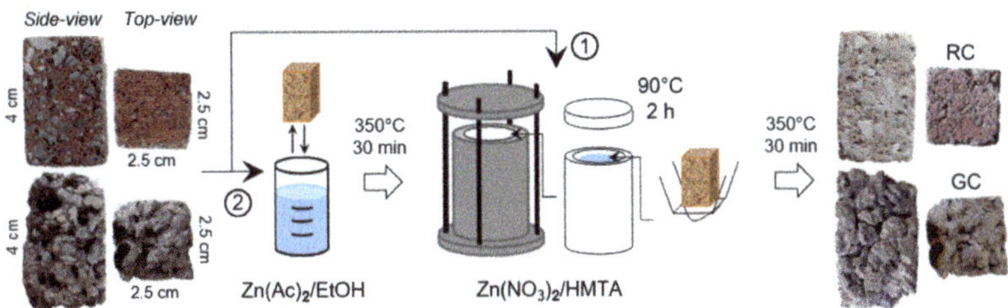

Figure 1. Schematic routes of concrete functionalization by classical hydrothermal growth in two steps (2) and seedless hydrothermal growth (1), and optical photographs of the grey concrete (GC) and red concrete (RC) blocks before and after the ZnO NS growth.

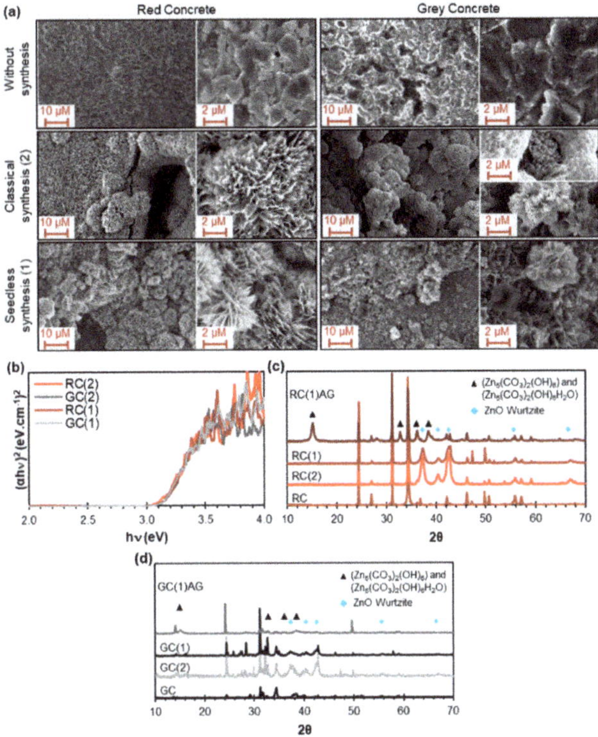

Figure 2. (**a**) SEM top-view pictures of bare concrete surfaces after 2 h of hydrothermal growth by the two-step hydrothermal method or the one-step seedless hydrothermal method; (**b**) UV-visible spectral plot with Tauc-Lorentz model of concrete surfaces after 2 h of hydrothermal growth by the two-step hydrothermal method or the one-step seedless hydrothermal method; (**c**,**d**) XRD pattern of concrete surfaces after 2 h of hydrothermal growth by the two-step hydrothermal method or the one-step seedless hydrothermal method.

However, in accordance with our previous results [5], a strong influence of the concrete complex substrate on the ZnO nanostructure growth was observed on all samples leading to complex ZnO nanostructures, such as nanosheets (NSHs), nanosheets self-assembled spheres (NSHSs), and hierarchical aggregates (HAs), instead of ZnO NWs commonly grown by this synthesis method. Indeed, on the tiling surface functionalized by route (2), ZnO NWs with an average diameter of 72 ± 10 nm and a density of 50 ± 5 NWs/µm² were obtained [5]. No information on ZnO NWs' length was measured on tiling, due to the impossibility of employing the classical cleavage method for SEM cross-section observation. The ZnO NWs on concrete were only observed punctually and were larger than those obtained from tiling. This radical morphological modification was attributed to:

- The turbulence and obstruction caused by the concrete surface complex microstructure, leading to the formation of aggregates and fusion of the ZnO NSs;
- The concrete basic pH property and the chemical composition with different adjuvants used, which could act as structure-modifying agents during the hydrothermal growth [5]. Hydroxyl ions (OH^-) excess and structure-modifying agent, such as trisodium citrate ($Na_3C_6H_5O_7$), were indeed assigned to the NSHs, NSHSs, and HAs formation by being selectively adsorbed on the positively charged zinc (001) plane, leading to a total or partial suppression of growth along the (002) direction, thus favoring the growth in other directions [5,15–17].

It should be noted that no trisodium citrate was used in the formulation of GC samples [18], which, although presenting a larger diversity of ZnO morphologies, presented relatively similar ZnO NSs to RC samples (Figure 2a). Therefore, it is possible to make the following assumption: the basic pH of concrete is the main drivers in the formation of dense ZnO NSs grown as NSHs, NSHSs, and HAs instead ZnO NWs. Nevertheless, the chemical composition, the turbulence, and the obstruction effect does not have to be set aside to explain the diversity of the ZnO NS morphology.

The basic pH of concrete was also considered as the main factor allowing the ZnO NS growth on the concrete surface without needing seed layer deposition and/or metal thin film deposition, such as Au, Ag, Cu, and Sn, to assist the ZnO NSs growth [12,19–21]. The presence of excess OH^- due to the basic pH property of concrete may ensure the formation of a nucleus on the concrete surface rather than in the growth solution. It is effectively well known that OH^- plays a key role, not only in the ZnO growth mechanism by affecting the intermediate chemical reaction and facet growth direction, but also in the initial growth stage by determining the initial growth stage pathway [9,22]. At high pH, the initial growth stage pathway follows Equations (1)–(6) and is described by a quick crystal growth and a slow nucleation rate leading to several growth sites on a ZnO single nucleus, thus explaining the self-assembled structures observed on SEM images [22].

$$C_6H_{12}N_4 + 6\,H_2O \rightarrow 4\,NH_3 + 6\,HCHO \tag{1}$$

$$NH_3 + H_2O \leftrightarrow NH_4^+ + OH^- \tag{2}$$

$$[Zn(NH_3)_4]^{2+} + 2\,OH^- \rightarrow ZnO + 4\,NH_3 + H_2O \tag{3}$$

$$[Zn(NH_3)_4]^{2+} + 4\,OH^- \rightarrow [ZnO_2]^{2-} + 4\,NH_3 + H_2O \tag{4}$$

$$ZnO + 2\,OH^- \rightarrow [ZnO_2]^{2-} + H_2O \tag{5}$$

$$[ZnO_2]^{2-} + H_2O \rightarrow ZnO + 2\,OH^- \tag{6}$$

The second factor, which is assumed to be a minor factor, allowing the concrete functionalization by the seedless hydrothermal growth, is the composition of the hydrothermal growth solution. Indeed, the presence of a local OH^- excess could also be favored by the presence of NO_3^- from the zinc salt used and the presence of NH_4 from HMTA thermal decomposition. The presence of NO_3^- could accelerate the Portlandite ($Ca(OH)_2$) dissolution thanks to its affinity with Ca^{2+}, with which it will react first to form $Ca(NO_3)_2$ before being resolved in solution due to its high solubility. In the literature, concentrated ammonium nitrate in DI water (~6 mol/L) is indeed used to study the accelerated concrete leaching processes over several weeks instead of several years in reality [23]. Nevertheless, regarding the low concentration (0.025 M) and the short duration (2 h), this pathway may not be the main parameter allowing the seedless hydrothermal synthesis and may not significantly damage the concrete.

In addition to the local OH^- excess in the concrete surface caused by its basic property and its reaction with hydrothermal solution, the last factor, which can allow the ZnO NSs' direct growth without seed layer deposition, is the use of additives in the concrete formulation. The presence of O^- groups from additives in the concrete formulation could contribute to the formation of the nucleus on the concrete surface rather than in the growth solution by acting as the place of nucleation for the ZnO seedless hydrothermal growth thanks to a strong adsorption of the {0001} ZnO plane on these carboxylate groups [24].

2.1.2. Gap Measurement

Optical properties were measured at room temperature using UV-visible spectrophotometry. Then, spectra were plotted with the Tauc-Lorentz model (Figure 2b). As it can be observed, all the post-annealed samples revealed an intense absorption under 400 nm corresponding to the bandgap of ZnO. Therefore, the presence of ZnO on all the post-annealed samples was confirmed and the bandgap energy of ZnO NSs was estimated. The intercept of a straight line on this linear part with the hν abscissa is the bandgap. For all functionalized concrete samples, lower ZnO NS bandgap values (~3.13 ± 0.4 eV) than traditional ones of ZnO NWs grown on a silicon substrate [25], on tiling, and on rock samples [10] (usually from 3.20 to 3.23 eV) were recorded in line with previous results on other concrete blocks studied [5]. This variation caused by the morphology modification of ZnO NSs could be associated with their crystal quality, dislocations density, impurities, size, and thickness, for instance [5,7,22]. Indeed, NSHs are supposed to contain more oxygen defects, which could reduce the bandgap by acting as an indirect donor energy level below the conduction band [21,26,27].

Relatively reproducible ZnO bandgap values were obtained on annealed concrete samples with and without seed layer deposition (Figure 2b). This last observation suggests that the removal of seed layer deposition before the hydrothermal synthesis seemed to prevent neither ZnO NSs' growth nor ZnO NSs' good-quality synthesis onto concrete surfaces. No evident gap values were obtained from the Tauc-Lorentz method on sample RC(1)AG and GC(1)AG, due to the lack of an evident gap transition on the UV-visible absorption spectrum.

2.1.3. XRD Characterization

The concrete surface composition and crystalline phase were determined by XRD (Figure 2c,d). For RC(2), RC(1), GC(2), and GC(1), the resulting peaks at 37.2°, 40.3°, 42.5°, 55.6, 66.5°, and 74.2° are assigned to ZnO Würtzite phase (100), (002), (101), (102), (110), and (103), respectively (retaliated values from ICDD NO. 98-002-9272) [28,29]. These peaks were not initially recorded on RC and GC corresponding to the non-functionalized substrates, therefore testifying the ZnO NSs synthesis by the hydrothermal method.

As expected in light of the results achieved, XRD peak intensities indicate that the preferential orientation of ZnO NSs grown on the concrete surface was not along the (002) plane as traditionally required by the hydrothermal route without modifying agents. Once again, it demonstrates the participation of the concrete substrate in the ZnO NS growth mechanism, avoiding a growth along (002) benefiting (100) and (101) planes, which are characteristic of NSs observed by SEM.

By comparing the ZnO Würtzite peaks intensity between the sample functionalized by routes (1) or (2), it is possible to observe that the intensity decreases with the seed layer suppression step (route 1). This last result could explain the impression of a very modest decline in the density of ZnO NSs, which is not quantifiable, for samples grown by route (1).

It is important to note that there is little to no ZnO Würtzite phase on the XRD pattern of RC(1)AG and GC(1)AG. As known in the literature, it confirmed the importance of the last annealing in order to obtain a good crystallinity of ZnO after the growth thanks to the possible Zn(OH)x phase and ZnO amorphous phase conversion into the ZnO Würtzite phase. This hypothesis and the participation of concrete properties in the growth mechanism seem to be proved by the apparition of new peaks in both samples related to possible hydrozincite ($Zn_5(CO_3)_2(OH)_6$), ashoverithe ($Zn(OH)_2$), and zinc carbonate hydroxide hydrate ($Zn_5(CO_3)_2(OH)_6H_2O$) peaks. In view of these elements, the growth mechanism in the hydrothermal solution seems to involve OH^- local excess and organic carbon groups. Then, the post-annealing process serves to convert ZnO amorphous and intermediate to ZnO Würtzite. Indeed, it is well known that species such as zinc hydroxide carbonate could be converted to ZnO thanks to annealing up to 300 °C [30].

In the light of this characterization information, the seedless one-step hydrothermal synthesis allows a functionalization of concrete with ZnO NSs without significant modification compared to those grown by two-step hydrothermal synthesis. The basic pH nature and the presence of O^- groups from concrete may be the main drivers in the formation of ZnO nuclei on the concrete surface followed by a ZnO NS growth. The suppression of the last phase of annealing seems to have an effect on the amount of quality of ZnO nanostructures.

2.2. Photocatalytic Activity Evaluation

2.2.1. Methyl Orange Removal

In order to validate the photocatalytic efficiency of the ZnO NSs functionalized concrete by the one-step seedless hydrothermal synthesis, MO degradation (60 mL at 10 μM) under artificial solar light was carried out in the presence of the samples functionalized by route (1) for subsequent experimental cycles and compared to the results with the samples functionalized by route (2) considered as a reference. Indeed, as described in our previous work [4,5], ZnO NSs grown on concrete by the two-step hydrothermal synthesis (route (2)) already showed an excellent efficiency by decomposing MO under UV light and natural solar light. The irradiance of the light received by the sample was adjusted to be comparable to a sunny summer day in France (~3700 μW/cm²). MO degradation was monitored by UV-visible spectrophotometry every 30 min for 5 h and the degradation efficiency X(%) was estimated thanks to Equation (7). The same experiments were carried out with no concrete or no light to evaluate the photolysis and the adsorption effects, respectively.

$$X(\%) = \left(\frac{A_0 - A}{A_0} \right) \times 100 \tag{7}$$

where A_0 and A, respectively, stand for the initial and actual absorption peak values at the wavelength of the maximum absorption for the studied dye (λ_{max} = 464 nm for MO and 515 nm for AR14).

As shown in Figure 3, no strong photolysis under artificial solar light (<20%) and no strong adsorption (<35%) for RC(2) were recorded after the 5 h process. After a second cycle process, the MO adsorption rate decreased to ~8%. Therefore, the MO degradation of 94.6% observed in cycle 1 in the presence of RC(2) under solar light could be mainly attributed to the photocatalytic activity of ZnO NSs. The strong adsorption decreases in RC(2) between cycle 1 and cycle 2 could explain the photocatalytic kinetic difference. After 5 h in the presence of RC(2) under solar light, the MO degradation reached 94.6%, 95.6%, 94%, 95%, 87.4%, and 90% for cycles 1 to 6, respectively. The MO degradation kinetics could be affected by the placement of the concrete under solar light. Indeed, even a slight variation in placement could induce different illuminated surface quantities, impacting the photocatalytic degradation rate. In this regard, the kinetic difference is not discussed in depth in this article. Therefore, it is possible to consider that the RC(2) photocatalytic efficiency remains relatively stable after 6 cycles with only a slight decrease of 5% compared to cycle 1.

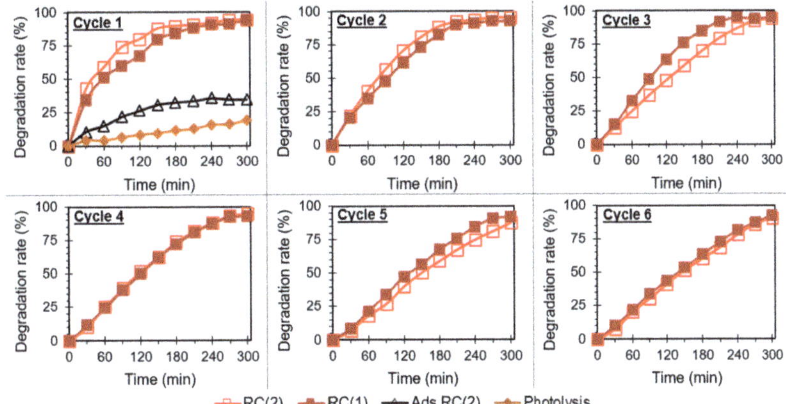

Figure 3. Photolysis degradation rate, adsorption rate in the dark (Ads), and photocatalysis degradation rate as a function of time for MO under artificial solar light in the presence of RC(2) or RC(1) for subsequent experiment cycles.

These results are promising as the stability is another important aspect in view of their application as an efficient urban depolluting surface over time. This good photocatalytic activity can be attributed to [6]:

- The fine texture of NSs and the presence of numerous boundaries increasing dye adsorption and light harvesting;
- The hierarchical structure, the various gains and interfaces facilitating the electron/hole (e^-/h^+) photogeneration, and the diffusion of e/h;
- The rich mesoporous structure helping dye adsorption and light-generated charge transfer.

By comparing the results of RC(2) with RC(1), the same trend was observed with only a small kinetic difference. After 5 h in the presence of RC(1) under solar light, the MO degradation reached 93.9%, 92.8%, 95.4%, 93%, 92%, and 92.6% for cycles 1 to 6, respectively. After several cycles, RC(1) seems slightly more stable and efficient than RC(2). Thus, it implies that the one-step seedless hydrothermal synthesis could be used to functionalize concrete surfaces with ZnO NSs as a more environmentally friendly approach to produce photocatalytic concrete for environmental pollution remediation.

To certify these observations and the universality of the synthesis process, the same experiments were carried out with the GC samples, which has a different composition and porosity than RC, as it can be seen in Figures 1 and 2c,d. In Figure 4, it is possible to note that the MO adsorption rate on GC(2) after 5 h is about 54% for the first use and about 6% for the second one, certifying that the degradation rate obtained under solar light is mainly due to the photocatalytic activity of ZnO NSs and justifying the difference in kinetics recorded between photocatalytic cycle 1 and 2. After 5 h in the presence of GC(2) under solar light, the MO degradation reached 95.7%, 94.1%, 93.7%, 91.1%, 82%, and 87.2% for cycles 1 to 6, respectively. Therefore, over 6 cycles, the GC(2) sample still provided a good photocatalytic activity comparable to RC(2) results and was relatively stable with only a slight decrease of ~9% between cycles 1 and 6.

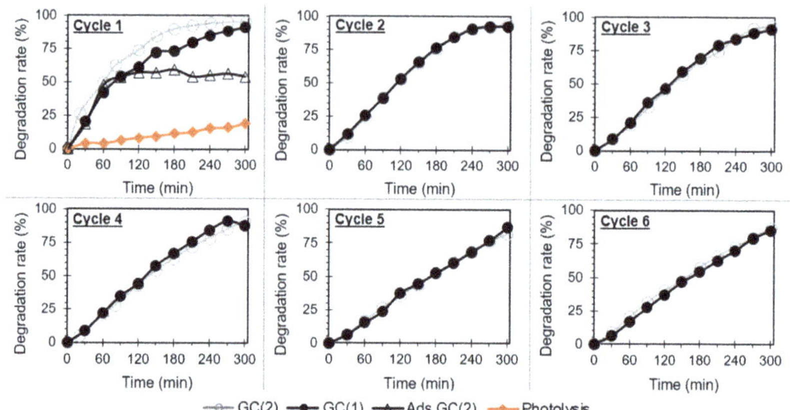

Figure 4. Photolysis degradation rate, adsorption rate in the dark (Ads), and photocatalysis degradation rate as a function of time for MO under artificial solar light in the presence of GC(2) or GC(1) for subsequent experiment cycles.

The small difference in the photocatalytic activity between GC(2) and RC(2) is likely due to the affinity and reactivity difference between MO, ZnO NSs, and the concrete. In fact, it is effectively well known that strong pollutant adsorption could be harmful for the photocatalysis depollution process by preventing water adsorption and, thus, hydroxyl radicals' creation. This variation could also be due to the porosity affecting the photocatalyst surface area, the pollutant, and light diffusion, which itself affects the photocatalytic efficiency. Indeed, photocatalytic activity could be impacted by the pollutants mass transfer from the environment to the ZnO NSs surface depending on the coupling of ZnO NSs/substrate [31].

A similar photocatalytic activity was recorded for GC(2) and GC(1) samples. After 5 h in the presence of GC(1) under solar light, the MO degradation reached 90.9%, 92.6%, 91.4%, 87.8%, 87%, and 85% for cycles 1 to 6, respectively. Therefore, the one-step seedless hydrothermal synthesis demonstrated once again its efficiency to produce more environmentally friendly ZnO-based photocatalytic concrete with similar activity and durability to those grown by two-step hydrothermal synthesis.

2.2.2. Acid Red 14 Removal

Regarding MO results, only RC(1) and GC(1) (samples obtained with seedless method), which are greener photocatalysts, were selected to conduct this study. The photocatalytic efficiency of RC(1) and GC(1) was, thus, evaluated for the degradation of another organic dye, AR14, in the same previous conditions in order to certify and compare their photocatalytic activity (Figure 5).

Their activity was proved by increasing the degradation rate ~30% after 5 h under artificial solar light compared to the photolysis degradation rate during cycle 1: 100% for RC(1), 96.3% for GC(1), and 69.2% for the photolysis. In this first cycle, it is possible to observe a sizeable difference between RC(1) and GC(1) photocatalytic kinetics, which disappeared after cycle 2. As discussed in the MO section, this initial difference could be due to the different affinity and reactivity of the pollutant with the samples and the different porosity of concrete substrates. The AR14 degradation rate after 5 h under artificial solar light for the second cycle was 100% for RC(1) and 95% for GC(1). Then, a similar slight decrease in kinetic activity both of RC(1) and GC(1) was recorded until cycle 3, leading to a degradation rate of 96.2% for RC(1) and 98.8% for GC(1) after 5 h under solar light. Beyond this cycle 3, the photocatalytic activity remained stable over time and the difference observed was mainly attributed to the error average between each cycle. The

AR14 degradation rates after 5 h under artificial solar light for cycles 3 to 5, respectively, were 99.4%, 100%, and 100% for RC(1), and 100%, 99.4%, and 97.6% for GC(1).

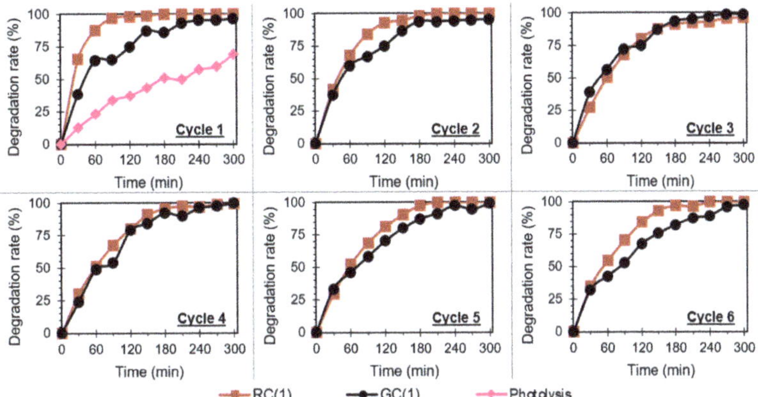

Figure 5. Photolysis degradation rate and photocatalysis degradation rate as a function of time for AR14 under artificial solar light in the presence of RC(1) or GC(1) for subsequent experiment cycles.

Despite the impossibility to make a direct comparison with previous results due to the non-identical experimental conditions, it is important to give an order of idea to the photocatalytic activity of other photocatalysts in the literature. An efficiency close to 70.4% of AR14 degradation was reached in 3.5 h with ZnO nanoparticles (60 ppm) in 50 mL of AR14-agitated solution at 20 ppm under UV-C light (30 W) [32]. A total degradation of 7 L of AR14 at 0.01 M was obtained thanks to an immobilized TiO_2 nanoparticle photocatalytic reactor, in which 7.5 mM of H_2O_2 was diffused, under a 15 W UV-C lamp in 1 h [33]. An amount of 48 mg of $ZnCo_2O_4/Co_3O_4$ eliminated 84.8% of AR14 in 65 min (1.2 mg of AR14 under 125 W of visible light) [34]. Considering photocatalytic concrete samples as fixed photocatalysts with weak weight, the results recorded are promising.

Therefore, the results from those functionalized concretes demonstrated an excellent photocatalytic activity and a relatively stable activity of samples obtained by the one-step seedless hydrothermal growth of ZnO NSs, thus illustrating that this faster and proper seedless hydrothermal growth could replace the two-step hydrothermal route for the ZnO-based photocatalytic concrete development at large scale. Nevertheless, even if this process is greener, at this stage, it still needs post-annealing at 350 °C, which may damage the substrate. For this reason and still with the objective to produce the most environmentally friendly photocatalytic concrete as possible, the post-grown annealing step of synthesis route (1) was deleted. As-grown obtained samples, RC(1)AG and GC(1)AG, were used for AR14 degradation for 3 subsequent cycles without annealing between experiments.

After 5 h under artificial solar light, the AR14 degradation rate reached 97.2%, 97%, and 96.3% for RC(1)AG from cycles 1 to 3 and 99.2%, 97.3%, and 96.2% for GC(1)AG from cycles 1 to 3 (Figure 6), respectively. The results showed that post-annealed samples and as-grown samples have similar trends and efficiency even if the amount of ZnO on the surface and its crystallinity are lower, as shown in Figure 2c,d. Only a slight drop in GC(1)AG activity could be observed and may be due to pollutants accumulation in the porous structure. It is important to note that as-grown samples were not annealed between each experiment, unlike other samples. Moreover, no other reactivation treatment, such as UV exposition, was used to replace the annealing reactivation. This last point underlines the promising application of this concrete in real life where the thermic regeneration is not possible.

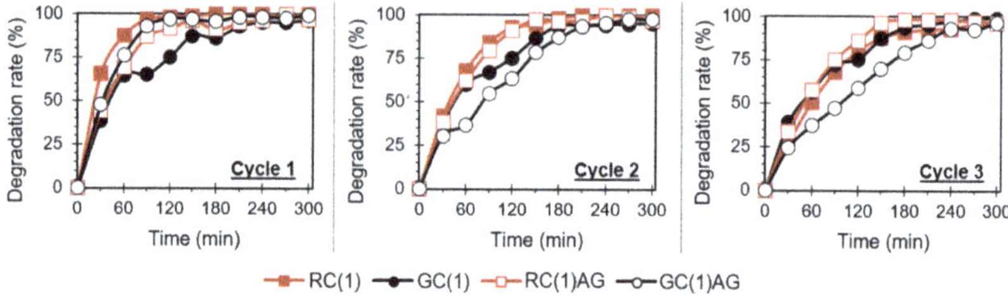

Figure 6. Photocatalysis degradation rate as a function of time for AR14 under artificial solar light in the presence of RC(1), GC(1), and RC(1)AG or GC(1)AG for 3 subsequent experimental cycles.

The results, particularly on RC samples, are very promising and put in light that it is possible to produce efficient photocatalytic concrete by only a short growth in hydrothermal solution with neither the seed layer deposition nor post-grown annealing. This method allows, therefore, a quick and low-chemical- and energy-consuming growth of ZnO NSs on concrete, which made this synthesis a very promising route for photocatalytic concrete daily life production with excellent photocatalytic activity.

3. Materials and Methods

3.1. Concrete Surface Functionalization

Red concrete blocks (purchased from DIY retailer Leroy Merlin, Lexovimes, France) and grey porous concrete blocks developed by Gustave Eiffel University for a permeable and removable urban road within the I-Street/CUD-SF project (Ademe, programmes d'investissement d'avenir (PIA), France) [12] cut into pieces of ~2.5 cm × 2.5 cm × 4 cm were used as substrates for the surface functionalization (Figure 1). Prior to any synthesis, a brief washing process was first carried out on all substrates by a rinse with DI water and a drying step at 100 °C during 30 min in order to remove all dust due to the cutting step. Then, the samples were functionalized by the two different routes described below before being carefully characterized by UV-visible spectroscopy (Maya2000 Pro from Ocean Optics, Duiven, Netherlands) and by field-emission scanning electron microscopy (Zeiss FE-SEM NEON 40, Oberkochen, Germany) to determine their bandgap value and to investigate the morphology of the obtained ZnO nanostructures. The samples were also characterized by X-ray diffraction (XRD, D8 Advance, Bruker France SAS, Champs sur Marne, France) using a cobalt anticathode (λ = 1.78897 Å, 35 kV, 40 mA).

3.1.1. Two-Step Hydrothermal Synthesis Method

The two-step hydrothermal synthesis known to be efficient for functionalizing concrete surfaces was used to produce reference samples (route (2) on the Figure 1) [4,6]. First, a ZnO seed layer was deposited on a substrate by an "horizontal impregnation" method with a solution of zinc acetate dihydrate ($Zn(Ac)_2 \cdot 2\ H_2O$, 98%, Sigma-Aldrich, St. Louis, MO, USA, CAS-No 5970-45-6) at 0.01 M in absolute ethanol (99.9%, Carlo Erba, Val-de-Reuil, France, CAS-No 64-17-5) followed by an annealing at 350 °C during 30 min. Then, a classical hydrothermal growth using 300 mL of equimolar aqueous solutions of hexamethylenetetramine (HMTA, $\geq$99%, VWR International, Radnor, PA, USA, CAS-No 100-97-0) and zinc nitrate hexahydrate ($Zn(NO_3)_2 \cdot 6\ H_2O$, 98%, Sigma-Aldrich, St. Louis, MO, USA, CAS-No 10196-18-6) at 0.025 M was carried out in an autoclave at 90 °C for 2 h. Finally, the as-synthesized samples were annealed for 30 min at 350 °C in order to remove all potential residues from the synthesis process and to improve the ZnO NS crystallinity. The samples were named RC(2) for the red concrete functionalized with ZnO NSs or GC(2) for the grey concrete, developed for a permeable removable urban road, functionalized with ZnO NSs (see the optical photographs on Figure 1).

3.1.2. One-Step Hydrothermal Synthesis Method

This method consists of a one-step seedless hydrothermal route directly functionalizing the concrete blocks substrates. Therefore, the same hydrothermal and annealing processes as described in Section 3.1.1 were applied to the concrete substrates directly after its washing process (route (1) in Figure 1). Then, the samples were named RC(1) for the red concrete functionalized with ZnO NSs or GC(1) for the grey concrete, developed for a permeable and removable urban road, functionalized with ZnO NSs. As-grown ZnO NSs without the final annealing step were also produced for comparison and named RC(1)AG and GC(1)AG.

3.2. Photocatalytic Activity Evaluation: Water Purification

The photocatalytic activity was evaluated by the degradation of MO (85%, Sigma-Aldrich, CAS-No 547-58-0) and AR14 (50%, Sigma-Aldrich, CAS-No 3567-69-9) with an initial concentration of 10 µM under artificial solar light irradiation (Sirius 300PU, Beijing, China; 200–2500 nm; P = 5–15 sun at 50 mm from the output). The sample was immerged in 60 mL of aqueous solution polluted by MO or AR14 dye and placed under the solar source with a fixed distance of 90 cm between the lamp and the sample surface. The UV light intensity received by the sample was around ~3700 $\mu W/cm^2$ corresponding to a sunny summer day in France. In order to estimate the photocatalytic efficiency, the photolysis (without sample) and adsorption experiments in the dark were also carried out. Experiments were monitored by UV-visible spectrophotometry every 30 min for 5 h and the degradation efficiency X(%) was estimated thanks to Equation (7).

4. Conclusions

In this work, the functionalization of two different concretes (different chemical formulation and porosity) by a seedless hydrothermal growth in 2 h was studied. Their photocatalytic activity for organic dye degradation in water was investigated and compared to the concretes functionalized by traditional two-step hydrothermal growth. The results of this study showed that the seedless hydrothermal growth in 2 h could be used as a novel, faster, less expensive, and more environmentally friendly process to produce ZnO NSs-based photocatalytic concrete. Similar gap values and nanostructures morphologies were recorded for all samples functionalized whatever the concrete substrate and the hydrothermal route used. The XRD pattern demonstrated the presence of a ZnO Würtzite phase on all functionalized post-annealed samples. It is important to note that there was little to no ZnO Würtzite phase in the XRD pattern of non-post-annealed samples. No strong ZnO NSs properties modifications were recorded. It was, therefore, assumed that the basic pH and the presence of O^- groups from concrete may be the main drivers in the formation of ZnO NSs at the concrete surface without the need of seed layer deposition and/or metal thin-film deposition. The universality of this route was proved for the two different kinds of concrete.

Photocatalytic concrete from this new approach demonstrated a similar activity with the same lifetime to photocatalytic concrete functionalized by the traditional two-step synthesis in our previous work. Amounts of 85–95% of MO degradation rate and 96–100% of AR14 degradation rate were reached after 5 h under artificial solar light in the presence of functionalized concrete even after 6 cycles of use. The obtained results also showed that the post-annealed samples and the as-grown samples have a similar trend and photocatalytic efficiency.

Therefore, short seedless hydrothermal growth in 2 h can function as a very beneficial and appropriate way to develop a photocatalytic concrete surface for environmental remediation. In terms of industrial upscaling for daily life production, this short and one-step synthesis route offers a real advantage for the fabrication of photocatalytic concrete.

Nevertheless, it is important to keep in mind that this study is still a primary investigation at the initial stage. Therefore, more research should be conducted prior to use of these functionalized blocks. The ZnO NSs should be tested for resistance in traffic abrasion, dust deposition, and other climate factors. The photocatalytic activity of aged samples should be evaluated. More in-depth research on the photocatalytic activity should be conducted to evaluate the photocatalytic activity in real conditions. The effect of the functionalization of concrete on its properties must be studied.

Author Contributions: The individual contributions was distributed as followed: conceptualization, M.L.P. and Y.L.-W.; methodology, M.L.P. and Y.L.-W.; experiments realization, M.L.P.; validation, M.L.P. and Y.L.-W.; formal analysis, M.L.P.; investigation, M.L.P.; resources, Y.L.-W.; data curation, M.L.P.; writing—original draft preparation, M.L.P.; writing—review and editing, M.L.P. and Y.L.-W.; visualization, M.L.P. and Y.L.-W.; supervision, Y.L.-W.; project administration, Y.L.-W.; funding acquisition, Y.L.-W. All authors have read and agreed to the published version of the manuscript.

Funding: This research was funded by The E3S project (2020, ANR via programmes d'investissement d'avenir (PIA), France, reference ANR-16-IDEX-0003) and the Smart Lab LABILITY of the University Gustave Eiffel (France), funded by the Region Île-de-France under Grant N°20012741.

Data Availability Statement: Not applicable.

Acknowledgments: The development of grey concrete for a permeable and removable urban road provided by Gustave Eiffel University was funded by the I-Street/CUD-SF project (Ademe, PIA). The financial and technical support linked to this project provided are gratefully acknowledged. The authors would like to thank M. Thierry Sedran and Julien Le Mouël from the University Gustave Eiffel for providing the concrete substrate and their technical help on concrete cutting. The authors would like to thank M. Nicolas Hautiére and Frédéric Bourquin from the University Gustave Eiffel for their invaluable help on this project. The authors would like to thank Myriam Duc from the University Gustave Eiffel for her helpful support to sample XRD characterization; and Stéphane Bastide from the East Paris Institute of Chemistry and Materials (ICMPE-CNRS) for his help on SEM characterization.

Conflicts of Interest: The authors declare no conflict of interest.

References

1. Liang, X.; Cui, S.; Li, H.; Abdelhady, A.; Wang, H.; Zhou, H. Removal effect on stormwater runoff pollution of porous concrete treated with nanometer titanium dioxide. *Trans. Res. Part D* **2019**, *73*, 34–45. [CrossRef]
2. Shen, S.; Burton, M.; Jobson, B.; Haselbach, L. Pervious concrete with titanium dioxide as photocatalyst compound for a greener urban road environment. *Const. Build. Mater.* **2012**, *35*, 874–883. [CrossRef]
3. Asadi, S.; Hassan, A.M.; Kevern, J.T.; Rupnow, T.D. Development of photocatalytic pervious concrete pavement for air and storm water improvements. *J. Trans. Res. Board* **2012**, *2290*, 161–167. [CrossRef]
4. Le Pivert, M.; Poupart, R.; Capochichi-Gnambodoe, M.; Martin, N.; Leprince-Wang, Y. Direct growth of ZnO nanowires on civil engineering materials: Smart materials for supported photodegradation. *Micro. Nanoengin.* **2019**, *5*, 57. [CrossRef]
5. Le Pivert, M.; Zerelli, B.; Martin, N.; Capochichi-Gnambodoe, M.; Leprince-Wang, Y. Smart ZnO decorated optimized engineering materials for water purification under natural sunlight. *Const. Build. Mater.* **2020**, *257*, 119592. [CrossRef]
6. Qiu, Y.; Wang, L.; Xu, L.; Shen, Y.; Wang, L.; Liu, Y. Shaped-controlled growth of sphere-like ZnO on modified polyester fabric in water bath. *Mater. Lett.* **2021**, *288*, 129342. [CrossRef]
7. Pastor, A.; Balbuena, J.; Cruz-Yusta, B.M.; Pavlovic, I.; Schànchez, L. ZnO on rice husk: A sustainable photocatalyst for urban air purification. *Chem. Eng. J.* **2019**, *368*, 659–667. [CrossRef]
8. Choudhary, S.; Sahu, K.; Bisht, A.; Satpati, B.; Mohapatra, S. Rapid synthesis of ZnO nanowires and nanoplates with highly enhanced photocatalytic performance. *App. Surf. Sci.* **2021**, *541*, 148484. [CrossRef]
9. Chandran, R.; Mallik, A. Facile, seedless and sufractant-free synthesis of ZnO nanostructures by wet chemical bath method and their characterization. *Appl. Nanosci.* **2018**, *8*, 1823–1830. [CrossRef]
10. Le Pivert, M.; Kerivel, O.; Zerelli, B.; Leprince-Wang, Y. ZnO nanostructures based innovative photocatalytic road for air purification. *J. Clean. Prod.* **2021**, *318*, 128447. [CrossRef]
11. Hossain, M.F.; Naka, S.; Okada, H. Fabrication of perovskite solar cells with ZnO nanostructures prepared on seedless ITO substrate. *J. Mater. Sci. Mater. Elec.* **2018**, *29*, 13864–13871. [CrossRef]
12. Hong, G.W.; Kim, J.; Lee, J.S.; Shin, K.; Jung, D.; Kim, J.H. A flexible tactile sensor using seedless hydrothermal growth of ZnO nanorods on fabrics. *J. Phys. Com.* **2020**, *4*, 045002. [CrossRef]
13. Zhang, Y.; Huang, X.; Yeom, J. A flotable piezo-photocatalytic platform based on semi-embedded ZnO nanowire array for high-performance water decontamination. *Nano-Micro Lett.* **2019**, *11*, 11. [CrossRef]
14. Habba, Y.G.; Capochichi-Gnambodoe, M.; Serairi, L.; Leprince-Wang, Y. Enhanced photocatalytic activity of ZnO nanostructure for water purification. *Phys. Status Solidi B* **2016**, *253*, 1480–1484. [CrossRef]

15. Shaban, M.; Zayed, M.; Hamdy, H. Nanostructured ZnO thin films for self-cleaning applications. *R. Soc. Chem.* **2017**, *7*, 617–631. [CrossRef]
16. Khoa, N.T.; Kim, S.W.; Thuan, D.V.; Yoo, D.H.; Kim, E.J.; Hahn, S.H. Hydrothermally controlled ZnO nanosheet self-assembled hollow spheres/hierarchical aggregates and their photocatalytic activities. *CrystEngComm* **2013**, *16*, 1344–1350. [CrossRef]
17. Luo, S.; Chen, R.; Xiang, L.; Wang, J. Hydrothermal synthesis of (001) facet highly exposed ZnO plates: A new insight into the effect of citrate. *Crystals* **2019**, *9*, 552. [CrossRef]
18. Sedran, T.; Gennesseaux, E.; Nguyen, M.-L.; Waligora, J.; Guyot, D.; Le Mouel, J. Development of a permeable removable urban pavement. In Proceedings of the International Conference on Concrete Roads, Krakow, Poland, 25–28 June 2023.
19. Wen, X.; Wu, W.; Ding, Y.; Wang, L.Z. Seedless synthesis of patterned ZnO nanowire arrays on metal thin films (Au, Ag, Cu, Sn) and their application for flexible electromechanical sensing. *J. Mater. Chem.* **2012**, *22*, 9469–9476. [CrossRef]
20. Mai, H.H.; Tran, D.H.; Janssens, E. Non-enzymatic fluorescent glucose sensor using vertically aligned ZnO nanotubes grown by a one-step, seedless hydrothermal method. *Microchim. Acta* **2019**, *186*, 245. [CrossRef]
21. Tian, J.H.; Hu, J.; Li, S.S.; Zhang, F.; Liu, J.; Shi, J.; Li, X.; Tian, Z.Q.; Chen, Y. Improved seedless hydrothermal synthesis of dense and ultralong ZnO nanowires. *Nanotechnology* **2011**, *22*, 245601. [CrossRef]
22. Adbulrahman, A.F.; Ahmed, S.M.; Hamad, S.M.; Almessiere, M.A.; Sadaji, S.M. Effect of different pH values on growth solutions for the ZnO nanostructures. *J. Phys.* **2021**, *71*, 175–189. [CrossRef]
23. Bilal, H.; Chen, T.; Ren, M.; Gao, X.; Su, A. Influence of silica fume, metakaolin & SBR latex on strength and durability performance of previous concrete. *Cons. Build. Mater.* **2021**, *275*, 122124. [CrossRef]
24. Winkler, N.; Edinger, S.; Kautek, W.; Dimopoulos, T. Mg-doped ZnO films prepared by chemical bath deposition. *J. Mater. Sci.* **2018**, *53*, 5159–5171. [CrossRef]
25. Leprince-Wang, Y.; Martin, N.; Ghozlane Habba, Y.; Le Pivert, M.; Capochichi-Gnambodoe, M. ZnO nanostructure based photocatalysis for water purification. *Nanoworld J.* **2020**, *6*, 1–6. [CrossRef]
26. Samadi, M.; Shivaee, H.A.; Pourjavadi, A.; Moshfegh, A.Z. Synergism of oxygen vacancy and carbonaceous species on enhanced photocatalytic activity of electrospun ZnO-carbon nanofibers: Charge carrier scavengers mechanism. *Appl. Catal. A Gen.* **2013**, *466*, 153–160. [CrossRef]
27. Yu, Z.; Moussa, H.; Liu, M.; Schneider, R.; Moliere, M.; Liao, H. Solution precursor plasma spray process as an alternative rapid one-step route for the development of hierarchical ZnO films for improved photocatalytic degradation. *Ceram. Int.* **2018**, *44*, 2085–2092. [CrossRef]
28. Yu, Z.; Moussa, H.; Chouchene, R.; Schneider, R.; Wang, W.; Moliere, M.; Liao, H. Tunable morphologies of ZnO films via the solution precursor plasma spray process for improved photocatalytic degradation performance. *Appl. Surf. Sci.* **2018**, *455*, 970–979. [CrossRef]
29. Joshi, S.; Jones, L.A.; Sabri, Y.M.; Bhargava, S.K.; Sunkara, M.V.; Ippolito, S.J. Facile conversion of zinc hydroxide carbonate to CaO-ZnO for selective CO_2 detection. *J. Colloid Interface Sci.* **2020**, *588*, 310–322. [CrossRef]
30. Dash, P.; Manna, A.; Mishra, N.C.; Varma, S. Synthesis and characterization of aligned ZnO nanorods for visible light photocatalysis. *Phys. E Low-Dimens. Sys. Nanostructures* **2019**, *107*, 38–46. [CrossRef]
31. Haghighat Mamaghani, A.; Haghighat, F.; Lee, C.S. Effect of titanium dioxide properties and support material on photocatalytic oxidation of indoor air pollutants. *Build. Environ.* **2021**, *189*, 107518. [CrossRef]
32. Daneshvar, N.; Salari, D.; Khataee, A.R. Photocatalytic degradation of azo dye acid red 14 in water on ZnO as an alternative catalyst to TiO_2. *J. Photochem. Photobiol. A Chem.* **2004**, *162*, 317–322. [CrossRef]
33. Mahmoodi, N.M.; Arami, M. Bulk phase degradation of acid red 14 by nanophotocatalysis using immobilize titanium(IV) oxide nanoparticles. *J. Photochem. Photobiol. A Chem.* **2006**, *182*, 60–66. [CrossRef]
34. Zinatloo-Ajabshir, S.; Heidari-Asil, S.A.; Salavati-Niasari, M. Recyclable magnetic $ZnCo_2O_4$-based ceramics nanostructure materials fabricated by simple sonochemical route for effective sunlight-driven photocatalytic degradation of organic pollution. *Ceram. Int.* **2021**, *47*, 8959–8979. [CrossRef]

 *catalysts*

Article

Enhanced Photoredox Activity of BiVO₄/Prussian Blue Nanocomposites for Efficient Pollutant Removal from Aqueous Media under Low-Cost LEDs Illumination

Abrar Ali Khan [1], Leonardo Marchiori [2], Elias Paiva Ferreira-Neto [3], Heberton Wender [4], Rashida Parveen [5], Mohammad Muneeb [1], Bianca Oliveira Mattos [6], Ubirajara Pereira Rodrigues-Filho [6], Sidney José Lima Ribeiro [2] and Sajjad Ullah [1,*]

[1] Institute of Chemical Sciences, University of Peshawar, Peshawar P.O. Box 25120, Pakistan
[2] Institute of Chemistry, São Paulo State University (UNESP), Araraquara 14800-060, SP, Brazil
[3] Department of Chemistry, Federal University of Santa Catarina (UFSC), Florianópolis 88040-900, SC, Brazil
[4] Nano&Photon Research Group, Laboratory of Nanomaterials and Applied Nanotechnology, Institute of Physics, Federal University of Mato Grosso do Sul, Campo Grande 79070-900, MS, Brazil
[5] Government Girls Degree College Dabgari, Peshawar 25000, KP, Pakistan
[6] Institute of Chemistry of São Carlos, University of São Paulo, São Carlos 13560-970, SP, Brazil
* Correspondence: sajjadullah@uop.edu.pk

Citation: Khan, A.A.; Marchiori, L.; Ferreira-Neto, E.P.; Wender, H.; Parveen, R.; Muneeb, M.; Mattos, B.O.; Rodrigues-Filho, U.P.; Ribeiro, S.J.L.; Ullah, S. Enhanced Photoredox Activity of BiVO₄/Prussian Blue Nanocomposites for Efficient Pollutant Removal from Aqueous Media under Low-Cost LEDs Illumination. *Catalysts* 2022, 12, 1612. https://doi.org/10.3390/catal12121612

Academic Editors: Gassan Hodaifa, Rafael Borja and Mha Albqmi

Received: 19 October 2022
Accepted: 4 December 2022
Published: 8 December 2022

Publisher's Note: MDPI stays neutral with regard to jurisdictional claims in published maps and institutional affiliations.

Abstract: Bismuth vanadate (BiVO₄, BV) is a widely explored photocatalyst for photo(electro)chemical applications, but its full photocatalytic potential is hindered by the fast recombination and low mobility of photogenerated charge carriers. Herein, we propose the photodeposition of different amounts of Prussian blue (PB) cocatalysts on the surface of monoclinic BV to obtain BV-PB composite photocatalysts with increased photoactivity. The as-prepared BV and BV-PB composites were characterized by an array of analytic techniques such scanning eletron microscopy (SEM), transmission eletron microscopy (TEM), X-day diffraction (XRD), and spectroscopic techniques including Fourier-transform infrared spectroscopy (FTIR), diffuse reflectance spectroscopy (DRS), electrochemical impedance spectroscopy (EIS), photoluminescence (PL), and Raman spectroscopy. The addition of PB not only increases the absorption of visible light, as indicated by DRS, but also improves the charge carriers' transfer across the photocatalysts/solution interface and hence reduces electron-hole (e^--h^+) recombination, as confirmed by EIS and PL measurements. Resultantly, the BV-PB composite photocatalysts with optimum PB loading exhibited enhanced Cr(VI) photoreduction efficiency as compared to pristine BV under visible light illumination from low-power blue light-emitting diodes (LEDs), thanks to the cocatalyst role of PB which mediates the transfer of photoexcited conduction band (CB) electrons from BV to Cr(VI) species in solution. Moreover, as compared to pristine BV and BV + H₂O₂, a drastic increase in the methylene blue (MB) photo-oxidation efficiency was observed for BV-PB in the presence of a minute quantity of H₂O₂ due to a synergic effect between the photocatalytic and Fenton-like processes. While pure BV photodegraded around 70% of MB dye within 120 min, the BV-PB/H₂O₂ and BV/H₂O₂ system could degrade almost 100% of the dye within 20 min ($k_{obs.}$ = 0.375 min⁻¹) and 40 min ($k_{obs.}$ = 0.055 min⁻¹), respectively. The practical approach employed in this work may pioneer new prospects for synthesizing new BV-based photocatalytic systems with low production costs and high photoredox efficiencies.

Keywords: photocatalysis; photo-fenton; BiVO₄; Prussian blue; cocatalyst; water purification

1. Introduction

With increasing industrialization and anthropogenic activities across the globe, freshwater bodies have been adversely affected by the uncontrolled contamination with many toxic substances, including both organic (dyes, pharmaceuticals, pesticides, etc.,) and inorganic contaminants (heavy metal ions). The presence of these toxic dyes and especially heavy metal toxins in aqueous environments has become a major global concern that needs to be addressed sooner [1–3]. For example, Cr(VI), which exists in different forms ($Cr_2O_7^{2-}$, $HCrO_4^-$, H_2CrO_4) in aqueous media [4], is highly toxic/carcinogenic, and even poses a threat of genetic modification in aquatic organisms and human beings [5] and is thus considered highly obnoxious and lethal by World Health Organization with a maximum permissible concentration of 0.05 mg/L in aqueous environments [6]. Hence, it is important to find out cost-effective methods to remove hexavalent chromium from water bodies. Consequently, various treatment methods have been introduced for Cr(VI) removal including adsorption [7,8], flocculation [9], chemical precipitation, reverse-osmosis [10,11], electro-coagulation [12], biological treatment [13], membrane separation [11,14], electro-chemical treatment [15,16], ion exchange [17] and photocatalytic reduction processes [18,19]. Similarly, the organic molecules present in aqueous media can be effectively oxidized using advanced oxidation processes (AOPs) which generate highly reactive oxidant species (such as superoxide $O_2^{\bullet-}$ and super hydroxyl $^\bullet OH$ radicals) for (photo)oxidation of organics [20].

Since Cr(III) is considered to be less toxic and since it can be easily removed as chromium hydroxide via precipitation using conventional techniques of water treatment, photocatalytic reduction of Cr(VI) to Cr(III) is considered a viable strategy to combat Cr(VI) pollution [1,18,19]. This process relies on the transfer of photogenerated electrons from the conduction band of photocatalysts to the Cr(VI) species solution, thereby reducing them to the less toxic Cr(III) form [18]. Unfortunately, the most studied wide-bandgap ($E_g > 3$ eV) semiconductor photocatalysts (such as ZnO and TiO_2) require UV light (which is less than 4% of the solar spectrum) for their photoexcitation and consequent photo(electro)chemical applications, and this practical limitation has attracted researcher's interest to develop visible light active photocatalysts able to use the 45% visible light, so visible light driven photocatalysts are essential to be studied for photocatalytic applications [21–25].

Among the visible light photocatalysts, bismuth vanadate (BV) has gained greater attention due to its narrow band gap (2.4–2.6 eV) and photo(chemical) stability [26]. BV has been widely studied for environmental pollutant degradation, water oxidation, water splitting, and biosensors, among other applications [27–30]. However, BV suffers from fast electron-hole (e^--h^+) recombination and high resistance to charge transfer to the target species in the solution. To overcome these limitations and bring efficacy in the photoelectrochemical applications of BV, several strategies have been employed, including the loading of metal nanoparticles or metal complexes on BV, hetero-junctions formation [23,26,31–35], metal-ions doping [36–38], and control of BV morphology [39–42].

Prussian blue (PB) is an ancient dye having chemical formulae of $Fe_4[Fe(CN)_6]_3$ with three Fe(II) in hexa-coordination with CN and the $Fe(CN)_6$ octahedral is then combined with four Fe(III) in three-dimensional extended Fe(III)$-$N$\equiv$C$-$Fe(II)$-$C$\equiv$N$-$Fe(III) linkages, leading to a unit cell with cubic lattice structure [43]. PB is a mixed-valence complex with different ligand-field-strength as a result of the ligand donor environment, where nitrogen-coordinated high-spin d^5 (S = 5/2) ferric sites and carbon-coordinated low-spin d^6 (S = 0) ferrous sites favor eminent catalytic and electro-magnetic properties [44]. PB-based nanomaterials have been previously explored for (photo)electrochemical applications in conjunction with semiconductor photocatalysts such as TiO_2 and g-C_3N_4 [18,44–49]. The often-enhanced photoactivity of such systems highlights the transfer of photo-generated electrons from the CB of semiconductor nanomaterials towards the deposited PB cocatalyst layer, thereby reducing PB to Prussian white (PW). The in-situ produced PW species may catalytically transfer electrons to the target species (metal-ions, organic molecules/pollutants) in the solution causing their photoreduction [18]. Similarly, Prussian blue analogues (PBA) have been found to improve the water oxidation efficiency of semiconductor photocatalysts [50–52].

Nevertheless, PB/semiconductor systems other than UV-active TiO_2-based systems and with high activity in the visible light range remain understudied for photocatalytic applications such as the photoreduction of Cr(VI) and oxidation of organic molecules. Therefore, we present, for the first time, a facile in-situ photodeposition methodology to obtain PB-decorated BV (BV-PB) photocatalysts employing low-cost blue-emitting LEDs (output 460 ± 10 nm (see Figure S1), total electric power = 1.26 W, light intensity output at 460 nm = 2.5 mW/cm^2) as illumination source [25]. LEDs have emerged as low-cost, durable, and safer alternatives to conventional lamps for use in photocatalytic applications [53–55], with inherent advantages such as high efficiency in terms of current-to-light emission at relatively lower operational temperature, longer life, small size, and tailorable shapes and, above all, their operation using direct current power supply [56]. The BV-PB system presented herein extended absorption in the visible region, reduced e$^-$-h$^+$ recombination, and improved photoactivity towards the mitigation of both organic (photooxidation of MB dye) and inorganic (photoreduction of Cr(VI)) pollutants. Finally, the mechanisms of charge transfer were discussed in detail.

2. Results and Discussion

2.1. Modification of BV with PB

The BV photocatalyst powders were surface modified with PB employing the photo-deposition methodology reported by Tada and co-workers [57]. This method utilizes $[Fe(CN)_6]^{3-}$ and Fe(III) salts as the molecular precursors and is based on the photo-reduction of $[Fe(CN)_6]^{3-}$ to $[Fe(CN)_6]^{4-}$ by the photo-excited electrons in the conduction band (CB) of photocatalyst (BV). The presence of Fe^{3+} cations facilitates the resulting reduced specie to initiate the selective nucleation/growth of PB ($Fe_4[Fe(CN)_6]_3$) over the surface BV particles, as depicted in the mechanism shown in Figure 1.

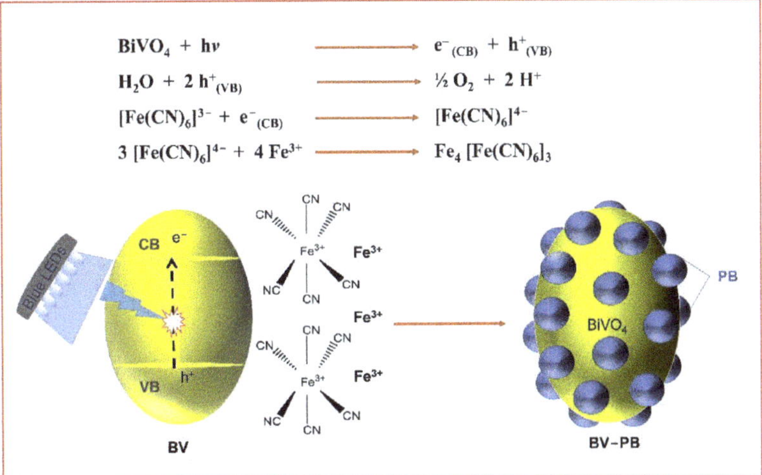

Figure 1. Infographic of the photoexcitation of BV with blue LEDs and the reactions involved in the photo-deposition of PB on the surface of BV.

The formation of PB over the surface of BV was confirmed by FTIR, EDX, XRD, and SEM-EDX analysis, as discussed below.

2.2. Morphological Analysis of BV and BV-PB

The comparison of the representative SEM images of BV and BV-PB3 in Figure 2 clearly shows that the bare BV consists of irregular-shaped hyper-branched BV particles with smooth surfaces (Figure 2a) and their surface roughness drastically increases after modification with PB (Figure 2b), confirming the successful photo-deposition of additional particles (PB) on the surface of BV particles. An analysis of the SEM images of the BV and BV-PB3 samples at different magnifications (Figure S2) confirms that the sample obtained after centrifugation does not contain isolated PB particles. One explanation is that the photo-deposition of PB requires the reduction of Fe^{3+} to Fe^{2+} by the conduction band electrons of

BV, and the photo-deposition most occurs on the surface of BV via heterogeneous nucleation. Furthermore, the SEM images of the different BV-PB samples at the same magnification show a gradual change in morphology, characterized by an increase in the amount of small photo-deposited PB particles as a function of the increasing PB loading (Figure S3).

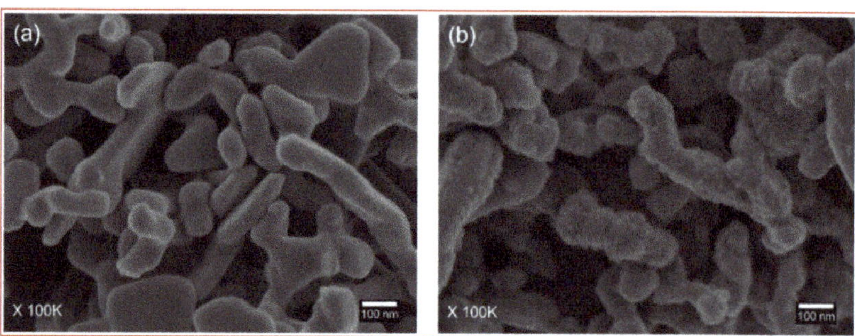

Figure 2. Representative FEG-SEM images of BV (**a**) and BV-PB3 (**b**) samples.

TEM analysis was performed to further study the structural features of BV-PB samples. A comparison of the TEM images of BV (Figure 3a) and BV-PB samples (Figure 3b) clearly confirms the formation of smaller than 10 nm PB crystals (average size 8 ± 1 nm) on the surface of BV in the later sample. The high-resolution TEM (HRTEM) image of the BV-PV sample (Figure 3c) shows that these small-surface deposited crystals exhibit interplanar distances of 0.22 nm, ~0.3 nm, and 0.38 nm which closely correspond to the (420), (311), and (220) crystal planes of PB, respectively (PDF n. 73-687).

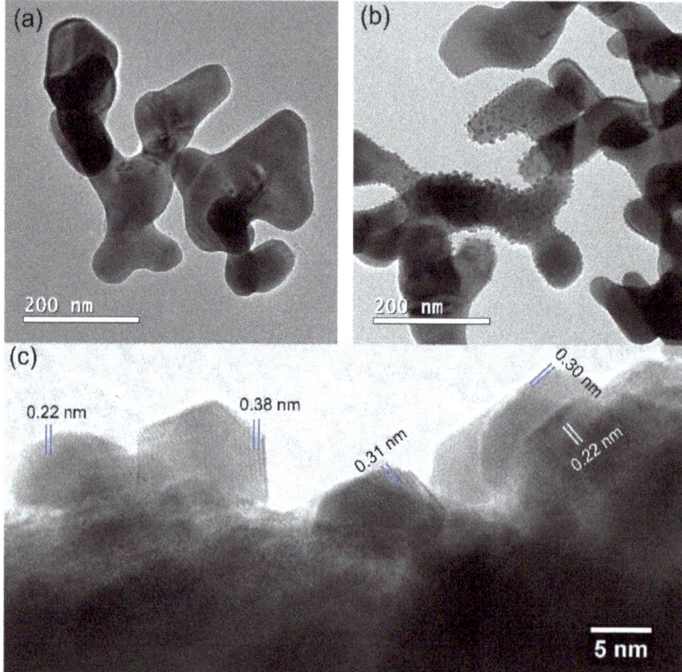

Figure 3. Representative TEM images of BV (**a**) and BV-PB2 (**b**) and the HRTEM image of the latter (**c**).

Qualitative nano-EDX analysis of the selected area of the BV-PB sample shown in the TEM image (Figure S4a) exhibits the X-ray emission lines of Bi, V, O, and Fe (Figure S4b), confirming the elemental identity of BV and the deposition of Fe. The elemental mapping (Figure S4c) performed by monitoring the X-ray lines from Bi (Lα1), V (Kα1), O (Kα1), and Fe (Kα1) clearly shows an overlap between the spatial distribution of these elements indicating that the BV-PB sample shown in Figure S4a consists of PB-loaded BV and that Fe (or PB) is present only on the surface of the BV particles.

The bulk elemental composition and PB loading (%Fe content) of the samples were then studied using SEM-EDX microanalysis (Figure 4). The EDX spectrum of pure BV exhibits the X-ray lines of Bi, V, and O with an atomic percentage of 17.7 ± 2.7, 17.6 ± 2.9, and 61.7 ± 7.5 %, respectively, giving a chemical composition $BiVO_{3.6}$ of the prepared BV sample. The EXD spectrum of the BV-PB sample (Figure 4a) shows additional X-ray emission lines at ~0.71 and 6.40 keV corresponding to Lα1 and Kα1 lines of Fe, respectively. The average Fe content (wt.%) and/or the Fe/Bi ratios obtained from triplicate measurements show a quasi-linear trend with the total nominal concentration of Fe precursors ($K_3[Fe(CN)_6]$ + $Fe(NO_3)_3$ or ($x + y$ mmol)) added during the reaction (Figure 4b), indicating efficient and controlled loading of PB over BV.

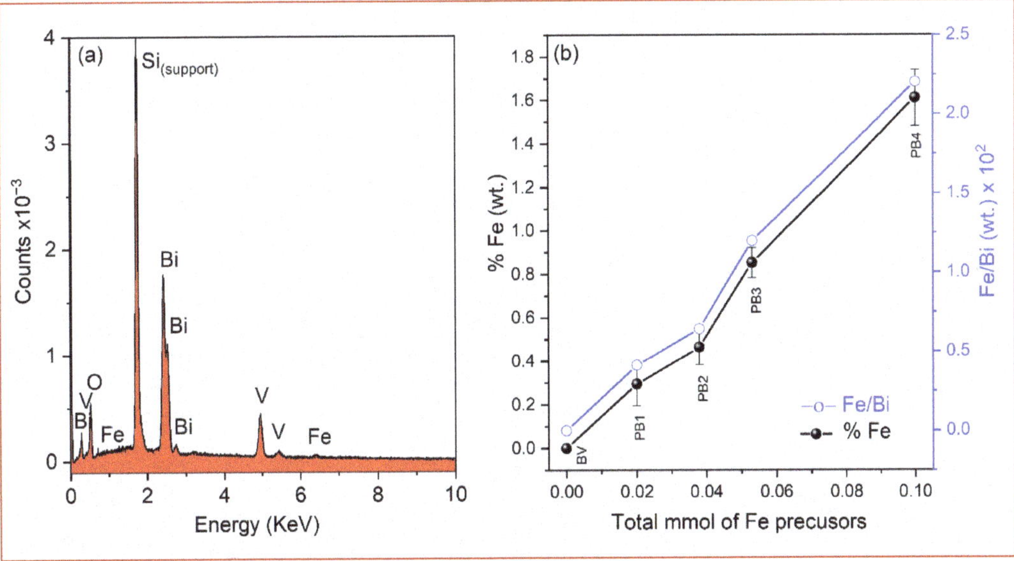

Figure 4. Representative EDX spectrum of BV-PB sample deposited as a thin layer on a silicon wafer (**a**) and the experimental Fe content (%Fe and Fe/Bi ratio) of different BV-BB samples measured by EDX analysis as a function of the total Fe precursor mmols ($x + y$) of $K_3[Fe(CN)_6]$ and $Fe(NO_3)_3$, added to in the reaction mixture during the photo-deposition process (**b**). The Si signal in the (**a**) comes from the Si wafer used as support.

2.3. Structural and Phase Analysis by XRD

The as-prepared BV and BV-PB samples were analyzed by XRD to further investigate their phase, crystallinity, and crystalline structure (Figure 5). The XRD pattern of the pristine BV sample matches perfectly with the standard diffraction patterns of monoclinic-scheelite BV (PDF no. 75–2480 and 14-0688). The XRD pattern of BV-PB is similar to that of the BV sample but a careful analysis shows the presence of small diffraction features around $2\theta = 17.5°$ (Figures 5 and S5), indicated by (*) and shown magnified in the inset of Figure 5, which corresponds to the most intense diffraction peak originated from the (200) planes of the cubic structure of PB (PDF n. 73-687) [18,58,59], confirming the formation of

PB, in agreement with HRTEM analysis (Figure 3c). The other weak diffraction features of PB expected around 35.4° (400) and 39.8 (420) could not be observed due to an overlap with the diffraction peak of BV (Figure S5) and/or the low PB content (Fe < 2%) (Figure 4b) [18], lower than the detection limit of XRD (2 > %) for mixed materials.

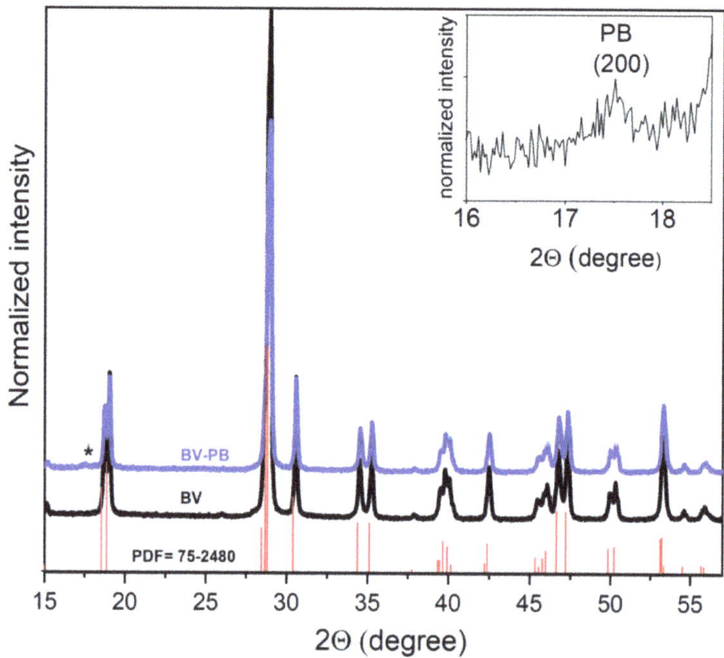

Figure 5. XRD patterns of BV (middle) and BV-PB3 (top) along with standard diffraction pattern of monoclinic BV (PDF 75–2480) (bottom). The inset shows a magnified view of the characteristic most intense diffraction feature of PB (indicated by an * in the diffractogram of BV-PB) at around 2θ = 17.5° corresponding to (200) planes of the cubic structure of PB.

2.4. Vibrational Spectroscopic (Raman and FTIR) Analysis

The formation of PB in BV-PB samples was further studied and confirmed by Raman spectroscopy (Figure 6a) and FTIR spectroscopy (Figure 6b) analysis. Raman spectroscopy is an appropriate methodology for the analysis of the local structure of materials. As shown in Figure 6a, the Raman analysis of unmodified BV shows Raman vibrational bands at 124 cm^{-1}, 209 cm^{-1}, 324 cm^{-1}, 367 cm^{-1}, 706 cm^{-1}, and 825 cm^{-1} (most predominant Raman band), all characteristic of the monoclinic phase of BV [23,60]. The vibrational modes at around 124 cm^{-1} and 209 cm^{-1} are related to the external modes, namely, translational (Ex_t), and rotational (Ex_r) twisting modes, respectively. Indeed, the band at 324 cm^{-1} and 367 cm^{-1} are produced from the asymmetric (δ_{as}) and symmetric (δ_s) bending vibration modes of the V–O bond in the VO_4^{3-} tetrahedra, sequentially. The low-intensity Raman shoulder at 706 cm^{-1} and high-intensity band at 825 cm^{-1} are respectively associated with asymmetric (ν_{as}) and symmetric (ν_s) stretching vibration modes of the V–O bonds, which are particularly sensitive to local-structural variations [50]. The Raman bands of BV-PB samples show the same vibration modes of BV but with a slight shift in band positions (to lower wavenumbers, cm^{-1}) and a decrease in band intensities and widths after PB deposition. For instance, the original stretching vibration modes of the V–O bonds 825 cm^{-1} (BV) shift to 810 cm^{-1} in BV-BP samples. Similarly, the low-intensity Raman shoulder at 706 cm^{-1} disappears and bands at 327 cm^{-1} and 367 cm^{-1} merge in BV-PB samples. Such spectral changes indicate significant interaction of BV with the photodeposited PB on its surface.

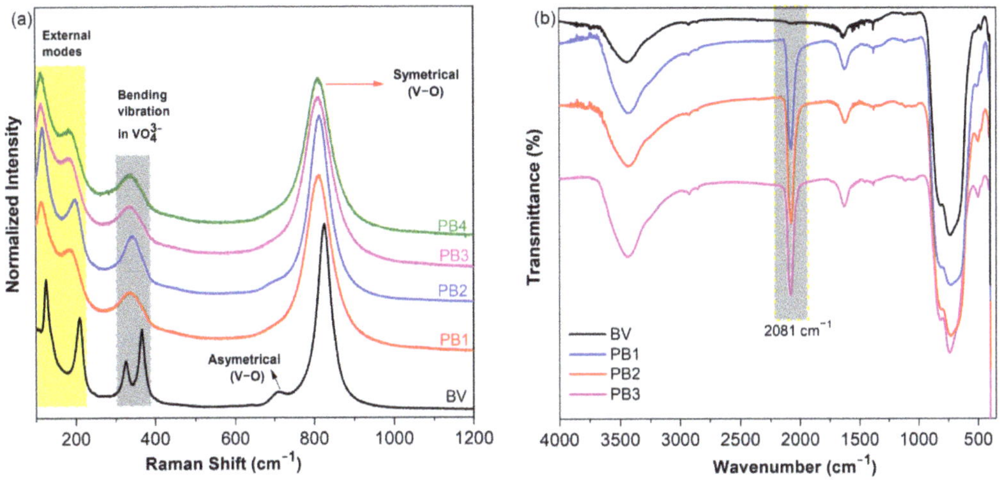

Figure 6. Raman spectra (**a**) and FTIR spectra (**b**) of BV and BV-PB samples. The shaded areas in Raman spectra correspond to external modes (yellow) and bending vibrations (gray), respectively. The shaded area in FTIR spectra corresponds to the signature C≡N stretching vibrations (2081 cm^{-1}) of PB.

The functional group's analysis of BV and BV-PB samples was performed using FTIR data (Figure 6b). The most important main bands in the 545–910 cm^{-1} region of the spectra arise from the symmetric stretching (ν_1) and antisymmetric (ν_3) vibration modes of metal-oxygen bond in BV (Figures 6b and S6). More clearly, the pure BV sample exhibits a characteristic strong vibrational band at 736 cm^{-1}, with a shoulder at around 822 cm^{-1}, ascribed to the asymmetric stretching vibration of metal oxide (V–O) groups in BV (see Figure S6) [61]. The bands at 3445 cm^{-1} and 1643 cm^{-1} are ascribed to the symmetric stretching and bending vibrations, respectively, of H-O-H molecules (atmospheric moisture) adsorbed on the surface of the photocatalyst. The small band at 413 cm^{-1} is ascribed to the presence of direct Bi–O linkages [62]. Importantly, the FTIR spectra of all BV-PB samples in Figure 6b exhibit a prominent FTIR band at 2081 cm^{-1}, characteristic of the stretching vibration of C≡N bonds (vibrational mode) in the cyanometallate network of PB [18]. The cyanide's vibrational modes at high wavenumbers (2089–2070 cm^{-1}) are in good agreement with the reported values for pure PB (2070 cm^{-1}) [63], thus confirming the formation of an extended network of PB synthesized via Fe(II) –CN–Fe(III) bridging linkages [18].

2.5. Optical Properties

The effect of PB deposition over BV and the optical characteristics of BV and BV-PB samples were studied using diffuse reflectance UV-visible spectroscopy (DRS) measurements (Figure 7). As shown in the digital images inserted in Figure 7a, the bright yellow BV sample visually turns greenish upon deposition of PB and BV starts to absorb light in the visible region at about 500–600 nm. Maximum absorbance is seen at wavelengths around 470 nm, a value that is comparable with the reported optical band-gap energy (E_g = 2.5–2.6 eV) of monoclinic BV [23,25,64]. Based on DRS data, the band gap energy (E_g) of BV was calculated by extrapolating the linear portion of the plot between $[F(R)h\upsilon]^2$ against $h\upsilon$ to 0 and was found to be 2.6 eV (Figure 7b), in agreement with literature data for monoclinic BV [23,64].

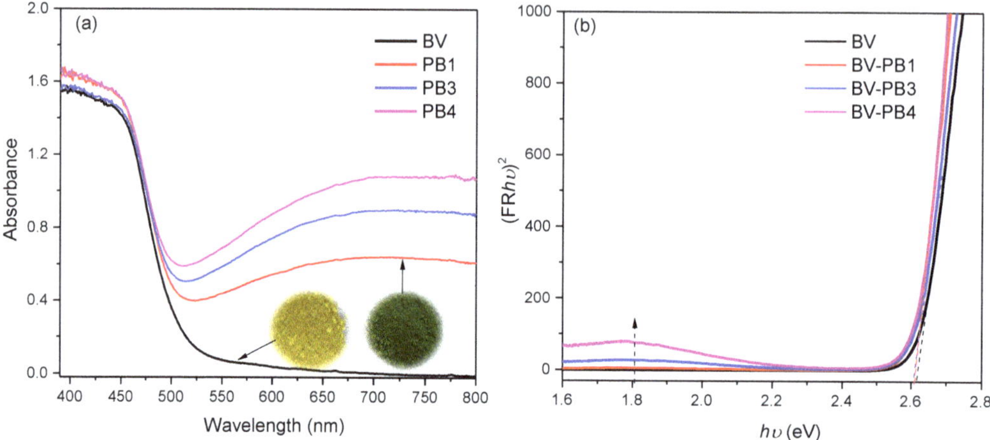

Figure 7. Absorbance spectra of BV and BV-PB samples (**a**) and the corresponding Kubelka-Munk plot for estimation of the band gap energy (E_g) of BV and BV-PB samples (**b**). The digital images of BV and BV-PB have been inserted in Figure 7a for visual comparison.

After the deposition of PB onto the BV surface, the visible light absorption of BV-PB samples strongly increases in comparison with BV, which is assigned to the metal-to-metal (Fe^{2+} to Fe^{3+}) intervalence charge transfer in PB. As discussed later, the improved visible-light absorbance of BV-PB is expected to enhance the photoactivity of the composite. Though PB loading increases visible light absorption by the nanocomposite, the absorption edge of BV and BV-PB samples were approximately the same (Figure 7b), with no considerable change in E_g (~2.6 eV), indicating only surface deposition of PB without structural modifications of BV.

2.6. Photocatalytic Properties

The photocatalytic properties of BV and BV-PB materials and the role of PB as cocatalyst were evaluated through the photodegradation of MB (Figure 8) and photoreduction of Cr(VI) (Figure 9) in aqueous media under blue LEDs irradiation. The difference in photoactivity of the samples, in terms of interfacial charge transfer at the photocatalysts/solution interface and electron-hole recombination, was then studied and verified using EIS and PL measurements (Figure 10), respectively, in addition to the DRS data presented above. Finally, a comprehensive photoactivity enhancement mechanism was proposed (Figure 11), as discussed at the end of this section.

2.6.1. Photooxidation of MB dye

We first studied the photodegradation of MB using BV and BV-PB samples as photocatalysts in the absence and presence of a minute quantity (0.2 mmol) of H_2O_2 in the reaction mixture under visible light illumination (Figure 8). The absorbance of MB dye decreases upon illumination with blue light from LEDs in the presence of photocatalysts (Figure S7), indicating photodegradation of the dye. While the pure BV could degrade around 70% of MB dye within 120 min (Figure 8b). However, the time for complete photodegradation of MB is reduced to 40 min in the presence of 0.2 mmol H_2O_2 in the reaction mixture (BV+H_2O_2, $k_{obs.}$ = 0.055 min^{-1}). Compared to pristine BV, the photocatalytic activity increases upon deposition of PB, and BV-PB could degrade around 85% of MB dye within 120 min. Importantly, a very prominent increase in photoactivity of BV-PB is observed in the presence of H_2O_2 and almost 100% degradation of MB is achieved within 20 min ($k_{obs.}$ = 0.375 min^{-1}) by the BV-PB/H_2O_2 system under conditions identical to those of BV + H_2O_2. Recyclability tests showed that the BV-PB/H_2O_2 system retains its photoactivity after repeated use and negligible loss in photoactivity was observed after five cycles of use (Figure 8b).

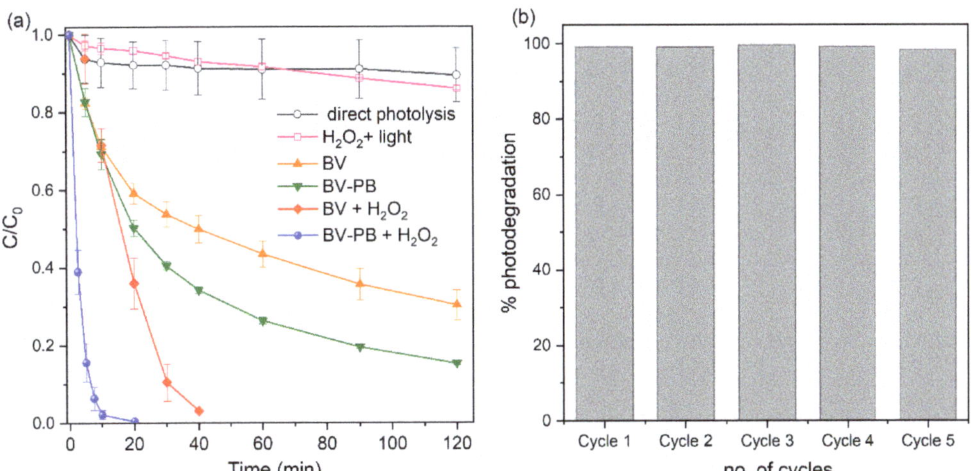

Figure 8. Kinetic profiles showing the photodegradation of MB as a function of visible light illumination time by the pristine BV or BV-PB samples in the presence and absence of a minute amount (0.2 mmol) H_2O_2 in the reaction mixture (**a**) and the MB photodegradation efficiency of the recycled BV-PB photocatalyst in the presence of H_2O_2 showing good recyclability for up to five cycles of use (**b**). Results of control experiments employing only LED illumination (direct photolysis), only H_2O_2 and H_2O_2 + LED illumination (emission centered at 460 ± 10 nm) are also reported in (**a**).

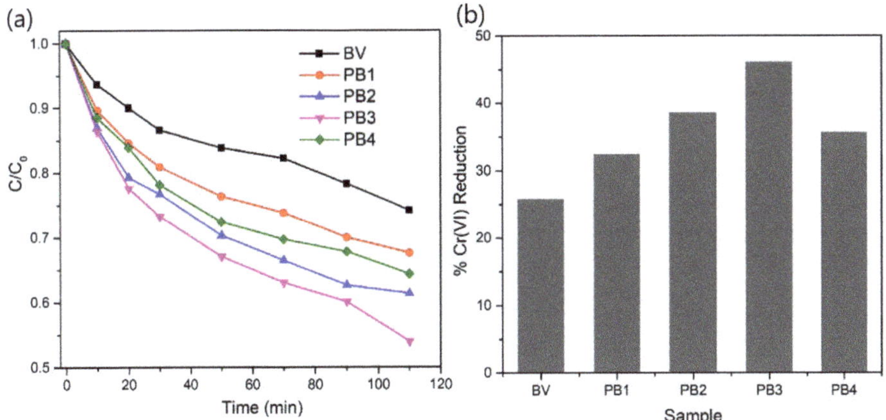

Figure 9. (**a**) Temporal changes in the concentration of Cr(VI) as a function of LEDs illumination time showing a comparison of the photo-catalytic activities of BV and BV-PB samples (**b**) A comparison of the Cr(VI) reduction efficiency of BV and different BV-PB samples. The PB layer acts as a cocatalyst, mediating the transfer of photoexcited conduction band electrons from BV to Cr(VI) (see also Figures 10 and 11). The experiments were performed without any pH adjustment (natural pH = 4.6) of the suspension and without adding any hole scavenger organic molecules. The experiments under light irradiation were performed using the LED reactor described above.

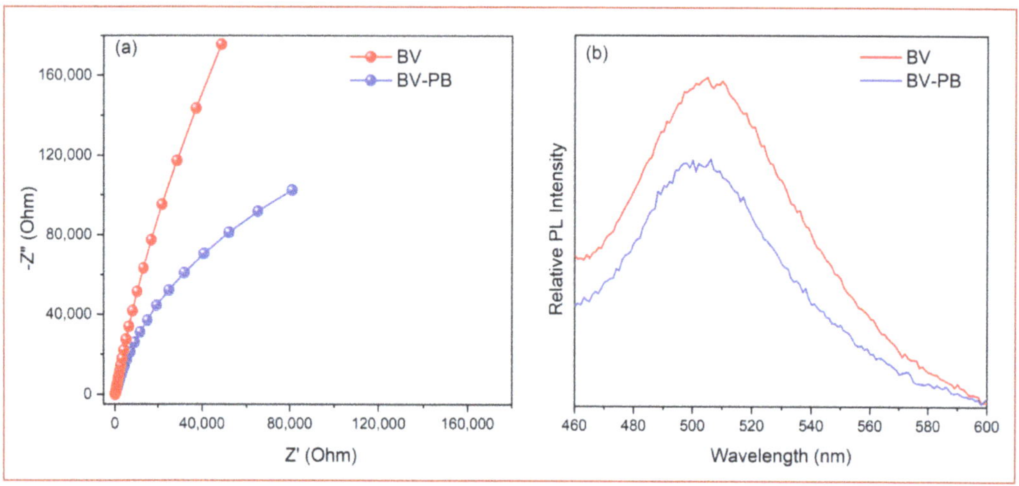

Figure 10. Nyquist plot obtained from EIS measurements under dark condition (**a**) and PL emission spectra arising from electron-hole recombination (**b**) of representative BV and BV-PB3 samples.

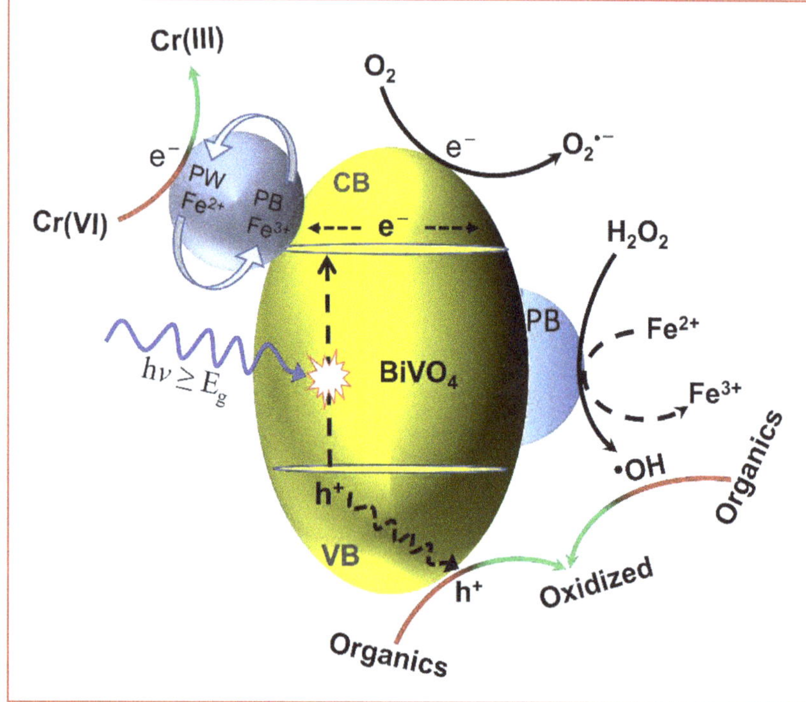

Figure 11. Schematics of the photoexcitation, charge transfer, and resulting photocatalytic processes in the BV/PB system. The photoexcited electrons from CB of BV transfer to PB and dissolved O$_2$ reducing them to PW and O$_2{}^{\bullet-}$, respectively. The PW, in turn, transfers electrons to Cr(VI) in a solution, itself becoming oxidized back to PB. Hydrogen peroxide is converted to $\bullet$OH radicals which, together with holes (h$^+$) in the valence band, can oxidize organic materials.

2.6.2. Photoreduction of Cr(VI)

Next, the photoreduction of toxic Cr(VI) to less toxic Cr(III) in aqueous media was studied under LED illumination apparatus and without any pH adjustment or addition of sacrificial organic molecules. Figure S8 shows the decrease in absorbance of the chromium (Cr)-DPC complex as a function of time under visible light illumination in the presence of BV-PB3 as representative photocatalysts, indicating photoreduction of Cr(VI) to Cr(III). All BV-PB samples exhibit better photoactivity than pristine BV (Figure 9) demonstrating that the PB assists in the photoreduction of Cr(VI), as explained below. The photoreduction efficiency of the samples follows the order BV-PB3 > BV-PB2 > BV-PB4 > BV-PB1 > BV.

2.7. Why BV-PB Shows Enhanced Photoactivity?

The above-noted superior photo-redox efficiency of the BV-PB samples, as compared to BV, could arise from:

1. Enhanced absorption of visible light (Figure 7),
2. The ability of Fe centers in PB to produce reactive oxygen species (ROS) in a Fenton-like derived process,
3. The role of PB as cocatalysts, lowering resistance to charge transfer and improving charge transfer ability at the photocatalysts/solution interface, thus promoting electron transfer from the conduction band (CB) of BV to the Cr(VI) and/or O_2 species in solution, and/or
4. Decreased electron-hole recombination due to lower charge transfer resistance.

While the enhanced visible light absorption (factor 1) is evident from Figure 7, an equally important PB-based Fenton-like process (factor 2) could occur at the Fe centers in PB in the presence of H_2O_2. This process generates additional ROS ($^\bullet$OH and HOO$^\bullet$) via oxidation of Fe^{2+} to Fe^{3+} (Equation (1)), and/or reduction of Fe^{3+} (Equation (2)).

$$Fe^{2+} + H_2O_2 \rightarrow Fe^{3+} + OH^- + {}^\bullet OH \tag{1}$$

$$Fe^{3+} + H_2O_2 \rightarrow Fe^{2+} + H^+ + HOO^\bullet \tag{2}$$

$$Fe^{3+} + e^-{}_{(CB)} \rightarrow Fe^{2+} \tag{3}$$

We have already observed that the photogenerated holes and ROS species ($^\bullet$OH and $O_2{}^{\bullet-}$) play a role in the photocatalytic degradation of dyes over $BiVO_4$ photocatalyst [23]. Since additional $^\bullet$OH species are expected to form in BV-PB material, the formation of $^\bullet$OH in the reaction mixture was confirmed by using terephthalic acid (TPA) as a fluorescent probe which reacts with $^\bullet$OH radicals to form 2-hydroxy terephthalic acid of relatively higher fluorescence intensity. The fluorescence intensity of TPA increases a little in the presence of only H_2O_2 but drastically in the presence of both BV-PB and H_2O_2 in the reaction mixture (Figure S9), confirming our hypothesis of the Fenton-like process in the BV-PB/H_2O_2 system. Thus, the excellent photoactivity of the BV-PB/H_2O_2 system can be assigned to a synergic effect between the BV-based photocatalytic (h^+ and $^\bullet$OH/$O_2{}^{\bullet-}$ generation) and PB-based Fenton-like processes (Figure 11, RHS), in addition to the other factors, such as lower electron-hole recombination and lower charge transfer resistance at photocatalysts/solution interface (*vide infra*). Furthermore, the improved interfacial charge transfer between BV and photo-deposited PB, as discussed below, may further increase the overall photo-Fenton process as it leads to the regeneration of Fe(II) sites (Equation (3)), which are known to be much more catalytic, active for hydroxyl radical formation in the presence of H_2O_2 [65].

The other two hypotheses (or factors 3 and 4) highlighting the role of PB as a cocatalyst) were evaluated using EIS measurements (Figure 10a) and PL measurements (Figure 10b), respectively. Factor 3 was evaluated by measuring the charge transferability of BV and BV-PB samples from the Nyquist plot using EIS measurements (Figure 10a). As compared to pure BV, the BV-PB samples show lower resistance to charge transfer and the charge transfer ability seems to improve the photoreduction of the target specie (Cr(VI) and/or O_2) in the solution [18]. Such improved charge transfer kinetics are expected to allow better charge separation and hence lower recombination (factor 4). The PL measurement of BV and BV-PB showed that the latter has lower PL intensity (that arises from electron-hole recombination) as compared to BV (Figure 10b), indicating reduced electron-hole recombination in BV-PB. The decrease in photoactivity at higher PB loading (BV-PB4 in Figure 9) may be related to the lower accessibility of photogenerated CB electron for Cr(VI) reduction in aqueous media due to a less effective transfer across relatively thicker BV/PB interface.

Considering the discussion above, it can be asserted that the blue LEDs illumination (λ output = 460 $\pm$ 10 nm which overlaps the absorption of BV (E_g = 2.6 eV)) can photoexcite electrons from the valance band (VB) of BV to its CB (Figure 11). The photoexcited electrons in the CB are subsequently transferred to molecular O_2 forming $O_2^{\bullet-}$ and to PB, partially reducing it ($Fe^{3+} + e^-_{(CB)} \rightarrow Fe^{2+}$) to Prussian white (PW). The resulting PW, in turn, transfers the extra electrons to Cr(VI) (and/or O_2) species in solution, reducing Cr(VI) to Cr(III) (and/or O_2 to $O_2^{\bullet-}$) and itself reverts back to PB for the next redox cycle (Figure 11, LHS). The mechanism proposed for charge transfers is represented in Figure 11. PB thus acts as a cocatalyst, mediating the transfer of electrons from BV to Cr(VI) (Figure 10a) and decreasing the electron-hole recombination (Figure 10b).

The practical approach employed in this work may guide new studies and pioneer new prospects for synthesizing new BV-based photocatalytic systems with low production costs and high photoredox efficiencies.

3. Materials and Methods

3.1. Reagents

Potassium hexacyanoferrate(III) ($K_3[Fe(CN)_6]$, 99%,), iron(III) nitrate ($Fe(NO_3)_3 \cdot 9H_2O$), and 1,5-diphenylcarbazide (DPC) were supplied by Sigma-Aldrich (São Paulo, Brazil). Ammonium metavanadate (NH_4VO_3, 99%, Merck, Darmstadt, Germany), $Bi(NO_3)_3 \cdot 5H_2O$ (>98%, Neon, São Paulo, Brazil), HNO_3 (Qhemis, São Paulo, Brazil), NH_4OH (28%, Synth, São Paulo, Brazil), acetone (Synth), H_2O_2 (50%, Synth), and potassium dichromate ($K_2Cr_2O_7$, Mallinckrodt, France) were used as received.

3.2. Preparation of BV and BV-PB Photocatalysts

Monoclinic BV nanoparticles were synthesized via the precipitation-hydrothermal route reported in our previous work [25]. Briefly, an aqueous solution of NH_4VO_3 (1 mmol) was drop-wise added to an equimolar aqueous solution of $Bi(NO_3)_3 \cdot 5H_2O$ under continuous stirring to give an orange-yellow solution, followed by pH adjustment (pH~ 6), and then microwave-assisted hydrothermal treatment (180 °C, 125 W, 275 psi, 10 min). The resulting bright yellow BV suspension (was centrifuged (6000 rpm), washed with deionized water twice, and dried at 80 °C for 24 h.

The BV-PB samples were prepared by a photo-assisted deposition method employing an array of home-made LEDs (output = 460 $\pm$ 10 nm (Figure S1), total electric power = 1.26 W, light intensity output at 460 nm = 2.5 mW/cm^2) as described elsewhere [25], using $K_3[Fe(CN)_6]$ and ($Fe(NO_3)_3 \cdot 9H_2O$) as the PB precursors. For this purpose, 150 mg of BV powder was dispersed in 100 mL water by sonication for 30 min, followed by the addition of different amounts (x mmol) of potassium $K_3[Fe(CN)_6]$ and stirring the mixture for 1h in dark (suspension A). Then, different amounts of $Fe(NO_3)_3 \cdot 9H_2O$ solution (30 mL solution in 0.03M HNO_3 containing y mmol of the precursor) were drop-wise added to suspension A under LED illumination and, after 1h illumination at 460 nm, the resulting suspension of BV-PB composite was separated via centrifugation, washed twice with deionized water and then dried in an oven at 80 °C. The molar concentrations of $K_3[Fe(CN)_6]$ (x mmol) and $Fe(NO_3)_3 \cdot 9H_2O$ (y mmol) were kept equal (1:1) for each set of samples and varied in a certain range ($x = y$ = 0.01, 0.019, 0.026, 0.05 mmol) to prepare different samples (BV-PB1, BV-PB2, BV-PB3, BV-PB4, respectively) with different PB loadings.

3.3. Characterization Techniques

Scanning electron microscopy (SEM) images were obtained using a FEG-SEM microscope (JSM-7200, JEOL, USA). The samples, deposited on silicon wafer pieces, were sputter-coated with 6 nm gold layer using BAL-TEC MED 020 (BAL-TEC, Balzers Liechtenstein) coating System (conditions: chamber pressure = 2.00×10^{-2} mbar; current = 60 mA; deposition rate 0.60 nm/s). Energy Dispersive X-ray Spectroscopy (EDX) microanalysis was performed using an XFlash$^{®}$ 6/60 detector (Bruker, Germany), employing ESPRIT 2.3 software, using a 15 kV electron beam, and using Cu standard for analytical calibration. For EDX analysis, the samples were pressed into a thick pellet and attached to the surface of conducting carbon tape. Three different regions (50 μm $\times$ 50 μm) of the same sample were analyzed to obtain an average elemental composition [23]. TEM analysis of the sample deposited from a dilute aqueous suspension onto carbon-coated copper grids was performed on a JEOL TEM (JEM-2100, USA) equipped with a LaB_6 electron source and operated at 200 kV electron accelerating voltage. X-ray diffraction (XRD) patterns of BV and BV-PB samples were measured using a D8 Advance X-ray diffractometer (Bruker, Germany) operating at 40 mA and 40 kV and employing Ni-filtered Cu Kα X-ray radiation (1.540 Å). The diffuse reflectance spectra (DRS) of the powder samples against a background of MgO powder (white standard) were obtained using a Cary 5000 UV-Vis-NIR spectrophotometer (Varian, Australia). Raman spectra (100–1200 cm^{-1}, acquisition time of 40 s, 2 cycles) were measured with a LabRAM HR 800 Raman spectrophotometer (Horiba Jobin Yvon) equipped with a He–Ne laser (632.81 nm). The photoluminescence (PL) emission spectra ($\lambda_{(excitation)}$= 375 nm) of the samples were acquired with a Horiba Jobin Yvon spectrofluorometer (Fluorolog-3 model FL3-122, USA) equipped with a Hamamatsu R-928 photomultiplier tube and a Xe lamp. Electrochemical impedance spectroscopy (EIS) measurements of the sample films on FTO glass were performed in a CorrTest potentiostat/galvanostat (model CS310, Wuhan China) using a three-electrode cell [23]. Fourier transform infrared spectroscopy (FTIR) spectra (400–4000 cm^{-1}) of the samples diluted with KBr and pressed into a pellet were collected with a NICOLET IS5 FTIR spectrophotometer (Thermo Scientific, Waltham, MA, USA) with a resolution of 2 cm^{-1} and averaged over 64 scans.

3.4. Evaluation of Photocatalytic Activity

The photocatalytic activity of BV and BV-PB samples was evaluated by photoreduction of Cr(IV) ions and photooxidation of MB dye as a model photocatalytic reaction.

3.4.1. Photocatalytic Reduction of Cr(VI)

The photoreduction of Cr(VI) was performed in order to evaluate the photoreduction efficiency of the BV and BV-PB and hence the role of PB in the composite photocatalysts. For this purpose, typically, 35 mg of the photocatalysts powder was dispersed in 35 mL of H_2O by sonication followed by the addition of 35 mL $K_2Cr_2O_7$ containing 20 $mg \cdot L^{-1}$ of Cr(VI). The resultant mixture (natural pH 4.6) was kept in the dark for 30 min and then illuminated with blue LEDs described above for different time intervals. The sample aliquots taken at various irradiation intervals were centrifuged to remove the suspended particle and 200 μL of the supernatant was then reacted with DPC (5 $g \cdot L^{-1}$ in acetone) as a selective colorimetric reagent [66] in the presence of H_2SO_4 in the reaction media (1 mL H_2O + 200 μL DPC + 100 μL H_2SO_4) to form Cr(VI)–DPC complex that shows maximum absorbance at 545 nm. During the reaction, Cr(VI) is reduced to Cr(III) and DPC is oxidized to 1,5-diphenylcabazone (DCPA). The decrease in the concentration of Cr(VI) was measured by observing the decrease in the absorbance (λ_{max} = 545 nm) of the Cr–DPC complex [18].

3.4.2. Photocatalytic Degradation of Methylene Blue (MB)

The photocatalytic activity of BV-PB and pure BV was also evaluated by the photodegradation of MB dye in the absence and presence of a minute amount of H_2O_2 in the reaction mixture. Typically, 25 mg of the photocatalyst was dispersed in 20 mL water by sonicated for 15 min, followed by the addition of 20 mL MB dye (15 $mg \cdot L^{-1}$) and allowing the dye-photocatalyst mixture to stir in dark for 30 min. The mixture was then irradiated with the blue LEDs photoreactor described above and sample aliquots were taken out, centrifuged, and electronic absorption spectra of the supernatant measured to follow the degradation of MB as a function of time. In the degradation studies performed in the presence of H_2O_2, 200 µL of 1 mol$\cdot$L^{-1} H_2O_2 solution was added to the reaction mixture just before illumination. A control experiment in the absence of photocatalysts in the reaction mixture was also performed under identical conditions to better evaluate the role of photocatalysts (BV, BV-PB), H_2O_2, and direct photolysis in the total MB removal efficiency of the photocatalytic systems studied. To compare the photoactivity of the samples towards MB degradation, the observed rate-constant (k_{obs}) values were calculated from the kinetic profiles ($\frac{C}{C_0}$ vs. time) using a first order exponential-function, ($\frac{C}{C0} = e^{-k_{obs}\cdot t}$).

The recyclability of the photocatalysts was tested by the same procedure just mentioned above, except that the used photocatalysts were recovered from the reaction mixture by centrifugation at 4500 rpm for 10 min, re-dispersed in 20 mL water by sonication, mixed with 20 mL MB (15 mg$\cdot$L^{-1}) and then employed in the next photocatalytic degradation cycles in the presence of H_2O_2 under LED illumination for 25 min.

To study the formation of $^\bullet$OH radicals by the BV-PB photocatalysts in the presence of H_2O_2, terephthalic acid (TPA, 4×10^{-4} mol$\cdot$L^{-1} solution in 2×10^{-3} mol$\cdot$L^{-1} NaOH) was used a fluorescence probe molecule [67]. The fluorescence spectra (340–600 nm, $\lambda_{max(emission)}$ = 425 nm) of TPA were recorded under 315 nm excitation [67] in the absence and presence of BV-PB and H_2O_2.

4. Conclusions

Addressing the inherent problems of $BiVO_4$ (BV) photocatalysts, including fast electron-hole recombination and slow charge transfer kinetics, we successfully prepared PB-loaded BV particles with enhanced photo-redox ability as investigated by photoreduction of Cr(VI) and photooxidation of MB. The PB was photo-deposited on the surface of hydrothermally synthesized monoclinic BV using low-cost commercial LEDs as the illumination source and the formation of BV-PB was confirmed by microscopic and spectroscopic analyses. The photo-deposited PB not only increases the absorption of visible light by the BV-PB composite photocatalyst, as indicated by DRS but also acts as cocatalysts, improving the charge carriers' transfer across the photocatalysts/solution interface and hence reducing their recombination, as confirmed by EIS and PL measurements, respectively. Consequently, the BV-PB composite photocatalysts with optimum PB loading exhibited enhanced Cr(VI) photoreduction efficiency as compared to pristine BV under visible light illumination from low-power LEDs, thanks to the cocatalyst role of PB which mediates the transfer of photoexcited conduction band electrons from BV to Cr(VI) species in solution. As compared to pristine BV (70% of MB degradation in 120 min), higher photoactivity was observed in the presence of a minute amount (0.2 mmol) of H_2O_2 in the reaction media for both BV (~97% photodegradation in 40 min, k_{obs} = 0.055 min^{-1}) and BV-PB (100% dye degradation within 20 min, k_{obs} = 0.375 min^{-1}) materials. Such high photo-oxidation efficiency of the BV-PB/H_2O_2 system is due to a synergic effect between the BV-based photocatalytic and PB-based Fenton-like processes, in addition to other factors including the role of PB as cocatalysts discussed above. The practical approach reported in this study may be extended to other photocatalytic systems with high photoredox efficiencies for photo(electro)chemical applications.

Supplementary Materials: The following supporting information can be downloaded at: https://www.mdpi.com/article/10.3390/catal12121612/s1, Figure S1: The emission spectrum of LEDs; Figures S2 and S3: SEM images of BV and BV-PB; Figure S4: TEM images and elemental mapping of BV-PB, Figure S5: XRD patterns of BV and BV-PB; Figure S6: FTIR Spectrum of pure BV; Figure S7: Electronic absorption spectra of MB dye as a function of LED illumination, Figure S8: Absorption spectra of Cr(VI)-DPC complex; Figure S9: Fluorescence spectra of pure terephthalic acid in the presence and absence of BV-PB2 and H_2O_2.

Author Contributions: Methodology, A.A.K. and L.M.; investigation, A.A.K., M.M., L.M. and B.O.M.; writing, review and editing, A.A.K., E.P.F.-N., H.W., R.P., U.P.R.-F. and S.J.L.R.; conceptualization, S.U. and E.P.F.-N.; funding acquisition and upervision, S.U., S.J.L.R. and U.P.R.-F. The manuscript was written through the contributions of all authors. All authors have read and agreed to the published version of the manuscript.

Funding: This research was funded by Higher Education Commission (HEC), Pakistan (Project No. 9286), the São Paulo Research Foundation (FAPESP, Brazil) (FAPESP Process 15/22828-6). H. Wender acknowledges financial support from CNPq (Projects 486342/2013-1, 427835/2016-0, 311798/2014-4, 313300/2020-8, and 310066/2017-4); FUNDECT (N. 099/2014, 106/2016, and 228/2022), and Coordenação de Aperfeiçoamento de Pessoal de Nível Superior—Brasil (CAPES)—Finance Code 001 and Capes-PrInt program (grant numbers 88881.311921/2018-01 and 88887.311920/2018-00). L. Marchiori also thanks CNPq for the Ph.D fellowship (grant number 140851/2021-6).

Data Availability Statement: The data supporting reported results can be shared on demand.

Acknowledgments: Sajjad Ullah also acknowledges the São Paulo State University (IQ-UNESP), Araraquara, SP, Brazil for the support and opportunity to work as a short-term visiting researcher.

Conflicts of Interest: The authors declare no conflict of interest.

References

1. Saiz, P.G.; Valverde, A.; Gonzalez-Navarrete, B.; Rosales, M.; Quintero, Y.M.; Fidalgo-Marijuan, A.; Orive, J.; Reizabal, A.; Larrea, E.S.; Arriortua, M.I.; et al. Modulation of the Bifunctional CrVI to CrIII Photoreduction and Adsorption Capacity in ZrIV and TiIV Benchmark Metal-Organic Frameworks. *Catalysts* **2021**, *11*, 51. [CrossRef]

2. Ukhurebor, K.E.; Aigbe, U.O.; Onyancha, R.B.; Nwankwo, W.; Osibote, O.A.; Paumo, H.K.; Ama, O.M.; Adetunji, C.O.; Siloko, I.U. Effect of Hexavalent Chromium on the Environment and Removal Techniques: A Review. *J. Environ. Manag.* **2021**, *280*, 111809. [CrossRef] [PubMed]

3. Liu, D.; Wang, Y.; Xu, X.; Xiang, Y.; Yang, Z.; Wang, P. Highly Efficient Photocatalytic Cr(VI) Reduction by Lead Molybdate Wrapped with D-A Conjugated Polymer under Visible Light. *Catalysts* **2021**, *11*, 106. [CrossRef]

4. Szabó, M.; Kalmár, J.; Ditrói, T.; Bellér, G.; Lente, G.; Simic, N.; Fábián, I. Equilibria and Kinetics of Chromium(VI) Speciation in Aqueous Solution—A Comprehensive Study from PH 2 to 11. *Inorganica Chim. Acta* **2018**, *472*, 295–301. [CrossRef]

5. Mishra, S.; Bharagava, R.N. Toxic and Genotoxic Effects of Hexavalent Chromium in Environment and Its Bioremediation Strategies. *J. Environ. Sci. Health Part C* **2016**, *34*, 1–32. [CrossRef]

6. WHO Expert Consultation for 2nd Addendum to the 3rd Edition of the Guidelines for Drinking-Water Quality: Geneva, 15–19 May 2006, WHO/SDE/WSH/06.05 v, 136 p. Available online: https://Apps.Who.Int/Iris/Handle/10665/69604 (accessed on 28 July 2022).

7. Ambika, S.; Kumar, M.; Pisharody, L.; Malhotra, M.; Kumar, G.; Sreedharan, V.; Singh, L.; Nidheesh, P.V.; Bhatnagar, A. Modified Biochar as a Green Adsorbent for Removal of Hexavalent Chromium from Various Environmental Matrices: Mechanisms, Methods, and Prospects. *Chem. Eng. J.* **2022**, *439*, 135716. [CrossRef]

8. Omer, A.M.; Abd El-Monaem, E.M.; Eltaweil, A.S. Novel Reusable Amine-Functionalized Cellulose Acetate Beads Impregnated Aminated Graphene Oxide for Adsorptive Removal of Hexavalent Chromium Ions. *Int. J. Biol. Macromol.* **2022**, *208*, 925–934. [CrossRef]

9. Du, J.; Shang, X.; Shi, J.; Guan, Y. Removal of Chromium from Industrial Wastewater by Magnetic Flocculation Treatment: Experimental Studies and PSO-BP Modelling. *J. Water Process Eng.* **2022**, *47*, 102822. [CrossRef]

10. Salman, R.H.; Hassan, H.A.; Abed, K.M.; Al-Alawy, A.F.; Tuama, D.A.; Hussein, K.M.; Jabir, H.A. Removal of Chromium Ions from a Real Wastewater of Leather Industry Using Electrocoagulation and Reverse Osmosis Processes. *AIP Conf. Proc.* **2020**, *2213*, 020186. [CrossRef]

11. Hu, W.; Chen, Y.; Dong, X.; Meng, Q.W.; Ge, Q. Positively Charged Membranes Constructed via Complexation for Chromium Removal through Micellar-Enhanced Forward Osmosis. *Chem. Eng. J.* **2021**, *420*, 129837. [CrossRef]

12. Zhang, C.; Li, Q.; Chen, Q. Electrochemical Treatment Of Landfill Leachate To Remove Chromium (Vi) Using Ni_3n And Nio Nps Anodes. *Int. J. Electrochem. Sci.* **2021**, *16*, 210710. [CrossRef]

13. Zhao, Y.; Gao, J.; Zhou, X.; Li, Z.; Zhao, C.; Jia, X.; Ji, M. Bio-Immobilization and Recovery of Chromium Using a Denitrifying Biofilm System: Identification of Reaction Zone, Binding Forms and End Products. *J. Environ. Sci.* **2023**, *126*, 70–80. [CrossRef]

14. Rathna, T.; PonnanEttiyappan, J.B.; RubenSudhakar, D. Fabrication of Visible-Light Assisted TiO_2-WO_3-PANI Membrane for Effective Reduction of Chromium (VI) in Photocatalytic Membrane Reactor. *Environ. Technol. Innov.* **2021**, *24*, 102023. [CrossRef]

15. Haji Ali, B.; Baghdadi, M.; Torabian, A. Application of Nickel Foam as an Effective Electrode for the Electrochemical Treatment of Liquid Hazardous Wastes of COD Analysis Containing Mercury, Silver, and Chromium (VI). *Environ. Technol. Innov.* **2021**, *23*, 101617. [CrossRef]

16. Thamaraiselvan, C.; Thakur, A.K.; Gupta, A.; Arnusch, C.J. Electrochemical Removal of Organic and Inorganic Pollutants Using Robust Laser-Induced Graphene Membranes. *ACS Appl. Mater. Interfaces* **2021**, *13*, 1452–1462. [CrossRef]

17. Peng, H.; Guo, J. Removal of Chromium from Wastewater by Membrane Filtration, Chemical Precipitation, Ion Exchange, Adsorption Electrocoagulation, Electrochemical Reduction, Electrodialysis, Electrodeionization, Photocatalysis and Nanotechnology: A Review. *Environ. Chem. Lett.* **2020**, *18*, 2055–2068. [CrossRef]
18. Ferreira-Neto, E.P.; Ullah, S.; Perissinotto, A.P.; de Vicente, F.S.; Ribeiro, S.J.L.; Worsley, M.A.; Rodrigues-Filho, U.P. Prussian Blue as a Co-Catalyst for Enhanced Cr(vi) Photocatalytic Reduction Promoted by Titania-Based Nanoparticles and Aerogels. *New J. Chem.* **2021**, *45*, 10217–10231. [CrossRef]
19. Yu, M.; Shang, J.; Kuang, Y. Efficient Photocatalytic Reduction of Chromium (VI) Using Photoreduced Graphene Oxide as Photocatalyst under Visible Light Irradiation. *J. Mater. Sci Technol.* **2021**, *91*, 17–27. [CrossRef]
20. Wang, J.L.; Xu, L.J. Advanced Oxidation Processes for Wastewater Treatment: Formation of Hydroxyl Radical and Application. *Crit. Rev. Environ. Sci. Technol.* **2012**, *42*, 251–325. [CrossRef]
21. Serge-correales, Y.E.; Ullah, S.; Ferreira-Neto, E.P.; Rojas-mantilla, H.D.; Hazra, C.; Ribeiro, S.J.L. A UV-Visible-NIR Active Smart Photocatalytic System Based on NaYbF4:Tm^{3+} Upconverting Particles and Ag$_3$PO$_4$/H$_2$O$_2$ for Photocatalytic Processes under Light on/Light off Conditions. *Mater. Adv.* **2022**, *3*, 2706–2715. [CrossRef]
22. Amirulsyafiee, A.; Khan, M.M.; Khan, M.Y.; Khan, A.; Harunsani, M.H. Visible Light Active La-Doped Ag$_3$PO$_4$ for Photocatalytic Degradation of Dyes and Reduction of Cr(VI). *Solid State Sci.* **2022**, *131*, 106950. [CrossRef]
23. Ullah, S.; Fayeza; Khan, A.A.; Jan, A.; Aain, S.Q.; Neto, E.P.F.; Serge-Correales, Y.E.; Parveen, R.; Wender, H.; Rodrigues-Filho, U.P.; et al. Enhanced Photoactivity of BiVO$_4$/Ag/Ag$_2$O Z-Scheme Photocatalyst for Efficient Environmental Remediation under Natural Sunlight and Low-Cost LED Illumination. *Colloids Surf. A Physicochem. Eng. Asp.* **2020**, *600*, 124946. [CrossRef]
24. Gomes, L.E.; Nogueira, A.C.; da Silva, M.F.; Plaça, L.F.; Maia, L.J.Q.; Gonçalves, R.V.; Ullah, S.; Khan, S.; Wender, H. Enhanced Photocatalytic Activity of BiVO$_4$/Pt/PtOx Photocatalyst: The Role of Pt Oxidation State. *Appl. Surf. Sci.* **2021**, *567*, 150773. [CrossRef]
25. Ullah, S.; Ferreira-Neto, E.P.P.; Hazra, C.; Parveen, R.; Rojas-Mantilla, H.D.D.; Calegaro, M.L.L.; Serge-Correales, Y.E.E.; Rodrigues-Filho, U.P.P.; Ribeiro, S.J.L.J.L. Broad Spectrum Photocatalytic System Based on BiVO$_4$ and NaYbF$_4$:Tm$_{3+}$ Upconversion Particles for Environmental Remediation under UV-Vis-NIR Illumination. *Appl. Catal. B* **2019**, *243*, 121–135. [CrossRef]
26. Shahzad, K.; Tahir, M.B.; Sagir, M.; Kabli, M.R. Role of CuCo2S4 in Z-Scheme MoSe$_2$/BiVO$_4$ Composite for Efficient Photocatalytic Reduction of Heavy Metals. *Ceram. Int.* **2019**, *45*, 23225–23232. [CrossRef]
27. Yin, G.; Liu, C.; Shi, T.; Ji, D.; Yao, Y.; Chen, Z. Porous BiVO4 Coupled with CuFeO$_2$ and NiFe Layered Double Hydroxide as Highly-Efficient Photoanode toward Boosted Photoelectrochemical Water Oxidation. *J. Photochem. Photobiol. A Chem.* **2022**, *426*, 113742. [CrossRef]
28. Li, S.; Xing, Z.; Feng, J.; Yan, L.; Wei, D.; Wang, H.; Wu, D.; Ma, H.; Fan, D.; Wei, Q. A Sensitive Biosensor of CdS Sensitized BiVO$_4$/GaON Composite for the Photoelectrochemical Immunoassay of Procalcitonin. *Sens. Actuators B Chem.* **2021**, *329*, 129244. [CrossRef]
29. Wei, Y.Y.; Wu, S.H.; Wang, Q.M.; Sun, J.J. A Novel Split-Type Photoelectrochemical Biosensor Based on Double-Strand Specific Nuclease-Assisted Cycle Amplification Coupled with Plasmonic Ag Enhanced BiVO4 Performance for Sensitive Detection of MicroRNA-155. *Sens. Actuators B Chem.* **2022**, *369*, 132251. [CrossRef]
30. Ho-Kimura, S.; Soontornchaiyakul, W.; Yamaguchi, Y.; Kudo, A. Preparation of Nanoparticle Porous-Structured BiVO$_4$ Photoanodes by a New Two-Step Electrochemical Deposition Method for Water Splitting. *Catalysts* **2021**, *11*, 136. [CrossRef]
31. Chen, M.; Zhao, J.; Huang, X.; Wang, Y.; Xu, Y. Improved Performance of BiVO$_4$ via Surface-Deposited Magnetic CuFe$_2$O$_4$ for Phenol Oxidation and O$_2$ Reduction and Evolution under Visible Light. *ACS Appl. Mater. Interfaces* **2019**, *11*, 45776–45784. [CrossRef]
32. Alhaddad, M.; Amin, M.S.; Zaki, Z.I. Novel BiVO$_4$/ZnO Heterojunction for Amended Photoreduction of Mercury (II) Ions. *Opt. Mater.* **2022**, *127*, 112251. [CrossRef]
33. Qin, N.; Zhang, S.; He, J.; Long, F.; Wang, L. In Situ Synthesis of BiVO$_4$/BiOBr Microsphere Heterojunction with Enhanced Photocatalytic Performance. *J. Alloys Compd.* **2022**, *927*, 166661. [CrossRef]
34. Yang, Z.; Li, H.; Cui, X.; Zhu, J.; Li, Y.; Zhang, P.; Li, J. Direct Z-Scheme Nanoporous BiVO$_4$/CdS Quantum Dots Heterojunction Composites as Photoanodes for Photocathodic Protection of 316 Stainless Steel under Visible Light. *Appl. Surf. Sci.* **2022**, *603*, 154394. [CrossRef]
35. Wang, S.; Zhao, L.; Gao, L.; Yang, D.; Wen, S.; Huang, W.; Sun, Z.; Guo, J.; Jiang, X.; Lu, C. Fabrication of Ternary Dual Z-Scheme AgI/ZnIn$_2$S$_4$/BiVO$_4$ Heterojunction Photocatalyst with Enhanced Photocatalytic Degradation of Tetracycline under Visible Light. *Arab. J. Chem.* **2022**, *15*, 104159. [CrossRef]
36. Pai, H.; Kuo, T.R.; Chung, R.J.; Kubendhiran, S.; Yougbaré, S.; Lin, L.Y. Enhanced Photocurrent Density for Photoelectrochemical Catalyzing Water Oxidation Using Novel W-Doped BiVO$_4$ and Metal Organic Framework Composites. *J. Colloid Interface Sci.* **2022**, *624*, 515–526. [CrossRef] [PubMed]
37. Maheskumar, V.; Jiang, Z.; Lin, Y.; Vidhya, B. The Structural and Optical Properties of Ag/Cu Co-Doped BiVO$_4$ Material: A Density Functional Study. *Mater. Lett.* **2022**, *315*, 131289. [CrossRef]
38. Prabhavathy, S.; Arivuoli, D. Visible Light-Induced Silver and Lanthanum Co-Doped BiVO$_4$ Nanoparticles for Photocatalytic Dye Degradation of Organic Pollutants. *Inorg. Chem. Commun.* **2022**, *141*, 109483. [CrossRef]
39. Sun, W.; Dong, Y.; Zhai, X.; Zhang, M.; Li, K.; Wang, Q.; Ding, Y. Crystal Facet Engineering of BiVO$_4$/CQDs/TPP with Improved Charge Transfer Efficiency for Photocatalytic Water Oxidation. *Chem. Eng. J.* **2022**, *430*, 132872. [CrossRef]

40. Srinivasan, N.; Anbuchezhiyan, M.; Harish, S.; Ponnusamy, S. Efficient Catalytic Activity of BiVO$_4$ Nanostructures by Crystal Facet Regulation for Environmental Remediation. *Chemosphere* **2022**, *289*, 133097. [CrossRef]
41. Kadam, A.N.; Babu, B.; Lee, S.-W.; Kim, J.; Yoo, K. Morphological Guided Sphere to Dendrite BiVO$_4$ for Highly Efficient Organic Pollutant Removal and Photoelectrochemical Performance under Solar Light. *Chemosphere* **2022**, *305*, 135461. [CrossRef]
42. Maheskumar, V.; Lin, Y.M.; Jiang, Z.; Vidhya, B.; Ghosal, A. New Insights into the Structural, Optical, Electronic and Photocatalytic Properties of Sulfur Doped Bulk BiVO$_4$ and Surface BiVO4 on {0 1 0} and {1 1 0} via a Collective Theoretical and Experimental Investigation. *J. Photochem. Photobiol. A Chem.* **2022**, *426*, 113757. [CrossRef]
43. Fatima, H.; Azhar, M.R.; Khiadani, M.; Zhong, Y.; Wang, W.; Su, C.; Shao, Z. Prussian Blue-Conjugated ZnO Nanoparticles for near-Infrared Light-Responsive Photocatalysis. *Mater. Today Energy* **2022**, *23*, 100895. [CrossRef]
44. Tada, H.; Tsuji, S.; Ito, S. Photodeposition of Prussian Blue Films on TiO$_2$: Additive Effect of Methanol and Influence of the TiO$_2$ Crystal Form. *J. Colloid Interface Sci.* **2001**, *239*, 196–199. [CrossRef] [PubMed]
45. Lu, J.; Chen, C.; Qian, M.; Xiao, P.; Ge, P.; Shen, C.; Wu, X.L.; Chen, J. Hollow-Structured Amorphous Prussian Blue Decorated on Graphitic Carbon Nitride for Photo-Assisted Activation of Peroxymonosulfate. *J. Colloid Interface Sci.* **2021**, *603*, 856–863. [CrossRef]
46. Kim, H.; Kim, M.; Kim, W.; Lee, W.; Kim, S. Photocatalytic Enhancement of Cesium Removal by Prussian Blue-Deposited TiO$_2$. *J. Hazard. Mater.* **2018**, *357*, 449–456. [CrossRef]
47. Song, Y.Y.; Zhang, K.; Xia, X.H. Photosynthesis and Characterization of Prussian Blue Nanocubes on Surfaces of TiO$_2$ Colloids. *Appl. Phys. Lett.* **2006**, *88*, 053112. [CrossRef]
48. Maurin-Pasturel, G.; Long, J.; Guari, Y.; Godiard, F.; Willinger, M.G.; Guerin, C.; Larionova, J. Nanosized Heterostructures of Au@Prussian Blue Analogues: Towards Multifunctionality at the Nanoscale. *Angew. Chem. Int. Ed.* **2014**, *53*, 3872–3876. [CrossRef]
49. Uchida, H.; Sasaki, T.; Ogura, K. Dark Catalytic Reduction of CO$_2$ over Prussian Blue-Deposited TiO$_2$ and the Photo-Reactivation of the Catalyst. *J. Mol. Catal.* **1994**, *93*, 269–277. [CrossRef]
50. Ghobadi, T.G.U.; Ghobadi, A.; Soydan, M.C.; Vishlaghi, M.B.; Kaya, S.; Karadas, F.; Ozbay, E. Strong Light–Matter Interactions in Au Plasmonic Nanoantennas Coupled with Prussian Blue Catalyst on BiVO$_4$ for Photoelectrochemical Water Splitting. *ChemSusChem* **2020**, *13*, 2577–2588. [CrossRef]
51. Meng, X.; Xu, S.; Zhang, C.; Feng, P.; Li, R.; Guan, H.; Ding, Y. Prussian Blue Type Cocatalysts for Enhancing the Photocatalytic Water Oxidation Performance of BiVO$_4$. *Chem. A Eur. J.* **2022**, *28*, e202201407. [CrossRef]
52. Li, J.; Chu, Y.; Zhang, C.; Zhang, X.; Wu, C.; Xiong, X.; Zhou, L.; Wu, C.; Han, D. CoFe Prussian Blue Decorated BiVO$_4$ as Novel Photoanode for Continuous Photocathodic Protection of 304 Stainless Steel. *J. Alloy. Compd.* **2021**, *887*, 161279. [CrossRef]
53. Yuan, D.; Sun, M.; Tang, S.; Zhang, Y.; Wang, Z.; Qi, J.; Rao, Y.; Zhang, Q. All-Solid-State BiVO$_4$/ZnIn$_2$S$_4$ Z-Scheme Composite with Efficient Charge Separations for Improved Visible Light Photocatalytic Organics Degradation. *Chin. Chem. Lett.* **2020**, *31*, 547–550. [CrossRef]
54. Yuan, D.; Sun, M.; Zhao, M.; Tang, S.; Qi, J.; Zhang, X.; Wang, K.; Li, B. Persulfate Promoted ZnIn$_2$S$_4$ Visible Light Photocatalytic Dye Decomposition. *Int. J. Electrochem. Sci.* **2020**, *15*, 8761–8770. [CrossRef]
55. Yu, L.; Achari, G.; Langford, C.H. LED-Based Photocatalytic Treatment of Pesticides and Chlorophenols. *J. Environ. Eng.* **2014**, *139*, 1146–1151. [CrossRef]
56. Izadifard, M.; Achari, G.; Langford, C. Application of Photocatalysts and LED Light Sources in Drinking Water Treatment. *Catalysts* **2013**, *3*, 726–743. [CrossRef]
57. Tada, H.; Saito, Y.; Kawahara, H. Photodeposition of Prussian Blue on TiO$_2$ Particles. *J. Electrochem. Soc.* **1991**, *138*, 140–144. [CrossRef]
58. Jang, S.C.; Hong, S.B.; Yang, H.M.; Lee, K.W.; Moon, J.K.; Seo, B.K.; Huh, Y.S.; Roh, C. Removal of Radioactive Cesium Using Prussian Blue Magnetic Nanoparticles. *Nanomaterials* **2014**, *4*, 894–901. [CrossRef]
59. Baggio, B.F.; Vicente, C.; Pelegrini, S.; Cid, C.C.P.; Brandt, I.S.; Tumelero, M.A.; Pasa, A.A. Morphology and Structure of Electrodeposited Prussian Blue and Prussian White Thin Films. *Materials* **2019**, *12*, 1103. [CrossRef]
60. Yu, J.; Kudo, A. Effects of Structural Variation on the Photocatalytic Performance of Hydrothermally Synthesized BiVO$_4$. *Adv. Funct. Mater.* **2006**, *16*, 2163–2169. [CrossRef]
61. Packiaraj, R.; Venkatesh, K.S.; Devendran, P.; Bahadur, S.A.; Nallamuthu, N. Structural, Morphological and Electrochemical Studies of Nanostructured BiVO$_4$ for Supercapacitor Application. *Mater. Sci. Semicond. Process.* **2020**, *115*, 105122. [CrossRef]
62. Mousavi-Kamazani, M. Facile Hydrothermal Synthesis of Egg-like BiVO$_4$ Nanostructures for Photocatalytic Desulfurization of Thiophene under Visible Light Irradiation. *J. Mater. Sci. Mater. Electron.* **2019**, *30*, 17735–17740. [CrossRef]
63. Kulesza, P.J.; Malik, M.A.; Denca, A.; Strojek, J. In Situ FT-IR/ATR Spectroelectrochemistry of Prussian Blue in the Solid State. *Anal. Chem.* **1996**, *68*, 2442–2446. [CrossRef]
64. Zhou, B.; Qu, J.; Zhao, X.; Liu, H. Fabrication and Photoelectrocatalytic Properties of Nanocrystalline Monoclinic BiVO$_4$ Thin-Film Electrode. *J. Environ. Sci.* **2011**, *23*, 151–159. [CrossRef] [PubMed]

65. Li, X.; Wang, J.; Rykov, A.I.; Sharma, V.K.; Wei, H.; Jin, C.; Liu, X.; Li, M.; Yu, S.; Sun, C.; et al. Prussian Blue/TiO_2 Nanocomposites as a Heterogeneous Photo-Fenton Catalyst for Degradation of Organic Pollutants in Water Xuning. *Catal. Sci. Technol.* **2015**, *5*, 504–514. [CrossRef]

66. Cheng, K.L.; Ueno, K.; Imamura, T. *Handbook of Organic Analytical Reagents*, 2nd ed.; CRC Press: Boca Raton, FL, USA, 1992; ISBN 9781315140568.

67. Ishibashi, K.I.; Fujishima, A.; Watanabe, T.; Hashimoto, K. Detection of Active Oxidative Species in TiO_2 Photocatalysis Using the Fluorescence Technique. *Electrochem. Commun.* **2000**, *2*, 207–210. [CrossRef]

Article

Constructing Active Sites on Self-Supporting $Ti_3C_2T_x$ (T = OH) Nanosheets for Enhanced Photocatalytic CO_2 Reduction into Alcohols

Shuqu Zhang, Man Zhang, Wuwan Xiong, Jianfei Long, Yong Xu, Lixia Yang and Weili Dai *

Key Laboratory of Jiangxi Province for Persistent Pollutants Control and Resources Recycle,
Nanchang Hangkong University, Nanchang 330063, China
* Correspondence: wldai81@126.com or daiweili@nchu.edu.cn

Abstract: $Ti_3C_2T_x$ (T = OH) was first prepared from Ti_3AlC_2 by HF etching and applied into a photocatalytic CO_2 reduction. Then, the $Ti_3C_2T_x$ nanosheets present interbedded a self-supporting structure and extended interlayer spacing. Meanwhile, the $Ti_3C_2T_x$ nanosheets are decorated with abundant oxygen-containing functional groups in the process of etching, which not only serve as active sites but also show efficient charge migration and separation. Among $Ti_3C_2T_x$ materials prepared by etching for different times, $Ti_3C_2T_x$-36 (Etching time: 36 h) showed the best performance for photoreduction of CO_2 into alcohols (methanol and ethanol), giving total yield of 61 µmol g catal.$^{-1}$, which is 2.8 times than that of Ti_3AlC_2. Moreover, excellent cycling stability for CO_2 reduction is beneficial from the stable morphology and crystalline structure. This work provided novel sights into constructing surface active sites controllably.

Keywords: photocatalytic CO_2 reduction; active sites; controllable exfoliation; $Ti_3C_2T_x$

Citation: Zhang, S.; Zhang, M.; Xiong, W.; Long, J.; Xu, Y.; Yang, L.; Dai, W. Constructing Active Sites on Self-Supporting $Ti_3C_2T_x$ (T = OH) Nanosheets for Enhanced Photocatalytic CO_2 Reduction into Alcohols. *Catalysts* 2022, 12, 1594. https://doi.org/10.3390/catal12121594

Academic Editors: Gassan Hodaifa, Rafael Borja and Mha Albqmi

Received: 8 November 2022
Accepted: 4 December 2022
Published: 6 December 2022

Publisher's Note: MDPI stays neutral with regard to jurisdictional claims in published maps and institutional affiliations.

1. Introduction

The continuous CO_2 emissions due to the depletion of fossil fuels have caused emerging problems in the environment and energy sectors [1]. Solar-driven CO_2 reduction that can produce various carbon fuels is considered a desirable strategy to resolve these problems [2]. Nevertheless, the perfect photocatalytic reduction of CO_2 process needs to meet the enhanced and broaden light absorption, abundant active sites and efficient charges separation [3]. At present, the researchers devote themselves improving the efficiency for photocatalytic CO_2 reduction towards the abovementioned objectives.

Two-dimensional semiconductors are valuable materials for photocatalytic applications because of their larger surface area and excellent electron mobility [4–7]. As a surface catalytic reaction, the performance of photocatalytic CO_2 reduction is also seriously determined by the reactive sites on the surface of photocatalysts [8]. Therefore, it is still urgent for constructing active sites on the surface of two-dimensional semiconductor photocatalysts to further enhance photocatalytic performance [9]. MXenes is formulated as $M_{n+1}X_nT_x$, in which M is a transition metal such as Ti, X is C or N, and T is a surface termination group such as -O or -OH [10]. They can be obtained by removing element A (mostly Al) from a ternary parent MAX phase through liquid exfoliation [11,12]. It was known that MXenes acted as a cocatalyst in photocatalytic CO_2 reduction due to its huge surface and excellent electronic conductivity [13]. Its huge surface provides the anchored sites for CO_2, and its excellent electronic conductivity is beneficial for migration of photogenerated electrons. However, there are still no reports about MXenes as separate photocatalyst in CO_2 reduction. As reported, the terminal oxygen-containing functional group on the MXenes surface could be redox-active [14]. Therefore, it is necessary to prepare MXenes nanosheets with a large surface area and explore the role of terminal functional groups on the performance of photocatalytic CO_2 reduction.

In this work, $Ti_3C_2T_X$ nanosheets were prepared by controllable etching, and firstly applicated into a photocatalytic CO_2 reduction. The $Ti_3C_2T_X$ nanosheets were decorated with different types of the oxygen-containing functional group. The interbedded self-supporting structure of layered $Ti_3C_2T_X$ not only exposed more active sites and preserved the stability of morphology and crystalline structure, but also benefitted for charge migration and separation. Eventually, $Ti_3C_2T_X$ nanosheets delivered excellent performance and stability for photocatalytic CO_2 reduction.

2. Results and Discussion

The XRD pattern was shown to investigate the stacking property and layered structure (Figure 1a). Diffraction peaks of $Ti_3C_2T_X$ correspond to JCPDS No. 52-0875. Stacking peak {002} shifts to a lower angle compared with Ti_3AlC_2, indicating extended interlayer spacing in Figure 1b [15]. $Ti_3C_2T_X$-36 shows the largest specific surface area among all the photocatalysts samples, which means that the $Ti_3C_2T_X$-36 holds the largest interlayer spacing, showing the lowest {002} peak intensity. Raman spectra of different samples was shown in Figure S1. The enhanced peak intensity at 203 cm^{-1} suggests powerful Ti-C vibration in $Ti_3C_2T_X$ [16]. The peak at 273 cm^{-1} belonging to Ti-Al vibration in Ti_3AlC_2 disappeared after etching. The enhanced Ti-C vibration and disappeared Ti-Al vibration suggest the removal of the Al layer. "Eg vibration" corresponds to out-of-plane vibration of Raman scattering for two-dimensional nanosheets. Eg vibration presents enhanced Raman scattering at 628 cm^{-1} for decoration of -OH groups on the terminated C atom of Ti_3C_2 [17]. All indicate successful formation of $Ti_3C_2T_X$ and decoration of oxygen-containing functional groups on it.

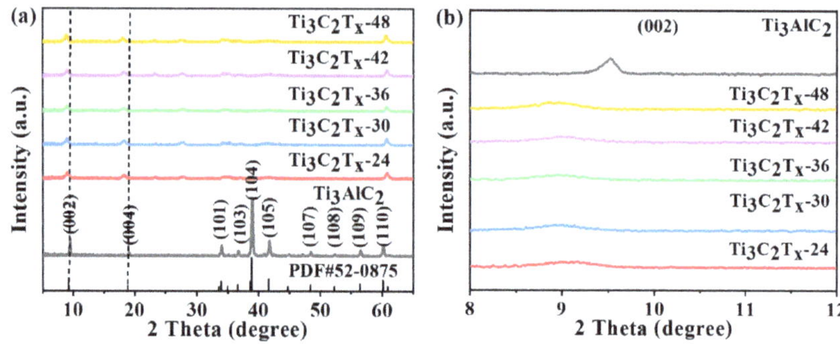

Figure 1. (a) XRD pattern and (b) (002) peak of Ti_3AlC_2, $Ti_3C_2T_X$-y nanosheets (y = 24, 30, 36, 42 and 48).

Figure 2 showed SEM images for $Ti_3C_2T_X$ samples with different etching time (24 h, 30 h, 36 h, 42 h and 48 h). $Ti_3C_2T_X$ samples show the obvious morphological features of layered structure with broadened interlayer spacing. It can be observed that $Ti_3C_2T_X$-36 shows uniform layers and a smooth surface. However, the $Ti_3C_2T_X$-42 and $Ti_3C_2T_X$-48 tended to aggregate and stack again with the increasing etching time. It is well accepted that the catalysis generally occurs on the active sites, while the active sites mostly exist in the edges, unsaturated steps, terraces, kinks, and/or corner atoms for layered structures [3,18].

The catalysts' surface holds a spot of active sites. The stacking, layered structure may cause the less active sites' exposure. $Ti_3C_2T_X$-36 shows a uniform layered structure (Figure S2a). Elemental mapping spectra presented Ti, C and O elements in Figure S2b,d. Oxygen-containing functional groups are decorated on the surface. The atomic structure of $Ti_3C_2T_X$ nanosheets is shown in Figure S2e. The side view for $Ti_3C_2T_X$ nanosheets shows a broadening layered structure (Figure S2f). TG analysis was conducted to inspect thermostability in Figure S3. $Ti_3C_2T_X$ nanosheets decorated with oxygen-containing functional groups shows the interbedded self-supporting structure, which also preserves morphological stability. Specifically, $Ti_3C_2T_X$-36 showed the best thermostability among all $Ti_3C_2T_X$

samples. The specific area of Ti_3AlC_2, $Ti_3C_2T_x$-24, $Ti_3C_2T_x$-30, $Ti_3C_2T_x$-36, $Ti_3C_2T_x$-42 and $Ti_3C_2T_x$-48 nanosheets are 0.56, 2.49, 3.27, 3.52, 2.97 and 1.41 m^2/g, respectively. $Ti_3C_2T_x$ shows a larger specific surface area compared with Ti_3AlC_2 form Nitrogen adsorption–desorption isotherms (Figure S4). $Ti_3C_2T_x$-36 holds the highest specific surface area and pore volume. The extended interlayer spacing means more surface is exposed and the stacking structure becomes open architecture. It is reported that the open architecture is beneficial for migration and diffusion of photogenerated carriers [19]. $Ti_3C_2T_x$-42 and $Ti_3C_2T_x$-48, with prolonged etching times, present the smaller specific surface area due to the stacking layers, which is in accord with the morphological features from Figure 1.

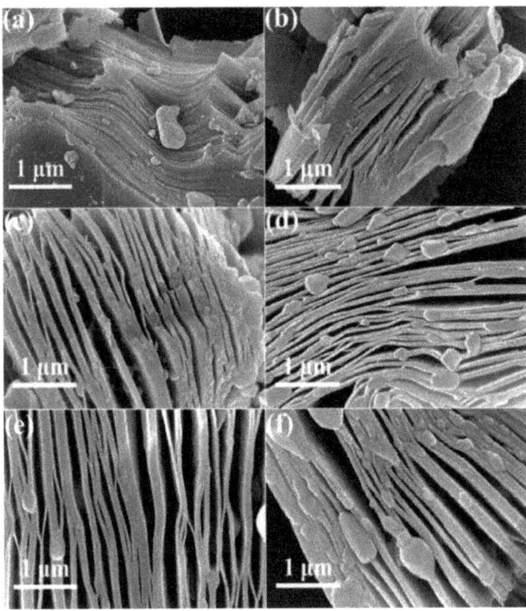

Figure 2. SEM images of (**a**) Ti_3AlC_2, (**b**) $Ti_3C_2T_X$-24, (**c**) $Ti_3C_2T_X$-30, (**d**) $Ti_3C_2T_X$-36, (**e**) $Ti_3C_2T_X$-42 and (**f**) $Ti_3C_2T_X$-48.

It is necessary to investigate the photoelectric property and identify performance of carrier separation. $Ti_3C_2T_x$ shows excellent UV-*vis* absorption ability (Figure S5a). The bandgap structure is not changed even though $Ti_3C_2T_x$ nanosheets are decorated with different oxygen-containing functional groups. The bandgap of $Ti_3C_2T_x$-24, $Ti_3C_2T_x$-30, $Ti_3C_2T_x$-36, $Ti_3C_2T_x$-42 and $Ti_3C_2T_x$-48 samples is 2.21, 2.14, 2.22, 2.38, 2.26 V, respectively, from the Kubelka–Munk function $(Ahv)^2$ vs. light energy (hv) in Figure S5b. The flat band potential (FB) of $Ti_3C_2T_x$ is -0.53 V vs. SCE by Mott-Schottky spectra in Figure S5c. The conduction band (CB) can be calculated as -0.39 V vs. NHE by the following equation:

$$E_{vs.\ NHE} = E_{FB} + E^0 + 0.059pH. \tag{1}$$

The valence band (VB) of $Ti_3C_2T_x$ is 1.82, 1.75, 1.83, 1.99, and 1.87 V vs. NHE (pH = 7) by the following equation:

$$(E_{VB} = E_{CB} + Eg) \tag{2}$$

It is reported that the oxidation potential is 0.82 V (vs. NHE, pH = 7) [20–22]. $Ti_3C_2T_x$ holds the ability to oxidize H_2O to provide H protons for a CO_2 reduction rection.

$Ti_3C_2T_x$-36 shows highest photocurrent, indicating efficient separation and transportation of photoinduced charge carriers (Figure 3a). In addition, $Ti_3C_2T_x$-36 shows a much smaller radius from electrochemical impedance spectroscopy (EIS) spectra (Figure 3b), demonstrating fast interfacial charge transfer. The efficient separation of photogenerated carriers and longer fluorescence (PL) lifetime imply that $Ti_3C_2T_x$-36 showed the best carrier generation and separation capability (Figure 3c,d). The enhanced photoelectric property is due to extended interlayer spacing and more introduced active sites.

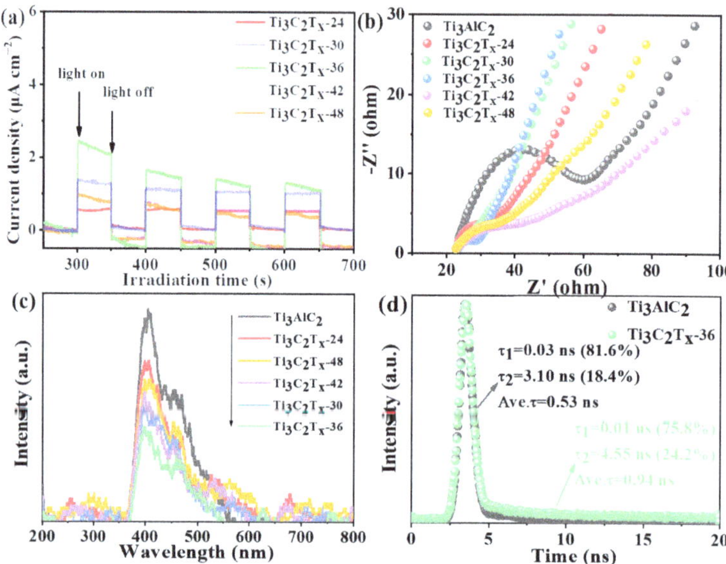

Figure 3. (**a**) Transient photocurrent responses, (**b**) EIS Nyquist plots, (**c**) PL spectra and (**d**) TRPL decay spectra of prepared photocatalysts.

The $Ti_3C_2T_x$ showed the better photoelectric property, therefore, it is needed to inspect the surface property of $Ti_3C_2T_x$. FTIR spectra was conducted to analyze the functional groups in Figure 4a. A length of 775–1237 cm^{-1} corresponds to various Ti-O vibrational modes [23]. A length of 1607 and 1631 cm^{-1} absorbs O vibration. A length of 2345 and 2372 cm^{-1} belongs to -OH groups vibration. A length of 3396 cm^{-1} corresponds to absorbed H_2O. The $Ti_3C_2T_x$ nanosheets were prepared by HF etching, therefore, there are few -F function groups linked with the C atom after Al removal (Figure S6a,b). The abundant oxygen-containing functional groups (i.e., -O, -OH) are decorated on the terminus of Ti_3C_2 after etching exfoliation from XPS measurement (Figure S6c). The binding energy at 527.15, 528.45 and 530.35 eV absorb O, -OH/Ox and H_2O, respectively [24]. The high-resolution XPS spectra of O 1s for the samples are analyzed to figure out the crucial oxygen-containing functional group in Figure 4b. The analysis results are listed in Figure 4c. The atomic O contents for $Ti_3C_2T_x$-24, $Ti_3C_2T_x$-30, $Ti_3C_2T_x$-36, $Ti_3C_2T_x$-42 and $Ti_3C_2T_x$-48 samples are 15.96%, 16.27%, 16.98%, 15.46% and 15.92%, respectively. $Ti_3C_2T_x$-36 shows the highest atomic O content because more oxygen-containing groups are decorated on a larger surface area. The stacking layers for $Ti_3C_2T_x$-42 and $Ti_3C_2T_x$-48 lead to the smaller O content. In $Ti_3C_2T_x$-36, the -OH/Ox and H_2O showed the highest content compared with other photocatalyst samples from the integral area of the corresponding peak, which means these two oxygen-containing groups play a crucial role for better photoelectric properties. As result, $Ti_3C_2T_x$-36 shows better dispersion from the morphological features of the layered structure in Figure S2g, which is due to the wider interlayer spacing for the decoration of -OH and -F functional groups. The selected area electron diffraction (SAED) pattern of $Ti_3C_2T_x$-36 suggests preservation of hexagonal basal structure derived from

Ti$_3$AlC$_2$ (Figure S2h) [25]. It was reported that this oxygen-containing functional group on the Ti$_3$C$_2$T$_x$ terminal could be redox-active, serving as adsorption active sites for CO$_2$ [26].

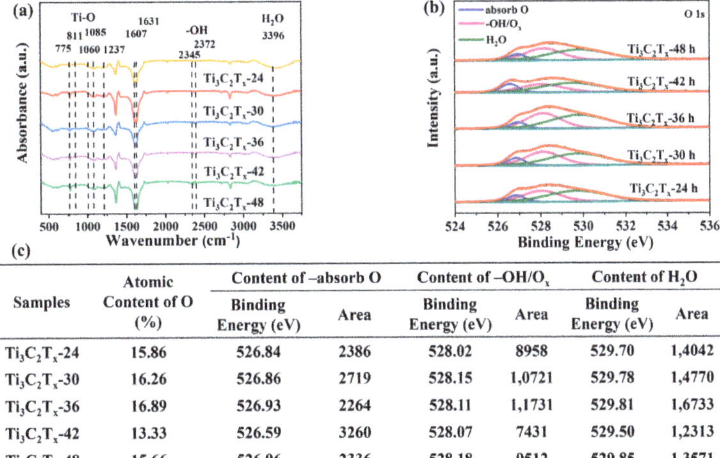

Samples	Atomic Content of O (%)	Content of –absorb O		Content of –OH/O$_x$		Content of H$_2$O	
		Binding Energy (eV)	Area	Binding Energy (eV)	Area	Binding Energy (eV)	Area
Ti$_3$C$_2$T$_x$-24	15.86	526.84	2386	528.02	8958	529.70	1,4042
Ti$_3$C$_2$T$_x$-30	16.26	526.86	2719	528.15	1,0721	529.78	1,4770
Ti$_3$C$_2$T$_x$-36	16.89	526.93	2264	528.11	1,1731	529.81	1,6733
Ti$_3$C$_2$T$_x$-42	13.33	526.59	3260	528.07	7431	529.50	1,2313
Ti$_3$C$_2$T$_x$-48	15.66	526.96	2336	528.18	9512	529.85	1,3571

Figure 4. (**a**) FTIR spectra, (**b**) High-resolution XPS spectra of O 1s, (**c**) Table of different O group comparison of Ti$_3$C$_2$T$_X$-24, Ti$_3$C$_2$T$_X$-30, Ti$_3$C$_2$T$_X$-36, Ti$_3$C$_2$T$_X$-42 and Ti$_3$C$_2$T$_X$-48.

The photocatalytic CO$_2$ reduction was proceeded to evaluate photocatalytic performance for photocatalysts. The products of photocatalytic CO$_2$ reduction are methanol and ethanol, and Ti$_3$C$_2$T$_x$ samples show enhanced photocatalytic performance with increasing irradiation time (Figure 5a,b). The produced rate for methanol and ethanol over Ti$_3$AlC$_2$ is 12.9 µmol g catal.$^{-1}$ and 8.7 µmol g catal.$^{-1}$. Among all samples, Ti$_3$C$_2$T$_x$-36 gives best methanol and ethanol yields, 38.1 µmol g catal.$^{-1}$ and 22.9 µmol g catal.$^{-1}$ after 4 h irradiation, respectively. The total yield for Ti$_3$C$_2$T$_x$-36 is 2.8 times than that of Ti$_3$AlC$_2$. The methanol and ethanol yields after 4 h irradiation of Ti$_3$C$_2$T$_x$-24, Ti$_3$C$_2$T$_x$-30, Ti$_3$C$_2$T$_x$-42 and Ti$_3$C$_2$T$_x$-48 are 19.34 and 17.7, 30.99 and 17.05, 28.85 and 18.11 and 19.1 and 14.34 µmol g catal.$^{-1}$, respectively. It is noted that Ti$_3$C$_2$T$_x$-42 and Ti$_3$C$_2$T$_x$-48 with less O contents show poorer photocatalytic performance of CO$_2$ reduction due to the restacking layers. The oxygen-containing content shows a positive correlation with production yields (Figures 4c and 5a,b). Table S1 showed the comparison of photocatalytic activity for CO$_2$ reduction (products: methanol and ethanol) by some photocatalyst systems. The enhanced performance for CO$_2$ reduction is due to more active sites constructed on Ti$_3$C$_2$T$_x$ surface and efficient carrier separation. ^{13}CO$_2$ was employed to replace ^{12}CO$_2$ to confirm carbon source of the produced methanol and ethanol with the corresponding MS spectra shown in Figure 5c,d. The intense signals of $m/z = 33$ and $m/z = 48$ are assigned to ^{13}CH$_3$OH and ^{13}C$_2$H$_5$OH, respectively. The nearby peaks belong to the fragment peaks. It verifies that CO$_2$ acts as the only carbon source of value-added alcohols over the Ti$_3$C$_2$T$_x$ photocatalyst. To further prove the water oxidation, O$_2$ amounts were detected during photocatalytic CO$_2$ reduction over Ti$_3$C$_2$T$_x$-36. The calibration curves of the relationships between peak area and O$_2$ volume was shown in Figure S7. The O$_2$ yield with 2, 5, 8, 13 µmol g catal.$^{-1}$ is increased during CO$_2$ reduction over Ti$_3$C$_2$T$_x$-36 in 4 h in Figure S8. It is true that self-supporting Ti$_3$C$_2$Tx nanosheets with constructed active sites could act as an efficient photocatalyst for CO$_2$ reduction and H$_2$O oxidation. The cycling stability of CO$_2$ reduction over Ti$_3$C$_2$T$_x$-36 was inspected in Figure 5e. The methanol and ethanol performance over Ti$_3$C$_2$T$_x$-36 in five cycles are 38.06 and 22.85, 32.1 and 28.64, 32.61 and 27.96, 31.5 and 27.52 and 30.49 and 27.42 µmol g catal.$^{-1}$, respectively. The performance over Ti$_3$C$_2$T$_x$-36 represents little decrease after five cycling runs, but crystal structure does not change (Figure 5f).

Interbedded self-supporting structures are responsible for excellent photocatalytic activity and stable morphology structure.

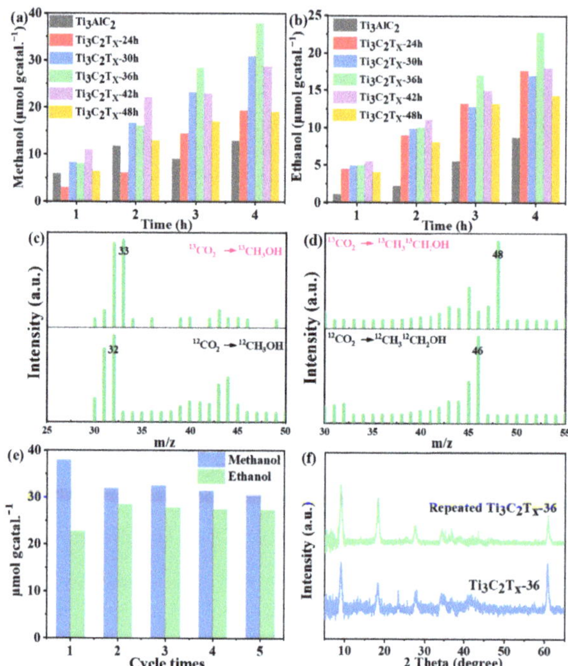

Figure 5. Yields of (**a**) methanol and (**b**) ethanol from photocatalytic CO_2 reduction under visible light. GC-MS spectra of (**c**) $^{13}CH_3OH$ and (**d**) $^{13}C_2H_5OH$ using $^{13}CO_2$ and $^{12}CO_2$ as the source of CO_2. (**e**) Cycling runs for CO_2 reduction over $Ti_3C_2T_x$-36. (**f**) XRD patterns of fresh and repeated $Ti_3C_2T_x$-36.

3. Materials and Methods

3.1. Preparation Methods

Ti_3AlC_2 powder (1 g) was dispersed in HF solution (10 mL) and vigorously stirred for different times at room temperature. The obtained powder was washed with deionized water until pH = 6, collected by centrifugation at 8000 rpm for 5 min and dried in the vacuum oven at 60 °C for 12 h. A series of $Ti_3C_2T_x$ were labeled as $Ti_3C_2T_x$-y (y = 24 h, 30 h, 36 h, 42 h and 48 h).

3.2. Materials Characterization

Crystal structure was analyzed by X-ray diffractometer with Cu Kα radiation (Bruker AXS-D8, Karlsruhe, Germany). Raman spectra of the samples were measured by Raman spectrophotometer (Horiba JY LabRAM HR800, Paris, France). Scanning electron microscope (SEM) images were obtained by Nova NanoSEM 450, Hillsboro, IL, USA, and transmission electron microscope (TEM) analyses were conducted with JEOL, JEM−2100F (HR, Tokyo, Japan). Specific surface area and pore property were collected by TriStar II 3020, Atlanta, GA, USA. Thermogravimetric (TG) analysis was obtained from SDT Q600 (TA Instruments, New Castle, DE, USA). Fourier transform infrared spectroscopy (FTIR) were conducted by Bruker VERTEX 70 (Bruker, Karlsruhe, Germany). X-ray photoelectron spectroscopy (XPS) was performed VG Escalab 250, Waltham, MA, USA spectrometer equipped with an Al anode. UV-*vis* diffuse reflectance spectra (DRS) were proceeded by Shimadzu UV-2450 (Tokyo, Japan) spectrophotometer. The photoluminescence (PL) spectra were measured to inspect charge recombination (F7000, Hitachi, Tokyo, Japan). A time-resolved fluorescence spectrofluorometer was used from Edinburgh, FS5.

3.3. Photoelectrochemical Measurements

Transient photocurrent response, electrochemical impedance spectroscopy and Mott–Schottky curves were carried out on the electrochemical workstation (CHI760E, Shanghai, China) in a standard three-electrode system with the Pt mesh as the counter electrode, the Saturated Calomel Electrode as the reference electrode, and the sample loaded electrodes as the working electrode in 0.1 M Na_2SO_4 aqueous solution (electrolyte solution) at room temperature. The distance between the counter electrode and the working electrode is 2 cm. Indium tin oxide (ITO) with a 1.0 cm $\times$ 1.0 cm area photocatalyst was used as the working electrode. The photocurrent measurement of the photocatalyst is measured by several switching cycles of light irradiated by a 300 W xenon lamp (using a 420 nm cut off filter).

3.4. Photocatalytic Reaction

The assessment for photocatalytic performance of CO_2 reduction as follows: 30 mg photocatalyst was dispersed in 30 mL deionized water and put into the reactor. The reactor was vacuumized. The saturated solution was obtained after admission with CO_2 (50 mL/min, 0.5 h). The reaction temperature was controlled at 4 °C. With the increasing irradiation time (light source: 300 W xenon lamp with a 420 nm cut offfilter, Perfect-Light, Beijing, China), the liquid reduction products were analyzed by gas chromatograph (GC7920-TF2A) equipped with a flame ionized detector (FID) and SE-54 capillary column. The isotope-labeled photocatalytic CO_2 reduction tests were performed by replacing $^{12}CO_2$ with $^{13}CO_2$ gas, while keeping the other reaction conditions unaltered. The obtained mixed gas was analyzed by gas chromatography-mass spectrometry (GC Model 6890 N/MS Model 5973, Agilent Technologies, Palo Alto, CA, USA).

4. Conclusions

In this work, $Ti_3C_2T_X$ with abundant oxygen-containing functional groups was successfully prepared and applicated into photocatalytic CO_2 reduction under visible light. The controllable content of oxygen-containing functional groups was achieved by tuning etching times as shown by the TG and XPS analysis. The exfoliation by extending the interlayer spacing exposed more active sites for generating more photo-induced carriers. The decorated oxygen-containing functional groups was beneficial for the charge migration and separation. The result was that the $Ti_3C_2T_X$-36 showed the best performance for photocatalytic CO_2 reduction (alcohols over production rate: 61 μmol g catal.$^{-1}$), which is 2.8 time than that of Ti_3AlC_2. The interbedded self-supporting structure of layered $Ti_3C_2T_X$ after successful exfoliation showed excellent stability of morphological structure, resulting in cycling stability for CO_2 reduction.

Supplementary Materials: The following supporting information can be downloaded at: https://www.mdpi.com/article/10.3390/catal12121594/s1, Figure S1. Raman spectra of Ti_3AlC_2, $Ti_3C_2T_x$-y nanosheets (y= 24, 30, 36, 42 and 48); Figure S2. SEM image of (a) $Ti_3C_2T_x$-36, (b-d) elemental mappings of (a). (e) Atomic structure of $Ti_3C_2T_x$, (f) STEM image, (g) TEM image and (h) selected area electron diffraction (SAED) pattern of $Ti_3C_2T_x$-36; Figure S3. TG analysis of (a) Ti_3AlC_2, (b) $Ti_3C_2T_x$-24, (c) $Ti_3C_2T_x$-30, (d) $Ti_3C_2T_x$-36, (e) $Ti_3C_2T_x$-42 and (f) $Ti_3C_2T_x$-48; Figure S4. (a) Nitrogen adsorption−desorption isotherms, (b) corresponding pore size distribution curves, and (c) information contrast of BET surface area, pore size and pore volume of $Ti_3C_2T_X$ with different etching times; Figure S5. (a) UV-vis absorption spectra, (b) Plots of transformed Kubelka–Munk function (Ahv)^2 *vs* light energy (hv) and (c) Mott-Schottky spectra of $Ti_3C_2T_X$-24, $Ti_3C_2T_X$-30, $Ti_3C_2T_X$-36, $Ti_3C_2T_X$-42, $Ti_3C_2T_X$-48; Figure S6. High-resolution XPS spectra of (a) Ti 2p, (b) C 1s and (c) O 1s over $Ti_3C_2T_X$-36; Figure S7. Calibration curves of the relationships between peak area and O_2 volume; Figure S8. O_2 evolution during photocatalytic CO_2 reduction over $Ti_3C_2T_x$-36; Table S1. The comparison of photocatalytic activity for CO_2 reduction (products: methanol and ethanol) by some photocatalyst systems [27–36].

Author Contributions: Conceptualization, S.Z.; Methodology, M.Z.; Methodology, W.X.; Formal analysis, J.L.; Investigation, Y.X.; Supervision, L.Y.; Writing—review and editing, W.D. All authors have read and agreed to the published version of the manuscript.

Funding: This work was financially supported by the National Natural Science Foundation of China (Grants 51908269, 52262037, 22006064, 52072165).

Data Availability Statement: In this manuscript, our characterizations were SEM, XRD, TEM, BET, and UV-*vis*. All data have been reported as the images.

Conflicts of Interest: The authors declare no conflict of interest.

References

1. Zhou, N.; Khanna, N.; Feng, W.; Ke, J.; Levine, M. Scenarios of energy efficiency and CO_2 emissions reduction potential in the buildings sector in China to year 2050. *Nat. Energy* **2018**, *3*, 978–984. [CrossRef]
2. He, J.; Janáky, C. Recent advances in solar-driven carbon dioxide conversion: Expectations versus reality. *ACS Energy Lett.* **2020**, *5*, 1996–2014. [CrossRef] [PubMed]
3. Luo, J.; Zhang, S.; Sun, M.; Yang, L.; Luo, S.; Crittenden, J. A critical review on energy conversion and environmental remediation of photocatalysts with remodeling crystal lattice, surface, and interface. *ACS Nano* **2019**, *13*, 9811–9840. [CrossRef]
4. Dai, W.; Yu, J.; Luo, S.; Hu, X.; Yang, L.; Zhang, S.; Li, B.; Luo, X.; Zou, J. WS_2 quantum dots seeding in Bi_2S_3 nanotubes: A novel Vis-NIR light sensitive photocatalyst with low-resistance junction interface for CO_2 reduction. *Chem. Eng. J.* **2020**, *389*, 123430. [CrossRef]
5. Dai, W.; Xiong, W.; Yu, J.; Zhang, S.; Li, B.; Yang, L.; Wang, T.; Luo, X.; Zou, J.; Luo, S. Bi_2MoO_6 Quantum Dots in Situ Grown on Reduced Graphene Oxide Layers: A Novel Electron-Rich Interface for Efficient CO_2 Reduction. *ACS Appl. Mater. Interfaces* **2020**, *12*, 25861–25874. [CrossRef] [PubMed]
6. Dai, W.; Long, J.; Yang, L.; Zhang, S.; Xu, Y.; Luo, X.; Zou, J.; Luo, S. Oxygen migration triggering molybdenum exposure in oxygen vacancy-rich ultra-thin Bi_2MoO_6 nanoflakes: Dual binding sites governing selective CO_2 reduction into liquid hydrocarbons. *J. Energy Chem.* **2021**, *61*, 281–289. [CrossRef]
7. Zhang, S.; Si, Y.; Li, B.; Yang, L.; Dai, W.; Luo, S. Atomic-Level and Modulated Interfaces of Photocatalyst Heterostructure Constructed by External Defect-Induced Strategy: A Critical Review. *Small* **2021**, *17*, 2004980. [CrossRef]
8. Habisreutinger, S.; Schmidt-Mende, L.; Stolarczyk, J. Photocatalytic reduction of CO_2 on TiO_2 and other semiconductors. *Angew. Chem. Int. Ed.* **2013**, *52*, 7372–7408. [CrossRef]
9. Liu, L.; Wang, S.; Huang, H.; Zhang, Y.; Ma, T. Surface sites engineering on semiconductors to boost photocatalytic CO_2 reduction. *Nano Energy* **2020**, *75*, 104959. [CrossRef]
10. Xiu, L.; Wang, Z.; Yu, M.; Wu, X.; Qiu, J. Aggregation-resistant 3D MXene-based architecture as efficient bifunctional electrocatalyst for overall water splitting. *ACS Nano* **2018**, *12*, 8017–8028. [CrossRef]
11. Li, T.; Yao, L.; Liu, Q.; Gu, J.; Luo, R.; Li, J.; Yan, X.; Wang, W.; Liu, P.; Chen, B.; et al. Fluorine-Free Synthesis of High-Purity $Ti_3C_2T_x$ (T = OH, O) via Alkali Treatment. *Angew. Chem. Int. Edit.* **2018**, *57*, 6115–6119. [CrossRef] [PubMed]
12. Natu, V.; Pai, R.; Sokol, M.; Carey, M.; Kalra, V.; Barsoum, M. 2D $Ti_3C_2T_z$ MXene synthesized by water-free etching of Ti_3AlC_2 in polar organic solvents. *Chem* **2020**, *6*, 616–630. [CrossRef]
13. He, F.; Zhu, B.; Cheng, B.; Yu, J.; Ho, W.; Macyk, W. 2D/2D/0D $TiO_2/C_3N_4/Ti_3C_2$ MXene composite S-scheme photocatalyst with enhanced CO_2 reduction activity. *Appl. Catal B-Environ.* **2020**, *272*, 119006. [CrossRef]
14. Lukatskaya, M.; Kota, S.; Lin, Z.; Zhao, M.-Q.; Shpigel, N.; Levi, M.; Halim, J.; Taberna, P.-L.; Barsoum, M.; Simon, P.; et al. Ultra-high-rate pseudocapacitive energy storage in two-dimensional transition metal carbides. *Nat. Energy* **2017**, *2*, 17105. [CrossRef]
15. Overbury, S.; Kolesnikov, A.; Brown, G.; Zhang, Z.; Nair, G.; Sacci, R.; Lotfi, R.; Duin, A.; Naguib, M. Complexity of intercalation in MXenes: Destabilization of urea by two-dimensional titanium carbide. *J. Am. Chem. Soc.* **2018**, *140*, 10305–10314. [CrossRef] [PubMed]
16. Li, M.; Han, M.; Zhou, J.; Deng, Q.; Zhou, X.; Xue, J.; Du, S.; Yin, X.; Huang, Q. Novel Scale-Like Structures of Graphite/TiC/Ti_3C_2 Hybrids for Electromagnetic Absorption. *Adv. Electron. Mater.* **2018**, *4*, 1700617. [CrossRef]
17. Sarycheva, A.; Makaryan, T.; Maleski, K.; Satheeshkumar, E.; Melikyan, A.; Minassian, H.; Yoshimura, M.; Gogotsi, Y. Two-dimensional titanium carbide (MXene) as surface-enhanced Raman scattering substrate. *J. Phys. Chem. C* **2017**, *121*, 19983–19988. [CrossRef]
18. Sun, Y.; Gao, S.; Lei, F.; Xie, Y. Atomically-thin two-dimensional sheets for understanding active sites in catalysis. *Chem. Soc. Rev.* **2015**, *44*, 623–636. [CrossRef]
19. Han, C.; Chen, Z.; Zhang, N.; Colmenares, J.; Xu, Y. Hierarchically CdS decorated 1D ZnO nanorods-2D graphene hybrids: Low temperature synthesis and enhanced photocatalytic performance. *Adv. Funct. Mater.* **2015**, *25*, 221–229. [CrossRef]

20. Ran, J.; Jaroniec, M.; Qiao, S. Cocatalysts in Semiconductor-based Photocatalytic CO_2 Reduction: Achievements, Challenges, and Opportunities. *Adv. Mater.* **2018**, *30*, 1704649. [CrossRef]

21. Li, Y.; Fan, S.; Tan, R.; Yao, H.; Peng, Y.; Liu, Q.; Li, Z. Selective Photocatalytic Reduction of CO_2 to CH_4 Modulated by Chloride Modification on Bi_2WO_6 Nanosheets. *ACS Appl. Mater. Interfaces* **2020**, *12*, 54507–54516. [CrossRef] [PubMed]

22. Zhang, S.; Xiong, W.; Long, J.; Si, Y.; Xu, Y.; Yang, L.; Zou, J.; Dai, W.; Luo, X.; Luo, S. High-throughput lateral and basal interface in $CeO_2@Ti_3C_2T_X$: Reverse and synergistic migration of carrier for enhanced photocatalytic CO_2 reduction. *J. Colloid Interfaces Sci.* **2022**, *615*, 716–724. [CrossRef] [PubMed]

23. Khan, A.; Tahir, M.; Zakaria, Z. Synergistic effect of anatase/rutile TiO_2 with exfoliated Ti_3C_2TR MXene multilayers composite for enhanced CO_2 photoreduction via dry and bi-reforming of methane under UV–visible light. *J. Environ. Chem. Eng.* **2021**, *9*, 105244. [CrossRef]

24. Kuang, P.; He, M.; Zhu, B.; Yu, J.; Fan, K.; Jaroniec, M. 0D/2D NiS_2/V-MXene composite for electrocatalytic H_2 evolution. *J. Catal.* **2019**, *375*, 8–20. [CrossRef]

25. He, P.; Cao, M.-S.; Shu, J.-C.; Cai, Y.; Wang, X.-X.; Zhao, Q.-L.; Yuan, J. Atomic layer tailoring titanium carbide MXene to tune transport and polarization for utilization of electromagnetic energy beyond solar and chemical energy. *ACS Appl. Mater. Interfaces* **2019**, *11*, 12535–12543. [CrossRef]

26. Li, W.; Jin, L.; Gao, F.; Wan, H.; Pu, Y.; Wei, X.; Chen, C.; Zou, W.; Zhu, C.; Dong, L. Advantageous roles of phosphate decorated octahedral $CeO_2\{111\}/g$-C_3N_4 in boosting photocatalytic CO_2 reduction: Charge transfer bridge and Lewis basic site. *Appl. Catal. B-Environ.* **2021**, *294*, 120257. [CrossRef]

27. Becerra, J.; Nguyen, D.-T.; Gopalakrishnan, V.-N.; Do, T.-O. Plasmonic Au nanoparticles incorporated in the zeolitic imidazolate framework (ZIF-67) for the efficient sunlight-driven photoreduction of CO_2. *ACS Appl. Energy Mater.* **2020**, *3*, 7659–7665. [CrossRef]

28. Lertthanaphol, N.; Prawiset, N.; Soontornapaluk, P.; Kitjanukit, N.; Neamsung, W.; Pienutsa, N.; Chusri, K.; Sornsuchat, T.; Chanthara, P.; Phadungbut, P.; et al. Soft template-assisted copper-doped sodium dititanate nanosheet/graphene oxide heterostructure for photoreduction of carbon dioxide to liquid fuels. *RSC Adv.* **2022**, *12*, 24362–24373. [CrossRef]

29. Wang, G.; He, C.-T.; Huang, R.; Mao, J.; Wang, D.; Li, Y. Photoinduction of Cu single atoms decorated on UiO-66-NH_2 for enhanced photocatalytic reduction of CO_2 to liquid fuels. *J. Am. Chem. Soc.* **2020**, *142*, 19339–19345. [CrossRef]

30. Maimaitizi, H.; Abulizi, A.; Kadeer, K.; Talifu, D.; Tursun, Y. In situ synthesis of Pt and N co-doped hollow hierarchical BiOCl microsphere as an efficient photocatalyst for organic pollutant degradation and photocatalytic CO_2 reduction. *Appl. Surf. Sci.* **2020**, *502*, 144083. [CrossRef]

31. Ma, M.; Huang, Z.; Doronkin, D.-E.; Fa, W.; Rao, Z.; Zou, Y.; Wang, R.; Zhong, Y.; Cao, Y.; Zhang, R.; et al. Ultrahigh surface density of Co-N_2C single-atom-sites for boosting photocatalytic CO_2 reduction to methanol. *Appl. Catal. B Environ.* **2022**, *300*, 120695. [CrossRef]

32. Wu, J.; Xie, Y.; Ling, Y.; Si, J.; Li, X.; Wang, J.; Ye, H.; Zhao, J.; Li, S.; Zhao, Q.; et al. One-step synthesis and Gd^{3+} decoration of BiOBr microspheres consisting of nanosheets toward improving photocatalytic reduction of CO_2 into hydrocarbon fuel. *Chem. Eng. J.* **2020**, *400*, 125944. [CrossRef]

33. Ribeiro, C.-S.; Lansarin, M.-A. Enhanced photocatalytic activity of Bi_2WO_6 with PVP addition for CO_2 reduction into ethanol under visible light. *Environ. Sci. Pollut. Res.* **2021**, *28*, 23667–23674. [CrossRef] [PubMed]

34. Seeharaj, P.; Kongmun, P.; Paiplod, P.; Prakobmit, S.; Sriwong, C.; Kim-Lohsoontorn, P.; Vittayakorn, N. Ultrasonically-assisted surface modified TiO_2/rGO/CeO_2 heterojunction photocatalysts for conversion of CO_2 to methanol and ethanol. *Ultrason. Sonochem.* **2019**, *58*, 104657. [CrossRef]

35. Ribeiro, S.-C.; Lansarin, A.-M. Facile solvo-hydrothermal synthesis of Bi_2MoO_6 for the photocatalytic reduction of CO_2 into ethanol in water under visible light. *React. Kinet. Mech. Catal.* **2019**, *127*, 1059–1071. [CrossRef]

36. Vu, N.-N.; Nguyen, C.-C.; Kaliaguine, S.; Do, T.-D. Reduced Cu/Pt–$HCa_2Ta_3O_{10}$ Perovskite Nanosheets for Sunlight-Driven Conversion of CO_2 into Valuable Fuels. *Adv. Sust. Syst.* **2017**, *1*, 1700048. [CrossRef]

Article

Photocatalytic Degradation of Tetracycline by Supramolecular Materials Constructed with Organic Cations and Silver Iodide

Xing-Xing Zhang, Xiao-Jia Wang and Yun-Yin Niu *

Green Catalysis Center, College of Chemistry, Zhengzhou University, Zhengzhou 450001, China
* Correspondence: niuyy@zzu.edu.cn

Abstract: Photocatalytic degradation, as a very significant advanced oxidation technology in the field of environmental purification, has attracted extensive attention in recent years. The design and synthesis of catalysts with high-intensity photocatalytic properties have been the focus of many researchers in recent years. In this contribution, two new supramolecular materials {[(L1)·(Ag$_4$I$_7$)]CH$_3$CN} (1), {[(L2)·(Ag$_4$I$_7$)]CH$_3$CN} (2) were synthesized by solution volatilization reaction of two cationic templates 1,3,5-Tris(4-aminopyridinylmethyl)-2,4,6-Trimethylphenyl bromide (L1) and 1,3,5-Tris(4-methyl pyridinyl methyl)-2,4,6-trimethylphenyl bromide (L2) with metal salt AgI at room temperature, respectively. The degradation effect of **1** and **2** as catalyst on tetracycline (TC) under visible light irradiation was studied. The results showed that the degradation of TC by **1** was better than that by **2** and both of them had good stability and cyclability. The effects of pH value, catalyst dosage, and anion in water on the photocatalytic performance were also investigated. The adsorption kinetics fit the quasi-first-order model best. After 180 min of irradiation with **1**, the degradation rate of TC can reach 97.91%. In addition, the trapping experiments showed that ·OH was the main active substance in the photocatalytic degradation of TC compared with ·O$_2$$^-$ and h$^+$. Because of its simple synthesis and high removal efficiency, catalyst **1** has potential value for the treatment of wastewater containing organic matter.

Keywords: AgI; supramolecular material; photocatalytic degradation (PDT); tetracycline (TC); catalyst

Citation: Zhang, X.-X.; Wang, X.-J.; Niu, Y.-Y. Photocatalytic Degradation of Tetracycline by Supramolecular Materials Constructed with Organic Cations and Silver Iodide. *Catalysts* **2022**, *12*, 1581. https://doi.org/10.3390/catal12121581

Academic Editors: Gassan Hodaifa, Rafael Borja and Mha Albqmi

Received: 1 November 2022
Accepted: 29 November 2022
Published: 5 December 2022

Publisher's Note: MDPI stays neutral with regard to jurisdictional claims in published maps and institutional affiliations.

1. Introduction

In recent years, with the continuous development of the economy and urbanization, chemicals have increasingly been synthesized to meet the growing needs of production and daily life. Unfortunately, various organic pollutants produced in the process of chemical production will cause great harm to water sources [1–3]. Tetracycline (TC) is one of the most widely used antibiotics in poultry and aquaculture [4]. Since most TC antibiotics cannot be metabolized and degraded in organisms, but are excreted as active drugs in feces and urine, they are often detected in the environment [5–8]. In addition, TC is a water-soluble antibiotic, which results in its mobility and long-term persistence in the environment. Therefore, the treatment of tetracycline in wastewater constitutes an important subject of research in environmental protection [9].

A variety of technologies have been proposed by researchers to remove organic pollutants from wastewater [10–16]. Among all technologies, physical adsorption and advanced oxidation processes (AOPs) are considered to be the most promising methods for TC removal because they are more eco-friendly [17–23]. However, adsorption technology can only remove pollutants from water, but not "destroy" pollutants completely. Therefore, the most commonly used method to remove pollutants is AOPs, including photocatalytic oxidation, ultrasonic oxidation, ozonation, wet oxidation, and various combination technologies (Figure 1), which can degrade pollutants by generating highly active, non-selective chemical oxidants (such as H$_2$O$_2$, ·OH, ·O$_2$$^-$, O$_3$) in situ [24–30]. Among them, photocatalytic oxidation, a method to convert solar energy into chemical energy, is the most sensible method and has a broad prospect in solving the problem of water pollution [31–33]. An important condition for achieving a good photocatalytic effect is that the photocatalyst has stronger activity; for photocatalytic degradation (PDT), the most important thing is to find an effective, stable, readily available, and excellent catalytic material [34,35].

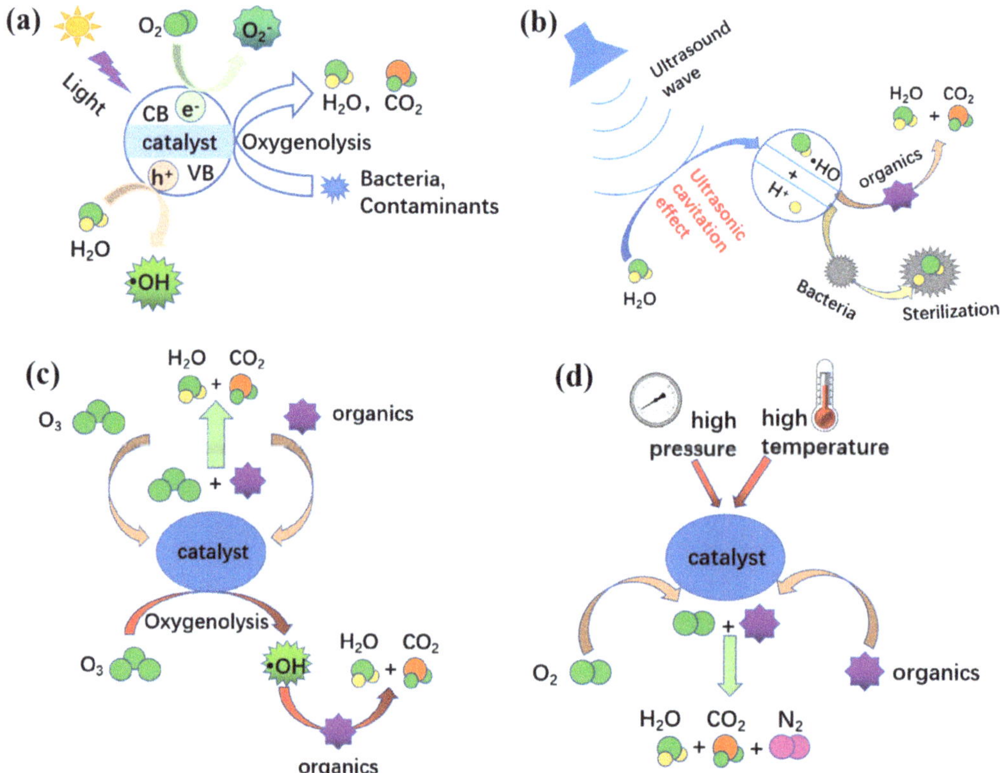

Figure 1. Four commonly used AOPs techniques: (**a**) photocatalytic oxidation technology; (**b**) ultrasonic oxidation technology; (**c**) ozonation technology; (**d**) wet oxidation technology.

Supramolecular chemistry, a complex and ordered combination of molecules with specific functions bonded by non-covalent weak interaction forces, is a sublimation of non-covalent molecular chemistry, known as "chemistry beyond molecular concept". It was first proposed by Jean-Marie Lehn, a French scientist, and was a fringe science composed of multiple disciplines [36,37]. Supramolecules are connected by two or more non-covalent bonding forces between chemical molecules, including hydrogen bonds, Van der Waals forces, electrostatic interactions, host-guest interactions, and π-π stacking [38–42]. Multivalent interactions that rely on noncovalent bonds are an important part of mediating biological processes as well as building complex (super) structures for material applications [43]; only by understanding and controlling non-covalent interactions one can assemble functional nanosystems [44] with a precision similar to nature, and develop highly complex chemical systems and advanced functional materials. Since the advent of supramolecular chemistry, the construction of multi-component supramolecular assemblies, known as supramolecular synthesis [45], has been a hot topic. It not only contains a fundamental understanding of self-assembly and molecular recognition processes related to the origin of life and evolution [46] but also provides support for the design of new materials in the future. Supramolecular compounds have also attracted the attention of researchers due to their advantages of structurally adjustable and functional diversity. Because the structure of nitrogen-containing heterocyclic ligands is easy to modify, various functional groups can be introduced, and then supramolecular compounds with diverse structures and rich functions can be induced and formed with them as templates, which can be further developed and applied in many fields, such as biomedicine [47,48], optoelectronics [49,50], functional materials [51,52], sensors [53], and so on. Especially in the field of photocatalysis, there have been some reports in recent years about whether supramolecular compounds can be used as photocatalysts. As shown in Table 1, compared with other supramolecular materials, catalysts synthesized by the organic cationic template and AgI have a higher degradation rate for organic pollutants, which offers a promising framework on which to build environmentally friendly materials with high reliability and excellent photocatalytic performance.

Table 1. Research results on the degradation of persistent organic pollutants by supramolecular materials.

Supramolecular Materials and Technologies	Various Research Results on the Degradation of Persistent Organic Pollutants
The supercapacitor and photocatalytic supermolecule materials constructed by $4'4$-pyridine and $\{PMo_{12}O_{40}\}$	Zhang et al. reported that the photocatalytic supramolecular material had a good photocatalytic degradation effect on methylene blue (95.8%) and rhodamine B (93.54%) [54].
Photocatalytically Active Supramolecular Organic–Inorganic Magnetic Composites	Sabina et al. prepared composites containing Zn-modified MgAl LDHs and Cu-phthalocyanine as a photosensitizer, which could remove up to 93% of β-lacamide antibiotics from water [55].
$AgI/BiPO_4$ n–n heterojunction photocatalyst	Zhou et al. prepared $AgI/BiPO_4$ n-n heterojunction hybrid by precipitation technology. As a catalyst, the degradation efficiency of this substance for tetracycline hydrochloride (TC) was 95% and methylene blue (MB) was 91%, respectively [56].
Supramolecular photocatalyst of perylene bisimide decorated with α-Fe_2O_3	Lu et al. prepared α-Fe_2O_3/PTCDI composite based on supramolecular photocatalyst (PTCDI) modified dinaphthalenediamine (PTCDI) by calcination of ferric nitrate. The degradation rates of phenol pollutants and coking wastewater reached 73% and 66.7%, respectively, by photofenton reaction [57].
Supramolecular Nanopumps with Chiral	Bao et al. prepared ph-responsive supramolecular nanopumps from porous tubules in the left hand to solid fibers in the right hand by self-assembly of aromatic amphiphiles with curved shapes. This superhydrophobic switching aromatic pore is ideal for effective removal and controlled release of organic pollutants from water through pulsating motion. It was found that the removal efficiency of the supramolecular compound was 78% for estradiol and 82% for bisphenol [58].
Two new supramolecular materials $\{[(L1)\cdot(Ag_4I_7)]CH_3CN\}$ (1), $\{[(L2)\cdot(Ag_4I_7)]CH_3CN\}$ (2) were synthesized by templatedself-assembly of tetracations L1 and L2 with AgI at room temperature	In this paper the degradation rate of TC by 1 can reach 97.91%.

2. Results and Discussion

*2.1. Description of Crystal Structures of $\{[(L1)\cdot(Ag_4I_7)]\cdot CH_3CN\}$ (**1**) and $\{[(L2)\cdot(Ag_4I_7)]\cdot CH_3CN\}$ (**2**)*

The crystal structures of compounds **1** and **2** are similar; both are supramolecular octa-nuclear Ag-I cluster structures belonging to the $P2_{1/n}$ space group in a monoclinic crystal system. The structural unit diagram of compound **1** is shown in Figure 2a. Metal Ag(I) in compound **1** has a distorted tetrahedral configuration. As shown in Figure 2b, two organic cations are sandwiched around an inorganic anion cluster $[Ag_8I_{14}]_6{}^-$, which interacts electrically to form a stable 2D supramolecular stacking structure by hydrogen bonding, electrostatic action, and intermolecular force action. The structural unit and packing diagram of catalyst **2** are shown in Figure 2c,d.

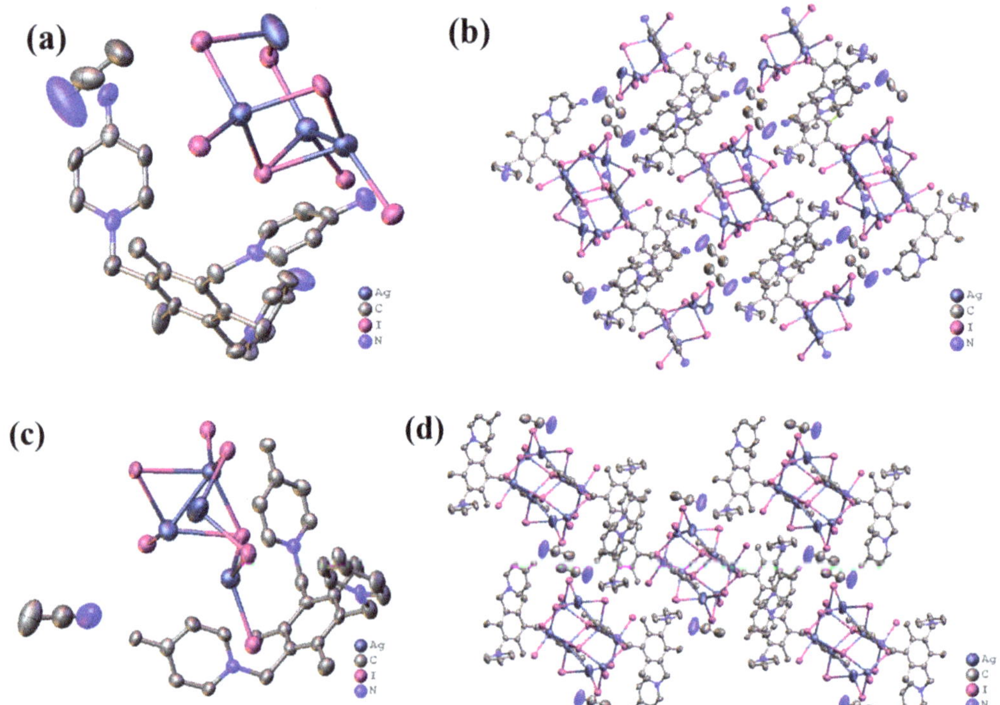

Figure 2. (**a**) The smallest structural unit of catalyst **1**; (**b**) the stacking diagram of catalyst **1**; (**c**) the structural unit of catalyst **2**; (**d**) the stacking diagram of catalyst **2** (with the H atom omitted).

2.2. TG–DTA Analysis

The thermal stability of compounds **1** and **2** was obtained by TG–TDA analysis. As shown in Figure 3a,b, the thermogravimetric degradation trends of compound 1 and compound 2 were similar. They were almost thermally stable between 50 °C and 250 °C; The mass loss between 250 °C and 400 °C may be attributed to the volatile decomposition of solvent molecules and organic cations in the molecule; the mass loss between 400 °C and 800 °C may be caused by the collapse of the inorganic anion skeleton.

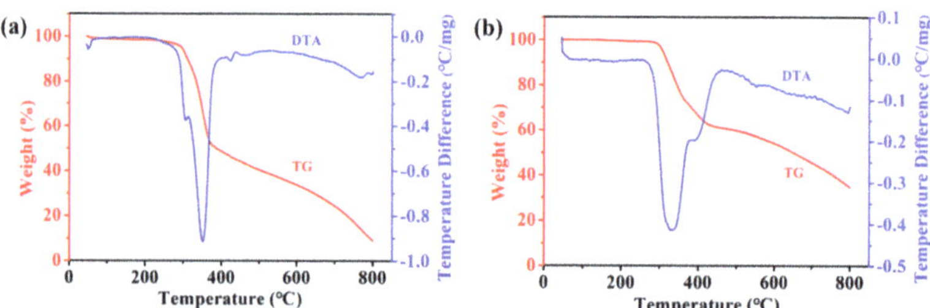

Figure 3. (**a**) TG–DTA curve of compound **1**; and (**b**) TG–DTA curve of compound **2**.

2.3. Semiconductor Properties of Compounds **1** and **2**

The absorption region of light is evaluated by the optical properties of the compounds. To evaluate the semiconductor performance of catalyst **1** and catalyst **2**, the Kubelka-Munk function $F = (1 - R)^2/2R$ [59] was used to calculate their band gap values, as shown in Figure 4a,b. The band gap values of catalyst **1** and catalyst **2** are 1.62 eV and 2.04 eV, respectively. These results indicate that catalysts **1** and **2** can be used as good semiconductor materials with excellent semiconductor properties and good photocatalytic activity under visible light.

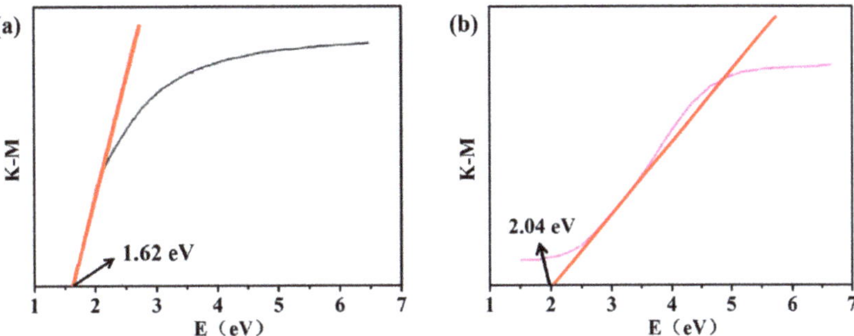

Figure 4. (**a**) Calculation diagram of band gap energy of catalyst **1** (The black line is the absorption curve of catalyst **1**, the red line is the tangent at the approximate linear in the absorption curve, and the intersection of the tangent line and the X-axis is the band gap value); and (**b**) calculation diagram of band gap energy of catalyst **2** (The pink line is the absorption curve of catalyst **2**, the red line is the tangent at the approximate linear in the absorption curve, and the intersection of the tangent line and the X-axis is the band gap value).

2.4. Catalytic Performance of the Degradation of TC with Compounds **1** and **2**

Tetracycline (TC) is an antibiotic widely used in poultry and aquaculture. Because TC cannot be metabolized and degraded in organisms, it is often excreted as raw drugs, which is very harmful to water production. To detect the photocatalytic properties of catalysts **1** and **2**, the organic contaminant TC is selected as the target for photocatalytic degradation. First, 10 mg of catalysts **1** and **2** were added to 20 mL TC solution (20 mg·L^{-1}, pH = 7), and the photocatalytic reaction was carried out in a photoreactor. To eliminate the effect of adsorption on the photocatalytic reaction, additional stirring for 30 min under dark conditions was performed. A 300 W xenon lamp was selected as the visible light source. Under the irradiation of visible light, 3 mL solution was sampled in intervals, separated by centrifugation, and the supernatant was taken. The absorbance of the solution was tested by UV-VIS spectrometer, and then the photocatalytic activity of catalysts **1** and **2** on TC was monitored. As shown in Figure 5a,b, the degradation of TC under visible light is negligible without adding the catalyst. After 180 min of irradiation with catalyst **1**, the degradation rate of TC can reach 97.91%. Furthermore, at 240 min of irradiation, the degradation rate of TC by catalyst **2** was 51.23%, indicating that the degradation efficiency of TC by catalyst **1** was much higher than that by catalyst **2**, which further confirmed the excellent photocatalytic activity of catalyst **1**. Table 2 describes the highest catalytic degradation efficiency of TC in a short time and with different types of catalysts through different treatment systems [60–66]. By comparing the above degradation with the catalyst prepared by us, it can be seen that catalyst 1 has a higher TC removal rate, and its preparation process is safer, easy to operate, and recyclable, and the catalyst is efficiently recovered. Catalyst **1** also had a higher removal rate of TC compared with the above MOFs and hybrid materials. With the increase in reaction time, the removal efficiency of catalysts **1** and **2** reaches equilibrium. In addition, the first-order and second-order kinetics models were used to fit the photocatalytic results, which are shown in Figure 5c,d. It can be seen from the figure that, compared with the second-order kinetics R^2, the first-order kinetics R^2 > 0.99, which indicates that first-order kinetics has the best fitting effect. Therefore, the photocatalytic degradation of TC by catalysts **1** and **2** is in line with the photocatalytic process controlled by the first-order kinetics.

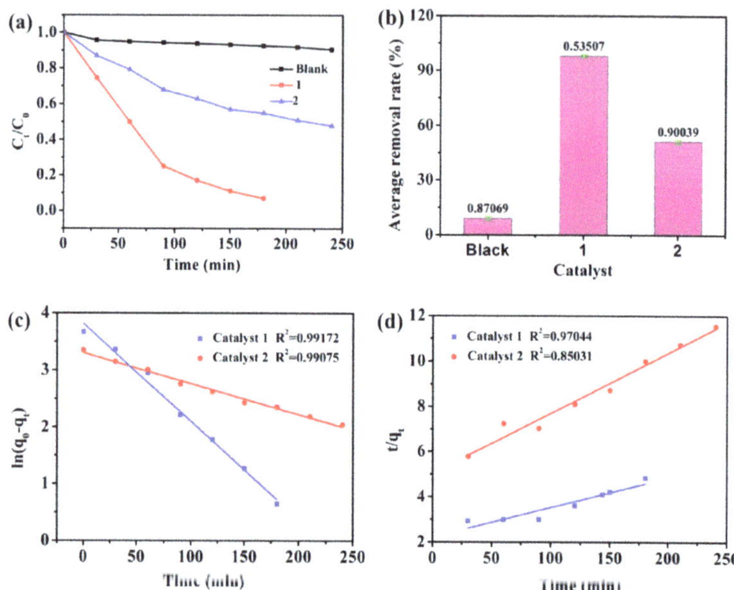

Figure 5. (**a**) The catalytic efficiency of catalysts **1** and **2** on TCs. (**b**) The average removal rate of TC. (**c**) The first-order kinetics of tetracycline degradation by catalysts **1** and **2**. (**d**) The second-order kinetics of tetracycline degradation by catalysts **1** and **2**.

Table 2. Degradation effect of different catalysts on TC.

Treatment System	Catalytic Efficacy of the Catalysts on the TC
Floatable cellulose acetate beads embedded with flower-like zwitterionic binary MOF/PDA	Cao et al. combined SnS_2 nanoparticles (SnSPs), a semiconductor material used for the photodegradation of antibiotics, onto the surface of a zirconia-based organic framework (UiO-66) to obtain a novel photocatalytic material with a degradation rate of 90% for TC [60].
TiO_2/Ag_3PO_4 nanojunctions on carbon fiber cloth	Eman et al. prepared the zwitterionic UiO-66/ZIF-8 binary MOF/polydopamine@cellulose acetate composite (UiO-66/ZIF-8/PDA@CA), the maximum removal rate of TC was 67% [61].
MXene-Based Photocatalysts	Zhang et al. constructed TiO_2/Ag_3PO_4 nanojunction on carbon fiber cloth, and the removal rate of TC from mobile wastewater reached 94.2% [62].
Ag_3PO_4/GO	Wu et al. prepared Ag_3PO_4/GO catalyst by ultrasonic reaction, and after 30 min irradiation, the degradation rate of Ag_3PO_4/5wt%GO to TC was 77.7% [63].
Enhanced sonochemical degradation of tetracycline by sulfate radicals.	Eslami et al. investigated the degradation of antibiotic TC in aqueous solution by silver-activated persulfate ($Na_2S_2O_8$) in the presence of ultrasonic radiation. Under optimal conditions, a TC removal rate of more than 83% was achieved in 120 min [64].
Facet-controlled activation of persulfate by magnetite nanoparticles	Hu et al. synthesized three magnetite nanoparticles (MNPs) with different bare crystal faces by hydrothermal method, and under heterogeneous activation of persulfate (PS), the degradation efficiency of the PS/MNPs {1 1 1} system for TC reached 74.38% within 4 h, which was much higher than that of the PS/MNPs {1 1 0} (19.29%) or PS/MNPs {1 0 0} (33.79%) systems [65].
A yolk-shell Bi@void@SnO$_2$ photocatalyst	Wu et al. synthesized the yolk-shell Bi@void@SnO$_2$ photocatalyst by a step-by-step method, and its catalytic degradation efficiency of TC reached 81.81% [66].
Two new supramolecular materials synthesized by templatedself-assembly of tetracations L1 and L2 with AgI	In this paper the degradation rate of TC by 1 can reach 97.91%.

2.5. Effect of pH Solution on TC Degradation

The pH value in the reaction system also plays an important role in the photocatalytic degradation of TC. In this paper, the following experiments were set up: put 20 mL TC solution (20 mg·L^{-1}), and 10 mg catalyst into the photocatalytic reactor, with hydrochloric acid and sodium hydroxide to adjust the pH value of the solution system to 3, 5, 7, 9, 11 respectively. As shown in Figure 6a,b, the TC degradation efficiencies of catalyst **1** at pH = 3–11 were 53.32%, 89.91%, 97.91%, 63.82%, and 52.65%, respectively. This indicated that the catalytic efficiency increased first and then decreased with the increase in pH value, and the removal efficiency of tetracycline was the highest when the pH value was 7. As shown in Figure 6c,d, the TC degradation efficiencies of catalyst **2** at pH = 3–11 were 37.76%, 41.72%, 67.32%, 70.28%, and 33.65%, respectively. With the increase of pH in the solution, the removal rate of tetracycline also increases at first. When pH = 9, the catalytic effect is the best, and when pH = 11, the degradation rate of photocatalytic tetracycline becomes smaller. The results show that the degradation efficiency at different pH values is related to the difference between the TC charge and the active site of the catalyst. As shown in Figure 6e, most TCs are positively charged (TCH_3^+) at pH < 3; zero charged (TCH_2^0) at neutral, and negatively charged (TCH^- or TC^{2-}) at pH > 7.5. Under acidic conditions, the degradation efficiency of TC by catalysts **1** and **2** at pH = 5 is better than that at pH = 3, mainly because the positively charged TCH_3^+ in the solution will compete with H^+ in the solution for the active site on the catalyst surface. For catalyst **1**, the degradation efficiency at pH = 9 is better than that at pH = 11 under alkaline conditions, mainly because the amino group in catalyst **1** may compete with TCH^- or TC^{2-} for the active site on the surface of the catalyst. When pH is 7, electrostatic force exists, so the removal efficiency of TC reaches the highest. For catalyst **2**, when the pH value changes from 7 to 9, the degradation efficiency of TC continues to improve, because TC becomes negatively charged at this stage and generates electrostatic attraction to catalyst **2**, resulting in an upward trend of degradation efficiency. When pH > 9, the degradation efficiency gradually decreases, possibly because the main form of TC is TC^{2-} exists; it is in contact with the catalyst without electrostatic interaction.

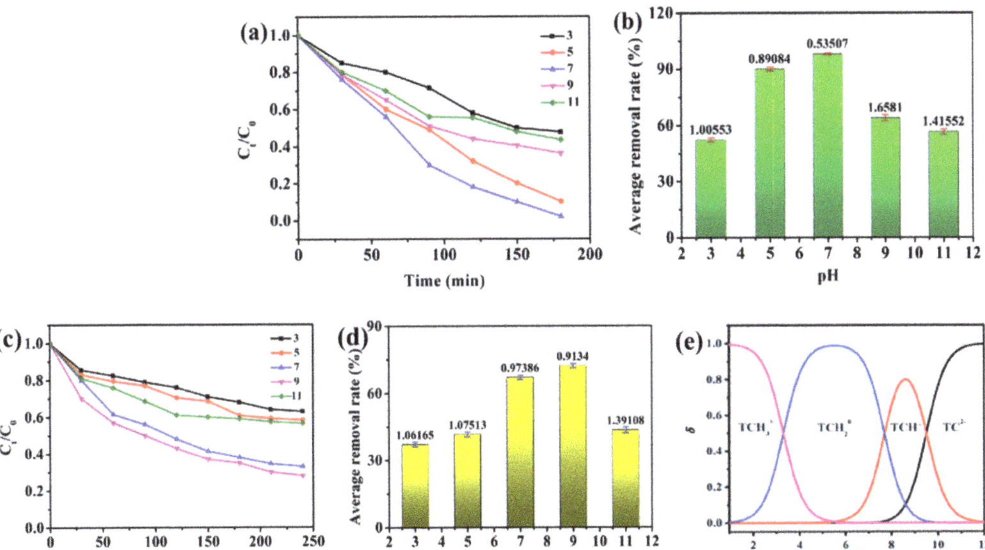

Figure 6. (**a**) Effect of catalyst **1** on TC degradation at different pH conditions. (**b**) Average removal rate of catalyst **1** to TC. (**c**) Effect of catalyst **2** on TC degradation at different pH conditions. (**d**) Average removal rate of catalyst **2** to TC. (**e**) The existence of tetracycline at different pH.

2.6. Effect of the Amount of Catalyst on TC Degradation

In the process of photocatalysis, the dosage of the photocatalyst also plays a vital role. To explore this effect of catalysts **1** and **2** on photocatalytic degradation of TC, 20 mL TC solution with an initial concentration of 20 mg·L^{-1} was taken, and the pH value of the solution was adjusted to 7 (catalyst **1**) or 9 (catalyst **2**); the catalyst dosage was 5 mg, 10 mg, and 20 mg for photocatalytic reaction. The results are shown in Figure 7a,b. When the amount of catalyst **1** was changed from 5 mg to 20 mg, the degradation rates of TC by it were 74.87%, 97.91%, and 61.27%; the degradation efficiency first increased and then decreased, possibly because with the increase of catalyst acting as adsorbent, the active site provided for photocatalytic reaction increased, and the collision probability between the active site and tetracycline also increased, so the concentration of adsorbent in solution increased. However, the accumulation of excessive catalyst particles will hinder and inhibit the light scattering and transport in the solution, and the organic matter adsorbed on the photocatalyst will reduce the light utilization rate. As shown in Figure 7c,d, the degradation efficiency of TC by catalyst **2** increased from 32.19% to 72.14%, which was due to the increase in the number of active centers with the increase in catalytic dose.

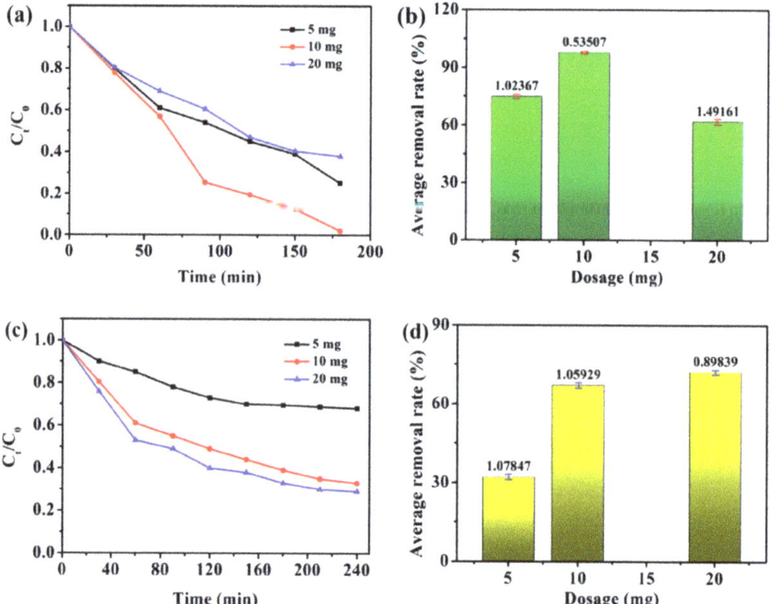

Figure 7. (**a**) Effect of different catalyst **1** dosages on TC degradation. (**b**) Average removal rate of TC degraded by catalyst **1**. (**c**) Effect of different catalyst **2** dosages on TC degradation. (**d**) Average removal rate of TC degraded by catalyst **2**.

2.7. Effects of Anions on Photocatalysis and Catalytic Mechanism

By optimizing the pH value of the TC solution and the number of catalysts, catalyst **1** with higher catalytic efficiency was selected to evaluate the influence of anions and catalytic mechanisms in the photocatalysis process. In this experiment, inorganic anions such as Cl^-, HCO_3^-, NO_3^-, and SO_4^{2-}, which are abundant in antibiotic wastewater in daily life, were selected as experimental subjects. The dosage of the photocatalyst was maintained at 10 mg, the pH of the solution was 7, and the initial concentration of tetracycline solution was 20 mg·L^{-1}. The experimental results are shown in Figure 8a, which clearly shows that the addition of these anions will inhibit the photocatalytic degradation of TC by the catalyst. This is because Cl^-, HCO_3^-, NO_3^- and SO_4^{2-} will combine with $\cdot OH$, H^+ or O_2^- to form other free radicals with lower activity, such as $NO_3\cdot$, $SO_4^-\cdot$, $Cl\cdot$, $PO_4^{2-}\cdot$ and $HCO_3\cdot/CO_3^-\cdot$. To have a deeper understanding of the active free radicals that play a major role in the photocatalytic degradation of TC, three different free radical trapping agents, including isopropanol (IPA, 1 mmol·L^{-1}, $\cdot OH$ scavenger), ethylenediamine tetraacetic acid disodium

(EDTA–2Na, 1 mmol·L^{-1}, h$^+$ scavenger), and 1,4–p–benzoquinone (PBQ, 1 mmol·L^{-1}, ·O$_2^-$ scavenger) were used to explore the photocatalytic mechanism of the catalyst **1**. The experimental results are shown in Figure 8b. After adding h$^+$ and ·O$_2^-$ capture agent, the photocatalytic degradation efficiency of TC had little change, which indicates that h$^+$ and ·O$_2^-$ hardly contribute to the photocatalytic process, but after adding ·OH capture agent, the photocatalytic efficiency decreased significantly. It is proved that ·OH plays an important role in the photocatalytic degradation of TC. Based on the above experimental results, the following possible photodegradation mechanisms can be proposed. Under the action of visible light, the catalyst composed of two staggered types of semiconductor, L1, and AgI, can induce the excitation of a small negative electron e$^-$ and a small positive hole h$^+$, respectively. After the generation of electron hole pairs, the electrons can transfer from L1 to AgI, and the photogenerated holes transfer in the opposite direction, thus achieving the spatial transfer of photogenerated electron holes, resulting in the photogenerated electrons at the conduction of L1 and the holes at the valence of AgI. The electron holes of L1 and AgI are retained to participate in the REDOX reaction, respectively. The oxygen in the air will produce hydrogen peroxide under the action of the photogenerated electron e$^-$ in the semiconductor L1 and then split into ·OH free radicals. H$_2$O also produces ·OH free radicals under the action of hole h$^+$ in semiconductor AgI. TC molecules can undergo deprotonation, dehydroxylation, demethylation, addition, carboxylation, and benzene ring cleavage under the action of ·OH radical. However, all intermediates eventually mineralize into smaller inorganic molecules, such as NH$_4^+$, NO$_3^-$, H$_2$O, CO$_2$, etc. [67]. The system can not only realize the spatial separation of REDOX sites but also ensure that the photocatalyst can maintain the appropriate valence band position, to maintain a strong REDOX reaction capacity. It was further shown that the added anions interact with the ·OH radical with the highest oxidation potential to form a less active free radical, resulting in reduced pollutant removal efficiency (Figure 8c).

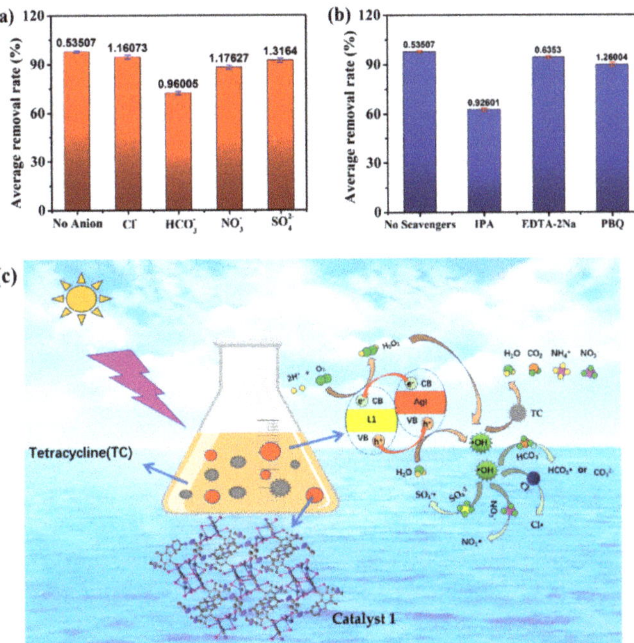

Figure 8. (**a**) Effect of different anions on TC degradation by catalyst **1**. (**b**) Effect of different free radical scavengers on TC degradation by catalyst **1**. (**c**) Photocatalytic degradation mechanism of catalyst **1** under the influence of anions.

2.8. Photocatalyst Recyclability

The reuse and stability of photocatalysts are important criteria for judging the applicable value of the catalyst. The higher the reuse rate, the higher the possibility to put into production and life in the future, which requires the catalyst to maintain excellent catalytic performance in the cycle process. To explore the reusability of catalysts **1** and **2**, cycle experiments were designed to test the photocatalytic efficiency during each cycle. As shown in Figure 9, catalysts **1** and **2** still reached 90.23% and 63.79% in the third round (97.91% and 71.28% in the first round, respectively), indicating that the photocatalytic degradation efficiency of these two catalysts did not decrease significantly. It is proven that catalysts **1** and **2** have easy recovery and good stability.

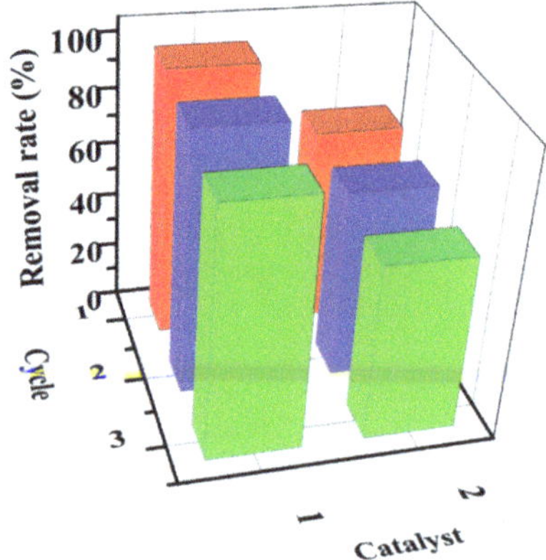

Figure 9. The effect diagram of the TC/catalysts **1** and **2** system for three cycles.

3. Materials and Methods

3.1. Materials

All the reagents of AR grade were used without any further purification. The cationic templates 1,3,5–Tris(4–aminopyridinylmethyl)–2,4,6–trimethylphenyl bromide (L1) [68] and 1,3,5–Tris(4–methylpyridinylmethyl)–2,4,6–trimethylphenyl bromide (L2) [69] were synthesized according to the methods described in the literature (Figure 10).

(a)

(b)

Figure 10. (a) The cationic template L1; and **(b)** the cationic template L2.

3.2. Synthesis of Compounds

3.2.1. The Synthesis of {[(L1)·(Ag$_4$I$_7$)]·CH$_3$CN} 1

L1 (0.0068 g, 0.01 mmol) in MeCN (3 mL) was added to AgI (0.0024 g, 0.01 mmol) in MeCN (2 mL) in the presence of an appropriate amount of KI, and then the mixed solution was stirred for 10 min and filtrated. The filtrate was volatilized at room temperature for 3 days, and colorless transparent columnar crystals were harvested. The yield was 15%. IR

(KBr, cm^{-1}): 3295.49 (s), 3179.36 (s), 2974.37 (s), 2926.54 (s), 1660.81 (s), 1558.41 (m), 1508.76 (s), 1456.24 (m), 1376.24 (s), 1291.79 (m), 1170.65 (s), 049.76 (s), 880.89 (w), 833.86 (s), 521.54 (w), 499.19 (w). Elemental Anal. Calc. For $C_{29}H_{36}N_7Ag_4I_7$ (1802.45): C, 19.32; H, 2.01; N, 5.44%. Found: C, 19.42; H, 2.48; N, 5.56%.

3.2.2. The Synthesis of {[(L2)·(Ag$_4$I$_7$)]·CH$_3$CN} 2

L2 (0.0068 g, 0.01 mmol) in MeCN (3 mL) was added to AgI (0.0024 g, 0.01 mmol) in DMF (3 mL) in the presence of an appropriate amount of KI, and then the mixed solution was stirred for 10 min and filtrated. Finally, the filtrate was volatilized at room temperature for 2 weeks, and colorless transparent crystals were harvested. The yield was 10%. IR (KBr, cm^{-1}): 3440 (s), 2923 (m), 2853 (s), 1635(s), 1463 (w), 1403(w), 1262(w), 936 (s), 801 (s), 516 (w). Elemental Anal. Calc. for $C_{32}H_{39}N_4Ag_4I_7$ (1799): C, 21.35; H, 2.18; N, 3.11. Found: C, 21.38 H,2.23; N, 3.15.

3.3. Methods for Characterizing Compounds

The KBr tablet pressing method and the Bruker TENSOR 27 infrared spectrometer were used to test the infrared spectrum (IR) in the range of 400 to 4000 cm^{-1} (Bruker, Shanghai, China). The NETZSCHTG209 thermogravimetric analyzer (Netzsch, Shanghai, China) was used to record changes in sample mass during a temperature rise from room temperature to 800 °C in a nitrogen atmosphere at a temperature rate of 5 °C·min^{-1}. Elemental analyses (C, H, and N) were performed on a Perkin-Elmer 240 elemental analyzer (PerkinElmer, Shanghai, China). UV-visible spectra were studied with a UV-5500 PC spectrophotometer in the scanning range of 200–800 nm. Using a graphite monochrome Cu-Kα light source (λ = 0.15418 nm) (XRD-6100, Shimadzu, Kyoto, Japan), setting the scan step to 2θ, the angular velocity of 4°·min^{-1}, and scanning range 5–50°, Powder X-ray diffraction (PXRD) data were tested with Philips X-pert X-ray diffraction (PXRD) with an X′Celerator detector. Simulation of the PXRD spectra was carried out by the single-crystal data and diffraction-crystal module of the Mercury (Hg) program. For compounds **1** and **2**, the synthesized PXRD plot is very close to the simulated single crystal diffraction pattern, indicating that the product is pure; their PXRD patterns were shown in Figure 11a,b. The compounds used Mo-Kα rays (λ = 0.71073 Å) for X-ray detection and crystallography data collection at the Bruker SMART CCD X-single crystal diffractometer. The structure was refined with full-matrix least-squares techniques on F^2 using the OLEX2 program package.

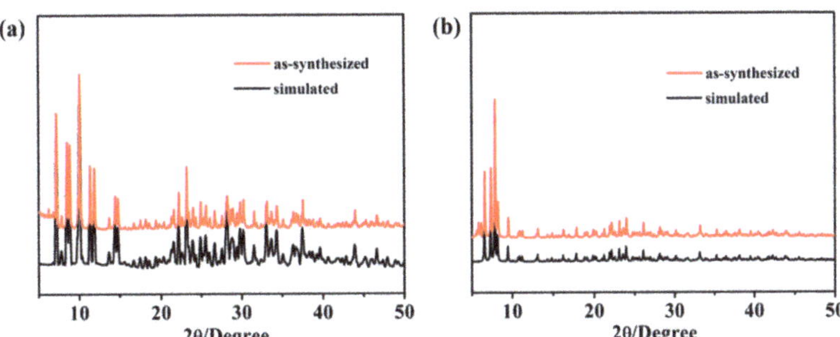

Figure 11. (**a**) PXRD pattern of catalyst **1**; and (**b**) PXRD pattern of catalyst **2**.

4. Conclusions

In this chapter, two supramolecular materials **1** and **2** were synthesized by solution volatilization at room temperature. Through a series of characterizations, it was proved that they had good catalytic potential, and the optimal conditions of photocatalytic degradation of TC were explored. The experimental results showed that for TC solution with the initial concentration of 20 mg·L^{-1}, the degradation rate of TC reached 97.91% after 180 min at pH = 7 with catalyst **1** of 10 mg. When the catalytic dose of catalyst **2** was 20 mg at pH = 9. The degradation rate of tetracycline reached 71.28% after 240 min of reaction. The photocatalytic degradation effect of catalyst **1** on TC was better than that of catalyst **2**. The inorganic anions in the system (Cl^-, $NO_3{}^-$, $SO_4{}^{2-}$, and $HCO_3{}^-$, etc.) can inhibit the photocatalytic degradation of the TC, because the ionic radicals (such as $Cl\cdot$, $NO_3\cdot$, $SO_4{}^-\cdot$, and $HCO_3\cdot/CO_3{}^-\cdot$) with lower activity than $\cdot OH$ or $\cdot O_2{}^-$ are generated. By adding free radical trapping agents it is shown that the $\cdot OH$ active free radicals play a major role in the photocatalytic reaction process. Through the photocatalytic cycle experiment, it was found that catalysts **1** and **2** still had high photocatalytic efficiency after three cycles, which fully proved the stability and repeatability of catalysts **1** and **2**. This study is helpful to solve the problem of water contamination by antibiotics. In subsequent studies, the detailed HRETM and XPS, together with the PL/RAMAN spectroscopy analysis can be conducted to further to provide information such as the lattice fridge, chemical state and electronic state of the elements, or charge carrier trapping, immigration, and transfer within the systems of catalysts 1 and 2.

Author Contributions: Conceptualization, X.-X.Z., X.-J.W. and Y.-Y.N.; methodology, X.-X.Z., X.-J.W. and Y.-Y.N.; software, X.-X.Z.; validation, X.-X.Z., X.-J.W. and Y.-Y.N.; formal analysis, Y.-Y.N.; investigation, X.-X.Z., X.-J.W. and Y.-Y.N.; resources, X.-J.W.; data curation, X.-X.Z.; writing—original draft preparation, X.-X.Z. and Y.-Y.N.; writing—review and editing, X.-X.Z. and Y.-Y.N.; visualization, X.-X.Z. and Y.-Y.N. All authors have read and agreed to the published version of the manuscript.

Funding: This research received no external funding.

Acknowledgments: The author would like to thank Yunyin Niu of College of Chemistry, Zhengzhou University for his guidance. All individuals included in this section have consented to the acknowledgement.

Conflicts of Interest: The authors declare no conflict of interest.

References

1. Salipira, K.L.; Mamba, B.B.; Krause, R.W.; Malefetse, T.J.; Durbach, S.H. Carbon nanotubes and cyclodextrin polymers for removing organic pollutants from water. *Environ. Chem. Lett.* **2006**, *5*, 13–17. [CrossRef]
2. Wojtyla, S.; Baran, T. Insight on doped ZnS and its activity towards photocatalytic removing of Cr(VI) from wastewater in the presence of organic pollutants. *Mater. Chem. Phys.* **2018**, *212*, 103–112. [CrossRef]
3. Zhuang, H.M.; Zhang, W.M.; Wang, L.; Zhu, Y.Y.; Xi, Y.Y.; Lin, X.F. Vapor Deposition-Prepared MIL-100(Cr)- and MIL-101(Cr)-Supported Iron Catalysts for Effectively Removing Organic Pollutants from Water. *ACS Omega* **2021**, *63*, 25311–25322. [CrossRef]
4. Liu, J.; Deng, W.J.; Ying, G.G.; Tsang, E.P.K.; Hong, H.C. Occurrence and distribution of antibiotics in surface water. *Ecotoxicology* **2022**, *31*, 1111–1119. [CrossRef]
5. Couch, M.; Agga, G.E.; Kasumba, J.; Parekh, R.R.; Loughrin, J.H.; Conte, E.D. Abundances of Tetracycline Resistance Genes and Tetracycline Antibiotics during Anaerobic Digestion of Swine Waste. *J. Environ. Qual.* **2019**, *48*, 171–178. [CrossRef] [PubMed]
6. Li, P.Y.; Rao, D.M.; Wang, Y.M.; Hu, X.R. Adsorption characteristics of polythiophene for tetracyclines and determination of tetracyclines in fish and chicken manure by solid phase extraction-HPLC method. *Microchem. J.* **2021**, *173*, 106935. [CrossRef]
7. Du, L.F.; Liu, W.K. Occurrence, fate, and ecotoxicity of antibiotics in agro-ecosystems: A review. *Agron. Sustain. Dev.* **2011**, *32*, 309–327. [CrossRef]
8. Kuppusamy, S.; Kakarla, D.; Venkateswarlu, K.; Megharaj, M.; Yoon, Y.E.; Lee, Y.B. Veterinary antibiotics (VAs) contamination as a global agro-ecological issue: A critical view. *Agric. Ecosyst. Environ.* **2018**, *257*, 47–59. [CrossRef]
9. Zhou, Y.; Yang, Q.; Zhang, D.N.; Gan, N.; Li, Q.P.; Cuan, J. Detection and removal of antibiotic tetracycline in water with a highly stable luminescent MOF. *Sens. Actuators B Chem.* **2018**, *262*, 137–143. [CrossRef]
10. Li, J.; Wang, X.X.; Zhao, G.X.; Chen, C.L.; Chai, Z.F.; Alsaedi, A.; Hayat, T.; Wang, X.K. Metal–organic framework-based materials: Superior adsorbents for the capture of toxic and radioactive metal ions. *Chem. Soc. Rev.* **2018**, *47*, 2322–2356. [CrossRef]
11. Zhao, G.X.; Huang, X.B.; Tang, Z.W.; Huang, Q.F.; Niu, F.L.; Wang, X.K. Polymer-based nanocomposites for heavy metal ions removal from aqueous solution: A review. *Polym. Chem.* **2018**, *9*, 3562–3582. [CrossRef]
12. Fu, S.F.; Chen, K.Q.; Zou, H.; Xu, J.X.; Zheng, Y.; Wang, Q.F. Using calcium peroxide (CaO_2) as a mediator to accelerate tetracycline removal and improve methane production during co-digestion of corn straw and chicken manure. *Energy Convers. Manag.* **2018**, *172*, 588–594. [CrossRef]
13. Oturan, N.; Wu, J.; Zhang, H.; Sharma, V.K.; Oturan, M.A. Electrocatalytic destruction of the antibiotic tetracycline in aqueous medium by electrochemical advanced oxidation processes: Effect of electrode materials. *Appl. Catal. B* **2013**, *140*, 92–97. [CrossRef]
14. Yu, L.L.; Cao, W.; Wu, S.C.; Yang, C.; Cheng, J.H. Removal of tetracycline from aqueous solution by MOF/graphite oxide pellets: Preparation, characteristic, adsorption performance and mechanism. *Ecotoxicol. Environ. Saf.* **2018**, *164*, 289–296. [CrossRef]

15. Park, J.-A.; Nam, A.; Kim, J.-H.; Yun, S.-T.; Choi, J.-W.; Lee, S.-H. Blend-electrospun graphene oxide/Poly(vinylidene fluoride) nanofibrous membranes with high flux, tetracycline removal and anti-fouling properties. *Chemosphere* **2018**, *207*, 347–356. [CrossRef] [PubMed]
16. Wang, X.J.; Qiao, X.Y.; Li, Z.Y.; Wei, D.H.; Niu, Y.Y. Influence of 4-cyanopyridinium multicationic isomers on the structure-property relationships of two-dimensional hybrid as photocatalyst for the degradation of organic dyes. *Inorg. Chem. Commun.* **2020**, *119*, 108126. [CrossRef]
17. Li, H.P.; Dou, Z.; Chen, S.Q.; Hu, M.; Li, S.; Sun, H.M.; Jiang, Y.C.; Zhai, Q.G. Design of a Multifunctional Indium–Organic Framework: Fluorescent Sensing of Nitro Compounds, Physical Adsorption, and Photocatalytic Degradation of Organic Dyes. *Inorg. Chem.* **2019**, *58*, 11220–11230. [CrossRef]
18. Deng, S.Q.; Mo, X.J.; Zheng, S.R.; Jin, X.; Gao, Y.; Cai, S.L.; Fan, J.; Zhang, W.G. Hydrolytically Stable Nanotubular Cationic Metal–Organic Framework for Rapid and Efficient Removal of Toxic Oxo-Anions and Dyes from Water. *Inorg. Chem.* **2019**, *58*, 2899–2909. [CrossRef]
19. Ma, H.F.; Liu, Q.Y.; Wang, Y.L.; Yin, S.G. A Water-Stable Anionic Metal–Organic Framework Constructed from Columnar Zinc-Adeninate Units for Highly Selective Light Hydrocarbon Separation and Efficient Separation of Organic Dyes. *Inorg. Chem.* **2017**, *56*, 2919–2925. [CrossRef]
20. Feng, L.Y.; Li, X.Y.; Chen, X.T.; Huang, Y.J.; Peng, K.S.; Huang, Y.X.; Yan, Y.Y.; Chen, Y.G. Pig manure-derived nitrogen-doped mesoporous carbon for adsorption and catalytic oxidation of tetracycline. *Sci. Total Environ.* **2019**, *708*, 135071. [CrossRef] [PubMed]
21. Giammarco, J.; Mochalin, V.N.; Haeckel, J.; Gogotsi, Y. The adsorption of tetracycline and vancomycin onto nanodiamond with controlled release. *J. Colloid Interface Sci.* **2016**, *468*, 253–261. [CrossRef]
22. Wang, D.; Jia, F.; Wang, H.; Chen, F.; Fang, Y.; Dong, W.; Zeng, G.; Li, X.; Yang, Q.; Yuan, X. Simultaneously efficient adsorption and photocatalytic degradation of tetracycline by Fe-based MOFs. *J. Colloid Interface Sci.* **2018**, *519*, 273–284. [CrossRef] [PubMed]
23. Zhang, S.P.; Li, Y.K.; Shi, C.H.; Guo, F.Y.; He, C.Z.; Cao, Z.; Hu, J.; Cui, C.Z.; Liu, H.L. Induced-fit adsorption of diol-based porous organic polymers for tetracycline removal. *Chemosphere* **2018**, *212*, 937–945. [CrossRef]
24. Dewil, R.; Mantzavinos, D.; Poulios, I.; Rodrigo, M.A. New perspectives for Advanced Oxidation Processes. *J. Environ. Manag.* **2017**, *195*, 93–99. [CrossRef]
25. Babu, S.G.; Ashokkumar, M.; Neppolian, B. The Role of Ultrasound on Advanced Oxidation Processes. *Top. Curr. Chem.* **2016**, *374*, 75. [CrossRef] [PubMed]
26. Bavasso, I.; Montanaro, D.; Petrucci, E. Ozone-based electrochemical advanced oxidation processes. *Curr. Opin. Electrochem.* **2022**, *34*, 101017. [CrossRef]
27. Wang, J.L.; Zhuan, R. Degradation of antibiotics by advanced oxidation processes: An overview. *Sci. Total Environ.* **2019**, *701*, 135023. [CrossRef] [PubMed]
28. De Filpo, G.; Pantuso, E.; Mashin, A.I.; Baratta, M.; Nicoletta, F.P. WO$_3$/Buckypaper Membranes for Advanced Oxidation Processes. *Membranes* **2020**, *45*, 157. [CrossRef]
29. Bethi, B.; Sonawane, S.H.; Bhanvase, B.A.; Gumfekar, S.P. Nanomaterials-based advanced oxidation processes for wastewater treatment: A review. *Chem. Eng. Process.* **2016**, *109*, 178–189. [CrossRef]
30. Dai, Q.Z.; Zhou, M.H.; Lei, L.C.; Zhang, X.W. A novel advanced oxidation process—Wet electro-catalytic oxidation for high concentrated organic wastewater treatment. *Sci. Bull.* **2007**, *52*, 1724–1727. [CrossRef]
31. Wang, X.Y.; Li, S.N.; Chen, P.; Li, F.X.; Hu, X.M.; Hua, T. Photocatalytic and antifouling properties of TiO$_2$-based photocatalytic membranes. *Mater. Today Chem.* **2021**, *23*, 100650. [CrossRef]
32. Koe, W.S.; Lee, J.W.; Chong, W.C.; Pang, Y.L.; Sim, L.C. An overview of photocatalytic degradation: Photocatalysts, mechanisms, and development of photocatalytic membrane. *Environ. Sci. Pollut. Res.* **2019**, *27*, 2522–2565. [CrossRef] [PubMed]
33. Chang, P.; Wang, Y.H.; Wang, Y.T.; Zhu, Y.Y. Current trends on In$_2$O$_3$ based heterojunction photocatalytic systems in photocatalytic application. *Chem. Eng. J.* **2022**, *450*, 137804. [CrossRef]
34. Qiao, X.Y.; Wang, C.H.; Niu, Y.Y. N-Benzyl HMTA induced self-assembly of organic-inorganic hybrid materials for efficient photocatalytic degradation of tetracycline. *J. Hazard. Mater.* **2020**, *391*, 122121. [CrossRef] [PubMed]
35. Wang, X.J.; Qiao, G.Y.; Zhu, G.H.; Li, J.; Guo, X.Y.; Liang, Y.; Niu, Y.Y. Preparation of 2D supramolecular material doping with TiO$_2$ for degradation of tetracycline. *Environ. Res.* **2021**, *202*, 111689. [CrossRef]
36. Zhou, C.H.; Gan, L.L.; Zhang, Y.Y.; Zhang, F.F.; Wang, G.Z.; Jin, L.; Geng, R.X. Review on supermolecules as chemical drugs. *Sci. China Chem.* **2009**, *52*, 415–458. [CrossRef]
37. Vicens, J.; Vicens, Q. Emergences of supramolecular chemistry: From supramolecular chemistry to supramolecular science. *J. Incl. Phenom. Macrocycl. Chem.* **2011**, *71*, 251–274. [CrossRef]
38. Lu, D.P.; Huang, Q.; Wang, S.D.; Wang, J.Y.; Huang, P.S.; Du, P.W. The Supramolecular Chemistry of Cycloparaphenylenes and Their Analogs. *Front. Chem.* **2019**, *7*, 668. [CrossRef]
39. Leclercq, L.; Douyère, G.; Nardello-Rataj, V. Supramolecular Chemistry and Self-Organization: A Veritable Playground for Catalysis. *Catalysts* **2019**, *9*, 163. [CrossRef]
40. Prins, L.J.; Reinhoudt, D.N.; Timmerman, P. Noncovalent Synthesis Using Hydrogen Bonding. *Angew. Chem. Int. Ed.* **2001**, *40*, 2382–2426. [CrossRef]

41. Adams, H.; Hunter, C.A.; Lawson, K.R.; Perkins, J.; Spey, S.E.; Urch, C.J.; Sanderson, J.M. A supramolecular system for quantifying aromatic stacking interactions. *Chem. Eur. J.* **2002**, *7*, 4863–4877. [CrossRef]

42. Müller-Dethlefs, K.; Hobza, P. Noncovalent interactions: A challenge for experiment and theory. *Chem. Rev.* **2001**, *100*, 143–168. [CrossRef]

43. Badjić, J.D.; Nelson, A.; Cantrill, S.J.; Turnbull, W.B.; Stoddart, J.F. Multivalency and cooperativity in supramolecular chemistry. *Acc. Chem. Res.* **2005**, *38*, 723–732. [CrossRef]

44. Philp, D.; Stoddart, J.F. Self-Assembly in Natural and Unnatural Systems. Angew. *Chem. Int. Ed.* **1996**, *35*, 1154–1196. [CrossRef]

45. Fyfe, M.C.T.; Stoddart, J.F. Synthetic Supramolecular Chemistry. *Acc. Chem. Res.* **1997**, *30*, 393–401. [CrossRef]

46. Elemans, J.A.A.W.; Rowan, A.E.; Nolte, R.J.M. Mastering molecular matter. Supramolecular architectures by hierarchical self-assembly. *J. Mater. Chem.* **2003**, *13*, 2661–2670. [CrossRef]

47. Yu, G.C.; Chen, X.Y. Host-Guest Chemistry in Supramolecular Theranostics. *Theranostics* **2019**, *9*, 3041–3074. [CrossRef] [PubMed]

48. Bickerton, L.E.; Johnson, T.G.; Kerckhoffs, A.; Langton, M.J. Supramolecular chemistry in lipid bilayer membranes. *Chem. Sci.* **2021**, *12*, 11252–11274. [CrossRef] [PubMed]

49. Kwon, T.; Choi, J.W.; Coskun, A. Prospect for Supramolecular Chemistry in High-Energy-Density Rechargeable Batteries. *Joule* **2019**, *3*, 662–682. [CrossRef]

50. Zhou, H.Y.; Yamada, T.; Kimizuka, N. Supramolecular Thermocells based on Thermo-Responsiveness of Host–Guest Chemistry. *Bull. Chem. Soc. Jpn.* **2021**, *94*, 1525–1546. [CrossRef]

51. Savyasachi, A.J.; Kotova, O.; Shanmugaraju, S.; Bradberry, S.J.; Ó'Máille, G.M.; Gunnlaugsson, T. Supramolecular Chemistry: A Toolkit for Soft Functional Materials and Organic Particles. *Chem* **2017**, *3*, 764–811. [CrossRef]

52. Düren, T.; Sarkisov, L.; Yaghi, O.M.; Snurr, R.Q. Design of new materials for methane storage. *Langmuir* **2005**, *20*, 2683–2689. [CrossRef] [PubMed]

53. Fukuhara, G. Analytical supramolecular chemistry: Colorimetric and fluorimetric chemosensors. *J. Photochem. Photobiol. C* **2020**, *42*, 100340. [CrossRef]

54. Zhang, L.-Y.; Zhao, X.-Y.; Wang, C.-M.; Yu, K.; Lv, J.-H.; Wang, C.-X.; Zhou, B.-B. The supercapacitor and photocatalytic supermolecule materials constructed by 4'4-pyridine and {PMo$_{12}$O$_{40}$}. *J. Solid State Chem.* **2022**, *312*, 123235. [CrossRef]

55. Sabina, G.I.; Octavian, D.P.; Nicolae, G.; Madalina, T.; Simona, M.C.; Vasile, I.P.; Bogdan, C.; Elisabeth, E.J. Use of Photocatalytically Active Supramolecular Organic–Inorganic Magnetic Composites as Efficient Route to Remove β-Lactam Antibiotics from Water. *Catalysts* **2022**, *12*, 1044.

56. Zhou, X.H.; Chen, Y.Y.; Wang, P.Y.; Xu, C.Y.; Yan, Q.S. Fabrication of AgI/BiPO$_4$ n–n heterojunction photocatalyst for efficient degradation of organic pollutants. *J. Mater. Sci. Mater. Electron.* **2020**, *31*, 12638–12648. [CrossRef]

57. Lu, J.R.; Yue, M.T.; Cui, W.Q.; Sun, C.H.; Liu, L. Supramolecular photocatalyst of perylene bisimide decorated with α-Fe$_2$O$_3$: Efficient photo-Fenton degradation of organic pollutants. *Colloids Surf. A Physicochem. Eng. Asp.* **2022**, *655*, 130222. [CrossRef]

58. Bao, S.H.; Wu, S.S.; Huang, L.P.; Xu, X.; Xu, R.; Li, Y.G.; Liang, Y.R.; Yang, M.Y.; Yoon, D.K.; Lee, M.; et al. Supramolecular Nanopumps with Chiral Recognition for Moving Organic Pollutants from Water. *ACS Appl. Mater. Interfaces* **2019**, *11*, 31220–31226. [CrossRef]

59. Wang, H.Y.; Hou, L.L.; Li, C.; Zhang, D.D.; Ma, P.T.; Wang, J.P.; Niu, J.Y. Synthesis, structure, and photocatalytic hydrogen evolution of a trimeric Nb/W addendum cluster. *RSC Adv.* **2017**, *7*, 36416–36420. [CrossRef]

60. Cao, P.Y.; Zhang, Y.P.; Gao, D.; Chen, H.X.; Zhou, M.L.; He, Y.F.; Song, P.F.; Wang, R.M. Constructing nano-heterojunction of MOFs with crystal regrowth for efficient degradation of tetracycline under visible light. *J. Alloys Compd.* **2022**, *904*, 164061. [CrossRef]

61. Abd El-Monaem, E.M.; Omer, A.M.; Khalifa, R.E.; Eltaweil, A.S. Floatable cellulose acetate beads embedded with flower-like zwitterionic binary MOF/PDA for efficient removal of tetracycline. *J. Colloid Interface Sci.* **2022**, *620*, 333–345. [CrossRef] [PubMed]

62. Zhang, Y.; Duoerkun, G.; Shi, Z.; Cao, W.; Liu, T.; Liu, J.; Zhang, L.; Li, M.; Chen, Z. Construction of TiO$_2$/Ag$_3$PO$_4$ nanojunctions on carbon fiber cloth for photocatalytically removing various organic pollutants in static or flowing wastewater. *J. Colloid Interface Sci.* **2020**, *571*, 213–221. [CrossRef] [PubMed]

63. Wu, F.; Zhou, F.; Zhu, Z.; Zhan, S.; He, Q. Enhanced photocatalytic activities of Ag$_3$PO$_4$/GO in tetracycline degradation. *Chem. Phys. Lett.* **2019**, *724*, 90–95. [CrossRef]

64. Eslami, A.; Bahrami, H.; Asadi, A.; Alinejad, A. Enhanced sonochemical degradation of tetracycline by sulfate radicals. *Water Sci. Technol.* **2016**, *287*, 131981. [CrossRef]

65. Hu, L.X.; Ren, X.H.; Yang, M.; Guo, W.L. Facet-controlled activation of persulfate by magnetite nanoparticles for the degradation of tetracycline. *Sep. Purif. Technol.* **2020**, *258*, 118014. [CrossRef]

66. Wu, X.F.; Wang, Y.J.; Song, L.J.; Su, J.Z.; Zhang, J.R.; Jia, Y.N.; Shang, J.L.; Nian, X.W.; Zhang, C.Y.; Sun, X.G. A yolk–shell Bi@void@SnO$_2$ photocatalyst with enhanced tetracycline degradation. *J. Mater. Sci. Mater. Electron.* **2019**, *30*, 14987–14994. [CrossRef]

67. Vladimir, A.; Anna, A.; Vadim, B.; Svetlana, K.; Varvara, V.; Daniil, K.; Madina, S.; Alexander, B.; Igor, F.; Roman, N.; et al. Fast Degradation of Tetracycline and Ciprofloxacin in Municipal Water under Hydrodynamic Cavitation/Plasma with CeO$_2$ Nanocatalyst. *Processes* **2022**, *10*, 2063.

68. Du, H.J.; Zhang, W.L.; Wang, C.H.; Niu, Y.Y.; Hou, H.W. A new nanocrystalline inorganic–organic hybrid exhibiting semiconducting properties and applications†. *Dalton Trans.* **2015**, *45*, 2624–2628. [CrossRef]
69. Beatríz, G.-A.; Ramón, M.-M.; José, V.R.-L.; Félix, S.; Juan, S. Discrimination between ω-amino acids with chromogenic acyclic tripodal receptors functionalized with stilbazolium dyes. *Tetrahedron Lett.* **2008**, *49*, 1997–2001.

Article

Study of Photocatalytic Oxidation of Micropollutants in Water and Intensification Case Study

Lucija Radetić [1], Jan Marčec [1], Ivan Brnardić [2], Tihana Čižmar [3] and Ivana Grčić [1,*]

[1] Faculty of Geotechnical Engineering, University of Zagreb, Hallerova Aleja 7, HR-42000 Varaždin, Croatia
[2] Faculty of Metallurgy, University of Zagreb, Aleja Narodnih Heroja 3, HR-44000 Sisak, Croatia
[3] Ruđer Bošković Institute, Bijenička 54, HR-10000 Zagreb, Croatia
* Correspondence: igrcic@gfv.hr or ivana.grcic@gfv.unizg.hr

Abstract: During the last decades, heterogenous photocatalysis has shown as the most promising advanced oxidation process for the removal of micropollutants due to degradation rate, sustainability, non-toxicity, and low-cost. Synergistic interaction of light irradiation, photocatalysts, and highly reactive species are used to break down pollutants toward inert products. Even though titanium dioxide (TiO_2) is the most researched photocatalyst, to overcome shortcomings, various modifications have been made to intensify photocatalytic activity in visible spectra range among which is modification with multiwalled carbon nanotubes (MWCNTs). Therefore, photocatalytic oxidation and its intensification by photocatalyst's modification was studied on the example of four micropollutants (diclofenac, DF; imidacloprid, IMI; 1-H benzotriazole, BT; methylene blue, MB) degradation. Compound parabolic collector (CPC) reactor was used as, nowadays, it has been considered the state-of-the-art system due to its usage of both direct and diffuse solar radiation and quantum efficiency. A commercially available TiO_2 P25 and nanocomposite of TiO_2 and MWCNT were immobilized on a glass fiber mesh by sol-gel method. Full-spectra solar lamps with appropriate UVB and UVA irradiation levels were used in all experiments. Photocatalytic degradation of DF, IMI, BT, and MB by immobilized TiO_2 and TiO_2/CNT photocatalysts was achieved. Mathematical modelling which included mass transfer and photon absorption was applied and intrinsic reaction rate constants were estimated: $k_{DF} = 3.56 \times 10^{-10} s^{-1} W^{-0.5} m^{1.5}$, $k_{IMI} = 8.90 \times 10^{-11} s^{-1} W^{-0.5} m^{1.5}$, $k_{BT} = 1.20 \times 10^{-9} s^{-1} W^{-0.5} m^{1.5}$, $k_{MB} = 1.62 \times 10^{-10} s^{-1} W^{-0.5} m^{1.5}$. Intensification of photocatalysis by TiO_2/CNT was observed for DF, IMI, and MB, while that was not the case for BT. The developed model can be effectively applied for different irradiation conditions which makes it extremely versatile and adaptable when predicting the degradation extents throughout the year using sunlight as the energy source at any location.

Keywords: micropollutants; TiO_2 films; TiO_2/CNT nanocomposites; photocatalysis kinetics

Citation: Radetić, L.; Marčec, J.; Brnardić, I.; Čižmar, T.; Grčić, I. Study of Photocatalytic Oxidation of Micropollutants in Water and Intensification Case Study. *Catalysts* **2022**, *12*, 1463. https://doi.org/10.3390/catal12111463

Academic Editors: Rafael Borja, Gassan Hodaifa and Mha Albqmi

Received: 9 October 2022
Accepted: 14 November 2022
Published: 18 November 2022

Publisher's Note: MDPI stays neutral with regard to jurisdictional claims in published maps and institutional affiliations.

1. Introduction

Advancement of technology has contributed to humankind progresses, but at the same time, there is the consequence of dealing with the environmental pollution and energy crisis [1]. Furthermore, advances in analytical methods have enabled detection of previously unknown compounds as well as their occurrence in the environment. Consequently, conventional wastewater treatment plants (WWTPs) were recognized as a key source of micropollutants since they are not designed for elimination of trace concentrations [2–4]. Pharmaceuticals, pesticides, personal care products, steroid hormones and other organic compounds were detected in effluents after WWTPs as well as in the environment at concentration levels ranging between $ng \cdot L^{-1}$ and $\mu g \cdot L^{-1}$ [5]. Even though removal of micropollutants is not mandatory in the European Union, up to now, three watch lists were established in order to monitor, collect data, and assess the potential risk for the environment and humans [4].

In this work, four micropollutants were chosen, each representing a field of application. Diclofenac (DF) is a substance commonly used as a pharmaceutical due to its anti-inflammatory properties. Therefore, it is one of the 12 most consumed drugs in the world. It has been detected in wastewaters, surface, and underground waters [6–8] since 65% of its initial oral dosage is eliminated as unmodified or metabolites through urine or feces [9], with removal rate of DF in the range of 21–40% after WWTPs [7,8]. At low concentrations, effects like cytotoxicity to kidney, liver and gill cells along with renal lesions were reported in literature [2]. Imidacloprid (IMI) is a widely used pesticide and belongs to neonicotinoid group of pesticides. It is used as an insecticide and has seed protection properties [10]. It is harmful for the pollinators [10], while its degradation products poses a risk to vertebrates, mammals, and humans [11]. Due to its wide application, IMI has been detected in surface waters and soil sediments, behaving as a hormone disruptor causing obesity and reproductive imbalance in organisms [12]. The importance of DF and IMI occurrence in surface waters was recognized as they were listed in first (both) and second (IMI) watch lists [4]. Another compound that is widely used in industrial activities and everyday life is 1H-benzotriazole (BT) [13]. Its application varies from corrosive inhibition in deicing fluids to application in bleaching, antifogging, and antifungal agents [14–17]. Moreover, it is formed during the synthesis of dyes, drugs, and fungicides as intermediates [13]. About 9000 tons of BT are produced annually [18,19] with airports being the hot spots of the BT releasement in the environment [13,20]. It has been estimated that approximately 30% of the total BT input into wastewaters are due to discharge from dishwashing machines [20]. Considering their wide application and yet low removal rate in WWTPs [21], and at same time chemical stability and high-water solubility, BT's residuals were detected mainly in aquatic environments in concentrations from nano to micro gram per liter [22–24]. Even though reported acute toxicity is generally low, considerable residuals levels were found in various organisms revealing a potential risk. In a lettuce leaf and strawberry, 153 and 44 $ng \cdot g^{-1}$ BT were found [25,26]. In the Yangtze and Pearl rivers in China, the maximum concentration of BT in wild fish was 2950 $ng \cdot L^{-1}$ [27–29]. In human urine, a maximum 24.5 $\mu g \cdot L^{-1}$ of BT and its derivates were determined [30]. Endocrine disrupting effects, hepatoxicity, and neurotoxicity are some of the effects related with the sublethal dosages and bioaccumulation effects [13,14]. Even though, research about impacts on human health are scarce [20], LC_{50} concentration of BT causing 50% mortality in the *Ceriodaphnia dubia* (48 h) and *Pimephales promelas* (96 h) were 86–120 and 38–75 $mg \cdot L^{-1}$, respectively [31]. Methylene blue (MB) is often used as a reference measurement to test photocatalytic activity of the prepared photocatalyst [32–40]. It's a very soluble synthetic dye with wide application in industry, such as paper, textile, chemical industry, as well in medical procedures [41]. It was reported that around 15% to 20% of the overall world production of dyes are lost during the dyeing process [42,43], which is significant since usually less than 1 ppm is sufficient to produce coloration of water and consequently reduce photosynthetic activity in the aquatic flora [43]. Since biodegradation process is slow, its removal is of interest [44].

Low-rate removal of contaminants from effluents and their occurrence in surface and underground waters have been directing efforts towards research of environmentally friendly techniques. Therefore, advanced oxidation processes (AOPs) are of considerable interest due to theirs feasible application in WWTPs [3,4,45]. Out of AOPs, heterogenic photocatalysis is one of the most promising. The basic principle can be described with the band gap model, according to which electrons from valence band are transferred into conduction band when photons (hν) induce the photocatalyst's surface with energy that is equal or higher than the photocatalyst's band gap energy (E_g), forming an electron–hole pair (h^+/e^-) [46] crucial for the initiation of redox processes. Prior to photocatalytic redox processes that occur on the photocatalyst surface, diffusion of pollutants and reactants takes place, followed by their adsorption on active sites. Simultaneously, desorption of the oxidized/reduced byproducts occurs and diffuses in the solution participating in new redox processes [6]. In those processes, powerful oxidizing species, hydroxyl radicals, are formed, which non-selectively mineralize pollutants to thermodynamically stable oxidation products such as CO_2, H_2O, and inorganic ions [3,47].

Semiconductors, especially, ZnO and TiO_2 have been recognized as the most effective photocatalysts [1]. The most researched one has been TiO_2 due to its cost, non-toxicity, and photochemical stability, with high photoactivity and mineralization efficiency. However, the drawback is its photocatalytic activity only in UV range ($\lambda \leq 390$ nm), and relatively high rate of electron–hole pairs recombination which reduces available charges for the redox reactions [3,48]. To overcome the shortcomings, various modifications have been made in order to intensify photocatalytic activity in the visible spectra range [1] which has huge potential from an environmental and economic point of view [49]. The aim of modifications is to form a localized state just above the conduction band or below the valence band; use semiconductor with low bandgap; color center formation in band gap or alter the surface. Frequently, it is done by doping with metal and/or nonmetal; codoping with diverse combination of donor and acceptor materials; forming composites; sensitization or substitution [49].

Modification of TiO_2 such as forming nanocomposites with carbon nanotubes (CNTs), contributes to boosting of photocatalytic activity. CNTs are chemically inert, stable, with high specific surface and charge mobility [50]. A nanocomposite system of TiO_2/CNT enables absorption of wider wavelengths due to formed C-O-Ti bonds [51]. It was noticed that increasing of CNT is proportional with the photocatalytic degradability of pollutants [52,53].

Even though suspended photocatalysts tend to be more reactive then immobilized, with appropriate design, mass and photon transfer limitations can be mastered [54]. The most suitable method used for synthesis of nanocomposites and its immobilization on inert carrier is the sol-gel method [51]. The method is considered as easy to use, does not require expensive equipment, and is non-destructive [55–57]. In current work, the mesh-like support was used to achieve a large and irregular photocatalytic layer by optimized sol-gel to overcome the mass and photon transfer limitations.

As a surface phenomenon, solar photocatalysis is highly dependent on the irradiation intensity. Hence, intensification of photocatalysis can be achieved not only by different photocatalyst's formulations, but also with an appropriate choice of reactor design and its usability of incident irradiation in reaction space. An optimal reactor, compound parabolic collector (CPC) reactor, is recommended [49,52] as state-of-the-art in design. Its application on a larger scale has been recognized as technically and economically feasible [58,59]. The usage varies from pilot scales to demonstration plants with collectors' surface areas from 3 to 150 m^2 [49].

Therefore, the aim of this work was to study the impact of solar photocatalysis intensification by using modified TiO_2/CNT photocatalyst in comparison to TiO_2 usage. Immobilized photocatalysts were placed in an optimal CPC reactor configuration under an artificial source of solar containing UVA and UVB light. Four widely used compounds (DF, IMI, BT, MB) were used as study micropollutants. The developed mathematical model [52] was applied in order to calculate intrinsic reaction rate constants (k_i) for studied micropollutants. In the model, the effect of the photocatalyst optical properties and incident irradiation under UVA and UVB light were defined, enabling the intrinsic reaction rate constant to be independent of the irradiation conditions and applied catalyst and applicable to other usage. Given settings enable the developed model to be effectively applied for different irradiation conditions.

2. Results and Discussion

2.1. Intensification by Photocatalyst Formulation

According to previous reporting in the paper [52], based on Micro-Raman spectroscopy results, indices of achieving the chemical bonds between composite TiO_2/CNT on a glass fiber mesh were observed. Pure samples of MWCNT showed characteristic peaks and bands at 1340, 1580, and 2680 cm^{-1}, while for TiO_2, characteristic bands at 148, 286, 399, and

633 cm^{-1}. In the composite sample of TiO$_2$/CNT, dominant bands were from TiO$_2$ along with the affected MWCNT bands which in composite appeared at 1290 and 1620 cm^{-1}. The indices about establishing chemical bonding between TiO$_2$ and MWCNT on glass fibers were additionally confirmed by SEM results. At higher magnifications (200,000× *g*), bonding of TiO$_2$ and MWCNT was clearly noticeable, despite the uneven immobilization along the fiber that can be noticed as aggregates at 1000× *g* magnifications (Figure 1).

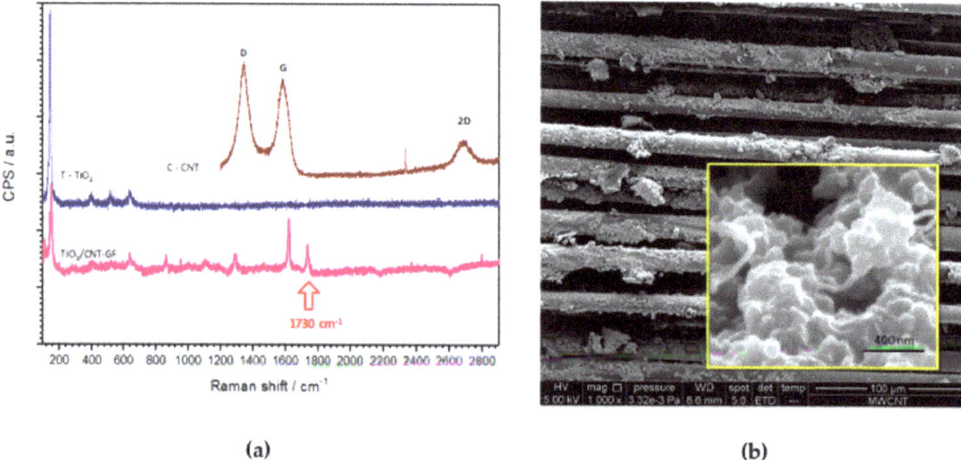

(a)　　　　　　　　　　　　　　(b)

Figure 1. (a) Raman spectra of the photocatalytic constituents and composite film on glass fiber mesh; (b) SEM image of TiO$_2$/CNT at 1000 magnification and 200,000× (given as embedded figure) [52].

According to the DRS spectra (Figure 2), all photocatalytic film samples exhibit photo-absorption in the range between 250 and 400 nm. To relatively enhance the absorption bands of different samples, the Kubelka–Munk function was applied, while the Tauc equation was applied to evaluate the type and extent of transition, along with the bandgap's extent. Therefore, while TiO$_2$ is without a doubt a semiconducting material with indirect allowed transitions and bandgap value near the 3.2 eV, the conventional calcium silicate glass, even though it is not a semiconducting material, may hinder or contribute to the semiconducting behavior of the composite with indirect allowed transitions and a bandgap value near the 3.85 eV. On the other hand, MWCNT as a conducting material shows direct allowed transitions with bandgap value between 2.9 to 3.6 eV depending on the treatment. For pure TiO$_2$ and TiO$_2$ immobilized on glass fibers, determined bandgap values were 3.22 and 3.48 eV, respectively. When it comes to the composite TiO$_2$/CNT, an average bandgap value of 3.12 eV can be expected while none of the transition types fully dominate. For example, if the TiO$_2$ is governing the composite behavior, the calculated bandgap value is 2.97 eV, while if the MWCNT is governing the composite behavior, the calculated bandgap value is 3.27 eV. Observed values suggest that a considerable synergetic photocatalytic effect was achieved.

In a research study [60], aging of photocatalysts were tested. Results have showed that depending on the preparation of the TiO$_2$/CNT photocatalyst during the immobilization process, uneven dispersion can impact on the lower photocatalytic activity compared to TiO$_2$. Regarding the durability of TiO$_2$/CNT and TiO$_2$ itself, it is observed that in a period of 90 days in water, photocatalyst's mass loss was 6% and 11% for TiO$_2$ and TiO$_2$/CNT, respectively. Reduced photocatalytic activity is observed as well, however with the aging of photocatalysts on the air, improvements of photocatalytic activity are achieved.

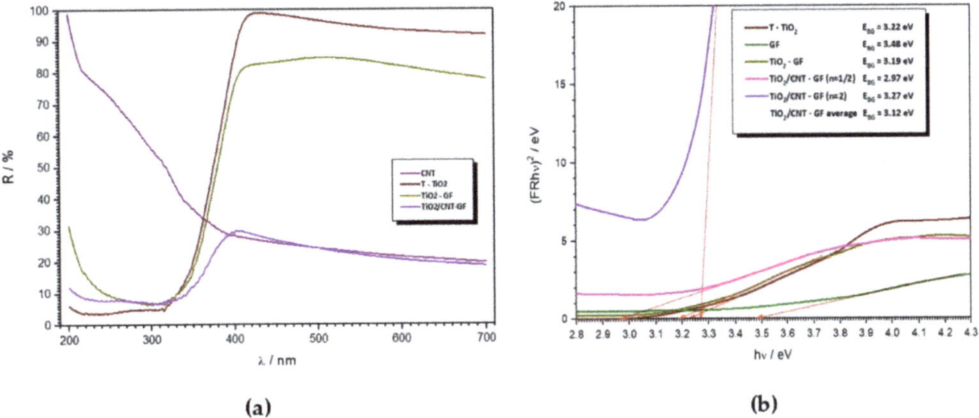

Figure 2. (**a**) DRS UV-vis spectra; (**b**) Tauc plot for the photocatalytic film constituents and photocatalytic composites on glass mesh [52].

2.2. Photolytic and Photocatalytic Degradation of Selected Pollutants

In order to discuss intensified degradation of selected pollutants, photolytic experiments were conducted under the same conditions as the photocatalytic one. Namely, once pollutants are in the environment, they are subjected to the photolysis. It is a process of chemical bond breakage due to photons absorption in an aqueous media initiated under the UV or visible irradiation spectra [5,61,62]. Its efficiency is impacted by the complexity of pollutant chemical structure and conditions in which experiments are conducted (models vs. real water samples). Therefore, photolysis alone, is usually not sufficient to achieve substantial effectiveness of pollutants' degradation [5].

However, the photolytic and photocatalytic degradation results were compared based on the assumption that degradation of selected pollutants follows the pseudo first-order kinetic model. According to the first-order kinetic model, basic kinetic Equation (1) with degradation rate $k > 0$ and initial pollutant's concentration $C(0) = C_0$ for time $t = 0$ can be written as the following Equations (2) and (3) [63]:

$$\text{r} = k[X]^n \tag{1}$$

$$C(t) = C_0 e^{-kt} \tag{2}$$

$$\ln(C(t)) = \ln(C_0) - kt \tag{3}$$

where $C(t)$ is concentration at time t (min), C_0 is initial concentration and k is degradation rate constant (min^{-1}).

Following the given Equations (2) and (3), degradation rate constants for photolysis, photocatalysis with TiO_2 and photocatalysis with TiO_2/CNT were obtained. Results are graphically presented on Figure 3, and further discussed.

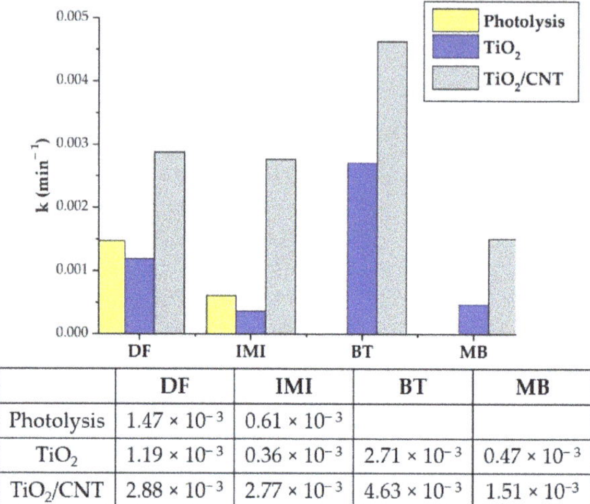

	DF	IMI	BT	MB
Photolysis	1.47×10^{-3}	0.61×10^{-3}		
TiO$_2$	1.19×10^{-3}	0.36×10^{-3}	2.71×10^{-3}	0.47×10^{-3}
TiO$_2$/CNT	2.88×10^{-3}	2.77×10^{-3}	4.63×10^{-3}	1.51×10^{-3}

Figure 3. Photolytic and photocatalytic (TiO$_2$, TiO$_2$/CNT) degradation rate constants k (min^{-1}) according to pseudo-first-order kinetic model for studied model pollutants (DF, IMI, BT, and MB).

Even though degradation of DF by hydroxyl radicals follows second-order kinetics, with approximation of steady-state concentration of hydroxyl radicals, reaction is usually treated as a pseudo-first-order kinetic model [7]. Furthermore, for highly diluted systems ($C_0 < 10^{-3}$ M), the reaction can be considered pseudo-first order [64]. In this paper, as it can be seen on Figure 3, degradation rates of both photolysis ($k_{DF} = 1.47 \times 10^{-3}$ min^{-1}, $R^2 = 0.99$) and photocatalysis ($k_{DF,TiO_2} = 1.19 \times 10^{-3}$ min^{-1}, $R^2 = 0 \times k_{DF,TiO_2/CNT} = 2.88 \times 10^{-3}$ min^{-1}, $R^2 = 0.99$) fit into pseudo-first order kinetic.

Furthermore, depending on experimental setup, photolytic degradation of DF can be described as negligible [45] or significantly better [2,7,65] in comparison with photocatalysis. In our study (Figure 3), photolytic degradation of DF is slightly better than photocatalytic degradation with TiO$_2$. Similar observations were made previously [2,65]. However, authors in their studies had worked with suspended TiO$_2$ in different reactor setup and irradiation values; CPC under sunlight [65] and batch reactor with UV/Vis spectra lamp [2], respectively. Nevertheless, as it was stated in [65], since DF absorbs UV light ($\lambda_{max} = 277$ nm), photolysis is a relevant process and it can be better than photocatalysis, thus, much less photons are available to generate electron–hole pairs due to interference with DF absorption of photons [2].

Regarding the photolysis studies in aqueous media, IMI has showed as readily degradable with a first-order rate kinetics [11,63]. As it can be seen on Figure 3, in this paper degradation rates of both photolysis ($k_{IMI} = 6.1 \times 10^{-4}$ min^{-1}, $R^2 = 0.74$) and photocatalysis ($k_{IMI,TiO_2} = 3.6 \times 10^{-4}$ min^{-1}, $R^2 = 0.87$; $k_{IMI,TiO_2/CNT} = 2.77 \times 10^{-3}$ min^{-1}, $R^2 = 0.99$) fit into the first order kinetic. It can be noticed that photolytic degradation of IMI is slightly faster than the photocatalytic degradation by TiO$_2$, even though the other way around would be expected. The same phenomenon was observed with DF photolytic and photocatalytic (by TiO$_2$) degradation rates. Therefore, since most pesticides absorb UV light ($\lambda_{IMI} = 277$ nm) [66], as well as formed intermediates [67], reduction of photons available to react with photocatalysts surface in reaction system occurs. As it was stated in the literature [11,63], different irradiation conditions in photolytic studies impact on the degradation rates of IMI. For instance, during the 2-h experiment under natural sunlight, 38% of IMI degradation was achieved [63]. On the other hand, 40-h were needed under UV lamps to achieve 95% of IMI degradation [11].

Although, BT is subjected to direct photolysis with first-order kinetic reaction mechanism under UV light [68], that was not the fact in our study. The most efficient degradation of BT is in UVC spectra since maximum absorption is detected at λ_{BT} = 254 nm, while in our study, experiments were done under lamps with UVA and UVB spectra. In their paper [68], authors have reported slow photolysis degradation under lamps with UVA and UVB light. However, DF and IMI absorb the UV light as well, but BT's relatively long persistence in the environment is due to the insensitivity to visible light [14]. Nevertheless, sunlight photolysis is a relevant process for BT degradation in surface waters [16] due to reaction with reactive transient species, though more toxic compounds can be formed if there is no mineralization [13]. In our study, BT photolysis was a negligible process as in [13], while photocatalysis were noticeable processes of degradation ($k_{BT,\ TiO_2} = 2.71 \times 10^{-3}$ min^{-1}, R^2 = 0.97; $k_{BT,TiO_2/CNT}$= 4.63$\times10^{-3}$ min^{-1}, R^2= 0.98.98).

The lack of photolysis and slower photocatalytic degradation than expected by TiO$_2$ can be explained due to the higher initial concentration (C$_{0,MB}$ = 25 ppm) used in experiments compared with literature (C$_{0,MB}$ = 10 ppm, [69]), as well as usage of TiO$_2$ in suspension, while in our case immobilized TiO$_2$ was used. However, in the study [70], photolysis of MB was as well reported as neglected, while photocatalysis with TiO$_2$ P25 achieved 46% during three hours of experiment. In their work [69], degradation rate of 3.4×10^{-3}min^{-1} was achieved with TiO$_2$ P25 in suspension, while we have observed 10 times slower degradation rate k_{MB,TiO_2}= 4.7$\times10^{-4}$ min^{-1} (R^2= 0.86). However, similar degradation rate was observed when a composite of 1%-MWCNT/TiO$_2$ was used; 4.7×10^{-3} min^{-1} versus $k_{MB,TiO_2/CNT}$= 1.51$\times10^{-3}$ min^{-1} (R^2= 0.98). The slight difference value is attributed to the form of photocatalyst's usage.

2.2.1. Intensification of Photocatalytic Degradation

For all studied pollutants model solutions, equilibrium was established in 30 min (Figure 4). Additionally, adsorption on the surface of both photocatalysts was observed, but it was negligible. In consistence with literature findings, [2] adsorption of DF on both photocatalysts was not significant; 1% and 6% on TiO$_2$ and TiO$_2$/CNT, respectively. Equilibrium was achieved during the 30 min in dark, and slightly higher adsorption (+5%) was observed with TiO$_2$/CNT photocatalyst. In the literature [66,71] during the 30 min of dark experiment, adsorption of IMI was negligible. However, in our study, 13% and 20% of adsorption on TiO$_2$ and TiO$_2$/CNT was observed, respectively. It was reported [67] that concentrations of IMI higher than 20 ppm inhibit the photocatalysis due to high adsorption rate of IMI on the photocatalyst's surface. Since in our study initial concentration was lower than stated (C$_0$ = 10 ppm), equilibrium was achieved. However, occupied sites on the photocatalysts surface by adsorbed IMI molecules could slow down the photocatalysis by TiO$_2$. In their work [69], when adsorption tests of MB on pure MWCNTs was conducted, strong adsorption capacity was noticed, 100% of MB adsorption was noticed during the 120 min. Nonetheless, despite adsorption, TiO$_2$ composites with CNT exhibit increased photocatalytic activity under visible light, and degrade MB more efficiently due to a simultaneous increase of active specific surface. In this study, 4% and 11% of MB adsorption on TiO$_2$ and TiO$_2$/CNT was observed, respectively. When it comes to BT adsorption, it was negligible, where 2% and 4% for TiO$_2$ and TiO$_2$/CNT were observed, respectively. Though, it was reported [72] that MWCNT can be used as a sorbent in solid phase extraction, and successfully used as a pretreatment to pre-concentrate BTs from real water samples.

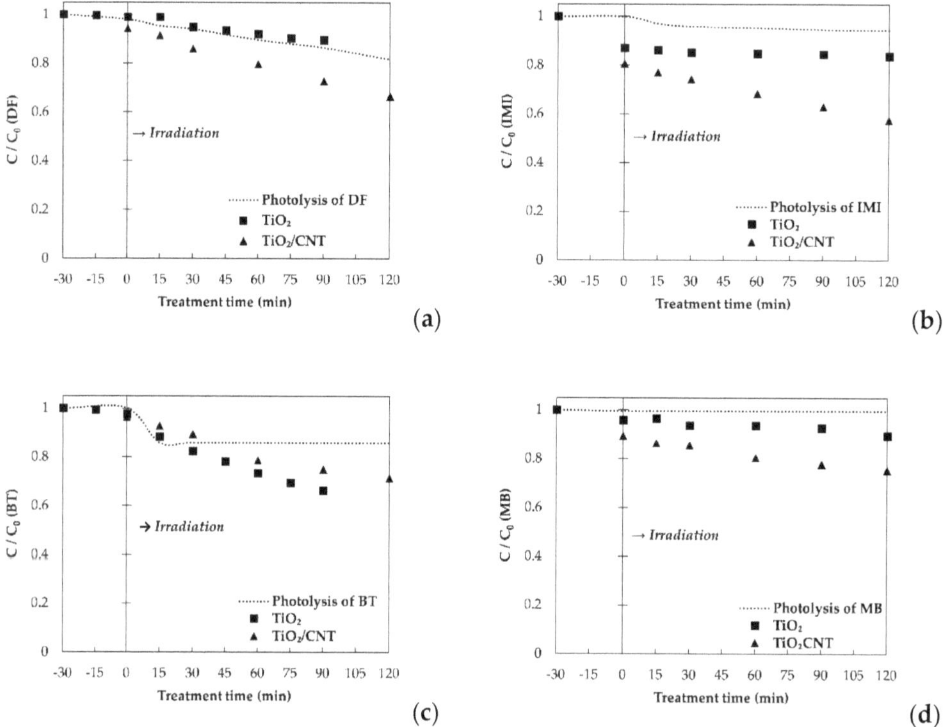

Figure 4. Experimental results of photolytic and photocatalytic degradation on TiO_2 and TiO_2/CNT of: (**a**) diclofenac (DF), (**b**) imidacloprid (IMI), (**c**) 1H-benzotriazole (BT), (**d**) methylene blue (MB). Presented results are obtained during the dark and irradiated experiments.

During the degradation process, to avoid formation of toxic derivates, an optimal treatment time is crucial to define. For instance, in the study [45], after 120 min of DF photocatalytic degradation, toxicity was not observed despite the formation of derivates as chloro and hydroxyl phenols radicals. Photolytic DF degradation rate of 28% was achieved in 4 h when demineralized water was used, while in the same time, 36% was achieved when fresh water was used [65], confirming that the presence of other compounds which can react as radicals contribute to the photolysis of DF [3]. In our study, during the 120 min, 17% of DF degradation was achieved. At the same time, by TiO_2 photocatalysis 10% and TiO_2/CNT 29% of DF degradation was achieved, respectively.

Even though in literature [9] 53.6% of DF degradation rate was achieved, the study was with TiO_2 in suspension. This confirms that the presence of DF molecules in the solution scavenge photons on their way to the immobilized photocatalysts surface. In systems with suspended photocatalysts, given phenomena is reduced due to the suspension form. It was reported that a vital role in DF degradation plays the concentration of dissolved oxygen [2], even though hydroxyl radicals have an important effect on the DF degradation as well [9].

The intensification of DF degradation by TiO_2/CNT photocatalysts in comparison with TiO_2 was observed in our study. For example, in their paper [2], the authors did not obtain significant photocatalytic degradation when composites of 10:100 (w/w) $MWCNT_{ox}$ and TiO_2 (anatase) were used. In their previous work [73], authors had given possible explanations, one of which is inhibition of electron–hole generation by interaction of TiO_2 and CNT. The stated does not apply to our study, since our experiments were conducted with immobilized photocatalysts, preventing interaction between nanoparticles, yet intensifying radicals' generation necessary for DF degradation.

Previous results [66] have demonstrated effective (68%) photocatalytic degradation of IMI by immobilized TiO_2 on a glass plate under the UVC irradiation in 180 min. It was observed that photocatalytic degradation by TiO_2 in batch reactors under UVA light is a relatively slow process as was the case with findings in our study. Photocatalytic degradation of IMI under UVA and UVB light by TiO_2 was 4%, however, by TiO_2 modification with CNT, IMI degradation increases to 29%.

The Intensification of IMI degradation by composites of TiO_2 and CNT was reported by [71]. During the 30-min dark experiments, a small increase in adsorption was observed, which had stabilized in given time. Under UV light, 32% and 53% IMI removal was achieved by TiO_2 and TiO_2/CNT photocatalysis in 3 h, respectively. Addition of CNT contributes to Ti-O-C bond formation, thus allowing the induction of e^- by photons to migrate into CNT and diminishing charge recombination, while e^- can attack H_2O to form hydroxyl radicals.

Consequently, usage of irradiated semiconductors has gained on the importance in the BT removal research. In their research [13], authors used TiO_2 in suspension form and demonstrated rapid removal of BT without formation of toxic byproducts. They proposed that any other AOP that use hydroxyl radicals as the main oxidants might be useful for BT removal. In our study, 32% of BT was degraded within 120 min by TiO_2. However, intensification of BT photocatalytic degradation by modification with CNT was not observed, only 26% BT removal was achieved. A similar trend was observed in the studies [73,74] where carbamazepine and DF photocatalytic degradation by TiO_2 and TiO_2/CNT were studied. Two hypotheses were offered; (1) high electrical conductivity of CNTs supplies TiO_2 with electrons and reduces separation of electron-hole pairs; (2) CNTs are excited to produce electron–hole pairs which are annihilating production of TiO_2 electron–hole pairs generation when conduction band of CNTs is higher than that of TiO_2.

Regarding the MB degradation, it has been reported [33] that about 80% of MB could be degraded under UVB light irradiation in 120 min, while in 60 min 90% of MB could be degraded under sunlight irradiation when immobilized TiO_2 was used. A higher removal rate of MB with suspended TiO_2 (47% in 180 min) was observed by [69]. However, only 6% of MB in 120 min by TiO_2 was degraded in our case. The lower removal efficiency of MB with TiO_2 was observed as well by [55], 13% in 100 min. Nonetheless, intensification of MB degradation when CNT were employed has been noticed. During the 180 min, 61.9% removal rate of MB was obtained with 1%-MWCNT/TiO_2 [14], while 76% of MB was degraded in 100 min by MWCNT:TiO_2 = 1:3 (w/w) [55]. In our study, immobilized TiO_2/CNT (10:1, w/w) was employed and 16% removal rate was observed.

2.2.2. Modeling of Photocatalytic Degradation

Concerning more detailed analysis of photocatalytic degradation, especially by TiO_2 modification with CNT, mathematical modeling with defined intrinsic parameters was applied. Description of model developing is given later in the text. Intrinsic degradation rate constants k_i, as well as intensification factors Y_{cat} for the studied pollutants were obtained and given in Table 1.

Table 1. Intrinsic degradation rate constants for DF, MI, BT and MB along with the intensification factors due to TiO_2 modification with CNT.

	K_i ($s^{-1}W^{-0.5}m^{1.5}$)	Y_{cat} (-)
DF	3.56×10^{-10}	2.35
IMI	8.90×10^{-11}	8.84
BT	1.20×10^{-09}	0.64
MB	1.62×10^{-10}	2.52

Photocatalytic degradation of studied pollutants modeled based on parameters given in Table 1, fit well with the experimental data as it can be seen in Figure 5. As it was already discussed, intensification of DF, IMI, and MB photocatalytic degradation was observed which can be noticed by the value of $Y_{cat} > 1$. The highest intensification contribution to a pollutant's degradation in comparison with TiO$_2$ can be given in the following order IMI (8.84) < MB (2.52) < DF (2.35). On the other hand, intensification of BT degradation by photocatalyst's modification was not observed and therefore value of $Y_{cat} < 1$.

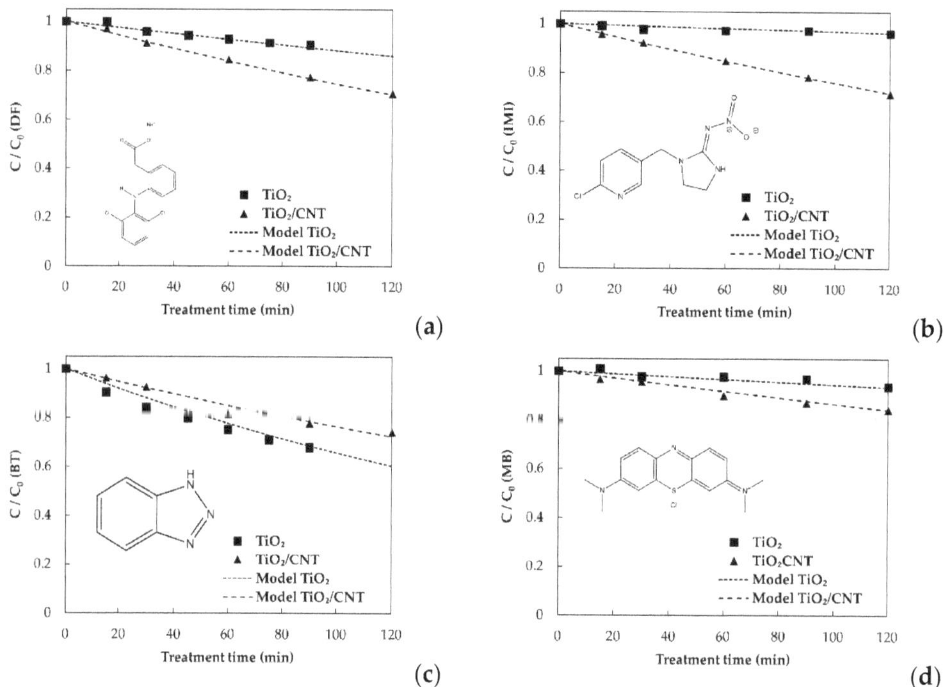

Figure 5. Photocatalytic degradation on TiO$_2$ and TiO$_2$/CNT of: (**a**) diclofenac (DF), (**b**) imidacloprid (IMI), (**c**) 1H-benzotriazole (BT), (**d**) methylene blue (MB). Fitting of mathematical modelling results along with the obtained experimental data.

Furthermore, by using a modeling approach, it is possible to predict time necessary for the pollutant's degradation. For instance, 90% of DF and BT degradation is possible to achieve within one day (Figure 6a,c), while degradation of IMI and MB depends on the used photocatalysts (Figure 6b,d). More elaborate data about the time degradation can be found in Table 2.

Table 2. Overview of time necessary to achieve 50% and 90% of efficient photocatalytic degradation for studied pollutants. Results are obtained based on developed model prediction.

	TiO$_2$				TiO$_2$/CNT			
	DF	IMI	BT	MB	DF	IMI	BT	MB
50%	9 h	37 h	3 h	21 h	4 h	4 h	4 h	8 h
90%	31 h	>84 h	9 h	67 h	13 h	14 h	15 h	27 h

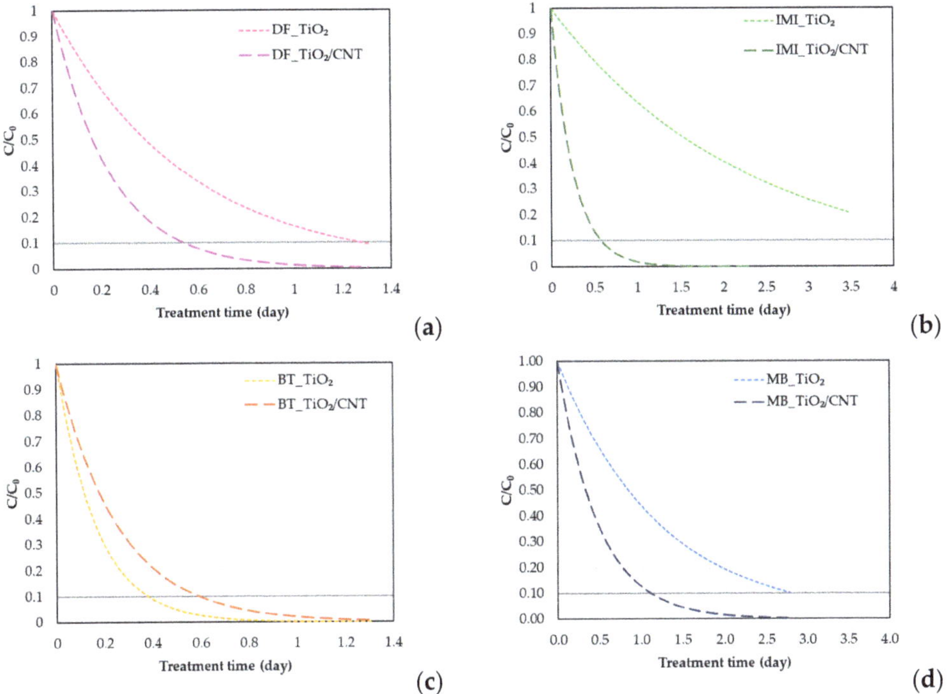

Figure 6. Modeled photocatalytic degradation on TiO_2 and TiO_2/CNT over period of time to achieve 90% efficiency of: (**a**) diclofenac (DF), (**b**) imidacloprid (IMI), (**c**) 1H-benzotriazole (BT), (**d**) methylene blue (MB).

3. Materials and Methods

3.1. Model Pollutants

In this work, photocatalytic degradation of 1H-benzotriazole (BT; 99%, Acros Organics, NJ, USA), imidacloprid (IMI; p.a., Sigma-Aldrich Chemie GmbH, Steinheim, Germany), and diclofenac (DF; p.a., Sigma-Aldrich Chemie GmbH, Steinheim, Germany) were studied. As a reference measurement, photocatalytic degradation of methylene blue (MB; VWR International, Leuven, Belgium) was studied as well. Model solutions were prepared with initial concentrations as follows: $C_0(BT) = 5$ ppm, $C_0(IMI) = 5$ ppm, $C_0(DF) = 10$ ppm and $C_0(MB) = 25$ ppm in accordance with the literature findings [7,67,75].

Concentrations of IMI and DF were determined by high-performance liquid chromatography (HPLC-UV, Agilent Technologies 1200 Series, Santa Clara, CA, USA). A mixture of 0.1% of formic acid and methanol was used as a mobile phase along with the flow rate 0.5 mL/min and 0.3 mL/min for IMI and DF, respectively. Detection was at 254 [76] and 258 nm, respectively [77].

Decrease of concentrations of BT and MB were determined by UV-Vis spectrophotometry (Avantes AvaLight-DH-S-Bal Spectrometer, Lafayette, USA) at 256 nm [78] and 660 nm [39,43,44,79–81], respectively. Regarding the MB, visible photodecolorization was observed as a result of degradation.

3.2. Photocatalysts Preparation

In the experiments, two photocatalysts were used: commercial TiO_2 (Evonik Operations GmbH, Aeroxid® TiO_2 P25, Hanau-Wolfgang, Germany, 30 nm, 56 m²/g, 75:25 anatas to rutil mass ratio) and TiO_2/CNT, a mixture of TiO_2 and CNT (MWCNT, Sigma-Aldrich Chemie GmbH, Steinheim, Germany, 50–90 nm, >95% carbon) in a ratio of 10:1

(*w:w*) (Figure 7). Both photocatalysts were immobilized on a glass fiber mesh (Kelteks, Karlovac, Croatia, density 480 g/m^2) of defined dimensions (480 × 250 mm) by modified sol-gel procedure [52]. According to the procedure, photocatalyst (TiO$_2$ or TiO$_2$/CNT) is added in a solution of distilled water and ethanol (p.a. 96%) with volume ratio 1:1 and stirred for 15 min during which the pH is adjusted to 1.5 with the addition of acetic acid (Kemika, Zagreb, Croatia). Homogenization of solution is then performed by ultrasonic probe for 2 min (30 W, 20 kHz). After sonication, tetraetoxysilane (TEOS) was added and immobilization solution was stirred for 1 h at 50 °C. Meanwhile, glass fiber mesh was cut to prepare supports for the immobilization. Supports were cleaned with ethanol, treated for 5 min with 10 M NaOH and rinsed with deionized water. Pretreated supports were 4 times in a row dipped in immobilization solution and then dried at 70 °C for 30 min. The immobilization procedure was finished after an additional week of immobilized photocatalysts drying at room temperature.

Figure 7. The image of photocatalysts: (**a**) TiO$_2$ photocatalytic film on glass fiber mesh, (**b**) TiO$_2$/CNT photocatalytic film on glass fiber mesh.

Characterization was performed by using Raman spectroscopy (HORIBA Jobin Yvone T64000 spectrometer, Bensheim, Germany) with a 532.5 nm, solid-state laser excitation); the scanning electron microscopy (FEG SEM Quanta250 FEI microscope, Hillsboro, Oregon, USA) and diffuse reflectance spectroscopy (DRS, Perkin-Elmer Lambda 35, Waltham, MA, USA). Additional information were described in more details in our previous paper [52].

3.3. Reaction Set Up

The experiments were performed in a compound parabolic collector (CPC) reactor (Figure 8) which represents state-of-the-art in reactor design. The CPC reactor set up consists of two parallel quartz tubes (L = 50 cm, D$_{outer}$ = 3 cm, D$_{inner}$ = 2.7 cm) connected with a PTFE U-tube of the same inner dimension to avoid changes in flows rates. Quartz tubes are placed in a compound parabolic mirror made of highly reflective alumina (JBL, Neuhofen, Germany, Solar Reflect 50).

As an irradiation source, a custom-made panel with three full spectra lamps (JBL, Neuhofen, Germany, Solar Ultra linear fluorescent lamps: Color, Tropic and Nature, T5, 145 cm, nominal power 80 W) and corresponding reflective mirrors (JBL, Neuhofen, Germany, Solar Reflect 146) was used. The UVB and UVA intensities were determined at the lamp wall (I$_w$) by UVX radiometer (UVP Products, Analytik Jena US LLC, Upland, CA) fitted with corresponding longwave UV-A UVX-36 (range 335–385 nm) and midrange UV-B UVX-31 (range 280–340 nm) sensors of ± 5% accuracy (Figure 9). The CPC reactor was inclined at 12° in correspondence with the inclination of a custom-made panel which was set up at 10 cm above the reactor.

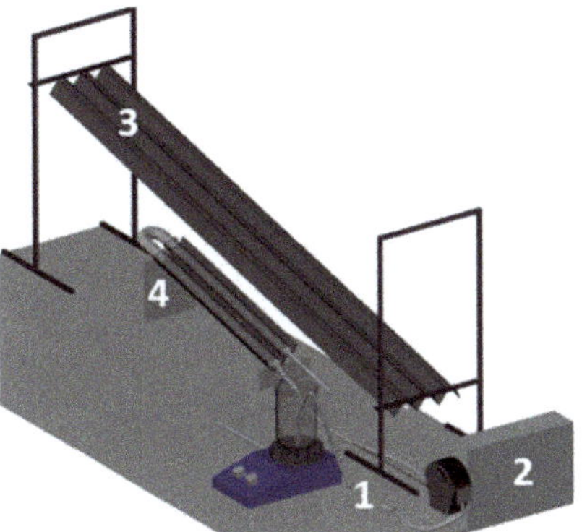

Figure 8. The set up for the CPC reactor. (1) Solution mixing and sampling; (2) Peristaltic pump; (3) Irradiation source; (4) CPC reactor.

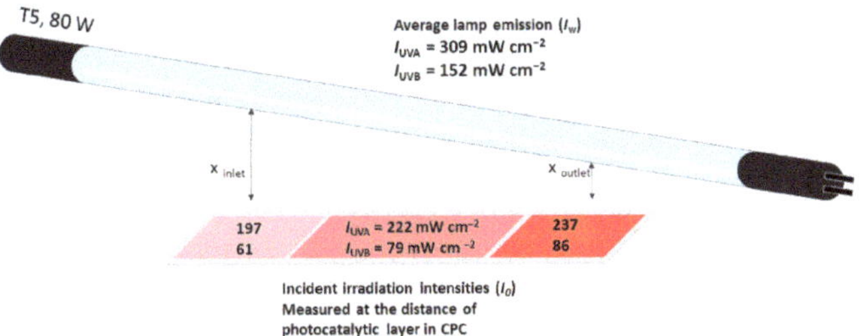

Figure 9. Average incident irradiation intensities on photocatalyst surface in CPC regarding lamps' positioning.

The CPC reactor was attached to a beaker of work volume 1.5 L with silicone tubes and peristaltic pump (Masterflex®) of workflow 26.5 cm³/s. The volume of CPC reactor with the tubes was 0.5 L. The beaker was placed on the magnetic stirrer in order to maintain the equilibrium in the solution. The CPC reactor was connected to the beaker in recirculation with the silicone tubes.

For the purposes of research, photolysis and photocatalysis experiments were conducted following the same methodology for all the model solutions. To achieve sorption equilibrium, the model solution was first recirculated for 30 min without irradiation. In the next 120 min, photolysis and photocatalysis were measured. To estimate the adsorption of model pollutants on the photocatalytic films, control experiments were conducted in 'dark' in the full length of 120 min. All experiments were conducted three times and average values were reported.

3.4. Mathematical Modeling of Intensification Factors

Determination of differences in the degradation rates of model pollutants due to different photocatalytic films requires proper modeling of kinetics parameters.

The basic kinetic Equation (1) was modified into Equation (4) to incorporate intrinsic parameters related with the photocatalytic degradation of selected pollutants over irradiated TiO₂ film.

$$r_i = -k_i((\mu I_0(L, W))_{UVB} + (\mu I_0(L, W))_{UVA})^m [X]^n \tag{4}$$

The $\mu(\text{m}^{-1})$ stands for the absorption coefficient averaged over the spectrum of incident irradiation (in UVB and UVA region), while the $I_0(L,W)$ (W m⁻²) stands for the incident photon flux at the film surface along its length, and m is the order of reaction with respect to irradiation absorption. By introducing the $(\mu I_0(L,W))^m$ into the kinetic model, resulting reaction rate constants became independent of irradiation condition and applied catalysts [82]. Therefore, k_i stands for the intrinsic degradation rate constant of selected pollutant (i).

By introducing the intensification index, Y_{cat}, Equation (4) was modified into Equation (5):

$$r_i = -k_i Y_{cat}((\mu I_0(L,W))_{UVB} + (\mu I_0(L,W))_{UVA})^m [X]^n \tag{5}$$

The Y_{cat} assumes the enhancement in light absorption by photocatalysts, both in the UVA and UVB region but also in the visible part of applied irradiation. It can be used for facile determination of total intensification for new and improved photocatalyst formulations.

The reaction kinetic model was further combined with the material balance for plug flow reactor in recirculation. The Reynolds number was estimated using the hydraulic diameter equal to the wetted perimeter and for maintained flow in CPC was 134.8. Despite the laminar flow, average velocity was used neglecting the radial and axial differences in the velocity profile. The averaged fluid velocity was estimated at $v = 4.99 \times 10^{-2}$ m/s in CPC, which was used in all related models. In CPC, outlet flow mixed with the reaction mixture in the recirculation tank led to different inlet concentrations in different reaction times. Material balance is given for the perfectly mixed reservoir tank:

$$V_{tank}\frac{d[X_i]^{out}_{tank}}{dt} = Q\left([X_i]^{in}_{tank} - [X_i]^{out}_{tank}\right) \tag{6}$$

where

$$v\frac{d[X_i]}{dL} = r_i(L,t) \tag{7}$$

The numerical simulation was performed dividing the reaction space along the length and width—i.e., L and W directions—in sufficiently small intervals. A small-time increment (Δt) equal to the reactor space time ($\tau = V_R/Q$) was introduced. The material balance in the reactor was solved at time t. The time step counter was increased, and the procedure repeated. All simulations were performed in VBA module (Excel). Reaction rate constants (k_i, s^{-1}W$^{-0.5}$m$^{1.5}$) were determined by the trial-and-error method fitting the experimental values into the model by minimizing the variance.

4. Conclusions

Photocatalytic degradation of DF, IMI, BT, and MB by immobilized TiO_2 and TiO_2/CNT photocatalysts were achieved. Mathematical modelling which included mass transfer and photon absorption was applied and intrinsic reaction rate constants were estimated: $k_{DF} = 3.56 \times 10^{-10}$s^{-1} W$^{-0.5}m^{1.5}$, $k_{IMI} = 8.90 \times 10^{-11}$s^{-1}W$^{-0.5}m^{1.5}$, $k_{BT} = 1.20 \times 10^{-9}$s^{-1} W$^{-0.5}m^{1.5}$, $k_{MB} = 1.62 \times 10^{-10}$s^{-1} W$^{-0.5}m^{1.5}$. In comparison with TiO_2 photocatalysis, intensification of photocatalysis by TiO_2/CNT was observed for DF (10% to 29%), IMI (4% to 29%) and MB (6% to 16%), while that was not the case for BT (32% vs. 26%.). Applied mathematical kinetic models can be effectively applied for different irradiation conditions which makes it extremely versatile and adaptable when predicting the degradation extents throughout the year using sunlight as the energy source at any location.

Author Contributions: Conceptualization, L.R., J.M. and I.G.; methodology, L.R. and J.M.; formal analysis, L.R.; investigation, L.R and J.M..; resources, L.R.; writing—original draft preparation, L.R. and J.M.; writing—review and editing, I.G., I.B. and T.Č.; visualization, J.M.; supervision, I.G.; project administration, I.G.; funding acquisition, I.G. All authors have read and agreed to the published version of the manuscript.

Funding: This work has been supported by the following projects "Waste & Sun for photocatalytic degradation of micropollutants in water" (OS-Mi), supported by European Regional Development Fund, KK.01.1.1.04.0006. and "PV-WALL" PZS-2019-02-1555 in the Research Cooperability Program of the Croatian Science Foundation funded by the European Union from the European Social Fund under the Operational Program Efficient Human Resources 2014–2020.

Data Availability Statement: Not applicable.

Acknowledgments: Special thanks to project VIRTULAB—Integrated laboratory for primary and secondary raw materials for the equipment usage supported by European Regional Development Fund KK.01.1.1.02.0022.

Conflicts of Interest: The authors declare no conflict of interest.

References

1. Ameta, R.; Solanki, M.S.; Benjamin, S.; Ameta, S.C. Photocatalysis. In *Advanced Oxidation Processes for Waste Water Treatment*; Elsevier: Amsterdam, The Netherlands, 2018; pp. 135–175, ISBN 9780128105252.
2. Martínez, C.; Canle, L.M.; Fernández, M.I.; Santaballa, J.A.; Faria, J. Aqueous degradation of diclofenac by heterogeneous photocatalysis using nanostructured materials. *Appl. Catal. B Environ.* **2011**, *107*, 110–118. [CrossRef]
3. Ribeiro, A.R.L.; Moreira, N.F.F.; Puma, G.L.; Silva, A.M.T. Impact of water matrix on the removal of micropollutants by advanced oxidation technologies. *Chem. Eng. J.* **2019**, *363*, 155–173. [CrossRef]
4. Kosek, K.; Luczkiewicz, A.; Fudala-Książek, S.; Jankowska, K.; Szopińska, M.; Svahn, O.; Tränckner, J.; Kaiser, A.; Langas, V.; Björklund, E. Implementation of advanced micropollutants removal technologies in wastewater treatment plants (WWTPs)— Examples and challenges based on selected EU countries. *Environ. Sci. Policy* **2020**, *112*, 213–226. [CrossRef]
5. Bustos, E.B.; Sandoval-González, A.; Martínez-Sánchez, C. Detection and Treatment of Persistent Pollutants in Water: General Review of Pharmaceutical Products. *ChemElectroChem* **2022**, *9*, e202200188. [CrossRef]
6. Espíndola, J.C.; Vilar, V.J.P. Innovative light-driven chemical/catalytic reactors towards contaminants of emerging concern mitigation: A review. *Chem. Eng. J.* **2020**, *394*, 124865. [CrossRef]
7. Arany, E.; Láng, J.; Somogyvári, D.; Láng, O.; Alapi, T.; Ilisz, I.; Gajda-Schrantz, K.; Dombi, A.; Kőhidai, L.; Hernádi, K. Vacuum ultraviolet photolysis of diclofenac and the effects of its treated aqueous solutions on the proliferation and migratory responses of Tetrahymena pyriformis. *Sci. Total Environ.* **2014**, *468–469*, 996–1006. [CrossRef] [PubMed]
8. Bi, L.; Chen, Z.; Li, L.; Kang, J.; Zhao, S.; Wang, B.; Yan, P.; Li, Y.; Zhang, X.; Shen, J. Selective adsorption and enhanced photodegradation of diclofenac in water by molecularly imprinted TiO_2. *J. Hazard. Mater.* **2021**, *407*, 124759. [CrossRef]
9. Diaz-Angulo, J.; Porras, J.; Mueses, M.; Torres-Palma, R.A.; Hernandez-Ramirez, A.; Machuca-Martinez, F. Coupling of heterogeneous photocatalysis and photosensitized oxidation for diclofenac degradation: Role of the oxidant species. *J. Photochem. Photobiol. A Chem.* **2019**, *383*, 112015. [CrossRef]
10. Náfrádi, M.; Hlogyik, T.; Farkas, L.; Alapi, T. Comparison of the heterogeneous photocatalysis of imidacloprid and thiacloprid— reaction mechanism, ecotoxicity, and the effect of matrices. *J. Environ. Chem. Eng.* **2021**, *9*, 106684. [CrossRef]
11. Ding, T.; Jacobs, D.; Lavine, B.K. Liquid chromatography-mass spectrometry identification of imidacloprid photolysis products. *Microchem. J.* **2011**, *99*, 535–541. [CrossRef]
12. Mikolić, A.; Karačonji, I.B. Imidacloprid as reproductive toxicant and endocrine disruptor: Investigations in laboratory animals. *Arch. Ind. Hyg. Toxicol.* **2018**, *69*, 103–108. [CrossRef] [PubMed]
13. Minella, M.; De Laurentiis, E.; Pellegrino, F.; Prozzi, M.; Dal Bello, F.; Maurino, V.; Minero, C. Photocatalytic Transformations of 1H-Benzotriazole and Benzotriazole Derivates. *Nanomaterials* **2020**, *10*, 1835. [CrossRef] [PubMed]
14. Shi, Z.-Q.; Liu, Y.-S.; Xiong, Q.; Cai, W.-W.; Ying, G.-G. Occurrence, toxicity and transformation of six typical benzotriazoles in the environment: A review. *Sci. Total Environ.* **2019**, *661*, 407–421. [CrossRef] [PubMed]
15. Lee, J.-E.; Kim, M.-K.; Lee, J.-Y.; Lee, Y.-M.; Zoh, K.-D. Degradation kinetics and pathway of 1H-benzotriazole during UV/chlorination process. *Chem. Eng. J.* **2019**, *359*, 1502–1508. [CrossRef]
16. Weidauer, C.; Davis, C.; Raeke, J.; Seiwert, B.; Reemtsma, T. Sunlight photolysis of benzotriazoles—Identification of transformation products and pathways. *Chemosphere* **2016**, *154*, 416–424. [CrossRef] [PubMed]
17. Bahnmüller, S.; Loi, C.H.; Linge, K.L.; von Gunten, U.; Canonica, S. Degradation rates of benzotriazoles and benzothiazoles under UV-C irradiation and the advanced oxidation process UV/H_2O_2. *Water Res.* **2015**, *74*, 143–154. [CrossRef] [PubMed]
18. Reemtsma, T.; Miehe, U.; Duennbier, U.; Jekel, M. Polar pollutants in municipal wastewater and the water cycle: Occurrence and removal of benzotriazoles. *Water Res.* **2010**, *44*, 596–604. [CrossRef]
19. Dai, Q.; Chen, W.; Luo, J.; Luo, X. Abatement kinetics of highly concentrated 1H-Benzotriazole in aqueous solution by ozonation. *Sep. Purif. Technol.* **2017**, *183*, 327–332. [CrossRef]
20. Alotaibi, M.D.; McKinley, A.J.; Patterson, B.M.; Reeder, A.Y. Benzotriazoles in the Aquatic Environment: A Review of Their Occurrence, Toxicity, Degradation and Analysis. *Water Air Soil Pollut.* **2015**, *226*, 226. [CrossRef]
21. Yin, W.; Shao, H.; Huo, Z.; Wang, S.; Zou, Q.; Xu, G. Degradation of anticorrosive agent benzotriazole by electron beam irradiation: Mechanisms, degradation pathway and toxicological analysis. *Chemosphere* **2022**, *287*, 132133. [CrossRef]

22. Cancilla, D.A.; Martinez, J.; van Aggelen, G.C. Detection of Aircraft Deicing/Antiicing Fluid Additives in a Perched Water Monitoring Well at an International Airport. *Environ. Sci. Technol.* **1998**, *32*, 3834–3835. [CrossRef]

23. Loos, R.; Locoro, G.; Comero, S.; Contini, S.; Schwesig, D.; Werres, F.; Balsaa, P.; Gans, O.; Weiss, S.; Blaha, L.; et al. Pan-European survey on the occurrence of selected polar organic persistent pollutants in ground water. *Water Res.* **2010**, *44*, 4115–4126. [CrossRef] [PubMed]

24. Loos, R.; Tavazzi, S.; Mariani, G.; Suurkuusk, G.; Paracchini, B.; Umlauf, G. Analysis of emerging organic contaminants in water, fish and suspended particulate matter (SPM) in the Joint Danube Survey using solid-phase extraction followed by UHPLC-MS-MS and GC–MS analysis. *Sci. Total Environ.* **2017**, *607–608*, 1201–1212. [CrossRef]

25. LeFevre, G.H.; Müller, C.E.; Li, R.J.; Luthy, R.G.; Sattely, E.S. Rapid Phytotransformation of Benzotriazole Generates Synthetic Tryptophan and Auxin Analogs in Arabidopsis. *Environ. Sci. Technol.* **2015**, *49*, 10959–10968. [CrossRef] [PubMed]

26. LeFevre, G.H.; Lipsky, A.; Hyland, K.C.; Blaine, A.C.; Higgins, C.P.; Luthy, R.G. Benzotriazole (BT) and BT plant metabolites in crops irrigated with recycled water. *Environ. Sci. Water Res. Technol.* **2017**, *3*, 213–223. [CrossRef]

27. Yao, L.; Zhao, J.-L.; Liu, Y.-S.; Yang, Y.-Y.; Liu, W.-R.; Ying, G.-G. Simultaneous determination of 24 personal care products in fish muscle and liver tissues using QuEChERS extraction coupled with ultra pressure liquid chromatography-tandem mass spectrometry and gas chromatography-mass spectrometer analyses. *Anal. Bioanal. Chem.* **2016**, *408*, 8177–8193. [CrossRef]

28. Yao, L.; Lv, Y.-Z.; Zhang, L.-J.; Liu, W.-R.; Zhao, J.-L.; Liu, Y.-S.; Zhang, Q.-Q.; Ying, G.-G. Determination of 24 personal care products in fish bile using hybrid solvent precipitation and dispersive solid phase extraction cleanup with ultrahigh performance liquid chromatography-tandem mass spectrometry and gas chromatography-mass spectrometry. *J. Chromatogr. A* **2018**, *1551*, 29–40. [CrossRef]

29. Yao, L.; Zhao, J.-L.; Liu, Y.-S.; Zhang, Q.-Q.; Jiang, Y.-X.; Liu, S.; Liu, W.-R.; Yang, Y.-Y.; Ying, G.-G. Personal care products in wild fish in two main Chinese rivers: Bioaccumulation potential and human health risks. *Sci. Total Environ.* **2018**, *621*, 1093–1102. [CrossRef]

30. Asimakopoulos, A.G.; Wang, L.; Thomaidis, N.S.; Kannan, K. Benzotriazoles and benzothiazoles in human urine from several countries: A perspective on occurrence, biotransformation, and human exposure. *Environ. Int.* **2013**, *59*, 274–281. [CrossRef]

31. Pillard, D.A.; Cornell, J.S.; DuFresne, D.L.; Hernandez, M.T. Toxicity of Benzotriazole and Benzotriazole Derivatives to Three Aquatic Species. *Water Res.* **2001**, *35*, 557–560. [CrossRef]

32. Samuel, J.J. Photocatalytic degradation of methylene blue under visible light by dye sensitized titania Photocatalytic degradation of methylene blue under visible light by dye sensitized titania. *Mater. Res. Express* **2020**, *7*, 015051. [CrossRef]

33. Islam, M.T.; Dominguez, A.; Turley, R.S.; Kim, H.; Sultana, K.A.; Shuvo, M.A.I.; Alvarado-Tenorio, B.; Montes, M.O.; Lin, Y.; Gardea-Torresdey, J.; et al. Development of photocatalytic paint based on TiO$_2$ and photopolymer resin for the degradation of organic pollutants in water. *Sci. Total Environ.* **2020**, *704*, 135406. [CrossRef] [PubMed]

34. Tschirch, J.; Dillert, R.; Bahnemann, D.; Proft, B.; Biedermann, A.; Goer, B. Photodegradation of methylene blue in water, a standard method to determine the activity of photocatalytic coatings? *Res. Chem. Intermed.* **2008**, *34*, 381–392. [CrossRef]

35. Gao, B.; Chen, G.Z.; Puma, G.L. Carbon nanotubes/titanium dioxide (CNTs/TiO$_2$) nanocomposites prepared by conventional and novel surfactant wrapping sol-gel methods exhibiting enhanced photocatalytic activity. *Appl. Catal. B Environ.* **2009**, *89*, 503–509. [CrossRef]

36. Kwon, C.H.; Shin, H.; Kim, J.H.; Choi, W.S.; Yoon, K.H. Degradation of methylene blue via photocatalysis of titanium dioxide. *Mater. Chem. Phys.* **2004**, *86*, 78–82. [CrossRef]

37. Ono, Y.; Rachi, T.; Okuda, T.; Yokouchi, M.; Kamimoto, Y.; Nakajima, A.; Okada, K. Kinetics study for photodegradation of methylene blue dye by titanium dioxide powder prepared by selective leaching method. *J. Phys. Chem. Solids* **2012**, *73*, 343–349. [CrossRef]

38. Ertuş, E.B.; Vakifahmetoglu, C.; Öztürk, A. Enhanced methylene blue removal efficiency of TiO$_2$ embedded porous glass. *J. Eur. Ceram. Soc.* **2020**, *41*, 1530–1536. [CrossRef]

39. Shidpour, R.; Simchi, A.; Ghanbari, F.; Vossoughi, M. Photo-degradation of organic dye by zinc oxide nanosystems with special defect structure: Effect of the morphology and annealing temperature. *Appl. Catal. A Gen.* **2014**, *472*, 198–204. [CrossRef]

40. Yang, C.; Dong, W.; Cui, G.; Zhao, Y.; Shi, X.; Xia, X.; Tang, B.; Wang, W. Highly efficient photocatalytic degradation of methylene blue by P2ABSA-modified TiO$_2$ nanocomposite due to the photosensitization synergetic effect of TiO$_2$ and P2ABSA. *RSC Adv.* **2017**, *7*, 23699–23708. [CrossRef]

41. National Center for Biotechnology Information PubChem Compound Summary for CID 6099, Methylene blue. Available online: https://pubchem.ncbi.nlm.nih.gov/compound/Methylene-blue#section=Use-and-Manufacturing (accessed on 15 November 2020).

42. Anwar, D.I.; Mulyadi, D. Synthesis of Fe-TiO$_2$ Composite as a Photocatalyst for Degradation of Methylene Blue. *Procedia Chem.* **2015**, *17*, 49–54. [CrossRef]

43. Savić, T.D.; Carević, M.V.; Mitrić, M.N.; Kuljanin-Jakovljević, J.Ž.; Abazović, N.D.; Čomor, M.I. Simulated solar light driven performance of nanosized ZnIn$_2$S$_4$/dye system: Decolourization vs. photodegradation. *J. Photochem. Photobiol. A Chem.* **2020**, *388*, 112154. [CrossRef]

44. Zhang, T.; Oyama, T.; Aoshima, A.; Hidaka, H.; Zhao, J.; Serpone, N. Photooxidative N-demethylation of methylene blue in aqueous TiO$_2$ dispersions under UV irradiation. *J. Photochem. Photobiol. A Chem.* **2001**, *140*, 163–172. [CrossRef]

45. Mahmoud, W.M.M.; Rastogi, T.; Kümmerer, K. Application of titanium dioxide nanoparticles as a photocatalyst for the removal of micropollutants such as pharmaceuticals from water. *Curr. Opin. Green Sustain. Chem.* **2017**, *6*, 1–10. [CrossRef]

46. Khan, M.M.; Pradhan, D.; Sohn, Y. Springer Series on Polymer and Composite Materials Nanocomposites for Visible Light-induced Photocatalysis. In *Nanocomposites Visible Light. Photocatal*; Springer: Cham, Switzerland, 2017; pp. 19–40. [CrossRef]

47. Grčić, I.; Marčec, J.; Radetić, L.; Radovan, A.-M.; Melnjak, I.; Jajčinović, I.; Brnardić, I. Ammonia and methane oxidation on TiO$_2$ supported on glass fiber mesh under artificial solar irradiation. *Environ. Sci. Pollut. Res.* **2020**, *28*, 18354–18367. [CrossRef]

48. Racovita, A.D. Titanium Dioxide: Structure, Impact, and Toxicity. *Int. J. Environ. Res. Public Health* **2022**, *19*, 5681. [CrossRef]

49. Casado, C.; García-Gil, Á.; van Grieken, R.; Marugán, J. Critical role of the light spectrum on the simulation of solar photocatalytic reactors. *Appl. Catal. B Environ.* **2019**, *252*, 1–9. [CrossRef]

50. Khalid, N.R.; Majid, A.; Tahir, M.B.; Niaz, N.A.; Khalid, S. Carbonaceous-TiO$_2$ nanomaterials for photocatalytic degradation of pollutants: A review. *Ceram. Int.* **2017**, *43*, 14552–14571. [CrossRef]

51. Woan, K.; Pyrgiotakis, G.; Sigmund, W. Photocatalytic carbon-nanotube-TiO$_2$ composites. *Adv. Mater.* **2009**, *21*, 2233–2239. [CrossRef]

52. Malinowski, S.; Presečki, I.; Jajčinović, I.; Brnardić, I.; Mandić, V.; Grčić, I. Intensification of Dihydroxybenzenes Degradation over Immobilized TiO$_2$ Based Photocatalysts under Simulated Solar Light. *Appl. Sci.* **2020**, *10*, 7571. [CrossRef]

53. Marques, R.R.N.; Sampaio, M.J.; Carrapiço, P.M.; Silva, C.G.; Morales-Torres, S.; Dražić, G.; Faria, J.L.; Silva, A.M.T. Photocatalytic degradation of caffeine: Developing solutions for emerging pollutants. *Catal. Today* **2013**, *209*, 108–115. [CrossRef]

54. Sundar, K.P.; Kanmani, S. Progression of Photocatalytic reactors and it's comparison: A Review. *Chem. Eng. Res. Des.* **2020**, *154*, 135–150. [CrossRef]

55. Zhao, D.; Yang, X.; Chen, C.; Wang, X. Enhanced photocatalytic degradation of methylene blue on multiwalled carbon nanotubes–TiO$_2$. *J. Colloid Interface Sci.* **2013**, *398*, 234–239. [CrossRef] [PubMed]

56. Pava-Gómez, B.; Vargas-Ramírez, X.; Díaz-Uribe, C.; Romero, H.; Duran, F. Evaluation of copper-doped TiO$_2$ film supported on glass and LDPE with the design of a pilot-scale solar photoreactor. *Sol. Energy* **2021**, *220*, 695–705. [CrossRef]

57. Mani, S.S.; Rajendran, S.; Nalajala, N.; Mathew, T.; Gopinath, C.S. Electronically Integrated Mesoporous Ag–TiO$_2$ Nanocomposite Thin films for Efficient Solar Hydrogen Production in Direct Sunlight. *Energy Technol.* **2022**, *10*, 2100356. [CrossRef]

58. Rodriguez-Acosta, J.W.; Mueses, M.Á.; Machuca-Martínez, F. Mixing Rules Formulation for a Kinetic Model of the Langmuir-Hinshelwood Semipredictive Type Applied to the Heterogeneous Photocatalytic Degradation of Multicomponent Mixtures. *Int. J. Photoenergy* **2014**, *2014*, 817538. [CrossRef]

59. Muñoz-Flores, P.; Poon, P.S.; Ania, C.O.; Matos, J. Performance of a C-containing Cu-based photocatalyst for the degradation of tartrazine: Comparison of performance in a slurry and CPC photoreactor under artificial and natural solar light. *J. Colloid Interface Sci.* **2022**, *623*, 646–659. [CrossRef] [PubMed]

60. Marić, T. Aging of Photocatalysis Based on Titanium (IV) Oxide and Carbon Nanotubes. Master's Thesis, University of Zagreb, Zagreb, Croatia, 2020.

61. Anisuzzaman, S.M.; Joseph, C.G.; Pang, C.K.; Affandi, N.A.; Maruja, S.N.; Vijayan, V. Current Trends in the Utilization of Photolysis and Photocatalysis Treatment Processes for the Remediation of Dye Wastewater: A Short Review. *ChemEngineering* **2022**, *6*, 58. [CrossRef]

62. Vaya, D.; Surolia, P.K. Semiconductor based photocatalytic degradation of pesticides: An overview. *Environ. Technol. Innov.* **2020**, *20*, 101128. [CrossRef]

63. Kurwadkar, S.; Evans, A.; DeWinne, D.; White, P.; Mitchell, F. Modeling photodegradation kinetics of three systemic neonicotinoids-dinotefuran, imidacloprid, and thiamethoxam-in aqueous and soil environment. *Environ. Toxicol. Chem.* **2016**, *35*, 1718–1726. [CrossRef]

64. Herrmann, J.-M. Photocatalysis. In *Kirk-Othmer Encyclopedia of Chemical Technology*; John Wiley & Sons, Inc.: Hoboken, NJ, USA, 2017; pp. 1–44, ISBN 9780128105252.

65. Pérez-Estrada, L.A.; Maldonado, M.I.; Gernjak, W.; Agüera, A.; Fernández-Alba, A.R.; Ballesteros, M.M.; Malato, S. Decomposition of diclofenac by solar driven photocatalysis at pilot plant scale. *Catal. Today* **2005**, *101*, 219–226. [CrossRef]

66. Shorgoli, A.A.; Shokri, M. Photocatalytic degradation of imidacloprid pesticide in aqueous solution by TiO$_2$ nanoparticles immobilized on the glass plate. *Chem. Eng. Commun.* **2017**, *204*, 1061–1069. [CrossRef]

67. Liang, R.; Tang, F.; Wang, J.; Yue, Y. Photo-degradation dynamics of five neonicotinoids: Bamboo vinegar as a synergistic agent for improved functional duration. *PLoS ONE* **2019**, *14*, e0223708. [CrossRef] [PubMed]

68. Ye, J.; Zhou, P.; Chen, Y.; Ou, H.; Liu, J.; Li, C.; Li, Q. Degradation of 1H-benzotriazole using ultraviolet activating persulfate: Mechanisms, products and toxicological analysis. *Chem. Eng. J.* **2018**, *334*, 1493–1501. [CrossRef]

69. Liu, X.; Wang, X.; Xing, X.; Li, Q.; Yang, J. Visible light photocatalytic activities of carbon nanotube/titanic acid nanotubes derived-TiO2 composites for the degradation of methylene blue. *Adv. Powder Technol.* **2015**, *26*, 8–13. [CrossRef]

70. Elghniji, K.; Ksibi, M.; Elaloui, E. Sol–gel reverse micelle preparation and characterization of N-doped TiO$_2$: Efficient photocatalytic degradation of methylene blue in water under visible light. *J. Ind. Eng. Chem.* **2012**, *18*, 178–182. [CrossRef]

71. de la Flor, M.P.; Camarillo, R.; Martínez, F.; Jiménez, C.; Quiles, R.; Rincón, J. Synthesis and characterization of TiO$_2$/CNT/Pd: An effective sunlight photocatalyst for neonicotinoids degradation. *J. Environ. Chem. Eng.* **2021**, *9*, 106278. [CrossRef]

72. Speltini, A.; Maraschi, F.; Sturini, M.; Contini, M.; Profumo, A. Dispersive multi-walled carbon nanotubes extraction of benzenesulfonamides, benzotriazoles, and benzothiazoles from environmental waters followed by microwave desorption and HPLC-HESI-MS/MS. *Anal. Bioanal. Chem.* **2017**, *409*, 6709–6718. [CrossRef]

73. Martínez, C.; Canle, L.M.; Fernández, M.I.; Santaballa, J.A.; Faria, J. Kinetics and mechanism of aqueous degradation of carbamazepine by heterogeneous photocatalysis using nanocrystalline TiO_2, ZnO and multi-walled carbon nanotubes–anatase composites. *Appl. Catal. B Environ.* **2011**, *102*, 563–571. [CrossRef]
74. Cao, Q.; Yu, Q.; Connell, D.W.; Yu, G. Titania/carbon nanotube composite (TiO_2/CNT) and its application for removal of organic pollutants. *Clean Technol. Environ. Policy* **2013**, *15*, 871–880. [CrossRef]
75. Yao, Y.; Luan, J. Preparation, Property Characterization of Gd_2YSbO_7/$ZnBiNbO_5$ Heterojunction Photocatalyst for Photocatalytic Degradation of Benzotriazole under Visible Light Irradiation. *Catalysts* **2022**, *12*, 159. [CrossRef]
76. Čižmar, T.; Panžić, I.; Capan, I.; Gajović, A. Nanostructured TiO_2 photocatalyst modified with Cu for improved imidacloprid degradation. *Appl. Surf. Sci.* **2021**, *569*, 151026. [CrossRef]
77. Bohač, M.; Čižmar, T.; Kojić, V.; Marčec, J.; Juraić, K.; Grčić, I.; Gajović, A. Novel, Simple and Low-Cost Preparation of Ba-Modified TiO_2 Nanotubes for Diclofenac Degradation under UV/Vis Radiation. *Nanomaterials* **2021**, *11*, 2714. [CrossRef] [PubMed]
78. Chen, Y.; Ye, J.; Li, C.; Zhou, P.; Liu, J.; Ou, H. Degradation of 1 H -benzotriazole by UV/H_2O_2 and UV/TiO_2: Kinetics, mechanisms, products and toxicology. *Environ. Sci. Water Res. Technol.* **2018**, *4*, 1282–1294. [CrossRef]
79. Nekouei, S.; Nekouei, F. Comparative procedure of photodegradation of methylene blue using N doped activated carbon loaded with hollow 3D flower like ZnS in two synergic phases of adsorption and catalytic. *J. Photochem. Photobiol. A Chem.* **2018**, *364*, 262–273. [CrossRef]
80. Salavati, H.; Kohestani, T. Preparation, characterization and photochemical degradation of dyes under UV light irradiation by inorganic–organic nanocomposite. *Mater. Sci. Semicond. Process.* **2013**, *16*, 1904–1911. [CrossRef]
81. Ji, K.H.; Jang, D.M.; Cho, Y.J.; Myung, Y.; Kim, H.S.; Kim, Y.; Park, J. Comparative Photocatalytic Ability of Nanocrystal-Carbon Nanotube and -TiO_2 Nanocrystal Hybrid Nanostructures. *J. Phys. Chem. C* **2009**, *113*, 19966–19972. [CrossRef]
82. Grčić, I.; Papić, S.; Brnardić, I. Photocatalytic Activity of TiO_2 Thin Films: Kinetic and Efficiency Study. *Int. J. Chem. React. Eng.* **2018**, *16*, 20160153. [CrossRef]

Article

Sepiolite-Supported WS$_2$ Nanosheets for Synergistically Promoting Photocatalytic Rhodamine B Degradation

Jiaxuan Bai [1,2], Kaibin Cui [1,2], Xinlei Xie [1,2], Baizeng Fang [3,*] and Fei Wang [1,2,*]

[1] Key Laboratory of Special Functional Materials for Ecological Environment and Information, Ministry of Education, Hebei University of Technology, Tianjin 300130, China
[2] Institute of Power Source and Ecomaterials Science, Hebei University of Technology, Tianjin 300130, China
[3] Department of Chemical & Biological Engineering, University of British Columbia, 2360 East Mall, Vancouver, BC V6T 1Z3, Canada
* Correspondence: baizengfang@163.com (B.F.); wangfei@hebut.edu.cn (F.W.)

Abstract: Pristine tungsten disulfide (WS$_2$) nanosheets are extremely prone to agglomeration, leading to blocked active sites and the decrease of catalytic activity. In this work, highly dispersed WS$_2$ nanosheets were fabricated via a one-step in situ solvothermal method, using sepiolite nanofibers as a functional carrier. The ammonium tetrathiotungstate was adopted as W and S precursors, and *N,N*-dimethylformamide could provide a neutral reaction environment. The electron microscope analysis revealed that the WS$_2$ nanosheets were stacked compactly in the shape of irregular plates, while they were uniformly grown on the surface of sepiolite nanofibers. Meanwhile, the BET measurement confirmed that the as-prepared composite has a larger specific surface area and is more mesoporous than the pure WS$_2$. Due to the improved dispersion of WS$_2$ and the synergistic effect between WS$_2$ and the mesoporous sepiolite mineral which significantly facilitated the mass transport, the WS$_2$/sepiolite composite exhibited ca. 2.6 times the photocatalytic efficiency of the pure WS$_2$ for rhodamine B degradation. This work provides a potential method for low-cost batch preparation of high-quality 2D materials via assembling on natural materials.

Keywords: photocatalysis; rhodamine B degradation; sepiolite nanofibers; catalyst support; WS$_2$ nanosheets

Citation: Bai, J.; Cui, K.; Xie, X.; Fang, B.; Wang, F. Sepiolite-Supported WS$_2$ Nanosheets for Synergistically Promoting Photocatalytic Rhodamine B Degradation. *Catalysts* **2022**, *12*, 1400. https://doi.org/10.3390/catal12111400

Academic Editors: Gassan Hodaifa, Rafael Borja and Mha Albqmi

Received: 28 September 2022
Accepted: 3 November 2022
Published: 9 November 2022

Publisher's Note: MDPI stays neutral with regard to jurisdictional claims in published maps and institutional affiliations.

1. Introduction

Since the last century, the environmental pollution caused by excessive discharge of organic pollutants in wastewater has led to the inadequacy of freshwater resources and threatened human health [1,2]. To solve these issues, photocatalysis could be regarded as a green and economical method. In fact, the rapid recombination rate of photogenerated electron-hole pairs and low solar light energy utilization efficiency significantly reduced the photocatalytic efficiency [3]. Compared with the prior photocatalysts containing ZnO and CdS, transition metal dichalcogenides (TMDCs) have received substantial interest due to their sandwich-like layered structure, strong photocatalytic activities, and favorable optical and electronic properties. WS$_2$ as a typical TMDC has been studied deeply due to its suitable bandgap, excellent stability, and nontoxic features [4,5]. However, pristine WS$_2$ has high specific surface energy and poor dispersion, and the spontaneous agglomeration of WS$_2$ with high specific surface energy always leads to a decrease in exposed active sites, which causes an unsatisfactory photocatalytic degradation effect. Thus, there has been little study on photocatalytic degradation of organic wastewater by pure WS$_2$, which has seriously hindered its practical application.

Fortunately, the development of supported WS$_2$ photocatalysts has become a feasible and promising approach. The active components dispersed on the surface of the support could increase the specific surface area and improve the catalytic efficiency of the active components per unit mass. A series of materials were combined with pure WS$_2$ to improve the photocatalytic performance. In previous reports, many carriers have been adopted to support WS_2, such as graphene [6], reduced graphene oxide [7], TiO_2 [8], CdS [9], ZnO [10], Bi_2MoO_6 [11], etc. However, the high fabrication cost and complex process made these carriers unsuitable for broad applications. Natural minerals have the advantages of low cost, environmental friendliness, special morphology, and unique properties, which make them a very promising candidate for carriers. To date, different mineral-based composites with highly dispersed active compounds have been successfully synthesized.

Mineral materials are widely available and have been reported as catalyst carriers. Sepiolite (Sep) is a kind of natural magnesium silicate clay mineral, which is abundant in nature. Ascribed to the 1D fibrous structure, it has a strong adsorption capacity and high specific surface area [12–15]. As a carrier, sepiolite can improve the dispersion of materials. In addition, the interface effect between sepiolite and the loaded material can also enhance the photocatalytic performance. Sep has attracted much attention in the field of adsorbent and functional carrier for wastewater treatment [16–19].

In our previous studies, we have successfully synthesized ultrathin MoS_2 nanosheets with the assistance of sepiolite and tourmaline [20,21]. Herein, a novel WS_2/Sep composite with few layered WS_2 nanosheets was fabricated via an environmentally friendly and simple solvothermal route. The ultrathin WS_2 nanosheets that uniformly grow on the surface of the Sep led to the exposure of more surface-active sites and the solution to agglomeration. Furthermore, the WS_2/Sep composite exhibited much higher photocatalytic efficiency toward rhodamine B (RhB) degradation than the pristine WS_2, and the synergistic effect between WS_2 and Sep was also noted. This work offers a new insight for low-cost preparation of dispersed 2D material using natural minerals.

2. Results and Discussion

2.1. Crystal Phase and Groups Analysis

XRD patterns were recorded to examine the phase of the prepared pure WS_2 and WS_2/Sep composite (Figure 1a). All of the diffraction peaks of both the pure WS_2 nanosheets and WS_2/Sep composite matched well with the 2H hexagonal phase (JCPDS Card No.87-2417) [22]. The XRD patterns showed that the main peaks of WS_2 of the WS_2/Sep were located at 2θ = 13.775°, 28.6989°, 34.248°, and 47.105°, corresponding to the (002), (004), (101), and (103) facets of WS_2, and the diffraction peaks at 7.423°, 11.982°, 20.785°, and 26.667° corresponded well to the (011), (031), (131), and (080) facets of the Sep, respectively. The d-spacing value of the composite corresponding to the peak at 13.775° was calculated to be 6.4 Å, which was close to that of the pure WS_2. The few broad bread-like peaks could be attributed to the partial crystallization of the WS_2 nanosheets. After the solvothermal treatment, WS_2 was evenly dispersed on the support, and the diffraction peaks corresponding to WS_2 were sharp and symmetric, further confirming the great crystallization of composite. Furthermore, the FT-IR was used to identify the functional groups of the WS_2/Sep sample (Figure 1b). The broad absorption peaks at around 3567 cm^{-1} and 1652 cm^{-1} were ascribed to the stretching vibration of the hydroxyl group of crystal water. The peaks at 1003, 975, 424, and 788 cm^{-1} were ascribed to the stretching of the Si-O band in the Si-O-Si groups of sepiolite. The peaks at 669, 688, and 645 cm^{-1} corresponded to the bending vibration of Mg_3OH. The peaks at 464 were attributed to Si-O-Al (octahedral) and Si-O-Si [23–25]. The two intensive peaks at around 2358 and 2119 cm^{-1} originated from the characteristic peaks of hexagonal WS_2. The results of XRD and FT-IR revealed the WS_2/Sep maintained the original crystal structure and surface functional groups of WS_2 and sepiolite.

2.2. Morphology and EDS Analysis

SEM, TEM, and HRTEM were used to observe the microstructure of the pure WS_2 and WS_2/Sep (Figure 2). The pure WS_2 with the shape of irregular plates consisted of WS_2 nanosheets stacked compactly (Figure 2a). In Figure 2b, one can observe the WS_2 nanosheets were uniformly anchored on the surface of Sep nanofibers, and a bark-like structure was formed intimately; the WS_2 nanosheets in the WS_2/Sep composite owned better dispersion. Compared with the pure WS_2 nanosheets, the mineral material (i.e., Sep) as a support of the composite could significantly reduce the agglomeration of the WS_2 nanosheets. Thus, more surface active sites of WS_2 could be exposed outside. The above results provided direct evidence that the WS_2 phase had been uniformly loaded on the Sep nanofibers.

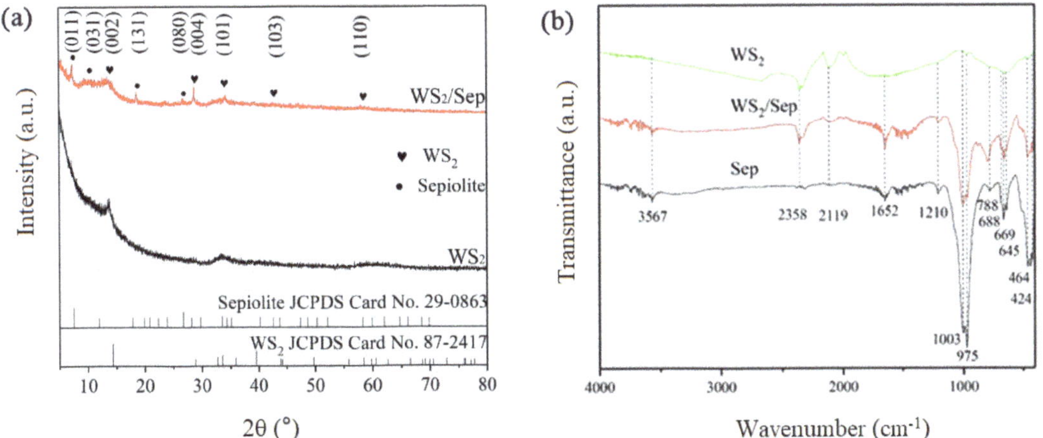

Figure 1. (**a**) XRD patterns of the pure WS_2 and WS_2/Sep nanocomposites obtained via solvothermal at 220 °C (**b**) and FT-IR spectra of the pure WS_2, sepiolite, and WS_2/Sep nanocomposites.

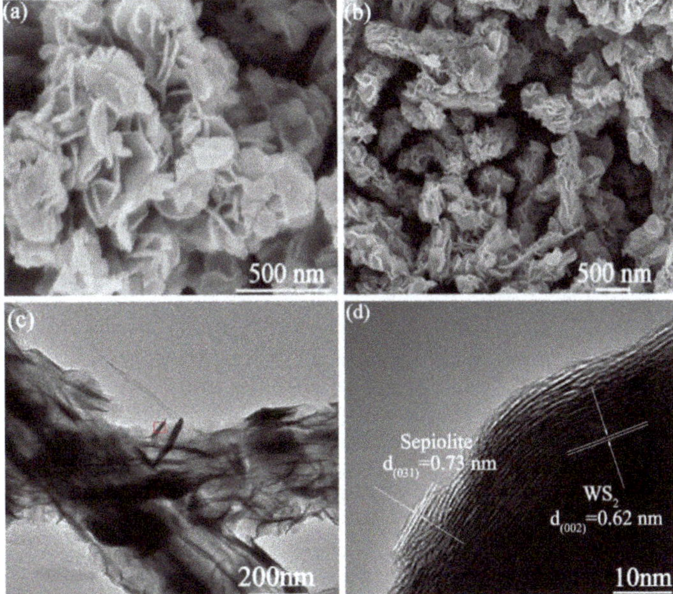

Figure 2. Morphology and structure of the pure WS_2 and WS_2/sepiolite nanocomposite. (**a,b**) SEM images of the pure WS_2 and WS_2/sepiolite nanocomposite obtained via solvothermal; (**c,d**) TEM and HRTEM images of WS_2/Sep.

TEM and HRTEM observations were adopted to deeply investigate the microstructure of the WS_2/Sep composite (Figure 2c,d), which further illustrated that the WS_2 phase had been loaded successfully on the surface of the Sep nanofibers. Moreover, the HRTEM images (Figure 2d) indicated a lattice space distance of 0.62 nm, which corresponded to the (002) facet of 2H-WS$_2$ [26,27]. Lattice spacing consistent with the d-spacing of Sep (031) was also detected. In addition, the elemental mapping images (Figure 3) also confirmed that WS$_2$ nanosheets were successfully assembled on the surface of Sep mineral.

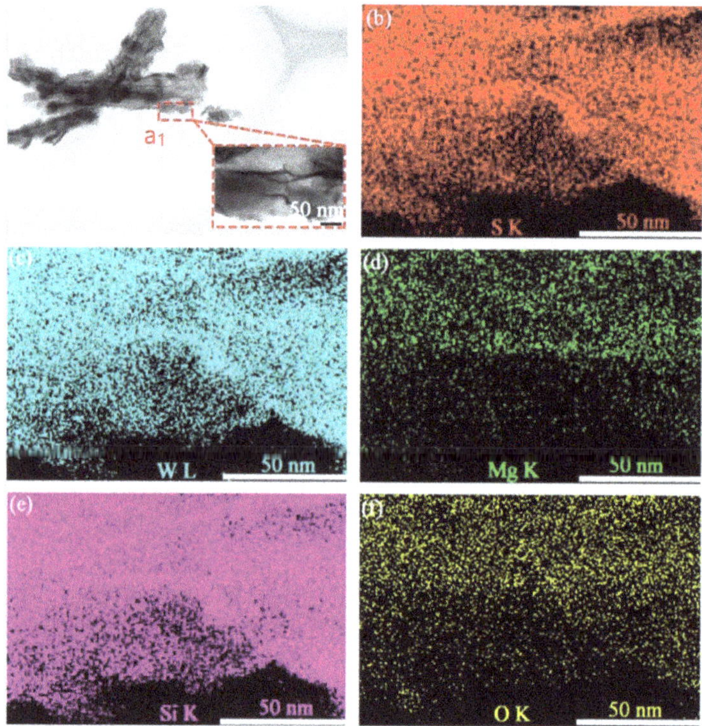

Figure 3. EDS mapping analysis of element distribution (**b**–**f**) in marked area (red rectangle in (**a**)) for the WS$_2$/Sep nanocomposite: S (**b**), W (**c**), Mg (**d**), Si (**e**), and O (**f**).

2.3. Nitrogen Adsorption-Desorption Analysis

Nitrogen adsorption-desorption isotherms and pore size distribution curves were adopted to further compare the surface area and pore structure between the WS$_2$/Sep composite and pure WS$_2$ (Figure 4). Both the samples presented Langmuir type IV isotherms with H3 hysteresis type loops, suggesting the existence of slit-like mesoporous structures because of the stacking of sheets. This was consistent with the electron microscopy images. The WS$_2$/Sep composite possessed a specific surface area of 45.9 m^2/g, which was much larger than that of the pure WS$_2$ (25.3 m^2/g). Meanwhile, the pore volume of the WS$_2$/Sep (0.107 cm^3/g) was also much larger than that of the WS$_2$ (0.048 cm^3/g). Consequently, the composite could expose more surface active sites and increase the number of mesopores, which tended toward improving catalytic activity [28].

2.4. Photocatalytic Performance of RhB Degradation

Comparative experiments were performed to evaluate the performance of the WS$_2$/Sep nanocomposite for RhB degradation, which is shown in Figure 5. It can be clearly seen that the WS$_2$/Sep composite exhibited much higher photocatalytic efficiency than the pure WS$_2$. After 150 min catalysis, the RhB degradation rate for the pure WS$_2$ only reached ~18%, while it achieved ca. 76% for the WS$_2$/Sep nanocomposites. The 4.2-fold improved activity could be mainly attributed to the better dispersion of the WS$_2$ nanosheets [20]. In addition, the excellent synergies between the WS$_2$ nanosheets and Sep nanofibers could also accelerate the photocatalysis, which will be further discussed in the following section.

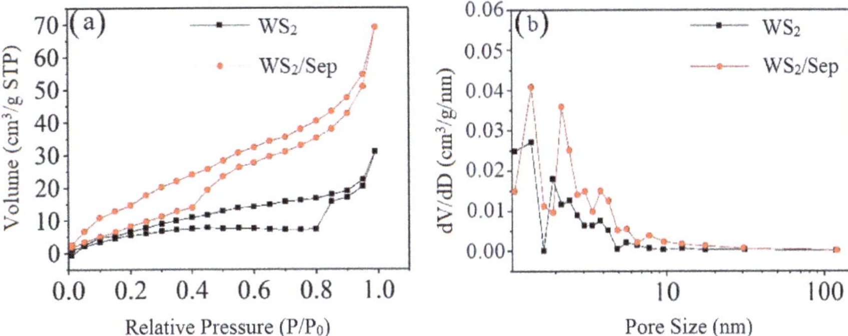

Figure 4. Nitrogen adsorption-desorption isotherms (**a**) and the corresponding pore size distribution curves (**b**) of WS$_2$ and WS$_2$/sepiolite composite.

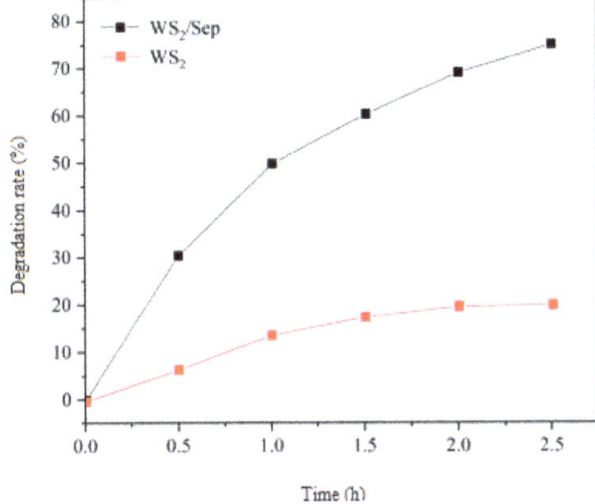

Figure 5. Photocatalytic activity of the WS$_2$/Sep composite and pure WS$_2$.

2.5. The Possible RhB Degradation Mechanism for the WS$_2$/Sep Composite

Figure 6 shows the possible process of the photocatalytic RhB degradation in which the synergic effect between the Sep and the grown WS$_2$ in the WS$_2$/sepiolite nanocomposite is illustrated. Under irradiation, the holes (h$^+$) and electrons (e$^-$) were produced. Then, the dissolved O$_2$ would react with e$^-$ and OH$-$ would react with h$^+$ to generate superoxide anion radical ($\cdot$O$_2{}^-$) and hydroxyl radical ($\cdot$OH), respectively [29,30]. Finally, the RhB molecules adsorbed on the surface-active sites of the WS$_2$ were degraded to CO$_2$ and H$_2$O [31,32]. During this process, the excellent adsorption of Sep was conductive to transferring the dissolved O$_2$ and RhB molecules to the active sites of the WS$_2$, and the abundant hydroxyl (–OH) groups on the Sep surface also benefited the production of $\cdot$OH [33–35]. Thus, in addition to the improved dispersion of WS$_2$, the synergic effects provided by the Sep during photocatalysis also enabled the WS$_2$/Sep composite to exhibit much higher photocatalytic activity than the pure WS$_2$ [36]. Furthermore, the presence of mesopores favored multilight scattering, resulting in enhanced harvesting of the exciting light, and accordingly, improved photocatalytic activity [37,38]. These mesopores also facilitated fast mass transport and thus enhanced the performance [39,40].

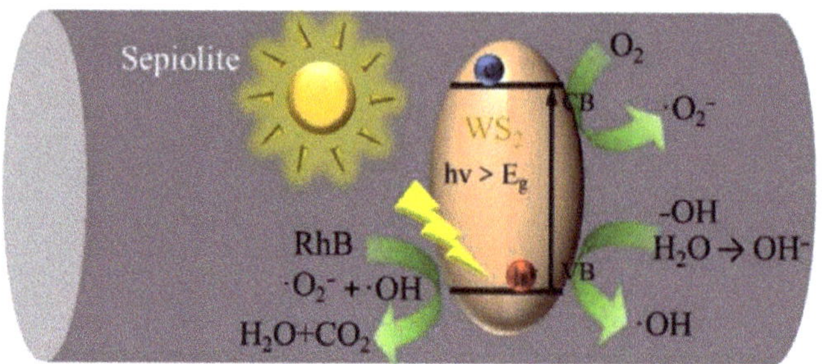

Figure 6. Possible RhB degradation mechanism of the WS$_2$/Sep composite.

3. Experiment

3.1. Chemicals and Reagents

The Sep was provided by LB Nanomaterials Technology Co., Ltd. (Henan, China), and the chemical analysis of the Sep was determined as SiO_2 of 53.56 wt%, MgO of 36.79 wt%, CaO of 5.53 wt%, Fe_2O_3 of 1.17 wt%, and other impurities of 2.95 wt%. The ammonium tetrathiotungstate ($H_8N_2S_4W$) and dimethylformamide (DMF) were supplied by Sigma-Aldrich Co., Ltd (Shanghai, China). The (RhB) was purchased from Kewei Chemical Group Co., Ltd. (Tianjin, China). All chemicals used in the experiments were of analytical grade. These materials were used without further purification. Deionized (DI) water was used in all experiments.

3.2. Synthesis of WS$_2$/Sep Nanocomposite

The WS$_2$/Sep nanocomposite was fabricated by a solvothermal method (Figure 7). A total of 30 mg of $H_8N_2S_4W$ was dissolved in 15 mL of *N,N*-dimethylformamide (DMF) and stirred for 0.5 h. Then, 10 mg of Sep powder was added into the solution and stirred continuously for 0.5 h. Next, the above mixture was sonicated for 10 min. Later, the above suspension was transferred into a 25 mL Teflon-sealed autoclave and heated to 220 °C for 24 h. After cooling down to room temperature, the final product was obtained by filtration, washed several times with DI water, and dried in a vacuum oven at 80 °C for 12 h. The preparation process of the pure WS$_2$ was similar to that of the WS$_2$/Sep nanocomposite but without the addition of Sep.

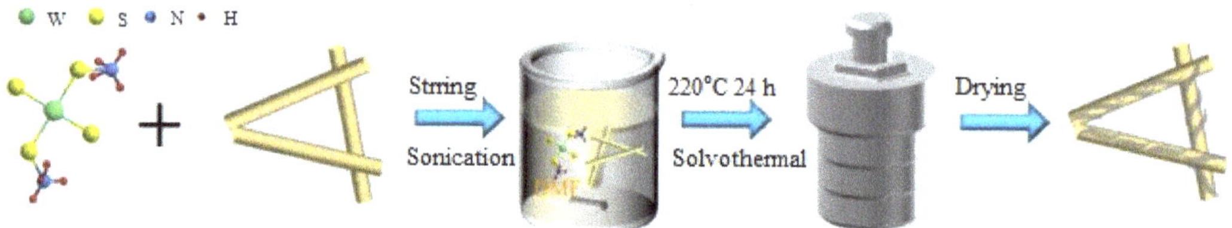

Figure 7. Schematic illustration for the synthesis process of WS$_2$/Sep composite.

3.3. Physicochemical Characterizations of the Synthesized WS$_2$/Sep Composite

X-ray powder diffraction (XRD) analysis was performed with the working conditions of Cu K$_\alpha$ radiation (λ = 1.54 Å), 40 mA, and 40 kV on a Smart Lab 9 KW X-ray diffractometer from Rigaku (Tokyo, Japan). Fourier-transform infrared spectroscopy (FTIR) spectra of the samples were recorded in a transmission mode from 400 to 4000 cm^{-1} on a Tensor II Fourier transform infrared spectrometer manufactured by Bruker (Saarbrucken, Germany). The morphologies of the as-synthesized samples were observed by using a S-4800 scanning electron microscope working at 5 kV, from JEOL (Tokyo, Japan). Transmission electron microscope (TEM) images, high-resolution TEM (HRTEM) images, and energy-dispersive

X-ray spectroscopy (EDS) were carried out on a JEM 2100F transmission electron microscope, from JEOL (Tokyo, Japan), with an accelerating voltage of 200 kV. The N_2 adsorption-desorption tests were conducted on Autosorb-iQ2, from Quantachrome (Boynton Beach, FL, USA). All samples were outgassed in nitrogen flow at 200 °C for 4 h before measurements. The specific surface area (SSA) was evaluated using the Brunauer-Emmett-Teller (BET) method, and the total pore volume was calculated at the relative pressure of approximately 0.99. The pore size distribution was computed using the Barrett-Joyner-Halenda (BJH) method.

3.4. Photocatalytic Performance Tests

Photocatalytic activity of the as-prepared samples was assessed by the reduction of RhB in aqueous solutions, using a 500 W Xe lamp equipped with a cut-off filter ($\lambda > 420$ nm). In a typical evaluation procedure, 20 mg of sample was added into 100 mL of RhB solution with a concentration of 20 mg/L at room temperature [41]. The suspension solution stirred vigorously in the dark for 0.5 h to reach the equilibrium of adsorption/desorption before visible light irradiation. At an interval of 0.5 h, 5.6 mL of the suspension was collected and centrifuged for absorbance analysis. The concentration of dye solution was analyzed by recording the UV–vis spectra at wavelength of 553 nm using a Shimadzu UV-1800 spectrophotometer from Shimadzu (Shimane, Japan).

4. Conclusions

In summary, a novel WS_2/Sep composite as a photocatalyst was successfully prepared via a facile solvothermal method. The physicochemical characterization results confirmed that the bark-like WS_2 nanosheets uniformly grew on the Sep nanofibers, which led to a larger surface area and more exposed active sites. In a typical photocatalytic application, the as-synthesized WS_2/Sep exhibited considerably improved photocatalytic performance for RhB degradation over the pure WS_2, which could be mainly attributed to the better dispersion of WS_2 nanosheets and the synergistic effect between the WS_2 nanosheets and the Sep nanofibers' support during the photocatalysis. This work is believed to provide new ideas for the low-cost batch preparation of high-quality two-dimensional materials based on natural minerals. However, further investigation of the as-developed composite photocatalyst with respect to reusability and the versatility for the degradation of other organic pollutants are still needed to see if there are any limitations to its practical applications.

Author Contributions: Conceptualization, F.W. and J.B.; methodology, F.W. and J.B.; software, J.B. and K.C.; validation, F.W. and J.B.; formal analysis, F.W. and J.B.; investigation, J.B. and K.C.; resources, F.W.; data curation, J.B. and F.W.; writing—original draft preparation, J.B.; K.C.; X.X., B.F.; and F.W.; writing—review and editing, F.W. and B.F.; visualization, F.W. and B.F.; supervision, F.W.; project administration, F.W.; funding acquisition, F.W. All authors have read and agreed to the published version of the manuscript.

Funding: This work was financially supported by the National Key R&D Program of China (No. 2021YFC1910605), the National Natural Science Foundation of China (No. 51874115), the Introduced Overseas Scholars Program of Hebei province, China (No. C201808), and the Excellent Young Scientist Foundation of Hebei province, China (No. E2018202241).

Data Availability Statement: Data will be available upon request from the corresponding authors.

Acknowledgments: Authors acknowledge the financial support by the National Key R&D Program of China (No. 2021YFC1910605), the National Natural Science Foundation of China (No. 51874115), the Introduced Overseas Scholars Program of Hebei province, China (No. C201808), and the Excellent Young Scientist Foundation of Hebei province, China (No. E2018202241).

Conflicts of Interest: The authors declare no conflict of interest.

References

1. Yahya, N.; Aziz, F.; Jamaludin, N.A.; Mutalib, M.A.; Ismail, A.F.; Salleh, W.N.W.; Jaafar, J.; Yusof, N.; Ludin, N.A. A review of integrated photocatalyst adsorbents for wastewater treatment. *J. Environ. Chem. Eng.* **2018**, *6*, 7411–7425. [CrossRef]
2. Yaqoob, A.A.; Parveen, T.; Umar, K.; Mohamad Ibrahim, M.N. Role of nanomaterials in the treatment of wastewater: A review. *Water* **2020**, *12*, 495. [CrossRef]
3. Liao, G.; Li, C.; Li, X.; Fang, B. Emerging polymeric carbon nitride Z-scheme systems for photocatalysis. *Cell Rep. Phys. Sci.* **2021**, *2*, 100355. [CrossRef]
4. Cong, C.; Shang, J.; Wang, Y.; Yu, T. Optical properties of 2D semiconductor WS_2. *Adv. Opt. Mater.* **2018**, *6*, 1700767. [CrossRef]
5. Manzeli, S.; Ovchinnikov, D.; Pasquier, D.; Yazyev, O.V.; Kis, A. 2D transition metal dichalcogenides. *Nat. Rev. Mater.* **2017**, *2*, 17033. [CrossRef]
6. Yuan, L.; Chung, T.F.; Kuc, A.; Wan, Y.; Xu, Y.; Chen, Y.P.; Heine, T.; Huang, L. Photocarrier generation from interlayer charge-transfer transitions in WS_2-graphene heterostructures. *Sci. Adv.* **2018**, *4*, e1700324. [CrossRef]
7. Wang, X.; Gu, D.; Li, X.; Lin, S.; Zhao, S.; Rumyantseva, M.; Gaskov, A. Reduced graphene oxide hybridized with WS_2 nanoflakes based heterojunctions for selective ammonia sensors at room temperature. *Sens. Actuators B Chem.* **2019**, *282*, 290–299. [CrossRef]
8. Lu, Z.; Cao, Z.; Hu, E.; Hu, K.; Hu, X. Preparation and tribological properties of WS_2 and WS_2/TiO_2 nanoparticles. *Tribol. Int.* **2019**, *130*, 308–316. [CrossRef]
9. Su, L.; Luo, L.; Song, H.; Wu, Z.; Tu, W.; Wang, Z.J.; Ye, J. Hemispherical shell-thin lamellar WS_2 porous structures composited with CdS photocatalysts for enhanced H_2 evolution. *Chem. Eng. J.* **2020**, *388*, 124346. [CrossRef]
10. Pataniya, P.M.; Late, D.; Sumesh, C.K. Photosensitive WS_2/ZnO nano-heterostructure-based electrocatalysts for hydrogen evolution reaction. *ACS Appl. Energy Mater.* **2021**, *4*, 755–762. [CrossRef]
11. Gao, J.; Liu, C.; Wang, F.; Jia, L.; Duan, K.; Liu, T. Facile synthesis of heterostructured WS_2/Bi_2MoO_6 as high-performance visible-light-driven photocatalysts. *Nanoscale Res. Lett.* **2017**, *12*, 377. [CrossRef] [PubMed]
12. Chen, Q.; Zhu, R.; Liu, S.; Wu, D.; Fu, H.; Zhu, J.; He, H. Self-templating synthesis of silicon nanorods from natural sepiolite for high-performance lithium-ion battery anodes. *J. Mater. Chem. A* **2018**, *6*, 6356–6362. [CrossRef]
13. Li, Z.; Gómez-Avilés, A.; Sellaoui, L.; Bedia, J.; Bonilla-Petriciolet, A.; Belver, C. Adsorption of ibuprofen on organo-sepiolite and on zeolite/sepiolite heterostructure: Synthesis, characterization and statistical physics modeling. *Chem. Eng. J.* **2019**, *371*, 868–875. [CrossRef]
14. Santos, S.C.R.; Boaventura, R.A.R. Adsorption of cationic and anionic azo dyes on sepiolite clay: Equilibrium and kinetic studies in batch mode. *J. Environ. Chem. Eng.* **2016**, *4*, 1473–1483. [CrossRef]
15. Zhang, J.; Yan, Z.; Ouyang, J.; Yang, H.; Chen, D. Highly dispersed sepiolite-based organic modified nanofibers for enhanced adsorption of Congo red. *Appl. Clay Sci.* **2018**, *157*, 76–85. [CrossRef]
16. Ma, Y.; Zhang, G. Sepiolite nanofiber-supported platinum nanoparticle catalysts toward the catalytic oxidation of formaldehyde at ambient temperature: Efficient and stable performance and mechanism. *Chem. Eng. J.* **2016**, *288*, 70–78. [CrossRef]
17. Selvitepe, N.; Balbay, A.; Saka, C. Optimisation of sepiolite clay with phosphoric acid treatment as support material for CoB catalyst and application to produce hydrogen from the $NaBH_4$ hydrolysis. *Int. J. Hydrogen Energy* **2019**, *44*, 16387–16399. [CrossRef]
18. Xu, X.; Chen, W.; Zong, S.; Ren, X.; Liu, D. Atrazine degradation using Fe_3O_4-sepiolite catalyzed persulfate: Reactivity, mechanism and stability. *J. Hazard. Mater.* **2019**, *377*, 62–69. [CrossRef]
19. Zhou, F.; Yan, C.; Sun, Q.; Komarneni, S. TiO_2/Sepiolite nanocomposites doped with rare earth ions: Preparation, characterization and visible light photocatalytic activity. *Micropor. Mesopor. Mater.* **2019**, *274*, 25–32. [CrossRef]
20. Wang, F.; Hao, M.; Liu, W.; Yan, P.; Fang, B.; Li, S.; Liang, J.; Zhu, M.; Cui, L. Low-cost fabrication of highly dispersed atomically-thin MoS_2 nanosheets with abundant active Mo-terminated edges. *Nano Mater. Sci.* **2021**, *3*, 205–212. [CrossRef]
21. Hao, M.; Li, H.; Cui, L.; Liu, W.; Fang, B.; Liang, J.; Xie, X.; Wang, D.; Wang, F. Higher photocatalytic removal of organic pollutants using pangolin-like composites made of 3–4 atomic layers of MoS_2 nanosheets deposited on tourmaline. *Environ. Chem. Lett.* **2021**, *19*, 3573–3582. [CrossRef]
22. Sun, Y.; Wu, J.; Zhang, L. Fabrication of Ag-WS_2 composites with preferentially oriented WS_2 and its anisotropic tribology behavior. *Mater. Lett.* **2020**, *260*, 126975. [CrossRef]
23. Arabkhani, P.; Javadian, H.; Asfaram, A.; Sadeghfar, F.; Sadegh, F. Synthesis of magnetic tungsten disulfide/carbon nanotubes nanocomposite (WS_2/Fe_3O_4/CNTs-NC) for highly efficient ultrasound-assisted rapid removal of amaranth and brilliant blue FCF hazardous dyes. *J. Hazard. Mater.* **2021**, *420*, 126644. [CrossRef] [PubMed]
24. Uğurlu, M.; Karaoğlu, M.H. TiO_2 supported on sepiolite: Preparation, structural and thermal characterization and catalytic behaviour in photocatalytic treatment of phenol and lignin from olive mill wastewater. *Chem. Eng. J.* **2011**, *166*, 859–867. [CrossRef]
25. Zhang, Y.; Wang, D.; Zhang, G. Photocatalytic degradation of organic contaminants by TiO_2/sepiolite composites prepared at low temperature. *Chem. Eng. J.* **2011**, *173*, 1–10. [CrossRef]
26. Anto Jeffery, A.; Nethravathi, C.; Rajamathi, M. Two-dimensional nanosheets and layered hybrids of MoS_2 and WS_2 through exfoliation of ammoniated MS_2 (M = Mo, W). *J. Phys. Chem. C* **2014**, *118*, 1386–1396. [CrossRef]

27. Yang, J.; Voiry, D.; Ahn, S.J.; Kang, D.; Kim, A.Y.; Chhowalla, M.; Shin, H.S. Two-dimensional hybrid nanosheets of tungsten disulfide and reduced graphene oxide as catalysts for enhanced hydrogen evolution. *Angew. Chem. Int. Ed.* **2013**, *52*, 13751–13754. [CrossRef]

28. Liu, C.; Chai, B.; Wang, C.; Yan, J.; Ren, Z. Solvothermal fabrication of MoS_2 anchored on $ZnIn_2S_4$ microspheres with boosted photocatalytic hydrogen evolution activity. *Int. J. Hydrogen Energy* **2018**, *43*, 6977–6986. [CrossRef]

29. Wu, Y.; Liu, Z.; Li, Y.; Chen, J.; Zhu, X.; Na, P. Construction of 2D-2D TiO_2 nanosheet/layered WS_2 heterojunctions with enhanced visible-light-responsive photocatalytic activity. *Chin. J. Catal.* **2019**, *40*, 60–69. [CrossRef]

30. Feng, Z.; Zeng, L.; Chen, Y.; Ma, Y.; Zhao, C.; Jin, R.; Lu, Y.; Wu, Y.; He, Y. In situ preparation of Z-scheme MoO_3/g-C_3N_4 composite with high performance in photocatalytic CO_2 reduction and RhB degradation. *J. Mater. Res.* **2017**, *32*, 3660–3668. [CrossRef]

31. Baral, A.; Das, D.; Minakshi, M.; Ghosh, M.; Padhi, D. Probing environmental remediation of RhB organic dye using α-MnO_2 under visible-light irradiation: Structural, photocatalytic and mineralization studies. *ChemistrySelect* **2016**, *1*, 4277–4285. [CrossRef]

32. Zhang, Q.; Tai, M.; Zhou, Y.; Zhou, Y.; Wei, Y.; Tan, C.; Wu, Z.; Li, J.; Lin, H. Enhanced Photocatalytic Property of γ-$CsPbI_3$ Perovskite Nanocrystals with WS_2. *ACS Sustain. Chem. Eng.* **2019**, *8*, 1219–1229. [CrossRef]

33. Elangovan, E.; Sivakumar, T.; Brindha, A.; Thamaraiselvi, K.; Sakthivel, K.; Kathiravan, K.; Aishwarya, S. Visible Active N-Doped TiO_2/WS_2 Heterojunction Nano Rods: Synthesis, Characterization and Photocatalytic Activity. *J. Nanosci. Nanotechnol.* **2019**, *19*, 4429–4437. [CrossRef] [PubMed]

34. Fu, S.; Yuan, W.; Liu, X.; Yan, Y.; Liu, H.; Li, L.; Zhao, F.; Zhou, J. A novel 0D/2D WS_2/BiOBr heterostructure with rich oxygen vacancies for enhanced broad-spectrum photocatalytic performance. *J. Colloid. Interface Sci.* **2020**, *569*, 150–163. [CrossRef]

35. Xiang, Q.; Cheng, F.; Lang, D. Hierarchical Layered WS_2/Graphene-Modified CdS Nanorods for Efficient Photocatalytic Hydrogen Evolution. *ChemSusChem* **2016**, *9*, 996–1002. [CrossRef] [PubMed]

36. Liu, X.; Xing, Z.; Zhang, Y.; Li, Z.; Wu, X.; Tan, S.; Yu, X.; Zhu, Q.; Zhou, W. Fabrication of 3D flower-like black N-TiO_{2-x}@MoS_2 for unprecedented-high visible-light-driven photocatalytic performance. *Appl. Catal. B Environ.* **2017**, *201*, 119–127. [CrossRef]

37. Liu, Y.; Shen, S.; Li, Z.; Ma, D.; Xu, G.; Fang, B. Mesoporous g-C_3N_4 nanosheets with improved photocatalytic performance for hydrogen evolution. *Mater. Charact.* **2021**, *174*, 111031. [CrossRef]

38. Liu, Y.; Xu, G.; Ma, D.; Li, Z.; Yan, Z.; Xu, A.; Zhong, W.; Fang, B. Synergistic effects of g-C_3N_4 three-dimensional inverse opals and Ag modification on high-efficiency photocatalytic H_2 evolution. *J. Clean. Prod.* **2021**, *328*, 129745. [CrossRef]

39. Fang, B.; Kim, J.; Kim, M.; Yu, J. Hierarchical nanostructured carbons with meso-macroporosity: Design, characterization and applications. *Acc. Chem. Res.* **2013**, *46*, 1397–1406. [CrossRef] [PubMed]

40. Fang, B.; Daniel, L.; Bonakdarpour, A.; Govindarajan, R.; Sharman, J.; Wilkinson, D. Dense Pt nanowire electrocatalysts for improved fuel cell performance using a graphitic carbon nitride-decorated hierarchical nanocarbon support. *Small* **2021**, *17*, 2102288. [CrossRef]

41. Ji, H.; Fei, T.; Zhang, L.; Yan, J.; Fan, Y.; Huang, J.; Song, Y.; Man, Y.; Tang, H.; Xu, H.; et al. Synergistic effects of MoO_2 nanosheets and graphene-like C_3N_4 for highly improved visible light photocatalytic activities. *Appl. Surf. Sci.* **2018**, *457*, 1142–1150. [CrossRef]

 *catalysts*

Review

Chalcogenides and Chalcogenide-Based Heterostructures as Photocatalysts for Water Splitting

Mohammad Mansoob Khan *[ID] and Ashmalina Rahman

Chemical Sciences, Faculty of Science, Universiti Brunei Darussalam, Jalan Tungku Link, Gadong BE1410, Brunei
* Correspondence: mmansoobkhan@yahoo.com or mansoob.khan@ubd.edu.bn

Abstract: Chalcogenides are essential in the conversion of solar energy into hydrogen fuel due to their narrow band gap energy. Hydrogen fuel could resolve future energy crises by substituting carbon fuels owing to zero-emission carbon-free gas and its eco-friendliness. The fabrication of different metal chalcogenide-based photocatalysts with enhanced photocatalytic water splitting have been summarized in this review. Different modifications of these chalcogenides, including coupling with another semiconductor, metal loading, and doping, are fabricated with different synthetic routes that can remarkably improve the photo-exciton separation and have been extensively investigated for photocatalytic hydrogen generation. In this direction, this review is undertaken to provide an overview of the enhanced photocatalytic performance of the binary and ternary chalcogenide heterostructures and their mechanisms for hydrogen production under irradiation of light.

Keywords: semiconductors; chalcogenides; photocatalysts; photocatalysis; water splitting

Citation: Khan, M.M.; Rahman, A. Chalcogenides and Chalcogenide-Based Heterostructures as Photocatalysts for Water Splitting. *Catalysts* **2022**, *12*, 1338. https://doi.org/10.3390/catal12111338

Academic Editors: Gassan Hodaifa, Rafael Borja, Mha Albqmi and Haralampos N. Miras

Received: 19 September 2022
Accepted: 25 October 2022
Published: 1 November 2022

Publisher's Note: MDPI stays neutral with regard to jurisdictional claims in published maps and institutional affiliations.

1. Introduction

Metal oxides and chalcogenides are semiconductors that possess conductivities between those of conductors and insulators and these materials are widely researched semiconductor materials. Due to their stability, resistance to photo-corrosion, non-toxicity, and inexpensive preparation, and the metal oxides such as TiO_2, ZnO, and SnO_2 are commonly applied as photocatalysts [1–4]. Despite these favorable characteristics, a major drawback of using these metal oxides as photocatalysts is their large band gap energies (≥ 3 eV) [5]. Such wide band gaps only enable them to absorb the ultraviolet portion of the solar spectrum (~4% of the total solar energy), which restricts widespread practical applications [2–4]. Many strategies have been devised to prepare metal oxides with narrower band gaps for improved harvesting of visible light [4,6,7]. However, the scope for band gap tuning by doping and other strategies is limited, and much larger decreases in band gap can be achieved by moving outside the oxide compositional space.

Chalcogenides, compounds comprising one or more electropositive elements and at least one chalcogen ion (S^{2-}, Se^{2-}, or Te^{2-}) [8], are well known for their narrow band gap energies [9,10]. These compounds continue to attract attention due to their numerous desirable properties, including narrow band gap energy, low toxicity, biocompatibility, low cost, and facile synthesis [11,12]. The utilization of chalcogenides and chalcogenide-based semiconductor materials in photocatalytic applications has been reported widely, mainly because of their narrow band gap energies that enable efficient harvesting of visible light [13].

Owing to the devastating impact of conventional fossil fuel use on the environment and the growing demand for energy, an increasing number of countries, companies, and researchers are considering hydrogen (H_2) energy as a potential solution to the pressing environmental and sustainability problems associated with the current global energy system [14]. H_2 is an alternative source of energy because they are clean, non-toxic, eco-friendly, and renewable. Unlike other alternative energy sources, H_2 has the advantage of producing only water as a by-product upon consumption, which does not cause any pollution or release greenhouse gases [15]. Moreover, the potential clean and green energy source H_2 as a replacement for fossil fuels can be produced in large quantities using simple and low-cost methods [16]. However, H_2 in nature exists in combination with other elements, such as oxygen in water, thus, methods to obtain pure hydrogen should be considered. The production of H_2 employing visible light is one of the most promising approaches for sustainable energy and environmental problems.

Water splitting is a process whereby water molecules are separated into oxygen and hydrogen. The equation for the chemical reaction is as follows:

$$2H_2O \rightarrow 2H_2 + O_2$$

Since pure water in nature does not absorb solar energy readily, a photocatalyst is required for the water-splitting process [17]. The reaction takes place in three steps as shown in Figure 1: step (1) to generate photoexcited electron-hole (e^-/h^+) pairs, the photocatalyst absorbs incoming photons with energies greater than the band gap of the material, step (2) the photogenerated carriers separate and migrate to the catalyst surface, and step (3) surface-adsorbed species (protons, H_2O molecules, and intermediates) are reduced to form hydrogen and oxidized to form oxygen by photogenerated electrons and holes, respectively [18]. According to Maeda and co-workers, the first and second steps depend on the electronic structure and properties of the semiconductor used. In general, high crystallinity promotes water-splitting activity because the density of defects acting as recombination centers is low [19]. In contrast, the third step occurs in the presence of a solid co-catalyst, which is usually a noble metal such as platinum or rhodium, or a metal oxide such as nickel oxide. The loading of the co-catalyst onto the photocatalyst surface provides active sites and lowers the activation energy required for H_2/O_2 gas evolution.

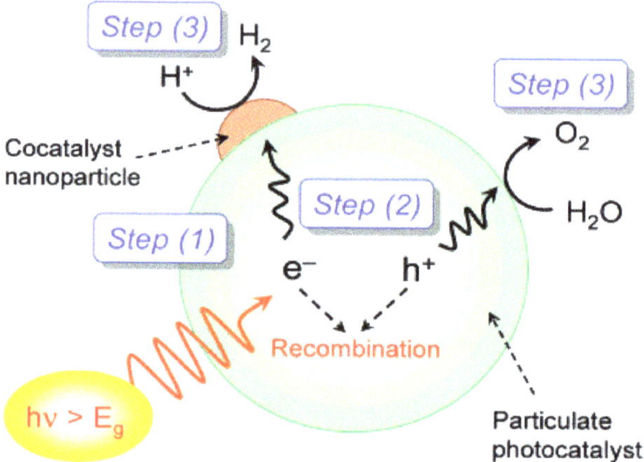

Figure 1. Overall steps involved in the photocatalytic water-splitting process [18].

There are several factors affecting the photocatalytic H_2O splitting as shown in Figure 2. The solar-to-hydrogen conversion efficiency of a material is mainly based on (1) light absorption, (2) e^-/h^+ photogeneration, and (3) separation and migration of e^-/h^+. There are a few principles to keep in mind, including the band gap energy, stability, the use of sacrificial reagents, solution pH, band structure and alignment, effective mass, carrier transport, crystallinity, and surface area of the material used.

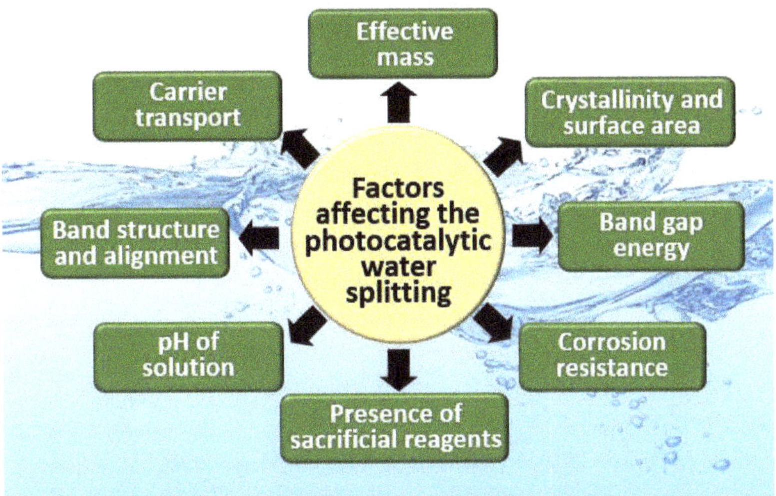

Figure 2. Factors affecting the photocatalytic water-splitting process.

According to Maeda and co-workers, photocatalysts for water splitting under visible-light irradiation must satisfy two requirements; the band gap energy should fall in the range of 1.23 eV (the thermodynamic minimum) to 3.0 eV (the edge of the visible spectrum), and the conduction and valence band edges should align with the H^+/H_2 and H_2O/O_2 redox potentials [19]. In practice, the lower end of the band gap range must be considerably larger than 1.23 eV, perhaps as high as ~2.4 eV, to account for entropic losses in the photocatalyst as well as concentration and kinetic over potential losses. Apart from that, the electron and hole-effective masses are among the key factors for efficient H_2O splitting. For a given mean free time between collisions, a smaller effective mass translates into larger carrier mobility and thus faster charge transport to the semiconductor–liquid interface. In general, the effective mass depends on the crystallographic direction, and therefore certain directions are more favorable for charge transport.

When selecting the material for H_2O splitting, the band edge position is crucial. Their VB should be more positive than the O_2 generation potential, whereas the CB should be more negative than the H_2 generation potential [20]. In addition to possessing a suitable band gap and band edge alignment, photocatalytic materials must have adequate light absorption in the visible region, which accounts for ~43% of the solar energy incident on the earth. The crystallinity and surface-active sites of the selected materials are also important for e^-/h^+ transfer and oxygen evolution reaction (OER)/hydrogen evolution reaction (HER), respectively. Materials with high crystallinity and few defects can reduce the e^-/h^+ recombination. The rate of OER/HER reactions should be larger than backward reactions, which is mostly depending on the number of active sites. Moreover, the material chosen should be stable either in acidic or basic electrolytes.

The present review provides an overview of recent literature on chalcogenides and chalcogenide-based heterostructures, with an emphasis on their synthesis and application for H_2 production by photocatalytic water splitting and the associated mechanisms. The main developments in binary and ternary chalcogenides of different metals including cadmium, copper, gallium, molybdenum, tin, titanium, vanadium, and zinc as well as their heterostructures are summarized. The relevant geometric structures and band structures of these chalcogenides as well as the key optical, electrical, and photocatalytic properties are also highlighted. Finally, the prospects of chalcogenide-based photocatalysts are discussed, and some general conclusions are drawn. The following sections cover the attributes of various chalcogenides and chalcogenide-based heterostructures that have recently been applied to photocatalytic water splitting [19,21].

2. General Synthesis Approaches of Chalcogenides

Although various synthesis methods of chalcogenides have been reported, synthesizing chalcogenides with high stability and to be photocatalytically active simultaneously is very challenging. Appropriate temperature, pressure, metal precursors, source of sulfur, and solvent combinations are important to synthesize these materials. Researchers have developed various synthesis routes for the fabrication of chalcogenides including hydrothermal [22], hot injection [23], solvothermal [24], and sol-gel [25]. Table 1 shows the different synthesis methods of chalcogenides. Gu et al. have synthesized $AgInS_2$ using a low-temperature liquid method and doped it with Mn^{2+} for H_2O splitting [26]. They yielded flower-like $AgInS_2$ spheres with sizes in the range of 200−800 nm. While the particles of Mn-doped $AgInS_2$ (1:100) are spherical with uniform particle sizes of about 300–400 nm. The calculated band gap energies of $AgInS_2$ and Mn-doped $AgInS_2$ (1:100) were found to be 1.63 and 1.52 eV, respectively. The H_2 production rate of $AgInS_2$ is 53 μmol $h^{-1}g^{-1}$. After Mn doping, the H_2 production rate of Mn-doped $AgInS_2$ (1:100) increases slightly, reaching 73 μmol $h^{-1}g^{-1}$. The hydrogen generation of Mn-doped $AgInS_2$ (1:100) was enhanced by loading multi-walled carbon nanotubes, reaching 105 μmol $h^{-1}g^{-1}$, which is ~2 times of $AgInS_2$ and ~1.4 times of Mn-doped $AgInS_2$ (1:100).

Hydrothermal involves a chemical reaction in an aqueous solution at elevated temperature and high pressure. For instance, Kannan et al. have hydrothermally synthesized $Cd_{0.5}Zn_{0.5}S$ using $Cu(NO_3)_2 \cdot 3H_2O$, $Na_2WO_4 \cdot 2H_2O$ and L-Cysteine as the starting materials at different temperatures (120, 140, 160, 180, and 200 °C) for 24 h [22]. Different morphologies, including flake-like and sponge-like examples of $Cd_{0.5}Zn_{0.5}S$, were observed. Moreover, they found that Cu_2WS_4 synthesized at the high temperature of 180 °C and exhibited excellent antibacterial activity against different bacterial strains. This shows that precise control over the reaction conditions, including pressure, time, pH value, concentration, and temperature are necessary for the successful synthesis of chalcogenides. In another study, Yang et al. have successfully fabricated NiS_2 using the hydrothermal method and $Ni(NO_3)_3 \cdot 6H_2O$ as Ni source and thioacetamide as the source of sulfur.

Solvothermal is similar to the hydrothermal method, but it uses a non-aqueous solvent (solvent other than water). For example, Wang et al. have used different $Cd(CH_3COO)_2 \cdot 2H_2O$ and sulfur powder ratios (1:1, 4:5, 2:3, 4:7, 1:2, and 1:3), precursor concentrations (1.5 and 6 mmol), temperatures (100, 140, 180, and 220 °C), and reaction times (4, 24, and 60 h) to yield CdS with different sizes and morphologies using solvothermal method [27]. The authors reported that the morphology, phase composition, and crystallinity of CdS were highly influenced by the precursor ratio, precursor concentration, temperature, and reaction time. The Cd concentration regulated the morphology of CdS morphology due to its effect on the formation of shape-determinant CdS nuclei. Nanorods, multipods, and triangular-like shape CdS nanocrystals were obtained with 1.5, 3, and 6 mmol of Cd, respectively. The arm diameter in CdS multipods increases from 10 to 60 nm with the sulfur concentration increasing when the cadmium concentration is maintained at 3 mmol.

Table 1. Different preparation methods of chalcogenides.

Synthesized Materials	Synthesis Method	Metal Precursor Used	Source of Sulfur	Solvent Used	Ref.
$AgInS_2$	Low-temperature liquid method	$AgNO_3$ $In(NO_3)_3 \cdot xH_2O$	Thioglycollic acid Thioacetamide	Water	[26]
$AgInS_2$	Microwave	$AgNO_3$ $In(NO_3)_3 \cdot 4.5H_2O$	Sulfur powder	Glycerol, Oleic acid, Oleylamine, 1-dodecanethiol and 1-octadecene	[28]
$Cd_{0.5}Zn_{0.5}S$	Hydrothermal	$Cd(CH_3COO)_2 \cdot 2H_2O$ $Zn(CH_3COO)_2 \cdot 2H_2O$	$Na_2S \cdot 9H_2O$	Water	[29]

Table 1. *Cont.*

Synthesized Materials	Synthesis Method	Metal Precursor Used	Source of Sulfur	Solvent Used	Ref.
CdS	Solvothermal	$Cd(CH_3COO)_2 \cdot 2H_2O$	Sulfur powder	Dodecylamine	[27]
Cu_2WS_4	Hydrothermal	$Cu(NO_3)_2 \cdot 3H_2O$ Na_2WO_4 $2H_2O$	L-Cysteine	Water	[22]
Cu_2WSe_4	Hot injection	$CuCl_2 \cdot 2H_2O$ WCl_4	Se powder	Oleylamine	[23]
Cu_2ZnSnS_4	Hot injection	$Cu(acac)_2$ $Zn(OAc)_2$ $\cdot 2H_2O$ $Sn(OAc)_4$	1-dodecylthiol and tert-dodecylthiol	1-octadecene	[30]
$CuCdS_2$	Solvothermal	Copper nitrate Cadmium acetate	Sodium thiosulphate pentahydrate	Ethylene glycol	[24]
CuS	Hydrothermal	Copper acetate dihydrate	Thiourea	Water	[31]
MoS_2	One-pot liquid-phase reaction	$(NH_4)_6Mo_7O_{24}$	Na_2S	Water	[32]
NiS_2	Hydrothermal	$Ni(NO_3)_3 \cdot 6H_2O$	Thioacetamide	Water	[33]
VS_2	Single-step chemical vapor deposition	VCl_3	Sulphur powder	-	[34]
$Zn_{0.5}Cd_{0.5}S$	Combustion method	$Zn(NO_3)_2 \cdot 4H_2O,$ $Cd(NO_3)_2 \cdot 6H_2O$	Thioacetamide	Water	[35]
ZnS	Co-precipitation	$Zn(NO_3)_2$	Na_2S	Water	[36]

Microwave-assisted synthesis is reported to be a clean, fast, and convenient synthetic route of chalcogenides. It works by applying microwave irradiation to chemical reactions which is based on efficient heating [5]. For example, Hu et al. have successfully synthesized $AgInS_2$ with tunable composition and optical properties using microwave-assisted synthesis [28]. In another study, the co-precipitation reaction between Na_2S solution and $Zn(NO_3)_2$ at room temperature produced ZnS [36]. The combustion method carried out by Tang et al. to produce $Zn_{0.5}Cd_{0.5}S$ using $Zn(NO_3)_2 \cdot 4H_2O$, $Cd(NO_3)_2 \cdot 6H_2O$ and thioacetamide [35]. Using water as the solvent, they heated the starting materials on the heating jacket until the gel-like precursor was formed, then the precursor was transferred into a Muffle furnace and heated at 300 °C for 15 min which resulted in the formation of yellow $Zn_{0.5}Cd_{0.5}S$ solid. Chemical vapor deposition (CVD) is a technique that uses thermally induced chemical reactions at the surface of a heated substrate. For example, Gopalakrishnan et al. have deposited VS_2 on the surface of vertically aligned Si nanowires using the CVD method [34]. In the typical synthesis, they used VCl_3 and sulfur powder as the precursors of V and S, respectively.

Based on the findings, it can be observed that the properties of chalcogenides can be tuned by controlling the reaction parameters during synthesis. Therefore, the optimization of these parameters is of crucial importance to fabricate the desired properties of chalcogenides with enhanced H_2O splitting ability.

3. Binary Chalcogenides and Their Photocatalytic Water-Splitting Activities

H_2 is a clean fuel, and its usage can address many of the issues caused by using fossil fuels. It has several uses, as shown in Figure 3; it is widely used as a feedstock in the chemical industry to produce ammonia, methanol, and various fuels like diesel, gasoline, etc. It is also used as a transport fuel. It has several other applications in the production of metals and agricultural industries. A cost-effective and long-lasting chalcogenide-based photocatalysts can make the H_2 generating process more economical and suitable. The use of binary chalcogenides and their modifications (compounds consisting of only one chalcogen and one electropositive atom [10]) for photocatalytic water splitting will be discussed in the following subsections.

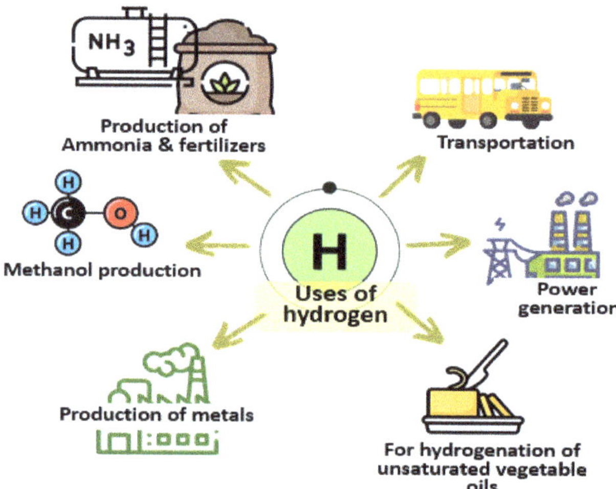

Figure 3. Several applications of H_2.

3.1. Cd-Based Chalcogenides

CdS and CdSe have both been used extensively in the fields of photocatalysis and photovoltaics, particularly in photoelectrochemical cells. Despite having a favorable bandgap for water splitting (~2.42 eV) [37,38] CdS is unstable in an aqueous solution under light irradiation due to photocorrosion. Therefore, in order to improve its photostability and photocatalytic activity, many studies have focused on synthesizing different morphologies of CdS together with the addition of co-catalysts and sacrificial agents [38]. Some of the different morphologies of CdS nanostructures that have been studied include hollow spheres [27], quasi-nanospheres [39], nanowires [39], and nanotubes [39], all of which can be synthesized by hydrothermal methods. Among these, CdS nanospheres were found to exhibit the highest H_2 evolution rate during photocatalytic testing in the presence of Na_2SO_3 and Na_2S hole scavengers.

In a recent study, one-dimensional CdS nanotubes displayed a remarkably high valence band edge compared with bulk CdS due to quantum confinement effects, and it was found that the photocatalytic efficiency of CdS nanotubes is higher than that of bulk CdS. Furthermore, the stabilized valence band, tubular structure, and strong p-d orbital hybridization led to a significant enhancement in photostability, making applications in aqueous solutions feasible [40].

In addition to the aforementioned studies focused on enhancing CdS through the preparation of different morphologies and sizes, another strategy to impart anti-photocorrosion properties to CdS is surface-modification with non-noble co-catalysts such as Ni, Ni_2P, and CuS-Ni_xP [41–43] as well as noble metals such as Pt and Au [44,45].

Another strategy to improve the photostability of CdS in an aqueous solution is to coat it with a thin layer of ZnO. Tso and co-workers have synthesized composite nanowires comprising CdS cores covered with thin ZnO shells, yielding CdS/ZnO core–shell nanowires [46]. These heterostructures provided improved stability compared with bare CdS nanowires and also led to reduced electron-hole recombination, causing an increase in H_2 evolution activity of over two orders of magnitude as shown in Figure 4. Investigations also revealed that nanowires with 10–30 nm-thick shells absorbed more incident light than nanowires with thinner ZnO shells and had a lower probability of electron-hole recombination, the combination of which led to higher H_2 evolution activity.

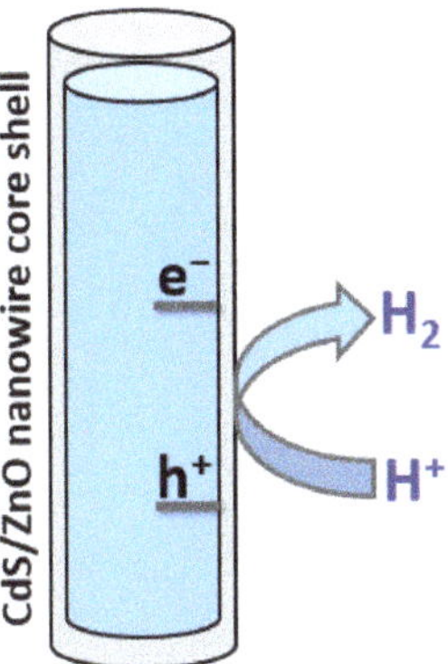

Figure 4. Schematic diagram of photocatalytic H$_2$ production using CdS/ZnO core–shell nanowire.

Several recent studies have examined photogenerated electron-hole pair separation in CdS/CuS nanocomposites [46]. Pan and co-workers found a high H$_2$ evolution rate of 561.7 µmol h^{-1} during photocatalytic water splitting using a CdS/CuS nanocomposite [47]. In another work, a CdS/CuS nanostructured heterojunction was synthesized [48]. The prepared material was applied as a water-splitting photocatalyst and showed that the nanocomposites exhibit an enhanced rate of photocatalytic H$_2$ evolution under visible-light irradiation. It can be concluded that CuS as a non-noble catalyst can improve the stability of CdS for photocatalytic H$_2$ production in an aqueous solution. The enhancement in stability found in the CdS/CuS system can be explained by the transfer of holes from CdS to CuS, which inhibits the photocorrosion of CdS (Figure 5) [49].

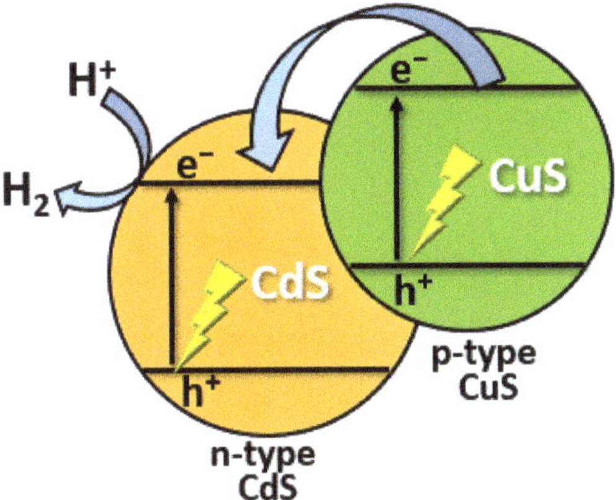

Figure 5. Mechanism of photocatalytic H$_2$ evolution using CdS/CuS nanocomposite.

Several authors have studied the factors affecting the performance of nanocrystalline CdSe quantum dots (QD) in photocatalytic water-splitting applications. Osterloh and co-workers showed a clear relationship between the extent of quantum size confinement and the photocatalytic water-splitting activity of suspended CdSe QDs. They found that the rate of H$_2$ evolution increased with QD size in nm as follows: 2.20 > 2.00 > 2.48 > 1.75 > 2.90 > 2.61 > 2.75 > 3.05 > 3.49 > 4.03 > 4.81. These results confirm that photocatalytic activity can be controlled by the extent of quantum confinement (Figure 6) [50].

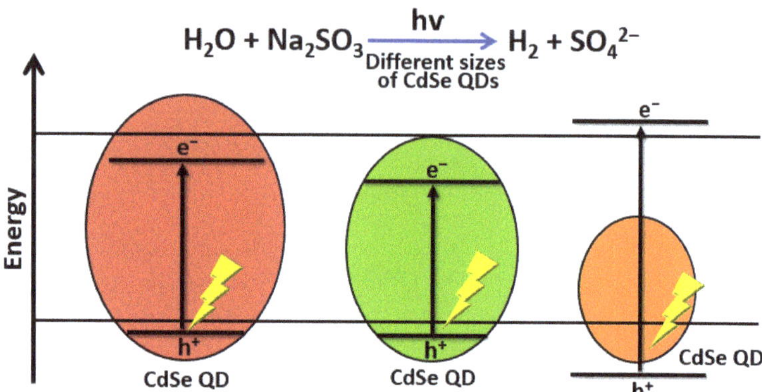

Figure 6. Schematic illustration of the effect of CdSe QD size in photocatalytic H$_2$ production.

Covalent organic frameworks (COFs) have drawn research interest owing to their stability, porous structure, large surface area, and tunable band structure [51]. It is an organic polymer arranged periodically with high crystallinity and high porosity, that is built from the integration of selected organic blocks via covalent bond linkage [52]. For instance, You et al. have synthesized COFs-p-phenylenediamine@CdS (COFs-Ph@CdS) using the hydrothermal method [53]. The authors reported that of the H$_2$ generation of bare CdS, COFs-Ph@CdS-x (x = 1, 2, 3, 4) photocatalysts were 16.4, 139.1, 17.5, 37.0, and 15.2 µmol, respectively. The photocatalytic H$_2$ production activity of solid COFs-Ph@CdS-1 exhibited 29 and eight times higher compared to those of pristine COFs and CdS, respectively. The enhanced activity may be attributed to a stable coordination bond between Cd^{2+} ions and N atoms in COFs-Ph@CdS was formed, which served as a migrated channel for photo-generated carriers and their unique hollow structure that enables the production of more photo-generated electrons and holes and absorb light more efficiently.

High rates of photocatalytic H$_2$ production were demonstrated by Das and co-workers using a surface ligand that binds strongly to CdSe QDs via tridentate coordination [54]. These authors also reported the effects of surface stabilizing and solubilizing agent dynamics. In more recent work, the influence of ligand exchange on the photocatalytic activity of CdSe QDs in solar water-splitting cells was studied [55]. It was found that CdSe QDs bearing shorter ligands showed better photocatalytic activity when compared to pristine CdSe QDs in a photoelectrochemical cell. These results should assist with the design and optimization of photochemical energy conversion systems employing nanocrystalline CdSe QD photocatalysts.

In another study, Zhou et al. have successfully prepared peanut-chocolate-ball-like CdS/Cd for H$_2$ evolution [56]. They reported that CdS/Cd exhibited a high photocatalytic H$_2$ production of 95.40 lmol h^{-1} which is about 32.3 times higher than bare CdS and displays exceptional photocatalytic stability over 205 h in comparison to CdS-based photocatalyst reported in previous studies. In recent studies, Li et al. 2% NiS/CdS present a high H$_2$ evolution rate of 18.9 mmol g^{-1}h^{-1} in comparison to pristine CdS [57]. The improved activity may be attributed to the presence of an interface between NiS and CdS that efficiently promote the charge separation and NiS nanoparticles serve as highly active H$_2$ evolution sites.

3.2. Cu-Based Chalcogenides

Visible-light-active CuS/Au nanostructures were obtained by decorating CuS particles with Au nanoparticles, which were prepared by reducing HAuCl$_4$ on the CuS surface [58]. The as-synthesized nanostructures displayed excellent catalytic performance in electrochemical H$_2$ evolution tests, with good durability under acidic conditions. It was found that the CuS/Au nanostructured catalyst could produce a current density of 10 mA/cm^2

upon the application of only 0.179 V versus the reversible hydrogen electrode (RHE). The CuS/Au catalyst also proved to be an efficient photocatalyst for the degradation of organic pollutants under visible-light irradiation.

The coupling of CuS with metal oxide has been shown to enhance its photocatalytic activities. For instance, Chandra and co-workers have synthesized CuS/TiO_2 heterostructures with varying TiO_2 contents using a simple hydrothermal method for the photocatalytic evolution of H_2 [59]. The highest rate of visible-light photocatalytic H_2 production was 1262 $\mu mol\ h^{-1}\ g^{-1}$, which is around 9.7 and 9.3 times faster in comparison to pristine TiO_2 nanospindles or CuS nanoflakes, respectively. The enhanced rate of H_2 evolution is attributable to increased light harvesting and more efficient charge separation when an optimum amount of CuS is deposited on TiO_2. Growing CuS nanoflakes on TiO_2 nanoparticles to form a CuS/TiO_2 heterostructure facilitates charge carrier separation at the heterostructure interface and thus enhances photocatalytic performance.

In another study, Dubale et al. fabricated nanocomposites comprising CuS and Cu_2O/CuO heterostructures with and without decoration by Pt nanoparticles using an in-situ growth method [60]. Cu_2O/CuO was prepared by straightforward anodization of Cu film, and a sputtering technique was used to deposit Pt nanoparticles onto the prepared $Cu_2O/CuO/CuS$ photocathodes. The prepared heterostructures were found to be very promising and highly stable photocathodes for H_2 evolution under visible-light irradiation, with optimized $Cu_2O/CuO/CuS$ photocathodes reaching photocurrent densities of up to 5.4 mA cm^2, some 2.5 times higher than that achieved by bare Cu_2O/CuO. Upon decorating the Cu_2O/CuO with both CuS and Pt, a further increase in photocurrent density to 5.7 mA/cm^2 was achieved due to the suppression of the charge carrier recombination.

Metal loading on the surface of Cu-based chalcogenides has been reported to improve the photocatalytic evolution of H_2. For instance, Ma and co-workers have reported that hydrothermally synthesized Au/CuSe/Pt nanoplates showed strong dual-plasmonic resonance and have been employed for photocatalytic H_2 generation under irradiation of visible and near-infrared [61]. The as-prepared Au/CuSe/Pt nanoplates exhibited outstanding H_2 generation activities of around 7.8 and 9.7 times those of Au/CuSe and Pt/CuSe composites, respectively. In the Au/CuSe/Pt system, light harvesting is enhanced by the plasmonic units (CuSe and Au), while Pt is mainly present as a co-catalyst promoting the H_2 evolution reaction as shown in Figure 7.

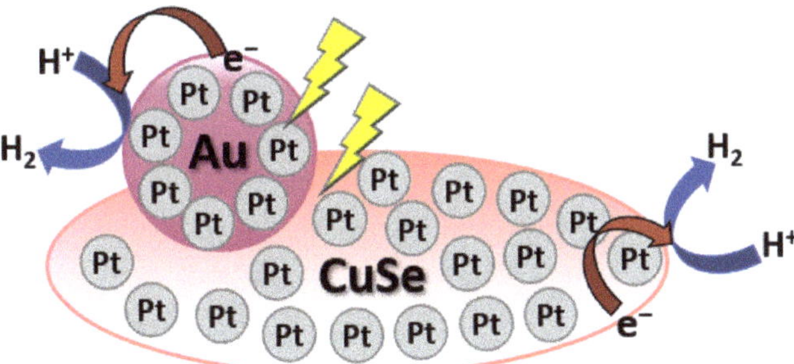

Figure 7. Schematic diagram of Au/CuSe/Pt ultrathin nanoplate hybrid for H_2 evolution under visible and near-infrared irradiation.

In a study conducted by Liang et al. CuS was deposited on the surface of CdS nanorods using the cation exchange method for generation of H_2 under the irradiation of light [62].

In a study by Hou et al., CuS has been used as a co-sensitizer along with CdS to improve photocatalytic H_2 production [63]. They decorated CdS–CuS on TiO_2 nanotube arrays using the hydrothermal method. The as-prepared material exhibited remarkably high photocatalytic H_2 evolution ability, and the photocatalytic H_2 production rate was about 62.02 $\mu mol\ cm^{-2}h^{-1}$. In a different study, Mandari and co-workers synthesized a stable $CuS/Ag_2O/g$-C_3N_4 material using hydrothermal and precipitation methods [31].

The as-synthesized material displayed a remarkable photocatalytic H_2 production of 1752 μmol h^{-1}g^{-1} which is considerably superior to those of the pristine CuS and g-C$_3$N$_4$. The enhanced H_2 production activity may be ascribed to the formation of heterojunctions which resulted in the increase in visible light harvested and suppressed the photogenerated charge carrier recombination.

3.3. Ga-Based Chalcogenides

Monolayers of GaS and GaSe have been studied as photocatalysts by several authors but their performance is limited by low optical absorption and inefficient electron-hole pair separation [64,65]. One strategy to overcome these limitations is the construction of van der Waals (vdW) heterostructures by pairing GaX (GaX, X = S, Se) with arsenene (two-dimensional arsenic), as demonstrated by Peng and co-workers using first-principle calculations [66]. Such GaX/As heterostructures possess band gaps and band alignment that satisfy the requirements for photocatalytic water splitting. In contrast to pristine GaX monolayers, the type of band gap in Se$_{0.5}$GaS$_{0.5}$/As and S$_{0.5}$GaSe$_{0.5}$/As changes favorably from indirect to direct as the interlayer distance is varied. Furthermore, these heterostructures exhibit transport anisotropy with high electron mobilities of up to $\sim$2000 cm^2 V^{-1} s^{-1}, which facilitates photogenerated electron-hole pair separation and migration. All of the GaX/As heterostructures studied also show enhanced visible-light absorption beyond that of pristine GaX monolayers. It seems likely that the exceptional properties of GaX/As heterostructures will render them competitive with existing photocatalysts for solar-driven water splitting [66].

3.4. Mo-Based Chalcogenides

Molybdenum disulfide (MoS_2) is a two-dimensional (2D) material, and it tends to exhibit more remarkable photocatalytic properties than its bulk counterparts [67,68]. Li and co-workers have investigated the photocatalytic properties of pristine MoS_2 nanosheets and Ag-modified MoS_2 nanosheets. They found that the rate of photocatalytic H_2 production by Ag-modified MoS_2 nanosheets could reach 2695 μmol h^{-1} g^{-1}. They also found that Ag-doped MoS_2 nanosheets exhibit superior stability and higher H_2 evolution efficiency compared with pristine MoS_2 nanosheets, which is mainly due to increased visible-light absorption with increasing Ag content (Figure 8) [69].

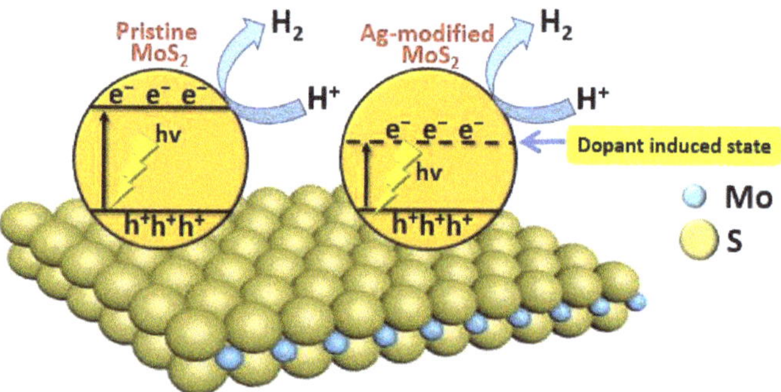

Figure 8. The mechanism of photocatalytic H_2 production using Ag-modified MoS_2 nanosheets.

Photocorrosion-resistant material, MoS_2/Zn$_{0.5}$Cd$_{0.5}$S/g-C$_3$N$_4$, has been successfully synthesized by Tang et al. to produce H_2 under irradiation of visible light [35]. The authors reported that the as-prepared materials exhibited a maximum H_2 production rate of 4914 μmol g^{-1}h^{-1} in Na$_2$S-Na$_2$SO$_3$ solutions which served as the sacrificial agent. The high evolution rate of H_2 was owing to the improved charge separation between the interfaces.

In another study, Yuan et al. designed 2D-2D MoS_2@Cu-ZnIn$_2$S$_4$ using the solvothermal method for photocatalytic production of H_2 [70]. The highest H_2 evolution rate of 5463 µmol g^{-1}h^{-1} was observed for 6% MoS_2@Cu-ZnIn$_2$S$_4$ under irradiation of visible light which is 72 times higher than that of pristine Cu-ZnIn$_2$S$_4$. The excellent activity may be ascribed to the high visible light absorption property and abundant active sites for the H_2 evolution reaction to take place. Moreover, the production rate of H_2 remain unchanged after three cycles which suggested the high durability of the synthesized MoS_2@Cu-ZnIn$_2$S$_4$ for photocatalytic H_2 production.

An enhanced photocatalytic H_2 evolution was performed by Yin and co-workers with the maximum evolution rate of 872.3 mol h^{-1} using MoS_2/C composites sensitized with Erythrosin B [32]. The enhanced performance was attributed to the efficient photogenerated charge transfer and separation between Erythrosin B and MoS_2 or C as well as the presence of more active sites for H_2 evolution. The photocatalytic evolution of H_2 could also be achieved through photoelectrochemical catalytic activities. For instance, Hassan and co-workers have reported the optimized MoS_2/GaN$_2$ exhibited significantly higher photoelectrochemical performance in comparison to the bare materials under the same visible conditions [71]. The as-prepared photoanode achieved efficient light harvesting with a photocurrent density of 5.2 mA cm^{-2} which is 2.6 times higher than bare GaN. Moreover, MoS_2/GaN$_2$ exhibited enhanced applied-bias-photon-to-current conversion efficiency of 0.91%, while the reference GaN yielded an efficiency of 0.32%. The authors reported that the decrease in charge transfer resistance between the electrolyte and semiconductor interface and the improved separation charge carriers in the MoS_2/GaN heterostructures were all attributed to significant improvements in photocurrent density and efficiency.

MoS_2-based nanomaterials have been found to be an efficient visible light-responsive material for H_2 production. Based on the reported work, revealed that the surface modification of the MoS_2 could lead to an improved photocatalytic activity owing to the effective charge transfer and separation of photo-generated e^-/h$^+$ pairs.

New strategies have been developed to use $MoSe_2$-based nanosheets as photocatalysts for the H_2 evolution reaction. Zhao and co-workers achieved an enhancement in the photocatalytic activity of $MoSe_2$ by surface modification and doping with Ni and Co [72]. Their study found that $Ni_{0.15}Mo_{0.85}Se_2$ nanosheets exhibited the highest photocatalytic activity in both alkaline and acidic media. These results provide an attractive alternative to Pt-based catalysts.

3.5. Sn-Based Chalcogenides

Recently, it has been discovered that monolayers of the Sn-based mono- and dichalcogenides SnX and SnX$_2$ (X = S or Se) are promising candidates for photocatalytic water splitting. These materials exhibit very good visible light absorption, low exciton binding energies, and high carrier mobility. However, in their pure and strain-free state, the valence band edges of Sn-monochalcogenide monolayers are predicted to be too high in energy for effective water oxidation. This issue can potentially be overcome by crystal engineering to induce a moderate tensile strain, which is predicted to lower the valence band edge to a level suitable for water oxidation [21]. Li and co-workers reported that monolayers of both SnX and SnX$_2$ are expected to exhibit good optical absorption in the visible region of the solar spectrum, and the predicted sequence of visible light absorption strength is as follows: SnSe > SnSe$_2$ > SnS > SnS$_2$. Moreover, SnX and SnX$_2$ monolayers also have excellent carrier mobility, which results in the fast migration of photogenerated carriers to the surface. Hence, the possibility of rapid redox reactions on the surface of the photocatalyst.

In another study, Lei et al. successfully prepared SnS/g-C$_3$N$_4$ nanosheets using a facile ultrasonic and microwave heating approach, which formed intimate interfacial contact and a suitable energy band structure [73]. The optimized sample displayed enhanced photocatalytic H_2 evolution from H_2O with the aid of Pt as a co-catalyst in comparison to pure g-C$_3$N$_4$. The stability of the photocatalysts was significantly improved after being loaded with MoO_3 particles due to the formation of a Z-scheme heterojunction. Among all as-prepared samples, the 10% SnS/g-C$_3$N$_4$ exhibits the highest photocatalytic rate of 818.93 mmol h^{-1}g^{-1} under AM1.5G irradiation, 2.90 times to pure g-C$_3$N$_4$, due to the matched energy band structure between g-C$_3$N$_4$ and SnS, which improves the separation efficiency of photo-generated carriers and hinders the recombination of hole-electron pairs. Additionally, SnS nanosheets have improved the light absorption efficiency of the prepared material and generated more catalytic active sites as well as shortening the carriers' transport path as well.

Liu and co-workers have synthesized flower-like nanostructure SnS$_2$ with generated sulfur vacancies using a new simple strategy of ion exchange [74]. The morphology of the flower-like structure provides shorter carrier diffusion lengths and more surface-active sites. While, the presence of S vacancies helps introduce defect levels in SnS$_2$, which leads to a reduction in work function, eventually improving carrier density and separation efficiency. The well-engineered sample 5%Cu/SnS$_{2-x}$ shows the highest photocatalytic activity with an H$_2$ yield of 1.37 mmol h^{-1}, which is about six times higher than that of pure SnS$_2$.

SnS$_2$/ZnIn$_2$S$_4$ composites were successfully fabricated by decorating SnS$_2$ nanosheets with ZnIn$_2$S$_4$ microspheres [75]. Geng and co-workers reported that the SnS$_2$/ZnIn$_2$S$_4$ composites showed superior photocatalytic properties for H$_2$ generation in comparison to pristine ZnIn$_2$S$_4$. Furthermore, a notable impact was observed on the photocatalytic activity of ZnIn$_2$S$_4$ when varying the mass ratio of SnS$_2$. The 2.5% SnS$_2$/ZnIn$_2$S$_4$ in particular exhibited the highest photocatalytic H$_2$ generation rate of 769 μmol g^{-1}h^{-1}, which was about 10.5 times the photocatalytic activity of pristine ZnIn$_2$S$_4$. The enhanced photocatalytic activity may be due to the formation of SnS$_2$/ZnIn$_2$S$_4$ heterojunction, which enabled highly active charge separation and transfers on the interface of SnS$_2$ and ZnIn$_2$S$_4$.

Li and co-workers have synthesized SnO$_2$/SnS$_2$ with excellent photocatalytic H$_2$ evolution performance under simulated light irradiation [76]. The material exhibited a high H$_2$ production rate of 50 μmol h^{-1} which is 4.2 times higher than that of pure SnO$_2$ under the same condition. A different study by Mangiri et al. has successfully synthesized a stable CdS/MoS$_2$-SnS$_2$ for water splitting via the solvothermal method [77]. The as-prepared CdS/MoS$_2$-SnS$_2$ (6 wt%) exhibited the highest H$_2$ production rate of 185.36 mmol h^{-1}g^{-1} which is much higher than pristine CdS (2.5 mmol h^{-1}g^{-1}) and 6 wt% of MoS$_2$-loaded CdS nanorods (123 mmol h^{-1}g^{-1}). The effective photocatalytic performance of MoS$_2$-SnS$_2$-loaded CdS may be ascribed to the ability of the material to harvest light effectively, better separation of e$^-$/h$^+$ pairs, formation of trapping sites, high active catalytic zones, migration of charge carriers towards the surface of a semiconductor, and suitable energy levels.

3.6. Ti-Based Chalcogenides

The tri-chalcogenide TiS$_3$ was found to be an effective photoanode for use in photoelectrochemical cells. Photocatalytic H$_2$ evolution rates of up to 1.80 ± 0.05 μmol min^{-1} were obtained, corresponding to a photoconversion efficiency of ~7% [78]. Furthermore, the ternary compounds Ti$_x$Nb$_{1-x}$S$_3$ (Nb-rich) and Nb$_x$Ti$_{1-x}$S$_3$ (Ti-rich) were tested as photoanodes and compared with the corresponding binary trisulfides. The binary compounds were polycrystalline with a nanoribbon morphology and adopted the monoclinic TiS$_3$ (P21/m) and triclinic NbS$_3$ (P2) crystal structures, respectively. H$_2$ generation experiments revealed that the Ti-rich phase was a superior photocatalyst compared with the Nb-rich phase, exhibiting H$_2$ production rates of up to 2.2 ± 0.1 mol/min cm^2 [79]. Ti-based chalcogenides are not widely explored for their photocatalytic H$_2$ production activity. Therefore, further studies and optimizations are required.

3.7. V-Based Chalcogenides

Certain transition metals located on the first row of the periodic table can participate in the formation of tetra-chalcogenides thanks to their ability to adopt +4 oxidation states (Figure 9). Among these metals is vanadium, which can form the binary tetra-chalcogenide VS$_4$. Nanocomposites comprising VS$_4$ and graphene were evaluated for solar-driven water splitting by Guo and co-workers. Excellent performance was obtained due to the formation of C–S bonds with a π-conjugated structure, which assists with transferring electrons from the S 2p orbital to graphene [80]. The ternary tetra-chalcogenides $Cu_3Nb_{1-x}V_xS_4$ and $Cu_3Ta_{1-x}V_xS_4$, which exist as solid solutions and adopt the sulvanite structure, were successfully synthesized in a solid-state reaction by Ikeda and co-workers [81]. The band gaps of the synthesized materials were estimated to fall in the range of 1.6–1.7 eV, and the band structure of the $Cu_3M_{1-x}V_xS_4$ system varies with composition. It was found that for most compositions of solid solution, superior photocatalytic activity was achieved compared with the ternary Cu_3MS_4 (M = V, Nb, Ta) compounds [81].

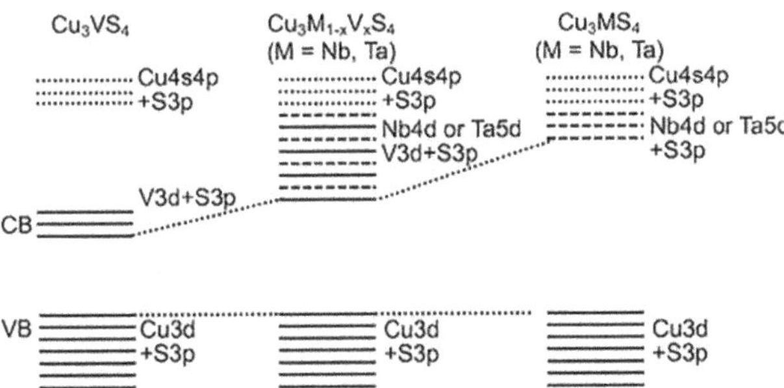

Figure 9. Proposed band structures for $Cu_3M_{1-x}V_xS_4$ (M = Nb and Ta) solid solutions.

Li et al. synthesized the $VS_2@C_3N_4$ heterojunction with S vacancies via the in situ supramolecular self-assembly method [82]. The as-synthesized material shows a remarkable synergistic H_2 production rate (9628 μmol g^{-1}h^{-1}) and wastewater degradation efficiency as well as stability, which is 16.0 times that of pristine C_3N_4. Based on the theoretical calculations and experiment findings, it shows that S vacancies have resulted in the formation of compact heterostructure and the reduction in the work function, which promotes interfacial carriers' transfer and surface properties. The core–shell structure improves the stability of S vacancies.

In another study, Gopalakrishnan et al. synthesized a core–shell heterostructures photocathode consisting of VS_2 deposited on a silicon nanowire surface via single-step chemical vapor deposition for H_2 evolution under solar irradiation [34]. The core–shell nanowire heterostructure photocathode displayed an outstanding solar-assisted water reduction performance at pH~7 over a pristine photocathode. The as-prepared material produces about 23 μmol cm^{-2} h^{-1} (at 0 V vs. RHE) of H_2 gas. Recently, Zhong et al. have doped Ni_3S_2 with nitrogen and coupled it with VS_2 using the hydrothermal method [83]. Based on their results, Ni_3S_2/VS_2 without nitrogen doping exhibited improved oxygen evolution reaction (OER) performance with an extremely low overpotential of 227 mV at 10 mA/cm^2 which may be attributed to the presence of high number of active sites and excellent interfaces. Moreover, Ni_3S_2/VS_2 with nitrogen doping shows high electrocatalytic hydrogen evolution reaction (HER) performance with a low HER overpotential (151 mV at 10 mA/cm^2) and this is due to the presence of nitrogen doping that significantly improves the conductivity and increases the number of catalytic active sites.

3.8. Zn-Based Chalcogenides

Kurnia and co-workers have investigated the performance of ZnS thin films during photoelectrochemical H_2 generation [84]. They found that ZnS is an inexpensive photocatalyst that exhibits activity under visible-light irradiation without the addition of a co-catalyst. The ZnS thin films were estimated to have band gaps of ~2.4 eV, and photocurrent densities exceeding 1.5 mA/cm^2 were achieved under visible-light irradiation ($\lambda \geq 435$ nm). These photocurrents are remarkably high for H_2 generation by undoped ZnS under visible-light irradiation.

A modification of ZnO nanosheets, using a thin layer of ZnS, was developed to enhance visible light absorption, leading to improved photoelectrochemical activity in water-splitting experiments [85,86]. Sánchez-Tovar and co-workers have prepared ZnO/ZnS heterostructures using anodization of Zn in glycerol/water/NH_4F electrolytes under controlled hydrodynamic conditions in the presence of various concentrations of Na_2S [85]. In another work, Zhang et al. synthesized ZnS/ZnO/ZnS sandwich nanosheets with an effective band gap of ~2.72 eV using thermal evaporation, and these materials resulted in potent visible-light photocatalytic activity [86]. These works show that coating ZnO with an ultrathin surface layer of ZnS is an interesting approach for enhancing its water-splitting activity under solar irradiation.

Arai and co-workers have synthesized hollow Cu-doped ZnS particles as visible-light photocatalysts for the decomposition of hydrogen sulfide under solar irradiation to generate H_2. The Cu-doped ZnS particles were synthesized starting from a ZnO precursor that was co-precipitated on Cu by exploiting the difference in reduction potential between Zn and Cu. The visible-light photoactivity of Cu-doped ZnS was six- and 130-times higher than the Cu-free "ZnS-shell" and pristine ZnS particles prepared using co-precipitation, respectively. A comparative study found that the "Cu-ZnS-shell" particles were highly active under visible-light irradiation (λ > 440 nm), and the rate of H_2 evolution was optimized with a Cu (2 wt%)/ZnS photocatalyst [36]. In another recent study, Cu/ZnS microspheres were prepared using a template-free microwave irradiation method for H_2 generation from an aqueous Na_2S solution under visible-light irradiation. The optimized H_2 evolution rate of 973.1 μmol h^{-1} g^{-1} was obtained for 2.0 mol% Cu-doped ZnS [87].

4. Ternary Chalcogenide Heterostructures for Water Splitting

Recently, ternary mixed-metal chalcogenides have shown excellent photocatalytic H_2 production activities, as shown in Table 2. For instance, $Zn_xIn_2S_{3+x}$ (x = 1–5) has attracted the attention of researchers as a new member of the semiconductor family. These compounds possess unique electronic structures, tunable optical properties, and, for certain compositions, band gaps suitable for visible-light absorption and energy bands that straddle the water redox potentials, making them ideal for use in photocatalysis and energy conversion reactions [88]. Wu and co-workers have successfully synthesized $Zn_xIn_2S_{3+x}$ with x ranging from 1 to 5 [88]. $Zn_xIn_2S_{3+x}$ samples display strong absorption of visible light due to excitation of the fundamental band gap transition, and the absorption edges of the samples move to a shorter wavelength as x is increased from 1 to 5 due to the widening of the band gap. $Zn_xIn_2S_{3+x}$ samples have band gap energies ranging from 2.65 eV to 2.84 eV, depending on chemical composition. $ZnIn_2S_4$ exhibited the lowest electron-hole recombination rate among all the compositions studied. One of the contributing factors leading to this finding is the presence of fewer composition faults in the $ZnIn_2S_4$ structure compared with the other samples. The existence of composition faults results in additional energy barriers to be overcome by charge carriers during transit to the particle surface, which leads to increased recombination and lower photocatalytic reaction rate. Based on the photoelectrochemical experiments conducted, it was found that $ZnIn_2S_4$ can generate more photocurrent than the other compositions. Electrochemical impedance spectroscopy results also suggest that $ZnIn_2S_4$ has the lowest resistance to interfacial electron transfer among the compositions studied, suggesting that $ZnIn_2S_4$ is the most efficient at interfacial charge transfer and consistent with the photocurrent trend. $ZnIn_2S_4$ shows the highest photocatalytic activity, followed by $Zn_2In_2S_5$, $Zn_3In_2S_6$, $Zn_4In_2S_7$, and, lastly, $Zn_4In_2S_8$, with H_2 production rates of 2.93, 2.86, 2.32, 2.15, and 2.05 mmol h^{-1}g^{-1}, respectively. Overall, it can be concluded that ternary metal chalcogenides with the formula $Zn_xIn_2S_{3+x}$ (x = 1–5) exhibit appropriate properties for use as photocatalysts, especially for water-splitting applications. In a different study, Fan et al. fabricated CuS@$ZnIn_2S_4$ hierarchical nanocages for H_2 evolution under visible light [89]. The presence of an interface between CuS and $ZnIn_2S_4$ improved the solar energy utilization and separation and transfer efficiency of photogenerated carriers. The as-prepared materials exhibited a photocatalytic H_2 evolution rate as high as 7910 μmol h^{-1}g^{-1}.

Table 2. Photocatalytic H_2 production activities using different ternary chalcogenides.

Photocatalyst Used	Amount of Photocatalyst Used	Electrolyte/ Solvent Used	Morphology	H_2 Produced	Ref.
$MoS_2/g\text{-}C_3N_4/ZnIn_2S_4$	50 mg	90 mL water and 10 mL triethanolamine	3D flower-like nanospheres	$ZnIn_2S_4$: 904 µmol g^{-1}h^{-1} $MoS_2/g\text{-}C_3N_4/ZnIn_2S_4$ loaded with with 3 wt% g-C$_3$N$_4$ and 1.5 wt% MoS$_2$: 6291 µmol g^{-1}h^{-1}	[90]
$ZnIn_2S_4@$ $CuInS_2$	50 mg	0.25 mol/L Na$_2$S and 0.35 mol/L Na$_2$SO$_3$ as a sacrificial agent in 100 mL of aqueous solution	Marigold-like spherical structure comprising numerous thin nanosheets	1168 µmol g^{-1}	[91]
Ti_3C_2 MXene@ $TiO_2/CuInS_2$	10 mg	100 mL water solution containing 20% methanol as sacrificial agent	Layered structure decorated with small CuInS$_2$ nanoparticles	356.27 µmol g^{-1}h^{-1}	[92]
$ZnIn_2S_4\text{-rGO-}CuInS_2$	15 mg	50 mL of aqueous Na$_2$S/Na$_2$SO$_3$ (10 vol%) solution	Dispersed marigold-like structured ZnIn$_2$S$_4$ and layer-structured CuInS$_2$ on reduced graphene oxide sheets	2531 µmol/g after 5 h	[93]
$CdS/CoMoS_4$	80 mg	8 mL of lactic acid was added into 80 mL of water	Nanorod structured	7.5 at.% CdS/ CoMoS$_4$: 208 µmol h^{-1}	[94]
Carbon nanotube modified $Zn_{0.83}Cd_{0.17}S$	1 mg	40 mL of an aqueous solution containing 0.1 M Na$_2$S/0.02 M Na$_2$SO$_3$	Irregular	0.25 wt% carbon nanotube modified Zn$_{0.83}$Cd$_{0.17}$S: 5.41 mmol h^{-1}g^{-1}	[95]

In another report by Hojamberdiev and co-workers, they have synthesized layered crystals of trigonal $ZnIn_2S_4$ by a flux method using a range of binary fluxes, including $CaCl_2$:$InCl_3$, $SrCl_2$:$InCl_3$, $BaCl_2$:$InCl_3$, $NaCl$:$InCl_3$, KCl:$InCl_3$, and $CsCl$:$InCl_3$, from waste containing ZnS from the mining industry [7]. Among them, KCl:$InCl_3$ was the most favorable for synthesizing phase-pure trigonal $ZnIn_2S_4$. The trigonal $ZnIn_2S_4$ crystals grown in this study using KCl:$InCl_3$ flux exhibited higher photocatalytic H_2 evolution activity ($132\ \mu mol\ h^{-1}$) in comparison to previously reported hexagonal $ZnIn_2S_4$ crystals prepared using the hydrothermal method. The superior performance obtained using the binary flux method can be ascribed to higher crystallinity and decreased defect density. The existence of secondary phases (namely ZnS and In_2S_3) in $ZnIn_2S_4$ crystals grown with $CaCl_2$:$InCl_3$ and $NaCl$:$InCl_3$ fluxes were found to have a positive impact on the photocatalytic activity, leading to H_2 evolution rates of 232 and 188 $\mu mol\ h^{-1}$, respectively. The enhanced H_2 production activity may be ascribed to efficient charge separation and interfacial charge transfer. Moreover, $Mn_{0.5}Cd_{0.5}S$/carbon black/CuS have been fabricated for H_2 production using sonochemical loading, and the subsequent in-situ deposition routes [96]. Lv et al. reported that the optimized $Mn_{0.5}Cd_{0.5}S$/carbon black/CuS exhibited the highest H_2 production of 819.9 $\mu mol\ h^{-1}$, which is 4.79-, 2.08-, and 1.47-fold increments more than pristine $Mn_{0.5}Cd_{0.5}S$, $Mn_{0.5}Cd_{0.5}S$/0.5 carbon black and $Mn_{0.5}Cd_{0.5}S$/CuS, respectively.

Group-III chalcogenide monolayers adopting two-faced 'Janus' structures are reported to be efficient photocatalysts for solar-driven water splitting. Several research groups have theoretically investigated the geometric structure, chemical stability, and electronic and optical properties of Janus group-III chalcogenide monolayers with layered ternary (XMMX′, where X, X′ = S, Se, or Te and M = Ga or In) [97,98] or quaternary composition (XMM′X′, where M and M′ are Ga and In) [99,100]. Among the ternary Janus monolayers, SGa_2Te, $SeGa_2Te$, SIn_2Te, and $SeIn_2Te$ are found to possess direct band gaps. Since direct gap semiconductors generally possess higher absorption coefficients than indirect gap semiconductors, these ternary group-III chalcogenide monolayers can potentially exhibit good light harvesting and efficient electron-hole pair generation [97]. However, most of these materials have not yet been synthesized, which will be an important next step in their photocatalytic evaluation. As reported by Bai and co-workers, group-III chalcogenide monolayers adopt a honeycomb-like geometric structure with a double M m (M = Ga or In) layer sandwiched between two chalcogen layers.

Depending on the combination of metal and chalcogen, both direct and indirect band gaps occur, with gap energies ranging from 1.54 eV to 2.91 eV. Although all these band gaps surpass the free energy change of the water-splitting reaction (1.23 eV per electron), photocatalytic water splitting is not guaranteed because the band edges must also straddle the water redox potentials. Figure 10 shows the calculated alignment between the band edges of the Janus XMMX′ monolayers and the water redox potentials. According to these calculations, all of the monolayers are capable of complete water splitting with the exception of SGa_2Te, which cannot drive the oxygen evolution reaction due to its valence band edge laying above the O_2/H_2O redox potential [97].

Although the pristine GaTe and InTe monolayers exhibit strong absorption at short wavelengths, they are inferior to the corresponding Janus monolayers at longer wavelengths. Among the Ga-based Janus chalcogenide monolayers, $SeGa_2Te$ and SGa_2Te possess the strongest optical absorption, while $SeIn_2Te$ and SIn_2Te exhibit the strongest absorption of all the Janus-type monolayers studied.

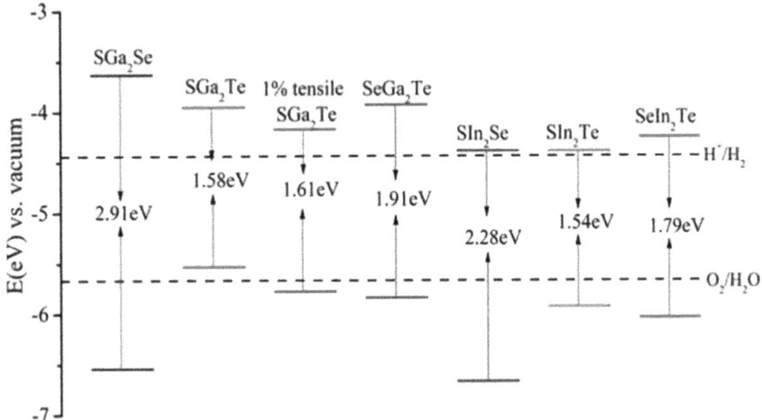

Figure 10. Calculated alignment of the band edges of the Janus XMMX′ monolayer with respect to the redox potentials of the water.

5. Prospects of Chalcogenides and Chalcogenide-Based Heterostructures

Even though steady progress has been made in utilizing chalcogenides as photocatalysts, a range of challenges still need to be addressed to promote the field, gain interest from other researchers, and achieve commercially exploitable results:

- In order to further boost the activity and stability of chalcogenides for water splitting, it is important to create precise methods for obtaining pure phase, active interface, exposed active surface, optimized electronic structure, and enhanced electronic conductivity.
- Recent research has shown that chalcogenides are very promising materials for H_2 evolution by photocatalytic water splitting. It is expected that, with further knowledge, controlled doping, surface engineering, and development of their performance can be further improved [101].
- Binary metal chalcogenides such as CdS and CdSe are unstable in acidic media and are also susceptible to photocorrosion. As such, potential replacements that are more stable in acidic media and that do not exhibit photocorrosion should be explored.
- Despite many recent studies on the use of ternary and quaternary chalcogenides as photocatalysts for H_2 production, the exact cause of photocorrosion in these materials is yet to be explored in detail and should be researched thoroughly in order to synthesize highly stable, multifunctional chalcogenides.
- Preparation of chalcogenide using low-cost methods while can produce large quantities of products also require more attention. Optimization of different parameters in the synthetic reactions such as precursors, temperature, pH, and reaction time should be studied for optimal yield to facilitate the production of these materials at a commercial scale.
- Detailed studies on extending the lifetime of photo-generated carriers and suppressing recombination are required to improve the photocatalytic activity of these materials for broader applications. Several approaches that could be employed include coupling with other semiconductors, loading of noble metals, and doping with metal or non-metal ions.

6. Conclusions

In summary, this review provides an overview of chalcogenide and chalcogenide-based heterostructures that have been extensively used for photocatalytic H_2O-splitting activities. This article summarizes the different modifications of chalcogenide materials that can improve their absorption of visible light ability, enhancing charge separation and reducing the recombination rate for water splitting. Moreover, this review also includes the mechanisms involved in the water splitting of binary, ternary, and chalcogenide-based heterostructures. Furthermore, combining different metal-based chalcogenides with other semiconductors has the potential to improve photocatalytic efficiency for generating H_2 and O_2 by water splitting.

Author Contributions: M.M.K.: Supervision, Conceptualization, Funding acquisition, Writing—review and editing. A.R.: Methodology, Data curation, Writing—original draft. All authors have read and agreed to the published version of the manuscript.

Funding: This research was funded by Universiti Brunei Darussalam, grant number UBD/RSCH/1.4/FICBF(b)/2021/035.

Acknowledgments: The authors would like to acknowledge the FIC block grant UBD/RSCH/1.4/FICBF(b)/2021/035 received from Universiti Brunei Darussalam, Brunei Darussalam.

Conflicts of Interest: The authors declare no conflict of interest.

References

1. Khan, M.M.; Ansari, S.A.; Pradhan, D.; Ansari, M.O.; Han, D.H.; Lee, J.; Cho, M.H. Band gap engineered TiO_2 nanoparticles for visible light induced photoelectrochemical and photocatalytic studies. *J. Mater. Chem. A* **2013**, *2*, 637–644. [CrossRef]
2. Rahman, A.; Harunsani, M.H.; Tan, A.L.; Khan, M.M. Zinc oxide and zinc oxide-based nanostructures: Biogenic and phytogenic synthesis, properties and applications. *Bioprocess Biosyst. Eng.* **2021**, *44*, 1333–1372. [CrossRef]
3. Matussin, S.; Harunsani, M.H.; Tan, A.L.; Khan, M.M. Plant-Extract-Mediated SnO_2 Nanoparticles: Synthesis and Applications. *ACS Sustain. Chem. Eng.* **2020**, *8*, 3040–3054. [CrossRef]
4. Khan, M.M.; Adil, S.F.; Al-Mayouf, A. Metal oxides as photocatalysts. *J. Saudi Chem. Soc.* **2015**, *19*, 462–464. [CrossRef]
5. Khan, M.M.; Rahman, A.; Matussin, S.N. Recent Progress of Metal-Organic Frameworks and Metal-Organic Frameworks-Based Heterostructures as Photocatalysts. *Nanomaterials* **2022**, *12*, 2820. [CrossRef]
6. Khan, M.M.; Pradhan, D.; Sohn, Y. *Nanocomposites for Visible Light-Induced Photocatalysis*; Springer: Berlin/Heidelberg, Germany, 2017. [CrossRef]
7. Hojamberdiev, M.; Cai, Y.; Vequizo, J.J.M.; Khan, M.M.; Vargas, R.; Yubuta, K.; Yamakata, A.; Teshima, K.; Hasegawa, M. Binary flux-promoted formation of trigonal $ZnIn_2S_4$ layered crystals using ZnS-containing industrial waste and their photocatalytic performance for H_2 production. *Green Chem.* **2018**, *20*, 3845–3856. [CrossRef]
8. Matussin, S.N.; Rahman, A.; Khan, M.M. Role of Anions in the Synthesis and Crystal Growth of Selected Semiconductors. *Front. Chem.* **2022**, *10*, 881518. [CrossRef]
9. Bouroushian, M. Chalcogens and Metal Chalcogenides. In *Electrochemistry of Metal Chalcogenides*; Monographs in Electrochemistry; Springer: Berlin/Heidelberg, Germany, 2010; pp. 1–56. [CrossRef]
10. Rahman, A.; Khan, M.M. Chalcogenides as photocatalysts. *New J. Chem.* **2021**, *45*, 19622–19635. [CrossRef]
11. Choi, Y.I.; Lee, S.; Kim, S.K.; Kim, Y.-I.; Cho, D.W.; Khan, M.M.; Sohn, Y. Fabrication of ZnO, ZnS, Ag-ZnS, and Au-ZnS microspheres for photocatalytic activities, CO oxidation and 2-hydroxyterephthalic acid synthesis. *J. Alloys Compd.* **2016**, *675*, 46–56. [CrossRef]
12. Khan, M.E.; Cho, M.H. CdS-graphene Nanocomposite for Efficient Visible-light-driven Photocatalytic and Photoelectrochemical Applications. *J. Colloid Interface Sci.* **2016**, *482*, 221–232. [CrossRef]
13. Khan, M.M. *Chalcogenide-Based Nanomaterials as Photocatalysts*; Elsevier: Amsterdam, The Netherlands, 2021. [CrossRef]
14. Tian, M.-W.; Yuen, H.-C.; Yan, S.-R.; Huang, W.-L. The multiple selections of fostering applications of hydrogen energy by integrating economic and industrial evaluation of different regions. *Int. J. Hydrogen Energy* **2019**, *44*, 29390–29398. [CrossRef]
15. Chen, Y.; Feng, X.; Liu, M.; Su, J.; Shen, S. Towards efficient solar-to-hydrogen conversion: Fundamentals and recent progress in copper-based chalcogenide photocathodes. *Nanophotonics* **2016**, *5*, 524–547. [CrossRef]
16. Revankar, S.T. Nuclear hydrogen production. In *Storage and Hybridization of Nuclear Energy: Techno-Economic Integration of Renewable and Nuclear Energy*; Elsevier: Amsterdam, The Netherlands, 2018; pp. 49–117. [CrossRef]
17. Yerga, R.M.N.; Alvarez-Galvan, M.C.; Del Valle, F.; De La Mano, J.A.V.; Fierro, J.L.G. Water Splitting on Semiconductor Catalysts under Visible-Light Irradiation. *ChemSusChem* **2009**, *2*, 471–485. [CrossRef] [PubMed]
18. Maeda, K.; Domen, K. New Non-Oxide Photocatalysts Designed for Overall Water Splitting under Visible Light. *J. Phys. Chem. C* **2007**, *111*, 7851–7861. [CrossRef]

19. Maeda, K.; Lu, D.; Domen, K. Direct Water Splitting into Hydrogen and Oxygen under Visible Light by using Modified TaON Photocatalysts with d^0 Electronic Configuration. *Chem. A Eur. J.* **2013**, *19*, 4986–4991. [CrossRef] [PubMed]
20. Walter, M.G.; Warren, E.L.; McKone, J.R.; Boettcher, S.W.; Mi, Q.; Santori, E.A.; Lewis, N.S. Solar Water Splitting Cells. *Chem. Rev.* **2010**, *110*, 6446–6473. [CrossRef]
21. Li, X.; Zuo, X.; Jiang, X.; Li, D.; Cui, B.; Liu, D. Enhanced photocatalysis for water splitting in layered tin chalcogenides with high carrier mobility. *Phys. Chem. Chem. Phys.* **2019**, *21*, 7559–7566. [CrossRef]
22. Kannan, S.; Vinitha, P.; Mohanraj, K.; Sivakumar, G. Antibacterial studies of novel Cu2WS4 ternary chalcogenide synthesized by hydrothermal process. *J. Solid State Chem.* **2018**, *258*, 376–382. [CrossRef]
23. Sarilmaz, A.; Can, M.; Ozel, F. Ternary copper tungsten selenide nanosheets synthesized by a facile hot-injection method. *J. Alloys Compd.* **2017**, *699*, 479–483. [CrossRef]
24. Saravanan, K.; Selladurai, S.; Ananthakumar, S.; Suriakarthick, R. Solvothermal synthesis of copper cadmium sulphide (CuCdS$_2$) nanoparticles and its structural, optical and morphological properties. *Mater. Sci. Semicond. Process.* **2019**, *93*, 345–356. [CrossRef]
25. Khalil, A.A.I.; El-Gawad, A.-S.H.A.; Gadallah, A.-S. Impact of silver dopants on structural, morphological, optical, and electrical properties of copper-zinc sulfide thin films prepared via sol-gel spin coating method. *Opt. Mater.* **2020**, *109*, 110250. [CrossRef]
26. Gu, X.; Tan, C.; He, L.; Guo, J.; Zhao, X.; Qi, K.; Yan, Y. Mn^{2+} doped AgInS$_2$ photocatalyst for formaldehyde degradation and hydrogen production from water splitting by carbon tube enhancement. *Chemosphere* **2022**, *304*, 135292. [CrossRef]
27. Wang, X.; Feng, Z.; Fan, D.; Fan, F.; Li, C. Shape-Controlled Synthesis of CdS Nanostructures via a Solvothermal Method. *Cryst. Growth Des.* **2010**, *10*, 5312–5318. [CrossRef]
28. Hu, Z.; Chen, T.; Xie, Z.; Guo, C.; Jiang, W.; Chen, Y.; Xu, Y. Emission tunable AgInS$_2$ quantum dots synthesized via microwave method for white light-emitting diodes application. *Opt. Mater.* **2022**, *124*, 111975. [CrossRef]
29. Li, S.; Cai, M.; Liu, Y.; Wang, C.; Yan, R.; Chen, X. Constructing Cd$_{0.5}$Zn$_{0.5}$S/Bi$_2$WO$_6$ S-scheme heterojunction for boosted photocatalytic antibiotic oxidation and Cr(VI) reduction. *Adv. Powder Mater.* **2023**, *2*, 100073. [CrossRef]
30. Thompson, M.J.; Ruberu, T.P.A.; Blakeney, K.J.; Torres, K.V.; Dilsaver, P.S.; Vela, J. Axial Composition Gradients and Phase Segregation Regulate the Aspect Ratio of Cu$_2$ZnSnS$_4$ Nanorods. *J. Phys. Chem. Lett.* **2013**, *4*, 3918–3923. [CrossRef]
31. Mandari, K.K.; Son, N.; Kang, M. CuS/Ag$_2$O nanoparticles on ultrathin g-C$_3$N$_4$ nanosheets to achieve high performance solar hydrogen evolution. *J. Colloid Interface Sci.* **2022**, *615*, 740–751. [CrossRef]
32. Yin, M.; Sun, J.; Li, Y.; Ye, Y.; Liang, K.; Fan, Y.; Li, Z. Efficient photocatalytic hydrogen evolution over MoS$_2$/activated carbon composite sensitized by Erythrosin B under LED light irradiation. *Catal. Commun.* **2020**, *142*, 106029. [CrossRef]
33. Yang, Z.; Xie, X.; Zhang, Z.; Yang, J.; Yu, C.; Dong, S.; Xiang, M.; Qin, H. NiS$_2$@V$_2$O$_5$/VS$_2$ ternary heterojunction for a high-performance electrocatalyst in overall water splitting. *Int. J. Hydrogen Energy* **2022**, *47*, 27338–27346. [CrossRef]
34. Gopalakrishnan, S.; Paulraj, G.; Eswaran, M.K.; Ray, A.; Singh, N.; Jeganathan, K. VS$_2$ wrapped Si nanowires as core-shell heterostructure photocathode for highly efficient photoelectrochemical water reduction performance. *Chemosphere* **2022**, *302*, 134708. [CrossRef]
35. Tang, Y.; Li, X.; Zhang, D.; Pu, X.; Ge, B.; Huang, Y. Noble metal-free ternary MoS$_2$/Zn$_{0.5}$Cd$_{0.5}$S/g-C$_3$N$_4$ heterojunction composite for highly efficient photocatalytic H$_2$ production. *Mater. Res. Bull.* **2018**, *110*, 214–222. [CrossRef]
36. Arai, T.; Senda, S.-I.; Sato, Y.; Takahashi, H.; Shinoda, K.; Jeyadevan, B.; Tohji, K. Cu-Doped ZnS Hollow Particle with High Activity for Hydrogen Generation from Alkaline Sulfide Solution under Visible Light. *Chem. Mater.* **2008**, *20*, 1997–2000. [CrossRef]
37. Oliva, A. Formation of the band gap energy on CdS thin films growth by two different techniques. *Thin Solid Films* **2001**, *391*, 28–35. [CrossRef]
38. Cheng, L.; Xiang, Q.; Liao, Y.; Zhang, H. CdS-Based photocatalysts. *Energy Environ. Sci.* **2018**, *11*, 1362–1391. [CrossRef]
39. Li, C.; Yuan, J.; Han, B.; Shangguan, W. Synthesis and photochemical performance of morphology-controlled CdS photocatalysts for hydrogen evolution under visible light. *Int. J. Hydrogen Energy* **2011**, *36*, 4271–4279. [CrossRef]
40. Garg, P.; Bhauriyal, P.; Mahata, A.; Rawat, K.S.; Pathak, B. Role of Dimensionality for Photocatalytic Water Splitting: CdS Nanotube versus Bulk Structure. *ChemPhysChem* **2018**, *20*, 383–391. [CrossRef]
41. Luo, M.; Liu, Y.; Hu, J.; Liu, H.; Li, J. One-Pot Synthesis of CdS and Ni-Doped CdS Hollow Spheres with Enhanced Photocatalytic Activity and Durability. *ACS Appl. Mater. Interfaces* **2012**, *4*, 1813–1821. [CrossRef]
42. Zhen, W.; Ning, X.; Yang, B.; Wu, Y.; Li, Z.; Lu, G. The enhancement of CdS photocatalytic activity for water splitting via anti-photocorrosion by coating Ni2P shell and removing nascent formed oxygen with artificial gill. *Appl. Catal. B Environ.* **2018**, *221*, 243–257. [CrossRef]
43. Li, Y.-H.; Qi, M.-Y.; Li, J.-Y.; Tang, Z.-R.; Xu, Y.-J. Noble metal free CdS@CuS-NixP hybrid with modulated charge transfer for enhanced photocatalytic performance. *Appl. Catal. B Environ.* **2019**, *257*, 117934. [CrossRef]
44. Yang, J.; Yan, H.; Wang, X.; Wen, F.; Wang, Z.; Fan, D.; Shi, J.; Li, C. Roles of cocatalysts in Pt–PdS/CdS with exceptionally high quantum efficiency for photocatalytic hydrogen production. *J. Catal.* **2012**, *290*, 151–157. [CrossRef]
45. Yang, T.-T.; Chen, W.-T.; Hsu, Y.-J.; Wei, K.-H.; Lin, T.-Y. Interfacial Charge Carrier Dynamics in Core–Shell Au-CdS Nanocrystals. *J. Phys. Chem. C* **2010**, *114*, 11414–11420. [CrossRef]
46. Tso, S.; Li, W.-S.; Wu, B.-H.; Chen, L.-J. Enhanced H$_2$ production in water splitting with CdS-ZnO core-shell nanowires. *Nano Energy* **2017**, *43*, 270–277. [CrossRef]
47. Zhang, F.; Zhuang, H.-Q.; Zhang, W.; Yin, J.; Cao, F.-H.; Pan, Y.-X. Noble-metal-free CuS/CdS photocatalyst for efficient visible-light-driven photocatalytic H$_2$ production from water. *Catal. Today* **2018**, *330*, 203–208. [CrossRef]

48. Cheng, F.; Xiang, Q. A solid-state approach to fabricate a CdS/CuS nano-heterojunction with promoted visible-light photocatalytic H_2-evolution activity. *RSC Adv.* **2016**, *6*, 76269–76272. [CrossRef]

49. Wu, Z.; Zhao, G.; Zhang, Y.-N.; Tian, H.; Li, D. Enhanced Photocurrent Responses and Antiphotocorrosion Performance of CdS Hybrid Derived from Triple Heterojunction. *J. Phys. Chem. C* **2012**, *116*, 12829–12835. [CrossRef]

50. Holmes, M.A.; Townsend, T.K.; Osterloh, F.E. Quantum confinement controlled photocatalytic water splitting by suspended CdSe nanocrystals. *Chem. Commun.* **2011**, *48*, 371–373. [CrossRef]

51. Yang, Q.; Luo, M.; Liu, K.; Cao, H.; Yan, H. Covalent organic frameworks for photocatalytic applications. *Appl. Catal. B Environ.* **2020**, *276*, 119174. [CrossRef]

52. You, J.; Zhao, Y.; Wang, L.; Bao, W. Recent developments in the photocatalytic applications of covalent organic frameworks: A review. *J. Clean. Prod.* **2021**, *291*, 125822. [CrossRef]

53. You, D.; Pan, Z.; Cheng, Q. COFs-Ph@CdS S-scheme heterojunctions with photocatalytic hydrogen evolution and efficient degradation properties. *J. Alloys Compd.* **2023**, *930*, 167069. [CrossRef]

54. Das, A.; Han, Z.; Haghighi, M.G.; Eisenberg, R. Photogeneration of hydrogen from water using CdSe nanocrystals demonstrating the importance of surface exchange. *Proc. Natl. Acad. Sci. USA* **2013**, *110*, 16716–16723. [CrossRef]

55. Park, Y.; Park, B. Effect of ligand exchange on photocurrent enhancement in cadmium selenide (CdSe) quantum dot water splitting cells. *Results Phys.* **2018**, *11*, 162–165. [CrossRef]

56. Zhou, W.; Li, F.; Yang, X.; Yang, W.; Wang, C.; Cao, R.; Zhou, C.; Tian, M. Peanut-chocolate-ball-inspired construction of the interface engineering between CdS and intergrown Cd: Boosting both the photocatalytic activity and photocorrosion resistance. *J. Energy Chem.* **2023**, *76*, 75–89. [CrossRef]

57. Li, K.; Pan, H.; Wang, F.; Zhang, Z.; Min, S. In-situ exsolved NiS nanoparticle-socketed CdS with strongly coupled interfaces as a superior visible-light-driven photocatalyst for hydrogen evolution. *Appl. Catal. B Environ.* **2023**, *321*, 122028. [CrossRef]

58. Basu, M.; Nazir, R.; Fageria, P.; Pande, S. Construction of CuS/Au Heterostructure through a Simple Photoreduction Route for Enhanced Electrochemical Hydrogen Evolution and Photocatalysis. *Sci. Rep.* **2016**, *6*, 34738. [CrossRef]

59. Chandra, M.; Bhunia, K.; Pradhan, D. Controlled Synthesis of CuS/TiO$_2$ Heterostructured Nanocomposites for Enhanced Photocatalytic Hydrogen Generation through Water Splitting. *Inorg. Chem.* **2018**, *57*, 4524–4533. [CrossRef]

60. Dubale, A.A.; Tamirat, A.G.; Chen, H.-M.; Berhe, T.A.; Pan, C.-J.; Su, W.-N.; Hwang, B.-J. A highly stable CuS and CuS–Pt modified Cu$_2$O/CuO heterostructure as an efficient photocathode for the hydrogen evolution reaction. *J. Mater. Chem. A* **2015**, *4*, 2205–2216. [CrossRef]

61. Ma, L.; Yang, D.-J.; Song, X.-P.; Li, H.-X.; Ding, S.-J.; Xiong, L.; Qin, P.-L.; Chen, X.-B. Pt Decorated (Au Nanosphere)/(CuSe Ultrathin Nanoplate) Tangential Hybrids for Efficient Photocatalytic Hydrogen Generation via Dual-Plasmon-Induced Strong VIS–NIR Light Absorption and Interfacial Electric Field Coupling. *Sol. RRL* **2019**, *4*, 1900376. [CrossRef]

62. Liang, H.; Mei, J.; Sun, H.; Cao, L. Enhanced photocatalytic hydrogen evolution of CdS@CuS core-shell nanorods under visible light. *Mater. Sci. Semicond. Process.* **2023**, *153*, 107105. [CrossRef]

63. Hou, J.; Huang, B.; Kong, L.; Xie, Y.; Liu, Y.; Chen, M.; Wang, Q. One-pot hydrothermal synthesis of CdS–CuS decorated TiO$_2$ NTs for improved photocatalytic dye degradation and hydrogen production. *Ceram. Int.* **2021**, *47*, 30860–30868. [CrossRef]

64. Zhuang, H.L.; Hennig, R.G. Single-Layer Group-III Monochalcogenide Photocatalysts for Water Splitting. *Chem. Mater.* **2013**, *25*, 3232–3238. [CrossRef]

65. Peng, Q.; Xiong, R.; Sa, B.; Zhou, J.; Wen, C.; Wu, B.; Anpo, M.; Sun, Z. Computational mining of photocatalysts for water splitting hydrogen production: Two-dimensional InSe-family monolayers. *Catal. Sci. Technol.* **2017**, *7*, 2744–2752. [CrossRef]

66. Peng, Q.; Guo, Z.; Sa, B.; Zhou, J.; Sun, Z. New gallium chalcogenides/arsenene van der Waals heterostructures promising for photocatalytic water splitting. *Int. J. Hydrogen Energy* **2018**, *43*, 15995–16004. [CrossRef]

67. Gupta, U.; Rao, C. Hydrogen generation by water splitting using MoS$_2$ and other transition metal dichalcogenides. *Nano Energy* **2017**, *41*, 49–65. [CrossRef]

68. Rahman, A.; Jennings, J.R.; Tan, A.L.; Khan, M.M. Molybdenum Disulfide-Based Nanomaterials for Visible-Light-Induced Photocatalysis. *ACS Omega* **2022**, *7*, 22089–22110. [CrossRef]

69. Li, M.; Cui, Z.; Li, E. Silver-modified MoS$_2$ nanosheets as a high-efficiency visible-light photocatalyst for water splitting. *Ceram. Int.* **2019**, *45*, 14449–14456. [CrossRef]

70. Yuan, Y.-J.; Chen, D.; Zhong, J.; Yang, L.-X.; Wang, J.; Liu, M.-J.; Tu, W.-G.; Yu, Z.-T.; Zou, Z.-G. Interface engineering of a noble-metal-free 2D–2D MoS$_2$/Cu-ZnIn$_2$S$_4$ photocatalyst for enhanced photocatalytic H$_2$ production. *J. Mater. Chem. A* **2017**, *5*, 15771–15779. [CrossRef]

71. Hassan, M.A.; Kim, M.-W.; Johar, M.A.; Waseem, A.; Kwon, M.-K.; Ryu, S.-W. Transferred monolayer MoS$_2$ onto GaN for heterostructure photoanode: Toward stable and efficient photoelectrochemical water splitting. *Sci. Rep.* **2019**, *9*, 20141. [CrossRef]

72. Zhao, G.; Wang, X.; Wang, S.; Rui, K.; Chen, Y.; Yu, H.; Ma, J.; Dou, S.X.; Sun, W. Heteroatom-doped MoSe$_2$ Nanosheets with Enhanced Hydrogen Evolution Kinetics for Alkaline Water Splitting. *Chem. Asian J.* **2018**, *14*, 301–306. [CrossRef]

73. Lei, W.; Wang, F.; Pan, X.; Ye, Z.; Lu, B. Z-scheme MoO$_3$-2D SnS nanosheets heterojunction assisted g-C$_3$N$_4$ composite for enhanced photocatalytic hydrogen evolutions. *Int. J. HydrogEN Energy* **2022**, *47*, 10877–10890. [CrossRef]

74. Liu, Y.; Zhou, Y.; Zhou, X.; Jin, X.; Li, B.; Liu, J.; Chen, G. Cu doped SnS$_2$ nanostructure induced sulfur vacancy towards boosted photocatalytic hydrogen evolution. *Chem. Eng. J.* **2020**, *407*, 127180. [CrossRef]

75. Geng, Y.; Zou, X.; Lu, Y.; Wang, L. Fabrication of the $SnS_2/ZnIn_2S_4$ heterojunction for highly efficient visible light photocatalytic H_2 evolution. *Int. J. Hydrogen Energy* **2022**, *47*, 11520–11527. [CrossRef]

76. Li, Y.-Y.; Wang, J.-G.; Sun, H.-H.; Hua, W.; Liu, X.-R. Heterostructured SnS_2/SnO_2 nanotubes with enhanced charge separation and excellent photocatalytic hydrogen production. *Int. J. Hydrogen Energy* **2018**, *43*, 14121–14129. [CrossRef]

77. Mangiri, R.; Kumar, K.S.; Subramanyam, K.; Ratnakaram, Y.; Sudharani, A.; Reddy, D.A.; Vijayalakshmi, R. Boosting solar driven hydrogen evolution rate of CdS nanorods adorned with MoS_2 and SnS_2 nanostructures. *Colloids Interface Sci. Commun.* **2021**, *43*, 100437. [CrossRef]

78. Barawi, M.; Flores, E.; Ferrer, I.J.; Ares, J.R.; Sánchez, C. Titanium trisulphide (TiS_3) nanoribbons for easy hydrogen photogeneration under visible light. *J. Mater. Chem. A* **2015**, *3*, 7959–7965. [CrossRef]

79. Flores, E.; Ares, J.; Sánchez, C.; Ferrer, I. Ternary transition titanium-niobium trisulfide as photoanode for assisted water splitting. *Catal. Today* **2018**, *321–322*, 107–112. [CrossRef]

80. Guo, W.; Wu, D. Facile synthesis of VS4/graphene nanocomposites and their visible-light-driven photocatalytic water splitting activities. *Int. J. Hydrogen Energy* **2014**, *39*, 16832–16840. [CrossRef]

81. Ikeda, S.; Aono, N.; Iwase, A.; Kobayashi, H.; Kudo, A. Cu_3MS_4 (M=V, Nb, Ta) and its Solid Solutions with Sulvanite Structure for Photocatalytic and Photoelectrochemical H_2 Evolution under Visible-Light Irradiation. *ChemSusChem* **2018**, *12*, 1977–1983. [CrossRef]

82. Li, G.; Deng, X.; Chen, P.; Wang, X.; Ma, J.; Liu, F.; Yin, S.-F. Sulphur vacancies-$VS_2@C_3N_4$ drived by in situ supramolecular self-assembly for synergistic photocatalytic degradation of real wastewater and H_2 production: Vacancies taming interfacial compact heterojunction and carriers transfer. *Chem. Eng. J.* **2022**, *433*, 134505. [CrossRef]

83. Zhong, X.; Tang, J.; Wang, J.; Shao, M.; Chai, J.; Wang, S.; Yang, M.; Yang, Y.; Wang, N.; Wang, S.; et al. 3D heterostructured pure and N-Doped Ni_3S_2/VS_2 nanosheets for high efficient overall water splitting. *Electrochim. Acta* **2018**, *269*, 55–61. [CrossRef]

84. Kurnia, F.; Ng, Y.H.; Amal, R.; Valanoor, N.; Hart, J.N. Defect engineering of ZnS thin films for photoelectrochemical water-splitting under visible light. *Sol. Energy Mater. Sol. Cells* **2016**, *153*, 179–185. [CrossRef]

85. Sánchez-Tovar, R.; Fernández-Domene, R.M.; Montañés, M.T.; Sanz-Marco, A.; Garcia-Antón, J. ZnO/ZnS heterostructures for hydrogen production by photoelectrochemical water splitting. *RSC Adv.* **2016**, *6*, 30425–30435. [CrossRef]

86. Zhang, X.; Zhou, Y.-Z.; Wu, D.-Y.; Liu, X.-H.; Zhang, R.; Liu, H.; Dong, C.-K.; Yang, J.; Kulinich, S.A.; Du, X.-W. ZnO nanosheets with atomically thin ZnS overlayers for photocatalytic water splitting. *J. Mater. Chem. A* **2018**, *6*, 9057–9063. [CrossRef]

87. Lee, G.-J.; Anandan, S.; Masten, S.J.; Wu, J.J. Photocatalytic hydrogen evolution from water splitting using Cu doped ZnS microspheres under visible light irradiation. *Renew. Energy* **2016**, *89*, 18–26. [CrossRef]

88. Wu, Y.; Wang, H.; Tu, W.; Wu, S.; Chew, J.W. Effects of composition faults in ternary metal chalcogenides (Zn In_2S_3+, x = 1 − 5) layered crystals for visible-light-driven catalytic hydrogen generation and carbon dioxide reduction. *Appl. Catal. B Environ.* **2019**, *256*, 117810. [CrossRef]

89. Fan, H.-T.; Wu, Z.; Liu, K.-C.; Liu, W.-S. Fabrication of 3D $CuS@ZnIn_2S_4$ hierarchical nanocages with 2D/2D nanosheet subunits p-n heterojunctions for improved photocatalytic hydrogen evolution. *Chem. Eng. J.* **2022**, *433*, 134474. [CrossRef]

90. Ni, T.; Yang, Z.; Cao, Y.; Lv, H.; Liu, Y. Rational design of $MoS_2/g\text{-}C_3N_4/ZnIn_2S_4$ hierarchical heterostructures with efficient charge transfer for significantly enhanced photocatalytic H_2 production. *Ceram. Int.* **2021**, *47*, 22985–22993. [CrossRef]

91. Guo, X.; Peng, Y.; Liu, G.; Xie, G.; Guo, Y.; Zhang, Y.; Yu, J. An Efficient $ZnIn_2S_4@CuInS_2$ Core–Shell p–n Heterojunction to Boost Visible-Light Photocatalytic Hydrogen Evolution. *J. Phys. Chem. C* **2020**, *124*, 5934–5943. [CrossRef]

92. Yang, W.; Ma, G.; Fu, Y.; Peng, K.; Yang, H.; Zhan, X.; Yang, W.; Wang, L.; Hou, H. Rationally designed Ti_3C_2 MXene$@TiO_2/CuInS_2$ Schottky/S-scheme integrated heterojunction for enhanced photocatalytic hydrogen evolution. *Chem. Eng. J.* **2021**, *429*, 132381. [CrossRef]

93. Raja, A.; Son, N.; Swaminathan, M.; Kang, M. Facile synthesis of sphere-like structured $ZnIn_2S_4$-rGO-$CuInS_2$ ternary heterojunction catalyst for efficient visible-active photocatalytic hydrogen evolution. *J. Colloid Interface Sci.* **2021**, *602*, 669–679. [CrossRef]

94. Li, Q.; Qiao, X.-Q.; Jia, Y.; Hou, D.; Li, D.-S. Amorphous $CoMoS_4$ Nanostructure for Photocatalytic H_2 Generation, Nitrophenol Reduction, and Methylene Blue Adsorption. *ACS Appl. Nano Mater.* **2019**, *3*, 68–76. [CrossRef]

95. Yao, Z.; Wang, L.; Zhang, Y.; Yu, Z.; Jiang, Z. Carbon nanotube modified $Zn_{0.83}Cd_{0.17}S$ nanocomposite photocatalyst and its hydrogen production under visible-light. *Int. J. Hydrogen Energy* **2014**, *39*, 15380–15386. [CrossRef]

96. Lv, H.; Kong, Y.; Gong, Z.; Zheng, J.; Liu, Y.; Wang, G. Engineering multifunctional carbon black interface over $Mn_{0.5}Cd_{0.5}S$ nanoparticles/CuS nanotubes heterojunction for boosting photocatalytic hydrogen generation activity. *Appl. Surf. Sci.* **2022**, *604*, 154513. [CrossRef]

97. Bai, Y.; Zhang, Q.; Xu, N.; Deng, K.; Kan, E. The Janus structures of group-III chalcogenide monolayers as promising photocatalysts for water splitting. *Appl. Surf. Sci.* **2019**, *478*, 522–531. [CrossRef]

98. Hu, L.; Wei, D. Janus Group-III Chalcogenide Monolayers and Derivative Type-II Heterojunctions as Water-Splitting Photocatalysts with Strong Visible-Light Absorbance. *J. Phys. Chem. C* **2018**, *122*, 27795–27802. [CrossRef]

99. Wang, P.; Zong, Y.; Liu, H.; Wen, H.; Wu, H.-B.; Xia, J.-B. Highly efficient photocatalytic water splitting and enhanced piezoelectric properties of 2D Janus group-III chalcogenides. *J. Mater. Chem. C* **2021**, *9*, 4989–4999. [CrossRef]

100. Ahmad, I.; Shahid, I.; Ali, A.; Gao, L.; Cai, J. Electronic, mechanical, optical and photocatalytic properties of two-dimensional Janus XGaInY (X, Y;= S, Se and Te) monolayers. *RSC Adv.* **2021**, *11*, 17230–17239. [CrossRef]
101. Chu, S.; Li, W.; Yan, Y.; Hamann, T.; Shih, I.; Wang, D.; Mi, Z. Roadmap on solar water splitting: Current status and future prospects. *Nano Futur.* **2017**, *1*, 022001. [CrossRef]

Ultrasound/Chlorine: A Novel Synergistic Sono-Hybrid Process for Allura Red AC Degradation

Oualid Hamdaoui [1,*], Slimane Merouani [2], Hadjer C. Benmahmoud [3], Meriem Ait Idir [3], Hamza Ferkous [3] and Abdulaziz Alghyamah [1]

1. Chemical Engineering Department, College of Engineering, King Saud University, P.O. Box 800, 11421 Riyadh, Saudi Arabia
2. Laboratory of Environmental Process Engineering, Department of Chemical Engineering, Faculty of Process Engineering, University Salah Boubnider Constantine 3, P.O. Box 72, Constantine 25000, Algeria
3. Process Engineering Department, Faculty of Technology, Badji Mokhtar Annaba University, P.O. Box 12, Annaba 23000, Algeria
* Correspondence: ohamdaoui@ksu.edu.sa or ohamdaoui@yahoo.fr

Abstract: Herein, we present an original report on chlorine activation by ultrasound (US: 600 kHz, 120 W) for intensifying the sonochemical treatment of hazardous organic materials. The coupling of US/chlorine produced synergy via the involvement of reactive chlorine species (RCSs: $Cl^{\bullet}$, $ClO^{\bullet}$ and $Cl_2^{\bullet-}$), resulting from the sono-activation of chlorine. The degradation of Allura Red AC (ARAC) textile dye, as a contaminant model, was drastically improved by the US/chlorine process as compared to the separated techniques. A synergy index of 1.74 was obtained by the US/chlorine process for the degradation of ARAC ($C_0 = 5$ mg·L^{-1}) at pH 5.5 and [chlorine]$_0$ = 250 mM. The synergistic index increased by up to 2.2 when chlorine concentration was 300 µM. Additionally, the synergetic effect was only obtained at pH 4–6, where HOCl is the sole chlorine species. Additionally, the effect of combining US and chlorine for ARAC degradation was additive for the argon atmosphere, synergistic for air and negative for N_2. An air atmosphere could provide the best synergy as it generates a relatively moderate concentration of reactive species as compared to argon, which marginalizes radical–radical reactions compared to radical–organic ones. Finally, the US/chlorine process was more synergistic for low pollutant concentrations ($C_0 \leq 10$ mg·L^{-1}); the coupling effect was additive for moderate concentrations (C_0~20–30 mg·L^{-1}) and negative for higher C_0 (>30 mg·L^{-1}). Consequently, the US/chlorine process was efficiently operable under typical water treatment conditions, although complete by-product analysis and toxicity assessment may still be necessary to establish process viability.

Keywords: ultrasound/chlorine process; reactive chlorine species (RCS); Allura Red AC (ARAC); degradation; synergy

Citation: Hamdaoui, O.; Merouani, S.; Benmahmoud, H.C.; Ait Idir, M.; Ferkous, H.; Alghyamah, A. Ultrasound/Chlorine: A Novel Synergistic Sono-Hybrid Process for Allura Red AC Degradation. *Catalysts* **2022**, *12*, 1171. https://doi.org/10.3390/catal12101171

Academic Editors: Gassan Hodaifa, Rafael Borja and Mha Albqmi

Received: 9 September 2022
Accepted: 30 September 2022
Published: 4 October 2022

Publisher's Note: MDPI stays neutral with regard to jurisdictional claims in published maps and institutional affiliations.

1. Introduction

Due to the high chemical stability and/or low biodegradability of most of water contaminants, one feasible option for removing organic pollutants from wastewater is the use of advanced oxidation processes (AOPs) [1]. These processes, e.g., Fe(II)/H_2O_2 (or $S_2O_8^{2-}$), UV/H_2O_2 (or $S_2O_8^{2-}$), UV/O_3, H_2O_2/O_3 and UV/TiO_2, have been widely recognized as highly effective treatments for recalcitrant wastewater or as a pretreatment to convert micropollutants into shorter chain substrates that can then be treated using conventional biological methods [2]. AOPs generate reactive free radicals, i.e., $^{\bullet}OH$ ($E^0 = 2.8$ V/NHE) or ($SO_4^{\bullet-}$, $E^0 = 2.6$ V/NHE), that are non-selective and highly reactive toward most organic pollutants [3,4].

High-power ultrasound (US: 100–1000 kHz) is one tool to generate $^{\bullet}OH$ radicals for water treatment applications [5]. Water sonolysis creates cavitation bubbles, which grow and collapse implosively in a time scale of microseconds, generating extreme temperatures and pressures within the bubbles (~5000 K and ~1000 atm) [6]. Radicals such as $^\bullet$OH, H$^\bullet$ and HO$_2$$^\bullet$, and atomic oxygen (O) are formed as a consequence of the pyrolytic reactions inside the bubble, under the extraordinary temperature developed at the final stage of the collapse [7–9]. The formation mechanism of these reactive moieties begins with the homolytic cleavage of water vapor and O$_2$ trapped inside the bubble to form $^\bullet$OH, H$^\bullet$ and O atoms (i.e., H$_2$O $\rightarrow$ $^\bullet$OH + H$^\bullet$ and O$_2$ $\rightarrow$ 2O) [10,11]. These primary species can react with H$_2$O or O$_2$ molecules to produce other reactive species such as HO$_2$$^\bullet$ (i.e., H$^\bullet$ + O$_2$ $\rightarrow$ HO$_2$$^\bullet$). Upon collapse, these radicals can diffuse into the bubble/solution interface to recombine or react with solutes present at the interfacial region. H$_2$O$_2$ is the main product of radicals' recombination at the bubble/solution interface, i.e., 2$^\bullet$OH $\rightarrow$ H$_2$O$_2$ and 2HO$_2$$^\bullet$ $\rightarrow$ H$_2$O$_2$ + O$_2$. Some radical reactions can also take place in the liquid bulk as ~10% of the formed radicals can reach the solution zone [12]. The hydroxyl radical is the most important radical implicated in the sonochemical treatment of hydrophobic/hydrophilic organic pollutants [5]. Parallel reaction pathways exist whereby volatile solutes may evaporate into the bubble and be pyrolyzed by the high core temperatures [13].

Recently, UV light and iron have been reported as efficient activators of chlorine for AOP-applications [14–16]. The UV/chlorine and Fe(II)/chlorine systems are emerging as attractive alternatives to UV/H$_2$O$_2$ and Fe(II)/H$_2$O$_2$ traditional AOPs. The interaction between UV light (<400 nm) or iron with chlorine can produce $^\bullet$OH and Cl$^\bullet$ as initial reactive species (Equations (1)–(4)) [16,17]. These radicals can then drive a reaction chain in which additional reactive chlorine species (RCSs), such as ClO$^\bullet$ and Cl$_2$$^{\bullet-}$, can be formed. ClO$^\bullet$ forms via a reaction of Cl$^\bullet$ or $^\bullet$OH with chlorine, while Cl$_2$$^{\bullet-}$ forms via a reaction of Cl$^\bullet$ with Cl$^-$. Interesting work on oxidant formation mechanisms in UV/chlorine and Fe(II)/chlorine systems and their subsequent reactions with organics has been delivered by many researchers [14–20].

$$HClO + h\nu \rightarrow {}^\bullet OH + Cl^\bullet \quad k_1 = \sim 1.3 \times 10^{-3} \text{ s}^{-1} \tag{1}$$

$$ClO^- + h\nu \rightarrow O^{\bullet-} + Cl^\bullet \quad k_2 = 9 \times 10^{-3} \text{ s}^{-1} \tag{2}$$

$$Fe(II) + HClO \rightarrow Fe(III) + {}^\bullet OH + Cl^- \quad k_3 = \sim 10^4 \text{ M}^{-1}\text{ s}^{-1} \tag{3}$$

$$Fe(II) + HClO \rightarrow Fe(III) + Cl^\bullet + OH^- \quad k_4 = \sim 10^4 \text{ M}^{-1}\text{ s}^{-1} \tag{4}$$

The generated reactive chlorine species (RCSs: Cl$^\bullet$, ClO$^\bullet$ and Cl$_2$$^{\bullet-}$) are of high redox potentials (2.43 V/NHE for Cl$^\bullet$, 2.13 V/NHE for Cl$_2$$^{\bullet-}$ and 1.5–1.8 V/NHE for ClO$^\bullet$) and are mostly implicated in the destruction of several water contaminants [21–32]. Unlike $^\bullet$OH, RCSs are selective oxidants that preferentially react with electron-rich moieties [33]. These radicals react with organic matter practically with the same mechanisms as those of $^\bullet$OH (i.e., H-atom abstraction, electron transfer or addition to unsaturated bands) [14]. The second-order rate constants for reactions involving RCS radicals varies in the order of ~10^8–10^{11} M^{-1}·s^{-1} for Cl$^\bullet$, ~10^7–10^9 M^{-1}·s^{-1} for ClO$^\bullet$ and 10^2–10^6 M^{-1}·s^{-1} for Cl$_2$$^{\bullet-}$ [14,15]. Additionally, RCSs have been characterized by a longer lifetime than that of $^\bullet$OH, i.e., 5 μs for Cl$^\bullet$ [34] and fractions of milliseconds for Cl$_2$$^{\bullet-}$ [34], making them available for longer in the solution. All these advantageous specifications of RCSs make them efficient oxidation agents that may create a parallel degradation pathway to that of $^\bullet$OH, thereby accelerating the degradation of micropollutants. Due to the significant implication of RCSs in the UV/chlorine system, this process was found to be more efficient in degrading several water contaminants than UV/H$_2$O$_2$ [25].

Chlorine is globally the most used chemical oxidant for drinking water disinfection and it led to the formation of trihalomethanes and haloacetic acids. Despite its low activity on microorganisms in biofilms, chlorine can lead to a significant removal of the majority of planktonic bacteria [35]. The combination of ultrasound with chlorine can improve the disinfection process, with the advantage of preventing the formation of disinfection by-products and reducing the chlorine dosage required for achieving admissible disinfection efficiencies [36]. Despite numerous reports of using acoustic cavitation to accelerate wastewater disinfection processes [36–38], the application of the US/chlorine system as an oxidation technique for the degradation of organic pollutants is, unexpectedly, very scarce [39]. The US/chlorine technique may present numerous preliminary advantages [39]:

- Chlorine is easy to handle, with more safety, as it is less harmful than other oxidants. Furthermore, the liquid phase of chlorine simplifies its use.
- Chlorine is more available and less expensive than other oxidants that require in situ production via a sophisticated expensive device. Consequently, the cost of the US/chlorine technique could be lower than other processes for similar experimental conditions.
- The US/chlorine treatment does not necessitate elimination of the residual chlorine as it is originally employed as a disinfectant.

In this work, chlorine activation by US as a new promising sono-hybrid advanced oxidation process is investigated, using Allura Red AC (ARAC) synthetic dye as a substrate model. ARAC is a very persistent textile dye with established carcinogenic and toxic effects [40–42]. The sonication runs were conducted at 600 kHz using a standing wave reactor operating in continuous mode. The aims of the work are: (i) to investigate the possible activation of chlorine by power ultrasound at 600 kHz and 120 W, (ii) to demonstrate the synergistic effect of the US/chlorine process toward the degradation of ARAC, (iii) to propose a reaction mechanism for the sono-activation of chlorine and (iv) to study the influence of various processing conditions, i.e., pH, chlorine and ARAC concentrations, and the nature of saturating gases on the synergistic effect of the US/chlorine sono-hybrid process. To the best of our knowledge, no previous research has been conducted to explore the synergistic effect of US/chlorine toward the oxidation of organic pollutants in aqueous media, except our recent paper [39].

2. Results and Discussion

2.1. Aqueous Chlorine Chemistry and ARAC Chlorination Tests

In water treatment, gaseous chlorine (Cl_2) or hypochlorite are commonly used for chlorination processes. Chlorine gas (Cl_2) hydrolyzes in water according to the reaction [14]:

$$Cl_2 + H_2O \rightleftharpoons HOCl + H^+ + Cl^- \quad K_{a1} = 3.94 \times 10^{-4} \text{ M}^2 \text{ at } 25\,°C \text{ and } I = 0 \tag{5}$$

For a temperature range of 0–25 °C, K_{a1} ranges from 1.3×10^{-4} to 5.1×10^{-4} M^2 [35]. Hypochlorous acid (HOCl) resulting from Equation (5) is a weak acid, which dissociates in aqueous solution according to the reaction from Equation (6) [14]:

$$HOCl \rightleftharpoons OCl^- + H^+ \quad pK_{a2} = 7.536 \quad K_{a2} = 2.9 \times 10^{-8} \text{ M at } 25\,°C \text{ and } I = 0 \tag{6}$$

The K_{a2} value also depends on temperature, it varies between 1.5×10^{-8} M at 0 °C and 2.9×10^{-8} M at 25 °C. The distribution of free chlorine species depends on chloride concentration, temperature and pH; of all, pH is the most impactful parameter. Figure S1a (Supplementary Materials) shows the calculated distribution of Cl_2, HOCl and ClO^- as a function of pH at 25 °C and for a chloride concentration of 2 mM. At 25 °C, Cl_2 is only present at low pH values (pH < 3). HOCl is the predominant free chlorine species at pH < 7.5, and ClO^- at pH > 7.5. More than 98% of free chlorine is present as HOCl in the pH range of 2.5–6, and as ClO^- at pH > 9. Therefore, under typical water treatment conditions in the pH range 6–9, hypochlorous acid and hypochlorite are the main chlorine species. In addition to these major chlorine species, other chlorine intermediates such as Cl_3^- and Cl_2O can also be formed [15], but their concentrations are very low. HOCl and ClO^- absorb UV light at wavelengths ranging from 200 to 375 nm (Figure S1b, Supplementary Materials). Absorption spectrums show maximum absorption bands centered at 236 nm for HOCl ($\varepsilon\sim102 \pm 2$ M^{-1} cm^{-1}) and at 294 nm for ClO^- ($\varepsilon\sim275 \pm 8$ M$^{-1}\cdot$cm^{-1}).

The effect of an initial solution pH in the range of 1–10 on the direct chlorination of ARAC at 25 °C when using 250 µM of chlorine in continuously stirred (300 rpm) solutions was investigated in our previous work [39]. Additionally, the effect of chlorine dosage (50–300 µM) was also investigated at pH 5.5 [39]. Fast chlorination rates of the dye were observed in strong acidic and basic mediums [39]. Removals of 80% for pH 1 and 45% for pH 8–10 were achieved after 10 min of reaction, while 97%, 69% and 90% of the initial ARAC concentrations were eliminated after 40 min for pH 1, 8 and 10, respectively. However, very low removals of about 10–15% were obtained at pH 3–6 for up to 40 min of reaction. The dye molecules kept the same molecular form at pH 1–10 as the pK_a of the dye was 11.4 [43]. Therefore, ARAC reacts efficiently with Cl_2 (pH 1) and OCl^- (pH 8 and 10), whereas the dye molecules showed strong persistence toward the reaction with HOCl (pH 3–6), even at varying chlorine dosages [39]. Such pH dependence of chlorine reactivity has been previously reported for several organic and inorganic micropollutants such as triclosan, estrogenic steroid hormones, bisphenol A, acetaminophen, 4-n-nonylphenol and ammonia [35]. The oxidation potential of chlorine species is as follows: $Cl_2 > HOCl > OCl^-$. However, for a given compound, HOCl and ClO^- reactivities are usually significantly varied [35]. In addition, different species of the pollutant can be present in solution (i.e., depending on the pK_a of substrates). Therefore, pH dependence of the second-order rate constant is typically reported for chlorination reactions [35]. An important review on chlorine species reactivity toward a number of water contaminants is given by Deborde and Gunten [35]. The review illustrates some reaction mechanisms that take place during the chlorination of micropollutants.

2.2. Chlorine Sonolysis

The sonication of 250 µM chlorine aqueous solutions at 600 kHz and 120 W was conducted for pH 5 and 9. The recorded absorption spectrums during the sonolytic runs are plotted in Figure 1a,b. At pH 5, the HClO spectrum increases with time and its intensity becomes more important at higher irradiation times, particularly for $\lambda < \lambda_{max} = 236$ nm. This trend reveals that the sonolysis of HClO may produce species that efficiently absorb UV light in the same band as that of HClO. These species may be characterized by higher molar absorption coefficients (ε) than that of HClO, which allows them to attain higher absorbance values even at lower concentration.

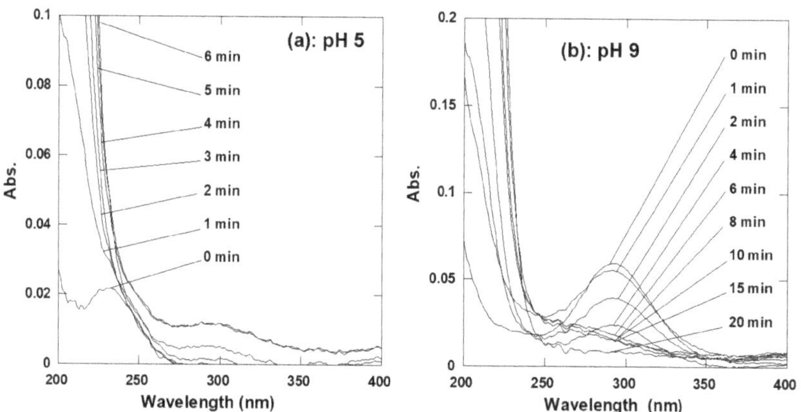

Figure 1. Chlorine sonolysis at pH 5 (**a**) and 9 (**b**) (conditions: [chlorine]$_0$ = 250 µM, V = 150 mL, temperature: 25 ± 1 °C, frequency: 600 kHz, power: 120 W).

However, chlorine decay upon sonolysis was clearly observed at pH 9 (Figure 1b). The pick intensity of OCl^- at 294 nm decreases progressively with time until it attains total disappearance at 20 min of irradiation; the trend was simultaneously accompanied by quick growth of the absorption band, for which $\lambda < 240$ nm, which can confirm the above suggestion concerning the high molar absorption coefficients (ε) of chlorine sono-products. Thus, it can be concluded that the product of ClO^- sonolysis does not absorb UV light in the same region as that of hypochlorite. Chlorine was presumably degraded via a radical pathway in which chlorine species initially react with the acoustically generated reactive species ($^{\bullet}OH$, $H^{\bullet}$, $HO_2^{\bullet}$ and H_2O_2). The following reaction mechanism (Equations (7)–(32)) including initiation [14,15,17,44], propagation [14,15,22,45] and termination reactions [14,46] (all involved in UV/chlorine AOP [14,15,17]) is postulated for a pH range of 1–10:

$$\text{Initiation: } HO_2^{\bullet} \rightarrow O_2^{\bullet-} + H^+ \quad pK_a = 4.7 \tag{7}$$

$$^{\bullet}OH + HClO \rightarrow ClO^{\bullet} + H_2O \quad k_8 = 2 \times 10^9 \text{ M}^{-1}\text{ s}^{-1} \tag{8}$$

$$^{\bullet}OH + ClO^- \rightarrow ClO^{\bullet} + OH^- \quad k_9 = 9.8 \times 10^9 \text{ M}^{-1}\text{ s}^{-1} \tag{9}$$

$$^{\bullet}H + HClO \rightarrow {}^{\bullet}OH + HCl \quad k_{10} = 5 \times 10^8 \text{ M}^{-1}\text{ s}^{-1} \tag{10}$$

$$^{\bullet}H + HClO \rightarrow Cl^{\bullet} + H_2O \quad k_{11} = 5 \times 10^8 \text{ M}^{-1}\text{ s}^{-1} \tag{11}$$

$$^{\bullet}H + HClO \rightarrow ClO^{\bullet} + H_2 \quad k_{12} = 5 \times 10^8 \text{ M}^{-1}\text{ s}^{-1} \tag{12}$$

$$HO_2^{\bullet}/O_2^{\bullet-} + HClO \rightarrow Cl^{\bullet} + O_2 + H_2O/OH^- \quad k_{13} = 7.5 \times 10^6 \text{ M}^{-1}\text{ s}^{-1} \tag{13}$$

$$H_2O_2 + HClO \rightarrow HCl + O_2 + H_2O \quad k_{14} = 1.1 \times 10^4 \text{ M}^{-1}\text{ s}^{-1} \tag{14}$$

$$H_2O_2 + ClO^- \rightarrow Cl^- + O_2 + H_2O \quad k_{15} = 1.7 \times 10^5 \text{ M}^{-1}\text{ s}^{-1} \tag{15}$$

$$\text{Propagation: } HCl \rightleftharpoons H^+ + Cl^- \quad pK_a = -6.3 \tag{16}$$

$$Cl^{\bullet} + HClO \rightarrow ClO^{\bullet} + H^+ + Cl^- \quad k_{17} = 3 \times 10^9 \text{ M}^{-1}\text{ s}^{-1} \tag{17}$$

$$Cl^{\bullet} + ClO^- \rightarrow ClO^{\bullet} + Cl^- \quad k_{18} = 8.2 \times 10^9 \text{ M}^{-1}\text{ s}^{-1} \tag{18}$$

$$Cl^{\bullet} + H_2O \rightarrow HClO^{\bullet-} + H^+ \quad k_{19} = 2.5 \times 10^5 \text{ s}^{-1} \tag{19}$$

$$Cl^{\bullet} + OH^- \rightarrow HClO^{\bullet-} \quad k_{20} = 1.8 \times 10^{10} \text{ M}^{-1}\text{ s}^{-1} \tag{20}$$

$$HClO^{\bullet-} \rightleftharpoons {}^{\bullet}OH + Cl^- \quad k_{21} = 6.9 \times 10^9 \text{ M}^{-1}\text{ s}^{-1} \tag{21}$$

$$HClO^{\bullet-} + H^+ \rightarrow Cl^{\bullet} + H_2O \quad k_{22} = 2.1 \times 10^{10} \text{ M}^{-1}\text{ s}^{-1} \tag{22}$$

$$HClO^{\bullet-} + Cl^- \rightarrow Cl_2^{\bullet-} + OH^- \quad k_{23} = 1 \times 10^5 \text{ M}^{-1}\text{ s}^{-1} \tag{23}$$

$$Cl^{\bullet} + Cl^- \rightleftharpoons Cl_2^{\bullet-} \quad k_{24} = 8.5 \times 10^9 \text{ M}^{-1}\text{ s}^{-1} \tag{24}$$

$$Cl_2^{\bullet-} + H_2O \rightarrow Cl^- + HClO^{\bullet-} + H^+ \quad k_{25} = 1 \times 10^5 \text{ M}^{-1}\text{ s}^{-1} \tag{25}$$

$$Cl_2^{\bullet-} + OH^- \rightarrow Cl^- + HClO^{\bullet-} \quad k_{26} = 1 \times 10^5 \text{ M}^{-1}\text{ s}^{-1} \tag{26}$$

$$\text{Termination: } Cl_2^{\bullet-} + {}^{\bullet}OH \rightarrow HClO + Cl^- \quad k_{27} = 1 \times 10^9 \text{ M}^{-1}\text{ s}^{-1} \tag{27}$$

$$Cl_2^{\bullet-} + Cl^{\bullet} \rightarrow Cl_2 + Cl^- \quad k_{28} = 2.1 \times 10^9 \text{ M}^{-1}\text{ s}^{-1} \tag{28}$$

$$^{\bullet}OH + {}^{\bullet}OH \rightarrow H_2O_2 \quad k_{29} = 5.5 \times 10^9 \text{ M}^{-1}\text{ s}^{-1} \tag{29}$$

$$Cl^{\bullet} + Cl^{\bullet} \rightarrow Cl_2 \quad k_{30} = 8.8 \times 10^7 \text{ M}^{-1}\text{ s}^{-1} \tag{30}$$

$$Cl_2^{\bullet-} + Cl_2^{\bullet-} \rightarrow Cl_2 + 2Cl^- \quad k_{31} = 6.3 \times 10^8 \text{ M}^{-1}\text{ s}^{-1} \tag{31}$$

$$ClO^{\bullet} + ClO^{\bullet} \rightarrow Cl_2O_2 \quad k_{32} = 7.5 \times 10^9 \text{ M}^{-1}\text{ s}^{-1} \tag{32}$$

Therefore, the sonolysis of chlorine can produce a number of highly reactive chlorine species (RCSs), i.e., mainly $Cl^{\bullet}$, $ClO^{\bullet}$ and $Cl_2^{\bullet-}$, which can be used for intensifying the degradation of water contaminants in a similar manner to that reported for UV/chlorine AOP.

2.3. Synergism of Coupling Ultrasound and Chlorine Treatments

Figure 2 shows the degradation kinetics of ARAC at 25 °C via the ultrasound (US: 600 kHz, 120 kHz), chlorine and US/chlorine processes for C_0 = 5 mg·L^{-1} (10 μM), [chlorine]$_0$ = 250 μM and a natural pH of 5.5. As clearly seen, removals of 10% and 50% were obtained after 10 min with, respectively, chlorine and US separately, whereas the US/chlorine combination ensured 92% removal at this time (i.e., 1.84- and 9.2-fold increases in chlorine and US yields separately). After 30 min, the ARAC was removed at 99%, compared to 15% and 70% for, respectively, chlorination alone and sonolysis lone. The initial rate of ARAC removal (r_0) was 0.8 mg·L^{-1}·min^{-1} for the US/chlorine treatment compared to 0.35 mg·L^{-1}·min^{-1} for US and 0.1 mg·L^{-1}·min^{-1} for chlorine oxidation, yielding an $r_{0,US/chlorine}/r_{0,US}$ ratio of 2.28 and a synergy index SI = $r_{0,US/chlorine}/(r_{0,US} + r_{0,chlorine})$ equal to 1.74.

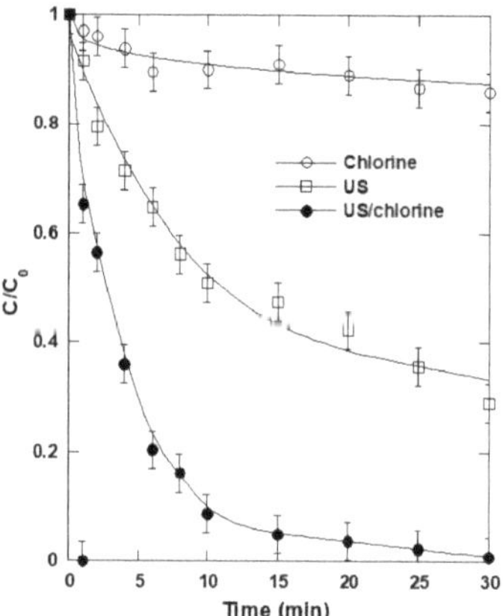

Figure 2. ARAC degradation kinetics via chlorine, ultrasound (US) and US/chlorine processes (conditions: C_0 = 5 mg·L^{-1} (10 μM), [chlorine]$_0$ = 250 μM, V = 150 mL, pH 5.5, temperature: 25 ± 1 °C, frequency: 600 kHz, power: 120 W).

The change in UV-Vis spectrums with time during the treatment of ARAC with the US and US/chlorine processes is shown in Figure 3a,b. For both processes, the abatement of the visible band of the chromophoric group is associated with (i) a decrease in the UV band at λ_{max} = 315 nm, which represents the absorptivity of the naphthenic group, and (ii) a rapid increase in the UV band for which λ_{max} < 250 nm (i.e., the absorption zone of the degradation by-products). This means that there is effective destruction of the dye molecules, and not only a decolorization process. However, the abatement rate of the UV-315 nm band is too rapid for the US/chlorine process compared to sole sonication. In fact, this band completely disappeared after 40 min under the sono-chlorination process compared to 100 min US. The initial rates of absorbance abatement at λ_{max} = 315 nm are 4.64×10^{-3} min^{-1} for US/chlorine and 1.21×10^{-3} min^{-1} for US, which provides a ratio of 3.83 (i.e., no change in absorbance at λ_{max} = 315 nm was recorded with the chlorine treatment). Thus, the synergistic effect of the US/chlorine system is more efficient for aromatic ring destruction than ARAC decolorization.

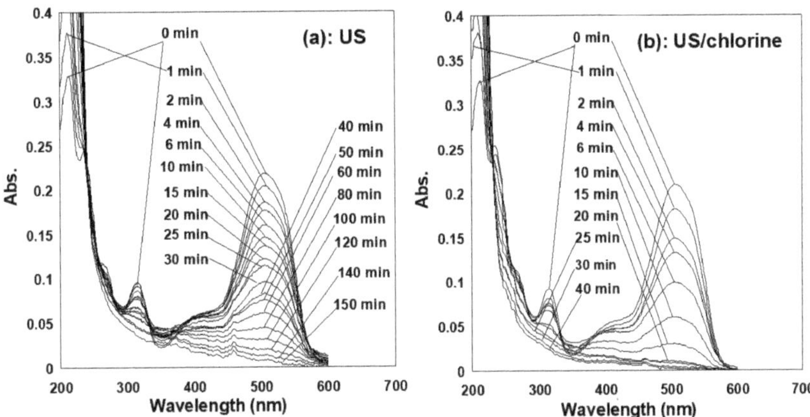

Figure 3. Changes in UV-Vis spectrums during the treatment of ARAC via ultrasound (US) (**a**) and US/chlorine (**b**) processes (conditions: C_0 = 5 mg·L^{-1} (10 µM), [chlorine]$_0$ = 250 µM, V = 150 mL, pH 5.5, temperature: 25 ± 1 °C, frequency: 600 kHz, power: 120 W).

2.4. Source of the Synergistic Effect

Firstly, ARAC is a highly hydrophilic water solute (solubility: 225 g·L^{-1}, log K$_{ow}$ = −0.55 [47]) of negligible volatility. Thus, ARAC cannot enter the bubbles during the sonolytic treatment but must be degraded outside the bubbles by a reaction with the $^\bullet$OH radical ejected from inside the bubbles. This degradation pathway has been confirmed by the addition of nitrobenzene (NB) as a selective scavenger of $^\bullet$OH ($k_{NB,\,^\bullet OH}$ = 3.9 × 10^9 [15]). ARAC removal was inhibited by more than 90% when NB was added at 1 mM [39]. Moreover, H$_2$O$_2$ analysis demonstrated the dominance of an $^\bullet$OH attack on ARAC molecules at the bubble/solution interface [39]. H$_2$O$_2$ is mainly formed at the bubble/solution interface via 2$^\bullet$OH → H$_2$O$_2$ (k = 5.5 × 10^9 M^{-1}·s^{-1}) [48,49]. The rate of hydrogen peroxide formation decreased from 5.6 µM·min^{-1} in pure water to 4.17 µM·min^{-1} in ARAC aqueous solution (5 mg·L^{-1}), meaning that ARAC molecules scavenge an appreciable portion of hydroxyl radicals located at the reactive interfacial region. Consequently, the sono-degradation of ARAC mainly takes place at the bubble/solution interface, but some degradation reactions can also occur in the bulk solution since NB addition in excess ([NB]$_0$/[ARAC]$_0$ = 100 at 1 mM NB) does not completely quench the dye degradation [39]. In fact, it is reported that ~10% of the formed $^\bullet$OH in the bubble can achieve the solution bulk (i.e., the concentration of radicals is higher at the bubble/solution interface) [12,50].

Therefore, the synergism resulting from the application of the US/chlorine process was attributed to the sonolytic activation of chlorine (Equations (7)–(32)). Acoustically generated reactive species ($^\bullet$OH, H$^\bullet$, HO$_2^\bullet$ and H$_2$O$_2$) can react with HClO/ClO$^-$ to produce RCSs (Cl$^\bullet$, ClO$^\bullet$ and Cl$_2^{\bullet-}$) that work together with $^\bullet$OH to quickly destroy the dye molecules. The overall degradation event mostly takes place at the bubble/solution interface where the maximum concentration of reactive species is present. However, RCSs were characterized by a longer lifetime than that of $^\bullet$OH, i.e., 5 µs for Cl$^\bullet$ [34] and fractions of milliseconds for Cl$_2^{\bullet-}$ [34], revealing that RCSs have enough time to diffuse far from the bubble interface towards the solution bulk and react with ARAC molecules.

Consequently, ARAC degradation in the US/chlorine system may also happen in the bulk of the solution. Thus, the US/chlorine process could be promising for the degradation of hydrophilic pollutants.

2.5. Synergism Dependence of Chlorine Dosage

Figure 4 shows the effect of an initial chlorine dosage in the range of 50–300 µM on the removal kinetics of ARAC (C_0 = 5 mg·L^{-1}) via the US/chlorine combination at 25 °C and pH 5.5. The removal rate of the dye increased rapidly with increasing [chlorine]$_0$. The ARAC removal efficiency after 6 min increased from 35% for US alone to 63%, 68%, 80% and 100% when chlorine was added at 100, 200, 250 and 300 µM, respectively. Chlorination alone removed at maximum 10–15% of ARAC for all investigated [chlorine]$_0$ [39]. The initial rate of ARAC removal (r_0) increased from 0.35 mg·L^{-1}·min^{-1} for US alone to 0.51, 0.70, 0.74, 0.80 and 1.01 mg·L^{-1}·min^{-1} for US/chlorine in the presence of 50, 100, 200, 250 and 300 µM of chlorine, respectively, yielding an increasing $r_{0,US/chlorine}/r_{0,US}$ ratio of 1.44 at [chlorine]$_0$ = 50 µM, 2 at [chlorine]$_0$ = 100 and 200 µM, 2.23 at [chlorine]$_0$ = 250 µM and 2.83 at [chlorine]$_0$ = 300 µM. Using a maximum elimination rate of 0.1 mg·L^{-1}·min^{-1} for the chlorination alone, the synergy index SI = $r_{0,US/chlorine}/(r_{0,US} + r_{0,chlorine})$ increased from 1.13 for [chlorine]$_0$ = 50 µM to 1.56 for [chlorine]$_0$ = 100 µM, 1.64 for [chlorine]$_0$ = 200 µM, 1.74 for [chlorine]$_0$ = 250 µM and 2.24 for [chlorine]$_0$ = 300 µM, respectively.

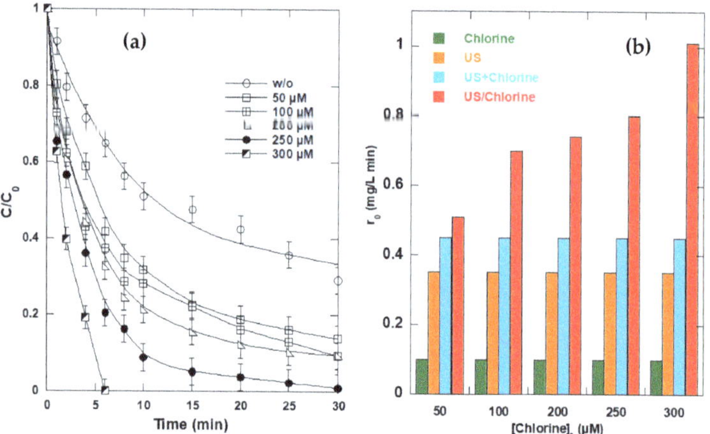

Figure 4. Effect of initial chlorine concentration on the sonochemical degradation of ARAC (**a**) (conditions: C_0 = 5 mg·L^{-1} (10 µM), [chlorine]$_0$ = 50–300 µM, V = 150 mL, pH 5.5, temperature: 25 ± 1 °C, frequency: 600 kHz, power: 120 W) and variation in initial ARAC removal rate (r_0) with respect to [chlorine]$_0$ for chlorination alone, US alone and US/chlorine combination (**b**) (the sum of the two processes separately, US + chlorine, was added for comparison with the combined process).

Therefore, the synergistic effect increases with increasing initial chlorine dosage without the observation of an optimum SI, as reported recently in the case of the UV/chlorine process [20,51,52]. It can therefore be concluded that increasing chlorine concentration in the solution could result in a higher concentration of RCSs (Cl$^\bullet$, ClO$^\bullet$ and Cl$_2^{\bullet-}$), thereby increasing the dye removal. The absence of an optimum was simply attributed to the fact that the required chlorine concentration, which quenches the beneficial effect of chlorine toward RCS generation and use, has not been attained.

2.6. Synergism Dependence of pH

The effect of varying the initial solution pH from 1 to 10 on the ARAC (C_0 = 5 mg·L^{-1}) removal kinetics via the US and US/chlorine processes is given in Figure 5 for an initial chlorine concentration of 250 µM. For both systems, the solution pH in the interval of 4 to 10 did not affect the degradation rate of the dye, but higher rates were obtained at pH 1 using ultrasound alone. However, the US/chlorine process ensured much higher degradation efficiency than sonication alone over the whole investigated range of pH values. A higher synergy index of 1.74 was always maintained for pH 4–6 (i.e., $r_{0,US/chlorine}$ ~ 0.8 mg·L^{-1}·min^{-1}, $r_{0,US}$~0.35 mg·L^{-1}·min^{-1}, $r_{0,chlorine}$~0.1 mg·L^{-1}·min^{-1}) where chlorination alone did not significantly affect the degradation of the dye [39]. At pH 1 and pH 8–10, ARAC chlorination happed at appreciable initial rates of 0.54 and 0.41 mg·L^{-1}·min^{-1}, respectively [39]. Additionally, the sonolytic degradation of ARAC in strong acidic medium (pH 1) was as high as that ensured by chlorination alone (0.832 mg·L^{-1}·min^{-1}), with the same r_0 obtained for the US/chlorine process. Therefore, the synergistic effect of the US/chlorine process was negative at pH 1 (SI = 0.6 < 1); the dye destruction in this case was predominately controlled by sonication alone rather than the coupled system. For pH 8 and 10, the effect of applying US/chlorine was additive as the synergistic index was equal to 1 (i.e., $r_{0,US/chlorine}$~0.8 mg·L^{-1}·min^{-1}, $r_{0,US}$~0.35 mg·L^{-1}·min^{-1}, $r_{0,chlorine}$~0.41 mg·L^{-1}·min^{-1}). Therefore, the synergism of applying US/chlorine treatment was obtained at pH 4–6, where HOCl is the sole chlorine species.

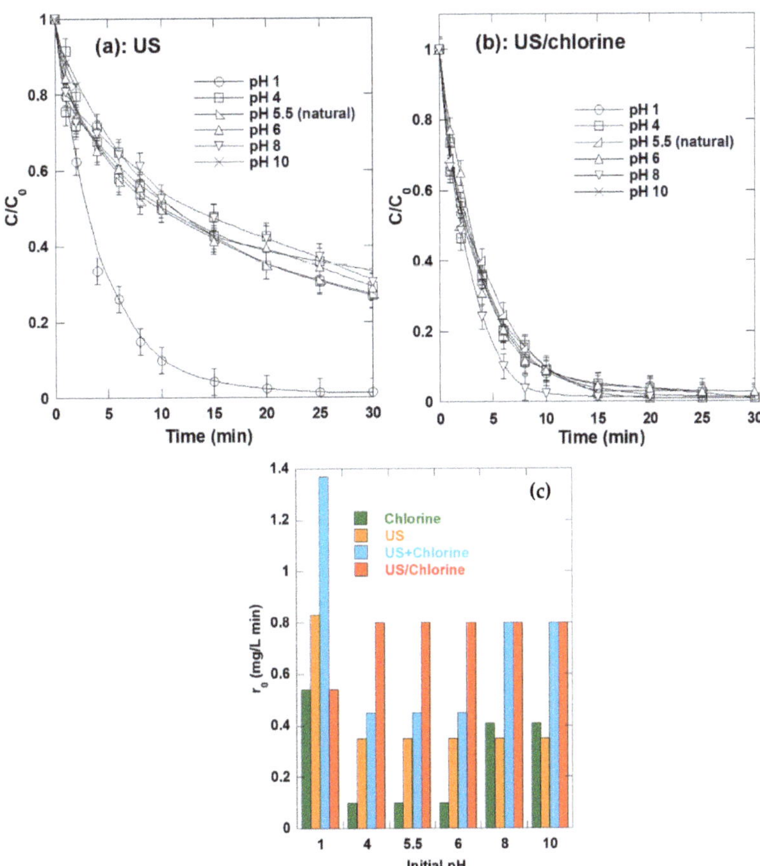

Figure 5. Effect of initial solution pH on the performance of US (**a**) and US/chlorine (**b**) processes toward the removal kinetics of ARAC (conditions: C_0 = 5 mg·L^{-1} (10 μM), [chlorine]$_0$ = 250 μM, V = 150 mL, pH 1–10, temperature: 25 ± 1 °C, frequency: 600 kHz, power: 120 W) and variation in ARAC initial removal rate (r_0) with respect to initial solution pH for chlorination alone, US alone and US/chlorine combination (**c**) (the sum of the two processes separately, US + chlorine, was added for comparison with the combined process).

In sonochemical treatments, the solution pH affects substrate ionization/protonation depending on their dissociation pK_a. Ionizable substrates are more hydrophilic than the protonated ones. They preferentially stay in the bulk solution, particularly at low concentrations, but their hydrophobicity increases when the protonated form is dominated. Because the pK_a of ARAC is 11.4 [43], the dye molecule can maintain the same form up to pH 10; therefore, dye sonolysis cannot be affected by pH elevation up to 10, as stated in Figure 5a. However, the strong acidic medium (pH 1) can protonate the two sulfonate groups ($-SO_3^-$) of the dye, which increases the dye hydrophobicity and accumulation at the reactive bubble/solution interface where the concentration of $^\bullet OH$ is at a higher level [53]. The degradation rate at pH 1 could therefore be more appreciable than under neutral and alkaline conditions. Additionally, the concentration of the dissolved CO_2 gas, i.e., from the atmosphere, in the solution is much higher at acidic conditions (i.e., at neutral and basic mediums, HCO_3^- and CO_3^{2-} are the most abundant acid carbonic forms). It has been reported that the injection of CO_2 at low concentration improves sonochemical treatment by increasing the number of active bubbles, i.e., since CO_2 could provide more nucleation sites for cavitation [54–56]. A detailed report on how CO_2 affects sonochemical efficiency was recently provided by Merouani et al. [57]. Therefore, the CO_2-improving cavitation event is another reason for the higher degradation extent of ARAC at pH 1. The same reason for the effect of pH 1 was maintained for the US/chlorine system, as the overall degradation rate is controlled by the sonolytic process, as stated earlier. For pH 4–6, the synergistic degradation rate is constant as HOCl is the sole chlorine species at pH 4–6 (Figure S1a, Supplementary Materials). For pH 8–10, the combined effect of US and chlorine is additive, and consequently, there is no need for it to be discussed.

2.7. Synergism Dependence of Saturating Gases

Figure 6 shows the effect of three saturating gases (argon, air and nitrogen) on ARAC degradation kinetics via US treatment and the US/chlorine process in the presence of 250 μM of chlorine. It is observed that the US/chlorine system provided the best degradation rates for the three gas atmospheres. For the US treatment, the degradation efficiency follows the order Ar > air > N_2, whereas the order for the US/chlorine system is air > Ar > N_2. Initial degradation rates (r_0) of 0.43 mg·L^{-1}·min^{-1} for Ar, 0.35 mg·L^{-1}·min^{-1} for air and 0.137 mg·L^{-1}·min^{-1} for N_2 were recorded for the ultrasonic treatment. Chlorine engenders an increase in the initial degradation rates by 30%, 128% and 37% for, respectively, for argon, air and N_2 saturations. The calculated synergy indexes are 1.1 for Ar, 1.74 for air and 0.8 for N_2 (i.e., for all gases, $r_{0,chlorine} = 0.1$ mg·L^{-1}·min^{-1}). Therefore, the effect of combining US and chlorine for ARAC degradation is additive for argon, synergistic for air and negative for N_2.

The obtained order of the saturating gases (Ar > air > N_2) for the US process is largely reported in the literature [58–61]. The measured accumulation rate of H_2O_2 in water, as $^\bullet OH$ quantifiers, were 6.4 μM·min^{-1} for Ar, 5.6 μM·min^{-1} for air and 3 μM·min^{-1} for N_2. Therefore, a higher production rate of hydroxyl radicals was associated with argon, then to air and, finally, to N_2. An interesting numerical investigation of how these gases affect the sonochemical activity was recently provided by Merouani et al. [9,62]. Overall, thanks to its beneficial physical properties (i.e., a greater polytropic ratio ($\gamma_{Ar} = 1.66$) and solubility ($x_{Ar} = 2.748 \times 10^{-5}$) and lower thermal conductivity ($\lambda_{Ar} = 0.018$ W m^{-2}·K^{-1}) than air and N_2 gases, which have the same γ and λ ($\gamma = 1.41$, $\lambda = 0.026$ W m^{-2}·K^{-1}) and a slight differences in their solubility ($x_{air} = 1.524 \times 10^{-5}$, $x_{N2} = 1.276 \times 10^{-5}$)), argon could produce the highest single-bubble yield ($^\bullet OH$ radical) and a greater number of bubbles than the other gases [60,63,64], allowing it to achieve the maximum sonochemical efficiency. The higher chemical efficiency in air-saturated solution than N_2 was mainly attributed to the internal bubble-chemistry [9,62]. The presence of a high N_2 concentration inside the bubble at the collapse decreased the production rate of radicals [9,62]. The reason for this trend was associated with the consumption of $^\bullet OH$ radicals through the reaction $NO + {}^\bullet OH + M \leftrightarrow HNO_2 + M$. Consequently, because oxidizing nitrogen NO is formed mainly, as found, through the reactions $N_2 + O \rightleftharpoons NO + N$ and $NO_2 + M \rightleftharpoons O + NO + M$, the higher the concentration of N_2 in the bubble, the higher the concentration of NO will be; this accelerates the consumption rate of $^{\bullet}OH$ radicals through the reaction $NO + {}^{\bullet}OH + M \rightleftharpoons HNO_2 + M$ [9,62]. Therefore, air could yield higher sonochemical efficiency than an N_2 atmosphere.

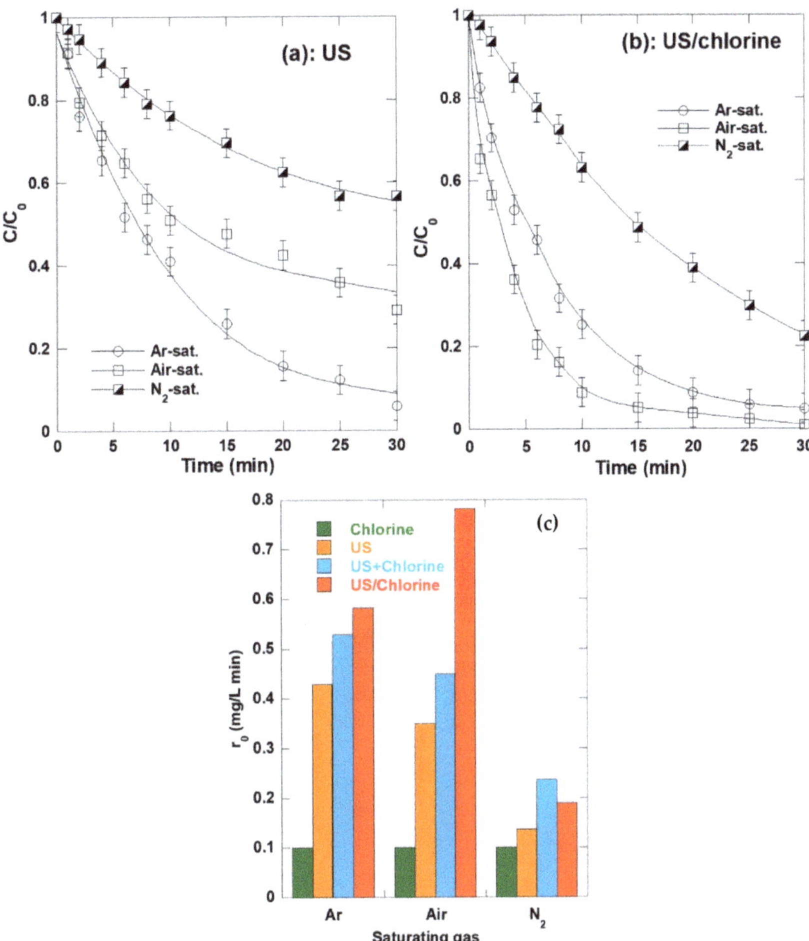

Figure 6. Effect of saturation gases on the performance of US (**a**) and US/chlorine (**b**) processes toward the removal kinetics of ARAC (conditions: $C_0 = 5$ mg·L^{-1} (10 µM), [chlorine]$_0 = 250$ µM, $V = 150$ mL, pH 5.5, temperature: 25 ± 1 °C, frequency: 600 kHz, power: 120 W) and variation in ARAC initial removal rate (r_0) with respect to saturating gas for chlorination alone, US alone and US/chlorine combination (**c**) (the sum of the two processes separately, US + chlorine, was added for comparison with the combined process).

For the US/chlorine system, the resulting negative synergy under a nitrogen saturation atmosphere was due to the poor production rate of the generated reactive species from the acoustic bubbles, as stated below. This could lower the production of RCSs, which are responsible for the synergistic effect. On the other hand, the absence of synergy under argon atmosphere was interpreted as follows: the radical–radical reactions, which are characterized by high second-order rate constants (Equations (27)–(32)), could always accompany the radical–organic reactions. The radical–radical reaction was classified as a parasite reaction for organic degradation because they reduce the radicals' availability in the solution [16,20,65]. As argon ensures a higher concentration of reactive species (as quantified by H_2O_2 dosage), it may be that the generated concentration of RCSs is too high, which favors radicals quenching by themselves (Reactions (27)–(32)) rather than their reaction with the organic pollutant. Such scenarios are widely reported in the literature for several cases of AOPs [16,20,65–67]. Therefore, an air atmosphere could provide the best synergy as it generates a relatively moderate concentration of hydroxyl radicals, and therefore RCSs, as compared to argon, which marginalizes radical–radical reactions compared to radical–organic ones.

2.8. Synergism Dependence of Initial ARAC Concentration

Figure 7 depicts the effect of initial dye concentration (C_0 = 5–40 mg·L^{-1}) on the performance of US treatment and the US/chlorine process (250 µM of chlorine) at pH 5.5. Higher removal rates were associated with the US/chlorine process as compared to US alone for the four tested concentrations of C_0. After 10 min of reaction, US/chlorine eliminated 92%, 73%, 53%, 43% and 34% for C_0 = 5, 10, 20, 30 and 40 mg L^{-1}, respectively, compared to 50%, 38%, 31%, 30% and 36% for US alone. The elimination ratio between the two processes decreased from 1.84 for C_0 = 5 mg·L^{-1} to 1.7 and 1.48 for C_0 = 20 and 40 mg·L^{-1}, respectively. Correspondingly, the calculated synergistic index SI = $r_{0,US/chlorine}$/ ($r_{0,US}$ + $r_{0,chlorine}$) decreased from 1.74 for C_0 = 5 mg·L^{-1} to 1.46 for C_0 = 10 mg·L^{-1}, 1 for C_0 = 20 and 30 mg·L^{-1} and 0.8 for C_0 = 40 mg·L^{-1}. Therefore, the US/chlorine process is more synergistic at low dye concentrations. For higher concentrations, the process is not synergistic.

In sonochemical treatment, increasing the pollutant concentration in the solution bulk could increase its concentration at the reactive interfacial region [68,69]. Therefore, the higher the C_0, the higher the initial degradation rate could be. In fact, initial rates of 0.35, 0.57, 0.85, 1 and 1.36 mg·L^{-1}·min^{-1} were recorded for C_0 = 5, 10, 20, 30 and 40 mg·L^{-1}, respectively, in the absence of chlorine. This means that the scavenging of the acoustically generated hydroxyl radicals at the bubble/solution interface could be more efficient at a higher pollutant concentration. When chlorine is present, it creates strong competition with the pollutant substrate to react with the cavitation-generated reactive species. Increasing C_0 could reduce the fraction of the reactive species scavenged by chlorine, thereby decreasing the concentration of RCSs responsible for the synergistic action. Therefore, it is preferable to operate the US/chlorine process at a low pollutant concentration to maintain higher synergistic level.

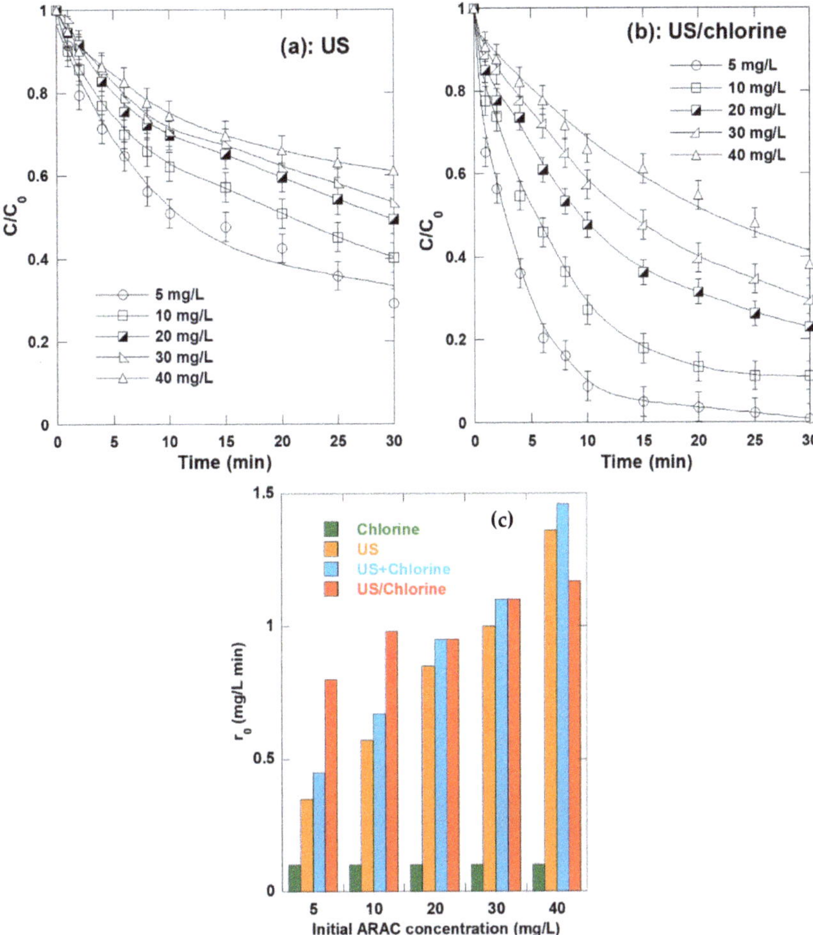

Figure 7. Effect of initial dye concentration on the performance of US (**a**) and US/chlorine (**b**) processes toward the removal kinetics of ARAC (conditions: C_0 = 5–40 mg·L^{-1} (10 µM), [chlorine]$_0$ = 250 µM, V = 150 mL, pH 5.5, temperature: 25 ± 1 °C, frequency: 600 kHz, power: 120 W) and variation in ARAC initial removal rate (r_0) with respect to initial de concentration for chlorination alone, US alone and US/chlorine combination (**c**) (the sum of the two processes separately, US + chlorine, was added for comparison with the combined process).

3. Materials and Methods

Throughout the study, ultrapure water was used for solution and sample preparation. Sodium hypochlorite solution (available active chlorine basis: ~16%) and Allura Red AC (abbreviation: ARAC; CAS number: 25956-17-6; chemical formula: $C_{18}H_{14}N_2Na_2O_8S_2$; molecular weight: 496,42 g·mol^{-1}) were supplied by Sigma-Aldrich (St. Louis, MO, USA). The molecular structure of ARAC is given in Figure S2. All other reagents (NaOH, H_2SO_4, KI, $(NH_4)_6Mo_7 \cdot 4H_2O$ and nitrobenzene) were commercial products of the purest grade available (Sigma-Aldrich).

Sonolytic runs were conducted using 150 mL of solution in the cylindrical water-jacketed glass reactor presented in Figure S3. The source of ultrasonic irradiation was a piezoelectric disc fixed on a stainless steel plate in the bottom of the reactor. US irradiation was emitted at a frequency of 600 kHz and at variable electric powers. For all experiments in this study, the electric power delivered from the generator was fixed at 120 W. The temperature of the irradiating liquid was controlled through the cooling jacket and displayed using a thermocouple immersed in solution. The acoustic energy dissipated in the solution (~23 W) was estimated using the calorimetric method [49].

Stock solutions of chlorine (100 mM, pH 5) and ARAC (500 mg·L^{-1}, pH ~ 5.5) were prepared and stored in the dark at 4 °C. Experiments were carried out under different conditions at ambient temperature (25 $\pm$ 1 °C). The pH of the solution was adjusted using NaOH or H$_2$SO$_4$ (0.1 M). For the test of gases, each gas was bubbled 20 min prior to start and until completion of experiments. Quantitative analysis of the dye concentration was performed using Biochrom WPA Lightwave II UV-Vis. spectrophotometer at λ_{max} = 504 nm. Note that the variation in pH in the range of 1–10 affected neither λ_{max} nor the initial absorbance at λ_{max}. During sonication, hydrogen peroxide concentrations were quantified according to the iodometric method [49]. To ensure reproducibility of the results, all runs were performed in triplicate and results were presented as averages. Error bars, reported in the relevant data, represent the deviation of means.

4. Conclusions

Based on the obtained results, it can be concluded that the US/chlorine process could be more suitable to quickly abate persistent organic pollutants under typical water treatment conditions (air-equilibrated solution, pH 4–6, ambient temperature and low pollutant concentration). The process can generate reactive chlorine species, i.e., via the sonochemical activation of chlorine, which greatly improve the degradation rate of the pollutant. A synergistic index of 1.74 was obtained via the US/chlorine process for the degradation of ARAC (C$_0$ = 5 mg·L^{-1}) at pH 5.5 and [chlorine]$_0$ = 250 mM. The synergy index increased with increasing initial chlorine dosage without the observation of an optimum. The absence of an optimum was simply attributed to the fact that the required chlorine concentration, which quenches the beneficial effect of chlorine toward RCS generation and use, was not reached. Additionally, the synergetic effect was only obtained at pH 4–6, where HOCl is the sole chlorine species. At pH 8–10, the combined effect of US and chlorine was additive. Furthermore, the influence of combining US and chlorine for ARAC removal was additive for argon, synergetic for air and negative for nitrogen. An air atmosphere could provide the best synergy as it produces a relatively moderate concentration of reactive species as compared to argon, which marginalizes radical–radical reactions compared to radical–organic ones. The US/chlorine process was more synergetic for low dye concentrations (C$_0$ $\leq$ 10 mg·L^{-1}); the coupling effect was additive for moderate concentrations (C$_0$~20–30 mg· L^{-1}) and negative for higher concentrations (>30 mg·L^{-1}).

The US/chlorine process could be involved as one novel AOP in wastewater treatment, although some specific analyses are still required. While the main objective of the current work was to explore the process performance in terms of pollutant removal, further investigations will be conducted for the sake of completeness, assessing the following issues:

- TOC, BOD$_5$ and toxicity evolutions;
- Radicals' identification and contribution to the overall degradation rate;
- The effect of processing conditions (e.g., pH, temperature, chlorine and pollutant concentrations) on radicals' distribution;
- The identification of degradation by-products because of the possible formation of toxic trihalomethanes;
- The reaction mechanism and scheme for ARAC degradation.

Supplementary Materials: The following supporting information can be downloaded at: https://www.mdpi.com/article/10.3390/catal12101171/s1. Figure S1. (a) Chlorine speciation in 0.5 mM total chlorine as function of pH and for a chloride concentration of 2 mM, and (b) molar absorption coefficients (ε) of HOCl (pH 5) and OCl$^-$ (pH 9) as a function of wavelength; Figure S2. Molecular structure of Allura Red AC (ARAC); Figure S3. Photograph of the sonochemical reactor used for the sonolytic experiments. (a) 600 kHz ultrasonic transducer, (b) cylindrical jacketed glass cell, (c) sonicated solution, (d) inlet cooling water, (e) outlet cooling water, (f) thermocouple.

Author Contributions: O.H.: supervision, conceptualization, methodology, formal analysis, project administration, resources, funding acquisition, investigation, validation and writing—review and editing; S.M.: conceptualization, methodology, validation, writing—original draft preparation and writing—review and editing; H.C.B.: investigation, validation and writing—review and editing; M.A.I.: investigation, validation and writing—review and editing; H.F.: validation, methodology, resources and writing—review and editing; A.A.: validation and writing—review and editing. All authors have read and agreed to the published version of the manuscript.

Funding: This research received no external funding.

Data Availability Statement: Not applicable.

Conflicts of Interest: The authors declare no conflict of interest.

References

1. Boczkaj, G.; Fernandes, A. Wastewater treatment by means of advanced oxidation processes at basic pH conditions: A review. *Chem. Eng. J.* **2017**, *320*, 608–633. [CrossRef]
2. Kanakaraju, D.; Glass, B.D.; Oelgemöller, M. Advanced oxidation process-mediated removal of pharmaceuticals from water: A review. *J. Environ. Manag.* **2018**, *219*, 189–207. [CrossRef] [PubMed]
3. Ameta, S.; Ameta, R. *Advanced Oxidation Processes for Wastewater Treatment: Emerging Green Chemical Technology*; Elsevier Science: London, UK, 2018; ISBN 9780128105252.
4. Stefan, M.I. *Advanced Oxidation Processes for Water Treatment: Fundamentals and Applications*; IWA Publishing: London, UK, 2017.
5. Pétrier, C. The use of power ultrasound for water treatment. In *Power Ultrasonics: Applications of High-Intensity Ultrasound*; Gallego-Juarez, J.A., Graff, K., Eds.; Elsevier: Cambridge, MA, USA, 2015; pp. 939–963.
6. Suslick, K.S.; Flannigan, D.J. Inside a Collapsing Bubble: Sonoluminescence and the Conditions During Cavitation. *Annu. Rev. Phys. Chem.* **2008**, *59*, 659–683. [CrossRef] [PubMed]
7. Yasui, K.; Tuziuti, T.; Lee, J.; Kozuka, T.; Towata, A.; Iida, Y. The range of ambient radius for an active bubble in sonoluminescence and sonochemical reactions. *J. Chem. Phys.* **2008**, *128*, 184705. [CrossRef]
8. Yasui, K.; Tuziuti, T.; Kozuka, T.; Towata, A.; Iida, Y. Relationship between the bubble temperature and main oxidant created inside an air bubble under ultrasound. *J. Chem. Phys.* **2007**, *127*, 154502. [CrossRef]
9. Merouani, S.; Hamdaoui, O.; Rezgui, Y.; Guemini, M. Sensitivity of free radicals production in acoustically driven bubble to the ultrasonic frequency and nature of dissolved gases. *Ultrason. Sonochem.* **2015**, *22*, 41–50. [CrossRef]
10. Makino, K.; Mossoba, M.M.; Riesz, P. Chemical effects of ultrasound on aqueous solutions. Evidence for $^\bullet$OH an $^\bullet$H by spin trapping. *J. Am. Chem. Soc.* **1982**, *104*, 3537–3539. [CrossRef]
11. Hart, E.J.; Henglein, A. Sonochemistry of aqueous solutions: H$_2$-O$_2$ combustion in cavitation bubbles. *J. Phys. Chem.* **1987**, *91*, 3654–3656. [CrossRef]
12. Henglein, A. Chemical effects of continuous and pulsed ultrasound in aqueous solutions. *Ultrason. Sonochem.* **1995**, *2*, 115–121. [CrossRef]
13. Thompson, L.H.; Doraiswamy, L.K. Sonochemistry: Science and Engineering. *Ind. Eng. Chem. Res.* **1999**, *38*, 1215–1249. [CrossRef]
14. Laat, J.D.E.; Stefan, M. UV/chlorine process. In *Advanced Oxidation Processes for Water Treatment*; Stefan, M.I., Ed.; IWA Publishing: London, UK, 2017; pp. 383–428.
15. Remucal, C.K.; Manley, D. Emerging investigators series: The efficacy of chlorine photolysis as an advanced oxidation process for drinking water treatment. *Environ. Sci. Water Res. Technol.* **2016**, *2*, 565–579. [CrossRef]
16. Meghlaoui, F.Z.; Merouani, S.; Hamdaoui, O.; Bouhelassa, M.; Ashokkumar, M. Rapid catalytic degradation of refractory textile dyes in Fe (II)/chlorine system at near neutral pH: Radical mechanism involving chlorine radical anion (Cl$_2^{\bullet-}$)-mediated transformation pathways and impact of environmental matrices. *Sep. Purif. Technol.* **2019**, *227*, 115685. [CrossRef]
17. Bulman, D.M.; Mezyk, S.P.; Remucal, C.K. The Impact of pH and Irradiation Wavelength on the Production of Reactive Oxidants during Chlorine Photolysis. *Environ. Sci. Technol.* **2019**, *53*, 4450–4459. [CrossRef] [PubMed]
18. Behin, J.; Akbari, A.; Mahmoudi, M.; Khajeh, M. Sodium hypochlorite as an alternative to hydrogen peroxide in Fenton process for industrial scale. *Water Res.* **2017**, *121*, 120–128. [CrossRef]

19. Meghlaoui, F.Z.; Merouani, S.; Hamdaoui, O.; Alghyamah, A.; Bouhelassa, M.; Ashokkumar, M. Fe(III)-catalyzed degradation of persistent textile dyes by chlorine at slightly acidic conditions: The crucial role of $Cl_2^{\bullet-}$ radical in the degradation process and impacts of mineral and organic competitors. *Asia-Pacific J. Chem. Eng.* **2020**, *16*, e2553. [CrossRef]
20. Belghit, A.; Merouani, S.; Hamdaoui, O.; Alghyamah, A.; Bouhelassa, M. Influence of processing conditions on the synergism between UV irradiation and chlorine toward the degradation of refractory organic pollutants in UV/chlorine advanced oxidation system. *Sci. Total Environ.* **2020**, *736*, 139623. [CrossRef]
21. Wang, W.L.; Wu, Q.Y.; Huang, N.; Wang, T.; Hu, H.Y. Synergistic effect between UV and chlorine (UV/chlorine) on the degradation of carbamazepine: Influence factors and radical species. *Water Res.* **2016**, *98*, 190–198. [CrossRef]
22. Fang, J.; Fu, Y.; Shang, C. The roles of reactive species in micropollutant degradation in the UV/free chlorine system. *Environ. Sci. Technol.* **2014**, *48*, 1859–1868. [CrossRef]
23. Wang, W.; Zhang, X.; Wu, Q.; Du, Y.; Hu, H. Degradation of natural organic matter by UV/chlorine oxidation: Molecular decomposition, formation of oxidation byproducts and cytotoxicity. *Water Res.* **2017**, *124*, 251–258. [CrossRef]
24. Wang, A.; Lin, Y.; Xu, B.; Hu, C.; Xia, S.; Zhang, T.; Chu, W. Kinetics and modeling of iodoform degradation during UV/chlorine advanced oxidation process. *Chem. Eng. J.* **2017**, *323*, 312–319. [CrossRef]
25. Kong, X.; Wu, Z.; Ren, Z.; Guo, K.; Hou, S.; Hua, Z.; Li, X.; Fang, J. Degradation of lipid regulators by the UV/chlorine process: Radical mechanisms, chlorine oxide radical ($ClO^{\bullet}$)-mediated transformation pathways and toxicity changes. *Water Res.* **2018**, *137*, 242–250. [CrossRef] [PubMed]
26. Xiang, Y.; Fang, J.; Shang, C. Kinetics and pathways of ibuprofen degradation by the UV/chlorine advanced oxidation process. *Water Res.* **2016**, *90*, 301–308. [CrossRef] [PubMed]
27. Guo, K.; Wu, Z.; Fang, J. UV-based advanced oxidation process for the treatment of pharmaceuticals and personal care products. In *Contaminants of Emerging Concern in Water and Wastewater*; Hernández-Maldonado, A.J., Blaney, L., Eds.; Elsevier Inc.: Amsterdam, The Netherlands, 2020; pp. 367–408. ISBN 9780128135617.
28. Wang, W.L.; Wu, Q.Y.; Li, Z.M.; Lu, Y.; Du, Y.; Wang, T.; Huang, N.; Hu, H.Y. Light-emitting diodes as an emerging UV source for UV/chlorine oxidation: Carbamazepine degradation and toxicity changes. *Chem. Eng. J.* **2017**, *310*, 148–156. [CrossRef]
29. Wu, Z.; Guo, K.; Fang, J.; Yang, X.; Xiao, H.; Hou, S.; Kong, X.; Shang, C.; Yang, X.; Meng, F.; et al. Factors affecting the roles of reactive species in the degradation of micropollutants by the UV/chlorine process. *Water Res.* **2017**, *126*, 351–360. [CrossRef] [PubMed]
30. Deng, J.; Wu, G.; Yuan, S.; Zhan, X.; Wang, W.; Hu, Z.H. Ciprofloxacin degradation in UV/chlorine advanced oxidation process: Influencing factors, mechanisms and degradation pathways. *J. Photochem. Photobiol. A Chem.* **2019**, *371*, 151–158. [CrossRef]
31. Guo, K.; Wu, Z.; Shang, C.; Yao, B.; Hou, S.; Yang, X.; Song, W.; Fang, J. Radical Chemistry and Structural Relationships of PPCP Degradation by UV/Chlorine Treatment in Simulated Drinking Water. *Environ. Sci. Technol.* **2017**, *51*, 10431–10439. [CrossRef]
32. Guo, Z.; Lin, Y.; Xu, B.; Huang, H.; Zhang, T.; Tian, F.; Gao, N. Degradation of chlortoluron during UV irradiation and UV/chlorine processes and formation of disinfection by-products in sequential chlorination. *Chem. Eng. J.* **2016**, *283*, 412–419. [CrossRef]
33. Dong, H.; Qiang, Z.; Hu, J.; Qu, J. Degradation of chloramphenicol by UV/chlorine treatment: Kinetics, mechanism and enhanced formation of halonitromethanes. *Water Res.* **2017**, *121*, 178–185. [CrossRef]
34. Alegre, M.L.; Geronees, M.; Rosso, J.A.; Bertolotti, S.G.; Braun, A.M.; Marrtire, D.O.; Gonzalez, M.C. Kinetic Study of the Reactions of Chlorine Atoms and $Cl_2^{\bullet-}$ Radical Anions in Aqueous Solutions. 1. Reaction with Benzene. *J. Phys. Chem. A* **2000**, *104*, 3117–3125. [CrossRef]
35. Deborde, M.; Gunten, U. Von Reactions of chlorine with inorganic and organic compounds during water treatment—Kinetics and mechanisms: A critical review. *Water Res.* **2008**, *42*, 13–51. [CrossRef]
36. Zou, H.; Tang, H. Comparison of different bacteria inactivation by a novel continuous-flow ultrasound/chlorination water treatment system in a pilot scale. *Water* **2019**, *11*, 258. [CrossRef]
37. Blume, T.; Neis, U. Improving chlorine disinfection of wastewater by ultrasound application. *Water Sci. Technol.* **2005**, *52*, 139–144. [CrossRef] [PubMed]
38. Lambert, N.; Rediers, H.; Hulsmans, A.; Joris, K.; Declerck, P.; De Laedt, Y.; Liers, S. Evaluation of ultrasound technology for the disinfection of process water and the prevention of biofilm formation in a pilot plant. *Water Sci. Technol.* **2010**, *61*, 1089–1096. [CrossRef] [PubMed]
39. Hamdaoui, O.; Merouani, S.; Ait Idir, M.; Benmahmoud, H.C.; Dehane, A.; Alghyamah, A. Ultrasound/chlorine sono-hybrid-advanced oxidation process: Impact of dissolved organic matter and mineral constituents. *Ultrason. Sonochem.* **2022**, *83*, 105918. [CrossRef] [PubMed]
40. Borzelleca, J.F.; Olson, J.W.; Reno, F.E. Lifetime toxicity/carcinogenicity study of FD & C Red No. 40 (Allura Red) in Sprague-Dawley rats. *Food Chem. Toxicol.* **1989**, *27*, 701–705. [CrossRef] [PubMed]
41. Garole, V.J.; Choudhary, B.C.; Tetgure, S.R.; Garole, D.J.; Borse, A.U. Detoxification of toxic dyes using biosynthesized iron nanoparticles by photo-Fenton processes. *Int. J. Environ. Sci. Technol.* **2018**, *15*, 1649–1656. [CrossRef]
42. Vorhees, C.V.; Butcher, R.E.; Brunner, R.L.; Wootten, V.; Sobotka, T.J. Development toxicity and psychotoxicity of FD and C red dye no. 40 (Allura red AC) in rates. *Toxicology* **1983**, *28*, 207–217. [CrossRef]
43. Sun, Q.; Yang, L.; Yang, J.; Liu, S.; Hu, X. Study on the interaction between Rhodamine dyes and Allura Red based on fluorescence spectra and its analytical application in soft Drinks. *Anal. Sci.* **2017**, *33*, 1181–1187. [CrossRef]
44. Vogt, R.; Schindler, R.N. Product channels in the photolysis of HOCl. *J. Photochem. Photobiol. A Chem.* **1992**, *66*, 133–140. [CrossRef]

45. Buxton, G.V.; Bydder, M.; Arthur Salmon, G. Reactivity of chlorine atoms in aqueous solution Part 1. The equilibrium $Cl^{MNsbd}+Cl^-Cl_2^-$. *J. Chem. Soc. Faraday Trans.* **1998**, *94*, 653–657. [CrossRef]
46. Buxton, G.V.; Greenstock, C.L.; Helman, W.P.; Ross, A.B. Critical review of rate constants for reactions of hydrated Electrons, hydrogen atoms and hydroxyl radicals ($^\bullet OH/O^-$) in aqueous solution. *J. Phys. Chem. Ref. Data* **1988**, *17*, 515–886. [CrossRef]
47. Pubchem. Allura Red AC. 2020. Available online: https://pubchem.ncbi.nlm.nih.gov/compound/Allura-Red-AC (accessed on 9 September 2022).
48. Pétrier, C.; Francony, A. Ultrasonic waste-water treatment: Incidence of ultrasonic frequency on the rate of phenol and carbon tetrachloride degradation. *Ultrason. Sonochem.* **1997**, *4*, 295–300. [CrossRef]
49. Merouani, S.; Hamdaoui, O.; Saoudi, F.; Chiha, M. Influence of experimental parameters on sonochemistry dosimetries: KI oxidation, Fricke reaction and H_2O_2 production. *J. Hazard. Mater.* **2010**, *178*, 1007–1014. [CrossRef] [PubMed]
50. Tauber, A.; Mark, G.; Schuchmann, H.-P.; von Sonntag, C. Sonolysis of tert-butyl alcohol in aqueous solution. *J. Chem. Soc. Perkin Trans. 2* **1999**, *2*, 1129–1136. [CrossRef]
51. Kong, X.; Jiang, J.; Ma, J.; Yang, Y.; Liu, W.; Liu, Y. Degradation of atrazine by UV/chlorine: Efficiency, influencing factors, and products. *Water Res.* **2016**, *90*, 15–23. [CrossRef]
52. Huang, N.; Wang, T.; Wang, W.; Wu, Q.; Li, A.; Hu, H. UV/chlorine as an advanced oxidation process for the degradation of benzalkonium chloride: Synergistic effect, transformation products and toxicity evaluation. *Water Res.* **2017**, *114*, 246–253. [CrossRef]
53. Chadi, N.E.; Merouani, S.; Hamdaoui, O.; Bouhelassa, M. New aspect of the effect of liquid temperature on sonochemical degradation of nonvolatile organic pollutants in aqueous media. *Sep. Purif. Technol.* **2018**, *200*, 68–74. [CrossRef]
54. Henglein, A. Sonolysis of carbon dioxide, nitrous oxide and methane in aqueous solution. *Z. Naturforsch. B* **1985**, *40*, 100–107. [CrossRef]
55. Harada, H.; Ono, Y. Improvement of the rate of sono-oxidation in the presence of CO_2. *Jpn. J. Appl. Phys.* **2015**, *54*, 52–55. [CrossRef]
56. Rooze, J. *Cavitation in Gas-Saturated Liquids*; Technische Universiteit Eindhoven: Eindhoven, The Netherlands, 2017. [CrossRef]
57. Merouani, S.; Hamdaoui, O.; Al-Zahrani, S.M. Toward understanding the mechanism of pure CO_2-quenching sonochemical processes. *J. Chem. Technol. Biotechnol.* **2020**, *95*, 553–566. [CrossRef]
58. Gao, Y.Q.; Gao, N.Y.; Deng, Y.; Gu, J.S.; Gu, Y.L.; Zhang, D. Factors affecting sonolytic degradation of sulfamethazine in water. *Ultrason. Sonochem.* **2013**, *20*, 1401–1407. [CrossRef] [PubMed]
59. Ferkous, H.; Hamdaoui, O.; Merouani, S. Sonochemical degradation of naphthol blue black in water: Effect of operating parameters. *Ultrason. Sonochem.* **2015**, *26*, 40–47. [CrossRef] [PubMed]
60. Boutamine, Z.; Merouani, S.; Hamdaoui, O. Sonochemical degradation of Basic Red 29 in aqueous media. *Turk. J. Chem.* **2017**, *41*, 99–115. [CrossRef]
61. Hamdaoui, O.; Merouani, S. Improvement of Sonochemical Degradation of Brilliant Blue R in Water Using Periodate Ions: Implication of Iodine Radicals in the Oxidation Process. *Ultrason. Sonochem.* **2017**, *37*, 344–350. [CrossRef] [PubMed]
62. Merouani, S.; Ferkous, H.; Hamdaoui, O.; Rezgui, Y.; Guemini, M. New interpretation of the effects of argon-saturating gas toward sonochemical reactions. *Ultrason. Sonochem.* **2015**, *23*, 37–45. [CrossRef] [PubMed]
63. Rooze, J.; Rebrov, E.V.; Schouten, J.C.; Keurentjes, J.T.F. Dissolved gas and ultrasonic cavitation—A review. *Ultrason. Sonochem.* **2013**, *20*, 1–11. [CrossRef]
64. Rooze, J.; Rebrov, E.V.; Schouten, J.C.; Keurentjes, J.T.F. Effect of resonance frequency, power input, and saturation gas type on the oxidation efficiency of an ultrasound horn. *Ultrason. Sonochem.* **2011**, *18*, 209–215. [CrossRef]
65. Chadi, N.E.; Merouani, S.; Hamdaoui, O.; Bouhelassa, M.; Ashokkumar, M. H2O2/Periodate (IO_4^-): A novel advanced oxidation technology for the degradation of refractory organic pollutants. *Environ. Sci. Water Res. Technol.* **2019**, *5*, 1113–1123. [CrossRef]
66. Ferkous, H.; Merouani, S.; Hamdaoui, O.; Pétrier, C. Ultrasonics Sonochemistry Persulfate-enhanced sonochemical degradation of naphthol blue black in water: Evidence of sulfate radical formation. *Ultrason. Sonochem.* **2017**, *34*, 580–587. [CrossRef]
67. Bekkouche, S.; Merouani, S.; Hamdaoui, O.; Bouhelassa, M. Efficient photocatalytic degradation of Safranin O by integrating solar-UV/TiO_2/persulfate treatment: Implication of sulfate radical in the oxidation process and effect of various water matrix components. *J. Photochem. Photobiol. A Chem.* **2017**, *345*, 80–91. [CrossRef]
68. Villaroel, E.; Silva-Agredo, J.; Petrier, C.; Taborda, G.; Torres-Palma, R.A. Ultrasonic degradation of acetaminophen in water: Effect of sonochemical parameters and water matrix. *Ultrason. Sonochem.* **2014**, *21*, 1763–1769. [CrossRef] [PubMed]
69. Merouani, S.; Hamdaoui, O.; Saoudi, F.; Chiha, M. Sonochemical degradation of Rhodamine B in aqueous phase: Effects of additives. *Chem. Eng. J.* **2010**, *158*, 550–557. [CrossRef]

Review

Recent Development in Non-Metal-Doped Titanium Dioxide Photocatalysts for Different Dyes Degradation and the Study of Their Strategic Factors: A Review

Parveen Akhter [1,*], Abdullah Arshad [1], Aimon Saleem [1] and Murid Hussain [2,*]

[1] Department of Chemistry, The University of Lahore, 1-Km Defence Road, Off Raiwind Road, Lahore 54000, Pakistan
[2] Department of Chemical Engineering, COMSATS University Islamabad, Lahore Campus, Defence Road, Off Raiwind Road, Lahore 54000, Pakistan
* Correspondence: parveen.akhter@chem.uol.edu.pk (P.A.); drmhussain@cuilahore.edu.pk (M.H.)

Citation: Akhter, P.; Arshad, A.; Saleem, A.; Hussain, M. Recent Development in Non-Metal-Doped Titanium Dioxide Photocatalysts for Different Dyes Degradation and the Study of Their Strategic Factors: A Review. *Catalysts* **2022**, *12*, 1331. https://doi.org/10.3390/catal12111331

Academic Editors: Rafael Borja, Gassan Hodaifa and Mha Albqmi

Received: 9 September 2022
Accepted: 26 October 2022
Published: 31 October 2022

Abstract: Semiconductor titanium dioxide in its basic form or doped with metals and non-metals is being extensively used in wastewater treatment by photocatalysis due to its versatile nature. Other numerous characteristics including being environmentally friendly, non-pernicious, economical, multi-phase, highly hydrophilic, versatile physio-chemical features, chemical stability, suitable band gap, and corrosion-resistance, along with its low price make TiO_2 the best candidate in the field of photocatalysis. Commercially, semiconductor and synthesized photocatalysts—which have been investigated for the last few decades owing to their wide band gap—and the doping of titania with p-block elements (non-metals) such as oxygen, sulfur, nitrogen, boron, carbon, phosphorus, and iodine enhances their photocatalytic efficiency under visible-light irradiation. This is because non-metals have a strong oxidizing ability. The key focus of this review is to discuss the various factors affecting the photocatalytic activity of non-metal-doped titania by decreasing its band gap. The working parameters discussed are the effect of pH, dyes concentration, photocatalyst's size and structure, pollutants concentration and types, the surface area of photocatalysts, the effect of light intensity and irradiation time, catalyst loading, the effect of temperature, and doping impact, etc. The mechanism of the photocatalytic action of several non-metallic dopants of titanium dioxide and composites is a promising approach for the exploration of photocatalysis activity. The various selected synthesis methods for non-metallic-doped TiO_2 have been reviewed in this study. Similarly, the effect of various conditions on the doping mode has been summarized in relation to several sorts of modified TiO_2.

Keywords: dyes degradation; non-metal-doped titania; photocatalysis; parameters; wastewater treatment

1. Introduction

Water is one of the most precious and non-replaceable commodities of mankind. The per capita usage of water varies between countries and its demand includes domestic, industrial, and public, but here lies a problem of the misuse of water on a larger scale. Obviously, photocatalysis is the most fundamental technique for the degradation of organic pollutants and their conversion into valuable products, the removal of industrial effluents, and energy utilization [1,2]. Appropriate water supplies are the most important and key parameters for human consumption, industrial, agricultural, and domestic purposes. Generally, a wide range of natural and artificial contaminations has deleteriously affected the environment directly or indirectly, in which textile dye removal is of major concern [3–5]. However, ample research has been carried out on TiO_2 as a paramount photocatalyst for a variety of applications, such as the degradation of organic pollutants, hydrogen production from water splitting, the purification of air and water, self-cleansing surfaces, food cosmetics, the paint industry, etc. Titania is extensively employed in energy-associated industries, specifically, in water splitting under visible-light irradiation [6]. However, different compounds have been used for the doping of TiO_2 with CdS [7]. SnO_2-doped TiO_2 [8–10], WO_3-doped TiO_2 [11], $ZrTiO_4$ [12–14], and are employed in the heterogeneous photocatalysis. Among these very endearing materials, non-metal-doped TiO_2 photocatalysts have gained significant attention due to their greater synergistic effect, which leads to the reduction in the band gap of titania from various metal-doped TiO_2 [14–16]. Many conventional techniques or pathways have been adopted for the wastewater treatment, but unfortunately, the problem remains unsolved. However, different measures such as adsorption, coagulation, flocculation, and oxidation have been employed to cast off dyes from wastewater. The most documented approaches for the removal of dyes from wastewater are precipitation, filtration, and electrochemical procedures. However, these methods also have disadvantages; moreover, some of them do not have the ability to break down the dye completely. Therefore, photocatalysis is now thought to be one of the best ways to remove dyes from contaminated water [17]. Additionally, to overcome this issue, the photocatalytic process of various dyes degradation has shown to be a more efficient, cost-effective, and convenient way to eliminate abundant contaminants and organic pollutants from water. This method has drawn the attention of researchers all over the world to the development of more effective techniques of wastewater treatment. For this mechanism, simple TiO_2, doped TiO_2, or composite of TiO_2 have been widely discussed in the literature [18–21]. Some modifications in titania can help in resolving the mistreatment of unadorned TiO_2. However, doping the TiO_2 would enhance the photocatalytic activity for a range of mechanisms, as well as reduce the band gap in TiO_2 materials [22]. The well-known polymorphs of TiO_2 are rutile, anatase, and brookite, which are naturally occurring. It has recently been demonstrated that photo-activity towards organic degradation depends on the phase composition and oxidizing agent; for example, when the performance of different crystalline forms was compared, it was discovered that rutile shows the highest photocatalytic activity with H_2O_2, whereas anatase shows the highest with O_2. A good photocatalyst should be photoactive, able to utilize visible and/or near UV light, and be biologically and chemically inert, photostable, inexpensive, and non-toxic. However, it was previously reported that the anatase phase of TiO_2 is more efficient owing to its excellent photocatalytic properties [19]. Rutile is stable and easy to fabricate in setting conditions, while brookite is metastable and the preparation is troublesome. Both rutile and anatase hold a tetragonal crystal structure with a band gap of approximately 3.0–3.2 eV, respectively [23]. Titania has many advantages; foremost important is its use as a photocatalyst, and it is innocuous and chemically stable, etc. However, it also has a few disadvantages, for instance, it has a higher energy gap around 3.2 eV, which is why its use is limited to UV light only. Moreover, it cannot be activated in visible light or sunlight; therefore, researchers have been working on other ways to use it. Few methods are being developed for this purpose; one of the methods is to dope non-metals or metals impurity, and the other is to introduce light-sensitive semiconductors such as WO_3 in order to enhance the photocatalytic activity of TiO_2 [24]. Below visible-light irradiation, metal doping enhances the performance of the doped TiO_2 by moving the absorption spectra into a low energy field. Metal doping, on the other hand, has significant disadvantages, including the thermal instability of doped TO_2, electron capture by metal centers, and the need for more expensive ion implants. Non-metallic doping increases the percentage of anatomical TiO_2, which slows down the formation of TiO_2 crystals and increases the specificity of titania [25]. The most important approach that is used widely is the sol-gel method, which produces high surface area TiO_2 by controlling the hydrolysis of titanium tetraisopropoxide (TTIP) [26]. Therefore, the re-integration of electron–hole pairs produced in the valence and conduction bands enhances the image performance of TiO_2 [27]. However, there is also provision for dye photosensitization and proper TiO_2 support with the help of an efficient TiO_2 modification method [28]. Likewise, mesoporous titania (mp-TiO_2) is considered the best due to its higher surface area, tunable pore-size distribution, and well-defined pore orientation [29]. Furthermore, nitrogen/fluorine (N/F) co-doped mp-TiO_2 has been shown to exhibit improved visible-light adsorption and photoactivity [30]. TiO_2 graphene (Gr) porous microspheres were prepared by ultrasonic spray pyrolysis [31]. However, N/F co-doped mp-TiO_2 showed the best photocatalytic activity compared with single-element doping. The solvothermal method was adopted for the synthesis of N/F-mp-TiO_2, using urea as a nitrogen source and ammonium fluoride as a source for fluorine. There has been a great change in band gaps by the doping of TiO_2 with non-metals constituents, which are reported as F-TiO_2, N-TiO_2, B/N-TiO_2 [32], and C-N-S [33,34]. Moreover, thin-film TiO_2 with a different nano size is being extensively used for solar cell, design sensors, displays, automobiles rearview mirrors, and many other gas purification applications [35–37].

In this review, we have unfolded all the possible prospects and aspects of TiO_2 photocatalyst doped with non-metals for light sensitivity. By the end of the paper, almost all the precarious applications, challenges, and future recommendations will have been proposed.

2. Effect of Non-Metal Doping

Different approaches are investigated to prevent the recombination of electron–hole pairs that are photogenerated by undoped TiO_2. Doping adjusts the electronic assembly of TiO_2, which boosts it from the UV region toward visible spectra. The doping of titania with any species enhances their ability to confer the chemistry of that particular material; for example, a high degree of crystallinity is mainly due to a large surface area and a large crystal size, which ultimately increases the photocatalytic activity and decreases the charge separation of the photogenerated e^-/h^+ pair recombination [38,39]. Nevertheless, the doping or modification of TiO_2 using non-metal/anion has been carefully adopted to overcome the use of maximum energy, high photocatalytic activity, reduced time span of charge separation, and higher efficiency of TiO_2, and the effect of various parameters has been discussed here. Moreover, non-metallic dopants such as carbon (C) [40–42], nitrogen (N) [43–47], sulfur (S) [47,48], boron (B) [49,50], iodine (I) [51], N/F-doped TiO_2 [30,52], L-amino acids (C–N co-doped or C–N–S tri-doped)-TiO_2, C-N-TiO_2 /CNT composites, etc., have also been discussed (Table 1). The compatibility of (N) was investigated, and the improved photocatalytic performance and morphology of TiO_2 was observed [23], as well as its composites [15]. Furthermore, GO/TiO_2 nanocomposite showed an upgraded photocatalytic efficiency [53]. Attention is being diverted to the treatment of contaminated water with porous mineral composite materials. The growth of non-metallic spongy minerals has resulted in a high specific surface area, strong adsorption capabilities, and tailored chemical accumulation [54]. Montmorillonite-supported-TiO_2 (MMT/TiO_2) will solve the cohesion problem of TiO_2 particles. Contaminants can be adsorbed on the surface of nano TiO_2 to improve the probability of ion exchange and contact between the catalysts and contaminants deterioration rate. Therefore, a porous MMT/TiO_2 complex system can improve its photodecomposition efficiency by reducing the agglomeration and intensifying the photocatalytic response. The comparatively improved absorbance of the composite material in visible light is (390–780 nm); therefore, MMT/TiO_2 also improves the optical absorption capacity from 70 to 87%, instead of using TiO_2 individually [55].

Table 1. Summary of non-metal-doped TiO_2 photocatalyst under varied circumstances.

Year of Study	Type of Non-Metal Dopants	Synthesis Route/Method	Type of Dye	Characterization Techniques	Ref.
2017	C-TiO_2	Hydrothermal	Methylene blue, Rhodamine B, p nitrophenol	XRD, SEM, TEM, STEM, XPS, UV-vis	[56]
2019	C-TiO_2	Hydrothermal	Methylene blue	XRD, FTIR, N_2 adsorption-desorption isotherm, SEM, UV-vis	[57]
2020	C-TiO_2	Sol-gel	Methylene blue	EDX, UV-vis DRS analysis, SEM	[58]

Table 1. *Cont.*

Year of Study	Type of Non-Metal Dopants	Synthesis Route/Method	Type of Dye	Characterization Techniques	Ref.
2020	C-TiO$_2$ double-layer hollow microsphere	Hydrolysis of thermal expandable microsphere	Rhodamine B	XRD, FTIR, TGA, SEM, Raman N$_2$ adsorption-desorption isotherm, XPS, UV-vis	[59]
2021	Carbon-doped TiO$_2$ nanoparticles	Sol-gel	Methylene blue	XRD, TEM, XPS, DRS,138	[60]
2022	C-TiO$_2$ nanoflakes (C-TNFs)	Facile hydrothermal	Methylene blue	XRD, FTIR, SEM, UV-vis	[40]
2007	N-TiO$_2$	Microemulsion-hydrothermal	Rhodamine B	XRD, Raman, XPS, PL emission spectra	[61]
2010	N-TiO$_2$	Sol-gel/acidic media	Lindane	XRD, SEM, TEM, Raman, XPS, GC-MS	[62]
2015	N and C-co-doped porous TiO$_2$ nanofibers	Electrospinning and calcination	Methylene blue	XRD, FESEM, TEM, XPS, DRS,	[63]
2017	N-TiO$_2$	Solvothermal	Rhodamine B	XRD, SEM, TEM, BET, XPS, UV-vis	[64]
2020	TiN/N-doped TiO$_2$ composites	Sputtering process	Methylene blue	Raman, XPS, UV-vis	[65]
2020	C-N-TiO$_2$ composite fibers	Hydrolysis and calcination	Rhodamine B	XRD, SEM, TEM, FTIR, Raman, XPS, UV-vis	[66]
2021	N-TiO$_2$ nanotubes	Hydrothermal	Methyl orange	XRD, SEM, XPS, UV-vis	[67]
2009	S-TiO$_2$	Hydrothermal	Methylene orange	XRD, TEM	[68]
2015	S-TiO$_2$	Wet-impregnation method.	Humic acid Humic acid	EDX, SEM, EEM fluorescence	[69]
2016	(S–TiO$_2$), (N–S–TiO$_2$)	Sol-gel	Phenol and MB	BET, FESEM, FTIR, XPS, DRS	[70]
2011	S-TiO$_2$, N-S-TiO$_2$	Sol-gel	Methyl orange	XRD, TEM, UV-vis DRS	[71]
2021	NS/TiO$_2$	Sol-gel	Methylene blue, methyl red	XRD, BET, SEM, FTIR, Raman, UV-vis	[21]
	P/TiO$_2$	Hydrothermal/sol-gel			[72]
	Ag-P/TiO$_2$ nanofibers	One-pot electrospinning	Methylene blue	XRD, XPS, FE-SEM, TEM, UV-vis	[73]
2022	P/TiO$_2$/MWCNTs	Sol-gel	Methylene blue	XRD, FE-SEM, FTIR, UV-vis	[74]
2017	Si/TiO$_2$	Solvothermal	Methyl orange	XRD, SEM, EDS, BET, XPS	[75]

3. Treatment Opportunities for Dyes

Dyes have been a constant source of contamination for decades; textile and industrial dyes in the wastewater are one of the major contributors of toxic organic pollutants. The ever-increasing production of dyes is because of rapid industrialization; therefore, it is urgently necessary to focus on the proper treatment of these dyes [76]. In industries, most of the fabrication and processing operations excrete dyes in the wastewaters; the range of dyes varies from 2% to 50% from basic to reactive colorants, respectively. Therefore, the toxins produced are harmful for ground as well as surface water, as most of the compounds present in these hazardous dyes are non-biodegradable and likely to cause cancer [77]. Now, the primary concern is the sufficient treatment of these dyes; for their removal, various chemical and physical methods are being reported. Some of these methods are: ozonation [78], activated carbon [79], bio-degradation [80], and photocatalytic degradation using TiO$_2$, etc. [81]. The removal of dyes from contaminated water has been accomplished through biological processes that include microbiological or enzymatic decomposition and environmental degradation. Anaerobic conditions also contribute to the deterioration of the azo bond, which leads to the removal of pigment, but also leads to an insufficient mineral formation of harmful and carcinogenic products. A variety of synthetic dyes have been extensively tested in recent years to develop a more promising method based on advanced oxidation processes (AOPs) that can release pollutants quickly and extensively. AOPs rely on the formation of highly active hydroxyl radicals (OH•), which can release any chemical present in a liquid matrix, usually with a reaction rate controlled by diffusion [82].

3.1. Types of Dyes

The method by which photo-degradation occurs is determined by the products generated, as these product molecules would be adsorbed on the surface of the semiconductor by modifying the layout of its electronic and active sites. In photo-catalytic degradation, it has been discovered that dyes with a positive charge (cationic) adsorb more on unaltered TiO_2 than dyes with a negative charge (anionic) [83]. Azo dyes (AzD) are attributed as the major (60–70%) class of synthetic dye stuff used in the textile, leather, oil, additives, and food industry, etc., and the resulting byproducts carry both the metal ion and the dyes. All the major classes of dyes are shown in Figure 1 and Table 2.

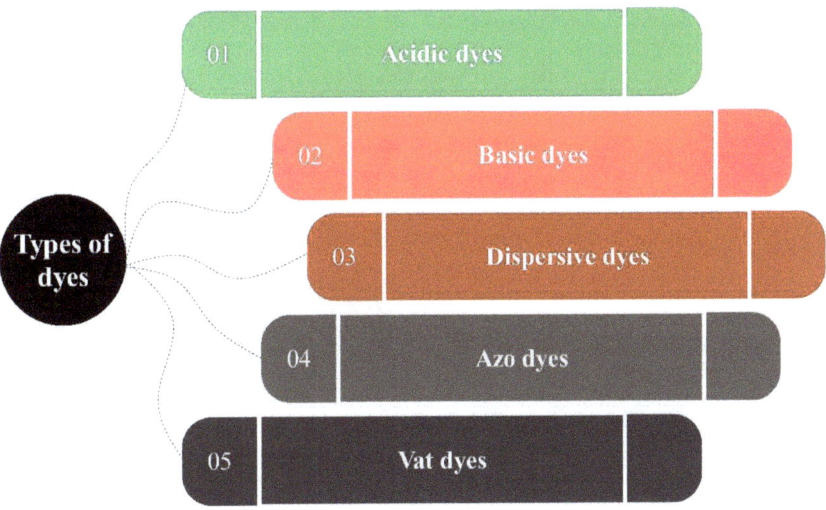

Figure 1. Major classes of dyes.

Table 2. Types and characteristics of Azo dyes [76].

Category of Dye	Features	Fiber	Pollutant	Dyes Fixation
Acidic	Water-soluble anionic compounds	Wool, nylon, cotton blends, acrylic, and protein fibers	Organic acids, unfixed dyes, color	80–93
Basic	Water-soluble, applied in weakly acidic dye baths, very bright dyes	Acrylic, cationic, polyester, nylon, cellulosic, and protein fibers	NA	97–98
Direct	Water-soluble, anionic compounds, applied without mordant	Cotton, rayon, and other cellulosic fibers	Surfactant, defoamer, leveling and retarding agents, finish, diluents	70–95

Table 2. *Cont.*

Category of Dye	Features	Fiber	Pollutant	Dyes Fixation
Dispersive	Insoluble in water	Polyester, acetate, modacrylic, nylon, polyester, triacetate, and olefin fibers	Phosphates, defoamer, lubricants, dispersants, diluents	80–92
Val	Oldest dyes, chemically complex, water-insoluble	Cotton, wool, and other cellulosic fibers	Alkali, oxidizing agents, reducing agents, color	60–70

These pigments comprise of a double bond between two nitrogens (–N=N–), which is further linked to two aromatic group moieties such as naphthalene/benzene rings [84]. AzDs are positively charged at pH 6.8 and negatively charged at a higher pH, which affects their adsorption on semiconductor surfaces. The pH of the effluent is not neutral, and the mixture of substances that would have dissolved in water varies the surface characteristics of the semiconductor. When charged species are present in solution, an electrical double layer forms, affecting electron–hole pair separation and dye adsorption properties on the semiconductor surface. pH and the amount of dye present influence the rate of photocatalytic dye abasement [85]. They have amphoteric properties. Depending on the pH of the medium, AzDs can be anionic (deprotonation in the acidic group), cationic (extension in the amino group), or non-ionic [86]. The most known AzDs are acid dyes, basic dyes (cationic dyes), direct dyes (substantive dyes), disperse dyes (non-ionic dyes), reactive dyes, vat dyes, and sulfur dyes. Acidic dyes acquire their name from the fact that they are frequently employed to dye nitrogenous textiles or fabrics in inorganic or organic acid solutions. Cations that are extensively utilized in the manufacturing of acrylic and modacrylic fibers are produced by basic dyes in the solution. Electrostatic forces are used to apply direct dyes to fibers/fabrics in an aqueous medium containing ionic salts and electrolytes [87]. Anaerobic conditions can also contribute to the deterioration of the azo bond, which leads to the removal of pigments, but also leads to an insufficient mineral formation of harmful and carcinogenic products [88].

3.2. Dye-Degradation Mechanisms

The degradation mechanism had been discussed in detail in the literature in various citations; only one is mentioned here [89]. When aqueous titania is subjected to the visible-light irradiation greater than 3.2 eV (band gap), the electrons of the conduction band (electrons $_{CB}^{-}$) and the holes of the valence band (holes $_{VB}^{+}$) are created. These light-generated electrons (e^-) can either react with e^- acceptor O_2 adsorbed on the surface of TiO_2, or they can reduce the dye; they can also dissolve in water, causing reduction and producing an anion which could be a superoxide radical—$O_2^{\bullet-}$. The holes that are generated have the tendency to oxidize any organic molecule to R^+, and they can also oxidize water or hydroxyl ion to $OH^\bullet$ radicals. All these radicals along with titania have the potential to decompose dyes photocatalytically. The series of reaction schemes is given below [90] by Scheme 1. The pictorial representation of the dye-degradation mechanism is represented in Figure 2.

$$\text{Photocatalyst} + hv \;\rightarrow\; (e^- CB + \mathbf{h}^+ VB) \dots\dots\dots\dots (1)$$

$$e^- CB + O_2 \;\rightarrow\; O_2^{\bullet -} \dots\dots\dots\dots\dots\dots\dots\dots (2)$$

$$H_2O + O_2^{\bullet -} \;\rightarrow\; OOH\bullet + OH^- \dots\dots\dots\dots (3)$$

$$2OOH\bullet \;\rightarrow\; O_2 + H_2O_2 \dots\dots\dots\dots\dots (4)$$

$$O_2^{\bullet -} + \text{dye} \;\rightarrow\; \text{dye}^- OO\bullet \dots\dots\dots\dots\dots (5)$$

$$OOH\bullet + H_2O + e^- CB \;\rightarrow\; H_2O_2 + OH^- \dots\dots\dots (6)$$

$$H_2O_2 + e^- CB \;\rightarrow\; OH\bullet + OH^- \dots\dots\dots\dots (7)$$

$$H_2O_2 + O_2^{\bullet -} \;\rightarrow\; OH\bullet + OH^- + O_2 \dots\dots\dots (8)$$

$$\text{Dye} + hv \rightarrow \text{dye}^* \dots\dots\dots\dots\dots\dots (9)$$

$$OH\bullet / O_2^{\bullet -} / \text{dye}^* + \text{photocatalyst} \rightarrow \text{Dye degradation} + \text{products} \dots\dots\dots (10)$$

Scheme 1. Reaction scheme for the degradation of dyes using TiO_2 [90].

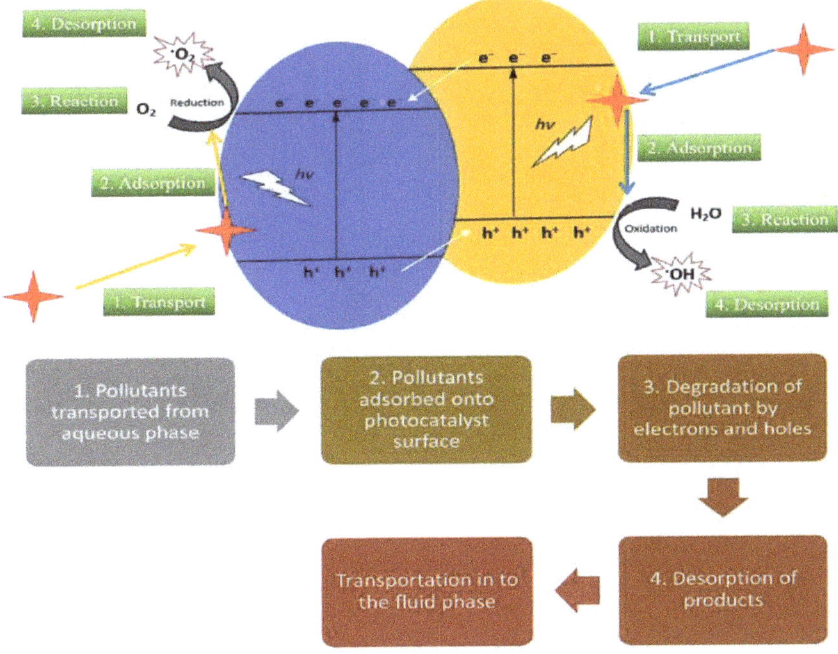

Figure 2. The process of dye-degradation mechanisms by photocatalysis, depicted graphically [91].

On the other hand, indirect methods can be used for the degradation mechanism, such as electron spin, resonance test scavenging, and radical trapping [92,93].

3.3. Technologies or Methods for the Removal of Dyes

There are innumerable dye-elimination techniques, which can be categorized as physical, chemical, and biological. The physical methods comprise of adsorption, ion exchange, and filtration/coagulation [94–100], whereas the biological include anaerobic deterioration, bio-sorption, and many more, while the chemical consist of ozonation, Fenton reagent, and photocatalytic processes, respectively [101,102]. The methodologies adopted for the removal of dyes is described in Figure 3.

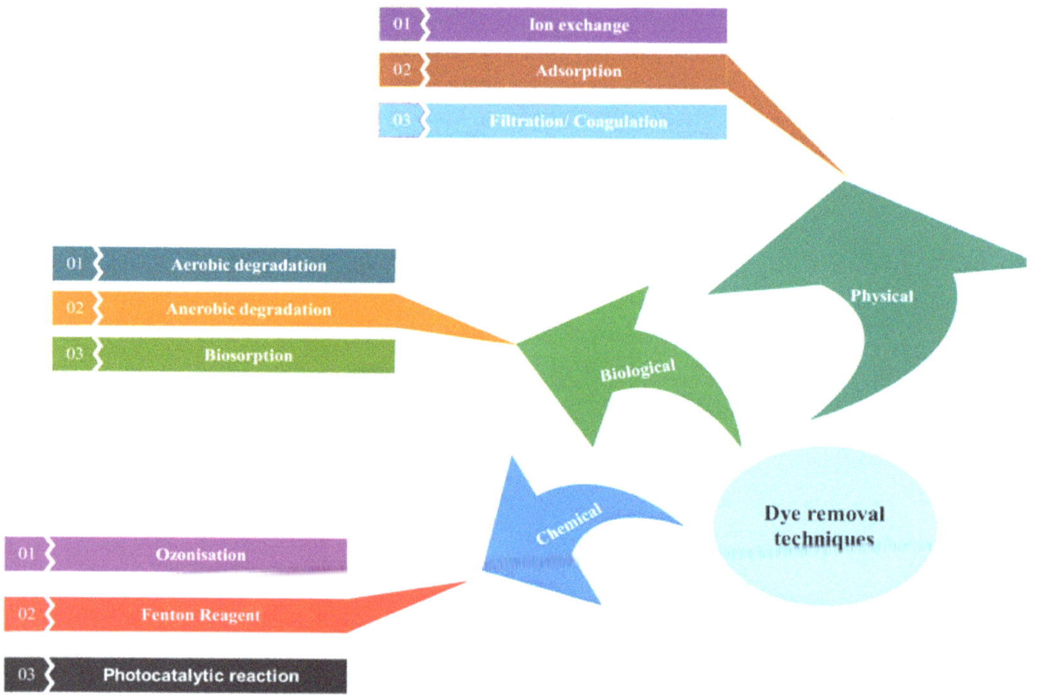

Figure 3. Different methodologies adopted for the removal of dyes.

In the physical method, different processes are used in the treatment of dyes, such as testing, mixing, precipitation, adsorption, and membrane filtration [97,103–105] (Table 3). Studies have shown that adsorption is the best method for the removal of dyes as it is simple, easily operative, highly efficient, cheap, and has been effective for toxic substances, whereas for membrane filtration, the contamination of the membranes and the cost of changing them are the biggest filtering limitations [106]. On the other hand, chemical treatments include coagulation-flocculation, oxidation, ozonation, Fenton oxidation, photocatalytic oxidation, irradiation, ion exchange, and electrochemical treatment [107,108]. These chemical methods are very effective in enabling the possibility of dyes removal, but excessive use of the chemicals causes difficulty in disposing of them into mud. Their disposal is too costly, and a reasonable amount of electrical power is required for these processes [109]. Biological analysis includes aerobic, anaerobic, and anaerobic–aerobic investigation, in which natural contamination is reversed into harmless and solid objects. The physicochemical and biological treatment of polluted water are the standard for dye-removal methods [110].

Table 3. Illustration of various methods for removal of organic dyes.

Strategies	Methods	Advantages	Disadvantages	Ref.
Chemical	Electro-Fenton reagent Ozonation Photocatalysis	Effective decolorization of soluble or insoluble dyes No sludge production initiates and accelerates Azo bond cleavages	No diminution of COD values by extra costs Sludge formation Formation of byproducts release of aromatic amines High costs	[108,111]

Table 3. *Cont.*

Strategies	Methods	Advantages	Disadvantages	Ref.
Physical	Ion exchange Adsorption Filtration/coagulation	Good removal of wide variety of dyes Regeneration No absorbent loss Good elimination of insoluble dyes Low-pressure process	Non-selective to absorbate Non-effective for all dyes High costs of sludge treatment Quality not high enough for re-using the flood	[103–105]
Biological	Enzymes Microbes Aerobic and anaerobic degradation Biosorption	Reduces the amount of waste that is delivered to landfills or incinerators Manufacturing requires less energy When it breaks down, it releases less hazardous compounds	Low biodegradability of dye Salt concentration stays constant	[98]

3.4. Stability of Non-Metal-Doped Titania

Non-metal-doped TiO_2 possesses higher catalytic activity in comparison to the undoped titania, as they have the enhanced capability of absorbing visible light, hence, more enhanced photocatalytic activity. Furthermore, the band gap becomes narrow because the lattice of titania is being substituted by oxygen. Almost all the non-metals reduce the re-connectivity of the e^-/h^+ duos but also decrease the band gap energy of titania via the formation of aperture energy. Therefore, the doping of non-metals in TiO_2 is an approach to increase the catalytic activity and the visible-light harvesting capacity. Therefore, it is safe to say that non-metal-doped titania can turn out to be the active catalyst for a visible-light catalyst for the degradation of different dyestuff [112]. Practical applications of nitrogen-doped TiO_2 for dye-sensitized solar cells were reported by Wei Guo and his co-workers, as N-doped titania dye-sensitized solar cells have shown 10.1% more conversion efficacy and are more stable because of the induction of N into the titania photo-electrode [113]. Carbon (C) and boron (B) co-doped photosensitizers have also been reported in the literature for efficient applications [114].

4. C-Doped TiO_2 (C/TiO_2)

The exploration of carbonaceous-doped materials with TiO_2 plays a significant role in photocatalysis, and they have shown spectacular growth due to their very simple method of synthesis [115]. Three dimensional cariogenic dot/TiO_2 nanoheterojunctions photocatalyst was synthesized by the hydrolysis process for the degradation of RhB, with different weight percent of C-dots for enhanced photocatalytic activity w [116]

The C/TiO_2 was immobilized with polyamide fibers with different weight percentages (1 wt.%, 2 wt.%, and 3 wt.%); the surface area of TiO_2 increased and the band gap energy (2.38 eV) decreased measurably after doping with non-metal for the degradation of MB [57]. The addition of C atoms to the TiO_2 structure can improve visible-light absorption by narrowing the band gap. However, organic chemicals, different dyes, and medicines were removed using C/TiO_2 photocatalyst matrices derived from aqueous [117]. The C/TiO_2 with several forms of carbon precursors was reported [118]. The most utilized carbon precursors are difficult to detect since the researchers use a variety of chemicals for titanium precursors, such as titanium carbide, titanium (IV) oxacetylacetonate, and titanium (IV) butoxide (TBOT) to act as TiO_2. The recombination of e^-/h^+ pairs is reduced by doping titania with carbon. Titanium dioxide (TiO_2) nanoparticles (NPs) were synthesized by sol-gel synthesis and doped with polydiallyldimethyl ammonium chloride (PDADMAC), as the carbon precursor caused a significant decrease in the band gap from 2.96 eV to 2.37 eV [41].

Recently, C/TiO_2 with anatase/rutile multi-phases coated on granular activated carbon was used under visible light for the removal of nonylphenol. A significant doping effect was observed in the band gap, which saw a decrease from 3.17 eV to 2.72 eV, respectively [119,120]. Indeed, carbon-based (nano) composites have improved photocatalytic activity due to the coupling effect from the incorporation of carbon species. However, several types of carbon–TiO_2 composites such as C-doped TiO_2, N–C-doped TiO_2, metal–C-doped TiO_2, and other co-doped C/TiO_2 composites have been reported (Table 4), which were synthesized by the solvothermal/HT method and sol-gel process [121].

Table 4. Methods and precursors used for the synthesis of C-doped TiO_2 photocatalysts.

Methods	TiO$_2$ Precursor	Carbon Source	References
Chemical bath deposition (CBD)	Titanium isopropoxide (TTIP)	Melamine borate	[122]
Sol-gel	Titanium isopropoxide (TTIP)	Microcrystalline cellulose (MCC)	[58]
Hydrothermal	TiC	-	[123]
Sol-gel	TTIP, TBOT, TiCl$_4$, TiCl$_3$	Ethanolamine (ETA), glycine, polyacrylonitrile (PAN), polystyrene (PS), starch, TBOT	[124,125]
Solvothermal treatment and calcinations	TiCl$_4$	Alcohols (benzyl alcohol and anhydrous ethanol)	[126]
Solvothermal	TTIP	Acetone	[127]
Electrospinning followed by heat treatment	TTIP	Acetic acid	[128]
Hydrolysis	TBOT	Glucose	[129]
Sol-gel	Titanium butoxide	-	[30]
Hydrothermal route	-	Various carbon sources	[42]

The C/TiO_2 photocatalyst exhibits enhanced photocatalytic activity in comparison to titania because the catalyst alters the crystal structure, lowers the pH, and narrows the band gap. C/TiO_2 supported by granular activated carbon for photocatalytic degradation of nonylphenol and anatase ratio is much better for the degradation in comparison to rutile [130]. Furthermore, it was discovered that annealing can increase the crystallinity of C/TiO_2 nanotubes, as shown in Figure 4 in the presence of Argon (Ar), instead of oxygen or nitrogen [131]. In addition, Ar or N_2 has proven beneficial in increasing the photocatalytic activity. The recombination of e^-/h^+ pairs is reduced by doping titania with carbon. TiO_2 nanoflakes (TNFs) and C/TiO_2 nanoflakes (CTNFs) were synthesized by the HT method, which is superior (92.7%) in the degradation of MB [40].

C/TiO_2 composites were further modified with nitrogen (as N–C-doped TiO_2 composites), and it was observed that N–C-doped TiO_2 composites exhibit improved photocatalytic activity as compared to only C/TiO_2 nano-formulations. High meso-porosity and a well-defined large surface area ($102 \text{ m}^2 \text{ g}^{-1}$) were obtained by N–C-doped TiO_2 composites synthesized by the solvothermal method with high photocatalytic evaluation [132–135]. In visible light, the catalytic image functions of non-metal-doping photocatalysts with different colors as model compounds are commonly used to analyze irradiation; however, this strategy has already been reported in the literature as an ineffective method [136]. This is due to the dye's ability to absorb visible light, which means that the photocatalytic process may not be driven solely by visible-light absorption. However, this is not only by the photocatalyst, but also by the dye's absorption of light (that is, dye sensitization). The most employed dyes in the literature are on visible-light active photocatalysts. In addition, Song and his colleagues reported a C-doped TiO_2/carbon nanofiber film (CTCNF) under visible-light irradiation for the breakdown of rhodamine B (RhB). The dye started to degrade, and its discoloration rate was 66.4–94.2% after 150 min of visible-light irradiation [137].

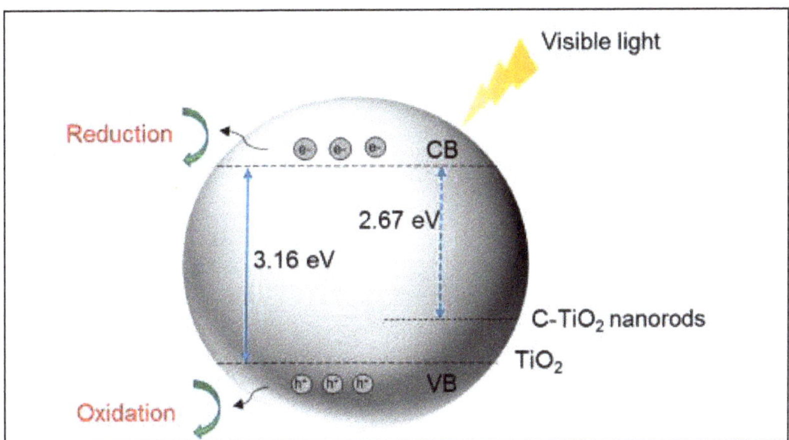

Figure 4. Photocatalysis mechanism of C-doped TiO$_2$ nanorods under visible light, adopted from [56].

In addition, the removal of other organic substances from pharmaceuticals, personal care items, and even microbes from water and hydrogen production are the best examples of compounds being removed with the help of a C/TiO$_2$ photocatalyst. C-TiO$_2$/rGO is used to form a hybrid nanocomposite that exhibits excellent photocatalytic activity for the better production of H$_2$ instead of pure TiO$_2$, as shown in Figure 5 [127].

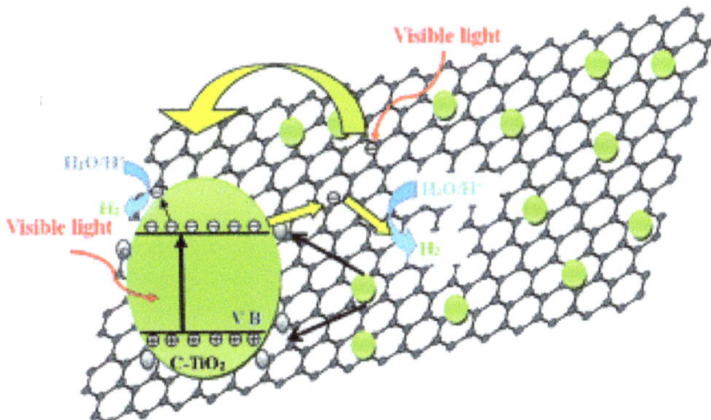

Figure 5. Photocatalytic mechanism of C-TiO$_2$/rGO adopted from [127].

The C/TiO$_2$ photocatalyst was prepared by the sol-gel process by using microcrystalline cellulose (MCC) as a carbon source material. In comparison to pure and C-TiO$_2$ a reduced band gap was observed. On the other hand, when Fe is co-doped with C-TiO$_2$ further reduction in the band gap exhibit stronger visible light absorption [138]. Some other carbon material, such as graphdiyne, plays a significant role in a variety of applications [139,140].

5. N-Doped TiO$_2$ (N/TiO$_2$)

Nitrogen is the most often-utilized non-metal dopant among all non-metal dopants [130]. N/TiO$_2$ has been reported (Table 5) for various photocatalytic applications [141–145]. Nitrogen (N) doping in TiO$_2$ nanotubes was prepared by the hydrothermal process for the degradation of dye and H$_2$ gas evolution. Urea was used as a N source, and optimized N/TiO$_2$ nanotubes (TiO$_2$ nanotubes vs. urea at 1:1 ratio) showed the highest degradation efficiency over methyl orange (MO) dye (~91% in 90 min) and manifested the highest rate of H_2 evolution (~19,848 μmol h$^{-1}\cdot$g^{-1}) under solar-light irradiation [67].

N/TiO$_2$ was synthesized by the direct hydrolysis of titanium tetraisopropoxide and ammonia as a nitrogen source for the photocatalytic degradation of organic dyes in liquid phase with visible light. A batch photo-reactor was used to study the photocatalytic evaluation of methylene blue (MB) dye degradation under the optimum operating parameters to obtain the maximum efficiency toward dye degradation. The band gap energy of titanium dioxide was shifted toward a lower band gap, i.e., in the visible range from 3.3 to 2.5 eV. This is because of its low ionization energy and the atomic size being comparable to the size of oxygen [146]. As proven within Table 5, quite a few nitrogen precursors were used in the synthesis of N-doped TiO$_2$.

The photocatalytic activity of N/TiO$_2$ and N-doped TiO$_2$ with transition metals (Fe, Cu) was reported for the degradation of MB under sunlight [147]. N/TiO$_2$ showed the highest activity among the doped TiO$_2$ nanoparticles (0.006 min^{-1}). Titanium nitride/nitrogen-doped titanium oxide (TiN/N-doped TiO$_2$) composite films were synthesized by the sputtering process. A Raman spectral analysis revealed the formation of TiO$_2$ with anatase, which later transformed to the rutile phase, but the results showed that the photodegradation efficiency of MB was higher in the case of titania anatase after exposure to visible light [65].

Table 5. Summary of N-doped TiO$_2$ photocatalysts synthesized by variety of methods and source precursor materials.

Year of Study	Method	TiO$_2$ Precursor	Nitrogen Source	Ref.
2017	Addition of N source to the TiO$_2$ precursor solution	TBOT	Tetra methyl-ethylene-diamine	[148]
2020	CVD	TICl$_4$	Tert-butylamine, benzyl amine	[149]
2017	Hydrothermal	TBOT	KNO$_3$	[150]
2019	Hydrolysis	TTIP	NH$_4$Cl, pyridine	[151]
2016	Electrochemical	Titania nanotubes	Diethylenetriamine, ethylenediamine, hydrazine	[152]
2019	Sol-gel	TTIP, TBOT, TiCl$_4$, Titanic acid	Urea, NH$_3$, nitro methane, n-butyl amine, N$_2$, hydrazine, HNO$_3$,	[153]

For the solar-driven photocatalysis over Ti^{3+} and N co-doped photo catalysts, Cao et al. [154] proposed a modified mechanism. The materials were made by reducing urea-modified mesoporous TiO$_2$ spheres with NaBH$_4$ at 350 °C in an Ar environment. The abovementioned N-doping of TiO$_2$ resulted in the emergence of a new impurity level above the VB. In addition, by introducing Ti^{3+} and O below the CB, an intermediate energy level was created. As a result, the band gap was decreased, which increased photocatalytic effectiveness in the visible light. The schematic diagram of N-doped TiO$_2$ is shown in Figure 6.

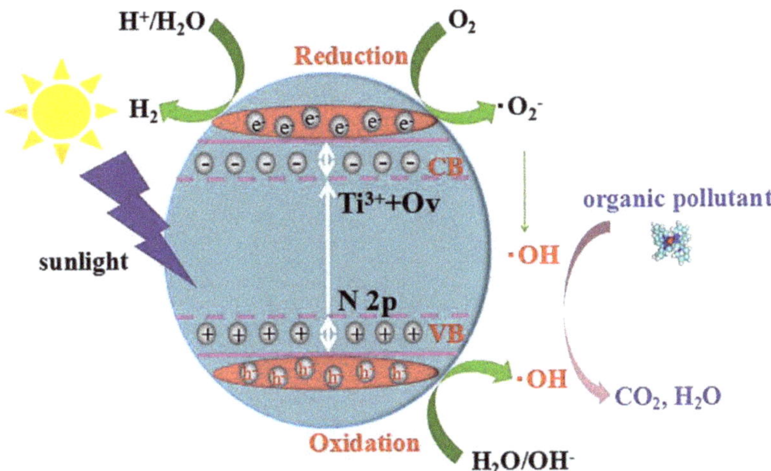

Figure 6. N-doped TiO$_2$ photocatalytic process driven under the sun is schematically represented [154].

The activity of N-doped TiO$_2$ as a photocatalyst has been examined in various studies. In general, N/TiO$_2$ photocatalysts have been utilized to remove organic chemicals from water, namely dyes and medicines, but phenol, furfural parabens, surfactants, and herbicides were among the other pollutants that were destroyed using doped TiO$_2$ with N [155]. It was also suggested that N/TiO$_2$ photocatalyst could be used to remove pollutants from the gaseous phase, such as acetaldehyde, benzene, ethylbenzene, NH$_3$, and NOx. Antibacterial characteristics were found in some of the photocatalysts, such as against Escherichia coli and oral cariogenic biofuels. Furthermore, the use of N-doped TiO$_2$ for human breast cancer diagnostics, and cancer treatment such as for melanoma, has been reported previously in Figure 7 [156]. N/F co-doped mp TiO$_2$ has been shown to have the highest adsorption and photocatalytic activity. The solvothermal method was adopted for the synthesis of N/F mp TiO$_2$ by using urea as a nitrogen source and ammonium fluoride as a fluorine source [29] The L-amino acid (C-N co-doped or C-N-S tri-doped) -TiO$_2$ photocatalyst was used for dyes removal under visible dye. The photodegradation of methyl orange and direct red 16 was studied by the first order kinetics, and the rate constant for DR16 photocatalytic removal using L-Arginine (1 wt.%)-TiO$_2$ was 2.9 and 4.3 times greater compared to those of L-Methionine (1.5 wt.%)-TiO$_2$ and L-Proline (2 wt.%)-TiO$_2$, respectively [157].

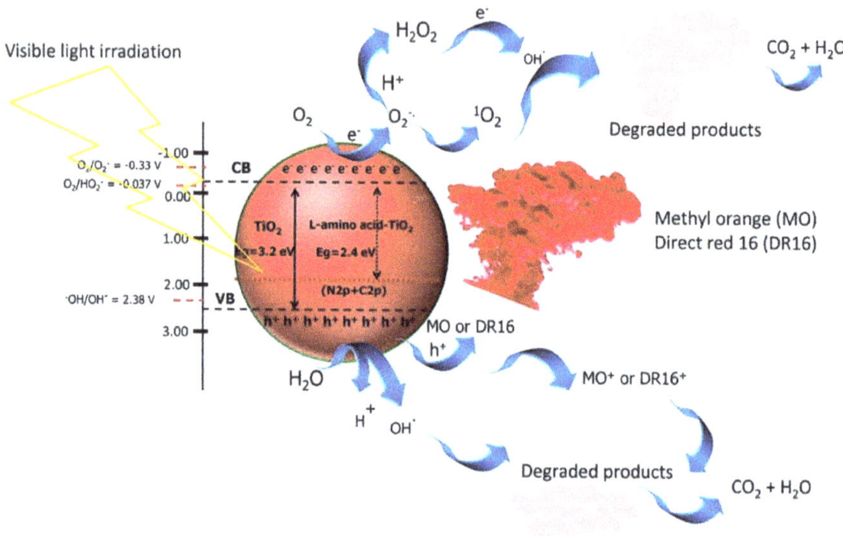

Figure 7. N-doped TiO_2 photocatalytic [157].

6. S-Doped TiO_2 (S/TiO_2 or SdT)

Amongst all the non-metals, sulfur would have been difficult to dope with titania because a large amount of formation energy is required for the replacement of S with O in titanium dioxide [158]. In contrast to the previous statement, S/TiO_2 (SdT) (anatase) was synthesized using TiS_2 (titanium disulfide) by the oxidative annealing. Therefore, this doping led to the red shift in the absorption band of SdT rather than the undoped titania. Furthermore, it was used for the photocatalytic degradation of methylene blue (MB). SdT instigated MB's irradiation quite efficiently in the visible-light region, as shown in Figure 8 [159]. SdT was also prepared by the heating of powder TiS_2 in the solution of HCl at 180 degrees, a comparatively lower temperature in comparison to the conventional methods of synthesizing SdT. Traditionally, it has been prepared by the thermal decomposition of thiourea (ThU) at elevated temperatures, whereas ThU is the source of sulfur. The formulated SdT at a demoted temperature was used for the desolation of 4-chlorophenol; it was concluded that the one-step hydrothermal process for the production of SdT was successful for the irradiation of 4-chlorophenol by visible-light photocatalytic activity [160].

Sulfur-doped and sulfate TiO_2 (SdST) were synthesized using the solvothermal method; potassium per sulfate (KPS) was taken as a source of sulfur and the irradiation rate of phenol was studied. As we know, titania showed lower photocatalytic activity in the visible light because it has a broad band gap energy, which limits the absorbance of UV light of less than 387 nm. However, the SdST catalyst showed higher degradation activity for phenol instead of pure TiO_2, especially in the range greater than 450 nm (longer visible-light range). Almost 51.3% of the containment (phenol) was degraded at a 0.5 ratio of the catalyst (SdST) for 10 h under the ranges, as mentioned in Figure 9 and Table 6 [161].

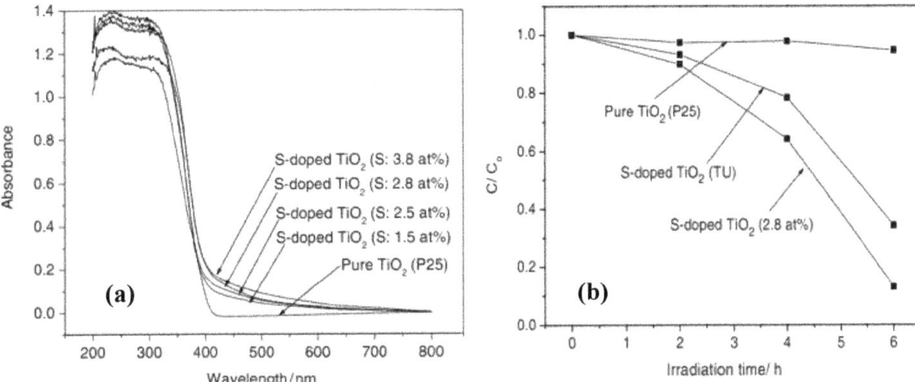

Figure 8. (**a**) Response of different concentrations of sulfur in S-doped TiO₂; (**b**) photocatalytic degradation of 4-chlorophenol by pure TiO₂, SdT prepared by ThU and SdT by hydrothermal process studied via UV-vis [160].

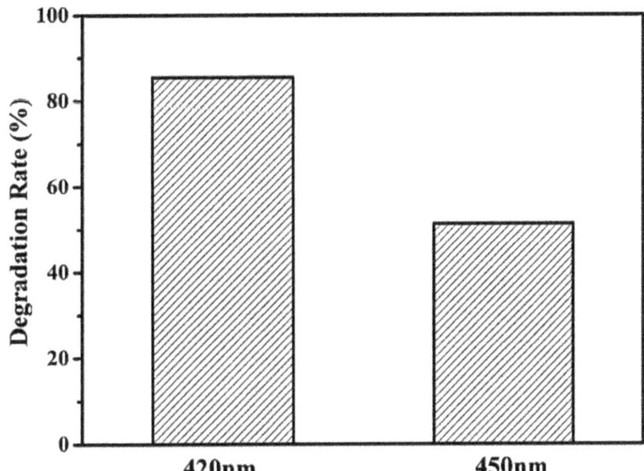

Figure 9. The photocatalytic degradation of phenol using SdST-0.5 at various wavelength irradiation [161].

Table 6. Summary of S-doped TiO₂ photocatalysts synthesized by variety of methods and source precursor materials.

Year of Study	Method	TiO₂ Precursor	Sulfur Source	Ref.
2003	Oxidative heating	Anatase	TiS₂	[159]
2006	Low-temp hydrothermal	Anatase	TiS₂ powder with HCl solution	[160]
2012	Solvothermal	TBOT	Potassium per sulfate	[161]
2016	Free oxidant peroxide method	Anatase	Thiourea (ThU)	[162]
2018	HT	Titanium sulfate (TiOSO₄)	TiOSO₄	[163]

7. P-Doped TiO₂ (P/TiO₂ or PdT)

In the recent years, P/TiO₂ (PdT) has gained attention because of their potential to increase the surface area, restrain the transformation of anatase to rutile, enhance the absorption of visible light and decrease grain growth [164] (Table 7). Therefore, many studies have been conducted for the generation of PdT and to study their properties regarding water splitting [165], dye-degradation, etc. [166]. Similarly, the synthesis of PdT nanoparticles using the sol-gel process was performed. The results showed a remarkable increased activity of the degradation of MB because of high visible-light pursuit, and the ESR spectroscopy (electron spin resonance) deduced its refined charge separation [167].

The fabrication of P doped over titania nanofibers (PdTNFs) was first reported by Zhu. Y. et al., from the method known as chemical vapor deposition (CVD), which showed remarkable results for electrochemical water splitting [168]. The doping of phosphorous decreases the band gap of titania and creates disproportion between O_2 and Ti^{4+} charges; therefore, the recombination of charge carriers is hindered. In TiO_2 (anatase type), the substitution of P^{3-} over O^{2-} is much greater than substituting P^{5+} onto Ti^{4+} (1.32 eV) because it requires high formation energy (15.48 eV). Hence, it shows that the incorporation of P^{5+} into titania lattice is achievable by forming the Ti-O-P bond rather than the Ti-P bond [169]. The separation of charges in the photocatalytic phenomenon is facilitated by the phosphate ions, which act as an electron-withdrawing species [170].

The photocatalytic activity of sulfamethazine (SMHZ) was investigated using mixed oxide novel-doped Fe_2O_3 and TiO_2. The weight percentage of 1.2 of mixed oxide had successfully degraded 30% of SMHZ; the percentage degraded is much higher than Fe_2O_3-TiO_2 or pure TiO_2 (Figure 10). [171].

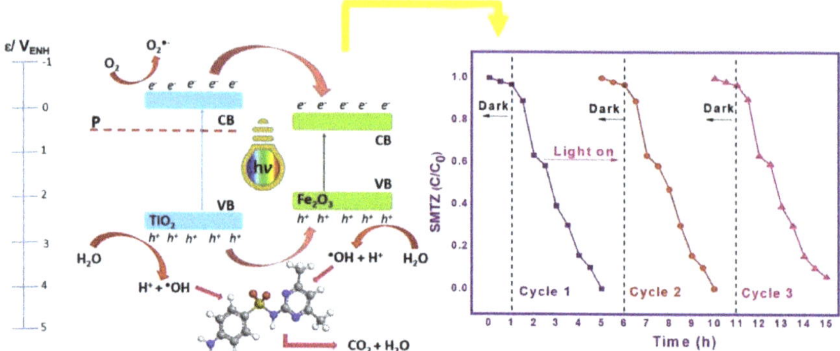

Figure 10. Photocatalytic mechanism of P-doped Fe_2O_3-TiO_2 against SMHZ [171].

Table 7. Summary of P/TiO_2 photocatalysts synthesized by variety of methods and source precursor materials.

Year of Study	Method	TiO₂ Precursor	Phosphorous Source	Ref.
2020	Chemical vapor deposition	Titanium (IV) butoxide (Ti(OC₄H₉)₄)	Red phosphorous	[171]
2014	Sol-gel method	TBO	Phosphorous acid	[172]
2011	Sol-gel process	Silicate/TiO₂ NPs	Phosphoric acid (H₃PO₄)	[164]
2009	Sol-gel	TBO	H₃PO₄	[173]

8. B-Doped TiO₂ (B/TiO₂ or BdT)

The non-metal doping of anatase TiO_2 was further doped with B, C, N, and F [174–176]. The B/TiO_2 hybrid hollow microspheres were synthesized by hydrothermal treatment. This boron-doped titania photocatalyst was used for MB degradation in aqueous solution and served as a target pollutant to evaluate the photocatalytic activity under sunlight. A higher photocatalytic activity of boron-doped hybrid hollow microspheres was observed (Table 8), which was comparatively much greater than undoped titania [177].

B/TiO$_2$-nanotubes (B/TiO$_2$NTs) were synthesized by the electrochemical anodization method with boron concentrations of 70, 140, 280, and 560 ppm and activated via UV-visible irradiation. The results showed that the dye-degradation rate of Acid Yellow1 (AY1) was twice greater at the doped electrode, which contained 280 ppm of B, and the activation of the electrode was maximum there, as observed from the UV-visible light. When the AY1 (100 ppm) was treated at the B/TiO$_2$NTs electrode for 120 min at +1.2 V, pH 2 and in 0.01 mol L^{-1} of sodium sulfate solution (Na$_2$SO$_4$), 100% discoloration of the dye was observed. Therefore, the synthesized amalgamation has the potential to become a stable electrochemical catalyst [178].

The synthesized BdT nanostructures was carried out using the sol-gel method. Studies have shown that, after the doping of boron, the band gap decreased from 2.98 eV of undoped titania to 2.95 eV of B/TiO$_2$ (7% boron content) (Figure 11a). Boron has the tendency to occupy the interstitial sites in the crystal lattice of TiO$_2$ and forms a Ti-O-B bond. Therefore, the degradation studies of 4-nitrophenol were studied using the said nanoformulations. The results showed that BdT (7%) displayed a 90% degradation efficacy compared to the undoped titania (79%) because the Ti-O-B linkage has a synergistic effect on supplementing the catalytic activity, as shown in Figure 11b [179].

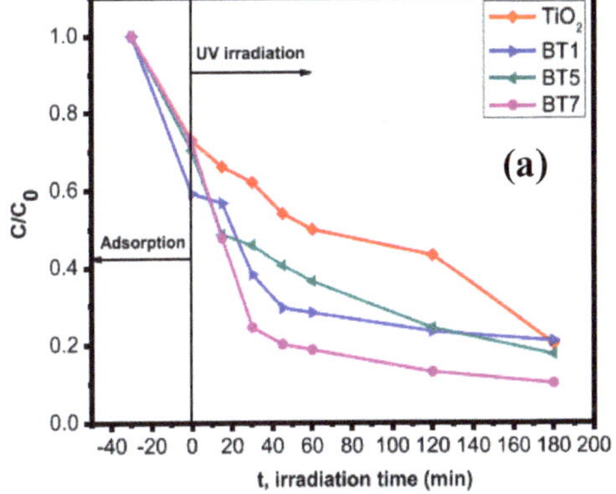
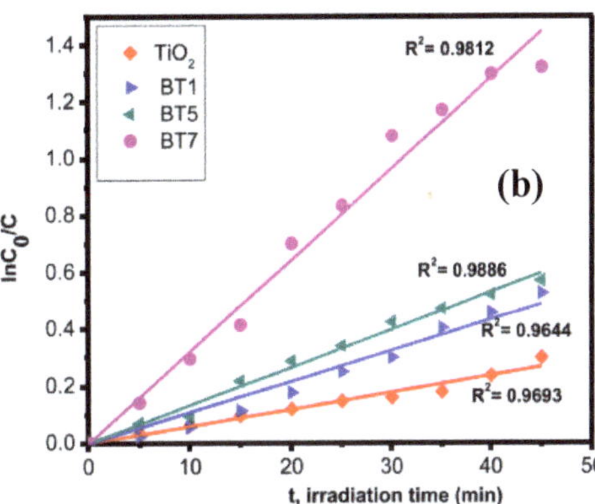

Figure 11. (**a**) Photocatalytic degradation of 4-nitrophenol using B/TiO$_2$ and undoped TiO$_2$. (**b**) First-order plot of photocatalytic degradation of 4-nitrophenol using UV-visible-light irradiation in the presence of BdT and undoped TiO$_2$ [171].

Table 8. Summary of B/TiO$_2$ photocatalysts synthesized by variety of methods and source precursor materials.

Year of Study	Method	TiO$_2$ Precursor	Boron Source	Ref.
2020	Sol-gel	Titanium isopropoxide	Boric acid	[179]
2015	Electrochemical anodization	Titanium sheets	NaBF$_4$	[178]
2011	HT	(NH$_4$)$_2$TiF$_6$	H$_3$BO$_3$	[177]

9. Halogens-Doped TiO$_2$ (X = F, Cl, Br, and I)

The hydrogenated F/TiO$_2$ (FdT) nanocrystals were synthesized using the physicochemical method. F-doping has the potential to increase the surface area of TiO$_2$; further, hydrogenation plays a pivotal role in forming the F-H and O-H bonds on the surface of titania and creating vacancies of Ti^{3+} and O^{2-}, which increases the range of absorption alongside the light utilization capacity of TiO$_2$. The bonds O-H and F-H can favor trapping the holes and can also react with water to produce an active species (OH$^\bullet$). Now, these hole and electron pairs can easily be separated to participate in the photocatalysis. Irradiation studies of MB were conducted and showed that the degradation rate constant of hydrogenated FdT is 0.146 min^{-1}, which is almost twice the rate of pure TiO$_2$ (0.063 min^{-1}) [180].

Hydrogen and fluoride co-doped TiO$_2$ nanostructures have been made from annealing. The doping of halogens decreases the band gap and bestows the oxide material a greater absorbance capacity. The results demonstrated that FdT is a better photocatalyst, as shown in Figure 12 and Table 9 [181].

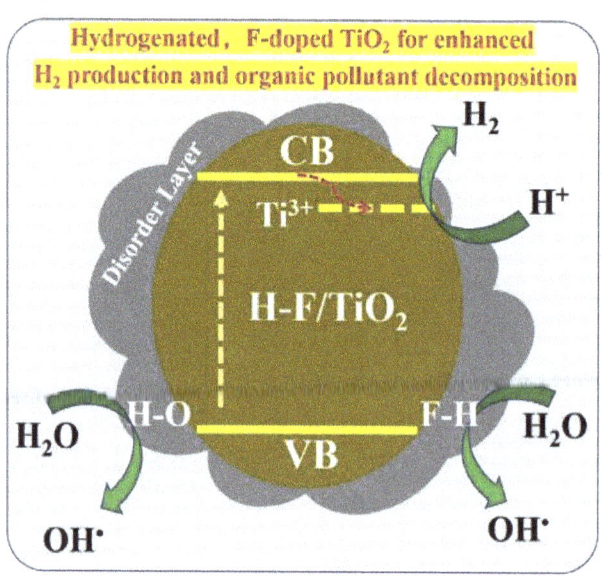

Figure 12. Suggested photocatalytic degradation for H$_2$ production and dye degradation [180].

Table 9. Summary of halogens-doped TiO$_2$ photocatalysts synthesized by variety of methods and source precursor materials.

Year of Study	Method	TiO$_2$ Precursor	Fluorine Source	Ref.
2020	Oxidative annealing	Titanium isopropoxide	NH$_4$F	[181]
2019	Physicochemical	TTIP	NH$_4$F	[180]
2017	Sol-gel	Titanium isopropoxide	Trifluroacetic acid	[182]
2014	Sol-gel	Tetrabutyl titanate	NH$_4$F	[183]
Year of Study	**Method**	**TiO$_2$ Precursor**	**Chloride Source**	**Ref.**
2020	Oxidative annealing	Titanium isopropoxide	NH$_4$Cl	[181]
2012	Sonochemical synthesis	Tetraisopropyl titanate	NaCl	[184]
2008	Hydrolysis	Tetrabutyl titanate	HCl	[185]
Year of Study	**Method**	**TiO$_2$ Precursor**	**Bromide Source**	**Ref.**
2017	HT	TBOT	NH$_4$Br	[186]
2009	Sol-gel	TBOT	Cetyl trimethyl Ammonium bromide (CTAB)	[187]
2004	HT	Titanium chloride	Hydrobromic acid	[188]
Year of Study	**Method**	**TiO$_2$ Precursor**	**Iodide Source**	**Ref.**
2017	Sol-gel	Titanium (IV) ter-butoxide	Potassium iodide	[189,190]

A Cl/TiO$_2$ (CldT) photocatalyst was prepared, which shows greater absorption in the range of visible light. The doping of chlorine decreases the band gap of titania; therefore, the absorption spectra are extended. CldT shows better degradation of phenol than the pure one that is 42.5% after 120 min [185].

The synthesized Br/TiO$_2$ (BrdT) hollow spheres are made by the hydrothermal method. Studies have revealed that the adsorption peak of BrdT hollow spheres is near 517 nm, which is greater than that of undoped titania. The band enhanced towards the visible-light range. The band gap of the fabricated spheres decreased (1.75 eV), whereas the band gap of undoped TiO$_2$ was around 2.85 eV; this illustrates that the doping of Br instigates the impurity level between the conduction and valence bands; therefore, the transition of electrons is promoted [186].

I/TiO$_2$ photocatalysts were prepared via the sol-gel method. They tend to have higher photo-degradation potential under direct-sunlight irradiation. Without catalysts, the photocatalytic degradation of RhB was not observed. The mechanism of the photocatalytic degradation of RhB using I/TiO$_2$ involves the following steps as shown in Scheme 2 [189]:

$$I/TiO_2 + h\nu \rightarrow I/TiO_2(e^- + h^+)$$

$$I/TiO_2(h^+) + H_2O \rightarrow I/TiO_2 + H^+ + \,^{\cdot}OH$$

$$I/TiO_2(h^+) + OH^- \rightarrow I/TiO_2 + \,^{\cdot}OH$$

$$I/TiO_2(e^-) + O_2 \rightarrow I/TiO_2 + O_2^{\cdot -}$$

$$RhB + OH^{\cdot} \rightarrow products$$

Scheme 2. Reaction scheme for the degradation of RhB dye using I/TiO$_2$.

10. Si-Doped TiO$_2$ (SidTiO$_2$)

Scheme 2 is based on nanotubes for MO. The results showed that 5% SidTiO$_2$ has a higher catalytic efficacy for MO degradation than the tubes synthesized with titania [75]. Similarly, in another study, SidTiO$_2$ was synthesized using the hydrothermal method for the photocatalytic degradation of organic pollutant, phenol. The results showed a nine-times higher percentage degradation with SidTiO$_2$ than undoped titania nanotubes [191] (Table 10).

Table 10. Summary of Si/TiO$_2$ photocatalysts synthesized by variety of methods and source precursor materials.

Year of Study	Method	TiO$_2$ Precursor	Silicon Source	Ref.
2019	Hydrothermal	Commercial TiO$_2$	SiO$_2$ commercial	[191]
2017	Solvothermal	Titanium oxysulfate	Tetraethoxysilane	[75]

11. Factors Affecting the Degradation of Photocatalytic Activity

The photocatalytic degradation of dyes is strongly affected by consideration of the following parameters: pH, dye concentration, the size and structure of the photocatalyst, pollutants concentration and types, light intensity and irradiation time, dopants' effect on dye concentration, etc. These factors and their impact on the dye-degradation performance are demonstrated with details in this section.

11.1. Effect of pH

pH is a pivotal parameter that impacts the photocatalytic activity. Some of the results on the effect of pH on dye degradation are presented in Table 11 pH fluctuations in combination with calcined non-metal-doped titania possess the best catalytic degradation results because of the synergistic effect of phase structure and crystallinity [192]. The results reported in Table 11 show that the facet TiO_2 can be deprotonated/protonated under alkaline/acidic conditions, respectively. Therefore, it can be concluded that pH alterations can result in the escalation efficacy of the dye degradation of titania without influencing the rate equation [193]. The effect of pH on the decomposition of MO was investigated by Guttai N et al., and it turned out to be the first-order reaction rate, which was 25 times higher at pH 2 than at pH 9 [194]. This also means that some types of dyestuff are preferentially photo-degraded on TiO_2 surfaces. Dyes can be degraded in three ways as a function of pH: directing hydroxyl radical, (ii) the direct involvement of a hole in the oxidation reaction, and (iii) a tête-à-têtes reduction in the participation of the activated electron in the steering band [195]. The effect of pH on the photo-degradation efficiency of the dyes must be considered in conjunction with several other parameters. The interaction of hydroxide ions can produce hydroxyl radicals. At a low pH, holes are the dominant forms of oxidation, and at a neutral or high pH, hydroxyl radicals are the dominant species, respectively.

Table 11. Different types of dyes under optimum pH.

Dye Type	Light Source	Photocatalyst	pH Range	Optimum pH	Ref.
Orange G (OG)	UV	$Sn/TiO_2/Ac$	1.0–12.0	2.0	[196]
(OG)	Visible	$N-TiO_2$	1.5–6.5	2.0	[197]
Bromo-cresol purple (BCP)	UV	TiO_2	4.5 & 8.0	4.5	[198,199]
Methyl Red (MR)	Visible	$3\%Ag+1.5\%Ni-TiO_2$	3–10	4	[199]
Malachite Green (MC)	Sun light	$Ni/MgFe_2O_4$	2.0–10.0	4	[200]
Indigo Carmine (IC)	UV	TiO_2	4.0–11.0	4	[201]
Textile dye (TD)	UV	TiO_2	3.0–7.0	5	[202]
Basic Yellow 28 (BY28)	UV	TiO_2	3.0–9.0	5	[203]
Methylene Blue (MB)	UV	TiO_2ZnO	1.0–6.0	2	[204]
Reactive Blue 4 (RB4)	UV	Anatase TiO_2	3.0–13.0	3–7	[205]
Procion Yellow (PY)	UV	TiO_2	2.0–10.0	7.8	[206]
Acid Orange (AO)	UV	WO_3-TiO_2	1.0–9.0	3	[84]
Methyl Orange (MO)	UV	TiO_2	2.0–10.0	8	[207]
Rhodamine B (RhB)	UV	ZnO	2.0–12.0	12	[208]
MO, RhB	UV	ZnO	2.0–10.0	Basic medium	[209]

Hydroxide (OH^-) is easy to produce in an alkaline solution with an oxidizing ion, which then penetrates through the semiconductor surface. Therefore, the efficiency of the process increases reasonably [210]. According to the findings, pH plays an important role in altering the charges on the dyes; for example, BCP dye degradation was better in acidic media than in alkaline media [198]. Azo dyes are positively charged at pH 6.8 and negatively charged at a higher pH, which affects their adsorption on semiconductor surfaces. When charged species are present in a solution an electrical double layer forms, which affects the electron–hole pair separation and adheres the adsorption properties of the dye on the surface of the semiconductor. pH and the amount of dye present influence the rate of photocatalytic activity [85].

11.2. Effect of Dye Concentration

The optimum concentration of the dye is very important for the photocatalytic reaction, as it is highly dependent on the type of dye being considered. Generally, by increasing the concentration of the colorant, the photocatalytic degradation efficiency of the dyestuff will decrease by suppressing the active sites, therefore, hindering the activity [211]. However, a small quantity of the dye is subjected to degradation and it only contributes to the process of photocatalysis. The initial concentration of the dye in the photocatalytic process is important to consider [81]. This is because as the dye concentration increases, more organic molecules are adsorbed on the surface of TiO_2, but fewer photons reach the catalyst surface, resulting in less OH production, and thus, lowering the percentage degradation [212]. On the other hand, the ionic behavior of dye molecules may result in either accelerating or slowing down the process of the degradation of dyestuff. Normally, metal ions are adsorbed on the surface of the photocatalyst and make them slightly positively charged. As this effect reduces the electrostatic repulsion for anionic dyes, they can be adsorbed and degraded readily in the presence of metal ions. On the contrary, a retarding effect can also be observed for cationic dyes due to the decrease in attraction between positively charged dyestuff and neutral or slightly positive catalyst surfaces [213].

11.3. Photocatalyst's Size and Structure

Surface morphology, such as particle size and agglomerate size, is an important component to address while discussing photocatalytic degradation, since there is a direct relationship between organic molecules and photocatalyst surface coverage. The speed of the reaction is determined by the number of photons striking the photocatalyst; therefore, it can be suggested that the reaction happens solely in the phase absorbed by photocatalysts [214]. Titania and Cu-TiO_2 nanocomposites were used to degrade MR. Studies have suggested that the Cu-modified titania shows more promising results for the catalytic degradation of MR in comparison to the titania because the morphology of the latter is different than the modified version of titania, which facilitates the advancement of organic molecules towards the catalyst [215].

11.4. Pollutant Concentrations and Types

The rate of the photocatalytic destruction of a pollutant is determined by the type of pollutant employed, along with its concentration and other chemicals present in the aqueous matrix. Many researchers have found that the rate of reaction of TiO_2 is influenced by the concentration contaminants present in water. High levels of water contaminants fill the TiO_2 site, reducing photonic efficiency, and moving the image of the catalyst towards malfunctioning [216]. The chemical structure of the target component, in addition to the concentrations of the pollutant, alters the photocatalytic degradation performance of the catalyst being used because of its conversion into its respective intermediates. For example, studies have shown that 4-chlorophenol takes longer to release irradiation than oxalic acid because it converts directly into carbon dioxide and water [217].

11.5. Surface Area of the Photocatalyst

The surface of TiO_2 is crucial in its employment as a photocatalyst because its reactions occur at the surface. This usually increases its availability either by utilizing it in the form of very fine particles, molding it into a porous sheet, or suspending it in some liquids [218]. Nanomaterials with a crystallite size/grain of less than 20 nm have gained particular interest among researchers because their physical characteristics differ significantly from their bulk counterparts. This has also opened new opportunities for their use as photographic catalysts in various fields. As the surface area of the image catalyst grows, more active sites become prominent [219].

11.6. Effect of the Intensity of Light and Irradiation Time

The intensity of light plays an important role in photocatalytic dye degradation, which results in the formation of less toxic byproducts of the dye under consideration. A large amount of research has been conducted on these specific parameters [220]. The effect of the intensity of light on the degradation and decolorization of RY14 is studied in [221]. The dye degradation is affected by both the strength of the light and the time it takes to expose it to degrade the dye or pollutants [193]. As the intensity of light significantly increases, the dye degradation rate enhances; this is because of an increase in the number of photons striking per unit time per unit area of the photocatalyst. Conversely, by further increasing the intensity of light, there might be a chance of higher thermal effects [173]. It was confirmed that N/TiO_2 showed higher dye degradation (almost 60% in 6 h) under visible light. On the contrary, under sunlight irradiation, N/TiO_2 unveiled degradation efficiency that was a little higher as compared to the non-doped TiO_2 efficiency of degradation. A negative light intensity effect on Congo Red was studied when the light intensity varied from 50 to 90 J·cm$^-$ [222].

11.7. Dopants' Impact on Dye Degradation

The main goal of doping is to cause a chromic change in the optical properties, which is defined as the reduction in the band gap or the introduction of intra-band gap conditions that leads to an increase in the absorption of visible light. Non-metal dopants' effects on photocatalytic activity is a challenging problem to tackle. The performance of TiO_2 can be improved by doping with non-metals to enhance the photocatalytic efficiency of titania. Depending on the type of dopants and its concentration, it can also allow light to be absorbed into the visible area at different levels. As a result, visible light can be used to stimulate photocatalysis on modified TiO_2. In mixed systems, dye deterioration is often much faster than systems alone, as the oxidation of dyes utilizes exciting holes quickly and effectively, reducing electron–hole regeneration [214]. The prime concentration of dyes for a photocatalytic reaction is a key parameter that is highly dependent on the type of dye employed. Typically, by increasing the pollutants concentration, the photocatalytic degradation efficiency deceases. The reason for this is that maximum or higher dye molecules compete for limited active sites along with turbidity increases [223].

11.8. Effect of Mass Loading on the Catalytic Activity

TiO_2 has been made to absorb lower-energy photons using a variety of approaches. Kaur et al., reported that under optimized conditions with the highest efficiency, the catalyst dosage for the maximum photocatalytic degradation of RR 198 is 0.3 g [224]. It is manifested that the photodegradation rate increases with the increase in the amount of photocatalyst and then decreases with the increase in the catalyst concentration [225]. In another study, an increase was reported in the weight of the catalyst from 1.0 to 4.0 g L^{-1}, which increases the dye decolorization sharply from 69.27% to 95.23% at 60 min and the dye degradation from 74.54% to 97.29% at 150 min. The optimum concentration of the catalyst for efficient solar photo decolorization and degradation is found to be 4 g/L [226].

12. Conclusions

This review focused on the comprehensive study of the fundamental aspects of non-metal-doped/TiO_2 nanoparticles. Titania is considered as the best visible-light-driven photocatalyst for the degradation of numerous dyes. Plentiful research has been carried out that encompasses the importance of non-metals-doped titania in comparison to the undoped TiO_2 for their better photocatalytic dye degradation of various anionic and cationic dyes, in order to overcome the environmental pollution issues via industrial effluents or other pollutants. We must incorporate most of the literature present on non-metals such as C, Si, N, P, B, halogens, etc. The effect of different parameters, for instance, pH, dyes concentration, photocatalyst's size and structure, pollutants concentration and types, the surface area of photocatalysts, the effect of light intensity along with its irradiation time, catalyst loading, variation in temperature, and doping (non-metals) impact with optimization has a strong correlation with pollutants degradation rate, which is also discussed in this review. The photocatalytic performance of non-metal-doped titania can be further improved with the design of efficient synthesis methodologies.

13. Opportunities, Challenges and Future Prospects

Titanium dioxide has a wide range of properties associated to it; it is a semiconductor [227] with varied properties such as being non-toxic [228], highly efficient [229], cost-effective [230], highly reactive [231], and eco-friendly. It has been accepted globally as a photocatalyst because of its high pore size [232], notable band gap [233], and large surface area [234]. Nevertheless, it has certain drawbacks, of which the following are related: the rejoining of the photo-generated charge bearers, low adsorption range, and the ineffectual visible-light utilization which requires improvement for the intensified photocatalytic activity. Numerous researchers have been regulated to undermine the limitations related to the photocatalysis of titania; the incorporation of nanoparticles and the doping of metals and non-metals have helped in enhancing its process to a larger extent. The amalgamation of NPs has been of utmost importance lately because they produce the desired results by amending the particle shape and size along with the physicochemical properties of TiO_2. Moreover, doping can strengthen the photocatalytic property by minimizing the reposting of charge carriers and decreasing the band-shift towards the region of visible light.

The photooxidation of organic effluent in wastewater and the elimination of nitrates and sulfates, along with acidic, basic, VAT, and azo dyes, can be achieved by the non-metallic doping of TiO_2. Moreover, the sensitization of dyes can be achieved by the doping of non-metals with titania; it can help in the breakdown of pesticides, the industrial discharge of dyes which have toxic and malignant effects, and the generation of photocatalytic hydrogen [235].

It is important to understand the prospects and future aspects on improving the photocatalytic activity of TiO_2 by different means, either by doping or co-doping. The studies have shown that both manifest remarkable results. The subject of high noticeability here is that whose process has produced more beneficial and long-lasting outcomes. More studies focusing on the engineered edges via doping for the improvement of the conduction and valence bands are required for enhancement of the absorption band of titania. Furthermore, the movability of charge carriers should be increased by introducing an impurity to improve the working efficiency of TiO_2. Therefore, the use of photocatalysis along with some other technologies would improve its application, which would benefit the environment for a longer period.

Moreover, a gradual deactivation of the photocatalytic materials concerns all the potential industrial applications. A periodical regeneration of the photocatalytic materials would be required, which would also increase the overall cost. Therefore, cost would obviously remain an essential issue for commercialization. Hence, to overcome this obstacle, extensive research would be necessary to develop both economical reactors and photocatalytic materials. Concomitant, concerted, and extensive research progress to achieve these goals is necessary for the practical implementation of this technology.

Author Contributions: Conceptualization, P.A. and M.H.; methodology, A.A.; software, A.S.; validation, A.A.; A.S.; formal analysis, A.A.; investigation, P.A.; resources, M.H.; data curation, A.A.; writing—original draft preparation, P.A.; writing—review and editing, M.H.; visualization, P.A.; supervision, P.A.; project administration, P.A.; funding acquisition, M.H. All authors have read and agreed to the published version of the manuscript.

Funding: This research received no external funding.

References

1. Lee, K.M.; Lai, C.W.; Ngai, K.S.; Juan, J.C. Recent developments of zinc oxide based photocatalyst in water treatment technology: A review. *Water Res.* **2016**, *88*, 428–448. [CrossRef] [PubMed]
2. Etman, A.S.; Abdelhamid, H.N.; Yuan, Y.; Wang, L.; Zou, X.; Sun, J. Facile Water-Based Strategy for Synthesizing MoO_{3-x} Nanosheets: Efficient Visible Light Photocatalysts for Dye Degradation. *ACS Omega* **2018**, *3*, 2193–2201. [CrossRef] [PubMed]
3. Mittal, A.; Mittal, A.; Mari, B.; Sharma, S.; Kumari, V.; Maken, S.; Kumari, K.; Kumar, N. Non-metal modified TiO_2: A step towards visible light photocatalysis. *J. Mater. Sci. Mater. Electron.* **2019**, *30*, 3186–3207. [CrossRef]

4.   Din, M.I.; Khalid, R.; Hussain, Z. Recent Research on Development and Modification of Nontoxic Semiconductor for Environmental Application. *Sep. Purif. Rev.* **2021**, *50*, 244–261. [CrossRef]

5.   Al-Mamun, M.R.; Kader, S.; Islam, M.S.; Khan, M.Z.H. Photocatalytic activity improvement and application of UV-TiO_2 photocatalysis in textile wastewater treatment: A review. *J. Environ. Chem. Eng.* **2019**, *7*, 103248. [CrossRef]

6.   Zhu, Z.; Kao, C.T.; Tang, B.H.; Chang, W.C.; Wu, R.J. Efficient hydrogen production by photocatalytic water-splitting using Pt-doped TiO_2 hollow spheres under visible light. *Ceram. Int.* **2016**, *42*, 6749–6754. [CrossRef]

7.   Makama, A.B.; Salmiaton, A.; Saion, E.B.; Choong, T.S.Y.; Abdullah, N. Synthesis of CdS Sensitized TiO_2 Photocatalysts: Methylene Blue Adsorption and Enhanced Photocatalytic Activities. *Int. J. Photoenergy* **2016**, *2016*, 2947510. [CrossRef]

8.   Huang, M.; Yu, S.; Li, B.; Lihui, D.; Zhang, F.; Fan, M.; Deng, C. Influence of preparation methods on the structure and catalytic performance of SnO_2-doped TiO_2 photocatalysts. *Ceram. Int.* **2014**, *40*, 13305–13312. [CrossRef]

9.   Hassan, S.M.; Ahmed, A.I.; Mannaa, M.A. Preparation and characterization of SnO_2 doped TiO_2 nanoparticles: Effect of phase changes on the photocatalytic and catalytic activity. *J. Sci. Adv. Mater. Devices* **2019**, *4*, 400–412. [CrossRef]

10. Dontsova, T.A.; Kutuzova, A.S.; Bila, K.O.; Kyrii, S.O.; Kosogina, I.V.; Nechyporuk, D.O. Enhanced Photocatalytic Activity of TiO_2/SnO_2 Binary Nanocomposites. *J. Nanomater.* **2020**, *2020*, 8349480. [CrossRef]

11. Wang, X.; Sun, M.; Murugananthan, M.; Zhang, Y.; Zhang, L. Electrochemically self-doped WO_3/TiO_2 nanotubes for photocatalytic degradation of volatile organic compounds. *Appl. Catal. B Environ.* **2020**, *260*, 118205. [CrossRef]

12. Amouhadi, E.; Aliyan, H.; Aghaei, H.; Fazaeli, R.; Richeson, D. Photodegradation and mineralization of metronidazole by a novel quadripartite $SnO_2@TiO_2/ZrTiO_4/ZrO_2$ photocatalyst: Comprehensive photocatalyst characterization and kinetic study. *Mater. Sci. Semicond. Process.* **2022**, *143*, 106560. [CrossRef]

13. Liu, C.; Li, X.; Wu, Y.; Sun, L.; Zhang, L.; Chang, X.; Wang, X. Enhanced photocatalytic activity by tailoring the interface in $TiO_2–ZrTiO_4$ heterostructure in $TiO_2–ZrTiO_4–SiO_2$ ternary system. *Ceram. Int.* **2019**, *45*, 17163–17172. [CrossRef]

14. Liu, C.; Li, X.; Wu, Y.; Zhang, L.; Chang, X.; Yuan, X.; Wang, X. Fabrication of multilayer porous structured $TiO_2–ZrTiO_4–SiO_2$ heterostructure towards enhanced photo-degradation activities. *Ceram. Int.* **2020**, *46*, 476–486. [CrossRef]

15. Xiang, Q.; Yu, J.; Jaroniec, M. Graphene-based semiconductor photocatalysts. *Chem. Soc. Rev.* **2012**, *41*, 782–796. [CrossRef]

16. Rommozzi, E.; Zannotti, M.; Giovannetti, R.; D'Amato, C.A.; Ferraro, S.; Minicucci, M.; Di Cicco, A. Reduced Graphene Oxide/TiO_2 Nanocomposite: From Synthesis to Characterization for Efficient Visible Light Photocatalytic Applications. *Catalysts* **2018**, *8*, 598. [CrossRef]

17. Rahman, M.M.; Choudhury, F.A.; Hossain, D.; Chowdhury, N.I.; Mohsin, S.; Hasan, M.; Uddin, F.; Sarker, N.C. A Comparative study on the photocatalytic degradation of industrial dyes using modified commercial and synthesized TiO_2 photocatalysts. *J. Chem. Eng.* **2014**, *27*, 65–71. [CrossRef]

18. Shehzad, N.; Tahir, M.; Johari, K.; Murugesan, T.; Hussain, M.A. Critical review on TiO_2 based photocatalytic CO_2 reduction system: Strategies to improve efficiency. *J. CO$_2$ Util.* **2018**, *26*, 98–122. [CrossRef]

19. Hussain, M.; Ceccarelli, R.; Marchisio, D.L.; Fino, D.; Russo, N.; Geobaldo, F. Synthesis, characterization, and photocatalytic application of novel TiO_2 nanoparticles. *Chem. Eng. J.* **2010**, *157*, 45–51. [CrossRef]

20. Siraj, Z.; Maafa, I.M.; Shafiq, I.; Shezad, N.; Akhter, P.; Yang, W.; Hussain, M. KIT-6 induced mesostructured TiO_2 for photocatalytic degradation of methyl blue. *Environ. Sci. Pollut. Res.* **2021**, *28*, 53340–53352. [CrossRef]

21. Rashid, R.; Shafiq, I.; Iqbal, M.J.; Shabir, M.; Akhter, P.; Hamayun, M.H.; Hussain, M. Synergistic effect of NS co-doped TiO_2 adsorbent for removal of cationic dyes. *J. Environ. Chem. Eng.* **2021**, *9*, 105480. [CrossRef]

22. Hlekelele, L.; Durbach, S.H.; Chauke, V.P.; Dziike, F.; Franklyn, P.J. Resin-gel incorporation of high concentrations of W^{6+} and Zn^{2+} into TiO_2-anatase crystal to form quaternary mixed-metal oxides: Effect on the a lattice parameter and photodegradation efficiency. *RSC Adv.* **2019**, *9*, 36875–36883. [CrossRef] [PubMed]

23. Thambiliyagodage, C. Activity enhanced TiO_2 nanomaterials for photodegradation of dyes—A review. *Environ. Nanotechnol. Monit. Manag.* **2021**, *16*, 100592. [CrossRef]

24. Tryba, B.; Piszcz, M.; Morawski, A.W. Photocatalytic activity of-composites. *Int. J. Photoenergy* **2009**, *2009*, 297319. [CrossRef]

25. Han, F.; Kambala, V.S.R.; Srinivasan, M.; Rajarathnam, D.; Naidu, R. Tailored titanium dioxide photocatalysts for the degradation of organic dyes in wastewater treatment: A review. *Appl. Catal. A Gen.* **2009**, *359*, 25–40. [CrossRef]

26. Pragati, T.; Roshan, K.N. Synthesis of sol-gel derived TiO_2 nanoparticles for the photocatalytic degradation of methyl orange dye. *Res. J. Chem. Environ.* **2011**, *15*, 145–149.

27. Deng, H.; He, H.; Sun, S.; Zhu, X.; Zhou, D.; Han, F.; Pan, X. Photocatalytic degradation of dye by Ag/TiO_2 nanoparticles prepared with different sol–gel crystallization in the presence of effluent organic matter. *Environ. Sci. Pollut. Res.* **2019**, *26*, 35900–35912. [CrossRef]

28. Al Jitan, S.; Palmisano, G.; Garlisi, C. Synthesis and surface modification of TiO_2-based photocatalysts for the conversion of CO_2. *Catalysts* **2020**, *10*, 227. [CrossRef]

29. Islam, S.Z.; Nagpure, S.; Kim, D.Y.; Rankin, S.E. Synthesis and Catalytic Applications of Non-Metal Doped Mesoporous Titania. *Inorganics* **2017**, *5*, 15. [CrossRef]

30. Wu, Y.; Xing, M.; Tian, B.; Zhang, J.; Chen, F. Preparation of nitrogen and fluorine co-doped mesoporous TiO_2 microsphere and photodegradation of acid orange 7 under visible light. *Chem. Eng. J.* **2010**, *162*, 710–717. [CrossRef]

31. Yang, J.; Zhang, X.; Li, B.; Liu, H.; Sun, P.; Wang, C.; Liu, Y. Photocatalytic activities of heterostructured TiO_2-graphene porous microspheres prepared by ultrasonic spray pyrolysis. *J. Alloys Compd.* **2014**, *584*, 180–184. [CrossRef]

32. Zhou, X.; Peng, F.; Wang, H.; Yu, H.; Yang, J. Effect of nitrogen-doping temperature on the structure and photocatalytic activity of the B, N-doped TiO_2. *J. Solid State Chem.* **2011**, *184*, 134–140. [CrossRef]

33. El-Sheikh, S.M.; Zhang, G.; El-Hosainy, H.M.; Ismail, A.A.; O'Shea, K.E.; Falaras, P.; Dionysiou, D.D. High performance sulfur, nitrogen and carbon doped mesoporous anatase–brookite TiO_2 photocatalyst for the removal of microcystin-LR under visible light irradiation. *J. Hazard. Mater.* **2014**, *280*, 723–733. [CrossRef]

34. Ao, Y.; Xu, J.; Fu, D.; Yuan, C. Synthesis of C, N, S-tridoped mesoporous titania with enhanced visible light-induced photocatalytic activity. *Microporous Mesoporous Mater.* **2009**, *122*, 1–6. [CrossRef]

35. Almaev, A.V.; Yakovlev, N.N.; Kushnarev, B.O.; Kopyev, V.V.; Novikov, V.A.; Zinoviev, M.M.; Yudin, N.N.; Podzivalov, S.N.; Erzakova, N.N.; Chikiryaka, A.V.; et al. Gas Sensitivity of IBSD Deposited TiO_2 Thin Films. *Coatings* **2022**, *12*, 1565. [CrossRef]

36. Mele, G.; del Sole, R.; Lü, X. 18—Applications of TiO_2 in sensor devices. In *Titanium Dioxide (TiO_2) and Its Applications*; Parrino, F., Palmisano, L., Eds.; Elsevier: Amsterdam, The Netherlands, 2021; pp. 527–581.

37. Gogova, D.; Iossifova, A.; Ivanova, T.; Dimitrova, Z.; Gesheva, K. Electrochromic behavior in CVD grown tungsten oxide films. *J. Cryst. Growth* **1999**, *198–199*, 1230–1234. [CrossRef]

38. Fan, X.; Chen, X.; Zhu, S.; Li, Z.; Yu, T.; Ye, J.; Zou, Z. The structural, physical and photocatalytic properties of the mesoporous Cr-doped TiO_2. *J. Mol. Catal. A Chem.* **2008**, *284*, 155–160. [CrossRef]

39. Murthy, N.S. Scattering techniques for structural analysis of biomaterials. In *Characterization of Biomaterials*; Elsevier: Amsterdam, The Netherlands, 2013; pp. 34–72.

40. Ghumro, S.S.; Lal, B.; Pirzada, T. Visible-Light-Driven Carbon-Doped TiO_2-Based Nanocatalysts for Enhanced Activity toward Microbes and Removal of Dye. *ACS Omega* **2022**, *7*, 4333–4341. [CrossRef]

41. Negi, C.; Kandwal, P.; Rawat, J.; Sharma, M.; Sharma, H.; Dalapati, G.; Dwivedi, C. Carbon-doped titanium dioxide nanoparticles for visible light driven photocatalytic activity. *Appl. Surf. Sci.* **2021**, *554*, 149553. [CrossRef]

42. Piątkowska, A.; Janus, M.; Szymański, K.; Mozia, S. C-, N- and S-Doped TiO_2 Photocatalysts: A Review. *Catalysts* **2021**, *11*, 144. [CrossRef]

43. Bergamonti, L.; Predieri, G.; Paz, Y.; Fornasini, L.; Lottici, P.P.; Bondioli, F. Enhanced self-cleaning properties of N-doped TiO_2 coating for Cultural Heritage. *Microchem. J.* **2017**, *133*, 1–12. [CrossRef]

44. Chen, X.; Kuo, D.-H.; Lu, D. N-doped mesoporous TiO_2 nanoparticles synthesized by using biological renewable nanocrystalline cellulose as template for the degradation of pollutants under visible and sun light. *Chem. Eng. J.* **2016**, *295*, 192–200. [CrossRef]

45. Hwang, Y.J.; Yang, S.; Lee, H. Surface analysis of N-doped TiO_2 nanorods and their enhanced photocatalytic oxidation activity. *Appl. Catal. B Environ.* **2017**, *204*, 209–215. [CrossRef]

46. Chung, K.-H.; Kim, B.J.; Park, Y.K.; Kim, S.C.; Jung, S.C. Photocatalytic Properties of Amorphous N-Doped TiO_2 Photocatalyst under Visible Light Irradiation. *Catalysts* **2021**, *11*, 1010. [CrossRef]

47. Shabir, M.; Shezad, N.; Shafiq, I.; Maafa, I.M.; Akhter, P.; Azam, K.; Hussain, M. Carbon nanotubes loaded N, S-codoped TiO_2: Heterojunction assembly for enhanced integrated adsorptive-photocatalytic performance. *J. Ind. Eng. Chem.* **2022**, *105*, 539–548. [CrossRef]

48. Li, T.; Abdelhaleem, A.; Chu, W.; Pu, S.; Qi, F.; Zou, J. S-doped TiO_2 photocatalyst for visible LED mediated oxone activation: Kinetics and mechanism study for the photocatalytic degradation of pyrimethanil fungicide. *Chem. Eng. J.* **2021**, *411*, 128450. [CrossRef]

49. Grabowska, E.; Zaleska, A.; Sobczak, J.W.; Gazda, M.; Hupka, J. Boron-doped TiO_2: Characteristics and photoactivity under visible light. *Procedia Chem.* **2009**, *1*, 1553–1559. [CrossRef]

50. Koli, V.B.; Ke, S.-C.; Dodamani, A.G.; Deshmukh, S.P.; Kim, J.-S. Boron-Doped TiO_2-CNT Nanocomposites with Improved Photocatalytic Efficiency toward Photodegradation of Toluene Gas and Photo-Inactivation of Escherichia coli. *Catalysts* **2020**, *10*, 632. [CrossRef]

51. Su, Y.; Xiao, Y.; Fu, X.; Deng, Y.; Zhang, F. Photocatalytic properties and electronic structures of iodine-doped TiO_2 nanotubes. *Mater. Res. Bull.* **2009**, *44*, 2169–2173. [CrossRef]

52. Xu, J.; Yang, B.; Wu, M.; Fu, Z.; Lv, Y.; Zhao, Y. Novel N−F-Codoped TiO_2 Inverse Opal with a Hierarchical Meso-/Macroporous Structure: Synthesis, Characterization, and Photocatalysis. *J. Phys. Chem. C* **2010**, *114*, 15251–15259. [CrossRef]

53. Morales-Torres, S.; Pastrana-Martínez, L.M.; Figueiredo, J.L.; Faria, J.L.; Silva, A.M. Design of graphene-based TiO_2 photocatalysts—A review. *Environ. Sci. Pollut. Res.* **2012**, *19*, 3676–3687. [CrossRef]

54. Suárez, S.; Jansson, I.; Ohtani, B.; Sánchez, B. From titania nanoparticles to decahedral anatase particles: Photocatalytic activity of TiO_2/zeolite hybrids for VOCs oxidation. *Catal. Today* **2019**, *326*, 2–7. [CrossRef]

55. Liang, H.; Wang, Z.; Liao, L.; Chen, L.; Li, Z.; Feng, J. High performance photocatalysts: Montmorillonite supported-nano TiO_2 composites. *Optik* **2017**, *136*, 44–51. [CrossRef]

56. Shao, J.; Sheng, W.; Wang, M.; Li, S.; Chen, J.; Zhang, Y.; Cao, S. In situ synthesis of carbon-doped TiO_2 single-crystal nanorods with a remarkably photocatalytic efficiency. *Appl. Catal. B Environ.* **2017**, *209*, 311–319. [CrossRef]

57. Saiful Amran, S.N.B.; Wongso, V.; Abdul Halim, N.S.; Husni, M.K.; Sambudi, N.S.; Wirzal, M.D.H. Immobilized carbon-doped TiO_2 in polyamide fibers for the degradation of methylene blue. *J. Asian Ceram. Soc.* **2019**, *7*, 321–330. [CrossRef]

58. Habibi, S.; Jamshidi, M. Sol–gel synthesis of carbon-doped TiO_2 nanoparticles based on microcrystalline cellulose for efficient photocatalytic degradation of methylene blue under visible light. *Environ. Technol.* **2020**, *41*, 3233–3247. [CrossRef]

59. Ji, L.; Liu, X.; Xu, T.; Gong, M.; Zhou, S. Preparation and photocatalytic properties of carbon/carbon-doped TiO_2 double-layer hollow microspheres. *J. Sol-Gel Sci. Technol.* **2020**, *93*, 380–390. [CrossRef]

60. Bhosale, R.R.; Pujari, S.R.; Muley, G.G.; Patil, S.H.; Patil, K.R.; Shaikh, M.F.; Gambhire, A.B. Solar photocatalytic degradation of methylene blue using doped TiO_2 nanoparticles. *J. Sol. Energy.* **2014**, *103*, 473–479. [CrossRef]

61. Cong, Y.; Zhang, J.; Chen, F.; Anpo, M. Synthesis and Characterization of Nitrogen-Doped TiO_2 Nanophotocatalyst with High Visible Light Activity. *J. Phys. Chem. C* **2007**, *111*, 6976–6982. [CrossRef]

62. Senthilnathan, J.; Philip, L. Photocatalytic degradation of lindane under UV and visible light using N-doped TiO_2. *Chem. Eng. J.* **2010**, *161*, 83–92. [CrossRef]

63. Liu, X.; Chen, Y.; Cao, C.; Xu, J.; Qian, Q.; Luo, Y.; Chen, Q. Electrospun nitrogen and carbon co-doped porous TiO_2 nanofibers with high visible light photocatalytic activity. *N. J. Chem.* **2015**, *39*, 6944–6950. [CrossRef]

64. Pu, X.; Hu, Y.; Cui, S.; Cheng, L.; Jiao, Z. Preparation of N-doped and oxygen-deficient TiO_2 microspheres via a novel electron beam-assisted method. *Solid State Sci.* **2017**, *70*, 66–73. [CrossRef]

65. Panghulan, G.R.; Vasquez Jr, M.R.; Edañol, Y.D.; Chanlek, N.; Payawan Jr, L.M. Synthesis of TiN/N-doped TiO_2 composite films as visible light active photocatalyst. *J. Vac. Sci. Technol. B* **2020**, *38*, 062203. [CrossRef]

66. Ji, L.; Zhou, S.; Liu, X.; Gong, M.; Xu, T. Synthesis of carbon- and nitrogen-doped TiO_2/carbon composite fibers by a surface-hydrolyzed PAN fiber and their photocatalytic property. *J. Mater. Sci.* **2020**, *55*, 2471–2481. [CrossRef]

67. Divyasri, Y.V.; Reddy, N.L.; Lee, K.; Sakar, M.; Rao, V.N.; Venkatramu, V.; Reddy, N.C. GOptimization of N doping in TiO_2 nanotubes for the enhanced solar light mediated photocatalytic H_2 production and dye degradation. *Environ. Pollut.* **2021**, *269*, 116170. [CrossRef]

68. Tian, H.; Ma, J.; Li, K.; Li, J. Hydrothermal synthesis of S-doped TiO_2 nanoparticles and their photocatalytic ability for degradation of methyl orange. *Ceram. Int.* **2009**, *35*, 1289–1292. [CrossRef]

69. Birben, N.C.; Uyguner-Demirel, C.S.; Sen-Kavurmaci, S.; Gurkan, Y.Y.; Turkten, N.; Cinar, Z.; Bekbolet, M. Comparative evaluation of anion doped photocatalysts on the mineralization and decolorization of natural organic matter. *Catal. Today* **2015**, *240*, 125–131. [CrossRef]

70. Syafiuddin, A.; Hadibarata, T.; Zon, N.F. Characterization of Titanium Dioxide Doped with Nitrogen and Sulfur and Its Photocatalytic Appraisal for Degradation of Phenol and Methylene Blue. *J. Chin. Chem. Soc.* **2017**, *64*, 1333–1339. [CrossRef]

71. Gao, H.-T.; Liu, Y.Y.; Ding, C.H.; Dai, D.M.; Liu, G.J. Synthesis, characterization, and theoretical study of N, S-codoped nano-TiO_2 with photocatalytic activities. *Int. J. Miner. Metall. Mater.* **2011**, *18*, 606. [CrossRef]

72. Raj, K.J.A.; Ramaswamy, A.; Viswanathan, B. Surface area, pore size, and particle size engineering of titania with seeding technique and phosphate modification. *J. Phys. Chem. C* **2009**, *113*, 13750–13757. [CrossRef]

73. Ghafoor, S.; Aftab, F.; Rauf, A.; Duran, H.; Kirchhoff, K.; Arshad, S.N. P-doped TiO_2 Nanofibers Decorated with Ag Nanoparticles for Enhanced Photocatalytic Activity under Simulated Solar Light. *ChemistrySelect* **2020**, *5*, 14078–14085. [CrossRef]

74. Sarker, D.R.; Uddin, M.N.; Elias, M.; Rahman, Z.; Paul, R.K.; Siddiquey, I.A.; Uddin, J. P-doped TiO_2-MWCNTs nanocomposite thin films with enhanced photocatalytic activity under visible light exposure. *Clean. Eng. Technol.* **2022**, *6*, 100364. [CrossRef]

75. Xiao, J.; Pan, Z.; Zhang, B.; Liu, G.; Zhang, H.; Song, X.; Zheng, Y. The research of photocatalytic activity on Si doped TiO_2 nanotubes. *Mater. Lett.* **2017**, *188*, 66–68. [CrossRef]

76. Ajmal, A.; Majeed, I.; Malik, R.N.; Idriss, H.; Nadeem, M.A. Principles and mechanisms of photocatalytic dye degradation on TiO_2 based photocatalysts: A comparative overview. *RSC Adv.* **2014**, *4*, 37003–37026. [CrossRef]

77. O'Neill, C.; Hawkes, F.R.; Hawkes, D.L.; Lourenço, N.D.; Pinheiro, H.M.; Delée, W. Colour in textile effluents–sources, measurement, discharge consents and simulation: A review. *J. Chem. Technol. Biotechnol. Int. Res. Process Environ. Clean Technol.* **1999**, *74*, 1009–1018. [CrossRef]

78. Tehrani-Bagha, A.; Mahmoodi, N.M.; Menger, F. Degradation of a persistent organic dye from colored textile wastewater by ozonation. *Desalination* **2010**, *260*, 34–38. [CrossRef]

79. Andriantsiferana, C.; Mohamed, E.F.; Delmas, H. Photocatalytic degradation of an azo-dye on TiO_2/activated carbon composite material. *Environ. Technol.* **2014**, *35*, 355–363. [CrossRef]

80. Gao, Y.; Yang, B.; Wang, Q. Biodegradation and decolorization of dye wastewater: A review. In *IOP Conference Series: Earth and Environmental Science*; IOP Publishing: Bristol, UK, 2018.

81. Reza, K.M.; Kurny, A.; Gulshan, F. Parameters affecting the photocatalytic degradation of dyes using TiO_2: A review. *Appl. Water Sci.* **2017**, *7*, 1569–1578. [CrossRef]

82. Madhavan, J.; Maruthamuthu, P.; Murugesan, S.; Anandan, S. Kinetic studies on visible light-assisted degradation of acid red 88 in presence of metal-ion coupled oxone reagent. *Appl. Catal. B Environ.* **2008**, *83*, 8–14. [CrossRef]

83. Baran, W.; Makowski, A.; Wardas, W. The influence of $FeCl_3$ on the photocatalytic degradation of dissolved azo dyes in aqueous TiO_2 suspensions. *Chemosphere* **2003**, *53*, 87–95. [CrossRef]

84. Rauf, M.; Meetani, M.; Hisaindee, S. An overview on the photocatalytic degradation of azo dyes in the presence of TiO_2 doped with selective transition metals. *Desalination* **2011**, *276*, 13–27. [CrossRef]

85. Zhang, T.; ki Oyama, T.; Horikoshi, S.; Hidaka, H.; Zhao, J.; Serpone, N. Photocatalyzed N-demethylation and degradation of methylene blue in titania dispersions exposed to concentrated sunlight. *Sol. Energy Mater. Sol. Cells* **2002**, *73*, 287–303. [CrossRef]

86. Suteu, D.; Malutan, T. Industrial cellolignin wastes as adsorbent for removal of methylene blue dye from aqueous solutions. *BioResources* **2013**, *8*, 427–446. [CrossRef]

87. Christie, R. *Carbonyl Dyes and Pigments*; RSC Publishing: London, UK, 2001.
88. Thung, W.-E.; Ong, S.A.; Ho, L.N.; Wong, Y.S.; Ridwan, F.; Oon, Y.L.; Lehl, H.K. A highly efficient single chambered up-flow membrane-less microbial fuel cell for treatment of azo dye Acid Orange 7-containing wastewater. *Bioresour. Technol.* **2015**, *197*, 284–288. [CrossRef] [PubMed]
89. Houas, A.; Lachheb, H.; Ksibi, M.; Elaloui, E.; Guillard, C.; Herrmann, J.M. Photocatalytic degradation pathway of methylene blue in water. *Appl. Catal. B Environ.* **2001**, *31*, 145–157. [CrossRef]
90. Konstantinou, I.K.; Albanis, T.A. TiO₂-assisted photocatalytic degradation of azo dyes in aqueous solution: Kinetic and mechanistic investigations: A review. *Appl. Catal. B Environ.* **2004**, *49*, 1–14. [CrossRef]
91. Anwer, H.; Mahmood, A.; Lee, J.; Kim, K.H.; Park, J.W.; Yip, A.C. Photocatalysts for degradation of dyes in industrial effluents: Opportunities and challenges. *Nano Res.* **2019**, *12*, 955–972. [CrossRef]
92. Chu, C.-Y.; Huang, M.H. Facet-dependent photocatalytic properties of Cu₂O crystals probed by using electron, hole and radical scavengers. *J. Mater. Chem. A* **2017**, *5*, 15116–15123. [CrossRef]
93. Meng, L.; Chen, Z.; Ma, Z.; He, S.; Hou, Y.; Li, H.H.; Long, J. Gold plasmon-induced photocatalytic dehydrogenative coupling of methane to ethane on polar oxide surfaces. *Energy Environ. Sci.* **2018**, *11*, 294–298. [CrossRef]
94. Wang, S.; Boyjoo, Y.; Choueib, A. A comparative study of dye removal using fly ash treated by different methods. *Chemosphere* **2005**, *60*, 1401–1407. [CrossRef]
95. Katheresan, V.; Kansedo, J.; Lau, S.Y. Efficiency of various recent wastewater dye removal methods: A review. *J. Environ. Chem. Eng.* **2018**, *6*, 4676–4697. [CrossRef]
96. Mondal, S. Methods of Dye Removal from Dye House Effluent—An Overview. *Environ. Eng. Sci.* **2008**, *25*, 383–396. [CrossRef]
97. Kaykhaii, M.; Sasani, M.; Marghzari, S. Removal of dyes from the environment by adsorption process. *Chem. Mater. Eng.* **2018**, *6*, 31–35. [CrossRef]
98. Piaskowski, K.; Świderska-Dąbrowska, R.; Zarzycki, P.K. Dye Removal from Water and Wastewater Using Various Physical, Chemical, and Biological Processes. *J. AOAC Int.* **2019**, *101*, 1371–1384. [CrossRef]
99. El-Desouky, M.G.; El-Bindary, M.A.; El-Bindary, A.A. Effective adsorptive removal of anionic dyes from aqueous solution. *Vietnam J. Chem.* **2021**, *59*, 341–361.
100. Alhujaily, A.; Yu, H.; Zhang, X.; Ma, F. Adsorptive removal of anionic dyes from aqueous solutions using spent mushroom waste. *Appl. Water Sci.* **2020**, *10*, 183. [CrossRef]
101. Pavithra, K.G.; Jaikumar, V. Removal of colorants from wastewater: A review on sources and treatment strategies. *J. Ind. Eng. Chem.* **2019**, *75*, 1–19. [CrossRef]
102. Zhao, X.; Wang, X.; Lou, T. Simultaneous adsorption for cationic and anionic dyes using chitosan/electrospun sodium alginate nanofiber composite sponges. *Carbohydr. Polym.* **2022**, *276*, 118728. [CrossRef]
103. Mallakpour, S.; Rashidimoghadam, S. 9—Carbon nanotubes for dyes removal. In *Composite Nanoadsorbents*; Kyzas, G.Z., Mitropoulos, A.C., Eds.; Elsevier: Amsterdam, The Netherlands, 2019; pp. 211–243.
104. Velusamy, S.; Roy, A.; Sundaram, S.; Kumar Mallick, T. A Review on Heavy Metal Ions and Containing Dyes Removal through Graphene Oxide-Based Adsorption Strategies for Textile Wastewater Treatment. *Chem. Rec.* **2021**, *21*, 1570–1610. [CrossRef]
105. Ling, C.; Yimin, D.; Qi, L.; Chengqian, F.; Zhiheng, W.; Yaqi, L.; Li, W. Novel High-efficiency adsorbent consisting of magnetic Cellulose-based ionic liquid for removal of anionic dyes. *J. Mol. Liq.* **2022**, *353*, 118723. [CrossRef]
106. Rahman, F.; Akter, M. Removal of dyes form textile wastewater by adsorption using shrimp shell. *Int. J. Waste Resour.* **2016**, *6*, 2–5.
107. Ledakowicz, S.; Paździor, K. Recent Achievements in Dyes Removal Focused on Advanced Oxidation Processes Integrated with Biological Methods. *Molecules* **2021**, *26*, 870. [CrossRef] [PubMed]
108. Bal, G.; Thakur, A. Distinct approaches of removal of dyes from wastewater: A review. *Mater. Today Proc.* **2022**, *50*, 1575–1579. [CrossRef]
109. Raval, N.P.; Shah, P.U.; Shah, N.K. Malachite green "a cationic dye" and its removal from aqueous solution by adsorption. *Appl. Water Sci.* **2017**, *7*, 3407–3445. [CrossRef]
110. Ruan, W.; Hu, J.; Qi, J.; Hou, Y.; Zhou, C.; Wei, X. Removal of dyes from wastewater by nanomaterials: A review. *Adv. Mater. Lett.* **2019**, *10*, 9–20. [CrossRef]
111. Sleiman, M.; Vildozo, D.; Ferronato, C.; Chovelon, J.M. Photocatalytic degradation of azo dye Metanil Yellow: Optimization and kinetic modeling using a chemometric approach. *Appl. Catal. B Environ.* **2007**, *77*, 1–11. [CrossRef]
112. Basavarajappa, P.S.; Patil, S.B.; Ganganagappa, N.; Reddy, K.R.; Raghu, A.V.; Reddy, C. Recent progress in metal-doped TiO₂, non-metal doped/codoped TiO₂ and TiO₂ nanostructured hybrids for enhanced photocatalysis. *Int. J. Hydrog. Energy* **2020**, *45*, 7764–7778. [CrossRef]
113. Guo, W.; Wu, L.; Chen, Z.; Boschloo, G.; Hagfeldt, A.; Ma, T. Highly efficient dye-sensitized solar cells based on nitrogen-doped titania with excellent stability. *J. Photochem. Photobiol. A Chem.* **2011**, *219*, 180–187. [CrossRef]
114. Wu, Y.; Xing, M.; Zhang, J. Gel-hydrothermal synthesis of carbon and boron co-doped TiO₂ and evaluating its photocatalytic activity. *J. Hazard. Mater.* **2011**, *192*, 368–373. [CrossRef]
115. Boikanyo, D.; Masheane, M.L.; Nthunya, L.N.; Mishra, S.B.; Mhlanga, S.D. Carbon-supported photocatalysts for organic dye photodegradation. In *New Polymer Nanocomposites for Environmental Remediation*; Elsevier: Amsterdam, The Netherlands, 2018; pp. 99–138.

116. Liu, J.; Zhu, W.; Yu, S.; Yan, X. Three dimensional carbogenic dots/TiO2 nanoheterojunctions with enhanced visible light-driven photocatalytic activity. *Carbon* **2014**, *79*, 369–379. [CrossRef]

117. Matos, J.; Miralles-Cuevas, S.; Ruíz-Delgado, A.; Oller, I.; Malato, S. Development of TiO$_2$-C photocatalysts for solar treatment of polluted water. *Carbon* **2017**, *122*, 361–373. [CrossRef]

118. Shi, Z.-J.; Ma, M.-G.; Zhu, J.-F. Recent Development of Photocatalysts Containing Carbon Species: A Review. *Catalysts* **2018**, *9*, 20. [CrossRef]

119. Noorimotlagh, Z.; Kazeminezhad, I.; Jaafarzadeh, N.; Ahmadi, M.; Ramezani, Z.; Martinez, S.S. The visible-light photodegradation of nonylphenol in the presence of carbon-doped TiO$_2$ with rutile/anatase ratio coated on GAC: Effect of parameters and degradation mechanism. *J. Hazard. Mater.* **2018**, *350*, 108–120. [CrossRef]

120. Shaban, Y.A.; El Maradny, A.A.; Al Farawati, R.K. Photocatalytic reduction of nitrate in seawater using C/TiO$_2$ nanoparticles. *J. Photochem. Photobiol. A Chem.* **2016**, *328*, 114–121. [CrossRef]

121. Ananpattarachai, J.; Seraphin, S.; Kajitvichyanukul, P. Formation of hydroxyl radicals and kinetic study of 2-chlorophenol photocatalytic oxidation using C-doped TiO$_2$, N-doped TiO$_2$, and C, N Co-doped TiO$_2$ under visible light. *Environ. Sci. Pollut. Res.* **2016**, *23*, 3884–3896. [CrossRef]

122. Rajamanickam, A.; Thirunavukkarasu, P.; Dhanakodi, K. A simple route to synthesis of carbon doped TiO$_2$ nanostructured thin film for enhanced visible-light photocatalytic activity. *J. Mater. Sci. Mater. Electron.* **2015**, *26*, 4038–4045. [CrossRef]

123. Shi, J.-W.; Liu, C.; He, C.; Li, J.; Xie, C.; Yang, S.; Chen, J.-W.; Li, S.; Niu, C. Carbon-doped titania nanoplates with exposed {001} facets: Facile synthesis, characterization and visible-light photocatalytic performance. *RSC Adv.* **2015**, *5*, 17667–17675. [CrossRef]

124. Shaban, Y.A.; Fallata, H.M. Sunlight-induced photocatalytic degradation of acetaminophen over efficient carbon doped TiO$_2$ (CTiO$_2$) nanoparticles. *Res. Chem. Intermed.* **2019**, *45*, 2529–2547. [CrossRef]

125. De Luna, M.D.G.; Chun-Te Lin, J.; Gotostos, M.J.N.; Lu, M.C. Photocatalytic oxidation of acetaminophen using carbon self-doped titanium dioxide. *Sustain. Environ. Res.* **2016**, *26*, 161–167. [CrossRef]

126. Yuan, Y.; Qian, X.; Han, H.; Chen, Y. Synthesis of carbon modified TiO$_2$ photocatalysts with high photocatalytic activity by a facile calcinations assisted solvothermal method. *J. Mater. Sci. Mater. Electron.* **2017**, *28*, 10028–10034. [CrossRef]

127. Kuang, L.; Zhang, W. Enhanced hydrogen production by carbon-doped TiO$_2$ decorated with reduced graphene oxide (rGO) under visible light irradiation. *RSC Adv.* **2016**, *6*, 2479–2488. [CrossRef]

128. An, N.; Ma, Y.; Liu, J.; Ma, H.; Yang, J.; Zhang, Q. Enhanced visible-light photocatalytic oxidation capability of carbon-doped TiO$_2$ via coupling with fly ash. *Chin. J. Catal.* **2018**, *39*, 1890–1900. [CrossRef]

129. Zhang, J.; Huang, G.F.; Li, D.; Zhou, B.X.; Chang, S.; Pan, A.; Huang, W.Q. Facile route to fabricate carbon-doped TiO$_2$ nanoparticles and its mechanism of enhanced visible light photocatalytic activity. *Appl. Phys. A* **2016**, *122*, 994. [CrossRef]

130. Noorimotlagh, Z.; Kazeminezhad, I.; Jaafarzadeh, N.; Ahmadi, M.; Ramezani, Z. Improved performance of immobilized TiO$_2$ under visible light for the commercial surfactant degradation: Role of carbon doped TiO$_{60}$ and anatase/rutile ratio. *Catal. Today* **2020**, *348*, 277–289. [CrossRef]

131. Saharudin, K.A.; Sreekantan, S.; Lai, C.W. Fabrication and photocatalysis of nanotubular C-doped TiO$_2$ arrays: Impact of annealing atmosphere on the degradation efficiency of methyl orange. *Mater. Sci. Semicond. Process.* **2014**, *20*, 1–6. [CrossRef]

132. Wang, X.; Lim, T.-T. Solvothermal synthesis of C–N codoped TiO$_2$ and photocatalytic evaluation for bisphenol A degradation using a visible-light irradiated LED photoreactor. *Appl. Catal. B Environ.* **2010**, *100*, 355–364. [CrossRef]

133. Wu, D.; Wang, L. Low-temperature synthesis of anatase C-N-TiO$_2$ photocatalyst with enhanced visible-light-induced photocatalytic activity. *Appl. Surf. Sci.* **2013**, *271*, 357–361. [CrossRef]

134. Wang, M.; Han, J.; Hu, Y.; Guo, R. Mesoporous C, N-codoped TiO$_2$ hybrid shells with enhanced visible light photocatalytic performance. *RSC Adv.* **2017**, *7*, 15513–15520. [CrossRef]

135. Zhang, X.; Zhang, Y.; Zhou, L.; Li, X.; Guo, X. In situ C,N-codoped mesoporous TiO$_2$ nanocrystallites with high surface areas and worm-like structure for efficient photocatalysis. *J. Porous Mater.* **2018**, *25*, 571–579. [CrossRef]

136. Ohtani, B. Photocatalysis A to Z—What we know and what we do not know in a scientific sense. *J. Photochem. Photobiol. C Photochem. Rev.* **2010**, *11*, 157–178. [CrossRef]

137. Song, L.; Jing, W.; Chen, J.; Zhang, S.; Zhu, Y.; Xiong, J. High reusability and durability of carbon-doped TiO$_2$/carbon nanofibrous film as visible-light-driven photocatalyst. *J. Mater. Sci.* **2019**, *54*, 3795–3804. [CrossRef]

138. Lavand, A.B.; Bhatu, M.N.; Malghe, Y.S. Visible light photocatalytic degradation of malachite green using modified titania. *J. Mater. Res. Technol.* **2019**, *8*, 299–308. [CrossRef]

139. Fang, Y.; Liu, Y.; Qi, L.; Xue, Y.; Li, Y. 2D graphdiyne: An emerging carbon material. *Chem. Soc. Rev.* **2022**, *51*, 2681–2709. [CrossRef]

140. Zheng, Z.; Xue, Y.; Li, Y. A new carbon allotrope: Graphdiyne. *Trends Chem.* **2022**, *4*, 754–768. [CrossRef]

141. Wang, W.; Chen, M.; Huang, D.; Zeng, G.; Zhang, C.; Lai, C.; Wang, Z. An overview on nitride and nitrogen-doped photocatalysts for energy and environmental applications. *Compos. Part B Eng.* **2019**, *172*, 704–723. [CrossRef]

142. Ansari, S.A.; Khan, M.M.; Ansari, M.O.; Cho, M.H. Nitrogen-doped titanium dioxide (N-doped TiO2) for visible light photocatalysis. *N. J. Chem.* **2016**, *40*, 3000–3009. [CrossRef]

143. Gomathi Devi, L.; Kavitha, R. Review on modified N–TiO$_2$ for green energy applications under UV/visible light: Selected results and reaction mechanisms. *RSC Adv.* **2014**, *4*, 28265–28299. [CrossRef]

144. Samokhvalov, A. Hydrogen by photocatalysis with nitrogen codoped titanium dioxide. *Renew. Sustain. Energy Rev.* **2017**, *72*, 981–1000. [CrossRef]
145. Peng, F.; Cai, L.; Yu, H.; Wang, H.; Yang, J. Synthesis and characterization of substitutional and interstitial nitrogen-doped titanium dioxides with visible light photocatalytic activity. *J. Solid State Chem.* **2008**, *181*, 130–136. [CrossRef]
146. Sacco, O.; Stoller, M.; Vaiano, V.; Ciambelli, P.; Chianese, A.; Sannino, D. Photocatalytic degradation of organic dyes under visible light on N-doped TiO$_2$ photocatalysts. *Int. J. Photoenergy* **2012**, *2012*, 626759. [CrossRef]
147. Thambiliyagodage, C.; Usgodaarachchi, L. Photocatalytic activity of N, Fe and Cu co-doped TiO$_2$ nanoparticles under sunlight. *Curr. Res. Green Sustain. Chem.* **2021**, *4*, 100186. [CrossRef]
148. Preethi, L.K.; Antony, R.P.; Mathews, T.; Walczak, L.; Gopinath, C.S. A study on doped heterojunctions in TiO$_2$ nanotubes: An efficient photocatalyst for solar water splitting. *Sci. Rep.* **2017**, *7*, 14314. [CrossRef] [PubMed]
149. Sirivallop, A.; Areerob, T.; Chiarakorn, S. Enhanced visible light photocatalytic activity of N and Ag doped and co-doped TiO$_2$ synthesized by using an in-situ solvothermal method for gas phase ammonia removal. *Catalysts* **2020**, *10*, 251. [CrossRef]
150. Xu, X.; Song, W. Enhanced H$_2$ production activity under solar irradiation over N-doped TiO$_2$ prepared using pyridine as a precursor: A typical sample of N-doped TiO$_2$ series. *Mater. Technol.* **2017**, *32*, 52–63. [CrossRef]
151. Ariza-Tarazona, M.C.; Villarreal-Chiu, J.F.; Barbieri, V.; Siligardi, C.; Cedillo-González, E.I. New strategy for microplastic degradation: Green photocatalysis using a protein-based porous N-TiO$_2$ semiconductor. *Ceram. Int.* **2019**, *45*, 9618–9624. [CrossRef]
152. Cheng, Z.-L.; Han, S. Preparation and photoelectrocatalytic performance of N-doped TiO$_2$/NaY zeolite membrane composite electrode material. *Water Sci. Technol.* **2016**, *73*, 486–492. [CrossRef]
153. Marques, J.; Gomes, T.D.; Forte, M.A.; Silva, R.F.; Tavares, C.J. A new route for the synthesis of highly-active N-doped TiO$_2$ nanoparticles for visible light photocatalysis using urea as nitrogen precursor. *Catal. Today* **2019**, *326*, 36–45. [CrossRef]
154. Cao, Y.; Xing, Z.; Shen, Y.; Li, Z.; Wu, X.; Yan, X.; Zhou, W. Mesoporous black Ti^{3+}/N-TiO$_2$ spheres for efficient visible-light-driven photocatalytic performance. *Chem. Eng. J.* **2017**, *325*, 199–207. [CrossRef]
155. Xing, X.; Du, Z.; Zhuang, J.; Wang, D. Removal of ciprofloxacin from water by nitrogen doped TiO$_2$ immobilized on glass spheres: Rapid screening of degradation products. *J. Photochem. Photobiol. A Chem.* **2018**, *359*, 23–32. [CrossRef]
156. Mohammadalipour, Z.; Rahmati, M.; Khataee, A.; Moosavi, M.A. Differential effects of N-TiO$_2$ nanoparticle and its photo-activated form on autophagy and necroptosis in human melanoma A375 cells. *J. Cell. Physiol.* **2020**, *235*, 8246–8259. [CrossRef]
157. Zangeneh, H.; Mousavi, S.A.; Eskandari, P. Comparison the visible photocatalytic activity and kinetic performance of amino acids (non-metal doped) TiO$_2$ for degradation of colored wastewater effluent. *Mater. Sci. Semicond. Process.* **2022**, *140*, 106383. [CrossRef]
158. Asahi, R.; Morikawa, T.; Ohwaki, T.; Aoki, K.; Taga, Y. Visible-light photocatalysis in nitrogen-doped titanium oxides. *Science* **2001**, *293*, 269–271. [CrossRef]
159. Umebayashi, T.; Yamaki, T.; Tanaka, S.; Asai, K. Visible light-induced degradation of methylene blue on S-doped TiO$_2$. *Chem. Lett.* **2003**, *32*, 330–331. [CrossRef]
160. Ho, W.; Jimmy, C.Y.; Lee, S. Low-temperature hydrothermal synthesis of S-doped TiO$_2$ with visible light photocatalytic activity. *J. Solid State Chem.* **2006**, *179*, 1171–1176. [CrossRef]
161. Niu, Y.; Xing, M.; Tian, B.; Zhang, J. Improving the visible light photocatalytic activity of nano-sized titanium dioxide via the synergistic effects between sulfur doping and sulfation. *Appl. Catal. B Environ.* **2012**, *115*, 253–260. [CrossRef]
162. Bakar, S.A.; Ribeiro, C. An insight toward the photocatalytic activity of S doped 1-D TiO$_2$ nanorods prepared via novel route: As promising platform for environmental leap. *J. Mol. Catal. A Chem.* **2016**, *412*, 78–92. [CrossRef]
163. Yan, X.; Yuan, K.; Lu, N.; Xu, H.; Zhang, S.; Takeuchi, N.; Kobayashi, H.; Li, R. The interplay of sulfur doping and surface hydroxyl in band gap engineering: Mesoporous sulfur-doped TiO$_2$ coupled with magnetite as a recyclable, efficient, visible light active photocatalyst for water purification. *Appl. Catal. B Environ.* **2017**, *218*, 20–23. [CrossRef]
164. Lv, Y.; Yuan, K.; Lu, N.; Xu, H.; Zhang, S.; Takeuchi, N.; Li, R. P-doped TiO$_2$ nanoparticles film coated on ground glass substrate and the repeated photodegradation of dye under solar light irradiation. *Appl. Surf. Sci.* **2011**, *257*, 5715–5719. [CrossRef]
165. Wu, J.; Zhang, Y.; Zhou, J.; Wang, K.; Zheng, Y.Z.; Tao, X. Uniformly assembling n-type metal oxide nanostructures (TiO$_2$ nanoparticles and SnO$_2$ nanowires) onto P doped g-C$_3$N$_4$ nanosheets for efficient photocatalytic water splitting. *Appl. Catal. B Environ.* **2020**, *278*, 119301. [CrossRef]
166. Zhang, G.; Ji, S.; Zhang, Y.; Wei, Y. Facile synthesis of pn heterojunction of phosphorus doped TiO$_2$ and BiOI with enhanced visible-light photocatalytic activity. *Solid State Commun.* **2017**, *259*, 34–39. [CrossRef]
167. Gopal, N.O.; Lo, H.H.; Ke, T.F.; Lee, C.H.; Chou, C.C.; Wu, J.D.; Ke, S.C. Visible light active phosphorus-doped TiO$_2$ nanoparticles: An EPR evidence for the enhanced charge separation. *J. Phys. Chem. C* **2012**, *116*, 16191–16197. [CrossRef]
168. Zhu, Y.; Li, J.; Dong, C.L.; Ren, J.; Huang, Y.C.; Zhao, D.; Yang, D. Red phosphorus decorated and doped TiO$_2$ nanofibers for efficient photocatalytic hydrogen evolution from pure water. *Appl. Catal. B Environ.* **2019**, *255*, 117764. [CrossRef]
169. Peng, Y.; He, J.; Liu, Q.; Sun, Z.; Yan, W.; Pan, Z.; Wei, S. Impurity concentration dependence of optical absorption for phosphorus-doped anatase TiO$_2$. *J. Phys. Chem. C* **2011**, *115*, 8184–8188. [CrossRef]
170. Olhero, S.M.; Ganesh, I.; Torres, P.M.; Ferreira, J.M. Surface passivation of MgAl$_2$O$_4$ spinel powder by chemisorbing H$_3$PO$_4$ for easy aqueous processing. *Langmuir* **2008**, *24*, 9525–9530. [CrossRef] [PubMed]

171. Mendiola-Alvarez, S.Y.; Hernández-Ramírez, A.; Guzmán-Mar, J.L.; Maya-Treviño, M.L.; Caballero-Quintero, A.; Hinojosa-Reyes, L. A novel P-doped Fe_2O_3-TiO_2 mixed oxide: Synthesis, characterization and photocatalytic activity under visible radiation. *Catal. Today.* **2019**, *328*, 91–98. [CrossRef]

172. Xia, Y.; Jiang, Y.; Li, F.; Xia, M.; Xue, B.; Li, Y. Effect of calcined atmosphere on the photocatalytic activity of P-doped TiO_2. *Appl. Surf. Sci.* **2014**, *289*, 306–315. [CrossRef]

173. Li, F.; Jiang, Y.; Xia, M.; Sun, M.; Xue, B.; Liu, D.; Zhang, X. Effect of the P/Ti ratio on the visible-light photocatalytic activity of P-doped TiO_2. *J. Phys. Chem. C* **2009**, *113*, 18134–18141. [CrossRef]

174. Valentin, C.; Pacchioni, G. Trends in non-metal doping of anatase TiO_2: B, C, N and F. *Catal. Today* **2013**, *206*, 12–18. [CrossRef]

175. Dozzi, M.V.; Selli, E. Doping TiO_2 with p-block elements: Effects on photocatalytic activity. *J. Photochem. Photobiol. C Photochem. Rev.* **2013**, *14*, 13–28. [CrossRef]

176. Kuo, C.-Y.; Jheng, H.-K.; Syu, S.-E. Effect of non-metal doping on the photocatalytic activity of titanium dioxide on the photodegradation of aqueous bisphenol A. *Environ. Technol.* **2021**, *42*, 1603–1611. [CrossRef]

177. Zheng, J.; Liu, Z.; Liu, X.; Yan, X.; Li, D.; Chu, W. Facile hydrothermal synthesis and characteristics of B-doped TiO_2 hybrid hollow microspheres with higher photo-catalytic activity. *J. Alloys Compd.* **2011**, *509*, 3771–3776. [CrossRef]

178. Bessegato, G.G.; Cardoso, J.C.; Zanoni, M.V.B. Enhanced photoelectrocatalytic degradation of an acid dye with boron-doped TiO_2 nanotube anodes. *Catal. Today* **2015**, *240*, 100–106. [CrossRef]

179. Yadav, V.; Verma, P.; Sharma, H.; Tripathy, S.; Saini, V.K. Photodegradation of 4-nitrophenol over B-doped TiO_2 nanostructure: Effect of dopant concentration, kinetics, and mechanism. *Environ. Sci. Pollut. Res.* **2020**, *27*, 10966–10980. [CrossRef]

180. Gao, Q.; Si, F.; Zhang, S.; Fang, Y.; Chen, X.; Yang, S. Hydrogenated F-doped TiO_2 for photocatalytic hydrogen evolution and pollutant degradation. *Int. J. Hydrogen Energy* **2019**, *44*, 8011–8019. [CrossRef]

181. Filippatos, P.-P.; Soultati, A.; Kelaidis, N.; Petaroudis, C.; Alivisatou, A.A.; Drivas, C.; Chroneos, A. Preparation of hydrogen, fluorine and chlorine doped and co-doped titanium dioxide photocatalysts: A theoretical and experimental approach. *Sci. Rep.* **2021**, *11*, 5700. [CrossRef]

182. Samsudin, E.M.; Hamid, S.B.A. Effect of band gap engineering in anionic-doped TiO_2 photocatalyst. *Appl. Surf. Sci.* **2017**, *391*, 326–336. [CrossRef]

183. Yu, W.; Liu, X.; Pan, L.; Li, J.; Liu, J.; Zhang, J.; Li, P.L.; Chen, C.; Sun, Z. Enhanced visible light photocatalytic degradation of methylene blue by F-doped TiO_2. *Appl. Surf. Sci.* **2014**, *319*, 107–112. [CrossRef]

184. Wang, X.-K.; Wang, C.; Jiang, W.Q.; Guo, W.L.; Wang, J.G. Sonochemical synthesis and characterization of Cl-doped TiO_2 and its application in the photodegradation of phthalate ester under visible light irradiation. *Chem. Eng. J.* **2012**, *189*, 288–294. [CrossRef]

185. Long, M.; Cai, W.; Chen, H.; Xu, J. Preparation, characterization and photocatalytic activity of visible light driven chlorine-doped TiO_2. *Front. Chem. China* **2007**, *2*, 278–282. [CrossRef]

186. Wang, Q.; Zhu, S.; Liang, Y.; Cui, Z.; Yang, X.; Liang, C.; Inoue, A. Synthesis of Br-doped TiO_2 hollow spheres with enhanced photocatalytic activity. *J. Nanoparticle Res.* **2017**, *19*, 72. [CrossRef]

187. Shen, Y.; Xiong, T.; Du, H.; Jin, H.; Shang, J.; Yang, K. Investigation of Br–N Co-doped TiO_2 photocatalysts: Preparation and photocatalytic activities under visible light. *J. Sol-Gel Sci. Technol.* **2009**, *52*, 41–48. [CrossRef]

188. Luo, H.; Takata, T.; Lee, Y.; Zhao, J.; Domen, K.; Yan, Y. Photocatalytic activity enhancing for titanium dioxide by co-doping with bromine and chlorine. *Chem. Mater.* **2004**, *16*, 846–849. [CrossRef]

189. Barkul, R.P.; Patil, M.K.; Patil, S.M.; Shevale, V.B.; Delekar, S.D. Sunlight-assisted photocatalytic degradation of textile effluent and Rhodamine B by using iodine doped TiO_2 nanoparticles. *J. Photochem. Photobiol. A Chem.* **2017**, *349*, 138–147. [CrossRef]

190. Gai, H.; Wang, H.; Liu, L.; Feng, B.; Xiao, M.; Tang, Y.; Qu, X.; Song, H.; Huang, T. Potassium and iodide codoped mesoporous titanium dioxide for enhancing photocatalytic degradation of phenolic compounds. *Chem. Phys. Lett.* **2021**, *767*, 138367. [CrossRef]

191. Van Viet, P.; Huy, T.H.; Sang, T.T.; Nguyet, H.M.; Thi, C.M. One-pot hydrothermal synthesis of Si doped TiO_2 nanotubes from commercial material sources for visible light-driven photocatalytic activity. *Mater. Res. Express* **2019**, *6*, 055006. [CrossRef]

192. Kim, M.G.; Kang, J.M.; Lee, J.E.; Kim, K.S.; Kim, K.H.; Cho, M.; Lee, S.G. Effects of calcination temperature on the phase composition, photocatalytic degradation, and virucidal activities of TiO_2 nanoparticles. *ACS Omega* **2021**, *6*, 10668–10678. [CrossRef]

193. Gaya, U.I.; Abdullah, A.H. Heterogeneous photocatalytic degradation of organic contaminants over titanium dioxide: A review of fundamentals, progress and problems. *J. Photochem. Photobiol. C Photochem. Rev.* **2008**, *9*, 1–12. [CrossRef]

194. Guettaï, N.; Amar, H.A. Photocatalytic oxidation of methyl orange in presence of titanium dioxide in aqueous suspension. Part I: Parametric study. *Desalination* **2005**, *185*, 427–437. [CrossRef]

195. Kinsinger, N.M.; Dudchenko, A.; Wong, A.; Kisailus, D. Synergistic effect of pH and phase in a nanocrystalline titania photocatalyst. *ACS Appl. Mater. Interfaces* **2013**, *5*, 6247–6254. [CrossRef]

196. Sun, J.; Wang, X.; Sun, J.; Sun, R.; Sun, S.; Qiao, L. Photocatalytic degradation and kinetics of Orange G using nano-sized Sn (IV)/TiO_2/AC photocatalyst. *J. Mol. Catal. A Chem.* **2006**, *260*, 241–246. [CrossRef]

197. Sun, J.; Qiao, L.; Sun, S.; Wang, G. Photocatalytic degradation of Orange G on nitrogen-doped TiO_2 catalysts under visible light and sunlight irradiation. *J. Hazard. Mater.* **2008**, *155*, 312–319. [CrossRef]

198. Baran, W.; Makowski, A.; Wardas, W. The effect of UV radiation absorption of cationic and anionic dye solutions on their photocatalytic degradation in the presence TiO_2. *Dye. Pigment.* **2008**, *76*, 226–230. [CrossRef]

199. Nhu, V.T.T.; QuangMinh, D.; Duy, N.N.; QuocHien, N. Photocatalytic Degradation of Azo Dye (Methyl Red) In Water under Visible Light Using AgNi/TiO$_2$ Sythesized by γ Irradiation Method. *Int. J. Environ. Agric. Biotechnol. (IJEAB)* **2017**, *2*, 529–538.

200. Kovalev, I.S. Stealth moths: The multi-plumed wings of the moth alucita hexadactyla may decrease the intensity of their echo to simulated bat echolocation cries. *Entomol. News* **2016**, *126*, 204–212. [CrossRef]

201. Ali, A.H. Study on the photocatalytic degradation of indigo carmine dye by TiO$_2$ photocatalyst. *J. Kerbala Univ.* **2013**, *11*, 145–153.

202. Viswanathan, B. Photocatalytic degradation of dyes: An overview. *Curr. Catal.* **2018**, *7*, 99–121. [CrossRef]

203. Đokić, V.R.; Vujović, J.; Marinković, A.; Petrović, R.; Janaćković, Đ.; Onjia, A.; Mijin, D. A study of the photocatalytic degradation of the textile dye CI Basic Yellow 28 in water using a P160 TiO$_2$-based catalyst. *J. Serb. Chem. Soc.* **2012**, *77*, 1747–1757. [CrossRef]

204. Mohabansi, N.; Patil, V.; Yenkie, N. A comparative study on photo degradation of methylene blue dye effluent by advanced oxidation process by using TiO$_2$/ZnO photo catalyst. *Rasayan J. Chem.* **2011**, *4*, 814–819.

205. Samsudin, E.M.; Goh, S.N.; Wu, T.Y.; Ling, T.T.; Hamid, S.A.; Juan, J.C. Evaluation on the photocatalytic degradation activity of reactive blue 4 using pure anatase nano-TiO$_2$. *Sains Malays.* **2015**, *44*, 1011–1019. [CrossRef]

206. Ram, C.; Pareek, R.K.; Singh, V. Photocatalytic degradation of textile dye by using titanium dioxide nanocatalyst. *Int. J. Theor. Appl. Sci.* **2012**, *4*, 82–88.

207. Meeti, M.; Sharma, T. Photo catalytic degradation of two commercial dyes in aqueous phase using photo catalyst TiO$_2$. *Adv. Appl. Sci. Res.* **2012**, *3*, 849–853.

208. Nagaraja, R.; Kottam, N.; Girija, C.R.; Nagabhushana, B.M. Photocatalytic degradation of Rhodamine B dye under UV/solar light using ZnO nanopowder synthesized by solution combustion route. *Powder Technol.* **2012**, *215*, 91–97. [CrossRef]

209. Choi, Y.; Koo, M.S.; Bokare, A.D.; Kim, D.H.; Bahnemann, D.W.; Choi, W. Sequential process combination of photocatalytic oxidation and dark reduction for the removal of organic pollutants and Cr (VI) using Ag/TiO$_2$. *Environ. Sci. Technol.* **2017**, *51*, 3973–3981. [CrossRef]

210. Guillard, C.; Disdier, J.; Monnet, C.; Dussaud, J.; Malato, S.; Blanco, J.; Maldonado, M.; Herrmann, J.-M. Solar efficiency of a new deposited titania photocatalyst: Chlorophenol, pesticide and dye removal applications. *Appl. Catal. B Environ.* **2003**, *46*, 319–332. [CrossRef]

211. Zhang, A.-Y.; Wang, W.K.; Pei, D.N.; Yu, H.Q. Degradation of refractory pollutants under solar light irradiation by a robust and self-protected ZnO/CdS/TiO$_2$ hybrid photocatalyst. *Water Res.* **2016**, *92*, 78–86. [CrossRef]

212. Azad, K.; Gajanan, P. Photodegradation of methyl orange in aqueous solution by the visible light active Co: La: TiO$_2$ nanocomposite. *Chem. Sci. J.* **2017**, *8*, 1000164.

213. Bhandari, S.; Vardia, J.; Malkani, R.K.; Ameta, S.C. Effect of transition metal ions on photocatalytic activity of ZnO in bleaching of some dyes. *Toxicol. Environ. Chem.* **2006**, *88*, 35–44. [CrossRef]

214. Kumar, A.; Pandey, G. A review on the factors affecting the photocatalytic degradation of hazardous materials. *Mater. Sci. Eng. Int. J.* **2017**, *1*, 106–114. [CrossRef]

215. Kumar, A.; Hitkari, G.; Gautam, M.; Singh, S.; Pandey, G. Synthesis, characterization and application of Cu-TiO$_2$ nanaocomposites in photodegradation of methyl red (MR). *Int. Adv. Res. J. Sci. Eng. Technol.* **2015**, *2*, 50–55. [CrossRef]

216. Hasanpour, M.; Hatami, M. Photocatalytic performance of aerogels for organic dyes removal from wastewaters: Review study. *J. Mol. Liq.* **2020**, *309*, 113094. [CrossRef]

217. Hinojosa–Reyes, M.; Camposeco–Solis, R.; Ruiz, F.; Rodríguez–González, V.; Moctezuma, E. Promotional effect of metal doping on nanostructured TiO$_2$ during the photocatalytic degradation of 4-chlorophenol and naproxen sodium as pollutants. *Mater. Sci. Semicond. Process.* **2019**, *100*, 130–139. [CrossRef]

218. Vorontsov, A.; Kabachkov, E.N.; Balikhin, I.L.; Kurkin, E.N.; Troitskii, V.N.; Smirniotis, P.G. Correlation of surface area with photocatalytic activity of TiO$_2$. *J. Adv. Oxid. Technol.* **2018**, *21*, 127–137. [CrossRef]

219. Ameen, S.; Seo, H.K.; Akhtar, M.S.; Shin, H.S. Novel graphene/polyaniline nanocomposites and its photocatalytic activity toward the degradation of rose Bengal dye. *Chem. Eng. J.* **2012**, *210*, 220–228. [CrossRef]

220. Cassano, A.E.; Alfano, O.M. Reaction engineering of suspended solid heterogeneous photocatalytic reactors. *Catal. Today* **2000**, *58*, 167–197. [CrossRef]

221. Muruganandham, M.; Swaminathan, M. TiO$_2$–UV photocatalytic oxidation of Reactive Yellow 14: Effect of operational parameters. *J. Hazard. Mater.* **2006**, *135*, 78–86. [CrossRef]

222. Elaziouti, A.; Ahmed, B. ZnO-assisted photocatalytic degradation of congo Red and benzopurpurine 4B in aqueous solution. *J. Chem. Eng. Process. Technol.* **2011**, *2*, 1000106.

223. Rafiq, A.; Ikram, M.; Ali, S.; Niaz, F.; Khan, M.; Khan, Q.; Maqbool, M. Photocatalytic degradation of dyes using semiconductor photocatalysts to clean industrial water pollution. *J. Ind. Eng. Chem.* **2021**, *97*, 111–128. [CrossRef]

224. Kaur, S.; Singh, V. TiO$_2$ mediated photocatalytic degradation studies of Reactive Red 198 by UV irradiation. *J. Hazard. Mater.* **2007**, *141*, 230–236. [CrossRef]

225. Shabir, M.; Yasin, M.; Hussain, M.; Shafiq, I.; Akhter, P.; Nizami, A.-S.; Jeon, B.-H.; Park, Y.-K. A review on recent advances in the treatment of dye-polluted wastewater. *J. Ind. Eng. Chem.* **2022**, *112*, 1–19. [CrossRef]

226. Muruganandham, M.; Swaminathan, M. Solar photocatalytic degradation of a reactive azo dye in TiO$_2$-suspension. *Solar Energy Mater. Sol. Cells* **2004**, *81*, 439–457. [CrossRef]

227. Litter, M.; Navio, J.A. Photocatalytic properties of iron-doped titania semiconductors. *J. Photochem. Photobiol. A Chem.* **1996**, *98*, 171–181. [CrossRef]

228. Leung, D.Y.; Fu, X.; Wang, C.; Ni, M.; Leung, M.K.; Wang, X.; Fu, X. Hydrogen production over titania-based photocatalysts. *ChemSusChem* **2010**, *3*, 681–694. [CrossRef] [PubMed]
229. Ma, T.; Akiyama, M.; Abe, E.; Imai, I. High-efficiency dye-sensitized solar cell based on a nitrogen-doped nanostructured titania electrode. *Nano Lett.* **2005**, *5*, 2543–2547. [CrossRef] [PubMed]
230. Mahmoud, M.; Ismail, A.A.; Sanad, M. Developing a cost-effective synthesis of active iron oxide doped titania photocatalysts loaded with palladium, platinum or silver nanoparticles. *Chem. Eng. J.* **2012**, *187*, 96–103. [CrossRef]
231. Han, X.; Kuang, Q.; Jin, M.; Xie, Z.; Zheng, L. Synthesis of titania nanosheets with a high percentage of exposed (001) facets and related photocatalytic properties. *J. Am. Chem. Soc.* **2009**, *131*, 3152–3153. [CrossRef]
232. Ikem, V.O.; Menner, A.; Bismarck, A. High-porosity macroporous polymers sythesized from titania-particle-stabilized medium and high internal phase emulsions. *Langmuir* **2010**, *26*, 8836–8841. [CrossRef]
233. Bartl, M.H.; Boettcher, S.W.; Frindell, K.L.; Stucky, G.D. 3-D molecular assembly of function in titania-based composite material systems. *Acc. Chem. Res.* **2005**, *38*, 263–271. [CrossRef]
234. Lindstrom, H.; Wootton, R.; Iles, A. High surface area titania photocatalytic microfluidic reactors. *AICHE J.* **2007**, *53*, 695–702. [CrossRef]
235. Peiris, S.; de Silva, H.B.; Ranasinghe, K.N.; Bandara, S.V.; Perera, I.R. Recent development and future prospects of TiO_2 photocatalysis. *J. Chin. Chem. Soc.* **2021**, *68*, 738–769. [CrossRef]

catalysts

Article

High-Temperature Abatement of N_2O over FeOx/CeO_2-Al_2O_3 Catalysts: The Effects of Oxygen Mobility

Larisa Pinaeva [1], Igor Prosvirin [1], Yuriy Chesalov [1] and Victor Atuchin [2,3,4,5,*]

[1] Boreskov Institute of Catalysis SB-RAS, 5, Lavrentiev ave, Novosibirsk 630090, Russia
[2] Laboratory of Optical Materials and Structures, Institute of Semiconductor Physics, SB-RAS, Novosibirsk 630090, Russia
[3] Research and Development Department, Kemerovo State University, Kemerovo 650000, Russia
[4] Department of Industrial Machinery Design, Novosibirsk State Technical University, Novosibirsk 630073, Russia
[5] R&D Center "Advanced Electronic Technologies", Tomsk State University, Tomsk 634034, Russia
[*] Correspondence: atuchin@isp.nsc.ru

Abstract: CeO_2-Al_2O_3 oxides prepared by co-precipitation (Ce+Al) or CeOx precipitation onto Al_2O_3 (Ce/Al) to obtain dispersed CeO_2 and samples with further supported FeOx (2.5–9.9 weight% in terms of Fe) were characterized by XRD, XPS, DDPA and Raman. Fe/Ce/Al samples with lower surface concentrations of Fe^{3+} were substantially more active in N_2O decomposition at 700–900 °C. It was related to higher oxygen mobility, as estimated from $^{16}O/^{18}O$ exchange experiments and provided by preferential exposing of (Fe-)Ce oxides. Stabilization of some Ce as isolated Ce^{3+} in Fe-Ce-Al mixed oxides dominating in the bulk and surface layers of Fe/(Ce + Al) samples retards the steps responsible for fast additional oxygen transfer to the sites of O_2 desorption.

Keywords: N_2O decomposition; Fe-Ce-Al mixed oxide; oxygen mobility

Citation: Pinaeva, L.; Prosvirin, I.; Chesalov, Y.; Atuchin, V. High-Temperature Abatement of N_2O over FeOx/CeO_2-Al_2O_3 Catalysts: The Effects of Oxygen Mobility. *Catalysts* **2022**, *12*, 938. https://doi.org/10.3390/catal12090938

Academic Editors: Gassan Hodaifa and Rafael Borja

Received: 4 August 2022
Accepted: 22 August 2022
Published: 24 August 2022

Publisher's Note: MDPI stays neutral with regard to jurisdictional claims in published maps and institutional affiliations.

1. Introduction

Nitric acid is a key chemical with annual production at above 65 million tons mainly driven by the fertilizer industry. Nitrous oxide (N_2O) is produced as an unwanted, but inevitable by-product at HNO_3 industrial manufacture by ammonia oxidation. Its emissions by nitric acid plants are growing from 5 to 12 kg N_2O/ton HNO_3 according to the tendency to high-pressure processes [1]. Meanwhile, nitrous oxide is a potent greenhouse gas characterized by about 300 times the Global Warming Potential of CO_2 and contributes to the greenhouse effect and ozone layer depletion. Therefore, 330–780 thousand tons of N_2O annually emitted at HNO_3 production are equivalent to 100–240 million tons/y CO_2e. It can be deleted by placing the catalysts in several points of the ammonia processing flowsheet [2], including that immediately downstream the Pt–Rh gauzes operating at 800–900 °C. A tendency of recent years to increase weight space velocity and reduce the charge of platinum-group metals in the reactor increased the risk of ammonia slipping at gauze ageing or damage. Therefore, efficient conversion of NH_3 to NO + NO_2, while N_2O is low is one of the necessary features for N_2O decomposition catalysts operating in these conditions as well.

Fe-Al-O-based catalytic compositions are efficient both in selective ammonia oxidation to NO and NO_2 [3–5] and in selective N_2O decomposition under the conditions of the nitric acid plant [4–8] without decrease in NOx yield [6,7]. The activity in both reactions was shown to increase by using CeO_2 as the support instead of alumina [4]. It was due to the appearance of the additional pathway of the oxygen supply through oxygen vacancies located in the near subsurface layers of ceria to Fe active sites facilitating the rate-determining step of O_2 desorption. An increase in both ceria and Co(Fe)O_x dispersion could additionally promote oxygen diffusion to Fe. In the first case it is because both distributions of oxygen vacancies and oxygen storage and release capability of CeO_2 are related to its particle size [9,10]. The positive effect of the dispersion of Co oxide spinel phase [11,12] or FeOx species over ceria and alumina [4,5,8] onto activity in both reactions was already shown and related to enhanced oxygen diffusion through the active component/support interface periphery [4,11,12]. However, for individual CeO_2 all positive effects resulting from fast oxygen transfer are restricted by its low surface area under the reaction conditions [10]. Several methods to stabilize dispersed CeO_2 particles over Al_2O_3 after calcination at above 800 °C were reported. They include the sol–gel method using aluminum tri-sec-butoxide and Ce(III) nitrates as precursors [13], impregnation of alumina with an aqueous solution of $(NH_4)_2[Ce(NO_3)_6]$ [14] or stabilization of a complex of the CeO_2 precursor in a homogeneous polymeric matrix (Pechini route) followed by supporting onto alumina [15,16].

In the current study, we used CeO_2 precipitation onto Al_2O_3 or CeO_2 and Al_2O_3 co-precipitation methods to increase CeO_2 dispersion in the final supports compared with individual CeO_2. The reasons for the principally different activity of FeOx/CeO_2-Al_2O_3 samples prepared using these supports have been elucidated.

2. Results and Discussion

2.1. Phase Composition (XRD)

XRD patterns of Ce/Al and (Ce + Al) samples revealed a mixture of CeO_2, θ-Al_2O_3 and γ-Al_2O_3 phases (Figure S1 in Supplementary Materials) Lattice parameters of CeO_2 in both samples were exactly the same (Table 1) and practically coincided with those for pure CeO_2 calcined at 900 °C (a = b = c = 5.412 Å), while the values of coherent scattering area d_{XRD} calculated by the Scherrer equation were lower (40 and 11 nm for Ce/Al and (Ce + Al), respectively, compared with 68 nm for pure CeO_2 [4]). Lattice parameters of $\gamma(+\theta)$-Al_2O_3 phases cannot be calculated correctly because of low intensity and substantial broadening of corresponding overlapping signals. Therefore, some penetration of Ce ions into their lattices cannot be excluded. At the same time, higher dispersion of alumina species in the (Ce + Al) sample follows from larger widths of the peaks at 45.5° and 67.0° (Figure S1).

Table 1. Structural characteristics of Ce/Al and (Ce + Al) based samples.

Sample	S_{BET} (m^2g^{-1})	Phase Composition (Lattice Parameters)	D_{XRD} (nm)
Ce/Al	63	CeO_2 (a = b = c = 5.411 Å)	40
		$(\theta + \gamma)$-Al_2O_3	-
3.8 Fe/Ce/Al	50	CeO_2 (a = b = c = 5.411 Å)	41
		$(\theta + \gamma)$-Al_2O_3	-
9.9 Fe/Ce/Al	35	CeO_2 (a = b = c = 5.411 Å)	46
		$(\theta + \gamma)$-Al_2O_3	-
		α-Fe_2O_3 (a = b = 5.035 Å, c = 13.741 Å)	32
(Ce + Al)	94	CeO_2 (a = b = c = 5.411 Å)	11
		$(\theta + \gamma)$-Al_2O_3	tr.
2.5 Fe/(Ce + Al)	70	CeO_2 (a = b = c = 5.411 Å)	12
		$(\theta + \gamma)$-Al_2O_3	tr.
6.6 Fe/(Ce + Al)	58	CeO_2 (a = b = c = 5.411 Å)	14
		$(\theta + \gamma)$-Al_2O_3	tr.
		α-Fe_2O_3 (a = b = 5.035 Å, c = 13.741 Å)	39

Fe deposition resulted in a decrease in S_{BET} value and a consistent increase in D_{XRD} of CeO_2 species; the value of effect depended on the quantity of Fe (Table 1). Although CeO_2 lattice parameters did not change for all Fe content ranges, the formation of Fe-Ce-O solid solution cannot be excluded. Ceria lattice shrinkage was measured in Au/Ce-Fe-O (Ce/Fe = 9) [17], Fe/CeO_2 (Ce/Fe = 4–9) [18] or CeO_2–Fe_2O_3 (Ce/Fe = 2.3) [19] samples prepared by procedures favoring homogeneous distribution of Fe and Ce in the samples before their high temperature treatment. It was due to the replacement of Ce^{4+} ions in the typical cubic fluorite lattice (radius 0.097–0.101 nm) by Fe^{3+} ions (0.049–0.078 nm, depending on coordination and spin state). However, at smaller Fe content in $Fe_xCe_{1-x}O_2$ ($x \leq 0.05$, or Ce/Fe ≥ 20) even a slight increase in lattice parameter compared with pure CeO_2 was observed [18] and related to partial Ce^{4+} reduction to the larger (0.123 nm) Ce^{3+} ion [20]. Although in our case Ce/Fe ratio varies within 1.1–6.2, these differently directed effects could compensate for each other because Fe was supported onto support already calcined at 900 °C diminishing the efficiency of its incorporation into CeO_2 lattice, while incorporation of some Fe cations into the alumina component is possible as well.

Nevertheless, neither relative intensity and dispersion of γ-,θ-Al_2O_3 changed (Figure S1, inserts), nor α-Al_2O_3 appeared in both Fe/Ce/Al and Fe/(Ce+Al) samples (Figure S1), unlike Fe/Al_2O_3 samples with reasonably close S_{BET} and Fe content values [4]. Hence, the formation of Fe-Al-O solid solutions in Ce-containing samples is hardly ever possible.

2.2. Catalytic Activity in N_2O Decomposition and NH_3 Oxidation

The activity of Ce/Al and (Ce + Al) based catalysts expressed in terms of N_2O conversion at 800 °C versus Fe content (Figure 1A) is typical for temperature range 700–900 °C (Figure S2). For every type of support, it increases with Fe content. O_2 and H_2O expectedly retard the reaction (Figure S3). Conversion values measured for Fe/(Ce + Al) samples are evidently lower than for Fe/Ce/Al. Therefore, better dispersion of CeO_2 species at close Ce content in the sample does not guarantee higher activity even at higher S_{BET} values (Table 1).

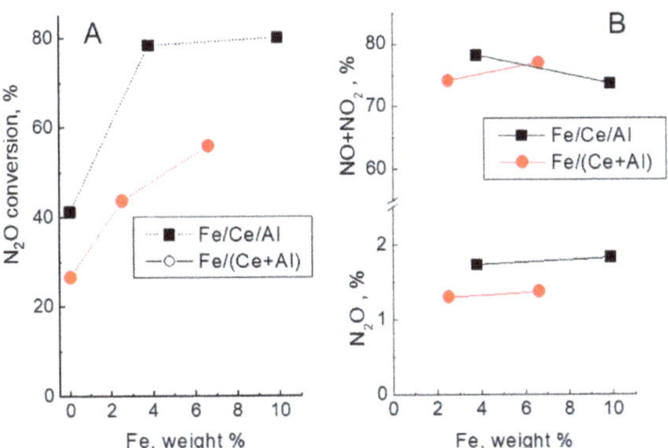

Figure 1. Dependence of N_2O conversion (**A**) and (NO + NO_2) and N_2O yield at NH_3 oxidation (**B**) on Fe content at 800 °C for Ce/Al and (Ce + Al) based samples.

For all samples, the conversion of NH_3 and yield of (NO + NO_2) (Y_{NO+NO2}) increase, while that of N_2O (Y_{N2O}) decreases with temperature; Y_{NO+NO2} = 74–80% and Y_{N2O} = 0.5–1.8% at T $\geq$ 800 °C (Figure S4A,B, and Figure 1B). High stability for 8 h of operation, at least, was shown as well (Figure S4C). We estimated that at an NH_3 slip of 1% (although such a value is practically impossible) the increment of N_2O concentration after the Fe-Ce-Al catalyst at 800 °C will be 65–90 ppm. At the average N_2O concentration downstream of the fresh gauzes of 1500 ppm [2], such an addition will diminish N_2O conversion only by 4–6%.

2.3. Microstructure and Morphology

2.3.1. DDPA

Detailed analysis of dissolution kinetics of Fe, Ce and Al in different samples has been presented in corresponding section of Supporting Information. Herein, we summarized and discussed the main facts that are important for understanding the composition and spatial distribution of different compounds in Fe/(Ce + Al) and Fe/Ce/Al samples.

Mixed oxides of different compositions dissolving consecutively in either HCl or HF (Figure S6) represent most of the 2.5 Fe/(Ce + Al) sample (Table 2). The most abundant compound is Fe-Ce-Al mixed oxide with $Fe_{0.03}Ce_{0.09}Al$ stoichiometry which (Figure 2A): encapsulates the particles of Fe-Ce mixed oxide ($Fe_{0.02}Ce$ stoichiometry); is partially screened by more disordered Fe-Ce-Al mixed oxide with higher content of Fe and Ce ($Fe_{0.05}Ce_{0.26}Al$ stoichiometry) and Fe-Al-O compounds dissolving in HCl.

Table 2. Stoichiometry/quantity (mmol/g, without account of for O) of compounds dissolved in HCl and HF flows in different samples. Quantities of insoluble Ce compounds were calculated from balance equations.

Sample	Ce + Al	2.5 Fe/(Ce + Al)	3.8 Fe/Ce/Al
HCl	Al/1.15	Fe/0.06 $Fe_{0.05}Al/1.30$ $Fe_{0.05}Ce_{0.26}Al/0.65$	Fe/0.07 $Fe_{0.29}Al/0.37$ $Fe_{0.19}Ce/0.12$
HF	$Ce_{0.05}Al/0.67$ $Ce_{0.05}Al/7.92$ Al/0.45 Ce/0.069	$Fe_{0.05}Ce_{0.26}Al/2.62$ $Fe_{0.03}Ce_{0.09}Al/8.14$ $Fe_{0.02}Ce/0.80$	Fe/0.03 $Fe_{0.08}Ce_{0.2}Al/2.93$ $Fe_{0.04}Al/7.47$
Insoluble Ce	2.4	0.7	2.0

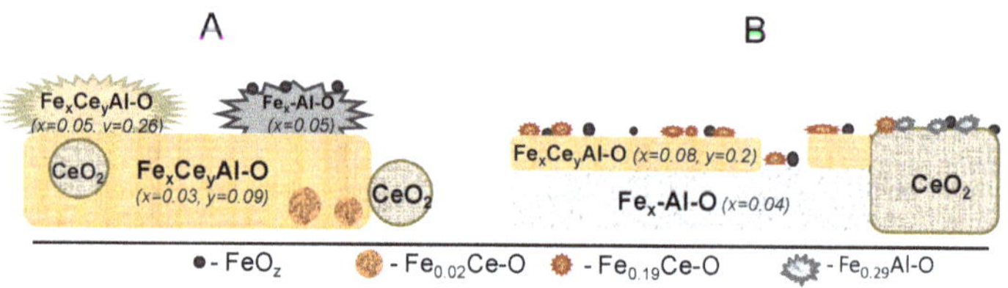

Figure 2. Structures formed in 2.5 Fe/(Ce + Al) (**A**) and 3.8 Fe/Ce/Al (**B**) samples.

The decreasing Fe/Al ratio in Fe-Al-O compounds points to the formation of mixed oxide with most disordered near-surface layers enriched by Fe^{n+} (Figure S6B). At the same time, the protocol of Fe supporting includes using of acidic solutions giving rise to the dissolution of some poorly crystallized alumina detected in the initial support (Figure S5A, Table 2). In this case, the formation of a γ-FeAlO$_3$-like compound with spinel structure is quite possible after following drying and calcination at T < 400 °C [21]. Since the mutual Fe_2O_3 and Al_2O_3 solubility in such structure decreased prominently with calcination temperature rise, its decomposition could happen in our case as well leading to the formation of highly dispersed FeO_x species (0.06 mmol/g) on the surface of Fe-Al mixed oxide with $Fe_{0.05}Al$ stoichiometry. Its quantity (1.3 mmole/g) is reasonably close to that of alumina dissolving in HCl in the initial support (1.15 mmole/g) (Table 2). Calculation showed that identified soluble compounds include all Al and Fe in the sample, while about 0.7 mmol/g of Ce, or 25% of its general content in the sample, represent insoluble oxides (Table S1 in Supplementary Materials, Table 2) which are, more probably, small sized CeO_2 species with unmodified fluorite lattice (Table 1). Their spatial location in the sample is not clear, but encapsulation by Ce-Al mixed oxide preventing Fe incorporation into the lattice is quite reasonable because the same was detected for the particles of Fe-Ce mixed oxide.

The overall quantity of soluble Ce-containing compounds in the 3.8 Fe/Ce/Al sample is substantially lower as compared with the 2.5 Fe/(Ce + A) sample (Table 2), which can be related to a larger size of CeO_2 particles (Table 1). Most of the dissolved part of the sample is represented by Fe-Al mixed oxide partially screened by Fe-Ce-Al one with $Fe_{0.04}Al$ and $Fe_{0.08}Ce_{0.2}Al$ stoichiometries, correspondingly (Figure S7C and Figure 2B, Table 2). Surface layers are composed of more dispersed/disordered binary Fe-Al and Fe-Ce oxides and some FeOx dissolved in HCl (Figure S7, Table 2). Completely independent kinetics of their dissolution can be in the case dispersed FeOx species are located on the surface of any of binary oxides and even non-dissolved large CeO_2 species with unmodified fluorite lattice including 2.0 mmol/g of Ce, or more than 70% of its content in the sample (Figure 2B, Table 2). A substantially higher quantity of Fe in the surface located Fe-Al-O compounds ($Fe_{0.29}Al$ stoichiometry) compared with that in 2.5 Fe/(Ce + Al) could be due to other mechanisms of their formation compared with that in 2.5 Fe/(Ce + Al) sample (see below).

2.3.2. Raman

The band at 462.9 cm^{-1} in the FT-Raman spectra of Ce/Al sample (Figure 3A) is typical for the F_{2g} vibration mode of the fluorite structure of CeO_2 [22,23]. Its width at ~24.5 cm^{-1} (Table 3) should correspond to a size of CeO_2 species at about 10 nm [23,24] which is substantially less than the estimated value of D_{XRD} at 40 nm (Table 1). At the same time, the particles of 10 nm in a size should be characterized by the Raman band at ~448 cm^{-1}, as follows from experimental dependence obtained for CeO_2 and CeO_2-Al_2O_3 composites [15]. Spanier et al. [24] showed that lattice strains in the smaller nanoparticles have the largest contribution to both Raman peak position shift and its broadening compared with other reasons (phonon confinement, particle size distribution, defects, phonon relaxation). Tsunekawa et al. [25] attributed the formation of additional strains to the reduction of Ce^{4+} ions to Ce^{3+} ions caused by an increasing molar fraction of oxygen vacancies. We supposed that in our case stabilization of additional Ce^{3+} within fluorite lattice and appearance of strains can be due to insertion of lower sized Al^{3+} cations into the interstitial sites of c-CeO_2 lattice resulting in charge redistribution, such as observed in c-CeO_2/α-Al_2O_3 composites with close Ce/Al ratio [26]. The intensity of the CeO_2 band in the spectra of Fe/Ce/Al samples does not decrease prominently (Table 3), while its position and especially the width become very close to those in the single crystals [23]. Such changes could be due to better ordering of CeO_2 structure resulting from Al^{3+} "extraction" from the lattice after impregnation by acidic Fe-containing solutions that proceed with temperature increase and, to a less extent, some enlargement of crystallized CeO_2 species (Table 1). It is the stabilization of small Fe-Al-O clusters in the points of Al insertion into CeO_2 lattice after their extraction, but not γ-$FeAlO_3$ phase decomposing at T > 400 °C, that can be responsible for the substantially higher content of Fe in the Fe-Al mixed oxide ($Fe_{0.29}Al$) identified by DDPA compared with that in 2.5 Fe/(Ce + Al) sample (Table 2). Bands at 231 (not shown), 301, 419 and 622 cm^{-1} in the spectrum of 9.9 Fe/Ce/Al sample are usually related to lattice vibrations in α-Fe_2O_3 which agrees with XRD data (Table 1).

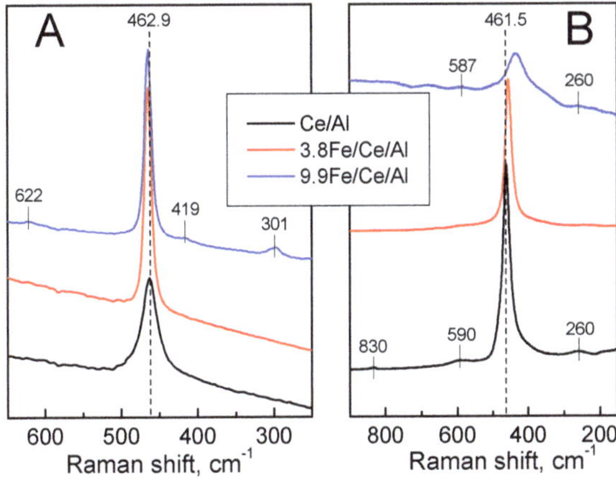

Figure 3. FT-Raman (**A**) and Raman (**B**) spectra of Ce/Al-based samples.

Table 3. Position, width (FWHM) and integral intensity of F_{2g} vibrational mode of CeO_2 in the FTRaman and Raman spectra.

Sample	Position, cm^{-1}		FWHM, cm^{-1}		Integral Intensity, arb.unit	
	FT-Raman	Raman	FT-Raman	Raman	FT-Raman	Raman
Ce/Al	462.9	461.5	24.5	21.2	0.101	30399
3.8 Fe/Ce/Al	465.1	455.8	10.6	21.8	0.087	27805
9.9 Fe/Ce/Al	464.9	433.0	10.8	66.3	0.079	26155
Ce + Al	464.1	462.7	18.3	17.0	0.219	266717
2.5 Fe/(Ce + Al)	464.0	455.8	17.3	23.4	0.0895	34159
6.6 Fe/(Ce + Al)	464.0	459.6	16.3	20.0	0.0327	19715

The absence of the band at ~377 cm^{-1} related to α-Al_2O_3 phases [27] in the Raman spectra giving information on near subsurface layers of all Ce/Al-based samples (Figure 3B) agrees with XRD and DDPA data. F_{2g} mode of CeO_2 at 461.5 cm^{-1} in the spectrum of support is supplemented by the bands at ~590, ~260 and ~830 cm^{-1} ascribed to the vacancy-interstitial (Frenkel-type) oxygen defects, surface mode, and peroxide ($O_2{}^{2-}$) stretching vibration at the defective ceria surfaces, respectively [28–30]. These bands disappear in the spectrum of the 3.8 Fe/Ce/Al sample because the doping element can annihilate oxygen defects by dopant interstitial compensation mechanism [31]. The low-frequency shift of the CeO_2 band and its broadening (Table 3) typical for smaller particles was detected in the spectra of Fe/Ce/Al samples which contradicts the slight increase in averaged sizes of crystalline CeO_2 particles (Table 1). Therefore, the formation of highly dispersed particles of Fe-Ce mixed oxide in the surface layers of the 3.8 Fe/Ce/Al samples was detected by DDPA (Table 2) with inhomogeneous strains due to Ce^{3+} ions therein [15,28] can be responsible for such changes. Asymmetrical broadening with a low-energy shoulder of CeO_2 band, especially prominent in the 9.9 Fe/Ce/Al sample, can result from a higher concentration of such compounds or higher content of α-Fe_2O_3 therein. In line with this, bands at ~260 and ~590 cm^{-1} appear again.

The band at 464 cm^{-1} which is responsible for F_{2g} mode of fluorite lattice attenuates drastically in Fe/(Ce + Al) samples compared with that in the initial support (Figure 4A, Table 3). It is, more probably, due to the noticeable additional inclusion of Ce into alumina or Ce-Al mixed oxides with low Ce content where it exists preferentially as isolated Ce^{3+} ions in the presence of Fe. Indeed, Ce/Al = 0.05 in Ce-Al mixed oxide detected by DDPA in the (Ce + Al) sample but increased up to 0.09–0.26 after Fe supporting (Table 2). The absolute value of mixed oxides increased as well, while the quantity of insoluble CeO_2 decreased from 2.4 mmol/g to 0.7 mmol/g (Table 2). At the same time, the position of CeO_2 F_{2g} mode in Fe/(Ce + Al) samples changed non-substantially pointing to the absence or low degree of ceria lattice modification by foreign atoms compared with the support. In this case, a slightly higher FWHM compared with that in Fe/Ce/Al samples is due to the particle size effect, which is true for crystallized particles, at least (Table 1) [15].

The band at ~377 cm^{-1} related to α-Al_2O_3 phases [27] was absent in the Raman spectra of near subsurface layers of (Ce + Al) based samples as well which agrees with XRD data (Figure S1, Table 1). After iron supporting, they undergo changes that are very similar to those in the bulk of the sample (Figure 4B, Table 3). An even more prominent drop in the intensity of the Ce band compared with that for the bulk can be related to the additional formation of Fe-Ce-Al mixed oxide with high Ce content ($Fe_{0.05}Ce_{0.26}Al$ stoichiometry) (Table 2). Nevertheless, the close CeO_2 band position and FWHM values in FT-Raman and Raman spectra point to a more uniform distribution of elements in the bulk and surface layers of these samples as compared to those in Ce/Al-based samples. This agrees with DDPA data on the preferential formation of Fe-Ce-Al-O mixed oxides in the 2.5 Fe/(Ce + Al) sample, while obviously spatially divided Fe-Al-O, Fe-Ce-O and more CeO_2 were found in the 3.8 Fe/Ce/Al sample.

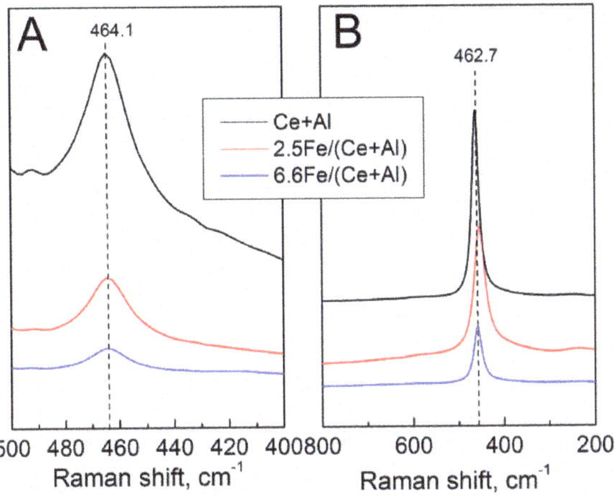

Figure 4. FT-Raman (**A**) and Raman (**B**) spectra of (Ce + Al) based samples.

2.4. Surface Composition (XPS)

The normal complex formed due to shake-down satellites from an O1s to Ce 4f electron transfer in the Ce 3d spectra of all CeO_2 containing samples (Figure 5A) was supplemented by the features marked as v′ and u′ due to the presence of Ce^{3+} [10]. An increase in Ce^{3+} fraction from 8% of total surface Ce concentration in CeO_2 (spectrum not shown for shortness) up to 20% in 0.86 Fe/CeO_2 sample was related to the formation of Fe-Ce–O solid solution or Fe-Ce mixed oxide [4,32]. Similar compounds of $Fe_{0.19}Ce$ stoichiometry have been detected in the 3.8 Fe/Ce/Al sample (Table 2, Figure 2). Therefore, there are two reasons that explain the further increase in Ce^{3+} fraction in 3.8 Fe/Ce/Al (25%) and especially Fe/(Ce + Al) (31-33%) samples (Table 4): stabilization of both isolated Ce cations in alumina-based structure of Fe-Ce-Al mixed oxides (Table 2) and Ce located on CeOx-AlOx boundary in the same oxides as Ce^{3+}; lower CeO_2 particle sizes (12 nm (2.5 Fe/(Ce + Al)) and 41 nm (3.8 Fe/Ce/Al), Table 1) compared with that in Fe/CeO_2 (55 nm [4]) favoring reduction in Ce^{4+} [25].

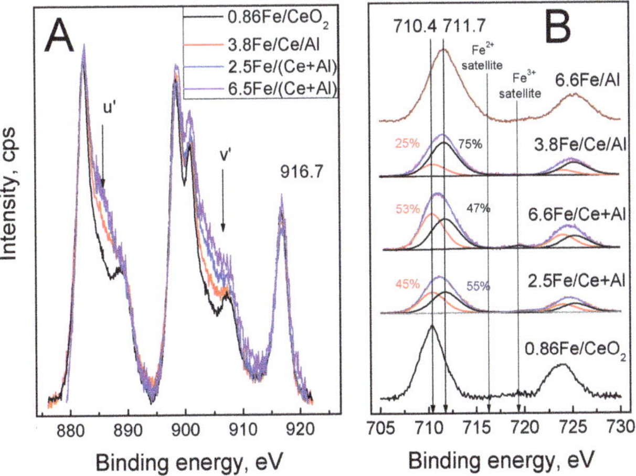

Figure 5. Ce 3d (**A**) Fe 2p (**B**) spectra of Fe-Ce-Al-O samples including contributions of Fe_2O_3 (710.4 eV) and FeOOH (711.7 eV)-like structures to overall Fe2p spectrum. 6.6 Fe/Al_2O_3 and 0.86 Fe/CeO_2 samples were studied by Pinaeva et al. [4].

Table 4. Surface (XPS) composition of different Fe-Ce-Al-O samples.

Sample	Ce^{3+}, % of Total	Ce_s, [a] at. %	Ce_{exp}, rel.units [b]	Ce/Al	Fe_s [a] at. % [a]	Fe_{exp}, rel.units [b]
3.8 Fe/Ce/Al	25	3.7	184	0.12	1.6	80
2.5 Fe/(Ce + Al)	31	2.4	171	0.075	1.7	117
6.6 Fe/(Ce + Al)	33	2.7	159	0.096	2.7	166

[a]: calculated supposing C-containing compounds are absent under reaction conditions; accounting of C does not change the order of values. [b]: calculated as $Fe(Ce)_{exp} = Fe(Ce)_s \cdot SSA$.

Surface Ce concentration (Ce_s) and Ce/Al ratios in all samples (Table 4) are substantially lower than should follow from its integral composition (Ce/Al $\approx$ 0.3, Table S1), which agrees in general with higher dispersion of Al-containing compounds. The highest Ce_s and Ce/Al values were measured in the 3.8 Fe/Ce/Al sample (Table 2). However, the concentration of exposed Ce (Ce_{exp}) in the Fe/Ce/Al and Fe/(Ce + Al) samples differ not so prominently because of higher S_{BET} values in the last (Table 1). It is problematic to estimate Ce_s in the Fe-Ce-Al-O compounds of the samples. This value should be slightly higher in 2.5 Fe/(Ce + Al) sample (Ce/Al = 0.26 instead of Ce/Al = 0.2 for 3.8 Fe/Ce/Al sample, Table 2). Surface areas of crystallized CeO_2 particles in Fe/(Ce + Al) and Fe/Ce/Al samples as formally estimated from their absolute content (Table S1) and sizes (Table 1) are quite close. However, encapsulation of about 29% of Ce composing Fe-Ce-O mixed oxides (Fe/Ce = 0.02, 0.8 mmol/g, Table 2) by Fe-Ce-Al mixed oxide was found for 2.5 Fe/(Ce + Al) sample (Figure 2A). This allows supposing bulk location of some CeO_2 species as well thus decreasing Ce_s value therein. In addition, preferential exposing of highly dispersed Fe-Ce-O compound with $Fe_{0.19}Ce$ stoichiometry can additionally contribute to higher Ce_s value in the 3.8 Fe/Ce/Al sample. Hence, a substantial part of Ce_s in the Fe/(Ce + Al) samples composes Fe-Ce-Al mixed oxides with a high concentration of Ce^+, while the fraction of CeO_2-like structures is higher and seems to be responsible for greater Ce_s value in Fe/Ce/Al samples.

The 3.8 Fe/Ce/Al sample is characterized by the lowest Fe surface concentration (Fe_s) and the quantity of exposed Fe as related to a weight unit (Fe_{exp}) (Table 4). BE values of the $Fe2p_{3/2}$ spectra ranging from 710.9 eV to 711.7 eV (Figure 5B) evidence presence of Fe^{3+} in oxyhydroxide environment (like in α- or γ-FeOOH), and in the oxide (like Fe_2O_3) structures [33–35], correspondingly dominating in 6.6 Fe/Al_2O_3 and 0.86 Fe/CeO_2 samples [4]. Although the positions of the Fe $2P_{3/2}$ peak for Fe^{2+} in $Fe_{0.94}O$, $2FeO \cdot SiO_2$ or $Fe_{0.01}Mg_{0.99}O(100)$ single crystal were characterized by the values between 709.0 and 710.4 eV, its absence or low content in our samples follows from the absence of distinct satellite at 714.6–716 eV [36,37]. FeOOH usually undergoes a dehydroxylation reaction at 250–300 °C resulting in Fe_2O_3 [38]. At the same time, tetrahedral Fe^{3+} cations coordinating both terminal and bridging OH groups were found on the surface of both Fe-Al-O solid solution and highly dispersed FeO_x species resulting from the decomposition of γ-$FeAlO_3$–like structures [21]. In accordance with this, the $Fe2p_{3/2}$ band at 711.7 eV (dominates in 3.8 Fe/Ce/Al sample) can be subscribed to: Fe^{3+} ions in Fe-Al mixed oxides, including those with $Fe_{0.29}Al$ stoichiometry (Table 2) formed by "extraction" of Al from CeO_2 lattice after Fe supporting; highly dispersed FeOx species resulted from the decomposition of Fe-Al-O.

The band at 710.4 eV can characterize FeOx species on the surface of CeO_2 or Fe^{3+} ions in the Fe-Ce-O compounds (Table 2, Figure 2). Its high abundance in Fe/(Ce + Al) samples (45–53% of total Fe_s) contradicts lower Ce_s values compared with that in 3.8 Fe/Ce/Al sample. However, the substantial part of the surface Fe in Fe/(Ce + Al) samples composes Fe-Ce-Al mixed oxides with reasonably high Ce content (Ce/Al = 0.26, Table 2). It resulted from the inclusion of additional Ce from CeO_2 into the surface layers of alumina or Ce-Al mixed oxide of the initial support with Fe participation. Therefore, preferential formation of Fe-Ce-O oxide-like compounds is quite possible in Fe-Ce-Al mixed oxides. The contribution of Fe-Ce-Al mixed oxides into the surface composition in the 3.8 Fe/Ce/Al sample is not high which follows from the preferential dissolution of FeOx, binary Fe-Al-O and Fe-Ce-O compounds in HCl (Table 2) and substantially less prominent decrease in integral intensity of F_{2g} vibrational mode of CeO_2 in the near-surface layers of Fe/Ce/Al samples compared with corresponding support (Table 3).

2.5. O_2 Adsorption/Desorption

O_2 transients recorded during the He/0.5 vol.% O_2 switch following 60 s stay in He flow revealed the delay of O_2 response compared with Ar (Figure 6) corresponding to adsorption of reasonably close quantities of oxygen (3×10^{19}–5×10^{19} oxygen atom/g) by reduced sites in the 2.5 Fe/(Ce + Al) and 3.8 Fe/Ce/Al samples.

Differently paired Fe^{2+} and Ce^{3+} ions resulting from the reduction in the oxygen absence are the most evident sites for O_2 dissociative adsorption as coordinatively unsaturated surface metal cations. They can form in the surface layers of FeO_x clusters, including those resulted from the decomposition of α-$FeAlO_3$; Fe-Al-O mixed oxide of $Fe_{0.29}Al$ stoichiometry formed in 3.8 Fe/Ce/Al sample by Al "extraction" from CeO_2 lattice during calcination; highly dispersed Fe-Ce mixed oxides ($Fe_{0.19}Ce$ stoichiometry (Table 2); Fe-Al-Ce-O mixed oxides with high Ce content ($Fe_{0.05}Ce_{0.26}Al$ or $Fe_{0.08}Ce_{0.2}Al$ stoichiometry).

Reduction of isolated Fe^{3+} in the Fe-Al-(Ce-)O mixed oxides with of $Fe_{0.03}Ce_{0.09}Al$ stoichiometry (most of Ce stabilized as Ce^{3+} and unlikely can participate in any redox processes) requires transport of second oxygen atom through the occasional vacancies on the surface or in the bulk of alumina and thus be slow.

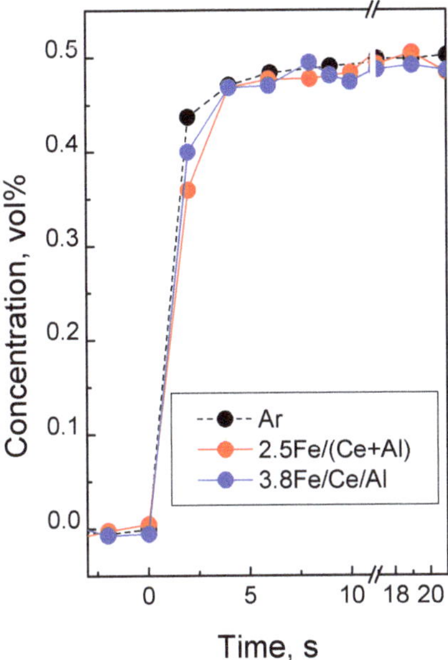

Figure 6. Responses of O_2 and inert label (Ar) during the switch from He to He + 0.5 vol.% O_2 flow over different samples. T = 800 °C.

2.6. Oxygen Mobility (^{18}O SSITKA)

The rates of oxygen exchange in Fe/Ce-Al-O samples in the period of 0–60 s after the $^{16}O/^{18}O$ switch can be ordered as 3.8 Fe/Ce/Al > 6.6 Fe/(Ce + Al)>2.5 Fe/(Ce + Al), as follows from the slope of $N_O(t)/g$ dependencies (Figure 7A). It does not correlate with S_{BET} values of the samples (Table 1) and Fe_s or Fe_{exp} values (Table 4). Moreover, the difference between 3.8 Fe/Ce/Al and Fe/(Ce + Al) samples becomes even more prominent when related to one surface Fe site (Figure 7B). Fast exchange proceeding with reasonably close to initial rate value in these samples during about 20–25 s includes a higher quantity of oxygen atoms ($N_O = 1 \times 10^{21}$–2×10^{21} O atoms/g) than can be related to Fe, even with an account of that located in the bulk (~1.0×10^{21} O atoms/g in the 6.6 Fe/(Ce + Al) sample with the highest Fe content). Therefore, efficient exchange in the near-surface layers of Ce-containing compounds contributes to N_O as well. Indeed, substantially more profound and faster oxygen exchange took place in Fe/CeO$_2$ samples compared with Fe/Al$_2$O$_3$ [4]. However, both Ce content and Ce$_{exp}$ values are reasonably close in all samples (Table 4). We consider that the higher rate of exchange in the 3.8 Fe/Ce/Al sample can be due to preferential exposure of Fe-Ce mixed oxide or crystallized CeO$_2$, including that bonded with clusters of FeOx or Fe-Al mixed oxides, while in 2.5 Fe/(Ce + Al) sample substantial part of Fe and Ce are included as preferentially isolated ions into Fe-Ce-Al mixed oxides that still retain alumina-like structure. Although Fe-Ce mixed oxide was detected in the 2.5 Fe/(Ce + Al) sample (Table 2, Figure S6), it obviously locates in bulk (Figure 2), like part of CeO$_2$, and thus does not contribute to the fast oxygen exchange.

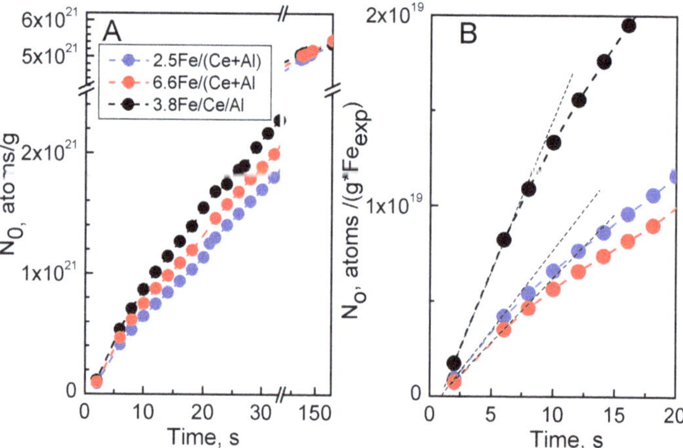

Figure 7. Time dependencies of the quantity of exchanged oxygen (No) as related to weight unit (**A**) or one surface Fe site (**B**) in ^{18}O SSITKA experiments for different samples. T = 800 °C.

2.7. Discussion

In accordance with the simplified redox scheme of N$_2$O decomposition proposed by Kapteijn et al. [39]:

$$N_2O + * \rightarrow N_2 + O*$$

$$2O* \leftrightarrow 2* + O_2$$

one active (reduced) site is sufficient for N$_2$ to evolve, but two neighboring O* are necessary for the desorption of O$_2$. Factually, two neighboring active sites for N$_2$O adsorption are desired for efficient reaction running, especially provided oxygen desorption from the surface is the rate-controlling step of this reaction [39]. Potential quantities of such reduced sites under reaction conditions in 2.5 Fe/(Ce + Al) and 3.8 Fe/Ce/Al samples are quite close (Figure 6). At the same time, the isotopic transient experiment ^{18}O$_2$/N$_2$^{16}O performed on LaMnO$_3$ at 900 °C revealed that there is direct oxygen transfer to the catalysts' bulk from N$_2$O molecule, and O$_2$ formed involves lattice oxygen [40]:

$$2N_2{}^{16}O + 2{}^{18}O_{latt} \rightarrow {}^{18}O_2 + 2{}^{16}O_{latt} + 2N_2 \qquad (1)$$

This means that catalytic activity should depend on oxygen exchange properties.

In CeO$_2$-containing compounds a simplified redox scheme realized in Fe oxides or FeOx/Al$_2$O$_3$ systems:

$$Fe^{2+} + N_2O \rightarrow N_2 + Fe^{3+}\text{-}O \qquad (2)$$

$$2Fe^{3+}\text{-}O \leftrightarrow 2Fe^{2+} + O_2 \tag{3}$$

is supplemented by analogous steps for Ce^{3+}/Ce^{4+} redox pair:

$$Ce^{3+} + N_2O \rightarrow N_2 + Ce^4\text{-}O \tag{4}$$

$$2Ce^{4+}\text{-}O \leftrightarrow 2Ce^{3+} + O_2 \tag{5}$$

And step corresponding to the additional pathway of oxygen species recombination provided by the presence of neighboring Fe^{3+} and Ce^{4+}:

$$Fe^{3+}O + Ce^{4+}O \leftrightarrow Fe^{2+} + Ce^{3+} + O_2 \tag{6}$$

In addition, reoxidation of Ce^{3+} due to fast oxygen diffusion from CeO_2 lattice (O_{lat}) by extended oxygen vacancies arising after insertion of Fe^{3+} ions into the fluorite lattice:

$$O_{lat} + Ce^{3+} \leftrightarrow Ce^{4+}O \tag{7}$$

Can become an alternative pathway for the supply of the second oxygen atom which is necessary for O_2 desorption and enhancing recombination of oxygen adspecies by steps 2B and 2C without adsorption of the second N_2O molecule. Pure CeO_2 revealed insubstantial activity as related per surface area unit compared with $FeOx/CeO_2$ and even Fe/Al_2O_3 samples [4]. Therefore, the contribution of this pathway is high in Fe/Ce/Al samples containing substantial quantities of both Fe-Ce-O mixed oxide and bulky CeO_2 bonded with FeO_x or Fe-Al-O clusters. Stabilization of a substantial fraction of Ce as isolated Ce^{3+} cations composing mixed Fe-Ce-Al oxides in the Fe/(Ce + Al) samples diminishes substantially the efficiency of oxygen transfer to Fe^3-O sites.

3. Materials and Methods

3.1. Catalysts Preparation

Two procedures were used to prepare the supports with a CeO_2:Al_2O_3 weight ratio of 1:1. In the first of them, the precursor of CeO_2 was deposited from 0.75 M $Ce(NO_3)_3 \cdot 6H_2O$ (99.0%, Vekton, Saint-Petersburg, Russia) water solution using 1 M $(NH_4)_2CO_3$ (99.3 %, Vekton, Saint-Petersburg, Russia) as a precipitation agent onto Al_2O_3 (the product of thermochemical activation of hydrargillite calcined at 500 °C) characterized by BET surface area at 210 m^2g^{-1}). After washing by H_2O unless the pH of the filtrate was 7 and drying at 60 °C overnight, the sample was calcined at 900 °C for 5 h and denoted below as Ce/Al. In the second procedure, the water solution of $(NH_4)_2CO_3$ was added dropwise to a water solution of $Al(NO_3)_3 \cdot 9H_2O$ (98.5%, Ecros, Saint-Petersburg, Russia) and $Ce(NO_3)_3 \cdot 6H_2O$ mixed in the necessary proportion unless (Ce + Al)/CO_3 (mole) = 2.5 value was reached. Thus, the obtained gel was dried at 60 °C for 22 h followed by calcination at 500 °C for 2 h and at 900 °C for 5 h resulting in a sample denoted below as (Ce + Al).

FeO_x was supported by incipient wetness impregnation of Ce/Al and (Ce + Al) by water solution of $Fe(NO_3)_3 \cdot 9H_2O$ (98.5%, Vekton, Saint-Petersburg, Russia) with necessary concentration with added citric acid (99.8%, Vekton, Saint-Petersburg, Russia) in 10 wt.% excess to the stoichiometric amount and ethyleneglycole (99.0%, Ecros, Saint-Petersburg, Russia), dried in air at 150 °C for 3 h and then calcined at 900 °C for 4 h. In the commonly used abbreviation n Fe/support, "n" corresponded to Fe weight % concentration in the sample.

3.2. Characterization

XRD patterns were recorded using D8 diffractometer (Bruker, Germany) with $CuK\alpha$ monochromatic radiation. Each sample was scanned in the range of 2θ from 10° to 70° with a step 0.05°. The surface composition of the samples was investigated by X-ray photoelectron spectroscopy (XPS) using spectrometer SPECS (SPECS, Germany) with Al $K\alpha$ irradiation ($h\nu$ = 1486.6 eV). The positions of the peaks of Au $4f_{7/2}$ (84.0 eV) and Cu $2p_{3/2}$

(932.67 eV) core levels were used for calibration of the binding energy (BE) scale. In Raman spectroscopy depth of light penetration depends on the wavelength of monochromatic radiation provided by different sources. We believe that green light (514.5 nm line of an Ar^+ laser with 2 mW power reaching the sample) provides information about the structure of preferentially near-surface layers of the samples, while data on the bulk composition are obtained with near-infrared (NIR) radiation at 1064 nm line (provided by an Nd-YAG laser with 100 mW power output) (Bruker Optik GmbH, Ettlingen, Germany). FT-Raman spectra (3700–100 cm^{-1}, 300 scans, resolution 4 cm^{-1}, 180° geometry) were collected using an RFS 100/S spectrometer (Bruker Optik GmbH, Ettlingen, Germany). Horiba Jobin Yvon T64000 spectrometer (HORIBA Scientific, Palaiseau, France) with micro-Raman setup and backscattering geometry for experimental spectra collection was used to measure the Raman spectra. The spectral resolution was not worse than 1.5 cm^{-1}. The detector was a silicon-based CCD matrix, cooled with liquid nitrogen. The band at 520.5 cm^{-1} of Si single crystal was used to calibrate the spectrometer.

The method of differential dissolution phase analysis (DDPA) [41] was used to reveal the composition and, in some cases, morphology and particle depth distribution of the compounds and phases (including X-Ray amorphous ones) formed in the samples. For this about 10 mg of the sample loaded in a quartz microreactor was dissolved in the flow (3.6 mL/min) of water-based solution with the composition changing from HCl (pH = 2) to 3M HCl (with continuous temperature increase from 20 °C to 90 °C), and finally to 3.6 M HF. Compounds dissolving in milder conditions (HCl (pH = 2) and 1–3M HCl) were reasonably supposed to form in the surface layers of the samples. Better crystallized structures, as a rule, dissolve in HF or even remain insoluble. Change of the outlet mixture composition in time was analyzed by ICP AES 262477-364A spectrometer (BAIRD, Zoeterwoude, The Netherlands) using the spectral lines at 238.2, 308.2, and 413.8 nm which are characteristic for Fe, Al and Ce, respectively.

3.3. Kinetic Measurements and Catalytic Tests

For $^{16}O/^{18}O$ exchange experiments, the sample was first heated to 800 °C in 0.58 vol %$^{16}O_2$ + He flow and kept at this temperature for 30 min. After, this gas mixture was replaced stepwise by the same one containing $^{18}O_2$ and Ar (1 vol.%) as an inert tracer. All responses were analyzed using QMS 200 gas analyzer (Stanford Research Systems, Sunnyvale, USA) as time variation of the ^{18}O atomic fraction in the gas phase $\alpha_g(t) = \frac{(^{16}O^{18}O + 2*^{18}O^{18}O)}{2*(^{16}O^{18}O + ^{18}O^{18}O + ^{16}O^{16}O)}$. Time dependencies of exchanged oxygen for different samples as related to the mass unit (N_O) were calculated using the formulae $N_o(t) = N_A * \frac{2*C_{O2}*U}{g} * \int_0^t (\alpha_g^{input} - \alpha_g)dt$, where α_g^{input}: isotope fraction in the inlet mixture (0.95), C_{O2}: inlet O_2 concentration (mol/mol), U: flow rate of the reaction mixture (mole/s), N_A: Avogadro number. Dynamics of oxygen adsorption/desorption at 800 °C was elucidated from the experiments on the stepwise replacement of He by 0.5 vol.% O_2 + 1 vol.% Ar + He mixture and vice versa flowing the reactor. (O_2 + He)/He switch was performed after at least 30 min sample stay in O_2 containing flow, while He/(O_2 + He) one followed in about 60 s. All kinetic measurements were performed with the sample (g = 0.025 g, particles of 250–500 μm in a size) loaded into a reactor (quartz tube, i.d. = 3 mm). The gas flow rates of all mixtures amounted to 16.7 cm^3/s.

The catalytic activity for samples with particles of 250–500 μm in a size was measured in a fixed-bed U-shaped reactor (3 mm i.d. quartz tube) at ambient pressure in the temperature range 700–900 °C. For NH_3 oxidation, mixture 1% NH_3 + 20% O_2 in N_2 was fed to the reactor charged by 0.015 g of the sample with a flow rate of 6.9 cm^3/s. Concentrations of NH_3 and NO_x (x = 0.5–2) in the outlet mixture were measured by infrared spectroscopy. For N_2O decomposition, a gas mixture of 0.15 vol% N_2O in He flowed the reactor charged with 0.038 g of the sample with a flow rate of 16.7 cm^3/s. In some experiments about 3 vol.% O_2 (+3 vol.% H_2O) were added to the inlet mixture. Outlet mixture composition was analyzed by gas chromatograph equipped with Porapack T (i.d. = 3mm, l = 3 m, for N_2O analysis) and NaX (i.d. = 3mm, l = 2 m, for N_2 analysis) columns. N_2O or NH_3

conversion ($X_{N2O(NH3)}$) and yield (Y_i) values calculated as $X_{N2O} = \frac{(C^o_{N_2O} - C_{N_2O})}{C^o_{N_2O}} * 100\%$, $X_{NH3} = \frac{(C^o_{NH_3} - C_{NH_3})}{C^o_{NH_3}} * 100\%$, and $Y_i = \frac{n*C_i}{C^o_i} * 100\%$, ($C_i$ and C^o_i: outlet and inlet concentrations of *i*th compound, n: number of N atoms in the *i*th molecule) were considered as a measure of samples activity.

4. Conclusions

FeOx (2.5–9.9 weight.% in terms of Fe) was supported by impregnation of mixed CeO_2-Al_2O_3 support prepared either by co-precipitation (Ce + Al) or CeO_2 precipitation onto Al_2O_3 (Ce/Al). Preferentially Fe-Ce-Al mixed oxides both in the surface layers and in the bulk were formed in Fe/(Ce + Al) samples, while the substantial spatial division of ceria and alumina-based compounds remains intact in Fe/Ce/Al samples. The stabilization of Ce^{3+} in Fe-Ce-Al mixed oxides was shown to inhibit oxygen mobility in the near-surface layers of Fe/(Ce + Al) samples. It results in retardation of the additional pathway of oxygen supply to the sites responsible for O_2 desorption and explains the lower activity of Fe/(Ce + Al) samples in N_2O decomposition.

Supplementary Materials: The following supporting information can be downloaded at: https://www.mdpi.com/article/10.3390/catal12090938/s1, Figure S1: XRD patterns of Ce/Al (A) and (Ce + Al) (B) based samples with different content of Fe, Figure S2: Dependence of N_2O conversion on the temperature for Ce/Al (A) and (Ce + Al) (B) based samples, Figure S3: Effect of O_2 and H_2O presence in the inlet flow on N_2O conversion over 3.8 Fe/Ce/Al and 2.5 Fe/(Ce + Al) samples. T = 800 °C, Figure S4: Temperature dependence of the conversion (A) yield of (NO + NO_2) and N2O (B) at NH_3 oxidation over different Fe/Ce-Al-O based samples. Stability tests for 9.9 Fe/Ce/Al sample (C) at 800 °C, Figure S5: Differential dissolution curves of Ce and Al (A) and identified compounds (B) at stepwise consecutive change of flow composition from HCl (pH = 2) to 1-3M HCl (1) and then to 3.6 M HF (2) over (Ce + Al) sample, Figure S6: Differential dissolution curves of Fe, Ce and Al (A) and identified compounds (B) at stepwise consecutive change of flow composition from HCl (pH = 2) to 1-3M HCl (1) and then to 3.6 M HF (2) over 2.5 Fe/(Ce + Al) sample, Figure S7: Differential dissolution curves of Fe, Ce and Al (A) and initially (B) and finally (C) identified Fe-Ce-Al compounds at stepwise consecutive change of flow composition from HCl (pH = 2) to 1–3M HCl (1) and then to 3.6 M HF (2) over 3.8 Fe/Ce/Al sample; Table S1: Quantity of Fe, Ce, Al (as prepared, dissolved and insoluble) in different samples (mmol/g). Quantities of insoluble compounds were calculated from balance equations.

Author Contributions: Conceptualization, L.P.; investigation, L.P., I.P., Y.C.; writing—original draft preparation, L.P.; visualization, L.P., I.P., Y.C; writing—review and editing L.P., V.A., supervision, L.P., V.A. All authors have read and agreed to the published version of the manuscript.

Funding: This work was supported by the Ministry of Science and Higher Education of the Russian Federation within the governmental order for Boreskov Institute of Catalysis (project AAAA-A21-121011390010-7), granted by the Government of the Russian Federation (075-15-2022-1132).

Acknowledgments: The authors appreciate the performing of DDPA measurements and discussion of the results by Dovlitova L.S.

Conflicts of Interest: The authors declare that they have no known competing interests or personal relationships that could have appeared to influence the work reported in this paper.

References

1. Groves, M.C.E.; Sasonow, A. Uhde EnviNOx® technology for NO_X and N_2O abatement: A contribution to reducing emissions from nitric acid plants. *J. Integr. Environ. Sci.* **2010**, *7*, 211–222. [CrossRef]
2. Pérez-Ramírez, J.; Kapteijn, F.; Schöffel, K.; Moulijn, J.A. Formation and control of N_2O in nitric acid production. Where do we stand today? *Appl. Catal. B Environ.* **2003**, *44*, 117–151. [CrossRef]
3. Sadykov, V.A.; Isupova, L.A.; Zolotarskii, I.A.; Bobrova, L.N.; Noskov, A.S.; Parmon, V.N.; Brushtein, E.A.; Telyatnikova, T.V.; Chernyshev, V.I.; Lunin, V.V. Oxide catalysts for ammonia oxidation in nitric acid production: Properties and perspectives. *Appl. Catal. A Chem.* **2000**, *204*, 59–87. [CrossRef]

4. Pinaeva, L.G.; Prosvirin, I.P.; Dovlitova, L.S.; Danilova, I.G.; Sadovskaya, E.M.; Isupova, L.A. MeO_x/Al_2O_3 and MeO_x/CeO_2 (Me = Fe, Co, Ni) catalysts for high temperature N_2O decomposition and NH_3 oxidation. *Catal. Sci. Technol.* **2016**, *6*, 2150–2161. [CrossRef]

5. Pinaeva, L.G.; Dovlitova, L.S.; Isupova, L.A. Monolithic FeO_x/Al_2O_3 Catalysts for Ammonia Oxidation and Nitrous Oxide Decomposition. *Kinet. Catal.* **2017**, *58*, 167–178. [CrossRef]

6. Giecko, G.; Borowiecki, T.; Gac, W.; Kruk, J. Fe_2O_3/Al_2O_3 catalysts for the N_2O decomposition in the nitric acid industry. *Catal. Today* **2008**, *137*, 403–409. [CrossRef]

7. Kruk, J.; Stołecki, K.; Michalska, K.; Konkol, M.; Kowalik, P. The influence of modifiers on the activity of Fe_2O_3 catalyst for high temperature N_2O decomposition (HT-deN_2O). *Catal. Today* **2012**, *191*, 125–128. [CrossRef]

8. Sádovská, G.; Tabor, E.; Bernauer, M.; Sazama, P.; Fíla, V.; Kmječ, T.; Kohout, J.; Závěta, K.; Tokarová, V.; Sobalík, Z. $FeOx/Al_2O_3$ catalysts for high-temperature decomposition of N_2O under conditions of NH_3 oxidation in nitric acid production. *Catal. Sci. Technol.* **2018**, *8*, 2841–2852. [CrossRef]

9. Imagawa, H.; Suda, A.; Yamamura, K.; Sun, S. Monodisperse CeO_2 Nanoparticles and Their Oxygen Storage and Release Properties. *J. Phys. Chem. C* **2011**, *115*, 1740–1745. [CrossRef]

10. Chen, L.; Fleming, P.; Morris, V.; Holmes, J.D.; Morris, M.A. Size-Related Lattice Parameter Changes and Surface Defects in Ceria Nanocrystals. *J. Phys. Chem. C* **2010**, *114*, 12909–12919. [CrossRef]

11. Iwanek, E.; Krawczyk, K.; Petryk, J.; Sobczak, J.W.; Kaszkur, Z. Direct nitrous oxide decomposition with $CoOx$-CeO_2 catalysts. *Appl. Catal. B Environ.* **2011**, *106*, 416–422. [CrossRef]

12. Grzybek, G.; Stelmachowski, P.; Gudyka, S.; Indyka, P.; Sojka, Z.; Guillén-Hurtado, N.; Rico-Pérez, V.; Bueno-López, A.; Kotarba, A. Strong dispersion effect of cobalt spinel active phase spread over ceria for catalytic N_2O decomposition: The role of the interface periphery. *Appl. Catal. B Environ.* **2016**, *180*, 622–629. [CrossRef]

13. Ferreira, A.P.; Zanchet, D.; Rinaldi, R.; Schuchardt, U.; Damyanova, S.; Bueno, J.M.C. Effect of the CeO_2 content on the surface and structural properties of CeO_2–Al_2O_3 mixed oxides prepared by sol–gel method. *Appl. Catal. A Chem.* **2010**, *388*, 45–56. [CrossRef]

14. Damyanova, S.; Perez, C.A.; Schmal, M.; Bueno, J.M.C. Characterization of ceria-coated alumina carrier. *Appl. Catal. A Chem.* **2002**, *234*, 271–282. [CrossRef]

15. Boullosa-Eiras, S.; Vanhaecke, E.; Zhao, T.; Chen, D.; Holmen, A. Raman spectroscopy and X-ray diffraction study of the phase transformation of ZrO_2–Al_2O_3 and CeO_2–Al_2O_3 nanocomposites. *Catal. Today* **2011**, *166*, 10–17. [CrossRef]

16. Ge, C.; Liu, L.; Liu, Z.; Yao, X.; Cao, Y.; Tang, C.; Gao, F.; Dong, L. Improving the dispersion of CeO_2 on γ-Al_2O_3 to enhance the catalytic performances of $CuO/CeO_2/\gamma$-Al_2O_3 catalysts for NO removal by CO. *Catal. Commun.* **2014**, *51*, 95–99. [CrossRef]

17. Laguna, O.H.; Romero Sarria, F.; Centeno, M.A.; Odriozola, J.A. Gold supported on metal-doped ceria catalysts (M = Zr, Zn and Fe) for the preferential oxidation of CO (PROX). *J. Catal.* **2010**, *276*, 360–370. [CrossRef]

18. Zhang, Z.; Han, D.; Wei, S.; Zhang, Y. Determination of active site densities and mechanisms for soot combustion with O_2 on Fe-doped CeO_2 mixed oxides. *J. Catal.* **2010**, *276*, 16–23. [CrossRef]

19. Li, K.; Wang, H.; Wei, Y.; Yan, D. Direct conversion of methane to synthesis gas using lattice oxygen of CeO_2–Fe_2O_3 complex oxides. *Chem. Eng. J.* **2010**, *156*, 512–518. [CrossRef]

20. Gupta, A.; Kumar, A.; Waghmare, U.V.; Hegde, M.S. Origin of activation of Lattice Oxygen and Synergistic Interaction in Bimetal-Ionic $Ce_{0.89}Fe_{0.1}Pd_{0.01}O_{2-\delta}$ Catalyst. *Chem. Mater.* **2009**, *21*, 4880–4891. [CrossRef]

21. Escribano, V.S.; Amores, J.M.G.; Finocchio, E.; Daturi, M.; Busca, G. Characterization of α-$(Fe,Al)_2O_3$ solid-solution powders. *J. Mater. Chem.* **1995**, *5*, 1943–1951. [CrossRef]

22. McBride, J.R.; Hass, K.C.; Poindexter, B.D.; Weber, W.H. Raman and x-ray studies of $Ce_{1-x}RE_xO_{2-y}$ where RE=La, Pr, Nd, Eu, Gd, and Tb. *J. Appl. Phys.* **1994**, *76*, 2435–2441. [CrossRef]

23. Kosacki, I.; Suzuki, T.; Anderson, H.U.; Colomban, P. Raman scattering and lattice defects in nanocrystalline CeO_2 thin films. *Solid State Ion.* **2002**, *149*, 99–105. [CrossRef]

24. Spanier, J.E.; Robinson, R.D.; Zhang, F.; Chan, S.W.; Herman, I.P. Size-dependent properties of CeO_{2-y} nanoparticles as studied by Raman scattering. *Phys. Rev. B* **2001**, *64*, 245407. [CrossRef]

25. Tsunekawa, S.; Ishikawa, K.; Li, Z.Q.; Kawazoe, Y.; Kasuya, Y. Origin of Anomalous Lattice Expansion in Oxide Nanoparticles. *Phys. Rev. Lett.* **2000**, *85*, 3440–3443. [CrossRef] [PubMed]

26. Ojha, A.K.; Ponnilavan, V.; Kannan, S. Structural, morphological and mechanical investigations of in situ synthesized c-CeO_2/α-Al_2O_3 composites. *Ceram. Int.* **2017**, *43*, 686–692. [CrossRef]

27. Porto, S.P.S.; Krishnan, R.S. Raman Effect of Corundum. *J. Chem. Phys.* **1967**, *47*, 1009–1012. [CrossRef]

28. Sudarsanama, P.; Malleshama, B.; Reddy, P.S.; Großmann, D.; Grünert, W.; Reddy, B.M. Nano-Au/CeO_2 catalysts for CO oxidation: Influence of dopants (Fe, La and Zr) on the physicochemical properties and catalytic activity. *Appl. Catal. B Environ.* **2014**, *144*, 900–908. [CrossRef]

29. Schilling, C.; Hofmann, A.; Hess, C.; Ganduglia-Pirovano, M.V. Raman Spectra of Polycrystalline CeO_2: A Density Functional Theory Study. *J. Phys. Chem. C* **2017**, *121*, 20834–20849. [CrossRef]

30. Reina, T.R.; Ivanova, S.; Centeno, M.A.; Odriozola, J.A. Boosting the activity of a $Au/CeO_2/Al_2O_3$ catalyst for the WGS reaction. *Catal. Today* **2015**, *253*, 149–154. [CrossRef]

31. Laguna, O.H.; Centeno, M.A.; Boutonnet, M.; Odriozola, J.A. Fe-doped ceria solids synthesized by the microemulsion method for CO oxidation reactions. *Appl. Catal. B Environ.* **2011**, *106*, 621–629. [CrossRef]

32. Perez-Alonso, F.J.; Melián-Cabrera, I.; López Granados, M.; Kapteijn, F.; Fierro, J.L.G. Synergy of $Fe_xCe_{1-x}O_2$ mixed oxides for N_2O decomposition. *J. Catal.* **2006**, *239*, 340–346. [CrossRef]

33. Suzuki, S.; Yanagihara, K.; Hirokawa, K. XPS study of oxides formed on the surface of high-purity iron exposed to air. *Surf. Interface Anal.* **2000**, *30*, 372–376. [CrossRef]

34. Grosvenor, A.P.; Kobe, B.A.; Biesinger, M.C.; McIntyre, N.S. Investigation of multiplet splitting of Fe 2p XPS spectra and bonding in iron compounds. *Surf. Interface Anal.* **2004**, *36*, 1564–1574. [CrossRef]

35. Atuchin, V.V.; Vinnik, D.A.; Gavrilova, T.A.; Gudkova, S.A.; Isaenko, L.I.; Jiang, X.; Pokrovsky, L.D.; Prosvirin, I.P.; Mashkovtseva, L.S.; Lin, Z. Flux Crystal Growth and the Electronic Structure of $BaFe_{12}O_{19}$ Hexaferrite. *J. Phys. Chem. C* **2016**, *120*, 5114–5123. [CrossRef]

36. Yamashita, T.; Hayes, P. Analysis of XPS spectra of Fe^{2+} and Fe^{3+} ions in oxide materials. *Appl. Surf. Sci.* **2008**, *254*, 2441–2449. [CrossRef]

37. Merte, L.R.; Gustafson, J.; Shipilin, M.; Zhang, C.; Lundgren, E. Redox behavior of iron at the surface of an $Fe_{0.01}Mg_{0.99}O(100)$ single crystal studied by ambient-pressure photoelectron spectroscopy. *Catal. Struct. React.* **2017**, *3*, 95–103.

38. Sayed, F.N.; Polshettiwar, V. Facile and sustainable synthesis of shaped iron oxide nanoparticles: Effect of iron precursor salts on the shapes of iron oxides. *Sci. Rep.* **2015**, *5*, 09733. [CrossRef]

39. Kapteijn, F.; Rodriguez-Mirasol, J.; Moulijn, J. Heterogeneous catalytic decomposition of nitrous oxide. *Appl. Catal. B Environ.* **1996**, *9*, 25–64. [CrossRef]

40. Ivanov, D.V.; Pinaeva, L.G.; Sadovskaya, E.M.; Isupova, L.A. Isotopic transient kinetic study of N_2O decomposition on $LaMnO_{3+\delta}$. *J. Mol. Catal. A Chem.* **2016**, *412*, 34–38. [CrossRef]

41. Malakhov, V.; Boldyreva, N.; Vlasov, A.; Dovlitova, L. Methodology and procedure of the stoichiographic analysis of solid inorganic substances and materials. *J. Analyt. Chem.* **2011**, *66*, 458–464. [CrossRef]

Review

Metallocavitins as Advanced Enzyme Mimics and Promising Chemical Catalysts

Albert A. Shteinman

Department of Kinetics and Catalysis, Institute of Problems of Chemical Physics of RAN, Prospect of Academic Semenov, 1, 142432 Chernogolovka, Russia; as237t@icp.ac.ru

Abstract: The supramolecular approach is becoming increasingly dominant in biomimetics and chemical catalysis due to the expansion of the enzyme active center idea, which now includes binding cavities (hydrophobic pockets), channels and canals for transporting substrates and products. For a long time, the mimetic strategy was mainly focused on the first coordination sphere of the metal ion. Understanding that a highly organized cavity-like enzymatic pocket plays a key role in the sophisticated functionality of enzymes and that the activity and selectivity of natural metalloenzymes are due to the effects of the second coordination sphere, created by the protein framework, opens up new perspectives in biomimetic chemistry and catalysis. There are two main goals of mimicking enzymatic catalysis: (1) scientific curiosity to gain insight into the mysterious nature of enzymes, and (2) practical tasks of mankind: to learn from nature and adopt from its many years of evolutionary experience. Understanding the chemistry within the enzyme nanocavity (confinement effect) requires the use of relatively simple model systems. The performance of the transition metal catalyst increases due to its retention in molecular nanocontainers (cavitins). Given the greater potential of chemical synthesis, it is hoped that these promising bioinspired catalysts will achieve catalytic efficiency and selectivity comparable to and even superior to the creations of nature. Now it is obvious that the cavity structure of molecular nanocontainers and the real possibility of modifying their cavities provide unlimited possibilities for simulating the active centers of metalloenzymes. This review will focus on how chemical reactivity is controlled in a well-defined cavitin nanospace. The author also intends to discuss advanced metal–cavitin catalysts related to the study of the main stages of artificial photosynthesis, including energy transfer and storage, water oxidation and proton reduction, as well as highlight the current challenges of activating small molecules, such as H_2O, CO_2, N_2, O_2, H_2, and CH_4.

Keywords: supramolecular chemistry; cavitins; biomimetics; metalloenzymes; metallocavitins; methane

Citation: Shteinman, A.A. Metallocavitins as Advanced Enzyme Mimics and Promising Chemical Catalysts. *Catalysts* **2023**, *13*, 415. https://doi.org/10.3390/catal13020415

Academic Editors: Gassan Hodaifa, Rafael Borja and Mha Albqmi

Received: 28 November 2022
Revised: 16 January 2023
Accepted: 17 January 2023
Published: 15 February 2023

1. Introduction

Biological catalysts–enzymes, usually demonstrate excellent selectivity and reactivity. Their mode of functioning is complex and far from completely understood. Metalloenzymes are ubiquitous and responsible for a wide range of challenging chemical transformations that proceed under mild conditions and with high chemo-, regio- and stereo-selectivity. Cavities and pores, being an integral feature of protein bodies in nature, are formed by folding and self-assembling of polypeptide helices through non-covalent and partially covalent interactions. In metalloenzymes they serve to accommodate the active sites for the delivery and bonding of substrates and the excretion of products (Figure 1a). The enzyme pocket–cavity in the protein body, plays a key role in the control of metal center nuclearity, substrate binding, substrate–catalyst–reactant pre-association, regio- and stereo-selectivity and substrate–product in/out exchanges [1]. It also provides a well-defined second co-ordination sphere for the activation and/or stabilization of intermediate reactive species and protects the metal center from undesired pathways. The local microenvironment of the metal center in the pocket differs substantially from the bulk solution. Cavities around the active sites of enzymes are of low symmetry and contain different chemical functionalities, such as recognition sites, catalytic groups and conformational switches. Chiral discrimination is one of the fundamental processes in enzymes. Functions of the metal complexes in the cavity are strongly dependent on its properties, including accessible spin states, oxidation potential and Lewis acidity. These properties are further fine-tuned by a well-defined first and second coordination sphere (Figure 1b).

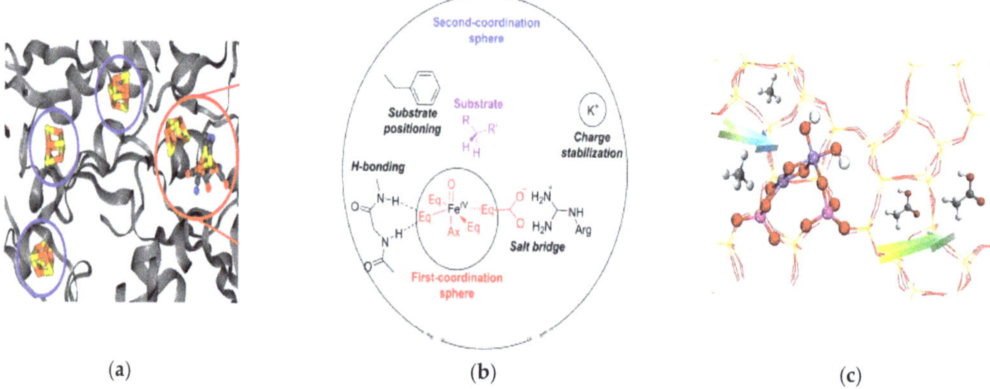

(a)   (b)   (c)

Figure 1. General performance mechanism of M-cavitin catalysts. (**a**) Crystal structure of [FeFe] hydrogenase (*Clostridium pasteurianum* CpI; PDB: 4XDC), showing [FeFe]-S(Cys)-[4Fe4S] active clusters included in its cavities. Adapted with permission from (Castner et al., 2021) [109]. Copyright 2021 American Chemical Society. (**b**) Cavity effects. Reprinted with permission from (Mukherjee et al., 2021) [29]. Copyright 2021 American Chemical Society. (**c**) Conversion of methane to CH_3COOH on $[Fe^{III}-(\mu O)_2-Fe^{III}]$-ZSM-5, the arrows show the transport of substrates in and products out. Reprinted with permission from (Wu et al., 2022) [72]. Copyright 2023 Elsevier.

The former is directly involved in metal coordination and usually consists of mixtures of different donor functionalities. In contrast, the second coordination sphere is not directly involved in metal binding but connected with the first by weak and reversible non-covalent interactions, such as hydrogen-bonding, electrostatic interactions, acid–base chemistry, van der Waals and hydrophobic forces. The second coordination sphere regulates the catalytic processes, proton or electron shuttling and substrate–product transport and determines the activity and selectivity of metalloenzymes. The pursuit of broadening our fundamental understanding of enzymatic catalysis has inspired scientists to develop and explore smaller synthetic complexes as enzyme mimics [2]. There are two main aims of mimicking enzyme catalysis: (1) curiosity, to gain insight into the nature of enzyme active sites and (2) practical tasks of mankind, to learn from nature and adopt from her long evolution experience. For a long time the traditional metalloenzyme modeling was mainly focused on the first coordination sphere of the metal ion and the second coordination sphere could be introduced directly only via the related chelate ligands of the first coordination sphere. In traditional homogeneous catalysis activity, the selectivity and stability of a transition metal catalyst was controlled by the ligands of the first coordination sphere. In biomimetic catalysis, a new direction for the research of advanced cavity-like models of metalloenzymes has gradually appeared and formed [3]. Many important aspects of enzymatic chemistry have been investigated by supramolecular chemistry, including molecular self-assembly, folding, molecular recognition, and host–guest chemistry on classical macrocyclic hosts provided by synthetic organic chemistry, such as cyclodextrins, crown ethers, cyclophanes, and calixarenes [4]. These gave rise to covalent cavitins [4,5]. On the other hand, supramolecular chemistry, built on weak and reversible non-covalent interactions, has emerged as a powerful and versatile strategy for the fabrication of coordination cavitins and has led to the creation of molecular containers or cages [5,6] and porous polymer molecules such as the metal-organic frameworks (MOF) and covalent-organic frameworks (COF) (Figure 2] [7,8].

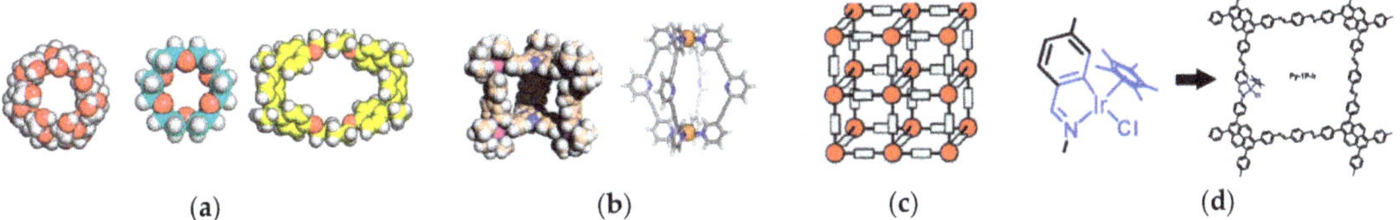

(a)   (b)   (c)   (d)

Figure 2. Diversity of cavitins. Discrete cavitins: (**a**) cyclodextrin, crown ether, cyclophane, and (**b**) metal-coordination cages. Extended cavitins: (**c**) MOF and (**d**) COF, the arrow symbolizes the incorporation of the M-complex into the COF cavity (Hu, J et al., 2021) [23].

The extended cavitins quickly gained recognition as industrial heterogeneous catalysis due to their outstanding characteristics: high-porosity and high-density catalytic metal centers, remarkable sorption properties, shape selectivity and easy syntheses in preparative quantities [9]. Metal catalysts can be encapsulated in various types of cavitins, providing the tools to control their activity and selectivity via the second coordination sphere. The supramolecular strategy has become more and more dominating in biomimetic chemistry over the last decade [10,11]. Due to their cavity-like structure and convenient modification, cavitins provide unlimited possibilities to mimic the active sites of natural enzymes.

2. The Diversity of Cavitins

Microporous compounds, such as charcoal or zeolites [12], have been used for a long time as carriers for metal ions in heterogeneous catalysts because they greatly increase the performance of the encapsulated transition metal. Cavity macrocycles, such as cyclodextrins, crown ethers, cyclophanes and calixarenes, have been studied as host molecules in the field of molecular recognition, which is key for the high catalytic efficiency and selectivity of natural enzymes [13]. To mimic the cavity and pores of natural enzymes, a number of polycycle molecular containers based on covalent bonding, such as carcerands, hemicarcerands, cryptophans, capsules and cages, have been prepared during the last decade. Among the classes of covalent cavitins the most popular are a derivatives of the cyclotriveratrylene (resorcinarene) [14] as well as cavitins formed by bonding resorcinarene units [15]. In the ocean of covalent and non-covalent cavitins there are two big classes: discrete individual molecule monocavitins (Figure 2a,b), such as cyclodextrins, calixarenes, cryptands, cucurbiturils, metal–organic cages (MOC), covalent organic cages (COC), helicates [16], and many others; and extended ones, polycavitins (Figure 2c,d), such as zeolites, MOF, COF, porous polymers, porous molecular crystals [17], hollow [18] and dynamic [19] MOF, and others which all together demonstrate a rich library of architectures varying in shape, size and geometry. H-bonded capsules based on resorcinarene units (Figure 3a) and H-organic frameworks (HOF) are evolving into novel and important classes of cavitins. Metal-COF (MCOF) are also emerging as a bridge between MOFs and COFs via integrating metal active sites into COFs (see Figure 2d) [20]. In recent years, there has been a growing interest in more exotic classes of cavitins, such as porous liquids and metal foams. A porous liquid (PL) is a liquid that combines the cavity of porous solids with the fluidity of liquids. The permanent pores endow PL unique physicochemical properties, interesting for catalysis, particularly for photocatalysis. A metal foam is a cellular structure consisting of a solid metal with pores comprising a large portion of its volume. They are considered as promising catalyst carriers due to their high porosity, large specific surface area, and satisfactory thermal and mechanical stability [21]. It is generally assumed that monocavins are more suitable for the modeling and academic study of enzyme active sites, while polycavitins are used for the fabrication of advanced heterogeneous catalysts. However, in the recent years the appearance of MOCs in catalysis has also increased. Using self-assembly and incorporating different functional groups, complex supramolecular hosts with diverse shapes, sizes and chemical environments of the cavity have been easily designed from relatively simple components [15,21].

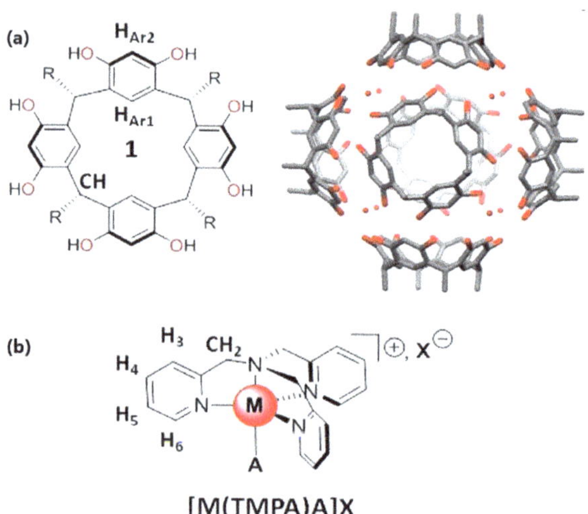

Figure 3. Encapsulation of a metallocomplex (**b**) in capsule (**a**). (**a**) The resorcin[4]arene **1** (R=C$_{11}$H$_{23}$) and structure of the H-bonded capsule. Reprinted with permission from (Zhang et al., 2021) [20]. Copyright 2021 Elsevier.

2.1. Metallocavitins

MOFs as well as MOCs are formed by coordination-driven self-assembly. They are composed of polydentate organic linkers and inorganic nodes containing metal ions or clusters known as secondary binding units (SBU). The metal in the coordination cavitins may serve not only in constructive goals but also as a coordinatively unsaturated catalytic center. Other approaches for the incorporation of metal active sites into cavitins include metallolinkers, non-covalent encapsulation of metal complexes and enzymes, templated metal–ligand assemblies and post synthetic metallation [3]. The most popular Schiff base COF possesses uniformly distributed imine linkages, which revealed a new metal binding mode. The imine linkage in COFs is the most tunable bond among all the currently employed reversible COF linkages and readily chelating transition metal via cyclometalation [22]. For example, the iridacycle B-decorated COF (Figure 2d) exhibited more than 10-fold efficiency than its molecular analog in photocatalytic hydrogen evolution from aqueous formate solution under mild conditions [23]. The robust porosity, stability, and chemical functionality of COF can be controlled by the reasonable selection of organic building blocks. Chirality is associated with the origin of life on Earth and plays a great role in the functioning of metalloenzymes. COFs are shown to be capable of inducing chiral molecular catalysts from non-enantioselective to highly enantioselective in organic reactions [24]. On the other hand, a method to synthesize chiral MOFs from achiral precursors by modifying the substituents utilizing chiral fragments was reported recently [25]. While the robust porosity of MOFs and COFs renders them as promising heterogeneous catalysts, they suffer from diffusion problems in mass transportation. Due to this, approach for the polymerization of soluble MOCs [26,27], and the preparation of semi-heterogeneous metal–enzyme-integrated catalysts using soluble porous imine molecular cages [23] were developed.

2.2. Design and Characterization

Design strategies have employed subcomponent self-assembly via the simultaneous formation of dynamic coordinative (N→metal), covalent (N=C), and other bonds. A key facet of metallo-supramolecular self-assembly is predicting the products of self-assembly based on constituent metal ion geometry and ligand conformation [28]. Binding selectivity, created by weak non-covalent interactions between the hosts and guests, is influenced by the size, shape and flexibility of cavitin [29]. A general design problem is that the linker units and SBUs should provide a unique flexibility/rigidity balance and directionality for the ligands to achieve the desired geometry and optimal host–guest interactions [30]. The right balance between the flexibility and rigidity of cavitin is favorable for binding substrates and releasing products [31]. Shape-persistent organic cavitins permit the precise control of their size, geometry, and the presence of functional groups in the interior of their cavities [32]. Adaptability is a hallmark of enzymes and flexible cavitins can mimics this via structural changes that accompany adsorption and desorption steps [33]. Host flexibility can greatly affect the cavity size and shape and lead to behaviors analogous to the induced fit of substrates within the active sites of enzymes. A little structural flexibility is inherent to some "rigid" metal–organic hosts, but torsional twisting of trigonal prismatic cages leads to a dramatic change in cavity size [34]. With advances in single-crystal X-ray diffraction and economic methods of computational structure optimization, cavity sizes can be readily determined. Practically very useful, simple rules, such as Rebek's 55% rule [35], fail to take into account structural flexibility that can allow hosts to significantly adapt their internal cavity [34]. Computational analysis offers a potential route to quantitatively examine the flexibility of metalorganic assemblies and may be used in the design of cavitins [36]. For example, a computational screening method able to predict new cavitins [37]. The Toolkit cgbind facilitates the characterization and prediction of functional metallocages [38]. A tight binding chemical method (GFN-xTB) has been developed specifically for geometry optimization in large molecular systems [39]. The volume calculations on empty cages and prospective guests with the online utility Voss Volume Voxelator confirmed that $Fe_4(Zn-L)_6$ has the appropriate size to accommodate a hydroformylation catalyst [39]. To explain the catalytic activity of the two dipalladium(II) cages the molecular dynamics simulation was explored for the evaluation of their conformational flexibility [38]. Hydrophobic MOFs have unique advantages as catalysts for various reactions: the hydrophobicity is beneficial for substrates to access the active sites and can improve the water stability of the MOF, they are also able to achieve spontaneous separation from the hydrophilic new products, thus improving selectivity [40]. Reversible bond formation is one of the prime prerequisites for the crystallization of cavitins. A general procedure to grow large single crystals of three-dimensional imine-based COFs was developed [41] using the principles of dynamic covalent chemistry [42,43]. In the design of complex cavitins it is necessary to take into account the balance between reversibility and robustness of the connecting bonds for enhanced crystal growth [44]. Many useful MOFs with enhanced catalytic performance possess varying degrees of chemical instability hampering their practical applications. The MOF/parylene-N hybrid not only imparts the chemical stability of an MOF without obviously impacting their inherent nature, but also broadens the scope of this catalysts in different aqueous environments [45]. A novel strategy for the synthesis of a highly crystalline and porous cyanurate-linked COF (CN-COF) by dynamic nucleophilic aromatic substitution was reported recently [46]. CN-COFs contain flexible backbones that exhibit unique AA′-stacking due to the interlayer H-bond interactions, exhibiting good stability. The complexity of cavitins has increased dramatically over the years. Heteroleptic, mixed-metal, hybrid and low symmetry assemblies are becoming more commonplace [47,48]. Improvements within SCXRD and the advancement of computational power allow the rapid and in-depth analysis of these systems [34]. Polyoxometalates (POMs) exhibit unique chemical properties that make them very attractive as catalysis. The ring-shaped lacunary POM comprises inorganic cavitins containing a large cavity useful for accumulating metal-cations. Recently an original approach was developed for the selective synthesis of multinuclear Cu-containing ring-shaped POM Cu4-Cu16 by the stepwise addition of four of copper(II) acetate equivalents (Figure 4) [49]. Insertion of POM into an MOF opens up new opportunities in heterogeneous catalysis [50].

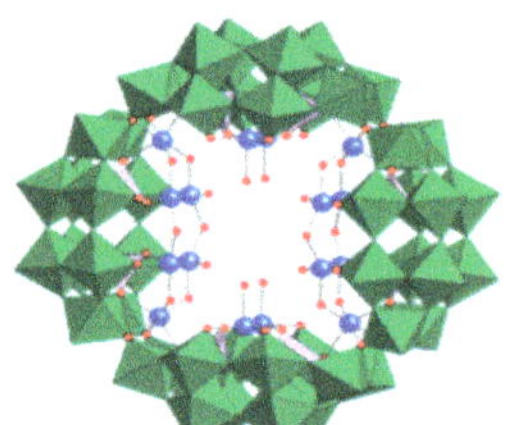

Figure 4. An example of an inorganic cavitin: multinuclear Cu-containing ring-shaped POM Cu16. Adapted with permission from (Koizumi et al., 2022) [49]. Copyright 2022 Wiley.

Perovkskites are another purely inorganic cavitin, which has recently received great attention as catalysts [51] due to their large surface area, low density, and high loading capacity. A typical perovskite oxide has the general formula ABO_3, in which A is a lanthanium or an alkaline earth metal and B is a transition metal. The enthusiasm for perovskite oxides is that they show a highly flexible elemental composition, with a large variation in properties that can be tailored by doping design. Progress in metallo-supramolecular chemistry has created the potential to synthesize metallocavitins with more than one function within the same assembly [47]. The inspiration has come from enzymes that congregate, for example, a substrate recognition site, an allosteric regulator element and a reaction center. The formation of heteroternary cucurbit[8]uril-viologen–naphthol complexes led to bifunctional photoredox catalysts for hydrogen generation [52]. Novel cubic cages with a different polarity to the peripheral environment surrounding the cage encapsulating catalytically active cobalt(II) meso-tetra(4-pyridyl) porphyrin were synthesized for study of polarity effects in cyclopropanation reactions [53]. The work [54] opens new perspectives for the synthesis of a more diverse library of coordination nanocages with innovative structures, metal ion composition and functionality.

3. Cavity Effects

Enzymes are a source of imagination and inspiration for chemists. They incorporate multiple functionalities in their substrate-binding cavities, in order to achieve high selectivity and activities. A good example of the cavity effect on a catalytically inactive binuclear iron complex was demonstrated in a paper [55] devoted to [FeFe]hydrogenase modeling. The complex $Fe_2[\mu\text{-}(SCH_2)_2NH](CN)_2(CO)_4{}^{2-}$ was shown to integrate into the inactive apo-form of [FeFe]-hydrogenases to yield a fully active enzyme. The cavities perform substrate encapsulation, molecular transformation, intermediate capturing, and product release, which facilitated the catalytic cycle. Cavity effects are based on entropy effects, cage-wall effects, absorption, desorption, shape- and size-selectivity [56] and include second sphere and hydrogen-bonding interactions, salt-bridges, and long-range allosteric effects from bound cations and anions (Figure 1b) and, at last, from component interactions. Basic atomic-molecular properties drastically change upon confinement within the catalyst framework, leading to effects such as increased excitation energy and lower polarizability, which can be explained using the "particle in the box" as a simplified model [57]. For larger molecules, spectroscopic evidence for the intrinsic decrease in the p-p* gap in aromatic hydrocarbons, such as naphthalene and anthracene, has been shown [57]. More recent studies, however, revealed that the pure effect of confinement rather leads to an increase in the HOMO–LUMO gap and point toward the specific properties of electrostatic stabilization within the active catalytic site [57]. The transition metal in the majority of enzymes is the center of their activity, but the cavity itself can perform some activation of the substrate, isolation of metal complexes to prevent aggregation or decomposition and increase the local concentration of catalyst and reaction partners. For example, the intermediary-sized cucurbit[7]uril in aqueous solution selects and effectively accelerates (in 4×10^5) the endo dimerization of cyclopentadiene. DFT calculations suggest that catalysis is due to an entropy dominated transition-state stabilization in the tightly packed ternary reaction complex [58]. In another example, the rigid, spherical cavity in spherical cavitin quantitatively encapsulates azobenzene and stilbene derivatives with 100% cis-selectivity in water and their cis-azo isomerization is suppressed due to the confinement effect [59]. The Narazov cyclization, which needs an acidic media, in the metallocage $[Ga_4L_6]^{12-}$ can proceed at pH 8. In this case, an acceleration comparable to some enzymes is observed, which is caused by the preorganization of the encapsulated substrate and stabilization of the transition state. The experimental results and quantum chemical calculations reveal that a $Ga_4L_6{}^{12-}$ cage accelerates the cyclization reactions of pentadienyl alcohols because of an increase in the basicity of the complexed alcohol [60,61]. The design of new catalytically effective cavitin is limited because of our poor understanding of cavity effects in depth. A simple and effective DFT protocol was suggested, which takes into account both the thermodynamic and kinetic aspects of catalysis permitting the elucidation of many effects on the molecular level [36]. The protection of the reactive functional groups favors reaction at the unprotected sites. Cavity effects alter the typical reaction pathway, switching reactions on and off [62], operating substrate selection based on size and shape, leading to unusual selectivity or inducing stereoselectivity through asymmetric scaffolds. Most importantly for the reaction rate are the proximity and orientation of the substrates and transition-state stabilization. The concepts of 'confinement effect' and 'second sphere effects', often used in modeling enzymes, are not clear enough and are often unjustifiably substituted or identified. In the active center of the enzyme there are a number of effects that affect the catalyzed reaction: purely geometric, such as a limited space, size and shape of the cavity, physicochemical, associated with the microenvironment, chemical, such as covalent and non-covalent bonds of the 'second sphere' functional groups with the reaction participants, long-term effects such as electrostatic [63], allosteric and the effects of inter-component interactions with other enzymes. As a result of these effects the confined molecules can fundamentally change their chemical and physical properties compared to those in bulk solution. Cavity effects (CE) are directed to the preorganization of catalysts and reactants, and confinement effects, usually identified with the second sphere, affect the reaction itself. Significant progress in our understanding of CE in cavitins was reached due to single molecule fluorescence microscopy imaging (SMS) over the last decade [64]. This technique was developed to directly monitor the behaviors of individual molecules in a confined space, thus enabling spatial and temporal visualization and a better understanding of molecular dynamics. Additionally, it was also observed that confinement induced reactivity change, on and off switching of reactions, substrate selection, stereoselectivity, regioselectivity, and product distribution variation.

3.1. Isolation from Bulky Solvent, Selective Incorporation and Stabilization of the M-Complex and Reactants

The separation of the complex from the bulk solvent and the control of the in/out exchange is reminiscent of the roles of the protein backbone in metalloenzymes. The cavity can change the structure of the active site and prevent the catalyst from decomposing. The neutral complex Ru(II) (Figure 5) was encapsulated inside a self-assembled hexameric host similar to (a) (Figure 3, [20]). Different spectral data and molecular dynamics simulations support the inclusion and motions of the complex inside the capsule. The embedded complex was assessed by the $NaIO_4$ catalytic oxidation arylmethyl alcohols into aldehydes, which is dependent on the substrates' size in the order benzyl > 4-phenyl-benzyl > 9-anthracenemethanol [65]. No discrimination between the substrates was observed in the absence of the cavitin.

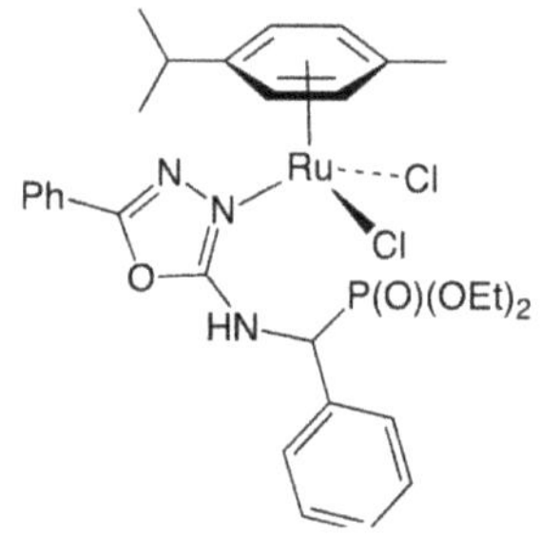

Figure 5. Neutral ruthenium(II) complex as a catalyst of arylmethyl alcohol oxidation (Hkiri S. et al., 2022) [65].

Usually cavitins control the nuclearity of the M-complex. For example, the encapsulation of metal complex (b) in monomer form has been demonstrated in a dynamic H-bonded capsule (a) (Figure 3) [20]. The position of the M-complex inside this capsule can be derived from NMR analysis and confirmed by docking simulations [19]. Assembling phenothiazine into a cavitin improves its photocatalytic performance and stability due to less aggregation inducing its quenching and also due to preorganization of the electron donor–acceptor complex within the cage [66]. Au$_{25}$ nanocluster encapsulated into MOF loses surface ligands and exhibits superior activity and stability in the oxidative esterification of furfural [67]. Enzymes bind reactants within their pockets and reduce the distance between the M-center and the substrates. Inspired by enzymatic behavior, cavities can be engineered to co-encapsulate metal catalysts with substrates in such a way that after M-complex encapsulation, some space remains vacant for the co-encapsulation of substrates. Substrates of appropriate size and shape react with the confined M-complex due to the thermodynamically favorable host–guest binding process of molecular recognition based on complementary physicochemical characteristics, which allows acquisition and orientation. Substrate selectivity is difficult to rationalize for small molecules, such as H$_2$, O$_2$, CO$_2$, and CH$_4$, that possess a range of physical characteristics too narrow to allow either precise positioning or discrimination between the reactants. Nevertheless, metalloenzymes have evolved to metabolize these substrates with high selectivity and efficiency due to small-molecule tunnels and gate-effects, for example, for the selective oxidation of methane [68]. Substrate may adopt a high-energy conformation that is structurally similar to the transition state, thus leading to a lowered activation energy. Conformational changes during substrate binding frequently appear in enzymes. Cavitins help to investigate this effect. The preferred conformation and orientation of the bound guest determines the molecular behavior in the cavity. The conformation is dependent upon the intrinsic encapsulation capability of the hosts. For example, for long chain alkanes or fatty acids, bent binding motifs are often observed when they are sequestrated. This can result in the close proximity of terminal reactive functional groups through enforced orientation of the guest. Guest encapsulation induces an entropic penalty that is often compensated for by favorable entropic and enthalpic gains from desolvation of the guest and the liberation of high-energy solvent molecules from the binding site [56]. This also involves the shielding of specific reactive groups by supramolecular encapsulation. For example, the electrophilic α-carbon on a [PhN$_2$]$^+$ ion can be selectively deactivated upon host–guest complexation with cucurbit[7]uril in aqueous media, achieving a 60-fold increase in the half-life of the carbocation. However, the electrophilic nitrogen of the encapsulated diazonium ion remains active towards diazo coupling with strong nucleophiles in water [69]. Electrostatic contributions are known as primary factors in enzyme catalysis. However, for a long time there were no models to study this mechanism. Positively charged hosts are able to attract negatively charged guests (and vice versa) and guest binding affinity, driven by electrostatic interactions, which can be modulated by different solvents [63]. Remarkable examples of reactive guest stabilization by confinement in synthetic molecular containers has been reported, involving the stabilization and detection of reaction intermediates through the formation of thermodynamically stable and covalent host–guest complexes with functionalized resorcin[4]arene cavitins [70]. A discrete nanocage of core−shell design with a hydrophilic interior and a hydrophobic exterior (Figure 6a) was able stabilize metal complexes (Figure 6b) with an uncommon oxidation state in organic solvents [71].

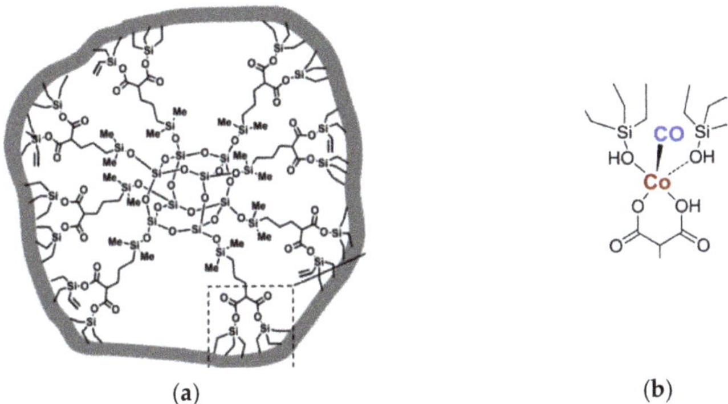

(a)   (b)

Figure 6. Stabilizing complex (**b**) with Co(I) oxidation state in a discrete nanocage (**a**) of core−shell design (Shen et al., 2014) [71].

Steric groups in the second coordination sphere define a corridor for the approach of substrates into the active sites. For instance, iron picket-fence porphyrin complexes have bulky amide substituents positioned in a single facial orientation, thereby constructing a cavity for small molecules, such as dioxygen [29]. The cavity effects on the catalytic reaction dynamics under variable nanopore morphologies, including pore length and diameter at the single-molecule level, were studied and were found to be dependent on the nanopore morphology [64].

3.2. Preorganization of the M-Catalyst and Reagents, Mutual Orientation and Shaping the Reaction Start Complex

The conversion of methane to acetic acid on Fe-ZSM-5 with ultrahigh selectivity has been attributed to the preorganization of the M-catalyst and reagents, the direct coupling of intermediate methyl radicals ($\bullet CH_3$) and the adsorbed CO* and OH* species on Fe site to form CH_3COOH (Figure 1c) [72]. Combining NMR analyses and molecular modelling showed significant differences in shape between the different complexes derived from α-, β- or γ-CD (Figure 7) [73].

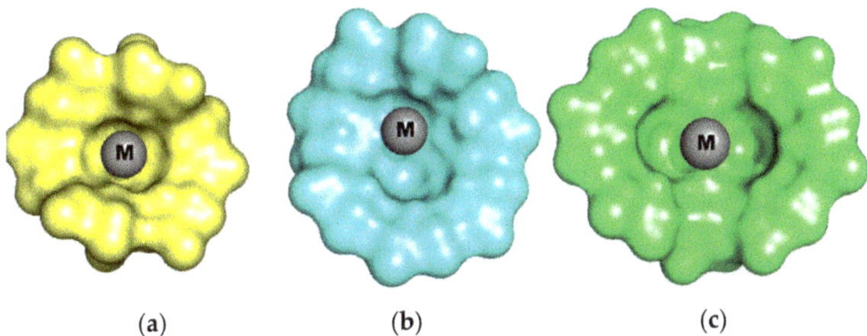

(a)   (b)   (c)

Figure 7. 3D structures of hybrid cyclodextrin–imidazolium M-cavitins: (**a**) α-ICD, (**b**) β-ICD and (**c**) γ-ICD. Reprinted with permission from (Roland et al., 2018) [74]. Copyright 2018 Wiley.

In the case of α-ICD a helical shape is apparent, when in the case of γ-ICD a symmetrical cavity shape is revealed [74]. The preorganization of a substrate in a higher energy conformation can accelerate the reaction and promote reactivity. Nanoenvironments allow (or enforce) the preorganization of substrates through conformational restrictions. Increasing the local concentrations, the mutual convergence and orientation of the catalyst and substrate enables the formation of the start reaction complex. For example, self-assembled nanospheres bearing guanidinium binding sites (Figure 8) bind sulfonate-functionalized ruthenium catalysts increasing the proximity of incoming water to the catalyst. This preorganization increases the reaction rate for electrochemical water oxidation in two-orders of magnitude comparable to the homogeneous system.

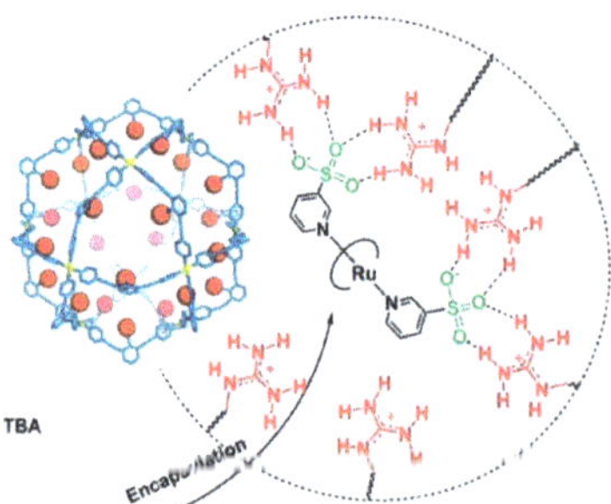

Figure 8. Ru-guanidinium nanosphere with incorporated ruthenium complex. Adapted with permission from (Yu et al., 2018) [90]. Copyright 2018 Wiley.

3.3. Transition State and Intermediate Stabilization

In cavitins the transition state of the target reaction can be stabilized more efficiently in comparison with bulk solutions. The Ir complex was incorporated into Zr-MOC-NH$_2$ with the formation Ir^{III}-MOC-NH$_2$. DFT calculations, mass spectrometry and in situ IR showed that the Ir(III) complex is the catalytic center, and −NH$_2$ in the cavity plays a synergistic role in the stabilization of the transition state and Ir·CO$_2$ intermediate [75]. The transition state or intermediate stabilization can not only lower the energy and enthalpy barrier of the reaction but can also alter the reaction mechanisms. The increased local concentration of reagents in the hydrophobic cage in the cobalt-catalyzed cyclopropanation of styrene, which involves radical intermediates and some shielding, reduces the number of unwanted side reactions of reactive radical intermediates and substantially improves performance [76]. The hexameric resorcinarene capsule (Figure 3a) can host trityl carbocation, which catalyzes Diels–Alder reactions between dienes and unsaturated aldehydes. The capsule promotes the formation of trityl carbocation from trityl chloride via the cleavage of the C-X bond promoted by OH/X H-bonding [77]. The labile imine and hemiaminal intermediates in the transformation of aldehydes to imines can be stabilized in water by hydrophobic cavitin containing a primary amine groups anchored in its cavity (Figure 9). The reaction favors the release of water from the hydrophobic microenvironment [78].

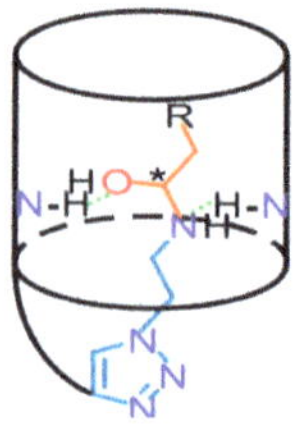

Figure 9. Stabilization of labile imine and hemiaminal intermediates through hydrogen bonding and hydrophobic effects by an endo-functionalized cavitin (Li et al., 2022) [78].

3.4. Second Sphere and Allosteric Effects

The interfacial interactions between substrate molecules and the cavity, such as the second sphere effect, can not only modulate the first sphere effect but also change the mass transport and adsorption–desorption equilibrium, thus significantly influencing the catalytic reaction activities and selectivities. For example, the electrostatic effects lower the activation energy of a reaction, and result in a rather large rate acceleration. Controlling the reactivity with the presence of acid–base residues and H-bonds is also very significant [20]. Among the eight water molecules embedded in the structure (Figure 3a), four feature a hydrogen atom hanging inside the cavity, which makes them good hydrogen-bond donors to azido ligands of $[Cu(TMPA)N_3]ClO_4$ serving as a reference probe of the second coordination sphere using IR spectroscopy. The presence of this hydrogen bonding was confirmed with docking experiments. The allosteric effects of central and peripheral interactions were extensively investigated. The accumulation of ions or neutral molecules at the periphery of a cavitin can results in a higher local concentration of substates inside of cavity. These peripheral interactions may be non-covalent or covalent in nature. The concentration of externally bound species decreases host flexibility and thus the guest exchange rate. The peripheral cage substituents control the activity of a caged cobalt-porphyrin-catalyst in cyclopropanation reactions. It was demonstrated that cage catalysts with non-polar external groups provided a higher activity compared to the free bulk catalyst and cages with no or polar exo-functionalization [76]. In some M-enzymes the substrate must diffuse through tunnel residues before binding to the active site. For example, the structure of cytochrome P450, which consists of a long hydrophobic tunnel, regulates substrate access and product release. The authors [76] declared that the cage serves as a mimic of the active site pocket of an enzyme whereas the periphery of the cage provides a synthetic equivalent of the substrate binding site tunnel. Mechanistic investigations into the role of the secondary coordination sphere and beyond on multi-electron electrocatalytic reactions showed that the introduction of additional interactions through the secondary coordination sphere beyond the active site, such as hydrogen-bonding or electrostatic interactions, also enables faster chemical steps in addition to its effects on the rate-limiting steps, examined earlier [79]. The cavity containing the M-complex confined in an asymmetric environment permits enantioselective catalytic reactions.

3.5. Changing the Reaction Course and Mechanism

The confinement can change the course of a reaction. The reactivity of the catalyst $[Re(^{C12}Anth-py_2)(CO)_3Br]$ was modulated by its encapsulation into a COF. The M-cavitin catalyzed either reductive etherification, oxidative esterification, or transfer hydrogenation depending on the local environment in the COF [80]. In conditions of alkyne hydration by NHC–Au complex the product is formed from intramolecular cyclisation induced by the confinement of the metal. Variations in product distribution were observed with (ICyD)AuCl complexes. A gold-catalyzed enyne cycloisomerization with an α-ICyD ligand gave a cyclopentenic product, while β-ICyD led a to six-membered ring. The outcome of the reaction depends on the conformation of the carbenic intermediate (Figure 10): inside the cavity of the α-ICyD conformation a was restricted while b fits better into β-ICyD [74]. Authors of this work have showed also that changing the size and shape of the cavity also changes the mechanism of reaction.

Figure 10. Stabilizing the conformation of a reactive intermediate, dependent on CD shape: (**a**) for α-CD and (**b**) for β-CD, determining the product distribution in a gold-catalyzed enyne cycloisomerization. Reprinted with permission from (Roland et al., 2018) [74]. Copyright 2018 Wiley.

3.6. Regioselectivity, Stereoselectivity and Product Selectivity

The chemical reaction rate and product selectivity are often dependent on the morphological properties of the cavity in both enzymes and cavitins. The size and shape of the cavity determines the selectivity. The use of ICyD ligands in the copper-catalyzed hydroboration of alkynes leads to an inversion of the regioselectivity that is controlled by a parallel or anti-parallel approach of the alkyne dependent on CD. While the β-ICyD ligand gives branched products, the smaller α-ICyD ligand gives linear vinyl boronates. A chiral second coordination sphere of the M-complex, confined in an asymmetric environment, permits enantioselective catalytic reactions. Such a scenario was demonstrated in enyne cycloisomerization for NHC-capped CD gold complexes (ICyD)AuCl encapsulated in CD in monomeric form. The stereoselectivity depends on the nature of the cyclodextrin and shows good yields and ee values (up to 80%). The selectivities were rationalized using the shapes of the cavities determined by NMR and modelling. While γ-ICyD does not afford enantioselecivity because of its symmetrical shape, α-ICyD and β-ICyD give the enantiomer for which the approach is the easiest according to their helical shape [73]. COFs are capable of inducing chiral molecular catalysts via preferential secondary interactions between the substrate and the framework that induce enantioselectivities not achievable in homogeneous systems. They catalyze the asymmetric acetalization of aromatic aldehydes and 2-aminobenzamide to generate products with up to 93% yield and 97% ee [24]. The highly selective synthesis of terpene compounds was demonstrated by the controlled conversion of (+)-limonene to terpinolene by kinetic suppression of overisomerization in a confined space of a porous metal–macrocycle framework in stark contrast to the acid-catalyzed reaction in bulk solution, which generally gives a complex mixture of thermodynamically favored isomers [81]. The microenvironment of the $\{Co_4^{II}O_4\}$ in the $Co_4@Ru_x$-Eu-MOF plays an important role in improving the performance and selectivity of CO_2 photoreduction and water splitting in syngas production. The H_2 and CO total yield can be improved by up to 2500 $\mu mol \cdot g^{-1}$ with the ratio of CO:H_2 ranging from 1:1 to 1:2 via changing the photosensitizer content in the confined space [82].

Thus, confinement of a metal complex is a promising way to induce their reactivity modulation and improve selectivity. Many of these effects are familiar from enzymology but need be studied in chemical models in order to provide deeper insight into the mechanisms of enzymatic efficiency and selectivity.

4. M-Cavitins as Advanced Chemical Models of Enzymes

Although considerable progress in study of enzyme catalysis has been realized by directly monitoring the catalytic processes of natural enzymes, the relationships between the supramolecular structures and the functionality of enzymes are still obscure. Based on their dynamic nature, supramolecular enzyme models with complex and hierarchical architectures have attracted considerable attention in the research area of mimicking the particular features of natural enzymes [83]. The design and development of enzyme mimics with supramolecular structures can help unravel the mystery features of enzyme catalysis. The earlier attempts to adequately model enzyme AC included metal complexes covalently attached to cyclodextrins [84]. Later metal complexes of porphyrin, salen and others were encapsulated within molecular cages [53]. In an effort to mimic the structure and functions of metalloenzymes discrete coordination and covalent metallocavitins were designed and synthesized either by embedding the active sites in the structures of the cage [85] or by the encapsulation of catalysts within the cage [53]. Being comparable to the AS of enzymes, monocavitins have intrinsic advantages as enzyme mimics. Their solubility in organic solvents permits their study in homogeneous systems [53]. This also facilitates the growing of single crystals that make it possible to reliably control their structure and functionality at the atomic level by using single-crystal X-ray diffraction (SCXRD) [30]. Many spectroscopic techniques, such as NMR, UV–Vis, and fluorescence spectroscopy, have been widely applied to monitor reactions catalyzed by monocavitins.

4.1. Stereoselectivity

Stereoselective binding is an essential feature for enzymatic catalysis. A variety of chiral cavitins have been constructed via covalent bonding or coordination assembly [86]. Over the past years, C3 symmetrical cages have emerged as an interesting class of supramolecular hosts that have been reported as efficient scaffolds for chirality dynamics: generation, control, and transfer [87]. Artificial metalloenzymes having a synthetic metal complex in its protein scaffold, selectively catalyze non-natural reactions and reactions inspired by nature in water under mild conditions. For example, the biotin-binding cavity of streptavidin can accommodate small coordination compounds to form the artificial enzyme for the enantioselective oxidation of prochiral sulfides, enhancing the activity and selectivity up to 93% ee for the sulfoxidation of methyl-2-naphthylsulfide in the presence of TBHP compared to the protein-free salt (Figure 11) [13]. Also the highest activity (82% ee with TON 2613) for the enantioselective dihydroxylation and epoxidation of styrene derivatives was obtained using a Ru complex linked with bovine serum albumin [62].

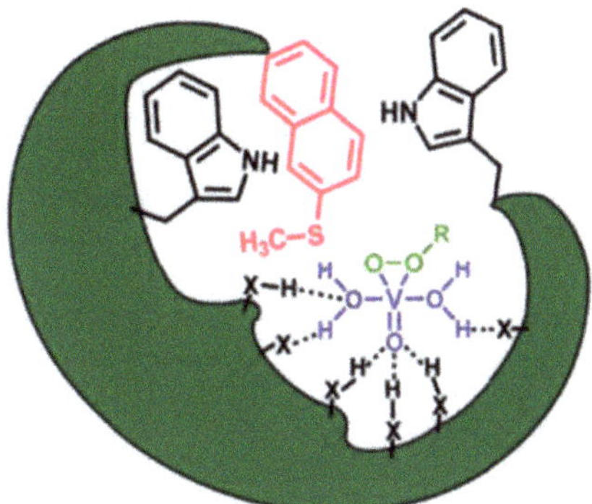

Figure 11. Vanadium-dependent artificial peroxidase for enantioselective sulfoxidation reactions. Reprinted with permission from (Dong et al., 2012) [13]. Copyright 2012 Royal Society of Chemistry.

4.2. Artificial Photosynthesis

Artificial photosynthesis, including photo-induced water oxidation and CO_2 reduction, has been widely studied in the past few years. Native photosynthesis involves three stages: light harvesting, charge separation and redox catalysis, and has special preorganization of chromophores and catalysts. Nature has evolved highly complex and well-organized supramolecular architectures, which can capture sunlight, and transform the solar energy with high efficiency. Cavitins are very suitable for modeling this process as follows from the previous sections. They provide a unique platform designing catalysts for photo-to-chemical energy conversion. Supramolecular chemistry is a powerful tool to achieve larger, more organized molecular structures with an increased level of complexity to optimize properties required for artificial photosynthesis. Like natural systems, perfect preorganization in cavitins leads to improved energy transfer processes, charge separation and redox catalysis. Porphyrins are synthetically accessible and stable analogues of nature's chlorophylls and therefore have been thoroughly studied as chromophores. Because of their excellent visible light harvesting ability and high electron transfer efficiency, ruthenium bipyridyl complexes are also classic photosensitizers. Thus, the active water oxidation catalyst *cis*-[Ru(bpy)(5,5′-dcbpy)(H$_2$O)$_2$]$^{2+}$ was incorporated into UIO-67 MOF using post-synthetic modification of the framework [88]. XAS, EPR, and Raman spectroscopy confirmed the formation of a highly active Ru^V=O key intermediate in M-cavitin [78]. Recently, MOC containing dinuclear and mononuclear Co active sites as well as a [Ru(bpy)$_3$]Cl$_2$ photo-sensitizer and a Na$_2$S$_2$O$_8$ electron scavenger was studied in photo-driven water oxidation (Figure 12) [89].

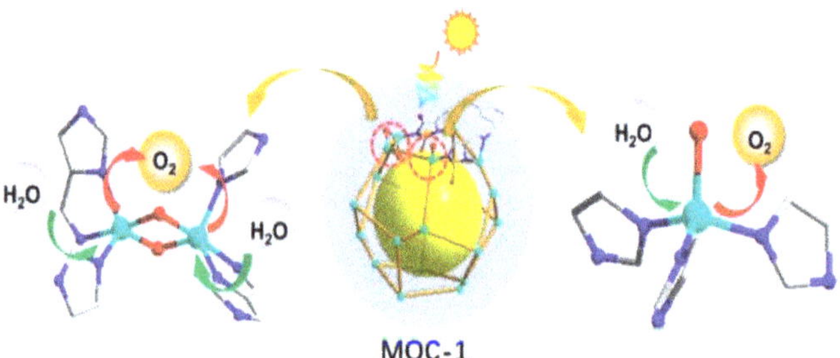

Figure 12. A Co MOC for photo-driven water oxidation. Reprinted with permission from (Chen et al., 2021) [89]. Copyright 2021 American Chemical Society.

This study revealed that photo-induced water oxidation initializes the electron transfer from the excited [Ru(bpy)$_3$]$^{2+*}$ to Na$_2$S$_2$O$_8$, and then, the bis(μ-oxo)dicobalt active sites further donate electrons to the oxidized [Ru(bpy)$_3$]$^{3+}$ to drive water oxidation [89]. Self-assembled nanospheres bearing guanidinium-binding sites can strongly bind sulfonate-functionalized ruthenium catalysts. Compared to the homogeneous system, the reaction rate for electrochemical water oxidation was enhanced by two-orders of magnitude by the preorganization of the ruthenium catalysts (Figure 8) [90]. There are more recent examples in Table 1. The study of artificial photosynthesis has great significance for future sustainable development, taking into account converting solar energy into chemical energy, including the production of H$_2$, water oxidation, carbon dioxide and nitrogen fixation, and fine organic syntheses (see next chapters).

Table 1. Photosynthesis mimics.

#	MC	Reaction + hv	Productivity, μmol g^{-1} (Time, h^{-1})	Select. %	Rate, μmol g^{-1}h^{-1} (TOF, h^{-1})	References
1	FDH@Rh-NU-1000	CO$_2$ + 2e2H$^+$ → HCOOH	144 M (24)	nd	(865)	[91]
2	UiO67-Ir-Cou6	CO$_2$ → H$_2$ → HCOOH	26,845 4808	95.5	nd	[92]
3	MIL-125-Py-Rh	CO$_2$ → HCOOH	9.5 mM (24)	nd	nd	[93]
4	CUST-804	CO$_2$ → CO	nd	82.8	2,710	[94]
5	Rh-MOP	CO$_2$ → HCOOH	76	nd	(60)	[95]
6	MAF-34-CoRu	CO$_2$ + H$_2$O → CO	nd	nd	11.2	[96]
7	PFC-58-30	CO$_2$ + H$_2$O → HCOOH	nd	nd	29.8	[97]

Table 1. *Cont.*

#	MC	Reaction + hν	Productivity, μmol g^{-1} (Time, h^{-1})	Select. %	Rate, μmol $g^{-1}h^{-1}$ (TOF, h^{-1})	References
8	POMs@INEP-20	$CO_2 \rightarrow CO$	nd	97.1	970 (2.43)	[98]
9	CABB@M-Ti	$CO_2 \rightarrow CH_4$	nd	88.7	32.9	[99]
10	CTF-Bpy-Co	$CO_2 \rightarrow CO$	120 μmol (10)	83.8	nd	[100]
11	SAS/Tr-COF	$CO_2 \rightarrow CO$		96.4	980.3	[101]
12	RuCOF-TPB	$H_2O \rightarrow H_2$	160	nd	20,308	[102]
13	CoP@ZnIn$_2$S$_4$	$H_2O \rightarrow H_2$	nd	nd	103	[103]
14	ZIF-67/CdS HS	$H_2O \rightarrow H_2$	nd	nd	1721	[104]
15	Ni-Py-COF	$H_2O \rightarrow H_2$	nd	nd	626	[105]
16	Co-Tz	$H_2O \rightarrow H_2$	9,320	nd	2,330	[106]
17	PTC-318	$H_2O \rightarrow H_2$	80	nd	nd	[107]
18	Ru(Bda)-COF	$H_2O \rightarrow O_2$	nd	nd	26,000	[108]

4.3. Models of Redox-Active Enzymes

The preparation and characterization of a redox-active MOF features both a biomimetic model of the hydrogenase active site as well as a redox-active linker that acts as an electron mediator, thereby mimicking the function of [4Fe4S] clusters in the enzyme [109]. Hydrogenase enzymes are highly efficient in reducing protons to hydrogen. Both major classes of hydrogenases, the [FeFe]- and [NiFe]-H$_2$ases, contain a bisthiolate-bridged dinuclear complex (FeFe or NiFe, respectively) in active site. The charge transport function of the [4Fe4S]-based enzymatic electron transport chain in redox enzymes is separated from the Fe2 catalytic function, but both these sites are linked as illustrated in Figure 1a for the [FeFe]-H$_2$ase. In the PCN-700 MOF mimic of [FeFe]-H$_2$ase the first function is modeled by an organic redox-active naphthalene diimide-based (NDI) linker, while the Fe2 subsite is modeled by a structurally related [FeFe](dcbdt)(CO)$_6$ (dcbdt = 1,4-dicarboxylbenzene-2,3-dithiolate) complex [109]. The two units reside in preorganized positions within the cavity and the NDI-to-Fe2 distance in the PCN-700/NDI/FeFe is nearly identical in [FeFe]-H$_2$ase. The simple encapsulation of a structural and functional model complex of [NiFe]-hydrogenase into the MOF cavities gives the advanced hydrogenase mimic NiFe@PCN-777 (Figure 13) [110].

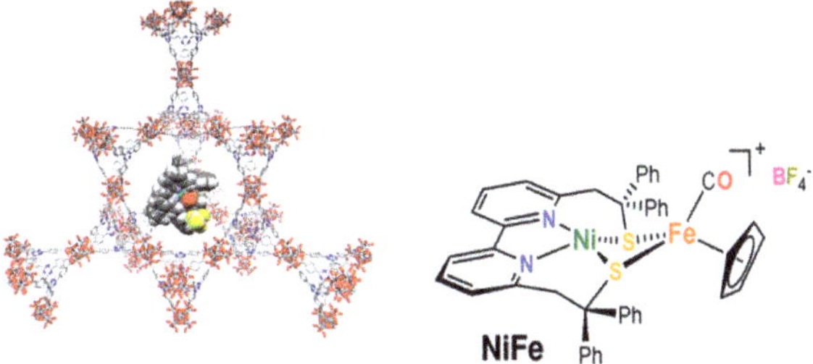

Figure 13. Advanced hydrogenase mimic NiFe@PCN-777. Adapted with permission from (Balestri et al., 2017) [110]. Copyright 2017 American Chemical Society.

MOF-818, containing trinuclear copper centers mimics the active sites of catechol oxidase, possessing efficient catechol oxidase activity with good specificity [111]. Other examples of redox-active enzymes may be found in Table 2.

Table 2. Models of redox-active enzymes.

#	MC	Enzyme	Reaction	Yield, % (μmol g^{-1})	Select, %	Rate, μmolg^{-1} h^{-1} (TOF, h^{-1})	Refs
1	Cu$_3$MOF-818	Catechol oxidase	DTBC + O$_2$ $\rightarrow$ DTBQ + H$_2$O$_2$	98	nd	nd	[111]
2	Cu$_2$MIL-125-Ti	Mono oxygenase	RH + O$_2$ $\rightarrow$ ROH $\rightarrow$ epoxide	94 92	nd	nd	[112]
3	Ce-AQ MOF	Mono oxygenase	C$_6$H$_{12}$ +O$_2$ $\rightarrow$ C$_6$H$_{10}$O	54.2	98.4	nd	[113]
4	Ce-UiO-Co	MMO	CH$_4$ +H$_2$O$_2$ $\rightarrow$ CH$_3$OH + H$_2$O	(2,166,000)	99	nd	[114]
5	Fe/Co-TFT	Hydrogenase	2H$^+$ + 2e$^-$ $\rightleftarrows$ H$_2$	nd	nd	(11,000)	[115]
6	UiO-MOF-Fe$_2$S$_2$	Hydrogenase	2H$^+$ + 2e$^-$ $\rightleftarrows$ H$_2$	35	nd	nd	[116]
7	[Pd$_{12}$(Fe$_2$BB)$_5$(BBNH$^+$)$_{19}$]$^{43+}$	Hydrogenase	2H$^+$ + 2e$^-$ $\rightleftarrows$ H$_2$	nd	nd	10,300 mol^{-1}s^{-1}	[117]
8	Mg$_3$(HiTP)$_2$	Reductase	O$_2$ + 2e,2H+ $\rightarrow$ H$_2$O$_2$	nd	90	nd	[118]
9	UiO66(SH)$_2$	Nitrogenase	N$_2$ + 6e,6H+ $\rightarrow$ 2NH$_3$	nd	nd	32.4	[119]
10	NFCO	Nitrogenase	N$_2$ + 6e,6H+ $\rightarrow$ 2NH$_3$	17,600	nd	126	[120]
11	MIL-101 (Fe^{II}/Fe^{III})	Nitrogenase	N$_2$ + 6e,6H+ $\rightarrow$ 2NH$_3$	nd	nd	466.8	[121]

4.4. MMO Mimics

The development of the direct low-temperature selective oxidation of methane to methanol has remained an active area of research over the last 50 years [122–124]. Native enzymes, the copper-containing particulate methane monooxygenase (pMMO) and the iron-containing soluble methane monooxygenase (sMMO), oxidize methane in ambient conditions by O$_2$ with two connected metal atoms selectively to methanol: CH$_4$ + O$_2$ + 2e$^-$ + 2H$^+$ $\rightarrow$ CH$_3$OH + H$_2$O (1), NADH $\rightarrow$ NAD+ + 2e$^-$ + 2H$^+$ (2). The exceptional MMO selectivity is controlled by a special regulatory mechanisms based on the AS hydrophobicity, substrate geometric dimensions, small-molecule tunnel, gate-mechanism and methane quantum tunneling, which together substantially accelerate the oxidation of methane compared to other substrates [68,125–127]. Catalyst encapsulation within a chemically stable porous COF provides a hydrophobic microenvironment around the active site. An anionic macrocyclic catalyst [Fe^{III}(Cl)bTAML]$^{2-}$ inside COF nanospheres (Figure 14) permits the oxyfunctionalization of hydrocarbons in water with an enhanced degree of selectivity using the catalyst-immobilized COF nanofilms [128].

It would be interesting to check with this cavitin a fine transformation of some Fe$_2$O$_2$ model complex of the MMO intermediate Q during its interaction with methane or its homologs taking into account following info. The combined structural and theoretical investigation of alkane uptake in a flexible MOF demonstrated accommodation of the C1–C4 alkanes, which are different in size and shape, and reveals that a turn stile mechanism facilitates their transport due to gate-opening [129]. A cavity-tailored MOC containing inward-facing ethyl groups selectively encapsulated methane, ethane, and ethylene at atmospheric pressures in acetonitrile and showed the strongest binding for methane [130]. MOFs bearing Fe(II) sites within Fe$_3$-μ_3-oxo nodes were active for the conversion of CH$_4$ + N$_2$O mixtures via Fe(IV)=O. On the basis of in situ IR spectroscopy and DFT calculations, it was demonstrated that methanol is protected within the MOF under reaction conditions as a methoxy group and its was concluded that there are steps beyond the radical-rebound mechanism to protect the desired CH$_3$OH product [131]. The synthesis of M-cavitin, modeling the coordination environment of the pMMO Cu_C site was reported recently. EPR analysis of the prepared Cu^{II} complex revealed striking similarities to the AS of pMMO. The similar Cu^I complex readily reacted with dioxygen and was capable of C-H bond oxidation (Figure 15) [132].

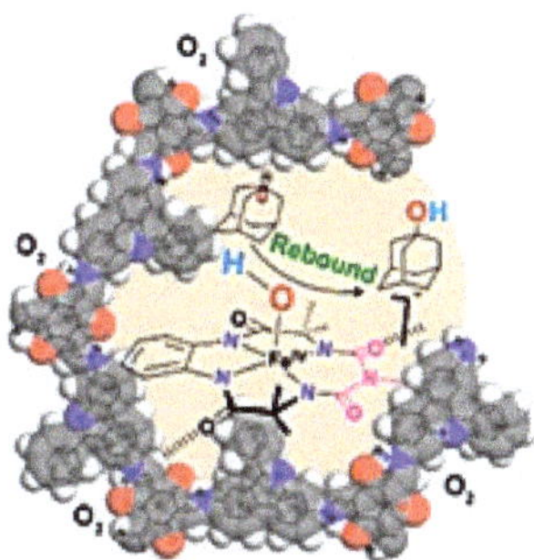

Figure 14. C-H functionalization in water via the stabilized Fe^{IV}=O intermediate in the cavity of a COF nanosphere. Adapted with permission from (Sasmal et al., 2021) [128]. Copyright 2021 American Chemical Society.

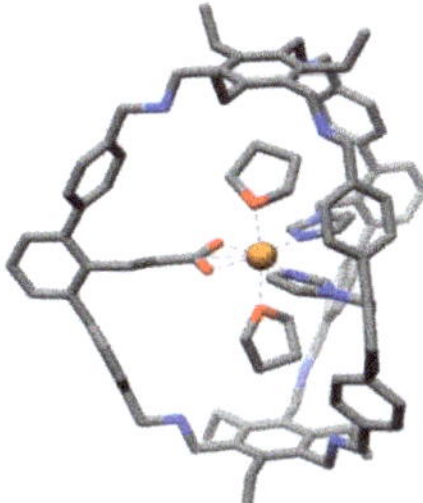

Figure 15. Structural and functional mimic of the pMMO Cu_C site. Adapted with permission from (Bete et al., 2022) [132]. Copyright 2022 Wiley.

An artificial binuclear copper monooxygenase Ti_8-Cu_2 was prepared by metalation of the SBUs in a Ti MOF. The closely spaced Cu^I pairs were oxidized by O_2 to afford the $Cu^{II}_2(\mu_2$-OH$)_2$ cofactor [112]. The SBU provided a precise binding pocket for the installation of binuclear Cu cofactors to cooperatively activate O_2. Ti_8-Cu_2 showed a turnover frequency at least 17 times higher than that of mononuclear Ti_8-Cu_1 [112]. Upon incorporation of mononuclear Fe^{II} tris(2-pyridylmetylamine) (FeTPA) into hemicriptophane, a Fe-cavitin was obtained which, in contrast to free Fe^{II}TPA, was able to oxidize methane by hydrogen peroxide under 60 °C and 30 bar to methanol [133]. The incorporation of the unstable Cu^I_3L complex into mesoporous silica gel allows one to obtain a catalyst which selectively transforms methane to methanol by hydrogen peroxide under room temperature with a conversion of 17.4% and TON 170 (Figure 16) [134].

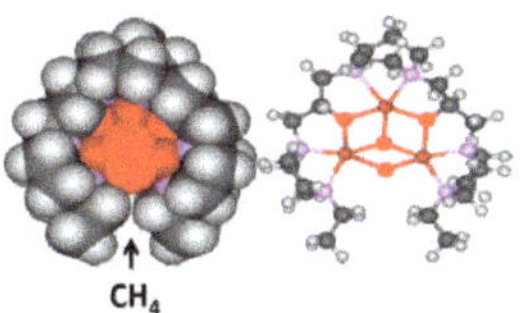

Figure 16. The Cu^I_3L complex in mesoporous silica gel for the transformation of methane to methanol using hydrogen peroxide under room temperature (Liu et al., 2016) [134].

The authors suggest that this effect is associated with the encapsulation of the complex into the hydrophobic cavity of the silica gel. Unfortunately, the paper does not contain enough evidence to support this suggestion. Nevertheless, the results are very interesting and very much encourage the continuation of this research despite the mechanism not being proven.

Thus, artificial enzymatic systems have been studying to mimic the structures and functions of their natural counterparts. However, there remains a significant gap between modeling and catalytic activity in these artificial systems. Unfortunately, the unlimited possibilities of organic chemistry and supramolecular chemistry have not yet been fully utilized for the biomimetic study of the complex structure of AS MMO and the mechanism of its functioning. It would be very useful to model the conformational effects of the polypeptide scaffold, dynamic changes in the coordination environment of the metal complex, and the sequential formation of intermediates in the multistage process of oxygen and methane activation. Of great interest in this connection is the reproduction of the MMOs catalytic cycle on the base of cavitins and, especially, a deeper penetration into the fine structure of the active intermediate Q and the peculiarities of its transformation during the interaction with methane, for example, using more adequate MMO model complexes [124,135] and advanced enzyme models of types demonstrated in Figures 11–15.

5. M-Cavitins in Fine Organic Synthesis

The primary field of MC application is fine organic synthesis and enantioselective catalysis.

Cavitins can induce enantioselectivities not achievable in homogeneous systems involving preferential secondary interactions with the included substrate. For example, they catalyze the asymmetric acetalization of aromatic aldehydes and 2-aminobenzamide with product yields up to 93% at 97% ee [24]. Additionally, the highest activity for the enantioselective dihydroxylation and epoxidation of styrene derivatives was obtained by using a Ru complex linked with a natural cavitin, bovine serum albumin Ru3-BSA-HA [62]. It is well known that special channels in enzymes facilitate the transport of substrates and products. The mesoporous MOF MnO_2@OMUiO-66(Ce), containing artificial substrate channels and MnO_2 attached to Ce-O clusters, was designed as a super-active artificial catalase [136]. MOF-818, containing trinuclear copper centers that mimic the active sites of catechol oxidase, shows efficient catechol oxidase activity. This artificial enzyme oxidizes o-diphenols to o-quinones with good substrate specificity [101]. The direct selective oxidations of the most difficult C-H bonds with O_2 are very challenging reactions and play an important role in fine organic synthesis [137]. Nature has created highly active and selective binuclear metal-containing monooxygenases working in participation with O_2 and a reducing agent to activate the most inert C-H bonds of alkanes involving methane. Recently the MOF-based artificial binuclear monooxygenase Ti_8-Cu_2 was prepared via the metalation of the SBU in a Ti-MOF (see 4.4 [112]). In the presence of coreductants, Ti_8-Cu_2 demonstrated excellent catalytic activity and selectivity in monooxygenation processes, including epoxidation, hydroxylation and sulfoxidation, with TOF, which is much higher than that of mononuclear Ti_8-Cu_1 (Figure 17) [112].

It would be interesting to check and develop this M-cavitin for alkane hydroxylation involving methane. While polycavitins are used for the fabrication of advanced heterogeneous catalysts, monocavins are more suitable for the modeling and academic study of enzyme active sites. However, the recent years MOC use in catalysis has also increased. Some examples are shown above. For the case of N-heterocyclic carbene-capped CD gold complexes (ICD)AuCl stereoselectivity in the enyne cycloisomerization depends on the nature of the cyclodextrin: α-ICD and β-ICD give the enantiomer for which the approach is the easiest according to their helical shape and γ-ICD does not afford enantioselecivity because of its symmetrical shape (Figure 10) [74]. The incorporation of iron porphyrin and L- or D-histidine endues chiral COF nanozymes with high activity and selectivity in the peroxidase oxidation of dopa enantiomers. This artificial peroxidase possesses 21.7 times higher activity than natural HRP [136]. The involvement of organic radical reactions in cavitins helps solve some of the problems connected with the high reactivity and reaction diversity of radicals via taming the reactivity, improving the selectivity or inducing new reaction outcomes [138]. Ti(IV)-based M-calixarene nanocage clusters exhibit extraordinary stability in concentrated acid–alkali solutions and can act as a stable photocatalyst for the oxidation of amines to imines [139]. Other interesting examples may be found in Table 3.

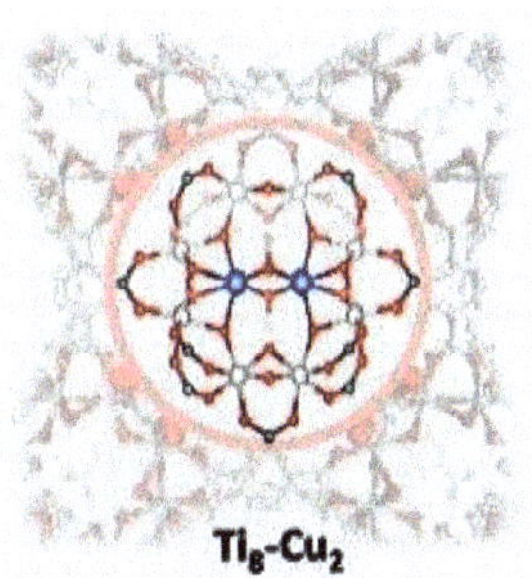

Figure 17. The artificial binuclear copper monooxygenase in a Ti MOF, the central circle shows the structure of the Cu bi-nuclear active site. Reprinted with permission from (Feng X. et al., 2021) [112]. Copyright 2021 American Chemical Society.

Table 3. Catalysts for fine organic synthesis.

#	MC	Reaction	Yield, %	Selectivity, %	Rate, mmol g^{-1}h^{-1} (TOF, h^{-1})	References
1	PdNPs/ZIF-8	$PhNO_2 + H_2 \rightarrow PhNH_2$	95	nd	nd	[140]
2	(Cu_1Pd_1)PCN-22(Co)	$2PhY + CO_2 \rightarrow Ph_2CO$	90	97	nd	[141]
3	Br-PMOF(Ir)	$PhEtNH + CO_2 + PhSiH_2 \rightarrow PhEtNCHO$	82	82	(507)	[142]
4	Zn-TACPA	$XC_6H_4NHCHCOOEt + PhCH=CH_2 \rightarrow$ 3a [143]	91	nd	nd	[143]
5	Co@Y	$MeCH=CH_2 + O_2 \rightarrow$ epoxyde	24.6	57	4.7	[144]
6	CoP@POC	$2PhCH_2NH_2 + O_2 \rightarrow PhCH_2N=CHPh$	93	99	(22,989)	[145]
7	CuHENU-8	$Me_2PhSiH + t\text{-}BuOOH \rightarrow Me_2PhSiOH$	89	95	132	[146]
8	Ni-SAPO-34	$C_6H_{10}=O + O_2 \rightarrow C_4H_{10}(COOH)_2$	30	87	nd	[147]
9	Prism1	$ArCH_2OH + O_2 \rightarrow ArCHO$	99.9	nd	nd	[148]
10	POM/MOF	$PhCH=CH_2 + H_2O_2 \rightarrow PhCHO$	96	99	nd	[149]
11	Zr-abtc	Carvone $+ H_2O_2 \rightarrow$ 1,2-epoxide	87	90	nd	[150]
12	SNNU-97-InV	Me-epoxide $+ CO_2 \rightarrow$ Me-c-carbonate	73.3	99	(24.2)	[151]

Table 3. *Cont.*

#	MC	Reaction	Yield, %	Selectivity, %	Rate, mmol g^{-1}h^{-1} (TOF, h^{-1})	References
13	NUC-45a	Ph-epoxide + CO_2 → Ph-carbonate	99	99	(316)	[152]
14	NUC-54a	PhCHO + CO_2 → Ph-carbonate	98	nd	(47)	[153]
15	MOF1	1-Et-2Ph-aziridine + CO_2 → oxazolidinone	99	97	nd	[139]
16	Ent-1(3b)	nerol → α-terpineol +limonen	73	70ee	nd	[154]
17	CuPMO	$NH_2C_6H_4OH + CH_2(COMe)_2$ → benzoazole	83	nd	nd	[155]
18	Ni-Ir@Tp-Bpy	$R^1R^2NH + RJ → R^1R^2NR$	94	nd	nd	[156]
19	Cu_2O@ZIF-8	$Me_2C(OH)$-C≡CH + CO_2 → α-alkylidenecarbonate	97	nd	(3.03)	[157]
20	(R)-CuTAPBP-COF	EtCHO+$PyCH_2Br$ (R)-MPP	98	95ee	nd	[158]
21	JLU-MOF-112	PhCHO + $CH_2(CN)_2$ → PhCH=C(CN)$_2$	98	nd	(198)	[159]
22	Cu-1D MOF	Pyrazole + PhJ → 1-Ph-1H-pyrazole	95	nd	nd	[160]
23	PdAg@ZIF-8	CH_2=$CHC_6H_4NO_2 + H_2$ → CH_2=$CHC_6H_4NH_2$	98	97.5	nd	[161]
24	JNM-4-Ns	$R_1C_6H_4$-C≡$CR_2 + B_2Pin_2$ → $R_1C_6H_4$-CH=CR_2BPin	90	nd	(41,734)	[162]
25	UiO-66-Gua$_{0.2}$	CO_2 + ECH → CPC	nd	nd	(110.3)	[163]
26	Bi_2S_3@quasi-Bi-MOF	4-NO_2PhOH + H_2 → 4-NH_2PhOH	nd	97	nd	[164]

It is worth to note an excellent reaction rate, high product selectivity and productivity outperforming most reported photocatalysts for MC #6 in Table 3 [145], in which the single Co atom organic cage CoP@POC demonstrates prominent photocatalytic efficiency for the oxidation of amines into imines in visible light.

6. M-Cavitins as Promising Industrial Catalysts

The growth of energy consumption and environmental problems have resulted in the search of catalysts for industrial energetically challenging processes with participation of small gas molecules involving innovative reactions, high selective to valuable products. The development of renewable and efficient energy conversion technologies is becoming extremely necessary. These technologies must be based on the principles of biomimetic chemistry and M-cavitin catalysis. For the realization of artificial photosynthesis, it is necessary to develop the design of fast and durable water oxidation catalysts that can be incorporated into future sunlight-to-chemical-fuel assemblies. The activation and transformation of small molecules, such as CO_2, N_2, O_2, CH_4 and H_2, into other products has always been central to endeavors of chemical science. Among various types of energy conversion, electrochemical CO_2 reduction (CO_2R) and water splitting (WS) have also been proven as promising strategies for their environmental benignancy and high efficiency. For the small molecules discussed here, the spatial and temporal control of protons and electrons delivery to/from the active site is crucial in maintaining product selectivity in these transformations [165]. The modern challenges of climate change, energy sustainability, and resource efficiency make the activation of small molecules more important than ever before.

6.1. H_2O

Water oxidation catalysis is of pivotal importance to progress the field of artificial photosynthesis. Water is an important renewable energy source and has the potential to meet the current energy crisis needs via photochemical, electrochemical, and photo-electrochemical splitting to produce oxygen and hydrogen green fuels. Water splitting is comprised of two half-cell reactions: an oxygen evolution reaction (OER) and a hydrogen evolution reaction (HER). The facile synthesis and electrocatalytic HER performance of SnTPPCOP was demonstrated recently (Figure 18) (Table 4, #16), which exhibited good HER activity with a low overpotential of 147 mV at 10 mA cm^{-2} due to its unique structural properties, ranking among the best new reports.

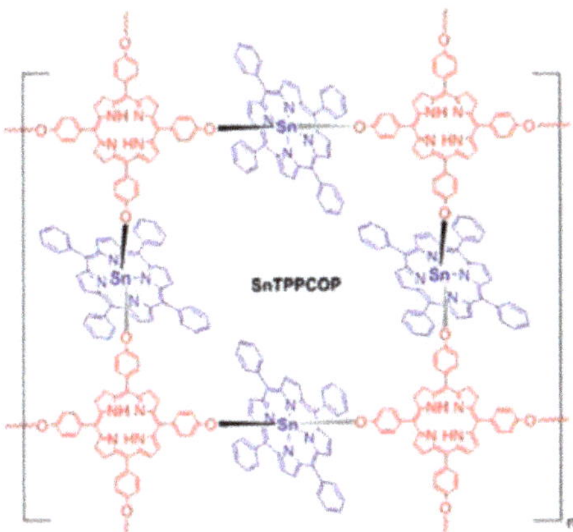

Figure 18. Tin porphyrin axially-coordinated 2D covalent organic polymer (SnTPPCOP) for HER activity (Wang Q. et al., 2022) [179].

Table 4. Activation of small molecules for sustainable development.

#	MC	Reaction	Yield, % (P, µmol g^{-1}/FE, %)	Selectivity, %	Rate, µmol g^{-1}h^{-1} (TOF h^{-1})	OP, mV (CD, mA sm^{-2})	Refs
1	Fe/Ni-MOF	$H_2O \rightarrow O_2$	nd	nd	(940)	239 (50)	[166]
2	CALF20	$CO_2 \rightarrow CO$	(nd/94)	nd	(1361)	(32.8)	[167]
3	HNTM-Ir/Pt	$H_2O \rightarrow H_2$	(600)	nd	201.9	nd	[168]
4	Ir^{III}-Uio-67-NH$_2$	$CO_2 \rightarrow CO$	nd (nd/6.71)	99.5	(120)	nd	[75]
5	TiO$_2$@ZIF-8	$H_2O \rightarrow H_2$	51 [1]	nd	262,000	nd	[18]
6	(NiCo)S$_2$/NCNF	$H_2O \rightarrow O_2 \rightarrow H_2$	nd	nd	nd	177 (10) 203 (10)	[169]
7	ZnO/Fe$_2$O$_3$ PN	$CH_4 \rightarrow CH_3OH$	(178.3/nd)	100	nd	nd	[170]
8	Mn$_1$Co$_1$/CN	$CO_2 \rightarrow CO$	nd	nd	47	nd	[171]
9	ZPMOF	$CO_2 \rightarrow CH_4$	nd	70	32	nd	[172]
10	T1-2Cu	$CO_2 \rightarrow CH_4$	nd	93	3.7	nd	[173]
11	NiFe-MOF/FF	$H_2O \rightarrow O_2$	nd	83.8	nd	216 (50)	[174]
12	Cu@FCN MOF/CF	$H_2O \rightarrow O_2$	nd	88.7	nd	290 (10)	[175]

Table 4. *Cont.*

#	MC	Reaction	Yield, % (P, μmol g^{-1}/FE, %)	Selectivity, %	Rate, μmol g^{-1}h^{-1} (TOF h^{-1})	OP, mV (CD, mA sm^{-2})	Refs
13	Fe$_3$-MOF-BDC-NH$_2$	$H_2O \rightarrow O_2$	nd	nd	nd	280 (10)	[176]
14	CoCu-MOF NBs	$H_2O \rightarrow O_2$	nd	nd	1084	271 (10)	[177]
15	CuNi-NKU-101	$H_2O \rightarrow H_2$	nd	100	nd	324 (10)	[178]
16	SnTPPCOP	$H_2O \rightarrow H_2$	nd	nd	nd	147 (10)	[179]
17	CoP/CNTHPs	$H_2O \rightarrow O_2 \rightarrow H_2$	nd	nd	nd	238 (10) 147 (10)	[180]
18	Ru/3DMNC	$H_2O \rightarrow O_2 \rightarrow H_2$	nd	nd	nd	217 (10) 51 (10)	[181]
19	MnZn-MUM-1/NF	$H_2O \rightarrow O_2$	nd	nd	(83.3)	253 (10)	[182]
20	Co-Fe-P	$H_2O \rightarrow O_2$	nd	nd	nd	240 (10)	[183]
21	Ce-Ni(OH)$_2$@Ni-MOF	$H_2O \rightarrow O_2$	nd	nd	(170)	272 (100)	[184]
22	FePc-pz	$N_2 \rightarrow NH_3$	(nd/31.9)	nd	33.6 [2]	nd	[185]
23	Fe$_1$S$_x$@TiO$_2$	$N_2 \rightarrow NH_3$	(nd/17.3)	nd	18.3 [2]	nd	[186]
24	UiO-66-H	$CH_4 \rightarrow CH_3OOH$	nd	100	350	nd	[187]

[1] quant eff; [2] μg μg h^{-1}.

Photo-induced WS into hydrogen and oxygen has been perceived as one of the most promising pathways for solving the energy crisis and environmental problems. The double-shelled TiO$_2$@ZIF-8 hollow spheres used for HER under illumination show efficient charge separation by electron injection from ZIF-8 to TiO$_2$, high photocatalytic quantum efficiency and a high HER rate, 3.5 times higher than bare TiO$_2$ (Table 4, #5) [18]. Donor–acceptor type imine-linked COFs can be produced, under visible light irradiation, upon protonation of their imine linkages. A significant redshift in light absorbance, largely improved charge separation efficiency, and an increase in hydrophilicity triggered by protonation of the Schiff-base moieties in the imine-linked COF are responsible for the improved photocatalytic performance [188]. Electrocatalytic WS has been regarded as one of the most promising approaches for producing hydrogen under mild conditions. Despite many progresses achieved in electrocatalytic WS, highly active and durable catalysts have to be developed to overcome the kinetic barriers in the water splitting process, especially for the OERs [189]. The theoretical basis for the design of new MOF electrocatalysts was recently elaborated through a study of the relationship between the structure and properties of trimetallic MOFs for efficient OERs. Fe$_3$-MOF-BDC-NH$_2$ exhibited an enhanced performance, superior to other reported catalysts (Table 4, #13) [176]. The multi-shelled hollow Mn/Fe-hexaiminobenzene MOF (Mn/Fe-HIB-MOF), featuring a conductive skeleton, was developed as an excellent bifunctional electrocatalyst for oxygen reduction reactions and OERs. It exhibited high OER performance, outperforming commercial RuO$_2$, Mn-HIB-MOF and Fe-HIB-MOF catalysts [190]. The FeNi-MOF showed remarkable electrocatalytic performance with a low overpotential of 266 mV at 100 mA cm^{-2}, and a high TOF value of 0.261 s^{-1} at an overpotential of 270 mV as well as superb long-term durability with a high current tolerance for water oxidation [166] (Figure 19).

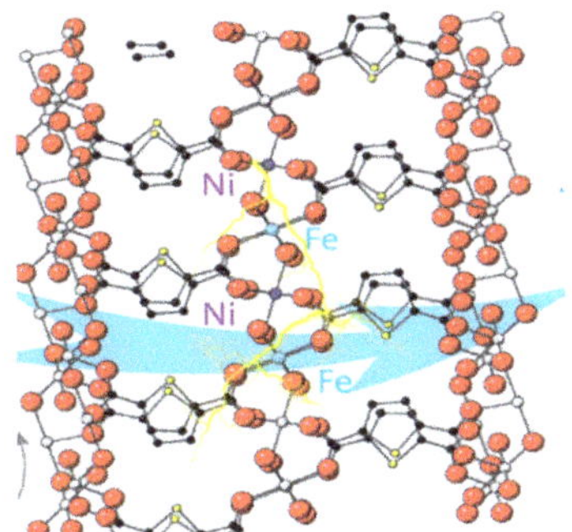

Figure 19. FeNi-MOF electrocatalyst for water oxidation. Reprinted with permission from (Wang CP et al., 2021) [166]. Copyright 2021 American Chemical Society.

Photo-induced water oxidation by an MOF has been widely studied in the past few years. The active water oxidation catalyst *cis*-[Ru(bpy)(5,5′-dcbpy)(H$_2$O)$_2$]$^{2+}$ was incorporated into a UIO-67 MOF using post-synthetic modification of the framework. OER was studied using an oxygraph with a Clark electrode at pH = 1. XAS, EPR, and Raman spectroscopy confirmed the formation of M-cavitin and the highly active Ru^V=O key intermediate [88]. MOCs based on cobalt ions and imidazolate ligands were studied on water photo-oxidation for the first time. These studies revealed that the reactions initialized via the electron transfer from the excited [Ru(bpy)$_3$]$^{2+*}$ to Na$_2$S$_2$O$_8$, and then, to the bis(μ-oxo)dicobalt active sites which further donated electrons to the oxidized [Ru(bpy)$_3$]$_3^+$ to drive water oxidation [89]. Recent advances in research of MOF nanoarchitectures for efficient electrochemical water splitting have reviewed [191]. Hierarchical bifunctional catalysts for WS are the most promising catalysts for energy transformation in future. For example, the bifunctional catalyst CoP/CNTHP, containing non-precious metals for efficient water splitting, has been shown to have outstanding catalytic activity and stability for overall WS [180].

6.2. CO$_2$

Because of the highly stacked layers, some COFs adopt semiconductive properties and exhibit promising catalytic performances in photo-CO$_2$R [192]. 3D flower-like SnS$_2$ with a sheet structure shows good performances for efficient CO$_2$ photoreduction under visible-light irradiation [193]. Under visible-light irradiation, the single Ir^{III}-MOC-NH$_2$ cage can convert CO$_2$ into CO with high selectivity and a TOF which is 3.4 times as much as bulk Ir^{III}-MOC-NH$_2$ and two orders of magnitude greater than that of the classical MOF counterpart, Ir^{III}-Uio-67-NH$_2$ [75] (Table 4, #4). The redox active In-MOF, In^{III}[Ni(C$_2$S$_2$(C$_6$H$_4$COO$_2$)$_2$], demonstrates the first example of a Ni-based MOF catalyst in electrocatalytic CO$_2$R, which opens promising prospects for designing novel and efficient non-noble metal-based, redox-active, biomimetic MOFs [194]. Conductive two-dimensional phthalocyanine-based MOF (NiPc-NiO$_4$) nanosheets linked by nickel-catecholate, are highly efficient electrocatalysts for CO$_2$R to CO electroreduction. The obtained NiPc-NiO$_4$ has good conductivity and exhibits a very high selectivity of 98.4% toward CO production and a large CO partial current density of 34.5 mA cm^{-2}, outperforming the reported MOF catalysts [195]. The MOF UiO-66 was used in tandem with its zirconium oxide nodes and incorporated ruthenium PNN pincer complex to hydrogenate carbon dioxide to methanol giving the highest reported turnover number (TON) (19,000) and turnover frequency (TOF) (9100 h^{-1}). Moreover, the reaction was readily recyclable, leading to a cumulative TON of 100,000 after 10 reaction cycles [196]. The neighboring Zn^{2+}-O-Zr^{4+} sites obtained by post-synthetic treatment of Zr$_6$(μ$_3$-O)$_4$(μ$_3$-OH)$_4$ nodes of MOF-808 by ZnEt$_2$ gave the MOF-808-Zn catalyst, which exhibited a >99% MeOH selectivity in CO$_2$ hydrogenation at 250 °C and good stability for at least 100 h [197] (Figure 20).

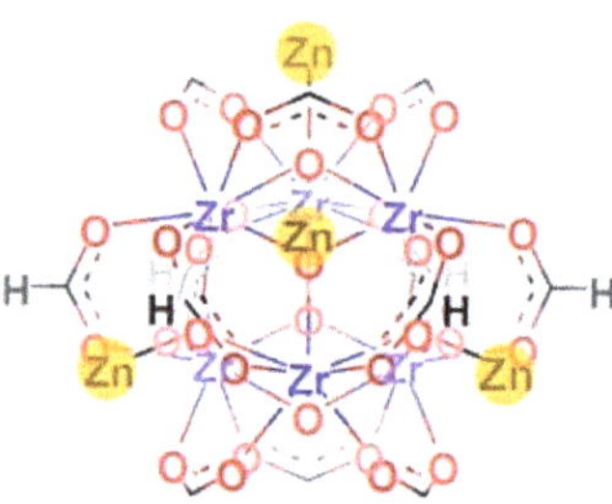

Figure 20. Zn-Zr cluster in the MOF-808-Zn catalyst for CO_2 hydrogenation (Zhang J. et al., 2021) [197].

Mechanistic investigations revealed that Zn^{2+} is responsible for H_2 activation and the $Zn^{2+}-O-Zr^{4+}$ site is critical for CO_2 adsorption and conversion.

Along with WS, the electrochemical reduction of carbon dioxide (CO_2R) has also been proven to be a promising strategy among various types of energy conversion. A desired electrocatalyst should have a high TON, a high TOF and a small overpotential. In cavitins the transition state of the target reaction can be stabilized more efficiently in comparison with bulk solution. The Ir complex was incorporated into Zr-MOC-NH$_2$ with the formation of Ir^{III}-MOC-NH$_2$. DFT calculations, mass spectrometry and in situ IR showed that the Ir(III) complex is the catalytic center, and $-NH_2$ in the cavity plays a synergistic role in the stabilization of the transition state and the Ir·CO_2 intermediate (Table 4, #4) [75]. Ni-MOF-derived catalysts for the light-driven methanation of CO_2 under UV–Vis IR irradiation displayed excellent recyclability without the loss of catalytic activity [198]. The high load of Zn-porphyrin in an anionic porous Zn-based PMOF has strong kinetic and thermodynamic advantages demonstrating a good performance in the photocatalytic CO_2-to-CH_4 conversion due to the atomically dispersed active catalytic sites and fast charge transfer (Table 4, #9) [172]. In another example, Cu^{2+} ions were dispersed in the crystal structure of the MOF matrix. The doping content of the Cu^{2+} ions and the photocatalytic performance displayed a volcanic relationship: the medium concentration (1Ti/2Cu) was optimal for the greatest performance for CH_4 and CO (3:1) (Table 4, #10) [173]. The d-UiO-66/MoS$_2$ composite facilitates the photo-catalytic conversion of CO_2 and H_2O to CH_3COOH under visible light [199]. Multiple Cu centers supported on the Ti-MOF catalyze CO_2 hydrogenation to ethylene and presents a new tandem route for CO_2-to-C_2H_4 conversion via CO_2 hydrogenation to ethanol followed by its dehydration [200]. Bifunctional MOFs containing tripyridyl complexes of Fe and Mn convert styrene into styrene carbonate via tandem epoxydation using O_2 and then CO_2 insertion. DFT calculations revealed the involvement of a high-spin Fe^{IV} (S = 2) center in the challenging oxidation of the sp^3 C-H bond [201]. A new porous copper−organic framework assembled from 12-nuclear [Cu12] nanocages with two types of nanotubular channels and a large specific surface area effectively catalyzed the cycloaddition of CO_2 to various epoxides under mild conditions [202]. Electrocatalytic N_2 reduction reactions (NNRs) at ambient conditions is a good way for sustainable NH_3 production, because the latter is a valuable raw material in organic synthesis and a significant clean energy carrier. The pyrazine-linked iron-phthalocyanine FePc-pz is an efficient electrocatalyst for simultaneously enhancing NRR activity and selectivity and is the best among the NNR electrocatalysts (Figure 21) (Table 4, #22) [185]. Inspired by the natural nitrogenase, the single-atom M-cavitin containing Fe$_1$S$_x$ in mesoporous TiO$_2$ appears as an excellent catalyst with a high rate and efficiency for NNR ((Table 4, #23) [186].

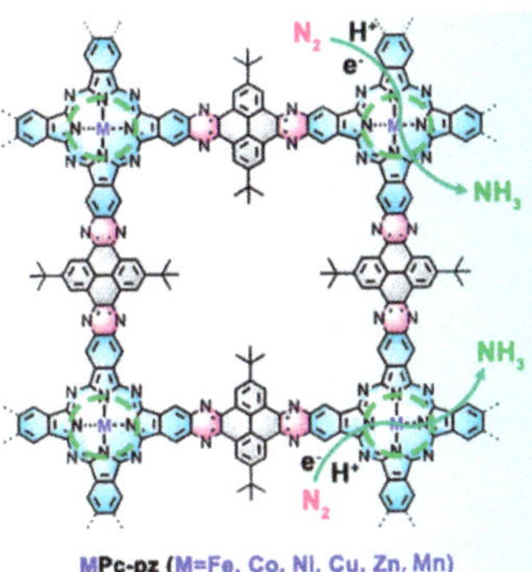

Figure 21. Pyrazine-linked iron-phthalocyanine FePc-pz for electrocatalytic N_2 reduction (Zhong H. et al., 2021) [185].

6.3. Methane

Compared to WS and CO_2 conversion, the effective and selective chemical routes to valorize the most abundant hydrocarbon on earth with the participation of cavitins are relatively less studied. Direct methane conversion has been carried out in the gas phase over Cu- and Fe-containing zeolites in the stepwise cyclical process that firstly involves interaction of the transition metal with O_2 or N_2O at 400–500 °C, forming an oxidative intermediate, then the methane reaction with this active intermediate at 200 °C, and finally the product extraction with water steam. However, the rate and productivity of these processes are still very low. A catalytic process with zeolite Fe-Cu-ZSM-5 with a selectivity of 20–80% was proposed for the hydroxylation of methane at 50 °C using aqueous H_2O_2 [203,204]. Such a Cu-Fe (2/0.1)/ZSM-5 catalyst is an efficient catalyst for the direct conversion of methane into methanol with excellent productivity and a methanol selectivity of 80% [204]. The mechanism based on the catalytic, spectroscopic and theoretical results was suggested: [204] the adjacent to the iron acid sites facilitates the formation of an active Fe(V)=O intermediate via the dehydration of the formed Fe-OOH in aqueous H_2O_2 solution, enabling the homolytic cleavage of the primary C-H by a radical-rebound mechanism to generate •CH_3 radicals that are quickly captured by •OH radicals to form CH_3OH. In contrast to Cu- and Fe-containing zeolites, the study of MOF-based MMO mimics is still in the early stages and suffers from the same problems, low productivity and rate and low methanol selectivity due to over-oxidation. The significant challenge for CH_4 photooxidation into CH_3OH are connected to activation of the inert C-H bond and inhibition of CH_3OH over-oxidation. In this connection several interesting works have recently appeared. Thus, it was shown that in reaction $CH_4 + H_2O_2 = CH_3OOH$, catalyzed by UiO-66-H, the high electronic density on Zr-oxo nodes facilitates coordination of methane and the formation of Zr-oxo/•OH intermediates which are competent to activate the methane C-H bond with 100% selectivity (Table 4, #24) [187]. It was suggested then from DFT calculations that in Zr coordination sphere the •CH_3 can quickly react with the •OOH to form CH_3OOH with extremely low energy barrier [187]. On one hand, the absolute selectivity is a very unbelievable result for direct methane oxidation, and the explanation is not convincing enough. Indeed, if this intermediate is competent in the reaction with methane [187], why is it not competent with CH_3OH [187]? On other hand, it was shown that two metal sites with different electronegativities can modulate the activity of CH4 activation and inhibit the overoxidation of CH_3OH. (Table 4, #7) [170]. Thus, the ZnO/Fe_2O_3 porous nanosheets efficiently performed CH_4 hydroxylation and suppressed CH_3OH overoxidation through strengthening its O-H bond. The experimental results and DFT calculations confirmed that the ZnO/Fe_2O_3 heterojunction results in a higher charge accumulation at the Fe sites through a charge transfer from the Zn sites, which favors the adsorption of CH4 molecules and further helps to lower the rate-limiting barrier of CH_3OH generation. These Fe sites endow the O-H bond of CH_3OH with a higher polarity through the transferring of electrons to the O atoms, this inhibits the homolytic cleavage of the O-H bond to generate highly reactive radicals [170]. In other work the electron donor–acceptor hybrid RhB/TiO_2 demonstrated the photocatalytic oxidation of CH_4 to CH_3OH with a rate 143 $\mu mol \cdot g^{-1} \cdot h^{-1}$ and a selectivity 94% in ambient conditions, utilizing visible light [205]. Other examples of small molecule activation may be found in Table 4.

7. Conclusions and Outlooks

Cavities and other holes are ubiquitous in the material word. Enzymes are natural cavitins. They have evolved over millions of years to provide extremely powerful catalysts toward a variety of reactions with excellent activities under mild conditions and exquisite substrate specificity and product selectivity. Our fundamental understanding of enzymatic catalysis has inspired scientists to develop and explore smaller synthetic complexes as enzyme mimics. Chemical cavitins have emerged due to the efforts of synthetic and supramolecular chemists. They represent an emerging class of molecules and supramolecular ensembles with intrinsic porosity. In-depth studies of these biomimetic artificial systems have provided important insights into natural enzymes [206]. However, there remains a significant gap between the structural modeling and catalytic activity in these artificial systems [112]. Due to advances in synthetic chemistry a huge diversity of cavitins inspired by enzymes has appeared during the last decade [8]. The ease of their synthesis has already provided us with a rich library of architectures. Their further functionalization to afford multifunctional assemblies is the challenge ahead [207]. Though significant advances have been achieved in recent years in cavitin chemistry [207], greater insights into the many subtle factors affecting their shape and size as well as cavity effects on catalysis are still required [208]. Compared to natural self-assembly, chemical cavitins often lack complexity, a feature highly desirable for enzyme mimics and advanced bioinspired catalysts [54,209,210]. Confined in a cavity, molecules can fundamentally change their chemical and physical properties compared to those in bulk solution [211]. Computational methods, advanced machine learning models, and direct and powerful techniques, such as in situ X-ray diffraction or single-crystal characterization, could be significant to better understanding cavity effects [212]. As demonstrated in recent studies, operando XAFS and FTIR techniques have been used as powerful tools for monitoring the evolution of the reactive centers at the molecular level. Although tremendous attention has been devoted to the development of single-atom catalysts [208], only a few reports are related to the construction of dinuclear and multi-nuclear metal species in cavitins. The exciting results were achieved during the last decade in the single-electron redox reactions [213]. At present, the greatest challenge to multielectron and proton photochemical transformations has appeared [214]. M-cavitins have opened up the possibility for chemistry to use the many principles developed in nature during evolution, expanded the tools of organic chemistry and have proven extremely important in fine organic synthesis and pharmaceutical chemistry, especially for enantioselective reactions. They not only improved the efficiency and selectivity of a number of reactions, but also allowed them to change their direction to new products. M-cavitins demonstrate high activities for energetically challenging reactions with the participation of small gas molecules and high selectivity to valuable products [215]. Great achievements have already been made in clean photocatalytic and electrochemical energy conversion using affordable and inexhaustible clean materials, such as water and carbon dioxide, which can produce valuable fuels and chemicals [216]. Using ubiquitous visible-light irradiation to reduce CO_2 to C-based products is an environmental and economic method which transforms solar energy in the form of chemical bonds. Looking forward, innovation efforts are still necessary to help solve the global energy crisis [217–219]. Achievements and solutions to this problem will be connected primarily with the development of fundamental scientific research in bioinspired catalysts [207]. These studies will require completely new approaches for the design and synthesis of a more diverse library of M-cavitins with innovative structures, metal ion composition and functionality. Photo- and electrocatalysis of other abundant resources, such as N_2 and methane, are now coming increasingly into focus. In particular, efforts to decipher the reaction mechanisms and extract fundamental insights are necessary to develop economically competitive routes using the direct methane oxidation to methanol [220]. The direct oxidation of methane in a laboratory setup with the participation of M-cavitins using mild conditions is still a challenging problem [221]. There is a huge gap between MMO and chemical catalysts based on M-cavitins, not only in activity and selectivity but also in the mechanism of direct oxidation of methane to methanol [220]. Though there still remains a considerable gap between the academic research and industrial applications [222], the numerous works covered in this review demonstrate a promising foundation for the future.

Funding: This research received no external funding.

Data Availability Statement: No new data were created or analyzed in this study. Data sharing is not applicable to this article.

Acknowledgments: The author would like to thank the Russian Ministry of Science and Education for supporting this work (via project no. AAAA-A19-119071190045-0 to the IPCP RAN), and Vera Kosareva for her help in the design of the figures.

Conflicts of Interest: The author declare no conflict of interest.

References

1. Li, W.L.; Head-Gordon, T. Catalytic Principles from Natural Enzymes and Translational Design Strategies for Synthetic Catalysts. *ACS Central Sci.* **2021**, *7*, 72–80. [CrossRef] [PubMed]
2. Rebilly, J.N.; Colasson, B.; Bistri, O.; Over, D.; Reinaud, O. Biomimetic cavity-based metal complexes. *Chem. Soc. Rev.* **2015**, *44*, 467–489. [CrossRef] [PubMed]
3. Leenders, S.H.A.M.; Gramage-Doria, R.; de Bruin, B.; Reek, J.N.H. Transition metal catalysis in confined spaces. *Chem. Soc. Rev.* **2015**, *44*, 433–448. [CrossRef]
4. Shang, J.; Liu, Y.; Pan, T. Macrocycles in Bioinspired Catalysis: From Molecules to Materials. *Front. Chem.* **2021**, *9*, 635315. [CrossRef]
5. Zarra, S.; Wood, D.M.; Roberts, D.A.; Nitschke, J.R. Molecular containers in complex chemical systems. *Chem. Soc. Rev.* **2015**, *44*, 419–432. [CrossRef]
6. Pappalardo, A.; Puglisi, R.; Sfrazzetto, T.G. Catalysis inside Supramolecular Capsules: Recent Developments. *Catalysts* **2019**, *9*, 630. [CrossRef]
7. Daliran, S.; Oveisi, A.R.; Peng, Y.; López-Magano, A.; Khajeh, M.; Mas-Ballesté, R.; Alemán, J.; Luque, R.; Garcia, H. Metal–organic framework (MOF)-, covalent-organic framework (COF)-, and porous-organic polymers (POP)-catalyzed selective C–H bond activation and functionalization reactions. *Chem. Soc. Rev.* **2022**, *51*, 7810–7882. [CrossRef] [PubMed]
8. Freund, R.; Canossa, S.; Cohen, S.M.; Yan, W.; Deng, H.; Guillerm, V.; Eddaoudi, M.; Madden, D.G.; Fairen-Jimenez, D.; Lyu, H.; et al. 25 Years of Reticular Chemistry. *Angew. Chem. Int. Ed.* **2021**, *60*, 23946–23974. [CrossRef] [PubMed]
9. Bavikina, A.; Kolobov, N.; Khan, I.S.; Bau, J.A.; Ramirez, A.; Gascon, J. Metal–Organic Frameworks in Heterogeneous Catalysis: Recent Progress, New Trends, and Future Perspectives. *Chem. Rev.* **2020**, *120*, 8468–8535. [CrossRef]
10. Wang, K.; Jordan, J.H.; Hu, X.Y.; Wang, L. Supramolecular Strategies for Controlling Reactivity within Confined Nanospaces. *Angew. Chem. Int. Ed.* **2020**, *59*, 13712–13721. [CrossRef]
11. Brown, C.L.; Toste, F.D.; Bergman, R.G.; Raymond, K.N. Supramolecular Catalysis in Metal–Ligand Cluster Hosts. *Chem. Rev.* **2015**, *115*, 3012–3035. [CrossRef] [PubMed]
12. Chai, Y.; Dai, W.; Wu, G.; Guan, N.; Li, L. Confinement in a Zeolite and Zeolite Catalysis. *Acc. Chem. Res.* **2021**, *54*, 2894–2904. [CrossRef] [PubMed]
13. Dong, Z.; Luo, Q.; Liu, J. Artificial enzymes based on supramolecular scaffolds. *Chem. Soc. Rev.* **2012**, *41*, 7890–7908. [CrossRef] [PubMed]
14. Zhang, D.; Martinez, A.; Dutasta, J.-P. Emergence of Hemicryptophanes: From Synthesis to Applications for Recognition, Molecular Machines, and Supramolecular Catalysis. *Chem. Rev.* **2017**, *117*, 4900–4942. [CrossRef]
15. Pei, W.-Y.; Lu, B.-B.; Yang, J.; Wang, T.; Ma, J.-F. Two new calix[4]resorcinarene-based coordination cages adjusted by metal ions for the Knoevenagel condensation reaction. *Dalton Trans.* **2021**, *50*, 9942–9948. [CrossRef]
16. Jiao, J.; Dong, J.; Li, Y.; Cui, Y. Fine-Tuning of Chiral Microenvironments within Triple-Stranded Helicates for Enhanced Enantioselectivity. *Angew. Chem. Int. Ed.* **2021**, *60*, 16568–16575. [CrossRef] [PubMed]
17. Yamagishi, H. Functions and fundamentals of porous molecular crystals sustained by labile bonds. *Chem. Commun.* **2022**, *58*, 11887–11897. [CrossRef] [PubMed]

18. Qiu, T.; Gao, S.; Liang, Z.; Wang, D.-G.; Tabassum, H.; Zhong, R.; Zou, R. Pristine Hollow Metal Organic Frameworks: Design, Synthesis and Application. *Angew. Chem. Int. Ed.* **2021**, *60*, 17314–17336. [CrossRef]
19. Dong, J.; Wee, V.; Peh, S.B.; Zhao, D. Molecular-Rotor-Driven Advanced Porous Materials. *Angew. Chem. Int. Ed.* **2021**, *60*, 16279–16292. [CrossRef]
20. Zhang, T.; Le Corre, L.; Reinaud, O.; Colasson, B. A promising approach for controlling the second coordination sphere of biomimetic metal complexes: Encapsulation in a dynamic hydrogen-bonded capsule. *Chem. Eur. J.* **2021**, *27*, 434–443. [CrossRef]
21. Li, R.-J.; Fadaei-Tirani, F.; Scopelliti, R.; Severin, K. Tuning the Size and Geometry of Heteroleptic Coordination Cages by Varying the Ligand Bent Angle. *Chem. Eur. J.* **2021**, *27*, 9439–9445. [CrossRef] [PubMed]
22. Gao, S.; Liu, Y.; Wang, L.; Wang, Z.; Liu, P.; Gao, J.; Jiang, Y. Incorporation of Metals and Enzymes with Porous Imine Molecule Cages for Highly Efficient Semiheterogeneous Chemoenzymatic Catalysis. *ACS Catal.* **2021**, *11*, 5544–5553. [CrossRef]
23. Hu, J.; Mehrabi, H.; Meng, Y.-S.; Taylor, M.; Zhan, J.-H.; Yan, Q.; Benamara, M.; Coridan, R.H.; Beyzavi, H. Probe metal binding mode of imine covalent organic frameworks: Cycloiridation for (photo)catalytic hydrogen evolution from formate. *Chem. Sci.* **2021**, *12*, 7930–7936. [CrossRef]
24. Hou, B.; Yang, S.; Yang, K.; Han, X.; Tang, X.; Liu, Y.; Jiang, J.; Cui, Y. Confinement-Driven Enantioselectivity in 3D Porous Chiral Covalent Organic Frameworks. *Angew. Chem. Int. Ed.* **2021**, *60*, 6086–6093. [CrossRef] [PubMed]
25. Gao, X.J.; Wu, T.T.; Ge, F.Y.; Lei, M.Y.; Zheng, H.G. Regulation of Chirality in Metal–Organic Frameworks (MOFs) Based on Achiral Precursors through Substituent Modification. *Inorg. Chem.* **2022**, *61*, 18335–18339. [CrossRef]
26. Schneider, M.L.; Markwell-Heys, A.W.; Linder-Patton, O.M.; Bloch, W.M. Assembly and Covalent Cross-Linking of an Amine-Functionalised Metal-Organic Cage. *Front. Chem.* **2021**, *9*, 696081. [CrossRef]
27. Jackson, N.; Vazquez, I.R.; Chen, Y.P.; Chen, Y.S.; Gao, W.Y. A porous supramolecular ionic solid. *Chem. Commun.* **2021**, *57*, 7248–7251. [CrossRef]
28. Zhang, D.; Ronson, T.K.; Nitschke, J.R. Functional Capsules via Subcomponent Self-Assembly. *Acc. Chem. Res.* **2018**, *51*, 2423–2436. [CrossRef]
29. Mukherjee, G.; Satpathy, J.K.; Bagha, U.K.; Qadri, M.; Mubarak, E.; Sastri, C.V.; de Visser, S.P. Inspiration from Nature: Influence of Engineered Ligand Scaffolds and Auxiliary Factors on the Reactivity of Biomimetic Oxidants. *ACS Catal.* **2021**, *11*, 9761–9797. [CrossRef]
30. Ivanova, S.; Kçster, E.; Holstein, J.J.; Keller, N.; Clever, G.H.; Bein, T.; Beuerle, F. Isoreticular Crystallization of Highly Porous Cubic Covalent Organic Cage Compounds. *Angew. Chem. Int. Ed.* **2021**, *60*, 17455–17463. [CrossRef]
31. Zhang, D.; Dutasta, J.-P.; Dufaud, V.; Guy, L.; Martinez, A. Sulfoxidation inside a C 3-Vanadium(V) Bowl-Shaped Catalyst. *ACS Catal.* **2017**, *7*, 7340–7345. [CrossRef]
32. Holsten, M.; Feierabend, S.; Elbert, S.M.; Rominger, F.; Oeser, T.; Mastalerz, M. Soluble Congeners of Prior Insoluble Shape-Persistent Imine Cages. *Chem. Eur. J.* **2021**, *27*, 9383–9390. [CrossRef]
33. Schoepff, L.; Kocher, L.; Durot, S.; Heitz, V. Chemically Induced Breathing of Flexible Porphyrinic Covalent Cages. *J. Org. Chem.* **2017**, *82*, 5845–5851. [CrossRef]
34. Díaz, A.E.M.; Lewis, J.E.M. Structural Flexibility in Metal-Organic Cages. *Front. Chem.* **2021**, *9*, 706462. [CrossRef]
35. Mecozzi, S.; Rebek, J., Jr. The 55% Solution: A Formula for Molecular Recognition in the Liquid State. *Chem. Eur. J.* **1998**, *4*, 1016–1022. [CrossRef]
36. Young, T.A.; Martí-Centelles, V.; Wang, J.; Lusby, P.J.; Duarte, F. Rationalizing the Activity of an "Artificial Diels-Alderase": Establishing Efficient and Accurate Protocols for Calculating Supramolecular Catalysis. *J. Am. Chem. Soc.* **2020**, *142*, 1300–1310. [CrossRef]
37. Kravchenko, O.; Varava, A.; Pokorny, F.T.; Devaurs, D.; Kavraki, L.E.; Kragic, D. A Robotics-Inspired Screening Algorithm for Molecular Caging Prediction. *J. Chem. Inf. Model.* **2020**, *60*, 1302–1316. [CrossRef]
38. Young, T.A.; Gheorghe, R.; Duarte, F. cgbind: A Python Module and Web App for Automated Metallocage Construction and Host-Guest Characterization. *J. Chem. Inf. Model.* **2020**, *60*, 3546–3557. [CrossRef]
39. Nurttila, S.S.; Brenner, W.; Mosquera, J.; van Vliet, K.M.; Nitschke, J.R.; Reek, J.N.H. Size-Selective Hydroformylation by a Rhodium Catalyst Confined in a Supramolecular Cage. *Chem. Eur. J.* **2019**, *25*, 609–620. [CrossRef]
40. Liu, L.; Tao, Z.P.; Chi, H.R.; Wang, B.; Wang, S.M.; Han, Z.B. The applications and prospects of hydrophobic metal–organic frameworks in catalysis. *Dalton Trans.* **2021**, *50*, 39–58. [CrossRef]
41. Ma, T.; Kapustin, E.A.; Yin, S.X.; Liang, L.; Zhou, Z.; Niu, J.; Li, L.-H.; Wang, Y.; Su, J.; Li, J.; et al. Single-crystal X-ray diffraction structures of covalent organic frameworks. *Science* **2018**, *361*, 48–52. [CrossRef]
42. Cusin, L.; Peng, H.; Ciesielski, A.; Samor, P. Chemical Conversion and Locking of the Imine Linkage: Enhancing the Functionality of Covalent Organic Frameworks. *Angew. Chem. Int. Ed.* **2021**, *60*, 14236–14250. [CrossRef]
43. Hu, J.; Gupta, S.K.; Ozdemir, J.; Beyzavi, M.H. Applications of Dynamic Covalent Chemistry Concept toward Tailored Covalent Organic Framework Nanomaterials: A Review. *ACS Appl. Nano Mater.* **2020**, *3*, 6239–6269. [CrossRef] [PubMed]
44. Hou, L.; Shan, C.; Song, Y.; Chen, S.; Wojtas, L.; Ma, S.; Sun, Q.; Zhang, L. Highly Stable Single Crystals of Three-Dimensional Porous Oligomer Frameworks Synthesized under Kinetic Conditions. *Angew. Chem. Int. Ed.* **2021**, *60*, 14664–14670. [CrossRef] [PubMed]
45. Zhang, B.; Zu, W.; Cui, X.; Zhou, J.; Fu, Y.; Chen, J. Preparation of Hydrophobic Metal–Organic Frameworks/Parylene Composites as a Platform for Enhanced Catalytic Performance. *Inorg. Chem.* **2022**, *61*, 18303–18310. [CrossRef] [PubMed]

46. Lei, Z. Cyanurate-Linked Covalent Organic Frameworks Enabled by Dynamic Nucleophilic Aromatic Substitution. *J. Am. Chem. Soc.* **2022**, *144*, 17737–17742. [CrossRef] [PubMed]
47. Pullen, S.; Tessarolo, J.; Clever, G.H. Increasing structural and functional complexity in self-assembled coordination cages. *Chem. Sci.* **2021**, *12*, 7269–7293. [CrossRef]
48. Mukoyoshi, M.; Kitagawa, H. Nanoparticle/metal–organic framework hybrid catalysts: Elucidating the role of the MOF. *Chem. Commun.* **2022**, *58*, 10757–10767. [CrossRef]
49. Koizumi, Y.; Yonesato, K.; Yamaguchi, K.; Suzuki, K. Ligand-Protecting Strategy for the Controlled Construction of Multinuclear Copper Cores within a Ring-Shaped Polyoxometalate. *Inorg. Chem.* **2022**, *61*, 9841–9848. [CrossRef]
50. Maru, K.; Kalla, S.; Jangir, R. MOF/POM hybrids as catalysts for organic Transformations. *Dalton Trans.* **2022**, *51*, 11952–11986. [CrossRef]
51. Wang, D.; Qi, J.; Yang, N.; Cui, W.; Wang, J.; Li, Q.; Zhang, Q.; Yu, X.; Gu, L.; Li, J.; et al. Dual-defects adjusted crystal field splitting of LaCo$_{1-x}$Ni$_x$O$_{3-\delta}$ hollow multishelled structures for efficient oxygen evolution. *Angew. Chem. Int. Ed* **2020**, *59*, 19691–19695. [CrossRef]
52. Lutz, F.; Lorenzo-Parodi, N.; Schmidt, T.C.; Niemeyer, J. Heteroternary cucurbit[8]uril complexes as supramolecular scaffolds for self-assembled bifunctional photoredoxcatalysts. *Chem. Commun.* **2021**, *57*, 2887–2890. [CrossRef]
53. Mouarrawis, V.; Bobylev, E.O.; de Bruin, B.; Reek, J.N.H. A Novel M8L6 Cubic Cage That Binds Tetrapyridyl Porphyrins: Cage and Solvent Effects in Cobalt-Porphyrin-Catalyzed Cyclopropanation Reactions. *Chem. Eur. J.* **2021**, *27*, 8390–8397. [CrossRef]
54. Hu, X.; Han, M.; Shao, L.; Zhang, C.; Zhang, L.; Kelley, S.P.; Zhang, C.; Lin, J.; Dalgarno, S.J.; Atwood, D.A.; et al. Self-Assembly of a Semiconductive and Photoactive Heterobimetallic Metal–Organic Capsule. *Angew. Chem. Int. Ed.* **2021**, *60*, 10516–10520. [CrossRef] [PubMed]
55. Esselborn, J.; Muraki, N.; Klein, K.; Engelbrecht, V.; Metzler-Nolte, N.; Apfel, U.P.; Hofmann, E.; Kurisu, G.; Happe, T. A structural view of synthetic cofactor integration into [FeFe]-hydrogenases. *Chem. Sci.* **2016**, *7*, 959–968. [CrossRef] [PubMed]
56. Mouarrawis, V.; Plessius, R.; van der Vlugt, J.; Reek, J.N.H. Confinement Effects in Catalysis Using Well-Defined Materials and Cages. *Front. Chem.* **2018**, *6*, 623. [CrossRef] [PubMed]
57. Mitschke, B.; Turberg, M.; List, B. Confinement as a unifying element in selective catalysis. *Chem* **2020**, *6*, 2515–2532. [CrossRef]
58. Tehrani, F.N.; Assaf, K.I.; Hein, R.; Jensen, C.M.E.; Nugent, T.C.; Nau, W.M. Supramolecular Catalysis of a Catalysis-Resistant Diels–Alder Reaction: Almost Theoretical Acceleration of Cyclopentadiene Dimerization inside Cucurbit[7]uril. *ACS Catal.* **2022**, *12*, 2261–2269. [CrossRef]
59. Yuasa, M.; Sumida, R.; Tanaka, Y.; Yoshizawa, M. Selective Encapsulation and Unusual Stabilization of cis-Isomers by a Spherical Polyaromatic Cavity. *Chem. Eur. J.* **2022**, *28*, e202104101. [CrossRef]
60. Nguyen, Q.N.N.; Xia, K.T.; Zhang, Y.; Chen, N.; Morimoto, M.; Pei, X.; Ha, Y.; Guo, J.; Yang, W.; Wang, L.P.; et al. Source of Rate Acceleration for Carbocation Cyclization in Biomimetic Supramolecular Cages. *J. Am. Chem. Soc.* **2022**, *144*, 11413–11424. [CrossRef]
61. Norjmaa, G.; Himo, F.; Maréchal, J.D.; Ujaque, G. Catalysis by [Ga4L6]$^{12-}$ Metallocage on the Nazarov Cyclization: The Basicity of Complexed Alcohol is Key. *Chem. Eur. J.* **2022**, *28*, e202201792. [CrossRef]
62. Leurs, M.; Dorn, B.; Wilhelm, S.; Manisegaran, M.; Tiller, J.C. Multicore Artificial Metalloenzymes Derived from Acylated Proteins as Catalysts for the Enantioselective Dihydroxylation and Epoxidation of Styrene Derivatives. *Chem. Eur. J.* **2018**, *24*, 10859–10867. [CrossRef]
63. Léonard, N.G.; Dhaoui, R.; Chantarojsiri, T.; Yang, J.Y. Electric Fields in Catalysis: From Enzymes to Molecular Catalysts. *ACS Catal.* **2021**, *11*, 10923–10932. [CrossRef] [PubMed]
64. Dong, B.; Mansour, N.; Huang, T.X.; Huang, W.; Fang, N. Single molecule fluorescence imaging of nanoconfinement in porous materials. *Chem. Soc. Rev.* **2021**, *50*, 6483–6506. [CrossRef] [PubMed]
65. Hkiri, S.; Steinmetz, M.; Schurhammer, R.; Sémeril, D. Encapsulated neutral ruthenium catalyst for substrate-selective oxidation of alcohols. *Chem. Eur. J.* **2022**, *28*, e202201887. [CrossRef]
66. Lin, H.; Xiao, Z.; Le, K.N.; Yan, T.; Cai, P.; Yang, Y.; Day, G.S.; Drake, H.F.; Xie, H.; Bose, R.; et al. Assembling Phenothiazine into a Porous Coordination Cage to Improve Its Photocatalytic Efficiency for Organic Transformations. *Angew. Chem. Int. Ed.* **2022**, *61*, e202214055. [CrossRef] [PubMed]
67. Wang, H.; Liu, X.; Yang, W.; Mao, G.; Meng, Z.; Wu, Z.; Jiang, H.L. Surface-Clean Au25 Nanoclusters in Modulated Microenvironment Enabled by Metal–Organic Frameworks for Enhanced Catalysis. *J. Am. Chem. Soc.* **2022**, *144*, 22008–22017. [CrossRef]
68. Banerjee, R.; Lipscomb, J.D. Small-Molecule Tunnels in Metalloenzymes Viewed as Extensions of the Active Site. *Acc. Chem. Res.* **2021**, *54*, 2185–2195. [CrossRef]
69. Moorthy, S.; Bonillo, A.C.; Lambert, H.; Kalenius, E.; Lee, T.C. Modulating the reaction pathway of phenyl diazonium ions using host–guest complexation with cucurbit[7]uril. *Chem. Commun.* **2022**, *58*, 3617–3620. [CrossRef]
70. Galan, A.; Ballester, P. Stabilization of reactive species by supramolecular encapsulation. *Chem. Soc. Rev.* **2016**, *45*, 1720–1737. [CrossRef]
71. Shen, J.M.; Kung, M.C.; Shen, Z.L.; Wang, Z.; Gunderson, W.A.; Hoffman, B.M.; Kung, H.H. Generating and Stabilizing Co(I) in a Nanocage Environment. *J. Am. Chem. Soc.* **2014**, *136*, 5185–5188. [CrossRef]
72. Wu, B.; Lin, T.; Lu, Z.; Yu, X.; Huang, M.; Yang, R.; Wang, C.; Tian, C.; Li, J.; Sun, Y.; et al. Fe binuclear sites convert methane to acetic acid with ultrahigh selectivity. *Chem* **2022**, *8*, 1658–1672. [CrossRef]

73. Zhang, P.; Tugny, C.; Suárez, J.M.; Guitet, M.; Derat, E.; Vanthuyne, N.; Zhang, Y.; Bistri, O.; Mouriès-Mansuy, V.; Ménand, M.; et al. Artificial chiral metallo-pockets Including a single metal serving as structural probe and catalytic center. *Chem* **2017**, *3*, 174–191. [CrossRef]

74. Roland, S.; Suarez, J.M.; Sollogoub, M. Confinement of metal–N-heterocyclic carbene complexes to control reactivity in catalytic reactions. *Chem. Eur. J.* **2018**, *24*, 12464–12473. [CrossRef]

75. Qi, X.; Zhong, R.; Chen, M.; Sun, C.; You, S.; Gu, J.; Shan, G.; Cui, D.; Wang, X.; Su, Z. Single Metal–Organic Cage Decorated with an Ir(III) Complex for CO_2 Photoreduction. *ACS Catal.* **2021**, *11*, 7241–7248. [CrossRef]

76. Mouarrawis, V.; Bobylev, E.O.; de Bruin, B.; Reek, J.N.H. Controlling the Activity of a Caged Cobalt-Porphyrin-Catalyst in Cyclopropanation Reactions with Peripheral Cage Substituents. *Eur. J. Inorg. Chem.* **2021**, *2021*, 2890–2898. [CrossRef]

77. De Rosa, M.; Gambaro, S.; Soriente, A.; Sala, P.D.; Gaeta, C.; Rescifina, A.; Neri, P. Carbocation catalysis in confined space: Activation of trityl chloride inside the hexameric resorcinarene capsule. *Chem. Sci.* **2022**, *13*, 8618–8625. [CrossRef]

78. Li, M.S.; Dong, Y.W.; Quan, M.; Jiang, W. Stabilization of Imines and Hemiaminals in Water by an Endo-Functionalized Container Molecule. *Angew. Chem. Int. Ed.* **2022**, *61*, e202208508. [CrossRef]

79. Wang, V.C.C. Beyond the Active Site: Mechanistic Investigations of the Role of the Secondary Coordination Sphere and Beyond on Multi-Electron Electrocatalytic Reactions and Their Relations Between Kinetics and Thermodynamics. *ACS Catal.* **2021**, *11*, 8292–8303. [CrossRef]

80. Goralski, S.T.; Cid-Seara, K.M.; Jarju, J.J.; Rodriguez-Lorenzo, L.; LaGrow, A.P.; Rose, M.J.; Salonen, L.M. Threefold reactivity of a COF-embedded rhenium catalyst: Reductive etherification, oxidative esterification or transfer hydrogenation. *Chem. Commun.* **2022**, *58*, 12074–12077. [CrossRef]

81. He, W.; Tashiro, S.; Shionoya, M. Highly selective acid-catalyzed olefin isomerization of limonene to terpinolene by kinetic suppression of overreactions in a confined space of porous metal–macrocycle frameworks. *Chem. Sci.* **2022**, *13*, 8752–8758. [CrossRef] [PubMed]

82. Yao, S.; Chang, L.P.; Guo, G.C.; Wang, Y.J.; Tian, Z.Y.; Guo, S.; Lu, T.B.; Zhang, Z.M. Microenvironment Regulation of {$Co_4^{II}O_4$} Cubane for Syngas Photosynthesis. *Inorg. Chem.* **2022**, *61*, 13058–13066. [CrossRef]

83. Lyu, Y.; Scrimin, P. Mimicking Enzymes: The Quest for Powerful Catalysts from Simple Molecules to Nanozymes. *ACS Catal.* **2021**, *11*, 11501–11509. [CrossRef]

84. Bellia, F.; La Mendola, D.; Pedone, C.; Rizzarelli, E.; Saviano, M.; Vecchio, G. Selectively functionalized cyclodextrins and their metal complexes. *Chem. Soc. Rev.* **2009**, *38*, 2756–2781. [CrossRef]

85. Hu, D.; Zhang, J.; Liu, M. Recent advances in the applications of porous organic cages. *Chem. Commun.* **2022**, *58*, 11333–11346. [CrossRef]

86. Liu, M.; Zhang, L.; Wang, T. Supramolecular chirality in self-assembled systems. *Chem. Rev.* **2015**, *115*, 7304–7397. [CrossRef] [PubMed]

87. Qiu, G.; Nava, P.; Colomban, C.; Martinez, A. Control and Transfer of Chirality Within Well-Defined Tripodal Supramolecular Cages. *Front. Chem.* **2020**, *8*, 599893. [CrossRef]

88. Ezhov, R.; Ravari, A.K.; Page, A.; Pushkar, Y. Water Oxidation Catalyst cis-[Ru(bpy)(5,5′-dcbpy)(H2O)$_2$]$^{2+}$ and Its Stabilization in Metal–Organic Framework. *ACS Catal.* **2020**, *10*, 5299–5308. [CrossRef]

89. Chen, Z.Y.; Long, Z.H.; Wang, X.Z.; Zhou, J.Y.; Wang, X.S.; Zhou, X.P.; Li, D. Cobalt-Based Metal–Organic Cages for Visible-Light-Driven Water Oxidation. *Inorg. Chem* **2021**, *60*, 10380–10386. [CrossRef]

90. Yu, F.; Poole, D.; Mathew, S.; Yan, N.; Hessels, J.; Orth, N.; Ivanović-Burmazović, I.; Reek, J.N.H. Control over electrochemical water oxidation catalysis by preorganization of molecular ruthenium catalysts in self-assembled nanospheres. *Angew. Chem. Int. Ed.* **2018**, *57*, 11247–11251. [CrossRef]

91. Chen, Y.; Li, P.; Zhou, J.; Buru, C.T.; Đorđević, L.; Li, P.; Zhang, X.; Cetin, M.M.; Stoddart, J.F.; Stupp, S.I.; et al. Integration of Enzymes and Photosensitizers in a Hierarchical Mesoporous Metal–Organic Framework for Light-Driven CO_2 Reduction. *J. Am. Chem. Soc.* **2020**, *142*, 1768–1773. [CrossRef]

92. Guo, S.; Kong, L.H.; Wang, P.; Yao, S.; Lu, T.B.; Zhang, Z.M. Switching Excited State Distribution of Metal–Organic Framework for Dramatically Boosting Photocatalysis. *Angew. Chem. Int. Ed.* **2022**, *61*, e202206193. [CrossRef]

93. Lin, G.; Zhang, Y.; Hua, Y.; Zhang, C.; Jia, C.; Ju, D.; Yu, C.; Li, P.; Liu, J. Bioinspired Metalation of the Metal-Organic Framework MIL-125-NH$_2$ for Photocatalytic NADH Regeneration and Gas-Liquid-Solid Three-Phase Enzymatic CO_2 Reduction. *Angew. Chem. Int. Ed.* **2022**, *61*, e202206283. [CrossRef]

94. Han, X.; Chu, Y.J.; Dong, M.; Chen, W.; Ding, G.; Wen, L.L.; Shao, K.Z.; Su, Z.; Zhang, M.; Wang, X.; et al. Copper-Based Metal–Organic Framework with a Tetraphenylethylene-Tetrazole Linker for Visible-Light-Driven CO_2 Photoconversion. *Inorg. Chem.* **2022**, *61*, 5869–5877. [CrossRef]

95. Ghosh, A.C.; Legrand, A.; Rajapaksha, R.; Craig, G.A.; Sassoye, C.; Balázs, G.; Farrusseng, D.; Furukawa, S.; Canivet, J.; Wisser, F.M. Rhodium-Based Metal–Organic Polyhedra Assemblies for Selective CO_2 Photoreduction. *J. Am. Chem. Soc.* **2022**, *144*, 3626–3636. [CrossRef]

96. Huang, N.Y.; Shen, J.Q.; Zhang, X.W.; Liao, P.Q.; Zhang, J.P.; Chen, X.M. Coupling Ruthenium Bipyridyl and Cobalt Imidazolate Units in a Metal–Organic Framework for an Efficient Photosynthetic Overall Reaction in Diluted CO_2. *J. Am. Chem. Soc.* **2022**, *144*, 8676–8682. [CrossRef] [PubMed]

97. Zhang, A.A.; Si, D.; Huang, H.; Xie, L.; Fang, Z.B.; Liu, T.F.; Cao, R. Partial Metalation of Porphyrin Moieties in Hydrogen-Bonded Organic Frameworks Provides Enhanced CO_2 Photoreduction Activity. *Angew. Chem. Int. Ed.* **2022**, *61*, e202203955. [CrossRef]

98. Huang, Z.W.; Hu, K.Q.; Mei, L.; Wang, D.G.; Wang, J.Y.; Wu, W.S.; Chai, Z.F.; Shi, W.Q. Encapsulation of Polymetallic Oxygen Clusters in a Mesoporous/Microporous Thorium-Based Porphyrin Metal–Organic Framework for Enhanced Photocatalytic CO_2 Reduction. *Inorg. Chem.* **2022**, *61*, 3368–3373. [CrossRef]

99. Sun, Q.M.; Xu, J.J.; Tao, F.F.; Ye, W.; Zhou, C.; He, J.H.; Lu, J.M. Boosted Inner Surface Charge Transfer in Perovskite Nanodots@Mesoporous Titania Frameworks for Efficient and Selective Photocatalytic CO_2 Reduction to Methane. *Angew. Chem. Int. Ed.* **2022**, *61*, e202200872. [CrossRef]

100. Hu, X.; Zheng, L.; Wang, S.; Wang, X.; Tan, B. Integrating single Co sites into crystalline covalent triazine frameworks for photoreduction of CO_2. *Chem. Commun.* **2022**, *58*, 8121–8124. [CrossRef] [PubMed]

101. Ran, L.; Li, Z.; Ran, B.; Cao, J.; Zhao, Y.; Shao, T.; Song, Y.; Leung, M.K.H.; Sun, L.; Hou, J. Engineering Single-Atom Active Sites on Covalent Organic Frameworks for Boosting CO_2 Photoreduction. *J. Am. Chem. Soc.* **2022**, *144*, 17097–17109. [CrossRef]

102. Han, W.K.; Liu, Y.; Yan, X.; Jiang, Y.; Zhang, J.; Gu, Z.G. Integrating Light-Harvesting Ruthenium(II)-based Units into Three-Dimensional Metal Covalent Organic Frameworks for Photocatalytic Hydrogen Evolution. *Angew. Chem. Int. Ed.* **2022**, *61*, e202208791. [CrossRef] [PubMed]

103. Wu, B.; Liu, N.; Lu, L.; Zhang, R.; Zhang, R.; Shi, W.; Cheng, P. A MOF-derived hierarchical CoP@ZnIn2S4 photocatalyst for visible light-driven hydrogen evolution. *Chem. Commun.* **2022**, *58*, 6622–6625. [CrossRef] [PubMed]

104. Ke, Y.; Zhang, J.; Liu, L.; Li, X.; Liang, Q.; Li, Z. Self-Assembled Zeolitic Imidazolate Framework/CdS Hollow heres with Efficient Charge Separation for Enhanced Photocatalytic Hydrogen Evolution. *Inorg. Chem.* **2022**, *61*, 10598–10608. [CrossRef]

105. Sun, L.; Lu, M.; Yang, Z.; Yu, Z.; Su, X.; Lan, Y.Q.; Chen, L. Nickel Glyoximate Based Metal–Covalent Organic Frameworks for Efficient Photocatalytic Hydrogen Evolution. *Angew. Chem. Int. Ed.* **2022**, *61*, e202204326. [CrossRef]

106. Pukdeejorhor, L.; Adpakpang, K.; Wannapaiboon, S.; Bureekaew, S. Co-based metal–organic framework for photocatalytic hydrogen generation. *Chem. Commun.* **2022**, *58*, 8194–8197. [CrossRef]

107. Teng, Q.; He, Y.P.; Chen, S.M.; Zhang, J. Synthesis of a Zr4(embonate)6–cobalt based superstructure for photocatalytic hydrogen production. *Dalton Trans.* **2022**, *51*, 11612–11616. [CrossRef]

108. Karak, S.; Stepanenko, V.; Addicoat, M.A.; Keßler, P.; Moser, S.; Beuerle, F.; Wurthner, F. A Covalent Organic Framework for Cooperative Water Oxidation. *J. Am. Chem. Soc.* **2022**, *144*, 17661–17670. [CrossRef]

109. Castner, A.T.; Johnson, B.A.; Cohen, S.M.; Ott, S. Mimicking the Electron Transport Chain and Active Site of [FeFe] Hydrogenases in One Metal–Organic Framework: Factors That Influence Charge Transport. *J. Am. Chem. Soc.* **2021**, *143*, 7991–7999. [CrossRef]

110. Balestri, D.; Roux, Y.; Mattarozzi, M.; Mucchino, C.; Heux, L.; Brazzolotto, D.; Artero, V.; Duboc, C.; Pelagatti, P.; Marchiò, L.; et al. Heterogenization of a [NiFe] Hydrogenase Mimic through Simple and Efficient Encapsulation into a Mesoporous MOF. *Inorg. Chem.* **2017**, *56*, 14801–14808. [CrossRef] [PubMed]

111. Li, M.; Chen, J.; Wu, W.; Fang, Y.; Dong, S. Oxidase-like MOF-818 Nanozyme with High Specificity for Catalysis of Catechol Oxidation. *J. Am. Chem. Soc.* **2020**, *142*, 15569–15574. [CrossRef]

112. Feng, X.; Song, Y.; Chen, J.S.; Xu, Z.; Dunn, S.J.; Lin, W. Rational Construction of an Artificial Binuclear Copper Monooxygenase in a Metal-Organic Framework. *J. Am. Chem. Soc.* **2021**, *143*, 1107–1118. [CrossRef]

113. Ji, G.; Zhao, L.; Wang, Y.; Tang, Y.; He, C.; Liu, S.; Duan, C. A Binuclear Cerium-Based Metal–Organic Framework as an Artificial Monooxygenase for the Saturated Hydrocarbon Aerobic Oxidation with High Efficiency and High Selectivity. *ACS Catal.* **2022**, *12*, 7821–7832. [CrossRef]

114. Antil, N.; Chauhan, M.; Akhtar, N.; Newar, R.; Begum, W.; Malik, J.; Manna, K. Metal–Organic Framework-Encaged Monomeric Cobalt(III) Hydroperoxides Enable Chemoselective Methane Oxidation to Methanol. *ACS Catal.* **2022**, *12*, 11159–11168. [CrossRef]

115. Jing, X.; He, C.; Yang, Y.; Duan, C. A Metal–Organic Tetrahedron as a Redox Vehicle to Encapsulate Organic Dyes for Photocatalytic Proton Reduction. *J. Am. Chem. Soc.* **2015**, *137*, 3967–3974. [CrossRef] [PubMed]

116. Wang, W.; Song, X.W.; Hong, Z.; Li, B.; Si, Y.; Ji, C.; Su, K.; Tan, Y.; Ju, Z.; Huang, Y.; et al. Incorporation of iron hydrogenase active sites into a stable photosensitizing metal-organic framework for enhanced hydrogen production. *Appl. Catal. B Environ.* **2019**, *258*, 117279. [CrossRef]

117. Zaffaroni, R.; Orth, N.; Ivanović-Burmazović, I.; Reek, J.N.H. Hydrogenase Mimics in $M_{12}L_{24}$ Nanospheres to Control Overpotential and Activity in Proton-Reduction Catalysis. *Angew. Chem. Int. Ed.* **2020**, *59*, 18485–18489. [CrossRef]

118. Dong, K.; Liang, J.; Wang, Y.; Zhang, L.; Xu, Z.; Sun, S.; Luo, Y.; Li, T.; Liu, Q.; Li, N.; et al. Conductive Two-Dimensional Magnesium Metal–Organic Frameworks for High-Efficiency O_2 Electroreduction to H_2O_2. *ACS Catal.* **2022**, *12*, 6092–6099. [CrossRef]

119. Guo, B.; Cheng, X.; Tang, Y.; Guo, W.; Deng, S.; Wu, L.; Fu, X. Dehydrated UiO-66(SH)$_2$: The Zr-O Cluster and Its Photocatalytic Role Mimicking the Biological Nitrogen Fixation. *Angew. Chem. Int. Ed.* **2022**, *61*, e202117244. [CrossRef]

120. Li, H.; Xia, M.; Chong, B.; Xiao, H.; Zhang, B.; Lin, B.; Yang, B.; Yang, G. Boosting Photocatalytic Nitrogen Fixation via Constructing Low-Oxidation-State Active Sites in the Nanoconfined Spinel Iron Cobalt Oxide. *ACS Catal.* **2022**, *12*, 10361–10372. [CrossRef]

121. Guo, L.; Li, F.; Liu, J.; Jia, Z.; Li, R.; Yu, Z.; Wanga, Y.; Fan, C. Improved visible light photocatalytic nitrogen fixation activity using a Fe^{II}-rich MIL-101(Fe): Breaking the scaling relationship by photoinduced Fe^{II}/Fe^{III} cycling. *Dalton Trans.* **2022**, *51*, 13085–13093. [CrossRef]

122. Friedlea, S.; Reisner, E.; Lippard, S.J. Current Challenges for Modeling Enzyme Active Sites by Biomimetic Synthetic Diiron Complexes. *Chem. Soc. Rev.* **2010**, *39*, 2768–2779. [CrossRef]
123. Freakley, S.J.; Dimitratos, N.; Willock, D.J.; Taylor, S.H.; Kiely, C.J.; Hutchings, G.J. Methane Oxidation to Methanol in Water. *Acc. Chem. Res.* **2021**, *54*, 2614–2623. [CrossRef] [PubMed]
124. Shteinman, A.A. Bioinspired oxidation of methane: From academic models of methane monooxygenases to direct conversion of methane to methanol. *Kinet. Catal.* **2020**, *61*, 339–359. [CrossRef]
125. Srinivas, V.; Banerjee, R.; Lebrette, H.; Jones, J.C.; Aurelius, O.; Kim, I.S.; Pham, C.C.; Gul, S.; Sutherlin, K.D.; Bhowmick, A.; et al. High-resolution XFEL structure of the soluble methane monooxygenase hydroxylase complex with its regulatory component at ambient temperature in two oxidation states. *J. Am. Chem. Soc.* **2020**, *142*, 14249–14266. [CrossRef] [PubMed]
126. Klinman, J.P.; Offenbacher, A.R. Understanding biological hydrogen transfer through the lens of temperature dependent kinetic isotope effects. *Acc. Chem. Res.* **2018**, *51*, 1966–1974. [CrossRef]
127. Lee, S.J.; McCormick, M.S.; Lippard, S.J.; Cho, U.S. Control of substrate access to the active site in methane monooxygenase. *Nature* **2013**, *494*, 380–384. [CrossRef] [PubMed]
128. Sasmal, H.S.; Bag, S.; Chandra, B.; Majumder, P.; Kuiry, H.; Karak, S.; Gupta, S.S.; Banerjee, R. Heterogeneous C−H Functionalization in Water via Porous Covalent Organic Framework Nanofilms: A Case of Catalytic Sphere Transmutation. *J. Am. Chem. Soc.* **2021**, *143*, 8426–8436. [CrossRef]
129. van Heerden, D.P.; Smith, V.J.; Aggarwal, H.; Barbour, L.J. High Pressure in Situ Single-Crystal X-ray Diffraction Reveals Turnstile Linker Rotation Upon Room-Temperature Stepped Uptake of Alkanes. *Angew. Chem. Int. Ed.* **2021**, *60*, 13430–13435. [CrossRef]
130. Zhu, J.L.; Zhang, D.; Ronson, T.K.; Wang, W.; Xu, L.; Yang, H.B.; Nitschke, J.R. A Cavity-Tailored Metal-Organic Cage Entraps Gases Selectively in Solution and the Amorphous Solid State. *Angew. Chem. Int. Ed.* **2021**, *60*, 11789–11792. [CrossRef]
131. Simons, M.C.; Prinslow, S.D.; Babucci, M.; Hoffman, A.S.; Hong, J.; Vitillo, J.G.; Bare, S.R.; Gates, B.C.; Lu, C.C.; Gagliardi, L.; et al. Beyond Radical Rebound: Methane Oxidation to Methanol Catalyzed by Iron Species in Metal–Organic Framework Nodes. *J. Am. Chem. Soc.* **2021**, *143*, 12165–12174. [CrossRef] [PubMed]
132. Bete, S.C.; May, L.K.; Woite, P.; Roemelt, M.; Otte, M. A Copper Cage-Complex as Mimic of the pMMO Cu$_C$ Site. *Angew. Chem. Int. Ed.* **2022**, *61*, e202206120. [CrossRef] [PubMed]
133. Ikbal, S.A.; Colomban, C.; Zhang, D.; Delecluse, M.; Brotin, T.; Dufaud, V.; Dutasta, J.-P.; Sorokin, A.B.; Martinez, A. Bioinspired Oxidation of Methane in the Confined Spaces of Molecular Cages. *Inorg. Chem.* **2019**, *58*, 7220–7228. [CrossRef] [PubMed]
134. Liu, C.C.; Mou, C.Y.; Yu, S.S.F.; Chan, S.I. Heterogeneous formulation of the tricopper complex for efficient catalytic conversion of methane into methanol at ambient temperature and pressure. *Energy Environ. Sci.* **2016**, *9*, 1361–1374. [CrossRef]
135. Jasniewski, A.J.; Que, L., Jr. Dioxygen activation by nonheme diiron enzymes: Diverse oxygen adducts, high-valent intermediates, and related model complexes. *Chem. Rev.* **2018**, *118*, 2554–2592. [CrossRef]
136. Zhou, Y.; Wei, Y.; Ren, J.; Qu, X. A chiral covalent organic frameworks (COFs) nanozyme with ultrahigh enzymatic activity. *Mater. Horiz.* **2020**, *7*, 3291–3297. [CrossRef]
137. Shteinman, A.A.; Mitra, M. Nonheme mono- and dinuclear iron complexes in bio-inspired C–H and C–C bond hydroxylation reactions: Mechanistic insight. *Inorg. Chim. Acta* **2021**, *523*, 120388. [CrossRef]
138. Wang, R.; Rebek, J.Y.Y. Organic radical reactions confined to containers in supramolecular systems. *Chem. Commun.* **2022**, *58*, 1828–1833. [CrossRef]
139. Tian, Y.Q.; Cui, Y.S.; Yu, W.D.; Xu, C.Q.; Yi, X.Y.; Yan, J.; Li, J.; Liu, C. An Ultrastable Ti-based metallocalixarene nanocage cluster with photocatalytic amine oxidation activity. *Chem. Commun.* **2022**, *58*, 6028–6031. [CrossRef]
140. Zhao, X.; Lu, X.; Chen, W.J.; Liu, Y.; Pan, X. Palladium decoration directed synthesis of ZIF-8 nanocubes with efficient catalytic activity for nitrobenzene hydrogenation. *Dalton Trans.* **2022**, *51*, 10847–10851. [CrossRef]
141. Fu, S.; Yao, S.; Guo, S.; Guo, G.C.; Yuan, W.; Lu, T.B.; Zhang, Z.M. Feeding Carbonylation with CO$_2$ via the Synergy of Single-Site/Nanocluster Catalysts in a Photosensitizing MOF. *J. Am. Chem. Soc.* **2021**, *143*, 20792–20801. [CrossRef]
142. Dong, J.; Mo, Q.; Wang, Y.; Jiang, L.; Zhang, L.; Su, C.Y. Ultrathin Two-Dimensional Metal–Organic Framework Nanosheets Based on a Halogen-Substituted Porphyrin Ligand: Synthesis and Catalytic Application in CO$_2$ Reductive Amination. *Chem. Eur. J.* **2022**, *28*, e202200555. [CrossRef]
143. Han, H.; Zheng, X.; Qiao, C.; Xia, Z.; Yang, Q.; Di, L.; Xing, Y.; Xie, G.; Zhou, C.; Wang, W.; et al. A Stable Zn-MOF for Photocatalytic C$_{sp3}$−H Oxidation: Vinyl Double Bonds Boosting Electron Transfer and Enhanced Oxygen Activation. *ACS Catal.* **2022**, *12*, 10668–10679. [CrossRef]
144. Li, W.; Wu, G.; Hu, W.; Dang, J.; Wang, C.; Weng, X.; da Silva, I.; Manuel, P.; Yang, S.; Guan, N.; et al. Direct Propylene Epoxidation with Molecular Oxygen over Cobalt-Containing Zeolites. *J. Am. Chem. Soc.* **2022**, *144*, 4260–4268. [CrossRef]
145. Zhang, F.; Ma, J.; Tan, Y.; Yu, G.; Qin, H.; Zheng, L.; Liu, H.; Li, R. Construction of Porphyrin Porous Organic Cage as a Support for Single Cobalt Atoms for Photocatalytic Oxidation in Visible Light. *ACS Catal.* **2022**, *12*, 5827–5833. [CrossRef]
146. Ma, X.; Jing, Z.; Li, K.; Chen, Y.; Li, D.; Ma, P.; Wang, J.; Niu, J. Copper-Containing Polyoxometalate-Based Metal–Organic Framework as a Catalyst for the Oxidation of Silanes: Effective Cooperative Catalysis by Metal Sites and POM Precursor. *Inorg. Chem.* **2022**, *61*, 4056–4061. [CrossRef] [PubMed]
147. She, M.; Gu, R.; Meng, D.; Yang, H.; Wen, Y.; Qian, X.; Guo, X.; Ding, W. Nanosheets of Ni-SAPO-34 Molecular Sieve for Selective Oxidation of Cyclohexanone to Adipic Acid. *Chem. Eur. J.* **2022**, *28*, e202200696. [CrossRef] [PubMed]

148. Lai, Y.L.; Wang, X.Z.; Zhou, X.C.; Dai, R.R.; Zhou, X.P.; Li, D. Self-assembly of a Mixed Valence Copper Triangular Prism and Transformation to Cage Triggered by an External Stimulus. *Inorg. Chem.* **2020**, *59*, 17374–17378. [CrossRef]

149. Cui, W.J.; Zhang, S.M.; Ma, Y.Y.; Wang, Y.; Miao, R.X.; Han, Z.G. Polyoxometalate-Incorporated Metal-Organic Network as a Heterogeneous Catalyst for Selective Oxidation of Aryl Alkenes. *Inorg. Chem.* **2022**, *61*, 9421–9432. [CrossRef]

150. Maksimchuk, N.V.; Ivanchikova, I.D.; Cho, K.H.; Zalomaeva, O.V.; Evtushok, V.Y.; Larionov, K.P.; Glazneva, T.S.; Chang, J.S.; Kholdeeva, O.A. Catalytic Performance of Zr-Based Metal–Organic Frameworks Zr-abtc and MIP-200 in Selective Oxidations with H_2O_2. *Chem. Eur. J.* **2021**, *27*, 6985–6992. [CrossRef]

151. Fan, S.C.; Chen, S.Q.; Wang, J.W.; Li, Y.P.; Zhang, P.; Wang, Y.; Yuan, W.; Zhai, Q.G. Precise Introduction of Single Vanadium Site into Indium−Organic Framework for CO_2 Capture and Photocatalytic Fixation. *Inorg. Chem.* **2022**, *61*, 14131–14139. [CrossRef] [PubMed]

152. Liu, S.; Chen, H.; Zhang, X. Bifunctional {Pb$_{10}$K$_2$}−Organic Framework for High Catalytic Activity in Cycloaddition of CO_2 with Epoxides and Knoevenagel Condensation. *ACS Catal.* **2022**, *12*, 10373–10383. [CrossRef]

153. Chen, H.; Zhang, T.; Liu, S.; Lv, H.; Fan, L.; Zhang, X. Fluorine-Functionalized NbO-Type {Cu$_2$}-Organic Framework: Enhanced Catalytic Performance on the Cycloaddition Reaction of CO_2 with Epoxides and Deacetalization-Knoevenagel Condensation. *Inorg. Chem.* **2022**, *61*, 11949–11958. [CrossRef] [PubMed]

154. Sokolova, D.; Piccini, G.; Tiefenbacher, K. Enantioselective Tail-to-Head Terpene Cyclizations by Optically Active Hexameric Resorcin[4]arene Capsule Derivatives. *Angew. Chem. Int. Ed.* **2022**, *61*, e202203384. [CrossRef] [PubMed]

155. Li, K.; Liu, Y.F.; Lin, X.L.; Yang, G.P. Copper-Containing Polyoxometalate-Based Metal–Organic Frameworks as Heterogeneous Catalysts for the Synthesis of N-Heterocycles. *Inorg. Chem.* **2022**, *61*, 6934–6942. [CrossRef] [PubMed]

156. Jati, A.; Dey, K.; Nurhuda, M.; Addicoat, M.A.; Banerjee, R.; Maji, B. Dual Metalation in a Two-Dimensional Covalent Organic Framework for Photocatalytic C−N Cross-Coupling Reactions. *J. Am. Chem. Soc.* **2022**, *144*, 7822–7833. [CrossRef]

157. Gu, A.L.; Zhang, Y.X.; Wu, Z.L.; Cui, H.Y.; Hu, T.D.; Zhao, B. Highly Efficient Conversion of Propargylic Alcohols and Propargylic Amines with CO_2 Activated by Noble-Metal-Free Catalyst Cu$_2$O@ZIF-8. *Angew. Chem. Int. Ed.* **2022**, *61*, e202114817. [CrossRef]

158. Ma, H.C.; Sun, Y.N.; Chen, G.J.; Dong, Y.B. A BINOL-phosphoric acid and metalloporphyrin derived chiral covalent organic framework for enantioselective α-benzylation of aldehydes. *Chem. Sci.* **2022**, *13*, 1906–1911. [CrossRef]

159. Qiao, J.; Zhang, B.; Yu, X.; Zou, X.; Liu, X.; Zhang, L.; Liu, Y. A Stable Y(III)-Based Amide Functionalized Metal–Organic Framework for Propane/Methane Separation and Knoevenagel Condensation. *Inorg. Chem.* **2022**, *61*, 3708–3715. [CrossRef]

160. Rasaily, S.; Sharma, D.; Pradhan, S.; Diyali, N.; Chettri, S.; Gurung, B.; Tamang, S.; Pariyar, A. Multifunctional Catalysis by a One-Dimensional Copper(II) Metal Organic Framework Containing Pre-existing Coordinatively Unsaturated Sites: Intermolecular C−N, C−O, and C−S Cross-Coupling; Stereoselective Intramolecular C−N Coupling; and Aziridination Reactions. *Inorg. Chem.* **2022**, *61*, 13685–13699. [CrossRef]

161. Li, L.; Li, Y.; Jiao, L.; Liu, X.; Ma, Z.; Zeng, Y.J.; Zheng, X.; Jiang, H.L. Light-Induced Selective Hydrogenation over PdAg Nanocages in Hollow MOF Microenvironment. *J. Am. Chem. Soc.* **2022**, *144*, 17075–17085. [CrossRef] [PubMed]

162. Wei, R.J.; You, P.Y.; Duan, H.; Xie, M.; Xia, R.Q.; Chen, X.; Zhao, X.; Ning, G.H.; Cooper, A.I.; Li, D. Ultrathin Metal−Organic Framework Nanosheets Exhibiting Exceptional Catalytic Activity. *J. Am. Chem. Soc.* **2022**, *144*, 17487–17495. [CrossRef] [PubMed]

163. Gao, A.; Li, F.; Xu, Z.; Ji, C.; Gu, J.; Zhou, Y.H. Guanidyl-implanted UiO-66 as an efficient catalyst for the enhanced conversion of carbon dioxide into cyclic carbonates. *Dalton Trans.* **2022**, *51*, 2567–2576. [CrossRef] [PubMed]

164. Jiang, J.; Wei, W.; Tang, Y.; Yang, S.; Wang, X.; Xu, Y.; Ai, L. In Situ Implantation of Bi2S3 Nanorods into Porous Quasi-Bi-MOF Architectures: Enabling Synergistic Dissociation of Borohydride for an Efficient and Fast Catalytic Reduction of 4-Nitrophenol. *Inorg. Chem.* **2022**, *61*, 19847–19856. [CrossRef] [PubMed]

165. Amanullah, S.; Saha, P.; Nayek, A.; Ahmed, M.E.; Dey, A. Biochemical and artificial pathways for the reduction of carbon dioxide, nitrite and the competing proton reduction: Effect of 2nd sphere interactions in catalysis. *Chem. Soc. Rev.* **2021**, *50*, 3755–3823. [CrossRef] [PubMed]

166. Wang, C.P.; Feng, Y.; Sun, H.; Wang, Y.; Yin, J.; Yao, Z.; Bu, X.-H.; Zhu, J. Self-Optimized Metal–Organic Framework Electrocatalysts with Structural Stability and High Current Tolerance for Water Oxidation. *ACS Catal.* **2021**, *11*, 7132–7143. [CrossRef]

167. Al-Attas, T.A.; Marei, N.N.; Yong, X.; Yasri, N.G.; Thangadurai, V.; Shimizu, G.; Siahrostami, S.; Kibria, G. Ligand-Engineered Metal–Organic Frameworks for Electrochemical Reduction of Carbon Dioxide to Carbon Monoxide. *ACS Catal.* **2021**, *11*, 7350–7357. [CrossRef]

168. He, T.; Chen, S.; Ni, B.; Gong, Y.; Wu, Z.; Song, L. Zirconium−Porphyrin-Based Metal-Organic Framework Hollow Nanotubes for Immobilization of Noble-Metal Single Atoms. *Angew. Chem. Int. Ed.* **2018**, *57*, 3493–3498. [CrossRef]

169. Zheng, Y.; Hu, H.; Zhu, Y.; Rong, J.; Zhang, T.; Yang, D.; Wen, Q.; Qiu, F. ZIF-67-Derived (NiCo)S2@NC Nanosheet Arrays Hybrid for Efficient Overall Water Splitting. *Inorg. Chem.* **2022**, *61*, 14436–14446. [CrossRef]

170. Zheng, K.; Wu, Y.; Zhu, J.; Wu, M.; Jiao, X.; Li, L.; Wang, S.; Fan, M.; Hu, J.; Yan, W.; et al. Room-Temperature Photooxidation of CH_4 to CH_3OH with Nearly 100% Selectivity over Hetero-ZnO/Fe_2O_3 Porous Nanosheets. *J. Am. Chem. Soc.* **2022**, *144*, 12357–12366. [CrossRef]

171. Ou, H.; Ning, S.; Zhu, P.; Chen, S.; Han, A.; Kang, Q.; Hu, Z.; Ye, J.; Wang, D.; Li, Y. Carbon Nitride Photocatalysts with Integrated Oxidation and Reduction Atomic Active Centers for Improved CO_2 Conversion. *Angew. Chem. Int. Ed.* **2022**, *61*, e202206579. [CrossRef] [PubMed]

172. Qin, J.H.; Xiao, Z.; Xu, P.; Li, Z.H.; Lu, X.; Yang, X.G.; Lu, W.; Ma, L.F.; Li, D.S. Anionic Porous Zn-Metalated Porphyrin Metal–Organic Framework with PtS Topology for Gas-Phase Photocatalytic CO_2 Reduction. *Inorg. Chem.* **2022**, *61*, 13234–13238. [CrossRef] [PubMed]
173. Zheng, Y.L.; Dai, M.D.; Yang, X.F.; Yin, H.J.; Zhang, Y.W. Copper(II)-Doped Two-Dimensional Titanium-Based Metal–Organic Frameworks toward Light-Driven CO_2 Reduction to Value-Added Products. *Inorg. Chem.* **2022**, *61*, 13981–13991. [CrossRef]
174. Jia, Y.; Xu, Z.; Li, L.; Lin, S. Formation of NiFe-MOF nanosheets on Fe foam to achieve advanced electrocatalytic oxygen evolution. *Dalton Trans.* **2022**, *51*, 5053–5060. [CrossRef]
175. Li, X.; Zheng, Y.; Yao, H.; Bai, J.; Yue, S.; Guo, X. Interface Synergistic Effect from Hierarchically Porous $Cu(OH)_2$@FCN MOF/CF Nanosheet Arrays Boosting Electrocatalytic Oxygen Evolution. *Catalysts* **2022**, *12*, 625. [CrossRef]
176. Kong, Y.; Lu, C.; Wang, J.; Ying, S.; Liu, T.; Ma, X.; Yi, F.Y. Molecular Regulation Based on Functional Trimetallic Metal–Organic Frameworks for Efficient Oxygen Evolution Reaction. *Inorg. Chem.* **2022**, *61*, 10934–10941. [CrossRef] [PubMed]
177. Cheng, W.; Wu, Z.P.; Luan, D.; Zang, S.Q.; Lou, X.W. Synergetic Cobalt-Copper-Based Bimetal–Organic Framework Nanoboxes toward Efficient Electrochemical Oxygen Evolution. *Angew. Chem. Int. Ed.* **2021**, *60*, 26397–26402. [CrossRef]
178. Zhang, R.Z.; Lu, L.L.; Chen, Z.H.; Zhang, X.; Wu, B.Y.; Shi, W.; Cheng, P. Bimetallic Cage-Based Metal–Organic Frameworks for Electrochemical Hydrogen Evolution Reaction with Enhanced Activity. *Chem. Eur. J.* **2022**, *28*, e202200401. [CrossRef]
179. Wang, Q.; Wang, A.; Dou, Y.; Shen, X.; Sudi, M.S.; Zhao, L.; Zhu, W.; Li, L. A tin porphyrin axially-coordinated twodimensional covalent organic polymer for efficient hydrogen evolution. *Chem. Commun.* **2022**, *58*, 7423–7426. [CrossRef]
180. Liao, W.; Tong, X.; Zhai, Y.; Dai, H.; Fu, Y.; Qian, M.; Wu, G.; Chen, T.; Yang, Q. ZIF-67-derived nanoframes as efficient bifunctional catalysts for overall water splitting in alkaline medium. *Dalton Trans.* **2022**, *51*, 7561–7570. [CrossRef]
181. Li, D.; Shi, X.; Sun, S.; Zheng, X.; Tian, D.; Jiang, D. Metal–Organic Framework-Derived Three-Dimensional Macropore Nitrogen-Doped Carbon Frameworks Decorated with Ultrafine Ru-Based Nanoparticles for Overall Water Splitting. *Inorg. Chem.* **2022**, *61*, 9685–9692. [CrossRef] [PubMed]
182. Abazari, R.; Amani-Ghadim, A.R.; Slawin, A M.Z.; Carpenter-Warren, C.L.; Kirillov, A.M. Non-Calcined Layer-Pillared $Mn_{0.5}Zn_{0.5}$ Bimetallic–Organic Framework as a Promising Electrocatalyst for Oxygen Evolution Reaction. *Inorg. Chem.* **2022**, *61*, 9514–9522. [CrossRef] [PubMed]
183. Xie, Y.; Huang, H.; Chen, Z.; He, Z.; Huang, Z.; Ning, S.; Fan, Y.; Barboiu, M.; Shi, J.-Y.; Wang, D.; et al. Co−Fe−P Nanosheet Arrays as a Highly Synergistic and Efficient Electrocatalyst for Oxygen Evolution Reaction. *Inorg. Chem.* **2022**, *61*, 8283–8290. [CrossRef]
184. Liu, D.; Zhao, Z.; Xu, Z.; Li, L.; Lin, S. Anchoring Ce-modified $Ni(OH)_2$ nanoparticles on Ni-MOF nanosheets to enhances the oxygen evolution performance. *Dalton Trans.* **2022**, *51*, 12839–12847. [CrossRef] [PubMed]
185. Zhong, H.; Wang, M.; Ghorbani-Asl, M.; Zhang, J.; Ly, K.H.; Liao, Z.; Chen, G.; Wei, Y.; Biswal, B.P.; Zschech, E.; et al. Boosting the Electrocatalytic Conversion of Nitrogen to Ammonia on Metal-Phthalocyanine-Based Two-Dimensional Conjugated Covalent Organic Frameworks. *J. Am. Chem. Soc.* **2021**, *143*, 19992–20000. [CrossRef]
186. Chen, J.; Kang, Y.; Zhang, W.; Zhang, Z.; Chen, Y.; Yang, Y.; Duan, L.; Li, Y.; Li, W. Lattice-Confined Single-Atom Fe_1S_x on Mesoporous TiO_2 for Boosting Ambient Electrocatalytic N_2 Reduction Reaction. *Angew. Chem. Int. Ed.* **2022**, *61*, e202203022. [CrossRef]
187. Fang, G.; Hu, J.N.; Tian, L.C.; Liang, J.X.; Li, L.; Zhu, C.; Wang, X. Zirconium-oxo Nodes of MOFs with Tunable Electronic Properties Provide Effective *OH Species for Enhanced Methane Hydroxylation. *Angew. Chem. Int. Ed.* **2022**, *61*, e202205077. [CrossRef]
188. Yang, J.; Acharjya, A.; Ye, M.Y.; Rabeah, J.; Li, S.; Kochovski, Z.; Youk, S.; Roeser, J.; Grüneberg, J.; Penschke, C.; et al. Protonated Imine-Linked Covalent Organic Frameworks for Photocatalytic Hydrogen Evolution. *Angew. Chem. Int. Ed.* **2021**, *60*, 19797–19803. [CrossRef]
189. Zhang, H.; Cheng, W.; Luan, D.; Lou, X.W.D. Atomically Dispersed Reactive Centers for Electrocatalytic CO_2 Reduction and Water Splitting. *Angew. Chem. Int. Ed.* **2021**, *60*, 13177–13196. [CrossRef]
190. Shinde, S.S.; Lee, C.H.; Jung, J.Y.; Wagh, N.K.; Kim, S.H.; Kim, D.H.; Lin, C.; Lee, S.U.; Lee, J.H. Unveiling dual-linkage 3D hexaiminobenzene metal–organic frameworks towards long-lasting advanced reversible Zn–air batteries. *Energy Environ. Sci.* **2019**, *12*, 727–738. [CrossRef]
191. Zhang, B.; Zheng, Y.; Ma, T.; Yang, C.; Peng, Y.; Zhou, Z.; Zhou, M.; Li, S.; Wang, Y.; Cheng, C. Designing MOF Nanoarchitectures for Electrochemical Water Splitting. *Adv. Mater.* **2021**, *33*, 2006042. [CrossRef] [PubMed]
192. Nguyen, H.L.; Alzamly, A. Covalent Organic Frameworks as Emerging Platforms for CO_2 Photoreduction. *ACS Catal.* **2021**, *11*, 9809–9824. [CrossRef]
193. You, F.; Hou, X.; Wei, P.; Qi, J. SnS_2 with Flower-like Structure for Efficient CO_2 Photoreduction under Visible-Light Irradiation. *Inorg. Chem.* **2021**, *60*, 18598–18602. [CrossRef] [PubMed]
194. Zhou, Y.; Liu, S.; Gu, Y.; Wen, G.H.; Ma, J.; Zuo, J.L.; Ding, M. In(III) Metal–Organic Framework Incorporated with Enzyme-Mimicking Nickel Bis(dithiolene) Ligand for Highly Selective CO_2 Electroreduction. *J. Am. Chem. Soc.* **2021**, *143*, 14071–14076. [CrossRef] [PubMed]
195. Yi, J.D.; Si, D.H.; Xie, R.; Yin, Q.; Zhang, M.D.; Wu, Q.; Chai, G.L.; Huang, Y.B.; Cao, R. Conductive Two-Dimensional Phthalocyanine-based Metal–Organic Framework Nanosheets for Efficient Electroreduction of CO_2. *Angew. Chem. Int. Ed.* **2021**, *60*, 17108–17114. [CrossRef]

196. Rayder, T.M.; Bensalah, A.T.; Li, B.; Byers, J.A.; Tsung, C.K. Engineering Second Sphere Interactions in a Host–Guest Multi-component Catalyst System for the Hydrogenation of Carbon Dioxide to Methanol. *J. Am. Chem. Soc.* **2021**, *143*, 1630–1640. [CrossRef]

197. Zhang, J.; An, B.; Li, Z.; Cao, Y.; Dai, Y.; Wang, W.; Zeng, L.; Lin, W.; Wang, C. Neighboring Zn–Zr Sites in a Metal–Organic Framework for CO_2 Hydrogenation. *J. Am. Chem. Soc.* **2021**, *143*, 8829–8837. [CrossRef]

198. Khan, I.S.; Mateo, D.; Shterk, G.; Shoinkhorova, T.; Poloneeva, D.; Garzón-Tovar, L.; Gascon, J. An Efficient Metal–Organic Framework-Derived Nickel Catalyst for the Light Driven Methanation of CO_2. *Angew. Chem. Int. Ed.* **2021**, *60*, 26476–26482. [CrossRef] [PubMed]

199. Yu, F.; Jing, X.; Wang, Y.; Sun, M.; Duan, C. Hierarchically Porous Metal–Organic Framework/MoS2 Interface for Selective Photocatalytic Conversion of CO_2 with H_2O into CH_3COOH. *Angew. Chem. Int. Ed.* **2021**, *133*, 25053–25057. [CrossRef]

200. Zeng, L.; Cao, Y.; Li, Z.; Dai, Y.; Wang, Y.; An, B.; Zhang, J.; Li, H.; Zhou, Y. Multiple Cuprous Centers Supported on a Titanium-Based Metal-Organic Framework Catalyze CO_2 Hydrogenation to Ethylene. *ACS Catal.* **2021**, *11*, 11696–11705. [CrossRef]

201. Shi, W.; Quan, Y.; Lan, G.; Ni, K.; Song, Y.; Jiang, X. Bifunctional Metal-Organic Layers for Tandem Catalytic Transformations Using Molecular Oxygen and Carbon Dioxide. *J. Am. Chem. Soc.* **2021**, *143*, 16718–16724. [CrossRef]

202. Wang, W.M.; Wang, W.T.; Wang, M.Y.; Gu, A.L.; Hu, T.D.; Zhang, Y.X.; Wu, Z.L. A Porous Copper–Organic Framework Assembled by $[Cu_{12}]$ Nanocages: Highly Efficient CO_2 Capture and Chemical Fixation and Theoretical DFT Calculations. *Inorg. Chem.* **2021**, *60*, 9122–9131. [CrossRef] [PubMed]

203. Hammond, C.; Forde, M.M.; Ab Rahim, M.H.; Thetford, A.; He, Q.; Jenkins, R.L.; Dimitratos, N.; Lopez-Sanchez, J.A.; Dummer, N.F.; Murphy, D.M.; et al. Direct catalytic conversion of methane to methanol in an aqueous medium by using copper-promoted Fe-ZSM-5. *Angew. Chem. Int. Ed.* **2012**, *51*, 5129–5133. [CrossRef]

204. Yu, T.; Li, Z.; Lin, L.; Chu, S.; Su, Y.; Song, W.; Wang, A.; Weckhuysen, B.M.; Luo, W. Highly Selective Oxidation of Methane into Methanol over Cu-Promoted Monomeric Fe/ZSM-5. *ACS Catal.* **2021**, *11*, 6684–6691. [CrossRef]

205. Wu, X.; Zeng, Y.; Liu, H.; Zhao, J.; Zhang, T.; Wang, S.L. Noble-metal-free dye-sensitized selective oxidation of methane to methanol with green light (550 nm). *Nano Res.* **2021**, *14*, 4584–4590. [CrossRef]

206. Stappen, C.V.; Deng, Y.; Liu, Y.; Heidari, H.; Wang, J.X.; Zhou, Y.; Ledray, A.P.; Lu, Y. Designing Artificial Metalloenzymes by Tuning of the Environment beyond the Primary Coordination Sphere. *Chem. Rev.* **2022**, *122*, 11974–12045. [CrossRef]

207. Lewis, J.E.M. Molecular engineering of confined space in metal–organic cages. *Chem. Commun.* **2022**, *58*, 13873–13886. [CrossRef] [PubMed]

208. Yu, Z.; Ji, N.; Li, X.; Zhang, R.; Qiao, Y.; Xiong, J.; Liu, J.; Lu, X. Kinetics Driven by Hollow Nanoreactors: An Opportunity for Controllable Catalysis. *Angew. Chem. Int. Ed.* **2023**, *62*, e202213612. [CrossRef] [PubMed]

209. McTernan, C.T.; Davies, J.A.; Nitschke, J.R. Beyond Platonic: How to Build Metal–Organic Polyhedra Capable of Binding Low-Symmetry, Information-Rich Molecular Cargoes. *Chem. Rev.* **2022**, *122*, 10393–10437. [CrossRef]

210. Pirro, F.; Gatta, S.L. A De Novo-Designed Type 3 Copper Protein Tunes Catechol Substrate Recognition and Reactivity. *Angew. Chem. Int. Ed.* **2023**, *62*, e202211552. [CrossRef] [PubMed]

211. Sicard, C. In Situ Enzyme Immobilization by Covalent Organic Frameworks. *Angew. Chem. Int. Ed.* **2023**, *62*, e202213405. [CrossRef] [PubMed]

212. Lin, H.Y.; Chen, X.; Dong, J.; Yang, J.F.; Xiao, H.; Ye, Y.; Li, L.H.; Zhan, C.G.; Yang, W.C.; Yang, G.F. Rational Redesign of Enzyme via the Combination of Quantum Mechanics/Molecular Mechanics, Molecular Dynamics, and Structural Biology Study. *J. Am. Chem. Soc.* **2021**, *143*, 15674–15687. [CrossRef]

213. Wang, L.; Wang, D.; Li, Y. Single-atom catalysis for carbon neutrality. *Carbon Energy* **2022**, *4*, 1021–1079. [CrossRef]

214. Meng, S.L.; Ye, C.; Li, X.B.; Tung, C.H.; Wu, L.Z. Photochemistry Journey to Multielectron and Multiproton Chemical Transformation. *J. Am. Chem. Soc.* **2022**, *144*, 16219–16231. [CrossRef] [PubMed]

215. Li, F.; Han, G.F.; Baek, J.B. Nanocatalytic Materials for Energy-Related Small-Molecules Conversions: Active Site Design, Identification and Structure–Performance Relationship Discovery. *Acc. Chem. Res.* **2022**, *55*, 110–120. [CrossRef]

216. Zhou, W.; Deng, Q.W.; He, H.J.; Yang, L.; Liu, T.Y.; Wang, X.; Zheng, D.Y.; Dai, Z.B.; Sun, L.; Liu, C.; et al. Heterogenization of Salen Metal Molecular Catalysts in Covalent Organic Frameworks for Photocatalytic Hydrogen Evolution. *Angew. Chem. Int. Ed.* **2023**, *62*, e202214143. [CrossRef]

217. Pan, Y.; Sanati, S.; Abazari, R.; Noveiri, V.N.; Gao, J.; Kirillov, A.M. Pillared-MOF@NiV-LDH Composite as a Remarkable Electrocatalyst for Water Oxidation. *Inorg. Chem.* **2022**, *61*, 20913–20922. [CrossRef]

218. Shi, Z.; Ge, Y.; Yun, Q.; Zhang, H. Two-Dimensional Nanomaterial-Templated Composites. *Acc. Chem. Res.* **2022**, *55*, 3581–3593. [CrossRef] [PubMed]

219. Jayaramulu, K.; Mukherjee, S.; Morales, D.M.; Dubal, D.P.; Nanjundan, A.K.; Schneemann, A.; Masa, J.; Kment, S.; Schuhmann, W.; Otyepka, M.; et al. Graphene-Based Metal–Organic Framework Hybrids for Applications in Catalysis, Environmental, and Energy Technologies. *Chem. Rev.* **2022**, *122*, 17241–17338. [CrossRef]

220. Takagaki, A.; Tsuji, Y.; Yamasaki, T.; Kim, S.; Shishido, T.S.; Ishihara, T.; Yoshizawa, K. Low-temperature selective oxidation of methane to methanol over a platinum oxide. *Chem. Commun.* **2023**, *59*, 286–289. [CrossRef] [PubMed]

221. Franz, R.; Uslamin, E.A.; Pidko, E.A. Challenges for the utilization of methane as a chemical feedstock. *Mendeleev Commun.* **2021**, *31*, 584–592. [CrossRef]
222. Liu, Q.; Pfriem, N.; Cheng, G.; Baráth, E.; Liu, Y.; Lercher, J.A. Maximum Impact of Ionic Strength on Acid-Catalyzed Reaction Rates Induced by a Zeolite Microporous Environment. *Angew. Chem. Int. Ed.* **2023**, *62*, e202208693. [CrossRef]

MDPI

St. Alban-Anlage 66

4052 Basel

Switzerland

Tel. +41 61 683 77 34

Fax +41 61 302 89 18

www.mdpi.com

Catalysts Editorial Office

E-mail: catalysts@mdpi.com

www.mdpi.com/journal/catalysts